— d ————————————————|————————— p

REPRESENTATIVE ELEMENTS   NOBLE GASES *inert*

non-reactive gases

				VII A	0
				halogens 1.0080 H[1] 1	4.00260 He[2] 2
III A	**IV A**	**V A**	**VI A**		
10.81 B[5] 3	12.011 C[6] 4	14.0067 N[7] 5	15.9994 O[8] 6	18.9984 F[9] 7	20.179 Ne[10] 8

—MENTS

	VIII		I B	II B	III A	IV A	V A	VI A	VII A	0
B					26.9815 Al[13] 3	28.086 Si[14] 4	30.9738 P[15] 5	32.06 S[16] 6	35.453 Cl[17] 7	39.948 Ar[18] 8

	VIII			I B	II B	III A	IV A	V A	VI A	VII A	0
30 ...5]	55.847 Fe[26] 14,2	58.9332 Co[27] 15,2	58.71 Ni[28] 16,2	63.545 Cu[29] 18,1	65.37 Zn[30] 18,2	69.72 Ga[31] 18,3	72.59 Ge[32] 18,4	74.9216 As[33] 18,5	78.96 Se[34] 18,6	79.90 Br[35] 18,7	83.80 Kr[36] 18,8
...3]	101.07 Ru[44] 15,1	102.906 Rh[45] 16,1	106.4 Pd[46] 18	107.868 Ag[47] 18,1	112.40 Cd[48] 18,2	114.82 In[49] 18,3	118.69 Sn[50] 18,4	121.75 Sb[51] 18,5	127.60 Te[52] 18,6	126.905 I[53] 18,7	131.30 Xe[54] 18,8
...5] 2	190.2 Os[76] 32,14,2	192.22 Ir[77] 32,15,2	195.09 Pt[78] 32,17,1	196.967 Au[79] 32,18,1	200.59 Hg[80] 32,18,2	204.37 Tl[81] 32,18,3	207.12 Pb[82] 32,18,4	208.981 Bi[83] 32,18,5	(209) Po[84] 32,18,6	(210) At[85] 32,18,7	(222) Rn[86] 32,18,8
	[108]										

——————— f —————————————————————→

150.4 Sm[62] 24,8,2	151.96 Eu[63] 25,8,2	157.25 Gd[64] 25,9,2	158.925 Tb[65] 27,8,2	162.50 Dy[66] 28,8,2	164.930 Ho[67] 29,8,2	167.26 Er[68] 30,8,2	168.934 Tm[69] 31,8,2	173.04 Yb[70] 32,8,2	174.97 Lu[71] 32,9,2
(244) Pu[94] 24,8,2	(243) Am[95] 25,8,2	(247) Cm[96] 25,9,2	(247) Bk[97] 26,9,2	(251) Cf[98] 28,8,2	(254) Es[99] 29,8,2	(253) Fm[100] 30,8,2	(256) Md[101] 31,8,2	(253) No[102] 32,8,2	(257) Lw[103] 32,9,2

stable isotopes

An Introduction to Physical Science

THIRD EDITION

James T. Shipman
Ohio University

Jerry L. Adams
Ohio University

Jerry D. Wilson
Lander College

D. C. HEATH AND COMPANY
Lexington, Massachusetts   Toronto

International Standard Book Number: 0-669-01720-5

Library of Congress Catalog Card Number: 78-55917

Cover Photographs

Cumulonimbus clouds developing an anvil shape on top—
Colorado. Grant Heilman;   Laboratory equipment at W. R.
Grace and Company's Washington Research Center. Courtesy
W. R. Grace and Company;   Friendship 7 launch, Cape
Kennedy, Florida. Courtesy NASA;   Solar flare. Courtesy
NASA;   Lava fountain, Kilauea, Hawaii. Courtesy U.S.
Geological Survey

Chapter Opening Photographs

1 Runk/Schoenberger from Grant Heilman **2** Ira
Kirschenbaum, Stock, Boston **3** Harold E. Edgerton
4 Courtesy Ohio University Publication Office **5** From
PSSC *Physics,* D. C. Heath and Company, Lexington, Mass.,
1965 **6** Manfred Kage from Peter Arnold Photo
Archives **7** Werner H. Muller from Peter Arnold Photo
Archives **8** Lick Observatory Photographs **9** A doughnut-
shaped device for achieving the temperature and pressure required
for nuclear fusion. Courtesy Los Alamos Scientific
Laboratory **10** Culver Pictures, Inc. **11** Diffraction pattern
produced by X-rays passing through a crystal. General Electric
Research & Development Center **12** Courtesy Fischer
Collection, Fischer Scientific Company **13** Manfred Kage from
Peter Arnold Photo Archives **14** Peter Vandermark
15 Courtesy James Crouse **16** Manfred Kage from
Peter Arnold Photo Archives **17** NASA **18** Mark
Harrington **19** NASA **20** The Whirlpool Galaxy. Mount
Wilson & Palomar Observatories **21** Alan Pitcairn from Grant
Heilman **22** Charles W. Flagg **23** Grant Heilman **24** Boston
Globe Photo **25** NCAR Photo **26** Grant Heilman **27** Peter
Southwick from Stock, Boston **28** Courtesy Standard Oil
Company **29** Courtesy U.S. Geological Survey **30** Courtesy
USDA-SCS **31** Courtesy USDA

Preface

The third edition of *An Introduction to Physical Science* is written to give college students a knowledge of the fundamentals of physical science and to help them understand the technology of modern society. Like its predecessors, this edition presents basic concepts in the five major areas of physical science—physics, chemistry, astronomy, meteorology, and geology. Organized in a logical rather than a historical fashion, the book first covers position and the science of measurement, followed by motion and the forces causing motion. These concepts are then used to explain the other phenomena of nature.

The treatment of the subject matter is both descriptive and quantitative, with the relative emphases of these two approaches left up to the instructor. The mathematical level has been reduced and as a result Chapters 1 through 5 have been revised extensively. However, the derivations of some equations remain in the Appendix for use by those who wish to know their origin.

All of the geology chapters (27 through 31) in this edition have been completely rewritten. Within the last few years there has been a revolution in the science of geology that gives us a better insight into the processes occurring near the earth's surface. The theory of plate tectonics, which considers the earth's surface to be made up of movable "plates," has surprisingly simple assumptions, but its consequences are far-reaching, giving a new understanding of the formation of mountains and ocean basins, volcanic activity, earthquakes, and the mechanisms of operation of many geologic processes. Its concepts are presented in Chapter 27 and are used to explain various geologic phenomena in the remaining chapters.

The text has been thoroughly revised and updated with the latest information in each of the major areas, and numerous practical applications of physical principles have been added. Learning objectives have been stated at the end of each chapter, and a complete glossary has also been added.

It is suggested that the material on physics and chemistry be taught first, followed by the sections on astronomy, meteorology, and geology. However, these topics are largely self-contained, and astronomy, meteorology, and geology can be presented before either physics or chemistry.

The book is planned for use with or without a laboratory course, and it has been used successfully under both the quarter and semester systems.

An instructor's manual has been prepared to supplement the instructor's background in the different physical science areas. The manual includes the following on a chapter-by-chapter basis:

1. An introduction to the chapter and suggestions for teaching it.
2. Answers to questions and worked-out solutions to all problems given at the end of the chapter.
3. A list of available films and possible demonstrations.
4. Questions of various types that can be used for examination purposes.
5. Selected references.

The program also includes a laboratory manual and student guide. The laboratory manual contains thirty-eight experiments in five physical science areas and is organized to follow and augment the textbook material. The student guide includes the following for each chapter of the text:

1. A review with essay questions and their answers.
2. A review with multiple-choice questions and their answers.
3. A programed summary consisting of a comprehensive treatment of questions and problems.
4. Problems with step-by-step solutions of the type found in the textbook chapter.

A review of basic mathematics and lists of fundamental constants and conversion factors are also included in the guide.

The authors wish to thank their colleagues and students for the many contributions they made to this edition of the text through discussions and classroom use of the second edition. The authors also express their gratitude to those individuals and organizations who contributed photographs, illustrations, and other information; in particular they wish to acknowledge the assistance of the U.S. Geological Survey, the U.S. Park Service, the Department of Energy, the American Institute of Physics, the Environmental Science Service Administration, the U.S. Air Force, the Ohio University Publications Office, the Field Enterprises Educational Corporation, James Crouse, and Paul H. Taft. A special note of thanks goes to Betty Craddock for her help in preparing the manuscript.

J. T. S.
J. L. A.
J. D. W.

Contents

Introduction

The world in which we live is changing continually, and people are continually seeking to discover and understand these changes. Work in this area is resulting in advances in technology related to all aspects of our daily life, making it most important that each individual knows and understands the physical concepts behind this advancing technology. It is only through an understanding of the basic physical concepts that the individual today can hope to remain abreast of this technology concerning the natural environment.

Physical science is the organized knowledge of our physical environment and the methods used to obtain it. The information acquired and applied by individuals over the past 7000 years through discovery and experimentation, fused with reflective thinking in the formation of theories and laws, has produced our modern civilization. Even after 7000 years of learning, however, there is still much we do not know. For example, the origin of human beings and how we came to be where we are today may never be known with certainty.

Much of our present knowledge about the environment is actually an approximation. The age of our planet is based on the age of the oldest known formation of rocks in North America, the Morton Gneiss of Minnesota, and therefore is only an approximation. Through the use of radioactive decay techniques, this formation of rocks has been dated back 3.5 billion years. Allowing another billion years for the cooling process to occur, assuming the planet started in the gaseous phase, the age of the earth is at least 4.5 billion years.

We do not know how life began, but the fact that we exist proves that the right ingredients, in the right amounts, and arranged in the correct way, did come together at some time in the past. The best approximation is that it happened at least one billion years ago.

The appearance of human beings is fairly recent with respect to the geological time scale. The earliest evidence indicates they have been in existence at least one million years. It has been suggested that human beings' ability to reason was first put to use during the Ice Age, when they were forced from the trees into a cold environment and "reasoned" that to survive they must hunt in groups. Possessing intelligence, they invented stone tools and developed a crude "language" in order to live together and to fight their enemies better. Early people fought not only the wild beasts and the forces of nature, but made war against their neighbors as well. They discovered fire and learned to control it. The need for refuge from the fear of certain natural phenomena gave rise to religious rites and the worship of gods. As the climate became warmer, the hunters became shepherds who, with the passing of time, became farmers. Through necessity the plow was invented, animals were domesticated, the use of clay was developed, and weaving wool and flax was learned. Copper was discovered and then tin, which when it was combined with copper made bronze. Iron and other metals were discovered and used in numerous ways. Industries were formed, roads were built, laws were made, medicine was practiced. These and many other progressions, with the passing of time, formed the complex groups of many thousands of people who built the cities that were the beginning of our early civilizations.

Our life on earth is probably just beginning (it could be ending, if we are to believe the views of some ecologists). A mere three hundred generations have lived since the early civilizations of the Egyptians and the Babylonians. Located within a few degrees of the 30° N parallel along the rich valley lands

Democritus

of the Nile, Indus, Yangtze, Euphrates, and Tigris Rivers, these early civilizations were the first to develop a knowledge of mathematics, astronomy, medicine, engineering, and agriculture and so were the originators of science. They invented writing and writing materials, thus providing a way of recording the knowledge acquired by one generation and a means of passing it on to the next.

By the middle of the fourth millennium B.C., the Egyptians in the Nile Valley and the Babylonians in the Tigris and Euphrates valleys had acquired a language, were able to write, and possessed technical skills in engineering and mathematics. They studied the stars and knew of the motion of the planets, sun, and moon, from which they fixed the solar year of 12 months.

These civilizations flourished until the fifteenth century B.C., at which time they were invaded by barbarian tribesmen from the north. Thereafter, they experienced a gradual decline.

The pursuit of scientific knowledge is an activity that serves both as a history of those who made important discoveries concerning nature and as a means for the development of new ideas. The following discussion includes some of the individuals whose discoveries or ideas have benefited the human race.

The development of physical science as we know it today began around 600 B.C. with the Greeks, who attempted to understand their immediate environ-

Aristotle

ment and the universe as a whole from a purely naturalistic point of view.

One of the earliest known scientists was a Greek named Thales of Miletus, who lived about 600 B.C. He was the first Greek mathematician and astronomer, and a number of geometric propositions are credited to him.

Pythagoras (580–500 B.C.), who is considered to be the founder of science, was also a Greek. He made many important discoveries in arithmetic, geometry, and astronomy. The idea of abstract numbers was established by Pythagoras, and he was the first to point out that science provided a means for a true understanding of our environment.

Anaxagoras (500–428 B.C.) is credited with the discovery of the true cause of eclipses and with the view that the moon shines by reflected light from the sun. Many historians also credit him with originating the atomic hypothesis, which was formulated by Democritus a generation later. According to Anaxagoras, matter is composed of an infinite number of so-called "seeds," which also exist in an infinite variety.

Empedocles (484–424 B.C.) chose four "primary substances" (fire, air, water, and earth), from which he contended all other substances were made. His famous experiment with the clepsydra (water clock) proved that air was something that occupied space and exerted force. Through his experiments, he proved that the true nature of the environment could be gained by analyzing nature at first hand.

Democritus (460–370 B.C.) supplied ancient scientists with the fundamental building block of matter. He expanded the atomic hypothesis of Anaxagoras by postulating the theory that the universe consists of a void (empty space) and an infinite number of invisible and indivisible particles, which are alike in substance but differ in form, size, shape, position, and arrangement. Democritus is also credited with the first thoughts concerning the conservation of matter. His ideas that matter cannot be created from nothing and that nothing which is can cease to exist were revolutionary.

Aristotle (384–322 B.C.), considered to be the greatest of the Greek philosophers, made contributions to many branches of science, including biology, logic, ethics, psychology, metaphysics, and natural science. Because his high social position prevented him from doing experiments, which were considered work for slaves, his scientific views originated only from reason and logic. As one would expect, many were incorrect. In the field of physical science, Aristotle contended that a heavy body dropped from a given height would reach the ground before a lighter one dropped from the same height at the same time. This was considered true until the seventeenth century, when Galileo proved it to be false by actual experiment. Aristotle was held in such high esteem and his authority was so great that his views concerning the structure of the universe, which placed the earth at rest at the center of the system, overruled the views of other scientists for nearly 2000 years.

Archimedes (287–212 B.C.) represents the present-day notion of the scientists. He experimented to establish the truth of a physical concept, and he combined the results with mathematics to formulate theories. A few of the concepts with which Archimedes is credited are density, specific weight, center of gravity, and the lever.

Aristarchus of Samos (310–230 B.C.), a Greek astronomer, was the first known person of authority to hold the theory that the earth revolves around the

Copernicus

The sixteenth century, however, was the beginning of an intellectual awakening. Known as the Renaissance, the era brought about advancements in the physical sciences and in other areas.

The new thinking in astronomy began with Nicholas Copernicus (1473–1543), a Polish astronomer who made the necessary mathematical calculations to prove the sun-centered, or heliocentric, theory put forth by Aristarchus and to disprove the theory of Aristotle.

Tycho Brahe (1546–1601), a Dane who was considered the greatest practical astronomer since the Greeks, spent his entire life observing and recording data on the planets and stars. His work might have gone unnoticed had not Johannes Kepler fallen heir to Brahe's large collection of data. Kepler (1571–1630), a German astronomer and mathematician, believed in Copernicus' heliocentric theory and, with Brahe's data, developed three laws of planetary motion which eliminated the geocentric or earth-centered theory and opened the way for our present views on the solar system. He discovered that the planets move in elliptical orbits around the sun.

sun. He was unable, however, to prove his theory, nor could he gain the support of his colleagues; thus it was that his idea was forgotten until the sixteenth century.

Eratosthenes (276–192 B.C.) of Alexandria, librarian and scientific writer, calculated the circumference of the earth and came close to the value we use today.

Hipparchus of Nicaea (second century A.D.), a Greek astronomer and mathematician, compiled the first star catalog and discovered the precession of the equinoxes.

Ptolemy of Alexandria (about 70–147 A.D.), another Greek astronomer, developed Aristotle's (incorrect) ideas that the sun, moon, and planets revolve in perfect circles about the earth, which he believed was the center of the universe.

Generally speaking, during the next 1200 years (200–1400 A.D.) after Ptolemy, very little scientific advancement was made. Roger Bacon (about 1214–1291 A.D.), the British philosopher, scientist, and educator, believed and taught that the true facts of nature could be revealed by experimentation and stressed the importance of mathematics in formulating scientific theories. Leonardo da Vinci (1452–1519 A.D.), the Italian painter whose work in the scientific field has only recently come to be known, also believed strongly in the experimental approach. These two scientists were largely responsible for what advancement in physical science there was during the years 1230–1520.

Brahe

Galileo Galilei (1564–1642), Italian astronomer, mathematician, and physicist, is regarded as the first true experimentalist and the father of modern physics. He was the first to view the moon, planets, sun, and stars through a telescope. He set forth the basic ideas of motion and laid the groundwork for the laws of motion to be formulated later by Newton; laws which are the foundations of the branch of physics known as mechanics.

Galileo designed and built his own telescope about 1609, after hearing about a Dutch lensmaker's success in magnifying distant objects with two lenses. His first telescope had a magnification of three, which he later improved to a power of more than 30. During the year 1610 he used his telescope to discover the craters and mountains of the moon, the spots on the sun which rotated every 27 days, the disc appearance of the planets, the four largest moons of Jupiter, the phase changes (the moon changes phase monthly) of Mercury and Venus, the rings of Saturn, and the multitude of stars in the Milky Way. His discoveries contributed more new information about the universe than those of any other single individual.

Galileo

Kepler

Sir Isaac Newton (1642–1727), the English physicist and mathematician regarded by many as the greatest scientist of all times, was born the year Galileo died. He lived at a time when a theory was needed to unify the discoveries made by Copernicus, Brahe, Kepler, and Galileo. Newton formulated such a theory and so advanced the branch of mechanics known as dynamics.

Newton was born on December 25, 1642 in a farmhouse near Grantham in Lincolnshire, England. He obtained his formal education at Trinity College, Cambridge University, receiving the M.A. degree in January 1665. The Great Plague closed the University in the fall of 1665, and Newton spent the next two years at home in Lincolnshire, at which time he developed the calculus, formulated his three laws of motion and the law of gravitation, and discovered the nature of white light. In 1667 Newton returned to teach at Cambridge University, where he spent most of his life.

Newton's theories were published many years after they were developed, and they might have gone unpublished had it not been for associates like Edmund Halley, who urged Newton to publish his work. Newton's *Mathematical Principles of Natural Philosophy,* or *The Principia* as it was called, appeared in 1687, over 20 years after he had formulated his laws of motion and gravitation. Following his recovery from a nervous breakdown in 1692, he did very little scientific work before he died in 1727.

The seventeenth century also produced Robert

Newton

leev and Meyer's periodic law of elements, and Graham's law of diffusion. Also, the wave theory of light was developed by Young; heat and thermodynamic theories were advanced by Kelvin, Joule, Rumford, Gibbs, and Carnot; additions to the theories of electricity and magnetism were made by Maxwell, Laplace, Green, and Poisson; and Ampere, Coulomb, Faraday, Henry, Oersted, Ohm, and Volta made fundamental electrical discoveries. Hamilton presented a new form of equation of motion to add to the mechanics field, radioactivity was discovered by Becquerel, X-rays by Roentgen, and electrons by Thomson.

The year 1900 proved to be the beginning of a completely new era in physics. It was in this year that Max Planck originated his quantum hypothesis. The word quantum means a discrete amount, and Planck's hypothesis stated that certain forms of energy could exist only in certain discrete amounts, or quanta, rather than in any amount chosen or desired. In 1905 Albert Einstein announced his theory of relativity, which completely altered our ideas of space and time. Einstein also related the mass of a particle to its total energy with his famous equation $E = mc^2$. In 1913 Niels Bohr, using a variation of Planck's hypothesis, developed a quantitatively correct model of

Boyle (1627–1691), an Englishman who enunciated the gas law that bears his name and who is considered to be one of the founders of chemistry; Christian Huygens (1629–1695), a Dutch physicist who developed the wave theory of light and helped formulate the law of conservation of momentum; and Robert Hooke (1635–1703), an English physicist who made many scientific contributions but is best known for his law of elasticity.

During the eighteenth century, theories on heat, light, electricity, and magnetism appeared. Advances were made in all fields of physical science by an increasing number of scientists. Fahrenheit developed the temperature scale which bears his name, Celsius developed the centigrade scale (called in recent years the Celsius scale), Black made the distinction between heat and temperature, Bernoulli and Euler discovered the law of angular momentum, Bradley elucidated the aberration of light, Priestly discovered oxygen, Lavoisier developed the analytical balance, J. L. Prout stated the law of definite proportions, Benjamin Franklin proved that lightning is electrical in nature, Cavendish proved the inverse square law of electrostatics, and Coulomb began quantitative research in electrostatics.

Included in the numerous achievements in the physical sciences during the nineteenth century are Dalton's atomic theory, Gay-Lussac's law of combining volumes, Avogadro's law for gases, Mende-

Avogadro

the simplest atom, hydrogen. It is evident that by this time, the science revolution had begun in earnest. The next big step was taken in 1925, when Louis de Broglie introduced the concept of matter waves. The notion of duality of light—that light acts sometimes as a wave and sometimes as a particle—was extended by de Broglie to the basic model of Bohr's atom. The electrons and protons in the atom were shown to act like waves as well as particles. Erwin Schrödinger in 1925 developed a wave equation which was obeyed by particles having mass. In the next seven years some of the greatest achievements in physics and chemistry were made. Schrödinger's equation was applied to atoms, and with the help of Wolfgang Pauli's exclusion principle, the periodic table of Mendeleev became understood from a submicroscopic point of view. Schrödinger's equation was applied to the structure of molecules and solids, and the modern theories of these structures began to take shape. The neutron was first found experimentally by James Chadwick in 1932, and the basic ideas of nuclear structure were then complete.

The impact of the discovery in 1939 by Otto Hahn and Fritz Strassman, German chemists, of the phenomenon of nuclear fission has not yet been measured. With this discovery the possibility of converting mass into great quantities of energy became very real. The explosion of the first atomic bomb on a lonely desert in New Mexico in 1945 ushered in the atomic age, the age of science and technology in which we now live.

As the principles of physics and chemistry developed, they were applied to the phenomena occurring in the environment. Efforts to understand weather phenomena and their causes gave rise to the science of meteorology. Probably no other branch of physical science is more readily observable and has a more direct effect on our daily activities than meteorology. From empirical weather observations and scientific principles has come a better understanding of the workings of our atmosphere. As a result, reasonably accurate predictions can be made of future weather conditions. The dynamics of weather is governed by physical laws. Although the physical principles are known, the vastness and the many variables of the atmospheric mass do not permit the age-old desire to completely understand and control the weather.

On an even larger scale, especially that of time, is the study of our earth—geology. Most of the fundamental geological principles that we use today were established in the seventeenth and eighteenth centuries. Very little knowledge of scientific value was known concerning the earth and its origin before this time. In 1669 Nicolaus Steno (1638–1687), Danish physician, published a small pamphlet concerning his observations and investigations of rocks and the fossils found in them. From his studies he concluded it should be possible to infer deductively the order in which sedimentary rocks are formed. His reasoning led to the Law of Sequences. In 1788 James Hutton (1726–1797), Scottish physician, lawyer, and geologist, published a paper entitled "Theory of the Earth." Hutton established the groundwork for the modern science of geology in his conclusion concerning the principle of uniformity of process or the principle of uniformitarianism. The

Chadwick

Hutton

principle states that "the present is the key to the past." The work of Hutton was advanced by Sir Charles Lyell (1797–1875), English geologist, who used Hutton's principle of uniformitarianism as the unifying theme in his classic publication *Principles of Geology.* Lyell is credited by many as having advanced the knowledge of geology more than any other person. The work of William Smith (1769–1839), English civil engineer and geological hobbyist, concerning the relationship between rock strata in different localities, led to the principle of faunal and floral succession. Smith noted that the rock layers could be identified by the fossil shells and skeletons they contain. This initiated the idea that major rock divisions as well as units of time might be based on the fossil animals and plants found in the rock. Perhaps the most important advance in our understanding of the geology of the earth during the twentieth century has been the development of the theory of continental drift and plate tectonics. Alfred Wegener, German meteorologist, in a publication in 1915, presented the concept that the present continents were at one time in the past one huge land mass or supercontinent, which he called Pangaea. This land mass broke up, drifted apart, and formed the continents in the positions we know today. Earth scientists have long regarded the earth as a rigid body, but the discoveries that the earth's interior may be slowly deformable led to the conclusion that the surface is mo-

bile. The results of geological research in the last 20 years supports Wegener's theory. But how could the large separation of the plates take place? In 1960 Hans H. Hess, of Princeton University, proposed that the sea floor cracks open along the crest of the mid-ocean ridges and that new sea floor forms there and spreads apart. The process is known as sea floor spreading. Today most geologists support the theory of continental drift, sea floor spreading, and plate tectonics.

It is evident that the development and advancement of scientific knowledge has been rapid since the sixteenth century. This accelerated growth was the result of many factors, not the least of which is the procedure scientists use in acquiring knowledge. Although scientists do not follow a given set of rules, they do follow a general plan for attacking a problem. Such a plan is referred to as the "scientific method" and includes the following, in general terms:

1. *The observation of phenomena and the recording of facts.* The "phenomena" are defined as whatever

Lyell

happens in the environment, and "facts" as accurate descriptions of what is observed.

2. *The formulation of physical laws from the generalization of the phenomena.* A "physical law" is a description of a certain behavior of nature that extends beyond what has been observed. A physical law is stated in general terms, often in symbol notation. Kepler's laws of planetary motion are examples.

3. *The development of a theory that is used to predict new phenomena.* A "theory" comprises a general scheme of thought, which correlates the known facts in such a manner that new thoughts and relations initiate the prediction of new phenomena. Einstein's theory of relativity and the kinetic theory of gases are examples.

A major factor in the development of physical science has been the formulation of concepts used to describe phenomena. A *concept* is a meaningful idea that can be used to describe phenomena. Concepts develop as a result of experimentation and are subject to change as new experimentation dictates. They

Smith

are fundamental in the sense that they form the rules for understanding nature, but they are also interconnected.

The meaning of a concept is established by a working definition. The definition may be stated in words, symbolic notation, or by means of a mathematical formula. In whatever way the definitions are stated, we must learn them, explain them, and interpret their meaning in respect to our environment.

An understanding of our physical environment begins with a study of fundamental concepts.

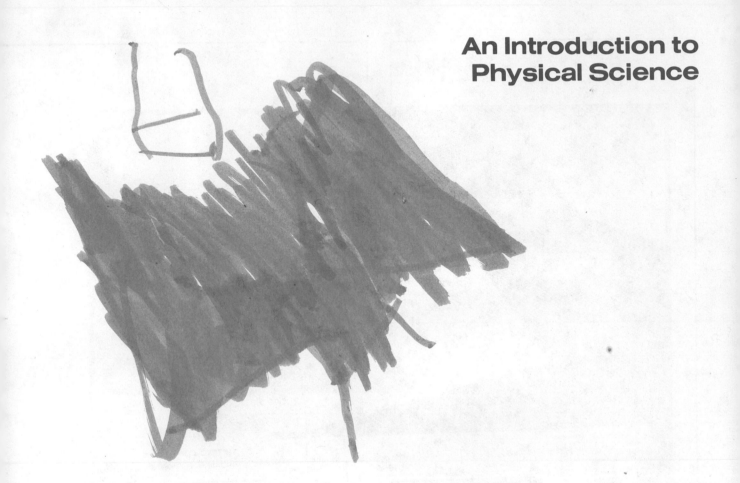

An Introduction to Physical Science

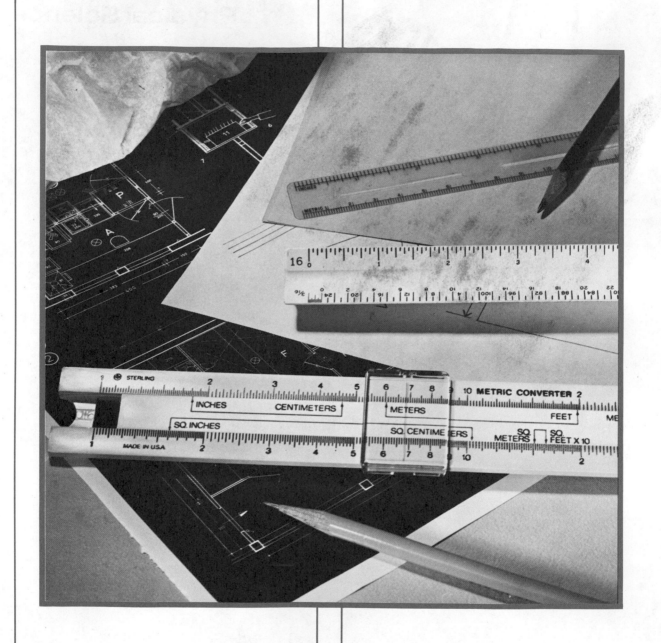

Chapter 1

MEASUREMENT

When you can measure what you are speaking about and express it in numbers, you know something about it; but when you cannot measure it, when you cannot express it in numbers, your knowledge is of meager and unsatisfactory kind.

Lord Kelvin

A first step in understanding our environment is to find out about the physical world through measurements. Over the centuries, human beings have developed increasingly sophisticated methods of measurements, and scientists make use of the most advanced of these. The student should gain both a workable knowledge of these methods and an awareness of their limitations.

We are continually making measurements in our daily life. Each day we plan our work, play, and rest schedules as a function of time. With watches and clocks we measure the time for events to take place. Every 10 years we take the census and determine (measure) the population. We count our money, our homes, our food, the minutes, hours, days, and years of our life. Some of us keep accurate measurements of food and drugs taken into the body because of illness. Many lives depend on accurate measurements being made by the medical doctor, laboratory technician, and pharmacist in the diagnosis and treatment of disease. Meteorologists measure the many elements (temperature, pressure, humidity, precipitation, wind) that make up the weather. This information is relayed to millions by the communications media, which must at all times measure all phases of their operation to stay within standards designed to protect the rights of others.

Human beings, in their efforts to understand their total environment, must measure the very small and the very large, probe farther inward to discover smaller and smaller particles, and explore outward to discover a larger and larger universe. At one time it was thought that all things could be measured with exact certainty. However, as we measured smaller and smaller objects, it became evident that the very act of measuring distorted the measurement. This uncertainty in making measurements of the very small will be discussed in detail in Chapter 11.

The ability of the scientist to know and predict is a function of accurate measurements. From accurate measurements taken on the moon's surface the geologist obtains new knowledge for understanding continental drift. Our conquest of the moon and planets is largely due to our ability to make accurate measurements on the hardware that makes up the space vehicle and to program computers that make accurate and continuous measurements of position, velocity, and the numerous other factors involved in space travel. Thus, we see relevance in all measurements and recognize the need to know the concepts of measurement as the first step in the understanding of our physical environment.

1-1
THE SENSES

Our environment stimulates our senses, either directly or indirectly. The five senses (sight, hearing, smell, touch, taste) make possible our knowledge of the environment. Therefore, they are a good starting point in the study and understanding of the physical world.

Most information about our environment comes through sight. This information is not always a true representation of the facts because the eyes, and therefore the mind, can be fooled. There are many well-known optical illusions, such as those in Fig. 1-1a. Many people are quite convinced that what they see in such drawings actually exists as they perceive it.

Hearing ranks second to sight in supplying the brain with information about the external world. The

Figure 1-1a *Some optical illusions. We can be deceived by what we see.*

senses of touch, taste, and smell, although very important for good health and happiness, rank well below sight and hearing in providing environmental information.

All the senses can be deceived and thus provide false information about our environment. Anyone who has gone to the beach to swim during the early morning hours when the air is cold knows how warm the water feels to the body. Later in the afternoon when the air becomes hot, the water, which has remained at practically the same temperature, feels cold. Thus, if we were asked to judge the temperature of the water, our answers would vary according to the temperature of the air.

Not only can the senses be deceived, but they also have their limitations. The naked eye is unable to distinguish the stars of our galaxy from the planets of our solar system. For centuries the planets were called wandering stars. If you were asked to distinguish between two identical sounds produced $\frac{1}{100}$ second apart, you would fail. The other senses have similar limitations. The handicaps of the senses can be conquered by close scrutiny of phenomena with measuring instruments. For example, in the diagram of optical illusions, diagonals (a) and (b) can be measured with a ruler, and their length accurately determined.

We extend our ability to measure our environment with many specialized instruments, or tools. These, too, have their limitations. Your wristwatch is a precision instrument, but it cannot be used to measure time intervals of less than $\frac{1}{100}$ second. Even the most precise instruments have limitations. More accurate information about our physical environment can be obtained by comparing the measurement of one instrument against another. Later in our study of the microscopic world we shall learn of other limitations concerning the measurement of physical quantities.

1-2
FUNDAMENTAL PROPERTIES

How would you describe your environment? As you begin to observe, you would ask questions like "Where is the book store?", "When does it open?", "How much do you want to buy?", and so forth. These questions of "Where?", "When?", and "How much?" refer to the basic concepts of space, time, and matter.

The description of space might refer to a location or to the size of an object. To measure locations and sizes we use the fundamental property of **length,** which is defined as the measurement of space in any direction. Space occupies three dimensions, each of

Figure 1-2a *Length, designated in the drawing by the lines labeled L, is the measurement of space in any direction. All components are reduced to three directions in this illustration.*

which can be measured by a length (see Fig. 1-2a). This can easily be seen by considering a rectangular object. It has a length, width, and height, and each of these dimensions can be measured as a length. A sphere such as the earth, has a radius, a diameter, and a circumference. Again, all of these are easily described by a length measurement.

Once we know where something is, we are frequently interested in what is happening to it. "Is the car moving?", "When will the next plane leave?", "What day will you be going home?"—all of these questions can be answered using the fundamental property of time. Each of us has an idea of what time is, but we would probably find it difficult to define or to explain it. Some terms that are often used refer to duration, period, or interval. **Time** may be defined as the continuous, forward flowing of events.

Without events or happenings of some sort, there would be no time (Fig. 1-2b). The mind has no direct experience of time, merely the awareness of events taking place in time. That is, we do not perceive time as such, only the events that take place in time.

Einstein, in his theory of relativity, has shown that space and time are inextricably linked together into what he calls space-time. In his theory time joins the three dimensions of space as a fourth dimension. His theory of relativity is the topic of Chapter 8. For the most part, however, we can use our intuitive

Figure 1-2b *Time can be described as the continuous forward flowing of events. Here the events are the dropping of sand particles in an hourglass.*

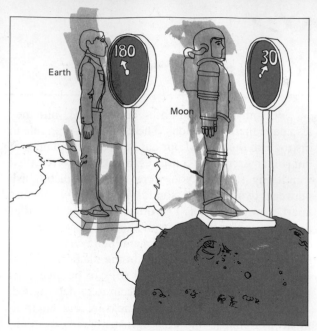

Figure 1-2c *The weight of an astronaut on the moon is $\frac{1}{6}$ what it is on the earth. The mass, however, is the same.*

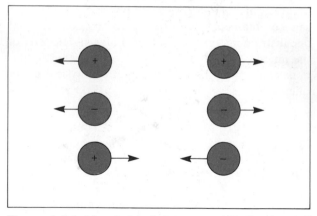

Figure 1-2d *Like electrical charges repel. Unlike electrical charges attract.*

ideas, which tend to regard space and time as separate fundamental properties.

When we ask questions concerning the quantity of matter, we need a third fundamental property known as mass. To define mass precisely, we need to understand the concepts of force and acceleration. These are discussed in Chapter 3. For now, let us simply say that **mass** is a measurement of the quantity of matter an object contains.

Since we live on earth, many of us tend to measure matter in terms of weight. However, weight is not a fundamental property. An astronaut who weighs 180 pounds on earth will weigh $\frac{1}{6}$ of that amount, or 30 pounds, on the moon (Fig. 1-2c). However, the astronaut's mass will be the same on the earth and the moon. Weight is related to the force of gravity, which changes depending on where we are in the universe. Mass, on the other hand, is a fundamental property, which remains the same throughout the universe. In Chapter 3 we will discuss in detail the relationship between mass and weight.

A fourth fundamental property is electric charge. There are two kinds of **electric charge,** called positive and negative. Two positive charges or two negative charges will repel each other, while opposite charges will attract (Fig. 1-2d). Electric charge is an important property of matter since all atoms are composed of

electrically charged particles and most matter is composed of atoms. Electric current is simply the flow of electrical charge. The concepts of electric charge and electric forces will be discussed in Chapter 6.

The four **fundamental properties** of nature are length, time, mass, and electric charge. These four properties and their various combinations are all that we need when we try to measure many of the complicated aspects of nature. Of course, there may be unknown fundamental properties as yet undiscovered, but for now, let us discuss how we measure these four.

1-3
STANDARD UNITS

In order to measure the fundamental properties and their various combinations, we need to refer them to reference or standard measurements. Each reference is called a standard unit. A **standard unit** has a fixed and reproducible value for the purpose of taking accurate measurements. There are several systems of measurement used around the world. Each system uses different standard units.

The United States is one of the few places in the world which uses the British engineering system of measurement. The **British system** uses the familiar unit of foot for length. It uses the pound as the unit of weight, and is thus called a gravitational system of measurement. The unit of mass in this system is the slug, which weighs approximately 32 pounds on the earth's surface. All systems of units use the second as the standard unit of time and the coulomb is usually used as the standard unit of charge.

Almost every country in the world except the United States uses the metric system of measurement.

This system is much simpler than the British system because converting from one unit to another is always done in the metric system by multiplying by factors of 10. For example, in the metric system, 1 kilometer is 1000 meters whereas in the British system, 1 mile is 5280 feet. Memorizing the various factors of 2, 4, 12, 16, 1760, 5280, etc., makes the British system unwieldy compared to the simplicity of the metric system.

There are actually two metric systems in common usage. One is called the mks system, the other is the cgs system. The letters **mks** stand for *m*eter, *k*ilogram, and *s*econd, while **cgs** stands for *c*entimeter *g*ram *s*econd. Table 1-3a lists the standard units for the mks, cgs, and British systems of measurement.

The standard unit of length in the meter-kilogram-second (mks) system is the **meter** (from the Greek, *metron*, to measure), which was originally intended to be one ten-millionth of the distance from the earth's equator to the geographic north pole. The unit was first adopted by the French in 1793 and is now used in scientific measurements of length throughout the world.

From 1889 to 1960 the standard meter was a platinum-iridium bar, 2 centimeters by 2 centimeters in cross section and slightly longer than 1 meter, kept in the vaults at the International Bureau of Weights and Measures near Paris, France. Two marks, one near either end of the bar, indicate the standard meter as the length between these two marks when the bar is at the temperature of melting ice.

Table 1-3a *Standard Units for the Metric and British Systems of Measurement*

Fundamental Property	Absolute Systems		Gravitational System
	Metric (mks)	Metric (cgs)	British
Length	Meter (m)	Centimeter (cm)	Foot (ft)
Mass	Kilogram (kg)	Gram (g)	Slug
Time	Second (s)	Second (s)	Second (s)
Electric charge	Coulomb (C)	Coulomb (C)	Coulomb (C)

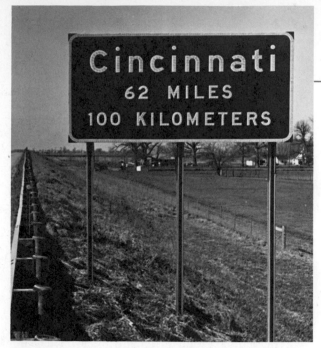

Figure 1-3a *The metric system is being used more in some states than in others. (Ohio State Department of Transportation.)*

Figure 1-3b *Prototype kilogram number 20, shown in photograph, is the United States standard unit of mass. The prototype is a platinum-iridium cylinder, 39 mm in diameter and 39 mm high. Prototype Number 20 was furnished to the United States by the International Bureau of Weights and Measures in accordance with the treaty of 1875. (Courtesy National Bureau of Standards.)*

Since the standard meter is a rigid body that could easily be destroyed by some disaster and the stability of the platinum-iridium bar was questioned by many scientists, the 11th General Conference of Weights and Measures meeting in Paris on October 14, 1960, adopted a new international standard unit of length, using the wavelength of the orange-red line in the spectrum of krypton-86. The new standard meter is defined as 1,650,763.73 wavelengths of the orange-red line.

From this basic standard, other units of length are defined. For example, the decimeter is defined as $\frac{1}{10}$ meter, the centimeter as $\frac{1}{100}$ meter, and the kilometer as 1000 meters. There is at present a worldwide move to adopt the metric system, because of its simplicity. In the United States, for example, some states have begun promoting the metric system by showing the relationship between it and the more familiar British system. Figure 1-3a shows what is being done in Ohio.

The standard unit of mass in the mks system is the kilogram, and it is associated with the standard unit of length, the meter. As originally conceived by the French in 1791, the standard unit of mass of 1 **kilogram** was the mass of 1 cubic decimeter (1 liter) of pure water at its maximum density. This choice was made on the assumption that 1 unit volume (1 cubic decimeter of water at its maximum density) could be easily reproduced in any laboratory in the world. As it turned out, 1 cubic decimeter of water of maximum density was difficult to obtain, so in the late 1880s the kilogram was redefined with reference to a platinum-iridium cylinder, with the intention that the mass of the new standard unit be equal to the old. A mistake was made, however, and the new standard turned out to have slightly less mass than the original. To avoid having two values for the kilogram, the liter was redefined as being 1000.028 cubic centimeters. The standard kilogram is kept at the International Bureau of Weights and Measures near Paris, with the United States prototype kept at the U.S. Bureau of Standards in Washington, D.C. See Fig. 1-3b.

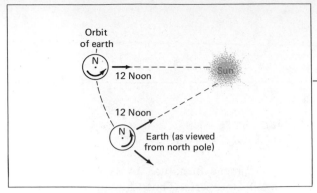

Figure 1-3c *Diagram illustrating one apparent solar day. One apparent solar day is the elapsed time between successive crossings of a meridian by the sun.*

The standard unit of time in both metric systems is the second. The word "second" was originated by the early Babylonian astronomers, whose observations were accurate enough to predict future eclipses. The Babylonians are also given credit for dividing the year into 12 months, the week into seven days, the day into 24 hours, the hour into 60 minutes, and the minute into 60 seconds.

The **second** was defined prior to 1956 as a fractional part (1/86,400) of the mean solar day. The elapsed time between two consecutive crossings of a meridian by the sun is called one apparent solar day. (A meridian is an imaginary line running from the earth's north geographic pole perpendicular to and through the equator to the south geographic pole.) Such a meridian, of course, passes through Washington, D.C. When the sun appears on the Washington meridian it will be 12 noon local solar time. When the next meridian crossing of the sun occurs at Washington, one apparent solar day will have elapsed. See Fig. 1-3c. The term "apparent" is used because the sun crossing the meridian is only apparent, not real. The earth's rotation produces the apparent movement of the sun. Since the earth's orbital velocity varies, the apparent solar days vary throughout the year. The mean solar day is the average time of all the apparent days computed over a period of one year.

The second was defined as 1/31,556,925.9747 of the 1900 tropical year in 1956. (The tropical year is defined as the time interval from one spring equinox to the next spring equinox.) Today we use an atomic definition of the second. The 12th General Conference of Weights and Measures, meeting in 1964 in Paris, adopted a definition of the second that is based on a certain frequency generated by the cesium-133 atom. (See Section 18-3 for a complete definition of the second.) The new standard is a constant of nature which can be reproduced with great accuracy in most laboratories throughout the world in a few minutes. The accuracy of time measurements with the new standard can be made to one part in 100 billion.

There are many units of measurement used in physical science. To make comprehension and exchange of ideas among people of different nations as simple as possible, an International System of Units (SI) has been established. The SI is an extension of the metric system and contains seven base units: the meter (m), the kilogram (kg), the second (s); the ampere (A), to measure rate of flow of electric charge; the kelvin (K), to measure temperature; the mole (mol), for measuring the amount of substance; and the candela (cd), to measure luminous intensity. A definition of each of these seven units is given in Appendix I. All other SI units are derived from the base units and *two* supplementary units (radian and steradian: a measure of plane and solid angle, respectively) by means of the established scientific laws relating the respective physical quantities.

An extensive search and study has been made concerning the origin and development of units. There still exists a great amount of uncertainty relating to the origin of many units, but the evidence collected by archaeologists, which has been the primary source of information, supports many of the views we have today. A detailed account of the origin and development of all or even several units of measurement cannot be given in this text, but a few of the more common units of length, mass, and time will be discussed briefly.

There is little doubt that the first units of length used by primitive people were the limbs of the human body. The units of weight most likely originated from the use of stones, grain, shells, or other objects found in the environment. The units of time originated from the measurement of the passage of the sun as it apparently crossed the sky from the eastern to the western horizon.

The foot, which was used by most early civilizations, was brought to Britain by the Romans, where it was modified and made a standard unit of length in the thirteenth century by King Edward I. Some historians believe he decreed that the yard should be defined as the distance from the tip of his nose to the tip of his index finger with his arm extended, the foot to be $\frac{1}{3}$ yard, and the inch to be $\frac{1}{36}$ yard. The decree was later legalized by the British parliament. The inch was originally the width of the human thumb. The Romans defined the inch as $\frac{1}{12}$ foot and introduced the term into England during the Roman occupation.

The land mile seems to have originated with the Romans and was defined as 1000 paces, or double steps. The pace, measured from the heel of one foot to the heel of the same foot in the next stride, was equal to 5 Roman feet; hence, the mile equaled 5000 feet. After being introduced into Britain by the Romans, the mile was changed to 5280 feet by a statute of Queen Elizabeth I (1533–1603) in order to make the furlong ($\frac{1}{8}$ mile) equal to 220 yards.

The units of weight—including the grain, stone, pound, and ton—seem to have originated from the weight of a certain volume of some particular item. The Roman pound, for example, was $\frac{1}{100}$ talent, which seems to have been the weight of an Egyptian cubic foot of water. There is strong evidence that the grain originated from the weight of a grain of wheat or some other grain product. For some reason, the pound and the grain were further divided into 16 and 12 equal parts, respectively, by the Romans, Greeks, and British. Equivalent measures for some of the more common metric and English units of measure are given in Appendix II.

1-4
DERIVED UNITS AND CONVERSION FACTORS

Most phenomena in nature involve more than one fundamental property. Thus, they are measured by combinations of the standard units. These combinations of standard units are called **derived units.** For example, the speed of an automobile may be measured in miles per hour (mi/h or mph), meters per sec (m/s), centimeters per second (cm/s), feet per second (ft/s), or any other combination of length per unit of time. All of these combinations of units are derived units. Some other simple examples are as follows:

Property	Derived Units
Area (length^2)	m^2, cm^2, ft^2, etc.
Volume (length^3)	m^3, cm^3, ft^3, etc.
Density (mass per volume)	kg/m^3, g/cm^3

The density of an object refers to how compacted a substance is with matter. More precisely, **density** is the mass per unit volume of an object and its formula can be written

$$\text{Density} = \frac{m}{V}$$

where m = mass of a substance,
V = volume of the substance.

Thus, something with a mass of 20 kg which occupies a volume of 5 m^3 would have a density of 20 kg/5 m^3 = 4 kg/m^3. In this example the mass was measured in kilograms and the volume in cubic meters. If mass is uniformly distributed throughout the volume, then the density of the matter will remain constant. Figure 1-4a shows that if you have a uniform substance, such as water, the density remains the same no matter how much of the substance is in question.

Figure 1-4a *Equal densities. Whether you have one, two, or three liters of water, having masses of one, two, or three kilograms, the density is the same. In this example, the density of water in each of the three containers is 1 kg/liter.*

The density of water is $1\ \text{g/cm}^3$ or 1 kg/liter. (Note that a liter is 1.056 quarts.) This is not a coincidence. The gram was originally defined as the mass of one cubic centimeter of water. If density is expressed in units of grams per cubic centimeter, it is easy to compare any density to that of water. For example, a rock might have a density of 3.3 g/cm^3, pure iron has a density of 7.86 g/cm^3, and the earth as a whole has a density of 5.5 g/cm^3.

The units on both sides of an equation must always be similar. For example, we cannot equate two quantities with fundamentally different units. We cannot equate something with units of velocity (e.g., centimeters per second) to something with units of area (e.g., cubic centimeters).

When a combination of units gets very complicated, we frequently give it a name of its own. Consider the following examples which will be discussed in later chapters:

$$\text{joule} = \text{kg} \times \text{m}^2/\text{s}^2$$
$$\text{newton} = \text{kg} \times \text{m}/\text{s}^2$$
$$\text{dyne} = \text{g} \times \text{cm}/\text{s}^2$$

$$\text{pound} = \text{slug} \times \text{ft}/\text{s}^2$$
$$\text{watt} = \text{kg} \times \text{m}^2/\text{s}^3$$

There are many more, but the point is that it is easi to talk about pounds than slug $\times$ ft/s^2. When a pa ticular combination of units gets to be important, it usually given its own name.

Frequently we want to convert from one syste of units to another in order to make comparisons. Fo instance, we frequently want to make comparison between the metric and the British systems. Many these conversion factors are listed in Appendix II. Fo instance,

$$1\ \text{in} = 2.54\ \text{cm}$$

This conversion factor is frequently used. For exam ple, if you want to know your height in centimeters and you are 5 ft 11 in tall, or 71 in, your height in centimeters is given by (see Fig. 1-4b)

$$71\ \text{in} = 71\ \text{in} \times 2.54\ \frac{\text{cm}}{\text{in}} = 180.3\ \text{cm}$$

Figure 1-4b *A person who is 5 ft 11 in tall is about 180.3 cm tall. (Photo courtesy James Crouse.)*

Figure 1-4c *A speedometer showing both miles per hour and feet per second. 60 mi/h is 88 ft/s.*

In the metric system your height would be 180.3 cm. If you ask someone what his or her height is and the reply is a number such as 165 or 180 or 190, this would imply so many centimeters. To convert these numbers to inches, you would have to divide by 2.54.

A similar exercise must be done when referring to weight or mass. The appropriate conversion factor is

$$1 \text{ kg mass} = 2.2 \text{ lb weight}$$

To find your mass in kilograms, simply divide your weight by 2.2 (actually 2.2 lb/kg). For instance, if you weigh 132 lb, then your mass in kilograms is

$$132 \text{ lb} \times \frac{1 \text{ kg}}{2.2 \text{ lb}} = 60 \text{ kg}$$

If a person's mass is known in kilograms, his equivalent weight on earth can be found by multiplying by 2.2 (actually 2.2 lb/kg). For example, a mass of 70 kg would be an equivalent weight of

$$70 \text{ kg} \times 2.2 \frac{\text{lb}}{\text{kg}} = 154 \text{ lb}$$

Often we are concerned with converting from miles per hour to feet per second. The conversion factor necessary to do this is

$$1 \frac{\text{mi}}{\text{h}} = 1.47 \frac{\text{ft}}{\text{s}}$$

To convert miles per hour to feet per second we simply multiply by 1.47 (actually $\frac{22}{15}$ ft/s ÷ mi/h). As an example, 60 mi/h can be converted as follows (see Fig. 1–4c):

$$60 \frac{\text{mi}}{\text{h}} \times 1.47 \frac{\text{ft/s}}{\text{mi/h}} = 88 \frac{\text{ft}}{\text{s}}$$

In Table 1-4a we summarize how to convert the quantities that we have discussed.

Table 1-4a *Conversion of Some Common Units*

To convert from inches to centimeters multiply by 2.54.
To convert from centimeters to inches divide by 2.54.
To convert from kilograms to pounds multiply by 2.2.
To convert from pounds to kilograms divided by 2.2.
To convert from miles per hour to feet per second multiply by 1.47 ($\frac{22}{15}$).
To convert from feet per second to miles per hour divide by 1.47 ($\frac{22}{15}$).

1-5
UNITS FOR A CIRCLE

The units of time, as mentioned earlier, probably originated with the Babylonians, who reckoned the year as 360 days. They were aware that, in any circle, a chord that is equal to the radius subtends an arc equal to 60 degrees (usually designated 60°), as shown in Fig. 1-5a.

This interesting property may have been the basis of their sexagesimal number system and perhaps accounts for the division of the **degree** into 60

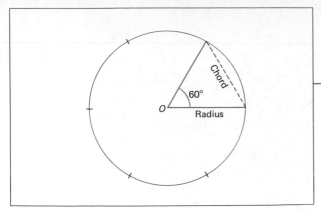

Figure 1-5a *A chord that is equal to the radius subtends an angle of 60°.*

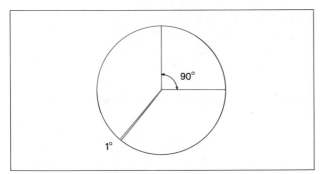

Figure 1-5b *A circle is divided into 360°. A right angle has 90°. Each degree is divided into 60 minutes of angle (designated 60'), and each minute of angle is divided into 60 seconds of angle (designated 60''). Thus, there are 3600 seconds of angle in 1 degree of angle.*

minutes, and minutes into 60 seconds (see Fig. 1-5b). Since the Babylonians used the apparent motion of the sun to tell the passing of daylight hours, their sexagesimal system may also be the basis for our present method for reckoning time, that is, 60 seconds equal 1 minute and 60 minutes equal 1 hour.

The early Greeks were extremely interested in geometry and the properties of circles. They were particularly interested in the ratio of the circumference of a circle to its diameter. This ratio they designated as π (pi). We can write **π** as

$$\pi = \frac{\text{Circumference of a circle}}{\text{Diameter of a circle}} \qquad \text{Eq. 1.1}$$

This ratio is always the same for every circle. The value of π is given by

$$\pi = 3.14159 \ldots \qquad \text{Eq. 1.2}$$

and is usually rounded off to 3.14 or $\frac{22}{7}$. From the definition of π, we find that the circumference of a circle is just $\pi \times$ diameter. But, since the diameter is twice the radius, we get

$$\text{Circumference} = 2\pi r \qquad \text{Eq. 1.3}$$

where $r =$ radius of the circle.

■ *Example* The earth goes around the sun in approximately a circular orbit with a radius of 93 million miles. What is the circumference of the earth's orbit?

□ *Solution* The circumference is given by

$$\text{Circumference} = 2 \times 3.14 \times 93,000,000 \text{ mi}$$
$$= 584,040,000 \text{ mi}$$

We see, then, that each year the earth travels over 500 million miles in its journey around the sun.

The quantity π is also used for angular measurement. If we divide Eq. 1.3 by r, the radius, we get

$$\frac{\text{Circumference}}{\text{Radius}} = 2\pi \qquad \text{Eq. 1.4}$$

We say that there are 2π radians in a circle. One **radian** is thus defined as the angle subtended at the center of any circle by an arc equal in length to the radius of that circle. Figure 1-5c shows the relation between the angle θ, the arc s, and the radius r of the circle.

Since 2π radians equals 360°, we get that 1 radian = 57.4°. Working with radians rather than degrees greatly simplifies many scientific formulas. For instance, if we use an object such as the earth moving in a circular path, we get the formula

$$\theta = \frac{s}{r} \qquad \text{Eq. 1.5}$$

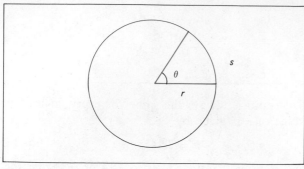

Figure 1-5c *the angle θ equals one radian when its arc s equals the radius r.*

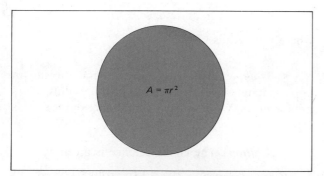

$A = \pi r^2$

Figure 1-5d *The area of a circle is given by $A = \pi r^2$.*

where θ = angle measured in radians,
  s = arc measured in length units,
  r = radius measured in length units,

when s equals the full circumference ($2\pi r$), we get

$$\theta = \frac{2\pi r}{r}$$

or

$$\theta = 2\pi$$

so we see that the total angle in one full circle is 2π radians.

The quantity π is also important when finding the area of a circle. The area of a circle is given by (see Fig. 1-5d)

$$A = \pi r^2$$

where A = area of the circle,
  r = radius of the circle.

1-6
EXPERIMENTAL ERROR

Making measurements of physical quantities requires the use of some device that will give a number denoting the ratio of the observed quantity to one of the standard units. A **measurement** is a comparison of the physical quantity with the standard unit. The process of taking any measurement by any known means always involves some uncertainty. This uncertainty is usually called **experimental error.** In taking measurements, an effort is made to keep the experimental error to a minimum.

Experimental errors are classified as either systematic or random. Systematic errors are always in the same direction; that is, the magnitude, or size, of the number obtained is always too small or too large, as with a watch that runs too fast or too slow, or a speedometer that shows too many or too few miles per hour. The observer may also be the cause of systematic errors, because of bad vision or other difficulties with his senses. Random errors, on the other hand, are in either direction and result from accidental variations of the observed physical quantity or of the measuring instruments. Random errors are caused by small variations in any direct or indirect physical quantity associated with the measurement. They can be the result of many causes such as temperature or pressure variations. An example of systematic experimental error is shown in Fig. 1-6a. Not all such errors are so apparent, however.

When random errors are small, the measurement is said to have high precision. **Precision** refers to the degree of reproducibility of a measurement, that is, to the maximum possible error of the measurement and may be expressed as a plus or minus correction. For example, if the length of the sample is expressed as 44.4 cm $\pm$ 0.1 cm (read plus or minus $\frac{1}{10}$ of 1 centimeter), we know the sample is somewhere between 44.3 and 44.5 cm long.

A measurement having high precision does not

Figure 1-6a *If the scientist does not realize that the zero point of his measuring instrument is incorrect, he will be making a systematic error. (Photo courtesy James Crouse.)*

necessarily have high accuracy. For example, one could determine the boiling point of a liquid with a precision of $\pm 0.01°C$ (read plus or minus $\frac{1}{100}$ of 1 degree Celsius), but any impurity in the liquid would prevent the accurate determination of the true boiling point of the liquid.

With experimental error always appearing in our measurements, how close an approximation can we obtain in making a measurement of any physical quantity; that is, how close can we come to the true value? The term **accuracy** refers to how close the measurement comes to the true value. What, then, is the true value? The true value is the value currently accepted by the best scientists in the field. This value is subject to change as measurements are made with better methods and better instruments.

It is obvious from the preceding discussion that the terms precision and accuracy should never be used interchangeably. They have different meanings and should be used only where they apply. The manufacturer of a measuring instrument states the precision of the product, that is, he states in the catalog the size of the maximum error. For example, he advertises a thermometer that will measure to the nearest hundredth of a degree Celsius. Thus, one can measure with a precision of ± 0.01 degree Celsius with the thermometer. Suppose an experimenter wants to determine the boiling point of pure water to the nearest 0.01 degree Celsius with this thermometer. If the experimenter has pure water, he can perform the experiment and obtain an accurate answer. If, however, the water is not pure, he cannot make an "accurate" measurement of the boiling point, even though he uses a high "precision" thermometer.

Since error is impossible to eliminate, methods have been devised to calculate the amount of error that exists in a given measurement. Calculation of the amount of error is done by either one of two methods, depending on the circumstances. If there is an accepted or true value of the physical quantity, then a calculation known as percentage error is made. **Percentage error** is defined as the ratio of the absolute difference* between the experimental and the accepted values to the accepted value, expressed as a percentage. This can be written as

Percentage error

$$= \frac{|\text{Exp. value} - \text{Accepted value}|}{\text{Accepted value}} \times 100$$

If there is no accepted value, then a percentage difference is obtained. **Percentage difference** is defined as the ratio of the absolute difference between

---

*"Absolute difference" is the result obtained when the smaller value is subtracted from the larger. It is designated by two vertical lines enclosing the subtraction.

the experimental values to an average of the experimental values, expressed as a percentage. This can be written as

Percentage difference

$$= \frac{|\text{1st exp. value} - \text{2nd exp. value}|}{\text{Average of the two values}} \times 100$$

A more useful value of measurement in experimental work is often the average value. For example, in finding the length of a table it would be better to take three separate measurements, find the sum of the three measurements, and then divide by three to obtain the average. When three or more measurements are made, the percentage difference is computed, using the average value and the measured values farthest from the average.

A better understanding of errors will be obtained when measurements are taken on real objects and the accuracy of the measurements determined. Generally, this kind of activity occurs in the laboratory.

1-7
POWERS OF 10 NOTATION

In physical science many numbers are very big or very small. In order to express very big or very small numbers, the **powers of 10** notation is frequently used. When the number 10 is squared or cubed we get

$$10^2 = 10 \times 10 = 100$$

$$10^3 = 10 \times 10 \times 10 = 1000$$

It is easy to see that the number of zeroes is just equal to the power of 10. As an example, 10^{23} is a one followed by 23 zeroes.

Negative powers of 10 can also be used. For example,

$$10^{-1} = \frac{1}{10} = 0.1$$

$$10^{-2} = \frac{1}{10^2} = \frac{1}{100} = 0.01$$

From this we see that if a number has a negative exponent we shift the decimal place to the left once for each power of 10. Thus, one micrometer, which is 10^{-6} m, equals 0.000001 m.

We can also multiply numbers by powers of 10. Table 1-7a shows a wealth of examples of various large and small numbers expressed in powers of 10 notation.

Table 1-7a *Numbers Expressed in Powers of 10 Notation*

Number	Powers of 10 Notation
0.025	2.5×10^{-2}
0.0000408	4.08×10^{-5}
0.0000001	1×10^{-7}
0.0000000000000000000016	1.6×10^{-19}
247	2.47×10^2
186,000	1.86×10^5
4,705,000	4.705×10^6
9,000,000,000	9×10^9
30,000,000,000	3×10^{10}
602,300,000,000,000,000,000,000	6.023×10^{23}

There are many standard prefixes which are used to represent powers of 10. These are listed in Table 1-7b. These prefixes are frequently used in expressing various units. The most important are **mega, kilo, milli,** and **micro.** Some examples are

1 *mega*ton = 10^6 tons
1 *kilo*meter = 10^3 meters
1 *milli*gram = 10^{-3} grams
1 *micro*second = 10^{-6} seconds

Table 1-7b *Prefixes Representing Powers of 10*

Multiple	Name	Abbreviation
10^{12}	Tera	T
10^9	Giga	G
10^6	Mega	M
10^3	Kilo	k
10^2	Hecto	h
10	Deka	da
10^{-1}	Deci	d
10^{-2}	Centi	c
10^{-3}	Milli	m
10^{-6}	Micro	μ
10^{-9}	Nano	n
10^{-12}	Pico	p
10^{-15}	Femto	f
10^{-18}	Atto	a

It is possible to represent a number in powers of 10 notation many different ways—all correct. For example, the distance from the earth to the sun is 93 million miles. This can be represented as

$$93,000,000 \text{ miles}$$

or 93×10^6 miles

or 9.3×10^7 miles

or 0.93×10^8 miles etc.

Scientists generally pick the power of 10 so that it is multiplied by a number between 1 and 10. Thus 9.3×10^7 miles would probably be chosen. However, any of the above representations of 93 million miles is correct.

Knowing how to express a measured quantity by using powers of 10 notation is not sufficient for solving problems in the physical sciences. One must know how to use the notation in the simple operations of addition, subtraction, multiplication, and division. See Appendix IV for rules and examples.

On the inside back cover of the book many examples of various powers of 10 involving length are shown. Along the horizontal axis very small things are represented in units of meters. The diameter of a proton is about 10^{-15} meters. This is close to the smallest thing that can be measured. Along the vertical axis some of the very large astronomical distances are represented. The distance to the farthest galaxy is approximately 10^{25} meters, and we may discover galaxies even further away. A close scrutiny of this chart will illustrate the extremely broad range of length which exists in our physical world. These tremendous differences of length are best expressed in powers of 10. This serves to illustrate another reason why powers of 10 are important in physical science.

IMPORTANT WORDS AND TERMS

Length	Kilogram	Experimental error
Time	Second	Precision
Mass	SI	Accuracy
Electric charge	Foot	Percentage error
Fundamental properties	Derived units	Percentage difference
Standard unit	Density	Powers of 10
British system	Degree	Mega
mks system	π	Kilo
cgs system	Radian	Milli
Meter	Measurement	Micro

LEARNING OBJECTIVES

After reading and studying this chapter, you should be able to do the following without referring to the text:

1. State four fundamental properties of nature.
2. State the units in which these fundamental properties are measured in three different systems of measurements.
3. Explain the concept of density and tell how it is different from mass.
4. State how the gram, meter, and second were originally defined.
5. Be able to convert from centimeters to inches, from pounds to kilograms, from miles per hour to feet per second or vice versa using Table 1-4a.
6. State the origin of π.
7. State the formulas for the circumference and area of a circle if the radius is known.
8. Distinguish between percentage error and percentage difference.
9. Express any number in powers of ten notation.
10. State the definitions of the prefixes mega, kilo, milli, and micro.

QUESTIONS

1. Why are measurements important?
2. Why are the senses important to measurement?
3. What are the four fundamental properties we seek to measure?
4. Would time exist if there were no motion?
5. What is space-time?
6. Is mass the same as weight? Explain.
7. What are the advantages of the metric system of measurement over the British system?
8. What is the origin of the (a) meter, (b) kilogram, (c) second?
9. Why have new standards of length and time been recently adopted?
10. Which is more dense, a kilogram of iron or a kilogram of feathers? Which has more mass?
11. What is the origin of π?
12. Explain the difference between precision and accuracy of a measurement.
13. Explain the difference between percentage error and percentage difference.
14. Why are powers of 10 used?

PROBLEMS

1. An astronaut has a mass of 60 kg.
   (a) What is his mass on the earth and on the moon?
   (b) What is his weight on the earth and on the moon in pounds?

2. Compute the height in feet and inches of a person who is 200 cm tall.

Answer: 6 ft 7 in

3. Compute the height in centimeters of a person who is 5 ft 4 in tall.

4. Compute the mass in kilograms of a person who weighs 150 lb.

Answer: 68.2 kg

5. Compute the speed in feet per second of an auto traveling at 40 mi/h.

6. Compute the speed in miles per hour of an auto traveling at 90 ft/s.

7. The moon is about 240,000 miles from the earth. Assuming the moon travels in a circle around the earth (this assumption is not quite correct),
   (a) What is the circumference of its path around the earth?
   (b) What is the area of this circle in square miles?
   (c) Express your answer to (b) in powers of 10 notation.

Answer: (c) 1.81×10^{11} mi^2

8. When Eratosthenes measured the circumference of the earth over 2000 years ago, he got an answer of 24,000 miles. What was his percentage error if the correct answer is 24,900 miles?

9. In powers of 10 notation give the following numbers:
   (a) 24 megawatts
   (b) 16 kilotons
   (c) 31 micrograms
   (d) 0.1 millimeters

Answer: (c) 31×10^{-6} g
or
3.1×10^{-5} g

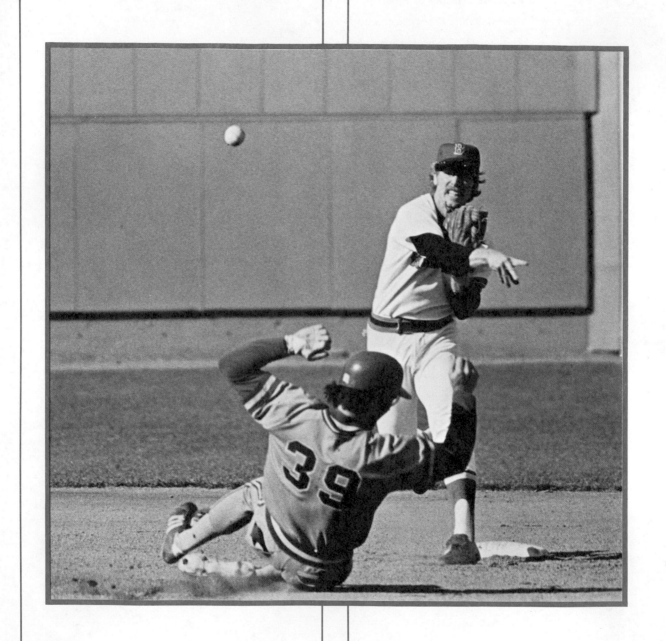

Chapter 2
KINEMATICS: THE STUDY OF MOTION

A physical law has been defined as a description of a certain behavior of nature extending beyond what has been observed. For example, the physicist is familiar with a law which states that an object at rest remains at rest ("at rest" means the object is not changing its position) unless acted upon by some outside force. The law seems simple enough, but the fact is that nowhere can we observe an object which is *truly* at rest. We can extend this statement by saying that we cannot mentally produce an object anywhere in the known universe that is truly at rest. True, we may think of an object at rest relative to another object, such as the Washington Monument, which is at rest relative to Washington, D.C.; but after a moment's thought we realize that the monument is moving because the earth on which it is resting is moving.

Motion is everywhere. We walk to class. We drive to the supermarket. Birds fly. The wind blows the trees. The clouds move. The rivers flow. Even the continents drift. The earth rotates and revolves. The sun and other stars are moving. The world in which we live is continually changing in all its physical aspects.

The basic discussion in this chapter revolves about the formulation of concepts and also about the terms used to describe the motion of objects observed in our environment. A study of motion in this manner (one that neglects the forces producing the motion) is called **kinematics.** We begin our study of kinematics

. . . *Scientific truth should be presented in different forms, and should be regarded as equally scientific whether it appears in the robust form and the vivid colouring of a physical illustration, or in the tenuity and paleness of a symbolical expression. . .*

James Clerk Maxwell

Figure 2-1a *Uniform rectilinear motion. Equal distances are traveled in equal periods of time by the dry ice puck.*

with a definition of the terms *position* and *motion*, and then develop the concepts of *speed*, *velocity*, and *acceleration*, which are used throughout the text to interpret the endless motion of all objects.

2-1
UNIFORM MOTION IN A STRAIGHT LINE

The term **position** refers to the location of an object.

To give the position of an object, another object or reference point must also be stated or implied.

The book is on the table. Cleveland is in Ohio. The window is there. In all these cases, a reference point is stated or implied. If an object located at one position is later found at another position, we say the object has moved, or changed positions.

When an object is undergoing a continuous change in position, we say the object is moving, or is in motion.

The term "continuous" should be stressed here, because to have motion the object must always be changing position.

As just stated, motion indicates a continuous change in position of an object. Since position is relative, motion also must be relative. The statement, "The student is walking 4 feet per second," would indicate that the student is changing position at the rate of 4 feet for each second of time, relative to the sidewalk, ground, or floor. Similarly, if we say, "The car is traveling at the rate of 40 miles per hour," we are using the road as the frame of reference for the car's motion.

The study of motion becomes complex if we try to describe too many of the different motions an object may be undergoing at any particular moment. Therefore, our concern in this section will be with rectilinear motion. Uniform rectilinear motion is motion in a straight line, with the visible parts of the

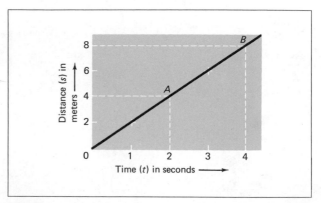

Figure 2-1b *Graph of data from Table 2-1a. Plot shows uniform rectilinear motion.*

object moving together, parallel to the reference frame.

An example of uniform rectilinear motion is that of a dry ice puck sliding across the flat surface of a table, where the table top is the reference frame. This motion is described using the fundamental units length and time. Suppose a dry ice puck moves across a table so that it travels two meters for each second of time. This can all be illustrated by a diagram, as shown in Fig. 2-1a and a graph, as shown in Fig. 2-1b which represents the data in Table 2-1a.

In Fig. 2-1b, the vertical axis represents the distance traveled in meters, and the horizontal axis represents the elapsed time in seconds. An examination of the graph will reveal that any distance-to-time relationship will be a constant value. For example, Points *A* and *B* on the graph represent a distance of 4 meters, and the time associated with the distance is

Figure 2-2a *Although the speed of light is 186,000 miles per second, it still takes over 8 minutes for light from the sun to reach us.*

Table 2-1a *Dry Ice Puck Data*

Distance (s) in meters	Time (t) in seconds
0	0
2	1
4	2
6	3
8	4

2 seconds. Taking the ratio of distance to time, we have

$$\frac{4 \text{ meters}}{2 \text{ seconds}} = 2 \frac{\text{meters}}{\text{second}}$$

2-2
SPEED AND VELOCITY

The terms speed and velocity are often used interchangeably by nonscientists. In physical science, however, the words speed and velocity have distinct meanings. The **average speed** of an object is given by the total distance traveled divided by the time it took to travel that distance. This can be written as

$$\text{Average speed} = \frac{\text{Total distance traveled}}{\text{Time to travel total distance}}$$

$$\text{Average speed} = \frac{d}{t} \qquad \text{Eq. 2.1}$$

where d = total distance traveled,
t = time taken.

As an example, consider the ice puck in Fig. 2-1a. It traveled 8 m in 4 s, so its average speed was 2 m/s. We can also rearrange Eq. 2.1 to read

$$t = \frac{d}{\text{average speed}}$$

For instance, if we know that the speed of light is 186,000 miles per second, we can calculate the time it takes for light from the sun to reach us using Eq. 2.2.

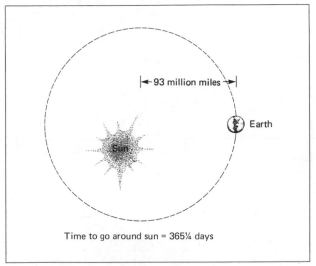

Figure 2-2b *The earth is 93 million miles from the sun. The total distance traveled around the sun is $2\pi \times 93$ million miles. It takes one year or $365\frac{1}{4}$ days to complete one revolution.*

Since the sun is 93 million miles away we get

$$t = \frac{93,000,000 \text{ mi}}{186,000 \text{ mi/s}}$$

$$t = 500 \text{ s}$$

$$t = 8.3 \text{ min}$$

From this we realize that although light travels very fast, it still takes 8.3 min for it to arrive at the earth after it leaves the sun. This is illustrated in Fig. 2-2a.

Referring back to Eq. 2.1, let us calculate the speed of the earth as it orbits the sun. We refer to Fig. 2-2b.

The speed of the earth on its journey around the sun can be computed by calculating the total number of miles traveled in one year. This is just the circumference of the circle in which it moves. (We assume it moves in a perfect circle. This is a good approximation.) We recall that the circumference of a circle is $2\pi r$, where r is the radius which is 93 million miles in our example. The time taken to travel this distance is one year or 365.25 days. Since there are 24 hours in a day, the total number of hours in one year is 365.25×24 hours. We can summarize as follows:

Distance traveled
in one year
$$= 2\pi \times 93{,}000{,}000 \text{ miles}$$
$$= 2 \times 3.14 \times 93{,}000{,}000 \text{ miles}$$
$$= 584{,}000{,}000 \text{ miles}$$
$$= 584 \times 10^6 \text{ miles}$$

$$\text{Time} = 365.25 \text{ days} \times 24 \text{ hours per day}$$
$$= 8766 \text{ hours}$$
$$= 8.766 \times 10^3 \text{ hours}$$

We can calculate the average speed of the earth around the sun from Eq. 2.1 as follows:

$$\text{Average speed} = \frac{584 \times 10^6 \text{ mi}}{8.766 \times 10^3 \text{ h}}$$
$$= 66.6 \times 10^3 \text{ mi/h}$$
$$= 66{,}600 \text{ mi/h}$$

This is a fantastic result! It says that we are hurtling through space at a speed of 66,600 miles per hour. Can you believe that? All of us are moving at an exceedingly high speed, and yet it doesn't seem to bother us. We will return to this question in Section 2-5.

One of the powerful aspects of physical science is that it is *quantitative.* Almost everyone is aware that the earth is going around the sun. This is a qualitative fact. The fact that we can calculate the speed of this motion as being 66,600 miles per hour quantifies our knowledge and gives a spectacular result. Thus, we begin to see the power and importance of going be-

Figure 2-2c *The speed indicated by the speedometer of an automobile is an example of instantaneous speed: the speed at that instant.* (*Courtesy J. D. Wilson,* Physics Concepts and Applications, *Heath, Lexington, Mass., 1977.*)

yond qualitative statements regarding our physical environment.

So far in our discussion of motion we have been discussing the concept of average speed. The speed at any instant of time may be different from the average speed. The **instantaneous speed** of an object is the speed of the object at any instant in time. A familiar example of instantaneous speed is the speed registered on an automobile speedometer such as shown in Fig. 2-2c. It is the speed at which the automobile is traveling right then, at that instant of time.

Speed is a scalar quantity. A **scalar** quantity is one that has magnitude (including units) only, but no direction. A **vector** quantity is one which has both magnitude (including units) and direction. Velocity is similar to speed, but it has a direction associated with it. The direction involved is frequently indicated by using arrows.

A good example of the difference between speed and velocity is found in the measurement of wind. The wind speed would be 10 mi/h. The wind velocity would also include the direction of the wind. A wind velocity would be 10 mi/h from the west.

The **average velocity** is defined in terms of displacement. **Displacement** is the straight-line distance

Figure 2-2d *The difference between distance and displacement is shown in this example. The displacement is the straight-line distance represented by path A. the distance traveled between the two points P_1 and P_2 depends on the path taken. If you took path B or C the distance would be greater than the displacement.*

between two points in space. It is defined by an arrow pointing from the original to the final position. The distinction between distance and displacement is best understood by referring to Fig. 2-2d.

In Fig. 2-2d the displacement is represented by path *A*. The distance traveled would vary depending on which path was taken, but the displacement between the points P_1 and P_2 would remain the same.

The **average velocity** can be written as the displacement divided by the time.

$$\text{Average velocity} = \frac{\text{Displacement}}{\text{Time}}$$

or
$$\bar{v} = \frac{s}{t} \qquad \text{Eq. 2.3}$$

where $\bar{v}$ = average velocity,
s = displacement,
t = elapsed time.

The student should note that the symbol *s* indicates a displacement and not speed. The displacement is measured in units of length and Eq. 2.3 is very similar to Eq. 2.1.

When indicating a velocity, Eq. 2.3 gives only the magnitude. To be complete, the direction of the velocity should also be given. For instance, a car might be traveling at a velocity of 40 miles per hour north. The **instantaneous velocity** is just the velocity at any instant of time. If we look at a very short time

interval Δt, we can say that

$$v = \frac{\Delta s}{\Delta t} \qquad \text{Eq. 2.4}$$

where v = velocity,
Δs = change in displacement,
Δt = change in time.

In uniform motion the instantaneous velocity v and the average velocity $\bar{v}$ are the same. In accelerated motion, discussed in Sections 2-3 and 2-4, we will be referring to Eq. 2.4 as the definition of velocity.

In straight-line motion, speed and velocity are very similar. Their magnitudes are the same, and the only distinction between them is that a direction must be specified when discussing velocity.

If the straight-line motion is not uniform, we can write a formula for the average velocity in terms of the original velocity v_o and the final velocity v_f. The formula for the average velocity is

$$\bar{v} = \frac{v_o + v_f}{2} \qquad \text{Eq. 2.5}$$

where $\bar{v}$ = average velocity,
v_o = original velocity,
v_f = final velocity.

Equation 2.5 holds true only if there is a uniform change in the velocity from v_o to v_f. In Fig. 2-2e, we have an example of a car uniformly accelerating from 30 miles per hour to 50 miles per hour. We can compute the average velocity as

$$\bar{v} = \frac{v_o + v_f}{2}$$

$$v_o = 30 \text{ mi/h}$$
$$v_f = 50 \text{ mi/h}$$

$$\bar{v} = \frac{30 \text{ mi/h} + 50 \text{ mi/h}}{2}$$

$$\bar{v} = 40 \text{ mi/h}$$

Figure 2-2e *A car which uniformly accelerates from 30 mi/h to 50 mi/h has an average velocity of 40 mi/h.*

There is nothing difficult about this. In fact, we can almost say intuitively that if a car accelerates from 30 mi/h to 50 mi/h, the average velocity was 40 mi/h. Equation 2.5 helps us to understand how we obtained the answer.

2-3

A mass that is moving with increasing or decreasing velocity is said to be accelerating.

Acceleration is defined as the time rate of change of velocity.

This can be written as

$$\text{Accel.} = \frac{\text{Final velocity} - \text{Original velocity}}{\text{Time for change to occur}}$$

or as

$$a = \frac{v_f - v_o}{t} = \frac{\Delta v}{t} \qquad \text{Eq. 2.6}$$

where a = acceleration,
v_f = final velocity,
v_o = original velocity,
t = time for velocity change to occur,
$\Delta v = v_f - v_o$.

As an example of Eq. 2.6, let us consider an automobile that accelerates from 66 ft/s (45 mi/h) to 88 ft/s (60 mi/h) in 10 s. The acceleration would be found by using Eq. 2.6.

$$a = \frac{v_f - v_o}{t}$$

$$a = \frac{88 \text{ ft/s} - 66 \text{ ft/s}}{10 \text{ s}}$$

$$a = \frac{22 \text{ ft/s}}{10 \text{ s}}$$

$$a = 2.2 \text{ ft/s}^2$$

The units of acceleration, such as ft/s² in the British system, may be confusing at first. It should be kept in mind that an acceleration is a measure of a

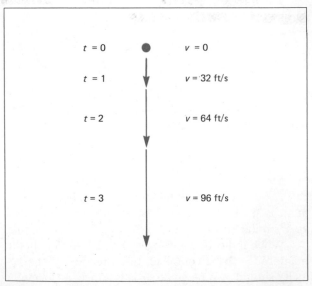

Figure 2-3a *For a constant acceleration of 32 ft/s², the velocity increases by 32 ft/s each second. The direction of the arrows indicates the direction of the velocity vector. The increasing length of the arrows indicates an increasing velocity.*

change in velocity during a given time period. Consider an acceleration of 32 ft/s². This means that the velocity increases by 32 ft/s each second. Thus, as the number of seconds increases, the velocity will go from 0 ft/s to 32 ft/s to 64 ft/s to 96 ft/s and so forth. This is illustrated in Fig. 2-3a.

We can also rewrite Eq. 2.6 to give a formula for the final velocity of an object if its original velocity and acceleration are known.

$$v_f - v_o = at$$

or $$v_f = v_o + at \qquad \text{Eq. 2.7}$$

This formula is useful for working problems in which the quantities a, v_o, and t are all known and we wish to find v_f. If the original velocity $v_o = 0$, then

$$v_f = at$$

Since velocity is a vector quantity, acceleration is also a vector quantity. In fact, we can represent each term in Eq. 2.7 as an arrow or vector. The length of the arrow represents the magnitude and the direction of the arrow represents the direction of the vector quantity.

Since t is a scalar quantity, the quantity at is a vector. Various illustrations of Eq. 2.7 are shown in Fig. 2-3b with positive, negative, and other types of acceleration.

In Fig. 2-3b, part (a) we have an illustration similar to Fig. 2-3a. Here both v_o and at are in the same direction and because of this, v_f is longer than v_o. Part (a) might represent a ball falling to the earth. In part (b) of Fig. 2-3b the acceleration vector points in the opposite direction of the velocity and serves to decelerate the object. Part (b) might represent a ball on its way up. Its velocity decreases because gravity tends to slow it down. Part (c) shows that even though the speed (or magnitude of the velocity) is a constant, an acceleration can change the direction of the velocity. To put it another way, an object moving at constant speed will be accelerating if its direction is changing. An example of this situation could be an object going in a circle at constant speed.

Figure 2-3b The equation $v_f = v_o + at$ is illustrated. The arrows v_o and at combine together to give the resultant arrow v_f. Note that v_f can be longer, shorter, or the same length as v_o.

2-4
ACCELERATION DUE TO GRAVITY

The great Italian physicist Galileo Galilei (1564–1642) was the first scientist to show that all objects will fall, that is, accelerate downward, at the same rate. Of course, this assumes that frictional effects are negligible. We can state this discovery of Galileo as follows:

If frictional effects can be disregarded, every freely falling object near the earth's surface accelerates downward at the same rate, regardless of the mass of the object.

Figure 2-4a *Astronaut Scott demonstrated that a feather and hammer fall at the same rate on the moon. There is no air on the moon to slow down the feather, so it accelerates at the same rate as the hammer.*

This can be easily demonstrated by taking a flat piece of tissue paper and a heavy object such as a book and dropping them at the same time. Air friction will prevent the tissue paper from falling as fast as the book. The correctness of Galileo's assertion can be seen by wadding up the tissue paper into a very small ball. This decreases the frictional effect to a negligible amount, and the wad of tissue will fall at the same rate as the book.

Another way to decrease the effects of friction is to perform an experiment on the moon. This was done by astronauts and is shown in Fig. 2-4a. A feather and hammer were dropped simultaneously. Since there is no atmosphere on the moon, the feather and hammer hit the moon's surface at the same time. This shows that Galileo's assertion applies on the moon as well as on the earth. Of course, on the moon all objects fall at a slower rate than do objects on the earth's surface. This is because the acceleration of gravity is less on the moon than on the earth.

The **acceleration of gravity** at the earth's surface is denoted by the letter g. Its value is

$$g = 32 \text{ ft/s}^2$$
$$g = 9.8 \text{ m/s}^2$$
$$g = 980 \text{ cm/s}^2$$

The acceleration of gravity varies slightly depending on such factors as how far you are from the equator and how high up you are. However, the variation is very small and for all practical purposes the acceleration due to gravity is constant everywhere on the earth's surface.

Figure 2-4b illustrates the motion of a freely falling object near the earth's surface. Each second the velocity increases by 32 ft/s. The average velocity during each second can be found using $\bar{v} = (v_o + v_f)/2$. Then the distance traveled during each second is found using $\Delta s = \bar{v}t$ and finally the total distance traveled is computed.

From Fig. 2-4b, we can see that the total distance s could have been computed using the formula

$$s = \tfrac{1}{2}gt^2$$

t	v	$\bar{v}$	Δs	s = total distance
0	0		0	0
1 s	32 ft/s	16 ft/s	16 ft	16 ft
2 s	64 ft/s	48 ft/s	48 ft	64 ft
3 s	96 ft/s	80 ft/s	80 ft	144 ft
4 s	128 ft/s	112 ft/s	112 ft	256 ft

Figure 2-4b *The velocity of a freely falling object at the end of each second. The average velocity $\bar{v}$ and the distance traveled are also computed from the formulas $\bar{v} = (v_o + v_f)/2$ and $s = \bar{v}t$.*

For instance, if $t = 3$ s, then s is given by

$$s = \frac{1}{2}\left(\frac{32 \text{ ft}}{\text{s}^2}\right)(3 \text{ s})^2$$

$$s = \frac{1}{2}\left(\frac{32 \text{ ft}}{\text{s}^2}\right)9 \text{ s}^2$$

$$s = 144 \text{ ft}$$

In Appendix V a general formula is derived showing that

$$s = v_o t + \tfrac{1}{2}at^2 \qquad\qquad \text{Eq. 2.8}$$

In our example the object was dropped so that the initial velocity $v_o = 0$. Then, since $a = g$ in our example, we obtain $s = \tfrac{1}{2}gt^2$.

Another formula is derived in Appendix V which states

$$v_f^2 = v_o^2 + 2as \qquad\qquad \text{Eq. 2.9}$$

This formula is useful if we want to know the velocity, but we don't know the time involved.

For instance, a major league baseball player once claimed that he could catch a baseball dropped from a 10-story building (100 ft). Let us calculate the velocity of the baseball and compare it to a major league fastball.

To do this calculation, we use $a = 32$ ft/s² and $s = 100$ ft. Since the baseball is dropped from rest, we have $v_o = 0$. Then we use Eq. 2.9 to obtain v_f.

$$v_f^2 = 0 + 2\left(32\,\frac{\text{ft}}{\text{s}^2}\right)(100 \text{ ft})$$

$$v_f^2 = 6400\,\frac{\text{ft}^2}{\text{s}^2}$$

so $\qquad\qquad v_f = 80\,\frac{\text{ft}}{\text{s}}$

This is not particularly fast since it corresponds to less than 60 mi/h. A good major league pitcher can throw a fastball nearly 100 mi/h. So the major league player who claimed he could catch a baseball dropped from 100 ft was correct.

2-5
ACCELERATION IN CIRCULAR MOTION

Whenever an object travels at a constant speed in a circle, its velocity vector is constantly changing direction. Because its velocity vector is changing, there is an acceleration. Consider a particle held by a string which is moving in a circle at constant speed. In Fig. 2-5a, the particle is at position P_1 and is directed to the right as shown. The velocity vector in Fig. 2-5a is drawn tangent to the circle at that point. If the string should break at this instant, the particle would fly off, tangent to the circle at that instant. Assuming the string holds, at some later instant the particle is at point P_2, moving with the same velocity v but in a direction tangent to the circle at point P_2. The two velocity vectors are drawn the same length to indicate that the velocity magnitude has remained the same. However, the velocity vectors have different directions and this indicates the *velocity* of the particle has changed in going from position P_1 to position P_2.

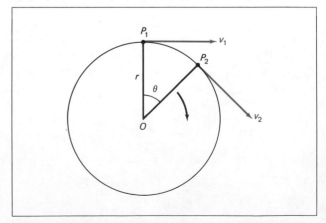

Figure 2-5a *The change in direction of a particle moving with constant speed in a circle. The direction of the velocity vector is continually changing.*

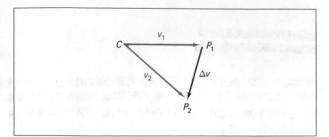

Figure 2-5b *Vector diagram for Fig. 2-5a. Δv represents the change in velocity due to a change in direction of v.*

Since the velocity of the particle has changed, it has therefore accelerated. What is the magnitude and direction of the acceleration? The answer to this question can be obtained from Fig. 2-5b. Lines CP_1 and CP_2 are drawn parallel to (and represent) velocities v_1 and v_2 in Fig. 2-5a at positions P_1 and P_2, respectively. Triangles $v_1 \Delta v v_2$ in Fig. 2-5b and $P_1 O P_2$ in Fig. 2-5a are similar, since they are isosceles triangles and have equal angles at C and O.

Therefore,
$$\frac{\Delta v}{v} = \frac{P_1 P_2}{OP_1}$$

and
$$\Delta v = \frac{P_1 P_2}{OP_1} v$$

If angle θ is very small, the distance from P_1 to P_2 is very small

or
$$\Delta v = \frac{s}{r} v$$

Dividing each side of this equation by t
$$\frac{\Delta v}{t} = \left(\frac{s}{t}\right)\frac{v}{r}$$

or
$$a = v \times \frac{v}{r}$$

Then,
$$a = \frac{v^2}{r} \qquad \text{(magnitude of acceleration)}$$

When the angle θ is very small the angle at C is very small, and the vector Δv is at right angles to the velocity v and is directed toward the center of the circle. Since the speed is constant, it is important to remember that the acceleration of a mass moving in a circle with uniform speed results from the changing direction of the mass.

As an example of acceleration of an object going in a circle, let us return to our example of the earth going around the sun. In Section 2-2, we calculated that the earth is revolving around the sun at a rate of 66,600 mi/h. This is equivalent to about 98,000 ft/s. The acceleration can then be found as follows:

$$a = \frac{v^2}{r}$$

$$a = \frac{(9.8 \times 10^4 \text{ ft/s})^2}{93 \times 10^6 \text{ mi} \times 5280 \text{ ft/mi}}$$

$$a = \frac{(9.8 \times 10^4 \text{ ft/s})^2}{4.91 \times 10^{11} \text{ ft}}$$

$$a = \frac{96.04 \times 10^8 \text{ ft}^2/\text{s}^2}{4.91 \times 10^{11} \text{ ft}}$$

$$a = 19.56 \times 10^{-3} \frac{\text{ft}}{\text{s}^2}$$

$$a = 0.01956 \frac{\text{ft}}{\text{s}^2}$$

Once again the power of quantitative calculations becomes evident. We see from the calculation that even though the earth's speed around the sun is 66,600 mi/h, the resultant acceleration is 0.02 ft/s² which is not even noticeable. For instance, compare this acceleration to g, the acceleration of gravity, which is 32 ft/s².

Another example we can consider is that of a car on a cloverleaf (see Fig. 2-5c). A good radius would

Figure 2-5c *An acceleration is necessary in order to go around a cloverleaf at a constant speed. (Photo courtesy James Crouse.)*

be 200 ft for the cloverleaf and a typical speed would be 30 ft/s, which is about 20 mi/h. The acceleration is then

$$a = \frac{v^2}{r}$$

$$= \frac{(30 \text{ ft/s})^2}{200 \text{ ft}}$$

$$= \frac{900 \text{ ft}^2/\text{s}^2}{200 \text{ ft}}$$

$$= 4.5 \frac{\text{ft}}{\text{s}^2}$$

This is not negligible compared to $g = 32 \text{ ft/s}^2$. In fact, we can feel the effects of this acceleration.

2-6
PROJECTILE MOTION

If a body is thrown out horizontally, it will follow the path shown in Fig. 2-6a. It can be seen from this photograph that an object which is thrown vertically will fall down at the same rate as an object which is dropped. We see from this that the horizontal and vertical directions are independent of each other.

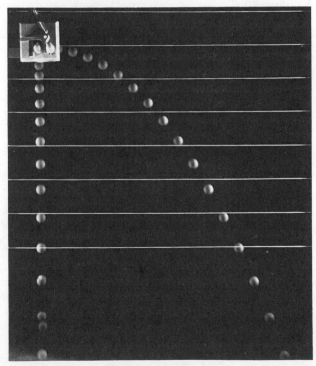

Figure 2-6a *A flash photograph of two golf balls, one projected horizontally at the same instant the other was dropped. The strings are 15 cm apart, and the interval between flashes was $\frac{1}{30}$ s. (From PSSC Physics, 3rd ed., Heath, Lexington, Mass., 1971.)*

That is, the velocity in the horizontal direction does not affect the velocity and acceleration in the vertical direction.

As an example, consider a person firing a rifle horizontally (see Fig. 2-6b). If a bullet is dropped simultaneously, the bullet fired from the rifle will hit the ground at the same time as the dropped bullet. Of course, this assumes that frictional effects are negligible and the ground is level. The bullets will hit simultaneously because the horizontal velocity of the fired bullet has no effect on its vertical motion. The vertical

Figure 2-6b *If a person fires a rifle horizontally and simultaneously drops a bullet, they will hit the ground at the same time.*

motion of the fired bullet is identical to that of the dropped bullet.

IMPORTANT WORDS AND TERMS

Kinematics *study of motion*
Position *location of an object*
Average speed
Instantaneous speed

Scalar
Vector
Displacement →
Average velocity →

Instantaneous velocity →
Acceleration *A in speed / A in time*
Acceleration of gravity
Motion *- an object undergoing a continuous change*

LEARNING OBJECTIVES

After reading and studying this chapter, you should be able to do the following without referring to the text:
1. Explain how you would find the speed of an object in terms of the distance traveled *Speed = dist / time* and the time.
2. Differentiate between scalars and vectors and give an example of each.
3. Differentiate between speed and velocity.
4. Differentiate between distance and displacement.
5. Differentiate between average velocity and instantaneous velocity and give an example of each.
6. Explain each term in the formula $v = \Delta s/\Delta t$.
7. Define the term "acceleration," and give an example referring to an automobile.
8. Give the units of acceleration in three different systems of units. *ft/s/s   m/h/h*
9. State Galileo's observation concerning freely falling objects.
10. Compute the speed of an object that has fallen for 1, 2, and 3 s ($g = 32$ ft/s^2).
11. Explain the meaning of each term in the equation $a = v^2/r$.
12. Give the mks units of each term in the equation $a = v^2/r$.
13. Explain how something can be traveling at a constant speed and still be accelerating.

QUESTIONS

1. Explain the difference between scalar and vector quantities.
2. Is velocity different from speed in rectilinear motion?
3. Is velocity different from speed in circular motion?

4. How is the earth moving in relation to the sun?
5. Is a speedometer reading a measurement of instantaneous speed, average speed, instantaneous velocity, or average velocity?
6. Can a car be moving at 30 mi/h and still be accelerating? Explain.
7. Can an object have an instantaneous velocity of zero and still be accelerating? Explain.
8. Can an object have an instantaneous velocity of 32 ft/s and simultaneously have an acceleration of 32 ft/s²? Explain.
9. Is the acceleration of gravity the same everywhere on the earth? Explain.
10. Is the earth accelerating as it goes around the sun?
11. What is the direction of the acceleration vector as the earth goes around the sun?

PROBLEMS

1. The moon is about 240,000 miles from the earth and an astronaut communicates using radio waves that travel at the speed of light, which is 186,000 miles per second. How long does it take for an astronaut's voice to reach the earth?

2. A car is driven for 2 h at 60 mi/h.
   (a) How far did the car go?
   The car is then driven for 2 h at 40 mi/h.
   (b) How far did the car go?
   (c) Find the average speed for the entire trip. *Answer: (c) 50 mi/h*

3. (a) A car goes 120 mi in 2 h. What is the average speed?
   (b) On the return trip it takes 3 h. What is the average speed?
   (c) The total distance traveled for the entire trip is 240 mi and the total time is 5 h. What is the average speed for the entire trip? (Note: This problem is different from problem 2 because more time is spent traveling at the slower speed.) *Answer: (c) 48 mi/h*

4. The earth's circumference is approximately 25,000 miles. A person on the equator spins with the earth and makes a circle 25,000 miles in its circumference in one day. Compute the person's speed in miles per hour. *Answer: 1042 mi/h*

5. A car traveling at 45 mi/h (66 ft/s) is accelerated at a rate of 2 ft/s² for 11 s. What is the speed of the car after 11 s?

6. A ball is thrown up at a velocity of 64 ft/s. It is being slowed down by the acceleration of gravity as it goes up.
   (a) What is the ball's velocity after 1 s?
   (b) What is the ball's velocity after 2 s? *Answer: (b) zero*
   (c) What happens next?

7. An object is dropped from a very tall building and hits the ground after 5 s. How many feet high is the building?

8. In problem 4 the speed of a person on the equator was computed as 1042 mi/h.
   (a) How many feet per second is this?
   (b) Calculate the acceleration on the person in feet per second. Use the radius of the earth as 4000 mi. *Answer: (b) 0.11 ft/s²*
   (c) Is this acceleration very big, small, or very small compared to 32 ft/s²?

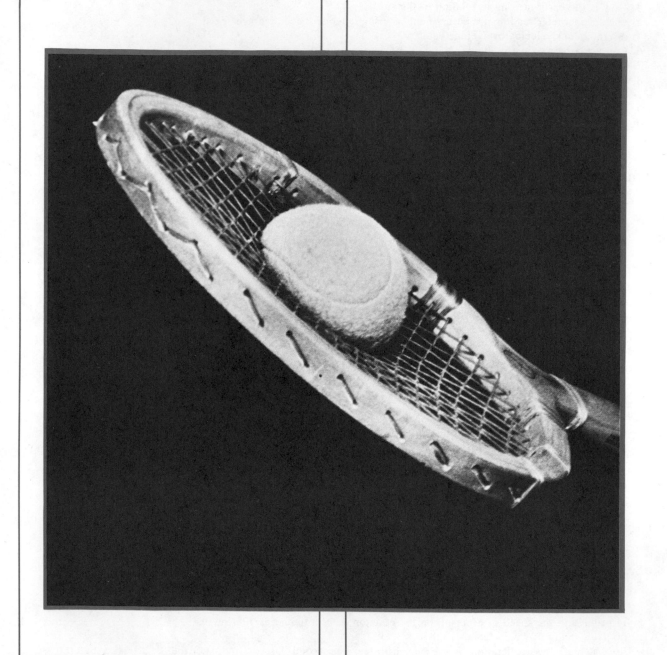

Chapter 3
FORCE AND MOTION

What is force? What is it that sets a moving object in motion? What stops it? These are fundamental questions to be dealt with in this chapter on force and motion. We will center our efforts on Newton's three laws of motion, his law of universal gravitation, and the laws of conservation of linear and angular momentum. Galileo (1546–1642) was the first scientist to make a formal statement concerning objects at rest and in motion. But it remained for Sir Isaac Newton, who was born the year Galileo died, to actually formulate the laws of motion mathematically and to explain the phenomena of moving objects on the surface of the earth plus the motion of the planets and other celestial bodies.

Newton was only 25 years old when he formulated his discoveries in mathematics and mechanics. His book *Mathematical Principles of Natural Philosophy* (commonly referred to as the *Principia*), published in 1687, is considered by many to be the most important publication in the history of physics. Certainly it established Newton as one of the greatest scientists of all times.

The whole burden of philosophy seems to consist in this—from the phenomena of motions to investigate the forces of nature, and from these forces to explain the other phenomena.

Sir Isaac Newton

3-1
NEWTON'S FIRST LAW OF MOTION

Before the time of Galileo and Newton, scientists asked themselves, What is the natural state of motion? The Aristotelian view of motion had held sway for

Figure 3-1a *Sir Isaac Newton (1642–1727). Along with Einstein, Newton is generally regarded as one of two greatest scientists who ever lived. A modest man, Newton wrote shortly before his death, "I seem to have been only a boy playing on the seashore and diverting myself in now and then finding a smoother pebble or a prettier shell than ordinary, whilst the great ocean of truth lay all undiscovered before me." A more accurate description of Newton's contribution to our knowledge was made by Alexander Pope who wrote, "Nature and nature's laws lay hid in the night. God said let Newton be, and all was light." (Courtesy AIP Niels Bohr Library, W. F. Meggers Collection.)*

centuries before Galileo and Newton showed it to be incorrect. Newton (Fig. 3-1a) answered the question by stating his **first law of motion.** This law can be stated in several different ways. One of these is as follows:

> A body will remain at rest or in uniform motion in a straight line unless acted upon by an external unbalanced force.

Uniform motion in a straight line means that the velocity is constant. Thus, another way to state Newton's first law is to say that the natural state of motion is at a constant velocity. If the constant velocity is equal to zero, we say that a body is at rest. From Newton's law we see that if objects are to speed up, slow down, go in a circle, or change direction, they must be acted upon by an external unbalanced force. For instance, a rocket in outer space will continue in uniform motion in a straight line. If it is heading in the correct direction, it doesn't have to burn fuel to continue on to its destination. Only if it needs to alter its course is it necessary to apply a force by burning fuel.

What is a "force"? We all have an intuitive feeling for the concept of force. A force is simply a push or pull. In terms of Newton's first law we see that a **force** is any quantity that is capable of producing a change in motion. This does not necessarily mean that a change in motion is produced, only that the capability is there.

In a tug-of-war (Fig. 3-1b, top), a lot of force is being applied but there is no motion. The forces in this example are large, but they are balanced. Movement only occurs when the forces are unbalanced (see Fig. 3-1b, bottom). For a change in motion or acceleration to occur, there must be an unbalanced or *net force* acting.

Forces act in particular directions as illustrated in Fig. 3-1b. Therefore, force is a *vector quantity*.

The property of matter to resist any change in motion is known as **inertia**. Inertia is a property that provides a means for measuring the quantity of matter present in an object. This quantity of matter is called **mass** and may be used as a measure of inertia. The greater the mass the greater the inertia, and vice versa.

You have experienced the effects of unbalanced forces and inertial masses when riding in an automobile. When the car is started quickly, you are thrown back against the seat; when it is stopped quickly, you are thrown forward. Your body (mass) resists the change in motion that is taking place. You are also thrown toward the outside of the car when the car is rounding a curve. Actually, in this case you are tending to travel in the original path of the car, but an unbalanced force is acting on the car from the road to turn the car from its straight path, and it is also acting on you to change your direction of motion.

Another example of inertia is shown in Fig. 3-1c. The stack of coins has an inertia which resists being moved when they are at rest. If the paper is quickly jerked, the inertia of the coins will prevent them from toppling.

Figure 3-1b *(Top) When forces are balanced, a change in motion is not produced. (Bottom) When F_2 is greater than F_1 the forces are unbalanced, and a change in motion occurs. (Courtesy J. D. Wilson,* Physics Concepts and Applications, *Heath, Lexington, Mass., 1977.)*

Figure 3-1c *A demonstration of inertia. If the strip of paper is given a quick jerk, the inertia of the stack of quarters prevents them from toppling. (Courtesy J. D. Wilson,* Physics Concepts and Applications, *Heath, Lexington, Mass., 1977.)*

3-2
NEWTON'S SECOND LAW OF MOTION

We defined acceleration in Chapter 2 as being the change in velocity divided by the change in time. This defines acceleration, but it doesn't tell us what causes acceleration. In Newton's first law of motion, we see that no acceleration will occur unless there is an external unbalanced force acting. From this, we see that force causes acceleration.

Newton realized that force causes acceleration, but he went much further. He quantified the relationship and found that acceleration is directly proportional to the unbalanced force. In Fig. 3-2a and Fig. 3-2b we see an example of this. In Fig. 3-2b the force is twice what it is in Fig. 3-2a. The acceleration will then be twice as great.

Acceleration of a mass does not depend on the force alone. It also depends on the mass. Mass is a means of measuring inertia. Inertia was defined as resistance to a change in motion. Thus, it seems reasonable to assume that the greater the mass, the smaller will be the change in motion. Another way to state this is to say that acceleration is inversely proportional to the mass: the smaller the mass the greater the acceleration. Comparing Fig. 3-2a and Fig. 3-2c we see that if an equal force is applied, the smaller mass will be accelerated more.

If we combine the effects of force and mass on acceleration and realize that both force and acceleration are vectors, we get **Newton's second law of motion:**

The acceleration produced by an unbalanced force acting on a mass is directly proportional to the unbalanced force, in the direction of the unbalanced force, and inversely proportional to the total mass being accelerated by the unbalanced force.

If we express Newton's second law mathematically, it is greatly simplified. Mathematically, we say

Figure 3-2a *An unbalanced force causes an acceleration. (Photo courtesy James Crouse.)*

Figure 3-2b *Twice the force causes double the acceleration. (Photo courtesy James Crouse.)*

Figure 3-2c *The same force as in Fig. 3-2a causes a greater acceleration since the mass here is smaller. (Photo courtesy James Crouse.)*

$$a \propto \frac{F}{m}$$

where a = the acceleration produced on the mass by the unbalanced force,

F = the unbalanced force acting on the mass,

m = the mass being accelerated.

When appropriate units are used, we can replace the proportionality sign with an equal sign and write

$$a = \frac{F}{m}$$

However, Newton's second law is easier to learn if we express it as

$$F = ma \qquad \text{Eq. 3.1}$$

Equation 3.1 is a mathematical statement of Newton's second law.

In examining Newton's first and second laws of motion we find that the first law is a special case of the second. This can be seen from the answer to a very easy question: What is the acceleration of a mass on which the unbalanced force is zero? The answer is zero; and if the acceleration is zero, then the mass must be moving with a constant velocity. But this is just Newton's first law: a mass that has no unbalanced force acting on it moves at a constant velocity.

Since units on both sides of an equation must be equivalent, the units of force must be the same as the units of mass times acceleration. In Table 3-2a we list the force units in the cgs, mks, and British systems of measurement.

The newton is the unit of force in the mks system. One **newton,** N, is equal to the force necessary to

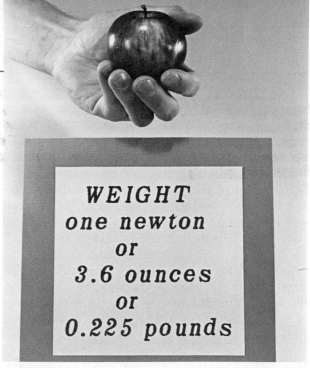

WEIGHT
one newton
or
3.6 ounces
or
0.225 pounds

Figure 3-2d *An average-sized apple weighs one newton, 3.6 ounces, or 0.225 pound. (Photo courtesy James Crouse.)*

accelerate one kilogram by one meter per second per second. A newton is equal to 0.225 pound or 3.6 ounces. In Fig. 3-2d, an apple is shown which weighs 3.6 ounces. Since the story is told (and probably true) that Isaac Newton was struck by an apple while meditating on the concept of gravity, it is easy to remember that an apple weighs approximately one newton.

The **dyne** is equal to the force necessary to accelerate one gram by one centimeter per second per second. This is a very small unit of force. A dyne is only 1/100,000 as large as a newton.

Table 3-2a *Units of Force*

	Mass Unit	Acceleration Unit	Force Unit		Special Name for Force Unit
cgs	g	cm/s^2	g $\times$ cm/s^2	=	dyne
mks	kg	m/s^2	kg $\times$ m/s^2	=	newton (N)
British	slug	ft/s^2	slug $\times$ ft/s^2	=	pound (lb)

The mass of an object is independent of the earth or any other object in the universe. Hence, a measure of inertial mass of a body is a quantity which that body alone possesses. This is not true of the **weight** of a body, which is the force of attraction between a body and the earth. This force of attraction varies, becoming smaller as the distance between the object and the earth increases. For example, a bowling ball whose mass is 0.5 slug weighs 16 pounds when placed on a spring scale located at the surface of the earth. If the ball is taken a billion miles from the earth, the gravitational attraction between the ball and the earth will, for all practical purposes, be zero; hence, the weight will be zero. However, since the quantity of matter has remained the same, the mass will still be 0.5 slug.

The cgs and the mks systems are known as absolute systems, because they are based upon mass (inertia), which does not depend upon gravity. The third force unit, the pound, is part of the British system of measurement, which is called a gravitational system because it is based on weight rather than mass.

This gravitational system is different from the two absolute systems. It is based upon the fundamental units of length, force, and time rather than length, mass, and time. Since we have a unit of force in the gravitational system called a pound, a mass unit needs to be defined in terms of the foot and pound. The force unit, **pound,** legally defined in respect to the international prototype kilogram, is the force of earth's gravity on 0.454 kg at a place where the acceleration due to gravity is 32.174 ft/s^2.

The unit of mass called the **slug** is defined as the mass of a body which will be accelerated 1 ft/s^2 when an unbalanced force of 1 pound acts upon the mass. Thus,

$$1 \text{ slug} = \frac{1 \text{ pound}}{1 \text{ foot/second}^2}$$

The dyne, newton, and slug are *derived units* as distinguished from the *fundamental units:* gram mass, kilogram mass, and pound force.

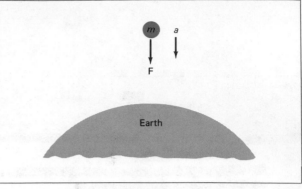

Figure 3-2e *The earth's force of gravity on an object causes the object to have weight. The force F on the object of mass m causes an acceleration a downward.*

The weight of an object has been defined as the force of attraction between the earth and the object. If we have an object, or mass, located near the surface of the earth, the force of attraction between the object and the earth will accelerate the object, providing the object is free to move. See Fig. 3-2e.

The relation $F = ma$ states that an unbalanced force acting on the mass m produces the acceleration a. But the force acting in the case of a falling object is the weight w of the object. Therefore, we can write

$$w = F$$
$$w = ma$$

Any mass falling freely over a short distance, near the surface of the earth, will accelerate downward due to gravity at a constant value of 980 cm/s^2, 9.8 m/s^2, or 32 feet/s^2. The acceleration due to gravity is a constant and is given the symbol g. Using this symbol for acceleration in the preceding relation, we obtain

$$w = ma = mg$$

or

$$w = mg \qquad\qquad \text{Eq. 3.2}$$

which states the relationship between mass and weight.

As an example of weight, let us compute the weight of a 1-kilogram mass. We use Eq. 3.2 to get

$$w = mg$$
$$w = 1 \text{ kg} \times 9.8 \text{ m/s}^2$$
$$w = 9.8 \text{ kg} \cdot \text{m/s}^2$$
$$w = 9.8 \text{ newtons}$$

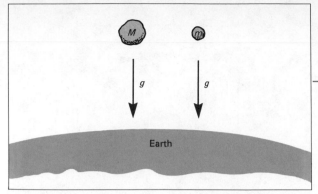

Figure 3-2f *Two rocks fall at the same acceleration. The rock that weighs more has more force on it. The small rock has less inertia.*

So we see that a 1-kilogram mass weighs 9.8 newtons.

A weight is a force and things that weigh more are more strongly attracted to the earth than are things that weigh less. If this is true, why do all things fall to the earth at the same rate? The answer can be seen by considering the mass of objects as well as the force on them. If you have two rocks as shown in Fig. 3-2f, the larger one has more inertia, and thus it is more difficult to move from rest. The smaller one is much easier to move from rest. Thus we have two factors involved that exactly cancel. The big rock weighs more, but it also has more inertia. The little rock weighs less, but it has less inertia. Since the two effects cancel, both rocks fall at the same rate. Mathematically we can write

$$a = \frac{F}{m}$$

or

$$a = \frac{w}{m}$$

or

$$a = \frac{mg}{m}$$

or

$$a = g$$

We see that the mass due to weight in the numerator exactly cancels the inertial mass in the denominator, and the acceleration of all falling objects is the same and equal to g. Remember that we are ignoring frictional effects.

3-3
NEWTON'S LAW OF GRAVITATION

The law governing the force of attraction between two bodies was formulated by Newton from his studies on planetary motion. Known as the **law of universal gravitation**, it is

Every mass in the universe attracts every other mass with a force which is directly proportional to the product of their masses and inversely proportional to the square of the distance between their centers.

Using the proportionality sign ($\propto$), this can be written as

$$\text{Force} \propto \frac{\text{First mass} \times \text{Second mass}}{\text{Sq. of dist. between centers}}$$

or as

$$F = G\left(\frac{m_1 m_2}{r^2}\right) \qquad \text{Eq. 3.3}$$

where F = the gravitational force of attraction,
  m_1 = the mass of one body,
  m_2 = the mass of the second body,
  r = the distance between the centers of the two masses,
  G = a proportionality constant known as the universal gravitational constant, the numerical value of which depends upon the units of F, m, and r.

The force of attraction between any object and the earth is what causes the object to fall to the earth when the object is released above the earth's surface. Newton said the same kind of force (gravitation) causes the planets to revolve around the sun, and the moon to revolve around the earth. Using his law of universal gravitation and his second law, he correctly calculated the time it would take for the moon to go around the earth. His calculation was done without knowing the value of G. In fact, the correct value of G was unknown during Newton's lifetime.

Table 3-3a *What is G?*

$G = 6.67 \times 10^{-8}$ dyne $\times$ cm^2/g^2
$G = 6.67 \times 10^{-11}$ N $\times$ m^2/kg^2
G is a very small quantity.
G is a universal constant. It is the same throughout the universe.
G is not g.
G is not a force.
G is not an acceleration.

The value for G was found experimentally in 1797, 70 years after Newton's death. Henry Cavendish (1731–1810), an English scientist, performed an experiment in which he succeeded in measuring the value of G. The value for G is very small and is given in Table 3-3a. This table tells us what G is and is not.

In Newton's law of universal gravitation the units on both sides must be equivalent. Thus, the units of G must be force units times length units squared divided by mass units squared. With these units for G, both sides of Eq. 3.3 have force units.

Note that G is very small. If we had two books, each with a mass of 1 kilogram (remember 1 kilogram weighs 2.2 pounds), and placed 1 meter apart, what would be the force between them? The solution is easy if we use Newton's law of universal gravitation (Eq. 3.3):

$$F = \frac{G m_1 m_2}{r^2}$$

$$F = 6.67 \times 10^{-11} \frac{\text{N} \times \text{m}^2}{\text{kg}^2} \frac{1\,\text{kg} \times 1\,\text{kg}}{(1\,\text{m})^2}$$

$$F = 6.67 \times 10^{-11}\,\text{N}$$

which is a very, very small number. According to Newton's law of universal gravitation, everything that has mass is attracted to everything else that has mass. However, since G is such a small number, at least one of the masses involved must be large before the gravitational force between two masses is appreciable. This is illustrated in Fig. 3-3a.

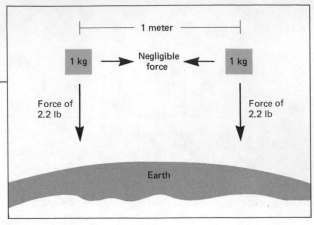

Figure 3-3a *Two masses of one kilogram each have a neglibible force of gravational attraction when they are one meter apart. However, because the earth's mass is quite large, they are both attracted to the earth with a force of 2.2 pounds. We call this force the weight of the masses.*

When a mass is attracted to the earth, we call the force involved the *weight of the mass*. Let us now work out the relationship between weight and the properties of the earth.

If a mass at the earth's surface is attracted to the earth, then the distance between the centers of the masses is approximately the radius of the earth. We will then find the force of attraction, using Newton's second law,

$$F = \frac{G m_{\text{object}}\, m_{\text{earth}}}{R^2_{\text{earth}}}$$

where R_{earth} = radius of the earth

This force is just what we call weight, so we can write

$$w = m_{\text{object}}\, g = \frac{G m_{\text{object}}\, m_{\text{earth}}}{R^2_{\text{earth}}}$$

Using this equation we get a relationship between g and G.

$$g = \frac{G m_{\text{earth}}}{R^2_{\text{earth}}}$$

This is a powerful relationship because it enables us to compute the mass of the earth if we know g, G, and R_{earth}. Since R_{earth} was known to be 6.37×10^6 m, the

Figure 3-3b *Objects on the moon have the same mass as on the earth, but they weigh only one-sixth as much. (Courtesy NASA)*

mass of the earth could be computed once G was determined. The calculations were done as follows:

$$m_{\text{earth}} = \frac{gR^2_{\text{earth}}}{G}$$

$$m_{\text{earth}} = \frac{(9.8 \text{ m/s}^2) \times (6.37 \times 10^6 \text{ m})^2}{6.67 \times 10^{-11} \text{ N(m}^2/\text{kg}^2)}$$

$$m_{\text{earth}} = \frac{(9.8 \text{ m/s}^2) \times 6.37 \times 6.37 \times 10^{12} \text{ m}^2}{6.67 \times 10^{-11} \text{ kg(m/s}^2)(\text{m}^2/\text{kg}^2)}$$

$$m_{\text{earth}} = 59.6 \times 10^{23} \text{ kg}$$

$$m_{\text{earth}} = 5.96 \times 10^{24} \text{ kg}$$

This shows that the mass of the earth is a very large number. The important thing is that we are able to compute it.

The acceleration of gravity on the moon can be computed by a formula similar to the one used to compute the acceleration of gravity on earth:

$$g_{\text{moon}} = \frac{Gm_{\text{moon}}}{R^2_{\text{moon}}}$$

From our knowledge of the moon's mass and radius we get

$$g_{\text{moon}} = \tfrac{1}{6} g$$

Thus, things on the moon fall at one-sixth the acceleration of the earth's gravity. Objects on the moon also weigh one-sixth as much although they will have the same mass (Fig. 3-3b). If you want to lose weight fast, go to the moon.

Why doesn't the moon fall down? Anything dropped seems to fall down, so why doesn't the moon? A good way to understand this problem is through examining Fig. 3-3c.

Let us suppose we are on top of a tall building throwing tennis balls out the window. At first they take path A in Fig. 3-3c, but when we throw harder they take path B. If we could throw them hard enough and there were no atmosphere, they would accelerate to earth with acceleration equal to g. However, they would take path C and never hit the earth. They would simply accelerate all the way around the earth, forming a circular path.

If we remember that anything going in a circle accelerates with $a = v^2/r$, we can understand that the moon is being constantly accelerated toward the earth. Newton's second law becomes

$$F_{\text{grav}} = m\frac{v^2}{r}$$

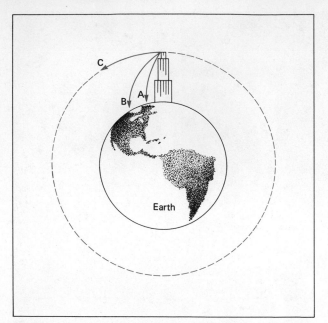

Figure 3-3c *An exaggerated drawing illustrating what happens when you throw something out of a building. If there were no atmosphere and you "threw" an object hard enough it would "fall around" the earth as shown in path C. The moon "falls around" the earth in a similar manner.*

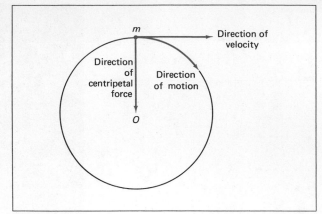

Figure 3-3d *The direction of motion, the velocity, and the centripetal force for a mass undergoing uniform circular motion.*

We understand this as follows: The earth's gravity pulls on the moon with a strong force. This force causes the moon to accelerate in a circle. There is only one force on the moon—the force of the earth's gravity. This unbalanced force causes an acceleration. This acceleration causes a circular motion rather than a straight-line motion. Newton was able to analyze this motion quantitatively, and he correctly calculated the time it takes for the moon to "fall around" the earth.

Whenever an object is going in circular motion, such as the moon going around the earth, the force that causes the circular motion is called the **centripetal force.** The word centripetal means "center-seeking." The centripetal force is always directed towards the center of the circle. The centripetal force and the acceleration it causes are both vector quantities and the direction for each is toward the center of the circle. Figure 3-3d is a diagram showing the direction of the velocity, the centripetal force, and the path of motion.

In our example of the moon going around the earth, the centripetal force was the force of gravity between the earth and the moon. When an automobile goes around a curve, the centripetal force is the friction between the tires and the pavement. If it is wet or icy, the friction may not be enough to provide the required centripetal force, and the automobile may go off the road.

3-4
NEWTON'S THIRD LAW OF MOTION

Newton's third law of motion is sometimes called the law of action and reaction. **Newton's third law** states:

> **For every action there is an equal and opposite reaction.**

In this statement the words "action" and "reaction" refer to forces. A more precise statement would be the following:

> Whenever one mass exerts a force upon a second mass, the second mass exerts an equal and opposite force upon the first mass.

Rocket motion is an example of Newton's third law. The exhaust gases are accelerated out of the base

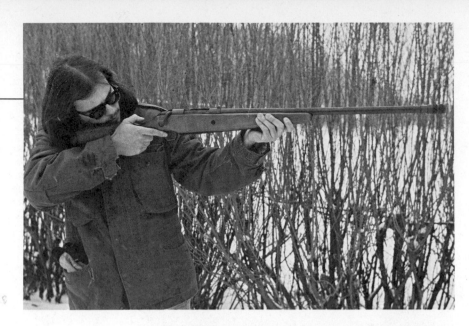

Figure 3-4a *When the bullet goes forward out of the rifle, the rifle recoils. (Photo courtesy James Crouse.)*

of the rocket, and the rocket accelerates in the opposite direction. A common misconception is that the exhaust gases push against the launch pad to accelerate the rocket. This is nonsense. If this were true, there would be no space travel since there is nothing to push against in space. The correct explanation is one of action (gases going out the back) and reaction (rocket propelled forward).

Figure 3-4a shows a person shooting a rifle. The shoulder pad is necessary because the rifle will "kick" when the bullet is shot. The action can be thought of as the bullet accelerating out of the gun. The reaction is the rifle recoiling backward. Another illustration was given by Newton: "If you press on a stone with your finger, the finger is also pressed upon by the stone." Figure 3-4b shows a person doing an isometric exercise based on Newton's third law.

Newton's third law can be expressed mathematically as

$$F_1 = -F_2 \qquad \text{Eq. 3.4}$$

where F_1 = force exerted on object 1 by object 2
F_2 = force exerted on object 2 by object 1

The minus sign indicates that F_2 is in the opposite direction from F_1. Using Newton's second law, we find

$$m_1 a_1 = -m_2 a_2 \qquad \text{Eq. 3.5}$$

Figure 3-4b *Isometrics are a good example of Newton's third law. During an isometric exercise, different muscles of the body work against each other. One muscle provides the action, another muscle provides the reaction. (Photo courtesy James Crouse.)*

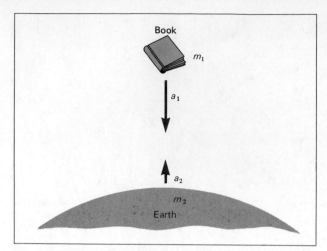

Figure 3-4c *As the book is accelerated downward, the earth is accelerated upward. Because the earth's mass is so large, the earth's acceleration is immeasurably small.*

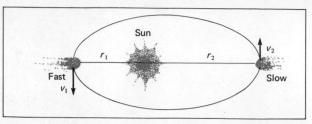

Figure 3-5a *The path of a comet going around the sun. The angular momentum of the comet is given by mv_1r_1 or mv_2r_2. Since the angular momentum is conserved, we get $mv_1r_1 = mv_2r_2$. As the comet gets closer to the sun, the radial distance r decreases so v must increase. The comet moves fastest when closest to the sun and slowest when furthest from the sun.*

From Eq. 3.5, we see that if m_2 is much bigger than m_1, then a_1 is much bigger than a_2.

As an example, consider dropping a book to the floor. As the book drops, it has a force acting on it (earth's gravity) that causes it to accelerate. What is the equal and opposite force? The equal and opposite force is the force of the book's gravitational pull on the earth. As illustrated in Fig. 3-4c the earth actually rises up to meet the book! However, since the earth's mass is so huge compared to the book, the earth's acceleration is miniscule and cannot be measured.

There are many other examples of action and reaction occurring in everyday life. In fact, from Newton's third law it is clear that every time there is a force, there must be a reaction force also.

3-5
CONSERVATION LAWS

Newton's second law of motion can be expressed as

$$F = m\frac{\Delta v}{\Delta t}$$

This is the well-known equation $F = ma$, with the definition of acceleration inserted. When a rocket accelerates, however, it loses mass; and a more general equation is necessary to take account of this mass change. The more general equation becomes

$$F = \frac{\Delta(mv)}{\Delta t}$$

The combination mv is very important and is called **linear momentum.** The linear momentum of a system is important because it is conserved: It does not change with time. This property makes it extremely important in analyzing the motion of various systems. The **law of conservation of linear momentum** can be stated as follows:

The total linear momentum of an isolated system of masses remains the same if there are no external forces acting on the system.

Both mass and velocity are involved in momentum. A small car and a large truck both traveling at 50 mi/h have the same velocity, but the truck has more momentum because it has a much larger mass.

Another important quantity which Newton found to be conserved is angular momentum. Angular momentum arises when objects go in paths around a fixed point. Consider a comet that goes around the sun as shown in Fig. 3-5a. The **angular momentum** is given by the product mvr, where m is the mass, v is the velocity, and r is the distance of the object from the center of motion which in this case is the sun.

The **law of conservation of angular momentum** states that the angular momentum of an object remains constant. That is,

$$mv_1r_1 = mv_2r_2 \qquad \text{Eq. 3.6}$$

In our example of a comet, the angular momentum,

Figure 3-5b *Conservation of angular momentum. (a) When the student holds the weights away from himself, the speed of rotation is slow. (b) When the weights are brought inward toward the body, the speed of rotation increases. The result is a fast, dizzy spin as indicated by the facial expression of the student. (Courtesy J. D. Wilson, Physics Concepts and Applications, Heath, Lexington, Mass., 1977.)*

mvr, remains the same. As the comet gets closer to the sun, r decreases so the speed v increases. For this reason, comets spend much less time close to the sun (where they can be seen more easily from the earth) and much longer periods of time far from the sun.

Another example of the conservation of angular momentum is demonstrated in Fig. 3-5b. Here a student holds weights on a spinning stool. With the weights held out, the spinning rate is slow because the radial distance r for the weight is large. When the weights are brought inward, r decreases, so the spinning rate must increase.

Ice skaters use the same principle to spin faster on the ice. The skater extends both arms and perhaps one leg and obtains a slow rotation, then the arms and leg are drawn inward and rapid angular velocity is gained because of the decrease in the radial distance of the mass.

The conservation of angular momentum law is applied throughout the wide range of physics—from the subatomic world to the astronomical universe. In all applications the law has proven true and extremely useful. For example, the earth in revolving around the sun does not travel in a true circular path. The actual path is the shape of an ellipse, which means the distance between the earth and sun is not the same at all times. Actually, the earth is closest to the sun in January and farthest away in July. Thus, the earth will be traveling slightly faster in its orbit about the sun in January and slightly slower in July, since angular momentum must be conserved.

The student can illustrate the law by performing the following experiment: Take a small object like a piece of chalk and tie a length of string about a foot long to the chalk. Hold the other end of the string by your forefinger and thumb and whirl the chalk around in a vertical circle. Extend your forefinger such that the string will wind itself around this finger and observe the increase in speed of the chalk as the distance between the chalk and finger becomes smaller.

The law of conservation of angular momentum would cause a helicopter with a single propeller to rotate. In order to conserve angular momentum, the body of the helicopter would have to rotate in the direction opposite that of the rotor. To prevent this, some helicopters have two oppositely rotating propellers. It is more common, however, for the helicopter to have a small propeller (Fig. 3-5c) on the tail which produces a rotation opposite to the rotation of

Figure 3-5c *The body of a helicopter with a single rotor would rotate in a direction opposite to that of the propeller. To prevent this, most helicopters have an auxiliary propeller on the tail. (Courtesy Bell Helicopter Company.)*

Figure 3-5d *A person driving a car twists the steering wheel if two equal and opposite forces are applied. Note that the net force is zero, but there is still a torque, or twisting effect.*

the body of the helicopter. With this configuration of propellers, the helicopter is stable.

The linear momentum of a system can be changed by the introduction of an external, unbalanced force; the angular momentum of a system can be changed by an external unbalanced torque. A **torque** is a twisting effect caused by one or more forces. For example, in Fig. 3-5d a torque on a steering wheel is caused by two equal and opposite forces acting on different parts of the steering wheel. This causes a twist, or torque, and causes the steering wheel to turn.

IMPORTANT WORDS AND TERMS

Newton's first law of motion
Force
Inertia
Mass
Newton's second law of motion $F = ma$
Newton
Dyne

Weight
Slug
Newton's law of universal $F = G\left(\frac{m_1 m_2}{r^2}\right)$
  gravitation
G — 6.67×10^{-8} dyne $\times 10^{-11}$ N
Centripetal force
Newton's third law of motion

Linear momentum $m \times v$
Law of conservation of
  linear momentum
Angular momentum mvr
Law of conservation of
  angular momentum
Torque

LEARNING OBJECTIVES

After reading and studying this chapter, you should be able to do the following without referring to the text:

1. State Newton's first law of motion and explain what the natural state of motion is.
2. Define the terms *force*, *inertia*, and *mass.*.
3. State Newton's second law of motion in words and with an equation.
4. Give the units of force in the cgs, mks, and British systems and give an everyday example of the mks unit.
5. Define the term *weight* and give its units in the cgs, mks, and British systems.
6. Explain why any two objects will drop at the same rate no matter what they weigh if air friction can be neglected.
7. Explain each term in the equation $F = Gm_1m_2/r^2$.
8. Explain the differences between G and g.
9. Explain why the moon doesn't fall down.
10. State and give an example of Newton's third law of motion.
11. Define *linear momentum* and *angular momentum*.
12. State and give an example of the law of conservation of angular momentum.

QUESTIONS

1. What is the natural state of motion?
2. Which of the following would change for a person who went to the moon? (a) weight (b) inertia (c) mass
3. Explain the relationship between (a) force and acceleration (b) mass and acceleration. $f = ma \quad a = \frac{m}{F}$
4. Is it possible for a body to be at rest if forces are being applied to it? Explain.
5. If no forces are acting on an object, can the object be in motion?
6. What is a newton?
7. If you dropped a 10-pound rock and a 1-pound rock, (a) why would you expect the 10-pound rock to hit first? (b) Why would you expect the 1-pound rock to hit first? (c) Explain what actually happens.
8. Discuss the quantities G and g and tell how they are different.
9. Explain why apples fall off trees, but the moon doesn't fall down.
10. Use Newton's third law to explain how a rocket blasts off.
11. Why are linear momentum and angular momentum important concepts?
12. Several keys on a string are whirled around your forefinger. As the string winds around your forefinger, explain what happens and why.

PROBLEMS

1. Determine the force necessary to give a 4-kg mass an acceleration of 6 m/s². $f = ma$ *Answer:* 24 newtons
2. A force of 42 N is applied to a mass of 7 kg. What will be the resultant acceleration? $a = \frac{F}{m}$
3. What is the force on a car moving at a constant velocity of 20 m/sec? 0
4. (a) What is the force in dynes acting on a 6-g mass that is falling? $980 \times 6g$ $f = ma$ *Answer:* (a) 5880 dynes
   (b) What is the weight of the same mass?
5. (a) Calculate your weight in newtons if your mass is 50 kg.
   (b) Calculate your weight in newtons if your weight is 110 lb.
6. Calculate the weight of a 1-slug mass.   *Answer:* 32 lb
7. Calculate the acceleration of gravity g in mks units. Use the formula $g = Gm_{earth}/R^2_{earth}$. (Hint: $m_{earth} = 5.96 \times 10^{24}$ kg and $R_{earth} = 6.37 \times 10^6$ m). $6.67 \times 10^{-11} \, N \times \frac{m^2}{kg^2}$
8. How much does a person weigh on the moon if the person's weight is 180 lb on the earth? $\frac{1}{6}$
9. Calculate the momentum of a truck of mass 15,000 kg which is traveling at 10 m/s. $m \times v$
10. A comet goes around the sun in a noncircular path. At its furthest point 600 million miles away it is traveling at a speed of 15,000 mi/h. How fast is it traveling when it is 100 million miles away from the sun? $m v_1 r_1 = m v_2 r_2$ *Answer:* 90,000 mi/h

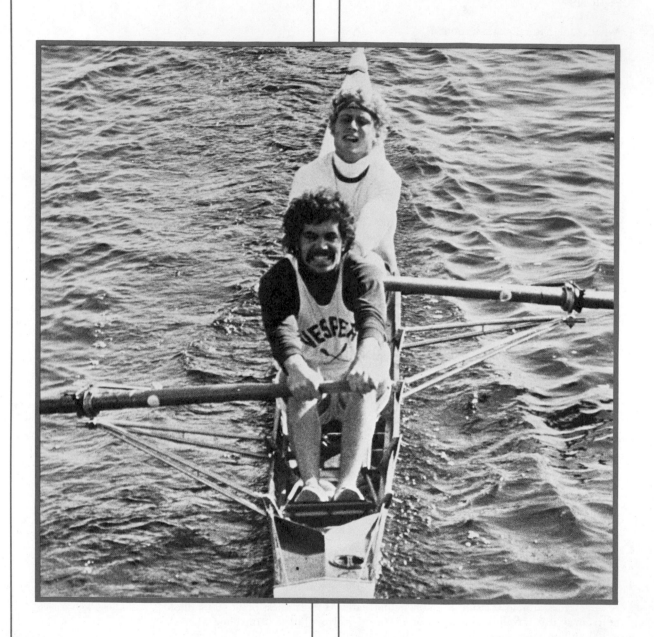

Chapter 4
WORK AND ENERGY

In the preceding chapter we learned that it is necessary to apply a force to an object in order to set it in motion, or to change its magnitude or its direction if the object is moving. Where do the forces necessary for producing motion come from? What happens to the forces when one moving object collides with another? In seeking the answer to such questions as these, Christian Huygens (1629–1695) formulated the first step for the concept of work and energy with his statement concerning the collision of two perfectly elastic spheres: the sum of the product of the masses of two perfectly elastic spheres times the square of their velocities will be the same before and after colliding.

The common meaning of the word "work" refers to the accomplishment of some task or job. When work is done, energy has been expended. Hence, work and energy are related. A student performs a certain amount of work during the day and becomes tired. He or she must obtain rest and food in order to continue the work. We know that rest alone is not sufficient to keep the student going; therefore, the food must serve as fuel to supply the necessary energy.

The technical meaning of the word "work" is quite different from the common meaning. A student standing at rest and holding several books is doing no work, although he or she will feel tired after a time. Technically speaking, work is accomplished only when a force acts through a distance. Energy has been defined as stored work, and our mastery of

Energy is a constant mathematical sum, independent of space, that takes many forms.

energy has produced today's modern civilization. From the control of fire to the mastery and control of nuclear energy, we have advanced our standard of living through our ability to use and control the flow of energy.

This aspect of the physical environment is probably the single most important concept we have developed in our search for physical knowledge. Energy takes many forms, such as mechanical, chemical, electrical, radiant, nuclear, and gravitational energy. These, as we shall learn, are classified in the more general categories of kinetic and potential energy.

Our main source of energy is the sun, which radiates into space each day an enormous amount of radiant energy; only a small portion of this energy is received by the planet earth. Radiant energy supports all life on our planet and provides us with the means for making nature work for the earth's inhabitants.

4-1
WORK

Work, as a physical concept, is defined as the product of the average applied force and the parallel distance through which the force acts. This can be written as

Work = Av. applied force × Parallel distance

or as
$$W = \bar{F}s$$
Eq. 4.1

where W = the work done,
  $\bar{F}$ = the average force acting,
  s = the parallel distance traveled by (or the displacement of) the object moved

Figures 4-1a, b, and c, illustrate the concept of work. In Fig. 4-1a, a force is being applied to the wall, but no work is being done since the wall is not moving through any distance. Figure 4-1b shows a mass being moved through a distance s by a force F, and the work done will be the product of the force and distance. The illustration also shows the force F and the distance s parallel to one another. When F and s

Figure 4-1a *Force is applied to wall but no work is done since s equals 0.*

Figure 4-1b *Force F is applied to mass M moving the mass through the distance s. Work equals the force times the distance the mass moves in the direction of the applied force.*

Figure 4-1c *Only the horizontal component F_H of the applied force F is used to do work in this example.*

Figure 4-1d *Person walking with constant velocity carrying mass M through some distance s. No work is done, since F is perpendicular to distance.*

are not parallel to one another, the component of the force F which is parallel to the distance s must be used to calculate work done. For example, when mowing the lawn with a push-type lawnmower only the force component parallel to the lawn (the horizontal force) is used to accomplish work. The vertical component of the force is doing no useful work, since this force is tending to push the lawnmower into the ground (Fig. 4-1c).

When the applied force is perpendicular to the distance through which the object is moved no work is done (Fig. 4-1d). The work done is zero, because there is no distance component parallel to the applied force.

One of the most important properties of work is that it is a scalar quantity. Both the force and the parallel distance have directions associated with them, but work does not. Work is expressed only as a magnitude with the proper units. It has no direction associated with it.

The units of work are listed in Table 4-1a. Since work is the product of a force and a distance, the units of work are the units of force times length.

Table 4-1a *Work Units (Energy Units)*

	Fundamental Units	Force × Distance Units	Name
cgs	g × cm²/s²	dyne × cm	erg
mks	kg × m²/s²	newton × m	joule
British	lb × ft	lb × ft	foot-pound

These units are given the special names erg, joule (pronounced jewel), and foot-pound. These work units are also energy units as we shall see later in the chapter.

The **joule** (J) is the unit of work or energy in the mks system. As seen from Table 4-1a, one joule is the amount of work you would do if you lifted an apple weighing one newton (N) to a height of one meter. The **erg** is the unit of work or energy in the cgs system. It is a very tiny amount of work or energy. One joule is 10^7 ergs. That is, an erg is one ten-millionth of a joule.

In the British system, the unit of work is called the **foot-pound** (ft · lb). One foot-pound is the work done by a force of one pound acting through a distance of one foot. It should be remembered that the pound referred to here is a force pound, not simply a mass unit, which in the British system would be the slug.

In the application of a force, according to Newton's third law of motion, we find that for every action there is an equal and opposite reaction exerted on the object causing the action. Therefore, when object A has work done on it by object B, then object B loses a quantity of work equal to that gained by object A.

Work can be done on objects in many different ways. The following are some of those ways:

(a) *Work against inertia.* We have learned that any object at rest remains at rest unless acted upon by some outside force, or, if the object is in motion, it will remain in motion at constant velocity in a straight line unless acted upon by some outside force. Newton referred to the properties of an object to stay at rest or in constant straight-line motion as *inertia.* When a force is applied to change the velocity or direction of an object, work will be done against inertia. Figure 4-1e illustrates this. A mass is shown at rest on a frictionless air table. In order to increase the velocity of the mass from 0 to 4 cm/s a force must be applied through the distance s, thus performing work. The work done has been against inertia.

W = some positive value

$v_0 = 0$

$v_f = 4$ cm/s

m

F

m

Air table

Floor

Figure 4-1e *Work done against inertia.* A force F is applied to move mass m from rest to a velocity of 4 cm/s. The air table provides an almost frictionless surface for the mass.

s

F

m

m

Force of friction

W = F s

Figure 4-1g *Work done against friction. The mass m is being moved at a constant velocity by the force F against the force of friction.*

s

F

W = F • s

Figure 4-1f *Work done against gravity. The force F is applied vertically upward, moving the dumbbell through a distance s.*

(b) *Work against gravity.* See Fig. 4-1f. A mass of 10 kilograms is lifted to a height of 1 meter. How much work is done? In order to use the formula for

work ($W = Fs$), we must first change the mass unit, kilograms, to force units by multiplying by g, acceleration due to gravity: ($10 \text{ kg} \times 9.8 \text{ m/s}^2 = 98 \text{ N}$).

$$W = \bar{F}s$$

$$W = 98 \text{ N} \times 1 \text{ m}$$

$$W = 98 \text{ N} \cdot \text{m}$$

$$W = 98 \text{ J}$$

(c) *Work against friction.* Examine Fig. 4-1g. A horizontal force of 10 newtons is needed to slide a 10-kilogram mass a distance of 1 meter along the top of a table. What is the work done?

$$W = \bar{F}s$$

$$W = 10 \text{ N} \times 1 \text{ m}$$

$$W = 10 \text{ N} \cdot \text{m}$$

$$W = 10 \text{ J}$$

You will notice it takes much less work to slide the 10-kilogram mass along the table 1 meter than to lift the same 10-kilogram mass to a height of 1 meter.

In Fig. 4-1g the force of friction is the force which acts to oppose the tendency of one surface to slide over the other. The force of friction is a constant for the two surfaces, and is always parallel to the surfaces which are in contact.

4-2
POWER

The rate of doing work is called **power.** It is calculated by dividing the work done by the time required to do it. This can be written as

$$\text{Power} = \frac{\text{Work}}{\text{Time}}$$

or as

$$P = \frac{W}{t}$$

Since work is the product of force and distance or displacement, the common formula used to find power is

$$P = \frac{\overline{F}s}{t}$$

Just as we have the cgs, mks, and British systems of units for measuring force and work, we also have units of power in these three systems.

In the cgs system, in which work is measured in ergs and time in seconds, power is measured in *ergs per second;* no other special name is given to it, and the designation usually is erg/s. In the mks system, in which work is measured in joules and time in seconds, power is measured in watts, one **watt** (W) equaling one joule per second.

In the British system, power is expressed in *foot-pounds per second* or *horsepower.* One **horsepower** (hp) is equal to 550 foot-pounds per second.

$$1 \text{ horsepower} = \frac{550 \text{ foot-pounds}}{\text{second}}$$

The term "horsepower" was originated by the British during the days when coal was mined by hand and brought to the surface with the use of horses. The average horse was capable of doing an amount of work equal to 550 foot-pounds per second. Horsepower is found by dividing power (as expressed in foot-pounds per second) by 550 foot-pounds per second, or 5.5×10^2 ft · lb/s, as it is often written.

The greater the power, the faster work is done. A 450-horsepower engine can accomplish work much faster than a 160-horsepower engine in a theoretical or laboratory situation. The smaller engine can do the same amount of work, but a longer period of time is required. For example, a student wishes to transfer books weighing 200 pounds from the first floor to the fourth floor of his dormitory. He could accomplish this task by making four trips carrying 50 pounds each trip. This would take perhaps 20 minutes of his time. If he has a willing roommate of equal strength, the same task could be performed in half the time. Thus, the greater the power, the faster the work can be done.

The following examples will serve to show how power is calculated:

■ *Example 1* A force of 50 newtons is required to push a student's stalled moped 20 meters along a flat road in 20 seconds. Calculate the work done in joules and the power in watts.

$$W = Fs$$

$$W = 50 \text{ N} \times 20 \text{ m}$$

$$W = 1000 \text{ J}$$

$$P = \frac{W}{t}$$

$$P = \frac{1000 \text{ J}}{20 \text{ s}} = \frac{50 \text{ J}}{\text{s}}$$

$$P = 50 \text{ W}$$

■ *Example 2* A student weighing 105 pounds and carrying books weighing 5 pounds climbs a stairway (a vertical height of 50 feet) in 20 seconds. Calculate the power in foot-pounds per second and in horsepower.

$$P = \frac{W}{t} = \frac{Fs}{t}$$

$$P = \frac{110 \text{ lb} \times 50 \text{ ft}}{20 \text{ s}}$$

$$P = \frac{5.5 \times 10^3 \text{ ft} \cdot \text{lb}}{20 \text{ s}}$$

$$P = 2.75 \times 10^2 \text{ ft} \cdot \text{lb/s}$$

$$P = \frac{2.75 \times 10^2 \text{ ft} \cdot \text{lb/s}}{5.5 \times 10^2 \text{ ft} \cdot \text{lb/s/hp}}$$

$$P = 0.5 \text{ hp}$$

Since power is work or energy expended per unit time, we can find the amount of energy used in terms of power.

$$P = \frac{W}{t}$$

so

$$W = P \times t$$

Since energy is closely related to work, we can write

$$E = P \times t$$

where E = energy consumed. This formula is useful in computing the amount of energy consumed in a home.

As an example, let us compute the energy consumed if we burn a 100-W light bulb for 80 hours.

$$E = P \times t$$

$$E = 100 \text{ W} \times 80 \text{ h}$$

$$E = 8000 \text{ W} \times \text{h}$$

$$E = 8 \text{ kW} \times \text{h}$$

$$E = 8 \text{ kWh}$$

We see that the total energy consumed by the bulb is 8 kWh. Electric companies use kilowatt-hours (abbreviated kWh) as a unit of energy. The price of electricity is figured at a certain rate per kilowatt-hour.

4-3
KINETIC ENERGY

The concept of energy is an abstract idea derived from a definition of work. Energy is often defined as the capacity to do work or simply as stored work. In mechanics, which is a branch of physics, energy is classified as *kinetic* or *potential* energy and work is defined as the product of force and distance. Starting with this concept of work, we derive in Appendix V an equation for work against inertia. This equation is

$$W = \frac{mv_f^2}{2} - \frac{mv_o^2}{2}$$

where m = mass of an object being moved,
v_o = original velocity of the object,
v_f = final velocity of the object.

In studying this equation we see that the right-hand side is the change in the quantity $mv^2/2$. So we can also write

$$W = \Delta\left(\frac{mv^2}{2}\right)$$

where the symbol Δ means "change in."

Thus, we see that the work on an object results in a change in the quantity $mv^2/2$. This very important combination of mass and velocity is given the name kinetic energy, which is the energy an object possesses due to its motion. In symbols, we write

$$E_k = \frac{mv^2}{2}$$

where E_k = kinetic energy.

We can now write the relationship between work and kinetic energy as

$$W = \Delta E_k \qquad \text{Eq. 4.2}$$

where W = work done,
Δ = change in,
E_k = kinetic energy.

Figure 4-3a *The work necessary to increase the velocity of a mass is equal to the increase in kinetic energy. (Assume no energy loss.)*

This is an extremely important formula. It relates work to energy of motion. Both the kinetic energy and work are scalar quantities. They must also have the same units, as was indicated in Table 4-1a.

The following examples show how work against inertia is related to kinetic energy:

■ *Example 3* A mass of 20 g at rest is given a velocity of 30 cm/s (Fig. 4-3a). Calculate the work done to accomplish this. (Assume no loss due to friction.)

m=20 g
v=30cm/s

$$W = \left(\frac{mv_f^2}{2}\right) - \left(\frac{mv_o^2}{2}\right)$$

W=mv²/2

Since $v_o = 0$,

$$W = \frac{mv_f^2}{2} = \frac{20 \text{ g}}{2} \times \frac{30^2 \text{ cm}^2}{\text{s}^2}$$

$$W = 10 \text{ g} \times 900 \text{ cm}^2/\text{s}^2$$

$$W = 9 \times 10^3 \text{ dyne cm}$$

or $\qquad W = 9 \times 10^3 \text{ ergs}$

Since the original velocity was zero, 9×10^3 ergs is also the magnitude of the change in the kinetic energy.

■ *Example 4* An automobile with a mass of 1600 kg is moving with velocity of 15 m/s. Calculate its kinetic energy at this velocity.

m = 1600 kg
v = 15 m/s
KE = mv²/2

$$E_k = \frac{mv^2}{2} = \frac{1600 \text{ kg} \times 15^2 \text{ m}^2/\text{s}^2}{2}$$

$$E_k = 1.8 \times 10^5 \text{ N} \cdot \text{m}$$

or $\qquad E_k = 1.8 \times 10^5 \text{ J}$

■ *Example 5* What is the kinetic energy of the same automobile (Example 4) at the time when its velocity is 30 m/s?

$$E_k = \frac{mv^2}{2} = \frac{1600 \text{ kg} \times 30^2 \text{ m}^2/\text{s}^2}{2}$$

$$E_k = 7.2 \times 10^5 \text{ N} \cdot \text{m}$$

or $\qquad E_k = 7.2 \times 10^5 \text{ J}$

(Note: Doubling the velocity increases the kinetic energy fourfold. Velocity used in the above formula is always instantaneous.)

4-4
POTENTIAL ENERGY

Potential energy is defined as the energy a body possesses due to its position in a force field. For example, if a book (mass 1 kg) at rest on the floor is lifted a height of 1 m to the top of the table, work is done. To find the amount of work done, the mass of 1 kg is first changed to force units by multiplying the mass by g, the acceleration due to gravity. (1 kg × 9.8 m/s² = 9.8 newtons.) Then,

$$W = \bar{F}s$$

$$W = 9.8 \text{ N} \times 1 \text{ m}$$

$$W = 9.8 \text{ N} \cdot \text{m}$$

or $\qquad W = 9.8 \text{ J}$

In this case, work has been done against the force of gravity. The book on the table now possesses energy and the ability to do work, due to its position in a gravitational force field. The energy is called **gravitational potential energy.** If the book were allowed to fall back to the floor, it would do work on the floor. From the equations

$$W = \bar{F}s$$

and

$$F = mg$$

the work it would do can be determined by

$$W = mgh$$

since the displacement s is the height h through which the book is lifted.

As noted previously, potential energy has been defined as the energy a body possesses due to its position in a force field; therefore, the work done is equal to the potential energy. This can be written

$$W = mgh = E_p \qquad \text{Eq. 4.3}$$

where W = work,
m = mass,
g = acceleration due to gravity,
h = height,
E_p = potential energy.

The work done on the book is independent of the path taken in lifting the book from the floor to the table top. The force necessary to lift the book is the force needed to overcome gravity, which is acting downward; therefore, the component of the applied force used to lift the book must be upward, no matter what path the book takes in arriving at the top of the table (Fig. 4-4a).

The potential energy a body possesses depends upon the reference level from which the height is measured, and the value of the energy may take on either a positive or negative value. For example, in Fig. 4-4b where surface O is taken as the reference level, the potential energy the mass m possesses at position T is equal to the work done in moving the mass from the reference level through the vertical

Figure 4-4a *Drawing illustrating the potential energy E_p gained by a mass transferred via three different paths from the floor to the top of the table. The E_p of m at the table top is independent of the path taken from the floor. The force acting downward is mg, and it acts vertically downward at all positions.*

Figure 4-4b *Diagram illustrating the idea that potential energy may be positive or negative. Whether the energy is positive or negative depends on the position of the reference level.*

height h_1. The magnitude equals mgh_1 and its value would be positive. When the mass is returned from position T to the reference level O, the mass releases energy equal to the amount received when raised through the height h_1.

When the mass is moved from the reference level O to position B through height h_2, the work required is equal to $-mgh_2$. The potential energy value will be negative. If we wish to return the mass to the reference level, work must be done on the mass equal to the amount it released in going from the reference level to position B.

Any reference level can be chosen. The energy difference between the levels will remain the same, and it is this *difference* which is useful for doing work.

The potential energy is a *scalar quantity*, as are work and kinetic energy. This makes it a very easy quantity to use because it is independent of the path taken as illustrated in Fig. 4-4a. Potential energy also has the same units as work and kinetic energy. These are given in Table 4-1a.

4-5
LAW OF CONSERVATION OF ENERGY

Although the meaning is always the same, the law of conservation of energy is stated in many different ways. Examples of this are as follows: *the total energy of a system remains constant; energy can be neither created nor destroyed; in changing from one form to another, energy is always conserved.*

The **law of conservation of energy,** as mentioned above, can be stated: The total energy of a system remains constant. This means that although energy may be changed from one form to another, energy is not lost.

In equation form we would state the law of conservation of energy as

(Total energy)$_{\text{time } 1}$ = (Total energy)$_{\text{time } 2}$ Eq. 4.4

That is, the total energy does not change with time. If the only forms of energy involved are kinetic and potential energy, we would write

$$(E_T)_{t_1} = (E_T)_{t_2}$$

where E_T = total energy,
         t_1 = time 1,
         t_2 = time 2, a later time.

$$(E_k + E_p)_{t_1} = (E_k + E_p)_{t_2}$$

or $(\tfrac{1}{2}mv^2 + mgh)_{t_1} = (\tfrac{1}{2}mv^2 + mgh)_{t_2}$

■   The law of conservation of energy can be illustrated by Susannah, who is sledding down

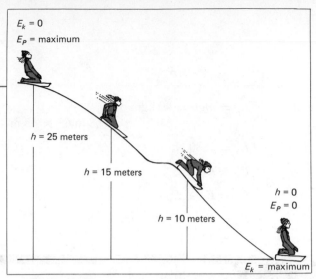

the snowy north side of Longview Heights (Fig. 4-5a). Susannah starts from rest at a height of 25 m above the lower level of the hill. What will be the kinetic and potential energies of Susannah plus the sled at the heights indicated on the drawing if she and the sled have a total mass of 40 kg? (Assume no losses due to friction.)

At the top of the hill, the velocity of the sled is zero.

$$E_T = E_k + E_p$$

Since $E_k = \dfrac{mv^2}{2} = 0$

then $E_T = E_p = mgh$
            $= 40 \text{ kg} \times 9.8 \text{ m/s}^2 \times 25 \text{ m}$

or $E_T = E_p = 9800 \text{ N} \cdot \text{m or joules}$

At $h = 15$ meters

$$E_T = E_k + E_p = 9800 \text{ J}$$

Figure 4-5a *The law of conservation of energy. All the available energy at the top of the hill is potential because the velocity of the sled and rider is zero. As the sled and rider move down the hill, the potential energy decreases and the kinetic energy increases. At the bottom of the hill all the available energy is kinetic, the velocity is maximum, and the potential energy is zero.*

In the figure: $E_k = 0$, E_p = maximum; $h = 25$ meters; $h = 15$ meters; $h = 10$ meters; $h = 0$, $E_p = 0$; E_k = maximum

and
$$E_p = mgh = 40 \text{ kgm} \times$$
$$9.8 \text{ m/s}^2 \times 15 \text{ m}$$

or
$$E_p = 5880 \text{ N} \cdot \text{m or joules}$$

From
$$E_T = E_k + E_p$$
$$E_k = E_T - E_p$$

or
$$E_k = E_T - E_p = 9800 - 5880$$
$$= 3920 \text{ J}$$

The loss in potential energy has been gained in kinetic energy. At $h = 10$ m

$$E_T = E_k + E_p = 9800 \text{ J}$$

and
$$E_p = mgh$$
$$= 40 \text{ kg} \times 9.8 \text{ m/s}^2 \times 10 \text{ m}$$

or
$$E_p = 3920 \text{ J}$$

from
$$E_T = E_k + E_p$$
$$E_k = E_T - E_p$$

or
$$E_k = 9800 - 3920$$
$$= 5880 \text{ J}$$

At $h = 0$
$$E_T = E_k + E_p = 9800 \text{ J}$$

since $h = 0$,
$$E_p = mgh = 0$$

from
$$E_T = E_k + E_p$$
$$E_k = E_T - E_p$$

or
$$E_k = 9800 \text{ J}$$

At any point along the slope of the hill, the sum of the potential and kinetic energies is equal to the total energy. Observe at the top, where the velocity is zero, the kinetic energy is zero, but h is a maximum, and the potential energy is maximum and equal to the total energy of the system. As the sled (and Susannah) travels down the hill, the potential energy becomes less because the height is decreasing. As the sled descends and loses potential energy, it gains in velocity and increases in kinetic energy. At the bottom, where h is zero, all the energy of the system is in the form of kinetic energy, and the velocity of the sled is at its maximum value.

Note that the shape of the hill makes no difference in these calculations (if we continue to assume that friction can be neglected). This fact makes the law of conservation of energy a very useful tool. For any system that starts out at rest and then changes its height by an amount h, we can write

$$(E_T)_\text{top} = (E_T)_\text{bottom}$$

or

$$(\tfrac{1}{2}mv^2 + mgh)_\text{top} = (\tfrac{1}{2}mv^2 + mgh)_\text{bottom}$$

but the velocity at the top is zero and the height at the bottom is zero, so our formula becomes

$$0 + mgh = \tfrac{1}{2}mv^2 + 0$$

or
$$\tfrac{1}{2}v^2 = gh$$

or
$$v^2 = 2gh$$

Remember in this formula that the symbol h stands for the *change* in the height.

The simple pendulum (Fig. 4-5b) also illustrates many of the same features of the law of conservation of energy. A **simple pendulum** consists of a heavy mass (called a bob) attached to a light string and supported so that the bob can swing back and forth, or oscillate, freely. The pendulum is called "simple" because most of the swinging mass is in the bob. The string, for all practical purposes, can be neglected. An object such as a freely swinging meter stick attached at one end is called a **compound pendulum.** The mass in this system is not concentrated at a point but distributed throughout the oscillating system.

The simple pendulum, as shown in Fig. 4-5b, can be set in motion by displacing the bob to one side (angle θ) and releasing it. The bob will swing back and forth with periodic motion, undergoing energy transformations. When the bob is at position a or c, it will be momentarily at rest. The velocity will be zero at these two positions, and the kinetic energy will be zero. The potential energy, however, will be maxi-

h is maximum
$E_p = mgh$ = maximum value
$v = 0$
$E_k = \dfrac{mv^2}{2} = 0$

b
v = maximum value
$E_k = \dfrac{mv^2}{2}$ = maximum value
$E_p = 0$

c
$v = 0$
h is maximum

d
v = maximum value
$h = 0$

Figure 4-5b *A simple pendulum. The pendulum swinging back and forth illustrates energy transformation.*

mum, because the mass is at its maximum height above the lowest part of the swing, or reference level. When the bob swings down from position *a*, the potential energy will decrease and the kinetic energy will increase correspondingly as the mass gains velocity. When the bob is at point *b*, it will have zero potential energy, because *h* will have become zero. The kinetic energy at this point will be maximum and the mass will be traveling at its maximum velocity. As the bob swings upward toward point *c*, the velocity will start to decrease, and the potential energy will begin to increase. As the bob swings higher, the velocity will continue to decrease until it stops momentarily at point *c*, where the kinetic energy will have decreased to zero and the potential energy will have reached the maximum. Thus, as the pendulum swings

back and forth, the energy is converted from potential to kinetic, back to potential.

The velocity of the bob at point *b* can be calculated by applying the law of conservation of energy as follows:

At points *a* or *c* $E_T = E_p$ because $E_k = 0$

At points *b* or *d* $E_T = E_k$ because $E_p = 0$

Therefore, E_k at the bottom of the swing $= E_p$ at the top of the swing

or
$$\frac{mv^2}{2} = mgh$$

Dividing through by m

$$\frac{v^2}{2} = gh$$

or
$$v^2 = 2\,gh$$

Taking the square root of both sides

$$v = \sqrt{2\,gh}$$

When h (the height above the reference level) is measured in feet, g will equal 32 ft/s²

then
$$v = \sqrt{2 \times 32 \times h}$$

or
$$v = 8\sqrt{h}\ \text{ft/s}$$

Again h is the difference in height between the top and bottom part of the pendulum's swing. This formula holds for a child's swing. If $h = 4$, the speed of the swing at the bottom will be 16 ft/s or 10.9 mi/h.

4-6
FORMS OF ENERGY

If we take the law of conservation of energy seriously, we must consider the examples of Section 4-5 in more detail. For instance, we know that both the sled and the pendulum will finally stop, and we

Figure 4-6a *We get our energy fuel from food. The body uses this energy to perform our daily tasks and to maintain our body temperature. Each of us radiates heat energy at about the same rate as a 100-watt light bulb. (Photo courtesy James Crouse.)*

Figure 4-6c *Petroleum products such as gasoline make up 44% of our nation's energy supply. Transportation needs comprise about one-fourth of our national energy needs. (Photo courtesy James Crouse.)*

Figure 4-6b *Our chief source of electrical energy is coal. (Photo courtesy James Crouse.)*

Figure 4-6d *Natural gas is widely used in home furnaces. Space heating needs comprise about one-fifth of our national energy needs. (Photo courtesy James Crouse.)*

must ask the question, Where did the energy go? Once again, friction is involved. In most practical situations the kinetic or potential energy of large objects eventually ends up as heat energy. Heat energy will be studied at some length in Chapter 7, but for now, let us say that heat energy is kinetic and potential energy on a microscopic level. As things get hot, the energies of the atoms and molecules that make up different substances increases.

There are many forms of energy. Besides the macroscopic kinetic and potential energy of objects, there are the following forms of energy: heat, radiant, sound, chemical, nuclear, and others. All of these are related to microscopic kinetic and potential energies which act together to give macroscopic effects. In addition we will see in Chapter 8 that mass itself is a form of energy.

The main unifying concept when considering energy is the law of conservation of energy. We cannot create or destroy the energy of a single system. However, we can transform it from one form to another. On a global scale, the source of practically all of our energy is the sun. Wood, coal, petroleum, and natural gas all come about because of the sun's radiant energy, which strikes the earth. This radiant energy provides us with an external source of energy that should be with us for billions of years. The earth's energy balance is discussed in more detail in Chapter 21.

As we go about our daily lives, each of us is constantly using energy and giving off energy in the form of body heat. The source of this energy is food (Fig. 4-6a). An average adult radiates heat energy at the same rate as a 100-watt light bulb. This explains why a crowded room soon gets hot. In winter, our clothing helps to keep this heat energy within our bodies. In summer, sweating helps us to give off the heat energy which our body produces.

The sources of our energy on a national scale are mainly coal, petroleum, and natural gas. Approximately 20% of our energy comes from coal, 44% from petroleum products, 34% from natural gas, and approximately 1% each from hydroelectric and nuclear power.

Coal is used primarily for the generation of electricity. Figure 4-6b shows a stockpile of coal for use at a power plant. Coal is very abundant in the United States and many national leaders have called for an increase in its usage.

Petroleum is our most widely used source of energy. It is used mainly in transportation (Fig. 4-6c), but it is also widely used in the residential, commercial, and industrial sectors of the economy. With the Arab oil boycott of 1973, the price of petroleum products increased greatly. This also caused other forms of energy to jump drastically in price. Not only is petroleum used as a fuel, it is also the starting material for the petrochemical industry. Plastics and many other synthetic materials, which are discussed in Chapter 16, are made from petroleum.

Natural gas is a cleanly burning fuel used in home furnaces (see Fig. 4-6d) as well as in business and industry. The burning of natural gas results in a minimum of pollution problems. It is vitally important to our nation's energy picture, and there are potentially vast reserves of this source of energy. In the winter of 1976–1977, a shortage of natural gas closed down industries and schools in several states. This caused a great deal of concern on both the state and national levels, and a national energy policy must deal with the natural gas question.

4-7
ENERGY CONSUMPTION

The importance of energy can be seen by comparing it to a country's gross national product or GNP. The GNP represents the dollar amount of goods and services produced by the country. It is closely related to a country's standard living. Figure 4-7a shows a plot of per capita GNP versus per capita energy

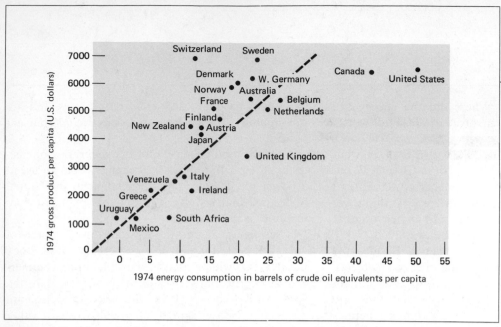

Figure 4-7a *A graph of per capita gross national product (GNP) versus per capita energy consumption shows that the two are clearly related. The per capita GNP is commonly referred to as the standard of living in a country. The United States seems to use far more energy to attain the same standard of living as Switzerland, Sweden, and Canada. (Source: U. N. Statistical Yearbook, 1975.)*

Table 4-7a *Energy Requirements in the U.S. According to End Use*

End Use	Percent of National Energy Requirements
Transportation (fuel only, excludes lubricants and greases)	24.9
Space heating (residential and commercial)	17.9
Process steam[a] (industrial)	16.7
Direct heat[a] (industrial)	11.5
Electric motors (industrial)	7.9
Feedstocks and raw materials (commercial, industrial, transportation)	5.5
Water heating (residential and commercial)	4.0
Air conditioning (residential and commercial)	2.5
Refrigeration (residential and commercial)	2.2
Lighting (residential and commercial)	1.5
Cooking (residential and commercial)	1.3
Electrolytic processes (industrial)	1.2
Other	2.9
Total	100.0

[a] Includes some uses for space heating; probably enough to bring the total for space heating to about 20%.

Source: Patterns of Energy Consumption in the United States, Office of Science and Technology, Executive Office of the President, January 1972, p. 7.

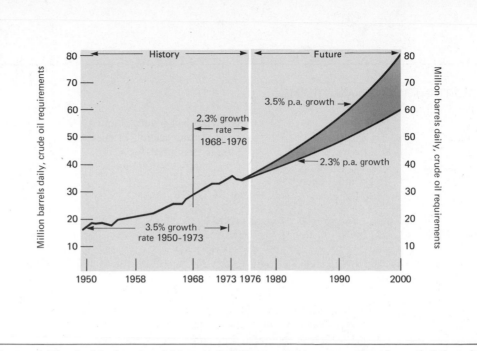

Figure 4-7b *A plot of past and future United States energy consumption shows that there is a considerable difference between a yearly (per annum) growth rate of 2.3% and 3.5%. (Source: U. S. Bureau of Mines).*

consumption for the year 1974 for various countries.

Several ideas are illustrated by Fig. 4-7a. First, it appears that the amount of energy consumed is related to a country's standard of living. This is represented by the dotted line. Second, it can be seen that the United States uses far more energy per person than do Switzerland and Sweden, yet these countries have a slightly higher standard of living. This tells us that we are using far more energy than is really necessary. Since our present energy supplies are slowly running out, it is important that we all try to conserve as much energy as possible.

The energy which we consume has constantly increased, partly due to an increase in population. During the years 1950–1973 the energy consumption in the United States increased at the rate of 3.5% per annum. However, this growth rate was less during the years 1968–1976 when it was 2.3% per annum. This lowered rate of increase was largely due to conservation efforts begun in 1973. If we can continue to conserve, we can drastically cut our future national energy needs. Figure 4-7b shows the amount of en-

ergy needed each day in order to supply our national energy needs. From this figure we can see that in 1973 there was actually a decrease in the amount of energy used in the United States. By keeping our growth rate at 2.3% per annum rather than 3.5%, we can cut 25% from our projected needs in the year 2000. Of course, predicting our energy needs in the future is a very risky business.

Where does our energy go that we consume each day? Table 4-7a shows the end use of our energy requirements. It can be seen from this table that transportation uses more energy than any other category. Space heating is the second highest use of energy. These two uses comprise almost 50% of the end use of all energy consumed in the United States. This gives us a good idea as to where our conservation efforts should be focused.

The energy problem will be with us for the rest of our lives. It is important that we as citizens learn to control and reduce our daily energy needs. If we all work together towards conservation efforts, the so-called energy crisis can be weathered.

IMPORTANT WORDS AND TERMS

Work *Fxd*
Joule *work in mks*
Erg *work in cgs*
Foot-pound *British work*
Power = *w/t*

Watt = *power in mks*
Horsepower *power in Brit.*
Energy = *stored work*
Kinetic energy = *motion*
Potential energy = *position*

Gravitational potential energy *work done against gravity*
Law of conservation of energy *can't be made constant*
Simple pendulum *Et remains constant no energy lost or destroyed*
Compound pendulum

LEARNING OBJECTIVES

After reading and studying this chapter, you should be able to do the following without referring to the text:

1. Define the term *work* and give its units in the cgs, mks, and British systems of units.
2. Define the term *power* and give its units in the cgs, mks, and British systems of units.
3. Define the terms *energy*, *kinetic energy*, and *potential energy* and give their units in the cgs, mks, and British systems of units.
4. State the law of conservation of energy, and explain why it is important.
5. Discuss the pendulum as an example of the law of conservation of energy.
6. List at least five different forms of energy.
7. List our main sources of energy.
8. State the relationship between standard of living and energy consumption.
9. State the two principal uses of energy in the United States.

QUESTIONS

1. Distinguish between work and energy.
2. Can you have a force being exerted with no work? Explain.
3. Which of the following are scalar quantities: force, parallel displacement, work, energy, power.
4. Distinguish between kinetic energy and potential energy.
5. Can you have energy without motion? Explain.
6. Can you have negative amounts of energy? Explain.
7. For a swinging pendulum, where is the kinetic energy the greatest? Where is it smallest?
8. What happens to the energy of a pendulum after it stops swinging?
9. When is the law of conservation of energy invalid?
10. Why does a crowded room get hot?
11. List five different forms of energy other than kinetic or potential energy.
12. What are the three principal sources of energy in the United States?
13. Which nation consumes the most energy per capita?
14. What are some of the principal uses of energy in the United States?

PROBLEMS

[handwritten: w = fd or mad or mgd]

1. A 10-kg mass is at rest on a table top.
   (a) Calculate the work required to lift the mass 2 m.
   (b) Calculate the work required to move the mass 3 m along the table if the force of friction is 12 newtons. *[handwritten: f=12N d=3m ...]*
   (c) If there is no friction, calculate the work required to give the mass a velocity of 5 m/s.

   Answer: (c) 125 joules

2. Referring to problem 1, what would be
   (a) The work required to lift the 10-kg mass a vertical distance of 2 m on the moon?
   (b) The work required on the moon to move the mass 3 m along the table if the force of friction is 12 N?
   (c) The work required on the moon to give the mass a velocity of 5 m/s, if there is no friction? *[handwritten: KE = \frac{mv^2}{2}]*

   Answer: (b) 36 joules

3. Calculate the kinetic energy of a rock weighing 0.5 kg moving at a velocity of 6 m/s.

4. For a comet orbiting the sun, answer the following questions:
   (a) Where is its *kinetic* energy the greatest?
   (b) Where is its *potential* energy the greatest?
   (c) Where is its *total* energy the greatest?

5. What is the potential energy of a child's swing of mass 3 kg that is stationary and 2 ft off the ground? Explain.

6. What is the horsepower necessary to carry 200 lb of books to a height of 22 ft in 40 s?

   Answer: 0.2 horsepower

7. A student weighing 600 N climbs a stairway (vertical height of 4 m) in 20 s.
   (a) How much work was done?
   (b) What was the power output during the 20 s?

8. A child is swinging on a rope which is 15 ft long. At the lowest point the child is 2 ft off the ground. At the highest point the child is 3 ft off the ground. What is the child's maximum speed?

   Answer: 8 ft/s

9. How many watts are generated in a crowded room with three 100-W light bulbs and 30 people? *[handwritten: 3300]*

10. A sled and rider with a combined weight of 700 N is at the top of a hill which is 5 m high.
    (a) What is their total energy at the top of the hill (use the bottom of the hill as a reference point)?
    (b) They slide down the hill and eventually slow down and stop. What happened to the energy? *[handwritten: transferred to heat unrecoverable]*

    Answer: (a) 3500 joules

Chapter 5
WAVE MOTION

Close observation and an accurate description of nature provides man with the true facts, with truth standing above all.

Since beginning our story of energy, much has been said about the forms of energy, the relation of energy to work, and the law of energy conservation—with many questions raised and answered. Other interesting questions remain, however. For example, how is energy changed from one form to another? How is energy transferred to the ball in a baseball game? How do our bodies obtain energy? How is energy transferred to a train, a car, a space ship, a particle of any size? How do we get energy from the sun?

A partial answer to these questions has been found in terms of particle collisions and wave motion, which are two important ways for transferring energy (see Fig. 5-1a). If a particle applies a force to another particle through a distance, then a transfer of energy has taken place through particle collision, one particle will gain energy, and the other will lose energy. If we are to believe the law of conservation of energy, and we know of no instance where the law has failed, then the increase of energy of one particle will be equal to the decrease of energy of the other, providing we neglect heat losses. In all cases, we find a transfer of energy takes place whenever two or more particles collide with one another.

When matter is disturbed, energy emanates from the disturbance; this emanation of energy is known as wave motion. For example, a stone dropped on the surface of a pond of water will disturb the water, and energy will be transferred outward from the disturbance as wave motion. Notice that only energy is transferred, not matter (the water). A similar situation can occur in a solid. For example, during an earthquake a disturbance takes place because of a slippage or other cause, and this disturbance is transmitted to

Figure 5-1a *Examples of transferring energy.*

all parts of the earth as wave motion. Again, this is a transfer of energy, not matter. The transfer of energy takes place with or without a medium. Sound waves in the air and waves upon stretched strings and steel wires are examples of energy transmissions that require a medium (air, strings, and wires in the above cases). The neighboring particles of the mediums react upon one another to transfer the disturbance. Electromagnetic waves, including radio, infrared, light, and X-rays, are transferred without a medium. We say these disturbances are radiated through space.

The data from which we learn about the planets, the sun, and other stars come to us by means of light. Later, when quantum mechanics is discussed, we shall study the dual nature of light. Strangely enough, light, which we consider to be a wave motion, has some of the properties of matter. This will be discussed in Chapter 10.

Our eyes and ears are two wave-detecting devices that serve to link us to our environment. Since a study of wave motion seems relevant to an understanding of our physical environment, a knowledge of wave motion is essential to the understanding of many scientific principles.

5-1
SIMPLE HARMONIC MOTION

The disturbance generating a wave motion is usually periodic; that is, the disturbance is repeated again and again at regular intervals. The plucking of a guitar string or the blowing of a whistle sets up periodic waves (Fig. 5-1b). The disturbance, and the resulting wave motion, does not necessarily have to be periodic; it may be a simple pulse, or shock wave, such as the one originating from a book hitting the floor, or from a jet plane passing through the sound barrier.

Several of the terms and concepts relating to particle disturbance are illustrated in Fig. 5-1c. At (a) is

Wooden ruler

Rubber band

Table

Steel or nylon string

Air column

Figure 5-1b *Some common methods used to generate wave motion.*

Figure 5-1c *Diagram illustrating how a mass m undergoes simple harmonic motion.*

shown a mass m attached to an elastic coiled spring supported at the top. The mass has stretched the spring slightly and is at rest in equilibrium. At (b) an external force has been applied downward, stretching the spring and giving it elastic potential energy. The spring and mass have been displaced through the distance s from the equilibrium position. The external force has set up a stress in the spring and a deformation has taken place; that is, the size and shape have changed from the original size and shape. A restoring force exists to return the spring to its original condition, and when the downward force is removed, the spring returns to its original equilibrium position, but due to the inertia of the system, the mass overshoots the original position, and the spring becomes compressed, (c). A restoring force, now acting downward, returns the spring and mass to their original equilibrium position, but again they pass through this position because of the property of inertia. Thus, the mass stretches and compresses the spring again and again; that is, the disturbance continues at periodic, or regular, intervals decreasing in maximum displacement with each cycle, until the energy originally supplied is dissipated as heat, and the mass and spring are at the equilibrium position, (d).

The motion illustrated by the mass and spring is known as **simple harmonic motion** (SHM). Simple harmonic motion is produced whenever a restoring force acting on an object is directly proportional to the displacement. The period, defined as the time required for one complete cycle to take place, can be found by the following equation:

$$T = 2\pi \sqrt{\frac{m}{k}}$$

where T = the time required to make one complete cycle,

m = mass of the object,

k = the force constant of the spring,

π = 3.1416 (a part of the equation, because we are dealing with cyclic motion).

5-2
WAVE PROPERTIES

A disturbance in an elastic body will set up wave motion. For example, in Fig. 5-2a a stretched spring is attached between two fixed points. When four or five coils at one end of the spring are compressed and released, the disturbance is propagated along the length of the coiled spring; that is, the disturbance moves with the velocity vector parallel to the direction in which the particles were displaced. The velocity is known as **wave velocity;** when the particle displacement and the wave motion are in the same direction the wave is called a **longitudinal wave.** Sound waves are of this type.

The term **transverse wave** is used to denote wave motion in which the individual particles are displaced perpendicular to the direction of the wave velocity

Figure 5-2a *A longitudinal wave in a stretched spring.*

Figure 5-2b *A transverse wave in a stretched cord.*

Figure 5-2c *The terms used to describe transverse wave motion.*

vector. Figure 5-2b is an illustration of a transverse wave in a stretched rubber cord. The cord is disturbed from its equilibrium position by moving the end of the cord up and down. Moving the cord end from side to side—or in any direction—will also produce a transverse wave. The wave motion in this example is to the right, as indicated by the velocity vector.

Transverse waves, unlike longitudinal waves, can be polarized, a physical property that is used to distinguish between longitudinal and transverse waves. A *polarized wave* is one in which the particle disturbance occurs in only one direction, perpendicular to the direction the wave is traveling. Polarization will be discussed in more detail when the properties of light are explained.

The transfer of energy by transverse waves can take place in the absence of a medium. All electromagnetic radiation is of this type of wave motion. The study of wave motion has resulted in certain terms which are used to explain the action of all waves (Fig. 5-2c). Velocity describes the direction and magnitude of the wave motion. The velocity of a longitudinal wave depends upon the properties of the medium. The **frequency** is the number of oscillations, or complete waves, that occur during a given period of time, usually one second. The units of frequency are cycles

per second. This unit is given the name hertz. One **hertz** (Hz) is one cycle per second. For example, if five complete wave crests pass a given spot in one second, the frequency would be five cycles per second or five hertz.

The **period** of a wave is the time it takes for one complete wave oscillation. If five crests pass by a given point in one second, one crest or complete cycle would pass in one-fifth of a second, and the period would be $\frac{1}{5}$ second. The frequency would be 5 cycles per second. From this example it is easy to see that

$$f = \frac{1}{T}$$

where f = the frequency in number of cycles per second, or hertz,
T = the period in seconds.

The wavelength of a wave is measured in units of length. One **wavelength** is the distance from any point on a wave to an identical point on the adjacent wave as shown in Fig. 5-2c.

A simple relation between wave velocity, wavelength, and period (or frequency) exists. This can be written as

$$v = \frac{\lambda}{T}$$

or

$$v = \lambda f$$

where v = wave velocity measured in meters per second, centimeters per second, feet per second, or any other velocity units,
λ = wavelength, measured in length units,
T = period of the wave, usually measured in seconds,
f = frequency of the wave, measured in cycles per second or hertz.

The **amplitude** of a wave refers to the maximum displacement of any part of the wave from its equilibrium condition. The amplitude of the wave does not affect the wave velocity. The amplitude of a wave squared is related to the energy transmitted by the wave.

5-3

ELECTROMAGNETIC WAVES

An electron at rest possesses an electric force field, the magnitude of which falls off rapidly with an increase in distance from the electron. Coulomb's equation* for the field indicates that the strength of the field varies inversely with the square of the distance. This electric force field, called the Coulomb field, is stationary in space; that is, there is no radiation of the field. When the electron is put in motion, a magnetic force field appears and is associated with the Coulomb field. The strength of the magnetic field also varies inversely with the square of the distance, and the field remains stationary in space, that is, there is no radiation of the field. Similar properties also exist with the proton. The proton is a very small particle with a mass about 2000 times that of the electron and a charge of equal magnitude but opposite sign.

When the electron is accelerated, work is done on the electron and energy is radiated away from the electron in the form of **electromagnetic waves;** that is, the electric and magnetic fields radiate outward with a velocity of 3×10^8 meters per second. This is always true of any accelerated charge. Many experiments have been performed with accelerated charges, and in all cases the propagation of energy takes place in the form of electromagnetic waves.

A drawing of an electromagnetic wave is shown in Fig. 5-3a. The drawing shows the wave traveling in a right-hand direction, slightly away from the plane of the page. The electric and magnetic vectors are at 90° to one another, and the velocity vector is at 90° to the vector fields.

Electromagnetic waves are produced when a charged particle is accelerated. Charged particles are accelerated in many different ways to produce electromagnetic waves of various frequencies. Waves with low frequencies, or long wavelengths, are known as

*Coulomb's equation is given in Section 6-1.

Figure 5-3a *Diagram of the vector components of electromagnetic waves. Electromagnetic waves consist of two force fields (electric and magnetic) perpendicular to one another and moving with a velocity vector that is perpendicular to both force fields.*

radio waves and are produced primarily by causing electrons to oscillate, or vibrate, in a resonant circuit. The frequency of oscillation is controlled by the physical dimensions and other properties of the tuned circuit.

The production of electromagnetic waves with frequencies greater than radio waves is accomplished by molecular excitation. In such cases, radiation occurs from the collision of molecules in hot gases and solids. Since the molecules carry charged particles which are greatly accelerated as the molecules vibrate, the particles will radiate electromagnetic waves ranging from 10^{12} cycles per second to 3.8×10^{14} cycles per second. This portion of the electromagnetic spectrum is called the *infrared* region. As the temperature of gases and solids is increased to higher and higher values, the atoms composing the molecules become more excited and electromagnetic radiation in the *visible* and *ultraviolet* regions of the spectrum is emitted. Still more energy applied to the atom will generate waves of higher frequencies, called *X-rays*, which range from 3×10^{17} to 3×10^{19} cycles per second. If sufficient energy is applied to the atom to disturb the nucleus, radiation known as *gamma rays* is emitted. The major portion of the electromagnetic spectrum is shown in Fig. 5-3b. The portion of the spectrum visible to the human eye is shown on color plate I.

Figure 5-3b *The electromagnetic spectrum.*

The word "light" is commonly given to visible electromagnetic radiation. However, only the frequency (or wavelength) distinguishes visible electromagnetic radiation from the other portions of the spectrum. Our human eyes are only sensitive to certain frequencies or wavelengths, but other instruments can detect other portions of the spectrum. For example, a radio receiver can detect radio waves.

Electromagnetic radiation consists of transverse waves. These waves can travel through a vacuum. For instance, radiation from the sun travels through the vacuum of space before arriving at the earth. All electromagnetic waves travel at the same speed in a vacuum. This speed is called the speed of light. The **speed of light** in a vacuum is designated by the letter c and is equal to the following:

$$c = 186,000 \text{ mi/s}$$
$$c = 3 \times 10^8 \text{ m/s}$$
$$c = 3 \times 10^{10} \text{ cm/s}$$

This is a very high speed; and, according to Einstein's theory of relativity, nothing can travel faster than c.

We can use the formula $v = \lambda f$ to find the wavelength of electromagnetic radiation if we know the frequency. Visible wavelengths of light are very short (about 10^{-5} cm) and are expressed in units called

angstroms, named for a Swedish scientist. One **angstrom** is equal to 10^{-8} cm. To find the wavelength if the frequency is known, we use the relation

$$\lambda = \frac{c}{f}$$

where λ = the wavelength in centimeters,
 f = the frequency in cycles per second,
 c = the velocity of electromagnetic radiation,
 or 3×10^{10} cm/s.

■ *Example 1* What is the wavelength of red light expressed in centimeters? In angstrom units?

We know from Fig. 5-3b that the frequency of red light is 3.8×10^{14} cycles/s. Hence,

$$\lambda = \frac{c}{f} = \frac{3 \times 10^{10} \text{ cm/s}}{3.8 \times 10^{14} \text{ cycles/s}}$$

$$\lambda = \frac{3 \times 10^{10} \times 10^{-14}}{3.8} \text{ cm}$$

$$\lambda = 7.89 \times 10^{-5} \text{ cm}$$

Since 1 angstrom (1 Å) = 10^{-8} cm

$$\frac{7.89 \times 10^{-5} \text{ cm}}{1} \times \frac{1 \text{ Å}}{10^{-8} \text{ cm}} = 7890 \text{ angstroms}$$

■ *Example 2* What is the wavelength in meters of radio waves produced by a station which has a frequency of 1200 kilohertz?

Since *kilo* means one thousand and *hertz* means cycles per second, we have

$$f = 1200 \times 10^3 \text{ cycles/s}$$
$$f = 1.2 \times 10^6 \text{ cycles/s}$$
$$\lambda = \frac{3 \times 10^8 \text{ m/s}}{1.2 \times 10^6 \text{ cycles/s}}$$
$$\lambda = 2.5 \times 10^2 \text{ m}$$
$$\lambda = 250 \text{ m}$$

We see from these two examples that radio waves have very much longer wavelengths than those of visible light.

Figure 5-4a *Illustrating the transmission of sound waves through air. Note the varying particle density along the radii, illustrating compressions and rarefactions.*

5-4
SOUND WAVES

Sound is a physical phenomenon originated by varying the degree of particle displacement and propagated by varying pressures and stresses in an elastic medium. Sound waves may originate from particle displacement in any kind of matter: solid, liquid, or gas. Some common sources include vibrating strings, tuning forks, ocean waves, tornadoes, and the human vocal cords. Transmission of sound through any medium is by longitudinal waves. Transmission of sound waves through air is illustrated in Fig. 5-4a.

The wave motion of sound in any medium consists of a series of wave compressions and rarefactions, displaced along the direction of the wave velocity. A **compression** is that portion of a longitudinal wave in which the vibrating particles are relatively close. A **rarefaction** is that portion of a longitudinal wave in which the vibrating particles are relatively far apart. The effect of these varying pressures and stresses on the eardrum and their subsequent interpretations by the brain is the sensation of sound.

Sound waves are generated with frequencies ranging from a few cycles per second up to frequencies in the millions of cycles per second. Frequencies audible to the human ear range from a minimum of about 20 to a maximum of about 20,000 cycles per second. The audible range is a function of the intensity of the sound as well as the frequency. The human ear is most sensitive to a frequency of about 3000 cycles per second. **Pitch** is the highness or lowness of a sound and is the consequence of the frequency, or the number of cycles per second, received by the ear.

The velocity of sound waves in air at 42°F is given by

$$v_{sound} = 1100 \text{ ft/s}$$

$$v_{sound} = 335 \text{ m/s}$$

At higher temperatures this velocity increases slightly and at lower temperatures it decreases. This velocity is much less than the speed of light.

Using the velocity and frequency of sound waves, we can compute their wavelengths.

■ *Example 3* What is the wavelength of a sound wave whose frequency is 2200 hertz?

We use the relation

$$\lambda = \frac{v_{sound}}{f}$$

$$\lambda = \frac{1100 \text{ ft/s}}{2200 \text{ cycles/s}}$$

$$\lambda = 0.5 \text{ ft}$$

From Example 3 and Example 1 we see that wavelengths of visible light (10^{-5} cm) are much shorter than wavelengths of audible sound.

The *loudness* of a sound wave relates to the amount of energy being transferred. Loudness is not related to the frequency or wavelength of a sound wave. Loudness is related to the sound wave's amplitude. When we speak of loudness, we are speaking of how our ear perceives a sound. The measurable quantity which corresponds to this is called intensity.

The **intensity** I of a sound wave is defined as the time rate of transfer of energy per unit area of the wave front. When the mks system of units is used, the rate of energy transfer per unit time is expressed in watts (joules per second) and the wave front area in square meters. The minimum intensity that can be detected by the average human ear is 10^{-12} W/m². The maximum intensity (sound becomes painful) is

Table 5-4a *Intensity Levels, in Decibels, of Sounds from Various Sources*

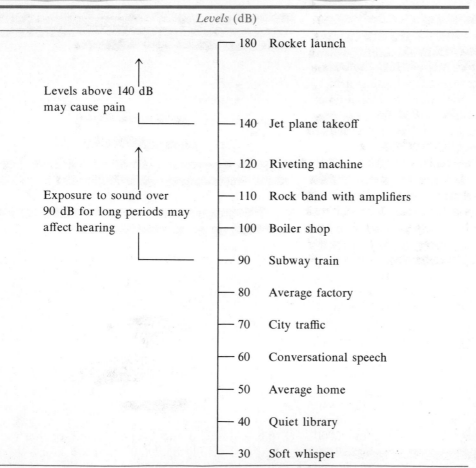

Levels (dB)

Level	Source
180	Rocket launch
140	Jet plane takeoff
120	Riveting machine
110	Rock band with amplifiers
100	Boiler shop
90	Subway train
80	Average factory
70	City traffic
60	Conversational speech
50	Average home
40	Quiet library
30	Soft whisper

Levels above 140 dB may cause pain

Exposure to sound over 90 dB for long periods may affect hearing

Note: An increase of 10 dB corresponds to an increase in the sound intensity level by a factor of 10. For example, an increase from 20 dB to 30 dB increases the sound intensity level from 100 times a standard intensity to 1000 times the same standard.

Figure 5-4b *A demonstration of resonance. The tuning fork on the right will vibrate when the tuning fork on the left is made to vibrate by striking it with a rubber hammer.*

about 10 W/m². This is a very wide range of intensities for the human ear to respond to. Because of this wide range of intensities and since the human ear responds, roughly, proportional to the logarithm* of the intensity, a better and more convenient method of measuring the intensity level of a sound uses the **decibel** scale, which is a logarithmic scale.

The intensity level of a sound of intensity I is defined as follows:

Intensity level in decibels (dB)
$$= 10 \text{ logarithm } (I/I_0)$$

where I = intensity of the sound in question
I_0 = reference intensity of 10^{-12} W/m² (minimum intensity detected by the human ear)

The decibel is equal to 0.1 bel. The bel is named after Alexander Graham Bell, inventor of the telephone. Table 5-4a gives the intensity levels in decibels of sounds from various sources.

Sound waves that come in contact with objects in the environment tend to produce vibrations in these objects. For example, the sonic boom of a jet plane flying at speeds greater than sound, the vibration of human vocal cords, a radio speaker, or other source of sound waves all cause our eardrums to vibrate. The reaction of a particular object to sound waves depends upon its own frequency of vibration. Objects similar in physical properties to the vibrating source tend to vibrate more noticeably than objects that are not similar (Fig. 5-4b). The two tuning forks in this figure are constructed to vibrate at the same frequency and are mounted a few meters apart, as shown. The tuning fork on the left is set in motion by striking it with a rubber hammer. This tuning fork will send out sound waves, which will be received by the tuning fork on the right. Since both forks have

*The logarithm of a number to a given base (the base 10 is used here) is the power to which the base must be raised to equal the number.

been constructed to vibrate at the same frequency, the fork on the right will start oscillating and continue to oscillate on its own. This condition of sympathetic vibration is known as **resonance,** which can be defined as a tuned condition existing whenever an external frequency corresponds to an object's natural frequency of vibration.

There are many instruments designed to use the principle of acoustic (sound) resonance. The structure of the throat and cavities of the head gives the human voice a better tone due to resonance. Resonance can produce a harmful effect at times. Examples of mechanical resonance occur when the wind causes large vibrations to build up in a steel bridge to the point where the bridge will buckle and collapse, or when parts of an automobile vibrate and sometimes break as a result of resonance, especially at particular velocities. Loudspeaker enclosures on some radios tend to vibrate at the lower frequencies because of the resonance effect, producing unwanted sounds.

There are many wave effects or phenomena which we observe quite frequently. A list of some important wave effects would include the following:

Resonance	Dispersion
Doppler effect	Diffraction
Reflection	Interference
Refraction	Polarization
Total internal reflection	

Resonance has already been discussed. We will now proceed to examine the others.

5-5
THE DOPPLER EFFECT

Whenever the source of a wave is moving relative to the observer, the frequency (and wavelength) which is observed changes. This is easily seen by referring to Fig. 5-5a. As the source moves along to the right, the waves at Observer A are closer together than normal while the waves perceived by Observer B are further apart. Observer A would perceive a higher frequency than the source is actually emitting. Observer B would hear a lower frequency.

If the source were stationary and the observer moved toward or away from it, a shift in frequency would also be observed. The **Doppler effect** is the apparent change in frequency resulting from the relative motion of the source or the observer.

The Doppler effect is a general effect that occurs for all kinds of waves, such as water waves, sound waves, and light waves. When a racing car with a loud engine noise approaches, a higher than usual frequency is heard. As it passes by, the frequency suddenly shifts lower and a "whoom" sound is heard. You may have heard a similar noise as a siren passes by.

When the Doppler effect occurs with visible light, the frequencies are shifted higher or toward the blue when the light source (such as a star) is approaching. Since blue is a high-frequency color, we say a *blue shift* has occurred. When a light source moves away from us, the frequency is shifted toward the red, and we say a **red shift** has occurred. The amount of blue shift or red shift will tell us how fast the light source, such as a distant star or galaxy, is moving toward or away from us. Red shifts are extremely important in astronomy because they tell us how fast distant galaxies are moving away from us. As we will see in Chapter 20, this speed is related to how far away the galaxy is from us.

Whenever the speed of the source exceeds the wave speed, a shock wave is formed. This is illustrated in Fig. 5-5b. When this happens with an air-

Figure 5-5a *If a wave source moves to the right at velocity v, the wave crests will be closer together on the right side and further apart on the left side. Observer A will hear a higher than usual frequency while Observer B will hear a lower than usual frequency.*

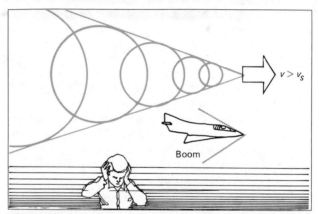

Figure 5-5b *When the velocity v exceeds the speed of sound v_s, a shock wave is formed which is heard as a "sonic boom" when it passes over an observer.*

plane exceeding the speed of sound, the resulting effect is a "sonic boom." A sonic boom occurs because the compressed air from the shock wave travels out and downward behind the plane. The "boom" travels across the ground, and an observer hears it only once. This effect is not observed for light waves in a vacuum, since nothing can travel faster than the speed of light in a vacuum. However, there are other common examples of shock waves, such as a speed boat that travels faster than the water waves.

5-6
REFLECTION, REFRACTION, AND DISPERSION

Waves travel through space in a straight line and will continue to do so unless forced to deviate from their original direction. A change in direction takes place when light strikes a surface or the boundary between two media. A change in direction by this method is called **reflection.** We see most of the objects in our environment because of reflected light. The reflection taking place from the surface of the paper of this text is called **diffuse reflection.** The rays (since light travels in a straight line, we can represent its motion by the use of a straight line called a ray) of light falling on the surface of the paper, when reflected, are not leaving the surface parallel to one another because the surface is not perfectly smooth. When the reflection takes place from a smooth surface, the reflected rays are parallel to one another, and this is called **regular reflection.** This occurs from mirrors and highly polished surfaces. See Fig. 5-6a for an illustration of the two types of reflection.

Figure 5-6b is an illustration of the two laws governing reflection which state: (1) the angle of reflection is equal to the angle of incidence; and (2) the reflected ray and the incident (or incoming) ray are in the same plane. The *angle of incidence* is defined as the angle between the incident ray of light and the normal, or perpendicular. The *reflected angle* is the angle between the reflected ray and the normal.

A diagram illustrating a method for determining the apparent location of an image formed by a mirror is shown in Fig. 5-6c. The image can be located by drawing any two rays of light emitted by the object and observing the laws of reflection. For a plane mirror, the image will be located behind the mirror the same distance as the object is in front of the mirror. Figure 5-6d is a ray diagram showing the light rays involved for the man to see the extremes of his image. Application of the laws of reflection reveals that for the man to see his total image in a plane

Figure 5-6a *Two types of reflected light.*

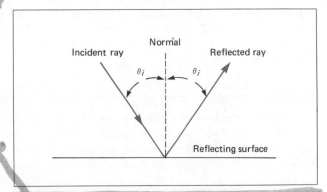

Figure 5-6b *When light is reflected from a smooth surface, the angle of reflection equals the angle of incidence θ_i.*

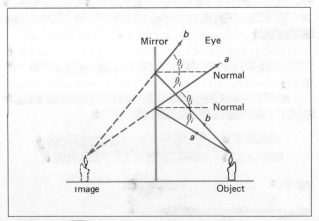

Figure 5-6c *Method used for locating the image formed by a plane mirror.*

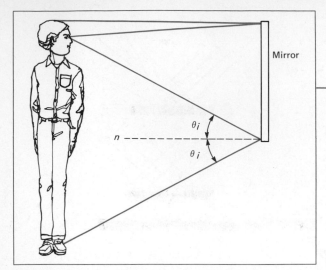

Figure 5-6d *Image formation in a plane mirror. Note that the vertical height of the mirror equals one-half the height of the man.*

mirror (distance from the mirror is not a factor) the mirror has to be only $\frac{1}{2}$ his height.

Another method of changing the direction of a wave is to produce a change in its velocity. This occurs whenever the wave penetrates from one medium, such as air, into another medium, such as water.

When light goes from one medium to another, its velocity changes. The change in velocity depends upon the two media involved and on the frequency of the light. Because of the change in velocity, the light wave is deviated from its original path. The bending of light waves because of a velocity change is called **refraction.** A pencil placed in a glass of water looks magnified and bent, because the light rays are refracted (that is, deflected from their normal directions) since their velocity is different in water than in air. The ratio of the velocity change, known as the **index of refraction,** can be written as

$$\frac{\text{Index of}}{\text{refraction}} = \frac{\text{Velocity of light in vacuum}}{\text{Velocity of light in medium}}$$

or as

$$n = \frac{c}{c_m}$$

where n = index of refraction,
c = velocity of light in vacuum,
c_m = velocity of light in medium.

Table 5-6a *Indices of Refraction of Some Common Substances*

Substance	n
Water	1.333
Crown glass	1.520
Diamond	2.420
Air (0°C, 1 atm)	1.00029

The index of refraction is a pure number, since c and c_m are measured in the same units. Indices of refraction for some common substances are shown in Table 5-6a.

When light enters a new medium, it is bent as shown in Fig. 5-6e. In this figure the refracted ray is bent towards the normal because the velocity of light in glass is less than the velocity of light in air. The amount of bending is due to the ratio of the indices of refraction of the two different media involved. Whenever refraction occurs, some reflection also occurs. This is designated by the dashed line in Fig. 5-6e.

The way in which bending of light is caused by a change in velocity can be seen by examining Fig. 5-6f. In Fig. 5-6f the rays of light are designated as a and b. Any line perpendicular to one of these rays, or any line perpendicular to both (as shown in this case with line OO') designates the wave front. As the wave front in this illustration approaches the glass, ray b arrives at point O' when ray a is at point O, the surface of the glass. If we start recording time at this instant, and continue until ray b arrives at the surface of the glass, the elapsed time will be Δt, and the distance traveled by ray b will be $c\,\Delta t$, which will equal $O'P'$. The time Δt will be the same for both rays; therefore, distance $O'P'$ will be greater than distance OP by a factor which is a function of the velocity of the ray in the new medium, c_m. This slowing down of ray a in the new medium, while ray b is still traveling in the air, and the subsequent slowing of the other rays between a and b, deviate the wave front toward the normal, or perpendicular.

Figure 5-6e *Diagram illustrating the deviation of a light ray from its original path as the ray passes from air into glass. A deviation of this type is called refraction. Some of the light is also reflected.*

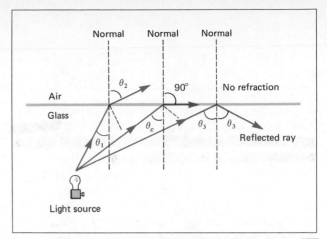

Figure 5-6g *Internal reflection. When light goes from glass to air, it is refracted away from the normal. At a certain critical angle θ_c, the angle of refraction is 90°. When the angle of incidence exceeds θ_c, all the light is reflected internally.*

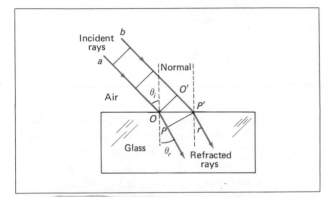

Figure 5-6f *The refraction of a wave front.*

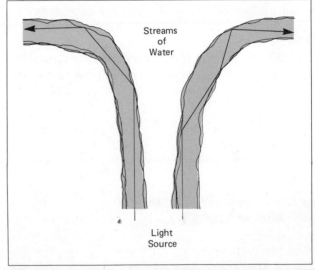

Figure 5-6h *When a light source illuminates a water fountain from below, the light is trapped inside the streams of water due to total internal reflection as shown.*

When light goes from a medium with a high index of refraction to one with a low index of refraction (e.g., from glass to air), an interesting phenomenon occurs. This is illustrated in Fig. 5-6g. If the angle is θ_1, then the light is refracted as shown. At the angle θ_c, called the *critical angle*, the refracted ray makes an angle of 90° with the normal. At angles greater than θ_c, the light is totally reflected, and none is refracted. This phenomenon is called **total internal reflection.**

Internal reflection enhances the brilliance of cut diamonds. A diamond is cut so that most of the light entering the diamond is internally reflected. The light then emerges from only certain portions with a brilliance far exceeding that of the incoming light. This effect creates the diamond's beautiful sparkle.

Another example of total internal reflection occurs when a fountain of water is illuminated from below. The light is totally reflected within the streams of water providing a spectacular effect. Figure 5-6h illustrates how this is done with a single stream.

Figure 5-7a *Diagram of the double-slit experiment illustrating the diffraction and interference of light waves.*

The angle of refraction of light depends somewhat on the light's frequency or wavelength. The fact that different frequencies are refracted at slightly different angles is called **dispersion.** When white light (electromagnetic radiation containing all wavelengths visible to the human eye) passes through a glass prism, the light rays are refracted on entering the glass and refracted again on leaving the glass. Since the velocity changes on entering (decreases) and on leaving (increases), the directions will change accordingly. The amount of refraction is a function of the wavelength, with the shortest wavelengths being deviated from their path by the greatest amount. Violet light has a shorter wavelength than red light. A prism used in this fashion produces the visible color spectrum, and we say the white light has been dispersed. The spectrum obtained by this method is shown in Plate I.

When sunlight is refracted, reflected, and refracted again from a water droplet, dispersion results and a rainbow is formed. Rainbows are discussed further in Chapter 23.

5-7
DIFFRACTION, INTERFERENCE, AND POLARIZATION

Another method by which waves can be deviated from a straight-line path is to let them pass through a narrow slit, where a small-scale interaction takes place with the edges of the slit to produce bright and dark fringes. The effect also occurs when light from a fairly bright source passes the edge of an opaque object. Close observation will show a shadow at the edges surrounded by narrow bands, or fringes. This phenomenon can easily be seen by holding two fingers close together to form a narrow slit and observing a bright light from a distance. If a fluorescent lamp is available, position the fingers parallel to the length of the lamp and try it. In all these cases, the deviation of light waves is referred to as **diffraction.**

Diffraction occurs with all types of wave motion. Sound is heard around corners due to diffraction, as well as to reflection and refraction. The directional property of the cheerleader's megaphone is an example of diffraction of sound waves. The diffraction of a wave as it passes through a small slit is shown in Figs. 5-7a and 5-7b. If light traveled in an exactly straight line as it passed through the slits, there would be only a point of light falling on the screen.

In our study of reflection and refraction, the location of a mirror image was done by drawing straight lines; that is, we solved the problem by using geometric methods. The study of optics using these methods is known as **geometrical optics.** An explanation of diffraction, however, cannot be made by using geometrical optics. We must return to the theory of wave motion to give a satisfactory explanation of the effects observed. Diffraction—and also interference—are the effects produced when a number of waves interact with one another because of a difference in "phase" and "amplitude." A study of optics from this point of view is known as **physical optics.**

Effects of physical optics occur when effects such as diffraction become important. The key ratio to a quantitative understanding of diffraction is

$$\frac{\lambda}{d}$$

where λ = wavelength of wave,
d = size of opening or obstacle causing diffraction.

Figure 5-7b *Straight waves passing through an opening. Note the waves curving around the ends of the barriers. (From* PSSC Physics, *3rd ed., Heath, Lexington, Mass., 1971.)*

If λ/d is much less than one, very little diffraction occurs. If λ/d is approximately one or greater, diffraction effects are easily observed. As an illustration, consider sound waves which have wavelengths of inches, feet, or more. These waves are easily diffracted around people, through doorways, and so on, because λ/d is approximately one or greater. Visible light waves, on the other hand, are of the order of 10^{-5} cm, and diffraction effects cannot be seen until the size of the slit opening is also approximately 10^{-5} cm. When you sit in a lecture room, a movie theater, or a concert hall, the sound will easily diffract around the people in front of you, but the light will not. This is due to the sizes of the wavelengths involved.

Interference effects are defined as changes in wave motion produced by phase and amplitude relations of two or more waves. The amplitude of a wave has been defined (Section 5-2) as the maximum displacement of the disturbance from its midposition. Two or more waves may reinforce one another when they are changing in the same direction, or tend to cancel one another when varying in opposite directions. When two or more waves have their displacement at all times in the same direction, they are said to be in phase with one another. If their displace-

ments at all times are in opposite directions, they are completely out of phase with one another. Figure 5-7c illustrates the concept of phase.

When interference occurs between waves, the two waves must remain exactly in phase for constructive interference (when waves reinforce each other) and 180° out of phase for destructive interference (when they completely cancel each other). For complete destructive interference, the two waves must have the same amplitude. They must also have the same wavelength and be traveling in the same direction.

Robert Young (1773–1829), a British physicist considered by many as the founder of the wave theory of light, demonstrated in 1801 the interference of light with apparatus similar to that shown in Fig. 5-7a. In order to meet the requirements set forth in the preceding paragraph, a monochromatic light source was used with a single frequency, which in turn was separated into two sources. The same amplitude was obtained. As the light waves travel outward from the two slits, the waves interfere with one another. The interaction of the waves, shown in Fig. 5-7a by the dotted lines, represents points where the waves aid one another (constructive interference) and

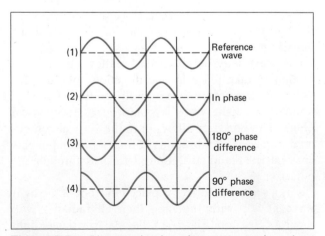

Figure 5-7c *Diagram showing when waves are in and out of phase with one another.*

a bright area is produced. Between these areas the waves cancel one another (destructive interference) and a dark band occurs.

A **diffraction grating,** which may be a piece of glass with many parallel slits uniformly scratched on it, can be used to produce a spectrum exactly like that produced by a prism. The grating is the heart of an instrument known as the diffraction grating spectrometer, which is used to analyze (measure accurately the wavelength) light from many sources, including that from distant stars.

In 1927 C. J. Davisson and L. H. Germer, two American physicists, demonstrated the diffraction of electrons, thus providing experimental evidence concerning the wave nature of particles. An electron diffraction pattern is shown in Figure 11.1b.

As you now know, the difference between transverse waves and longitudinal waves is the direction of the particle displacement in respect to the velocity vector. In a transverse wave the particle displacement is perpendicular to the wave motion. Viewing a transverse light wave from the front, we observe a pattern similar to Fig. 5-7d, which shows the displacement vectors in a plane 90° to the direction of travel. There are an infinite number of these vectors in the plane, all pointing in different directions. We call such a wave **unpolarized.** If we restrict the vibrations to one direction, then the wave is said to be **linearly polarized.**

A wave may be polarized by allowing the wave motion to take place in one direction only, using a Polaroid film. Figure 5-7e is an illustration of how an unpolarized light wave is polarized in the vertical plane. The polarizer allows only the vertical component to pass. The vibrations of the other components are restricted and not allowed to pass through the polarizer.

If a second polarizer is placed in front of the light wave at 90° with the first, then all radiation will theoretically stop (Fig. 5-7f). The polarization of light is strong experimental proof that light is a transverse

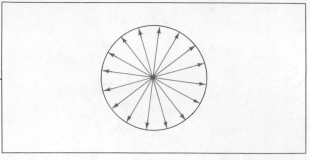

Figure 5-7d *A "head-on-view" of a beam of unpolarized light.*

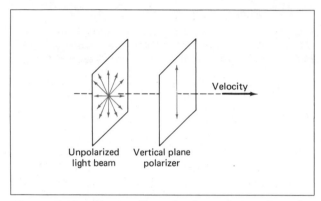

Figure 5-7e *Diagram illustrating how a beam of unpolarized light is polarized.*

Figure 5-7f *The polarization of light using Polaroid material. The left drawing illustrates the planes of polarization overlapping but parallel to one another. The right drawing illustrates the planes of polarization overlapping but perpendicular to one another. The black in the right drawing is achieved through complete absorption of the waves on the overlapping portions of the Polaroid material.*

Figure 5-7g *Polarized sunglasses consist of a single piece of Polaroid material. When two pieces of Polaroid are placed at right angles, no light can pass through.*

Figure 5-8a *Light rays from a concave spherical mirror when incoming light rays are parallel to the principal axis. Image is at focal point.*

wave motion. Longitudinal waves cannot be polarized.

Polarized sunglasses (Fig. 5-7g) consist of a single sheet of Polaroid material which cuts down the total amount of light passing through it. Only vertically polarized light can go through the sunglasses. When unpolarized sunlight falls upon a horizontal surface, such as a road, the reflected light is polarized in the horizontal direction. Since the Polaroid in the sunglasses only allows vertically polarized light to pass, the reflected glare is blocked out.

5-8
SPHERICAL MIRRORS

A spherical reflecting surface can be used to form real images from divergent light waves. The construction of such a mirror is illustrated in Fig. 5-8a. Some appropriate and important optical terms are given in the figure. The spherical surface of a **concave mirror** is a small inside section of the total curvature of the circle. (A **convex mirror** has a reflecting surface on the

outside of the spherical surface.) The center point of a concave mirror is called the *vertex v*, and a line drawn through *v* and the center of curvature of the mirror *R*, is called the *principal axis* of the mirror. For a concave mirror, rays of light coming in parallel to the principal axis will be reflected and brought to a focus on the principal axis at a point *f* known as the *focal point*. The distance from the focal point to the vertex is called the **focal length.** The geometric construction of the concave mirror makes the focal length one-half the value of the radius of curvature, where the radius of curvature is the radius of the spherical surface. Expressed in symbols

$$f = \frac{R}{2}$$

where f = the focal length,
R = the radius of curvature.

Rays of light that come in parallel to one another but not parallel to the principal axis are reflected and brought to a clear image in a *focal plane*, as shown in Fig. 5-8b.

The characteristics of the images formed by concave mirrors may be real or virtual; erect or inverted; larger, smaller, or the same size as the object; in front of or behind the mirror. A **real image** is defined as one that can be brought to a focus on a screen, and a **virtual image** is one that cannot. A virtual image is always behind the mirror, and one must look into the mirror to see the image.

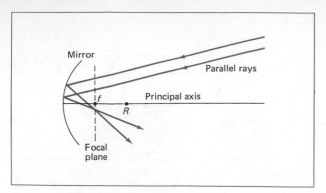

Figure 5-8b *Light rays from a concave spherical mirror when incoming light rays are not parallel to principal axis. The image is formed on the focal plane under these conditions.*

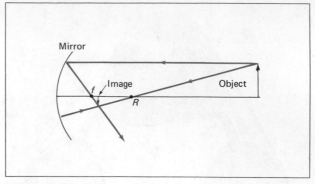

Figure 5-8c *The graphic solution for locating the image formed by a concave spherical mirror when the object is beyond the center of curvature R.*

The relationship between the object distance, the image distance, and the focal length is given by the equation

$$\frac{1}{D_o} + \frac{1}{D_i} = \frac{1}{f}$$

where D_o = object distance from vertex of mirror,
D_i = image distance from vertex of mirror,
f = focal length.

The values for the object distance D_o and the focal length f are always considered to be positive. If D_i is positive, then the image is real, inverted, and in front of the mirror. If D_i has a negative value, the image is virtual, erect, larger than the object, and behind the mirror.

Information concerning the location of the image can be obtained through the equation given above or by means of the graphic solutions shown in Figs. 5-8c, 5-8d, and 5-8e. In all three of these illustrations, the location of an image (represented by an arrow) can be determined by drawing two light rays, so that ray 1 is drawn parallel to the principal axis from the point on the object to the mirror, then reflected so that it passes through f, and ray 2 is drawn from the same point on the object through R and reflected on itself. The intersection of the reflected rays 1 and 2 will be the location of the image (represented by the broken-line arrows in the three figures). When the object is located between the mirror and the focal point, as in Fig. 5-8e, the reflected rays of light will

Figure 5-8d *The graphic solution for locating the image formed by a concave spherical mirror when the object is between f and R.*

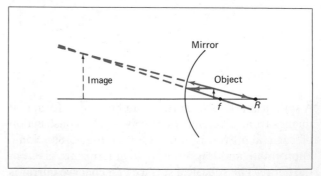

Figure 5-8e *The graphic solution for locating the image formed by a concave spherical mirror when the object is between the mirror and f.*

Figure 5-8f *The graphic solution for locating the image formed by a convex spherical mirror.*

diverge from the mirror, and they will never intersect. The image in this case can be located by extending rays 1 and 2 behind the mirror, as shown in the illustration. Their point of intersection behind the mirror is the location of the image.

When the image is formed behind the concave mirror, it is larger than the object. The equation expressing the **magnification** can be written

$$M = -\frac{D_i}{D_o}$$

where

M = magnification,
D_i = image distance from center of mirror,
D_o = object distance from center of mirror.

The minus sign in the equation indicates that D_i is negative.

If the spherical surface constructed in Fig. 5-8a had been silvered on the outside, then a convex mir-

ror would have been formed. The mirror equation can be used to determine the image position by making f a negative value. Geometrical constructions also can be used for image location (see Fig. 5-8f).

We see from Fig. 5-8f that images in a convex mirror are always virtual. They are always upright and smaller than the object. Of course, left and right are also reversed. An example of what is seen in a convex mirror is shown in Fig. 5-8g. Convex mirrors provide an expanded field of view. Consequently,

Figure 5-8g *A spherical convex mirror gives an expanded field of view. The images are smaller than the original objects and have right and left reversed. (Courtesy J. D. Wilson,* Physics Concepts and Applications, *Heath, Lexington, Mass., 1977.)*

they are often used as side mirrors for large trucks and for store monitoring as in Fig. 5-8g.

■ *Example 4* What is the magnification of an image in a convex mirror for an object 6 m away if the focal length of the mirror is 2 m?

Here we have

$$f = -2 \text{ m}$$
$$D_o = +6 \text{ m}$$

We use

$$\frac{1}{D_o} + \frac{1}{D_i} = \frac{1}{f}$$

to get

$$\frac{1}{6} + \frac{1}{D_i} = -\frac{1}{2}$$

so

$$\frac{1}{D_i} = -\frac{1}{2} - \frac{1}{6} = -\frac{2}{3}$$

$$D_i = -\frac{3}{2}$$

and

$$M = -\frac{D_i}{D_o} = -\frac{-\dfrac{3}{2}}{6} = +\frac{1}{4}$$

This shows that the image is not magnified, but is $\frac{1}{4}$ as small as the object. This can be seen in Fig. 5-8g.

5-9
CONVEX LENSES

There are real similarities between the effects produced by convex lenses and concave mirrors. The construction of a double convex lens is shown in Fig. 5-9a. The lens pictured is considered to be a thin lens, that is, the thickness may be neglected and the verti-

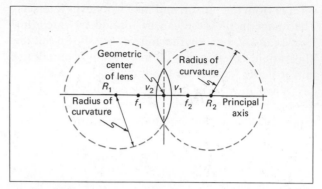

Figure 5-9a *Diagram illustrating the construction of a double-convex lens.*

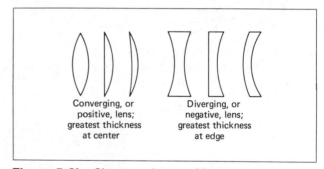

Figure 5-9b *Classes and types of lenses.*

ces of the two surfaces may be regarded as coinciding at the geometrical center of the lens. Although the radius of curvature for one face may differ from that for the other, the radii are the same in most cases, as illustrated by the diagram. Most lenses are made of ordinary glass, but some are constructed from materials like quartz, plastics, and fluorite.

Although the physical properties of only convex lenses will be discussed in this section, it should be pointed out that there are many different types of lenses. Figure 5-9b shows the two major classes of

Figure 5-9c *The graphical solution for locating the image formed by a double-convex lens.*

Figure 5-9d *Diagram illustrating the object's position and formation of the image by a simple magnifying glass.*

lenses with three different types in each class.

Information concerning the image formed by a convex lens can be obtained by the graphical method illustrated in Fig. 5-9c or analytically, using the following equation:

$$\frac{1}{D_o} + \frac{1}{D_i} = \frac{1}{f}$$

where D_o = object distance from center of lens,
D_i = image distance from center of lens,
f = focal length of lens.

D_o and f are always positive in value. When D_i is positive, the image will be a real image, that is, it can be brought to a focus on a screen; when D_i is negative, the image will be virtual and can be viewed only by looking through a lens. The following example will illustrate the use of the equation:

■ *Example 5* An object is placed 6 cm in front of a convex lens, the focal length of which is 10 cm. Calculate the image position.

Substituting the values into the lens equation given above

$$\frac{1}{6} + \frac{1}{D_i} = \frac{1}{10}$$

or

$$\frac{1}{D_i} = \frac{1}{10} - \frac{1}{6} = \frac{3}{30} - \frac{5}{30}$$

$$= -\frac{2}{30} = -\frac{1}{15}$$

so

$$D_i = -15 \text{ cm}$$

Since D_i has a negative value, the image is virtual, erect, larger than the object, and located on the same side of the lens as the object.

A simple magnifying glass is a convex lens and functions as illustrated in Example 5. The object to be viewed and magnified must be located between the focal point and the lens. See Fig. 5-9d. As with mirrors, the magnification is again given as $M = -D_i/D_o$. In Example 5 the magnification is $M = -(-15 \text{ cm})/6 \text{ cm} = +2.5$.

The optometrist or ophthalmologist specifies the power of a lens in diopters, written as the reciprocal of the focal length in meters.

Lens power (in diopters)

$$= \frac{1}{\text{Focal length (in meters)}}$$

Figure 5-9e *The human eye has a lens with an index of refraction which is greater than the surrounding parts of the eye. This convex lens focuses light onto the retina.*

Retina

Lens

The human eye can be thought of as a convex lens. In Fig. 5-9e the normal path of light is shown. The lens and other parts of the eye have different indices of refraction to give the effect of a convex lens. The light is focused on the **retina,** to which is attached the optic nerve that relays the visual signals to the brain.

Because the distance between the lens and the retina do not vary, we have a situation in which D_i is a constant. Since D_o varies, the focal length f of the eye lens must be varied. This is done by the ciliary muscles.

Defects of vision occur when the image does not fall on the retina. If the image falls in front of the retina, as in Fig. 5-9f, the eye is said to be **nearsighted.** As the object moves closer, the image distance moves back to the retina. Hence, near objects can be seen clearly by a nearsighted person. To correct this defect, glasses consisting of diverging lenses are worn.

The opposite effect occurs when the image falls behind the retina (see Fig. 5-9g). This effect causes the eye to be **farsighted,** and distant objects are seen clearly. To correct this defect, glasses consisting of converging lenses are worn.

Uncorrected

Corrected

Figure 5-9f *Nearsightedness is corrected by placing glasses with a diverging lens in front of the eye.*

Uncorrected

Corrected

Figure 5-9g *Farsightedness is corrected by placing glasses with a converging lens in front of the eye.*

IMPORTANT WORDS AND TERMS

Simple harmonic motion
Wave velocity
Longitudinal wave
Transverse wave
Frequency
Hertz
Period
Wavelength
Amplitude
Electromagnetic waves
Speed of light
Angstrom
Sound
Compression
Rarefaction

Pitch
Intensity
Decibel
Resonance
Doppler effect
Red shift
Reflection
Diffuse reflection
Regular reflection
Refraction
Index of refraction
Total internal reflection
Dispersion
Diffraction
Geometrical optics

Physical optics
Interference effects
Diffraction grating
Unpolarized
Linearly polarized
Concave mirror
Convex mirror
Focal length
Real image
Virtual image
Magnification
Retina
Nearsighted
Farsighted

LEARNING OBJECTIVES

After reading and studying this chapter, you should be able to do the following without referring to the text:

1. Define the two main types of waves and give an everyday example of each.
2. Define five properties of every wave. *reflect refract diffract dispers polariz interf*
3. List in order of increasing frequency the various types of electromagnetic waves.
4. State the speed of light and the speed of sound in the British system of units.
5. Differentiate between the wavelengths of visible light and audible sound.
6. State what properties of waves give rise to pitch, loudness, color, and brightness.
7. Define and give examples of resonance and the Doppler effect.
8. Define and give examples of refraction, total internal reflection, and dispersion.
9. Define and give examples of diffraction and polarization.
10. Discuss the difference between sound waves and light waves with regard to diffraction and polarization.
11. Discuss the differences and similarities between the images seen in a plane mirror and a convex mirror.
12. Sketch diagrams of a converging lens and a diverging lens.
13. Discuss briefly how the eye works and how nearsighted and farsighted vision is corrected.
14. Define all terms in the formula $\frac{1}{D_o} + \frac{1}{D_i} = \frac{1}{f}$ and state how a magnification is calculated.

QUESTIONS

1. How are sound waves fundamentally different from light waves?
2. Define the terms wave velocity, wavelength, frequency, period, and amplitude.
3. Give the mks system of units for each of the terms in question 2.
4. List in order of increasing frequency the various types of electromagnetic waves.
5. What wave property distinguishes between (a) different colors, (b) different sound pitches, and (c) different radio stations?
6. What is the approximate wavelength of (a) visible light, (b) a radio wave?
7. What wave property distinguishes (a) loudness, (b) brightness?
8. What happens to the frequency of light seen by an astronomer when he or she looks at a galaxy moving away from the earth at one-half the speed of light?
9. What causes a sonic boom?
10. What causes refraction?
11. Draw a diagram to show how total internal reflection is possible.
12. What is dispersion? What causes dispersion?
13. Why is diffraction of sound waves an everyday effect, but diffraction of light waves is rarely observed?
14. (a) What is unpolarized light? (b) What is polarized light? (c) Can you have polarized sound waves?
15. Describe the images in a spherical convex mirror.
16. How does a simple magnifying glass work?
17. (a) How does the human eye work? (b) What causes farsightedness and how can it be corrected?

PROBLEMS

1. Waves traveling on Lake Michigan have a speed of 8 m/s. The distance between adjacent crests is 3 m.
   (a) Determine the frequency of the waves.
   (b) What is the period?

2. The velocity of sound is about 1100 ft/s. How many miles per hour is this? *Answer:* 750 mi/h

3. What is the wavelength in cm of microwaves produced in an oven, if the waves have a frequency of 10^{10} hertz?

4. What is the wavelength in cm of an X-ray with a frequency of 10^{18} hertz? *Answer:* 3×10^{-8} cm or 3 Å

5. During a thunderstorm, 4 s elapses between a flash of lightning and the corresponding thunder. Approximately how far away was the lightning?

6. A subway train has a sound intensity of about 90 dB and a rock band has a sound intensity of about 110 dB. How many times louder is the rock band than the subway? *Answer:* 100 times louder

7. Determine the speed of light in water.

8. An object 3 cm high is placed 12 cm in front of a convex spherical mirror that has a focal length of 2 cm. Determine graphically the position and size of the image.

Answer: size $= \frac{3}{7}$ cm

9. An object 3 cm high is placed 12 cm in front of a convex spherical mirror that has a focal length of 2 cm. Calculate the image distance and image magnification.

Answer: $D_i = -12/7$ cm
$M = 1/7$

10. A simple magnifying glass has a focal length of 8 cm. An object is placed 6 cm in front of it.
   (a) Determine the image distance.
   (b) Determine the magnification.

Answer: (b) $M = 4$

Chapter 6

ELECTRICITY AND MAGNETISM

Like charges repel, and unlike charges attract each other, with a force that varies inversely with the square of the distance between them. . . . Frictional forces, wind forces, chemical bonds, viscosity, magnetism, the forces that make the wheels of industry go round—all these are nothing but Coulomb's law. . . .

J. R. Zacharias

The magnetic properties of the mineral lodestone (magnetite, Fe_3O_4) were known to the Greeks as early as 600 B.C. Thales of Miletus (640–546 B.C.), an early Greek mathematician and astronomer, was aware of the properties of attraction and repulsion of lodestone with similar pieces of lodestone; he also knew of the amber effect, that is, the attraction of bits of straw to an amber rod that had been rubbed with wool. The word "magnet" seems to have been derived from Magnesia, a province in Asia Minor, where the Greeks first discovered lodestone. The Chinese were probably the first to use the lodestone as a compass, both on land and sea. Early records indicate that ships sailing between Canton, China, and Sumatra as early as 1000 A.D. were navigated by the use of the magnetic compass.

In the thirteenth century, a Frenchman, Petrus P. de Maricount, described the magnetic compass in some detail and applied the term "pole" to the regions on the compass where the fields of influence were the strongest. The north-seeking pole he called north, and the south-seeking pole he called south. The attraction of unlike poles, the repulsion of like poles, and the formation of new unlike poles when a magnet was broken into two pieces were also described by de Maricount.

In 1600 Dr. William Gilbert (1540–1603), court physician to Queen Elizabeth, published his book on magnetism, *De Magnete*. The book contained all information then known about electricity and magnetism, plus experiments carried out by Gilbert. These experiments included information on the dip (the angle the earth's magnetic field makes with the earth's surface) and declination (the angle the compass needle deviates from the geographical north) of the compass, the loss of magnetism by a magnet when heated, and experiments with a sphere-shaped magnet, which led him to the conclusion that the earth acts like a huge magnet.

Gilbert was also aware of the amber effect. We now know that this effect is due to repulsion and attraction of electric charges. Gilbert carried out many experiments on the amber effect with an instrument he called a versorium (Latin, *verso*, to turn around). The versorium was nothing more than a slender arrow-type nonconducting material balanced on a pivot point so as to give a high degree of sensitivity to the force of attraction when an amber rod or other substances were placed in its vicinity. With his versorium he discovered that many substances possess the amber effect. Gilbert is responsible for the word "electron," which is very familiar today. He classified those substances possessing the amber effect as "electrics" (Greek, *electron*, amber).

A similar major study of electricity and magnetism took place in the eighteenth century, when Charles Coulomb (1736–1806) established the inverse square law of attraction and repulsion between electrostatic charges, and verified the same law to hold for magnetic poles.

In 1820 Hans Christian Oersted (1777–1851) found that a compass was deflected by an electric current-carrying wire. In 1831 Michael Faraday (1791–1865) and Joseph Henry (1797–1878) independently found that a magnet plunged into a coil of wire would induce an electric current. The idea that electricity and magnetism were intimately related gave rise to the *theory of electromagnetism*. James Clerk Maxwell (1831–1879) brilliantly incorporated all known electric and magnetic phenomena into a complete theory of electromagnetism in 1873. He also stated that light was made up of electromagnetic

waves and that other wavelengths were possible. It was due to this theory that radio waves were searched for and soon discovered.

A knowledge of the basic laws of electricity and magnetism is essential in understanding the forces that hold matter together. In our study of atomic physics, chemistry, meteorology, geology, and other subjects in physical science, we are concerned with electromagnetic forces. It is electromagnetic forces that are responsible for the structure of atoms, molecules, and hence, for matter in general. The facts that paper tears, rocks are hard, and countless other facts are due to the action of electromagnetic forces.

6-1
ELECTRIC CHARGE

All matter, according to modern theory, consists of very small particles called atoms which are composed in part of negatively charged particles called **electrons,** positively charged particles called **protons** and particles called **neutrons** which carry no electric charge. Table 6-1a summarizes the fundamental properties of these atomic particles.

As can be seen from the table, all three of these particles have mass, but only the electrons and protons possess electric charge. The magnitudes of electric charge on an electron and a proton are equal, but their signs are different. When we have the same number of electrons and protons, the total charge is zero, and we have a *neutral* situation. Hence, the name *neutron* for the small particle with zero electric charge. Neutrons have virtually no effect in atomic physics. They become important, however, when nuclear physics is discussed in Chapter 9.

The unit of electric charge is called the **coulomb,** named after Charles Coulomb, the French scientist. Electric charge is usually designated by the letter q. When electric charges move, they give rise to an

Table 6-1a *Some Properties of Fundamental Atomic Particles*

Particle	Symbol	Mass	Charge
Electron	e^-	9.11×10^{-31} kg	-1.6×10^{-19} coulomb
Proton	p^+	1.67×10^{-27} kg	$+1.6 \times 10^{-19}$ coulomb
Neutron	n^0	1.67×10^{-27} kg	0

electric current. **Current** is defined as the rate of flow of electric charge. Current is usually designated by the letter I. Current is measured in terms of amperes. One **ampere** is equal to one coulomb per second.

In symbol notation we can relate electric current and electric charge by the formula

$$I = \frac{q}{t} \qquad \text{Eq. 6.1}$$

where I = electric current, measured in amperes,
q = electric charge flowing past a given point, measured in coulombs,
t = time for the charge to move past the point, measured in seconds.

From Eq. 6.1, we can write

$$q = I \times t \qquad \text{Eq. 6.2}$$

or 1 coulomb = 1 ampere $\times$ 1 second.

This equation is important because the experimental quantity used as the fundamental unit is the ampere. To be precise, the *coulomb* is defined as the amount of charge which flows past a given point in one second when the current is one ampere.

Electric charge cannot be created or destroyed. All protons have the same charge, 1.6×10^{-19} coulomb, and all electrons have the same charge of -1.6×10^{-19} coulomb. So far, all particles found in nature have multiples of this amount of charge. For example, no one has ever found a charge of $\frac{3}{2}$ or 6.5. Many scientists are searching for particles called *quarks*, which are believed to have charges such as $+\frac{2}{3}$ or $-\frac{1}{3}$. Maybe these particles will one day be found.

It has long been known that like charges repel and unlike charges attract. It was Charles Coulomb who first quantified this attraction and repulsion in what is now known as Coulomb's law for electric charges.

Coulomb's law states:

The force of attraction or repulsion between two charged bodies is directly proportional to the product of the two charges and inversely proportional to the square of the distance between them.

This can be written as

$$F = k \frac{q_1 q_2}{r^2} \qquad \text{Eq. 6.3}$$

where F = the force of attraction or repulsion,
q_1 = magnitude of first charge,
q_2 = magnitude of second charge,
r = distance between charges,

k is a proportionality constant

$$k = 9 \times 10^9 \frac{\text{newton} \times \text{meter}^2}{\text{coulomb}^2}$$

Coulomb's law is very similar to Newton's law of universal gravitation. The similarity comes from the fact that both are dependent on the square of the distance between the centers of the objects. One obvious difference between the two laws is that Coulomb's law depends on charge while Newton's law depends on mass. There are two other important differences. One is that Coulomb's law can give rise to either an attractive or a repulsive force, depending on whether the two charges are different or the same. The force of gravitation, on the other hand, is *always* attractive. The other important difference is that the electric forces are generally much stronger than the gravitational forces for ordinary objects. This arises from the fact that the constant of proportionality k is much bigger than the constant of proportionality G in the law of gravitation. Of course, these two constants cannot be compared directly since they have different units, but observe that in mks units, $k = 9 \times 10^9$ whereas $G = 6.67 \times 10^{-11}$. From this we see that k is much bigger than G. This is reflected in the fact that in dealing with atoms and the properties of matter, the gravitational forces can be ignored compared to the electric forces of attraction and repulsion.

An object with an excess of electrons is said to be negatively charged, and an object with a deficiency of electrons is said to be positively charged. A negative charge can be placed on a rubber rod by stroking the rod with fur. In Fig. 6-1a, a rubber rod which has been rubbed with fur is shown suspended by a thin thread which allows the rod to swing freely. When a similar rubber rod which has been stroked with fur is brought close to the suspended rod, it will swing away from the second rubber rod; that is, the charged rods repel one another. The same procedure using two glass rods which have been stroked by silk will show similar results (Fig. 6-1b). Although the experiment shows repulsion in both cases, the charge on the glass rods is different from the charge on the rubber rod. This is shown in Fig. 6-1c, which indicates that the charges on the stroked rubber and glass rods attract one another. The charge placed on the glass rod is positive; on the rubber rod the charge is negative.

From Coulomb's law we see that as two charges get closer together, the force of attraction or repulsion increases. This important effect can give rise to nonzero electric forces as shown in Fig. 6-1d. When a negatively charged rubber comb approaches a small piece of paper, the charges within the paper arrange themselves so that the positive charges move toward the comb and the negative charges are repelled away.

Figure 6-1a *Two negatively charged bodies repel one another.*

Figure 6-1b *Two positively charged bodies repel one another.*

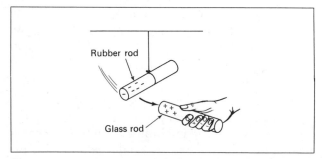

Figure 6-1c *A negatively charged body and a positively charged body attract one another.*

Since the positive charges are closer to the comb, the attractive forces are stronger than the repulsive forces. Thus, there is a net attraction between the comb and the piece of paper. Remember, however, that the paper is uncharged.

A similar effect occurs when a negatively charged rubber rod is brought close to a stream of water. The molecules of water arrange themselves so that the

Figure 6-1d *A negatively charged rubber comb is brought near a small piece of paper which has no net charge on it. The comb will attract the positive charges and repel the negative charges in the paper. The comb then exerts a net attractive force on the paper because the positive charges are closer to it.*

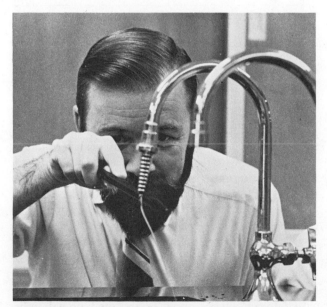

Figure 6-1e *A charged rubber rod brought close to a stream of water will bend the stream. (Courtesy J. D. Wilson,* Physics Concepts and Applications, *Heath, Lexington, Mass., 1977.)*

negative ends are closer than the positive ends. Then the water is attracted to the rod and the stream is bent as shown in Fig. 6-1e.

Since atoms are fundamental parts of matter, it follows that electrons and protons are always present in any material. The electrons, which are located in the outer part of the atom, move within the material with varying degrees of freedom. The protons located in the inner part of the atom are much heavier and are not free to move. In metals, such as copper, silver, and aluminum, some of the electrons can move easily. Such metals are called **conductors.** Materials, such as wood, paper, cork, glass, and plastics are made such that the electrons cannot move freely. These materials are called **insulators.** Materials that are neither good conductors nor good insulators are called *semiconductors*.

6-2
ELECTRICITY

The effects produced by moving charges give rise to what we call **electricity.** In order to get charges to move, they must be acted upon by positive or negative charges.

Consider the situation shown in Fig. 6-2a. We start out with some unseparated charges and then begin to separate them. It takes very little work to

Figure 6-2a *It takes work to separate positive and negative charges. When they are separated, a free negative charge such as an electron will move from the negative to the positive side.*

pull the first negative charge to the left and the first positive charge to the right. When the next negative charge is moved to the left, it is repelled by the negative charge already there, so more work is needed. Similarly, it takes more work to move the second positive charge to the right. As we separate more and more charges, it takes more and more work.

Since work is done in separating the charges, we are creating an **electric potential energy.** If we insert a negative charge in between the separated charges, it will move to the right as shown in Fig. 6-2a; that is, it will acquire a velocity and kinetic energy.

Instead of speaking of electric potential energy, we usually speak of a potential difference or voltage. The **voltage** is defined as the amount of work it would take to move a charge between two points, divided by the value of the charge. This can be written as

$$V = \frac{W}{q} \qquad \text{Eq. 6.4}$$

where W = work done in joules,
$\quad\ V$ = voltage in volts,
$\quad\ q$ = charge in coulombs.

The **volt** is the unit of voltage and is equal to one joule per coulomb. A voltage is caused by a separation of charge. Once the charges are separated, a current can be set up, and we get electricity.

Whenever a current flows in a wire, it meets with some opposition. This opposition to the flow of charge is called **resistance.** It is measured in units called ohms. There is a simple relationship between voltage, current, and resistance that was formulated by George Ohm (1787–1854), a German physicist. **Ohm's law** can best be stated as the equation

$$V = IR \qquad \text{Eq. 6.5}$$

where V = voltage in volts,
$\quad\ I$ = current in amperes,
$\quad\ R$ = resistance in ohms.

From this equation we see that one **ohm** is a volt per ampere.

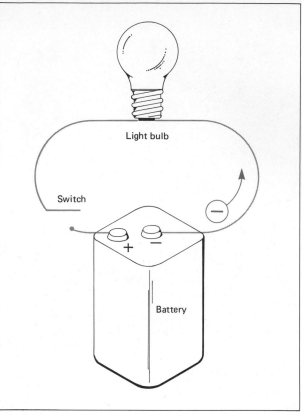

Figure 6-2b *A simple electric circuit. A battery provides the voltage. The electrons can be thought of as flowing from the negative to the positive terminals. The light bulb provides the resistance. When the switch is open, as shown, nothing happens. When the switch is closed, the electrons move and the bulb lights up.*

An example of a very simple electric circuit is shown in Fig. 6-2b. The battery provides the voltage. A conducting wire is placed between the positive and negative terminals, and electrons in the wire can move away from the negative end toward the positive end. If we put a light bulb along the path of the electrons, it provides resistance and will light up.

The switch in the circuit allows the path of the electrons to be closed or open. When the switch is open, as shown in Fig. 6-2b, the electrons will not move and nothing will happen. When the switch is closed, the circuit is completed and the bulb will light up.

Whenever current flows in a circuit, work is done to overcome the resistance of the circuit and power is expended. Recall that one definition of power was

$$P = \frac{W}{t}$$

If we substitute qV for W, we get

$$P = \frac{q}{t}V$$

We now recognize that $q/t = I$ to get

$$P = IV \qquad \text{Eq. 6.6}$$

If we use Ohm's law for V, we get

$$P = I(IR)$$

$$P = I^2R \qquad \text{Eq. 6.7}$$

The power that is dissipated in an electric circuit is in the form of heat. This heating effect is used in electric stoves, heaters, cooking ranges, hair dryers, etc. When a light bulb lights up, much of the power goes to produce heat as well as light. The units of power are watts and light bulbs are rated in watts (see Fig. 6-2c).

An electric company delivers electric energy. Our electric utility bill is computed on the number of kilowatt-hours (kWh) used. This is a unit of energy, since work produces energy and

$$W = P \times t \qquad \text{Eq. 6.8}$$

Hence, we pay a "power" company for energy, not power.

There are two principal forms of electric current. Figure 6-2b shows a simple circuit where a battery is used. This type of current is called **dc** or direct current. We use dc in battery-powered devices such as flashlights, radios, and automobiles. Much more common is **ac** or alternating current, which is produced by constantly changing the voltage from positive to negative to positive, etc. The frequency of changing from positive to negative is usually at the rate of 60 cycles per second or 60 hertz. Alternating current is produced by electric companies and is used in homes.

Figure 6-2c *Light bulbs are rated in units of power or watts. This tells us how much electric energy is being used per second to produce heat and light. (Courtesy J. D. Wilson, Physics Concepts and Applications, Heath, Lexington, Mass., 1977.)*

One of the principal reasons we use alternating current in our homes is that the voltages of alternating current can be easily increased or decreased by means of **transformers.** This is important because in transmitting electrical power, some of the electrical energy is lost as heat when the transmission wires heat up. The amount of energy lost as heat is inversely proportional to the square of the voltage; that is, the higher the voltage of the transmitted power, the lower the heating losses. These heating losses are minimized by transmitting the energy at very high voltages (as much as 230,000 volts). The high voltages are attained by using transformers to increase the voltage. This high voltage is then stepped down, again using transformers (Fig. 6-2d), until it reaches the home at approximately 110 to 120 volts. In Section 6-4 we will discuss the principles of transformers.

As the electricity is transmitted at these very high voltages, insulators are used to keep the wires away from the supporting towers and apart from each other. These insulators are shown in Fig. 6-2e.

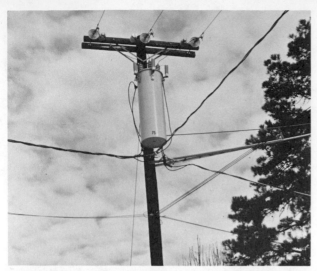

Figure 6-2d *The transformer shown decreases the voltage from the high-voltage power line to a lower voltage for residential use.* (*Courtesy J. D. Wilson,* Physics Concepts and Applications, *Heath, Lexington, Mass., 1977.*)

Figure 6-2f *A string of Christmas tree lights is an example of a series circuit. All the current passes through all the lights.*

Figure 6-2e *High-voltage electrical transmission lines are used to minimize heat losses as power is transmitted. Long insulators keep the lines away from metal support towers and insulating separators keep the wires from touching.* (*Courtesy J. D. Wilson,* Physics Concepts and Applications, *Heath, Lexington, Mass., 1977.*)

Once the electricity enters the home or business, it is used in electric circuits. There are two simple kinds of electric circuits which contain only resistance. In Fig. 6-2f, a string of Christmas tree lights illustrates a series circuit. In a **series circuit,** all of the current passes through all of the resistors, and the various resistances add up. The total voltage is given by

$$V = IR_1 + IR_2 + IR_3 + \cdots \qquad \text{Eq. 6.9}$$

where V = voltage of the circuit,

$\quad I$ = current passing through the various resistors,

$\quad R_1$ = first resistance,

$\quad R_2$ = second resistance, etc.

In our example of Christmas tree lights, the various voltages on the right will add up to 110 volts, since this is the total voltage that is supplied.

When we use Christmas tree lights in a series circuit, we find that if one bulb goes out, it is as if a switch has opened, and none of the bulbs will light. This can be a very frustrating experience when you

Figure 6-2g *Home electric circuit. All electric components are connected in parallel. Each appliance has 110 volts available to it.*

try to discover which light bulb doesn't work. Note that many Christmas tree lights are no longer made up in a series circuit.

Another common type of simple circuit is called a parallel circuit. An illustration is shown in Fig. 6-2g. In a **parallel circuit,** the voltage across each resistance is the same but the current through each resistance varies. The total current is equal to the sum of all the individual currents through the resistances. Home appliances are wired in parallel. There are two major advantages to the parallel circuit. First, the designed voltage for the appliance is available throughout the house. Second, if one appliance fails to operate, the others are not affected.

As more and more appliances are turned on, more and more current flows and the wires get hotter and hotter. The fuse shown in Fig. 6-2g is a safety device which prevents the wires from getting too hot. When too much current flows, the fuse gets so hot that it burns through and acts as an open switch. This prevents any current from flowing until the fuse is replaced.

An electric shock can be very dangerous and touching exposed electric wires should always be avoided. People are killed every year through electric shocks. The danger is proportional to the amount of electric current which goes through the body. The amount of current going through the body is given by Ohm's law as

$$I = \frac{V}{R_{body}} \qquad \text{Eq. 6.10}$$

where R_{body} = resistance of the body.

A current of 0.001 ampere can be felt as a shock, and a current as low as 0.05 ampere can be fatal. A current of 0.1 ampere is nearly always fatal.

The amount of current can be seen from Eq. 6.10 to be very dependent on the body's resistance. The body's resistance varies considerably. A dry body can have a resistance as high as 500,000 ohms and the

current from a 110-volt source will be 0.00022 ampere. The danger occurs when the skin is moist or wet. Then the resistance of the body can go as low as 100 ohms, and the current will rise to 1.1 amperes. Injuries and death from shocks usually occur when the skin is wet. For instance, touching an exposed household wire while sitting in a bathtub will frequently be fatal. You should avoid ever touching even a wall switch when your hands or feet are wet.

6-3
MAGNETISM

One of the first things we notice in examining a bar magnet is that it has two ends or poles. We call these two poles north, N, and south, S, because a compass needle was one of the earliest applications of magnetism. The north pole of a magnet, such as a compass, points north, and the south pole points south.

The basic principle of magnetic force has been mentioned already in the introduction to this chapter.

Like poles repel and unlike poles attract.

We say that north and south poles, N-S attract and N-N or S-S poles repel each other. The strength of the attraction or repulsion depends on the strength of the magnetic poles. Also, in a manner similar to Coulomb's law, the strength of the magnetic force is inversely proportional to the distance between the

Figure 6-3a *Small toy magnets show a repulsive effect that causes them to be suspended in air. (Photo courtesy James Crouse.)*

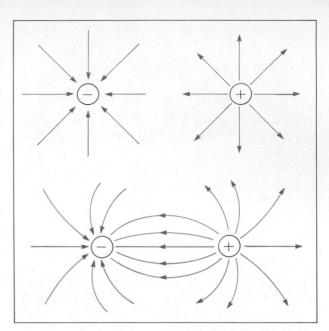

Figure 6-3c *Electric field lines for a single positive charge, a single negative charge, and an electric dipole which is a positive and negative charge together.*

Figure 6-3b *Magnetic field lines may be found by using a very small compass as shown. The north pole of the compass points in the direction of the field.*

magnetic poles. Figure 6-3a shows some toy magnets which seem to defy gravity due to their magnetic repulsion.

Every magnet produces a force on every other magnet. In order to discuss these force effects, it is common to introduce the concept of a magnetic field. A **magnetic field** is a set of imaginary lines that indicates the direction a small compass needle would point if it were placed at a particular spot. Figure 6-3b shows the magnetic field lines around a simple bar magnet. The arrows in the field lines indicate the direction a compass north pole would point. The closer together the field lines, the stronger the magnetic force on the imaginary compass.

The field concept also applies to the electric force. An **electric field** is a set of imaginary lines that indicates the direction a small positive charge would move, if it were placed at a particular spot. The electric fields are indicated in Fig. 6-3c for a single positive charge, a single negative charge, and a positive and negative charge together. This last configuration is called an *electric dipole* because there are two poles involved. A common magnet such as that shown in Fig. 6-3b would be called a *magnetic dipole* because it also has two poles. Note the similarity and difference between the fields for an electric dipole and a magnetic dipole.

Two facts about magnetic and electric fields are evident from Figs. 6-3b and 6-3c. First, field lines never cross. If they did, it would indicate that at the point of crossing the compass pointed in two directions or the positive charge would move in two directions. Second, electric fields begin and end on electric charges, whereas magnetic field lines are continuous. Magnetic field lines have no beginning or end. All magnets occur with two poles, i.e., as *di*poles. Electric charge can be thought of as isolated single poles or *mono*poles. There is nothing to prevent a magnetic monopole from existing, but so far none has been found. The discovery of a magnetic monopole would be an important fundamental development. A **magnetic monopole** would consist of a single north or south pole without the other. In 1975 a discovery of a magnetic monopole was claimed, but this claim has not been confirmed.

Electric and magnetic fields have directions associated with them, so they are *vector quantities*. Electromagnetic waves, discussed in Chapter 5, are caused by electric and magnetic fields that vary with time.

Magnetic field patterns are frequently generated by using small iron filings. The iron filings are magnetized and act like small compasses. The magnetic field produced in this way using two pieces of naturally magnetic lodestone (Fe_3O_4) is shown in Fig. 6-3d.

Electricity and magnetism are discussed together in this chapter because they are linked. In fact, the source of magnetism can be understood as being produced by moving and "spinning" electric charges. Oersted first discovered in 1820 that a compass needle was deflected by an electric current-carrying wire. To

Figure 6-3d *An iron filing pattern near two pieces of naturally magnetic lodestone.*

see this effect, it is necessary to have a fairly strong electric current.

Current-carrying wires cause different magnetic field configurations. Some of these are shown in Fig. 6-3e. A straight wire produces a field in a circular pattern around the wire. A single loop of wire acts like a small bar magnet, and a coil of wire creates a field similar to a bar magnet.

Individual electrons in an atom have a magnetic field due to what is best understood as a spinning motion. This causes all electrons to act like magnetic dipoles or very small bar magnets. The electrons also go in orbit around the atom's nucleus. This effect is similar to a single loop of wire and again creates a magnetic field similar to a small bar magnet. There are thus magnetic effects due to an electron's movement and due to its spinning motion. Most atoms have many electrons and the magnetic effects of all these electrons usually cancel out. So most materials

Figure 6-3e *Iron filings near current-carrying wires create magnetic field patterns for a straight wire (bottom), a single loop (middle) and a coiled wire (top). (From PSSC Physics, 3rd ed., Heath, Lexington, Mass., 1971.)*

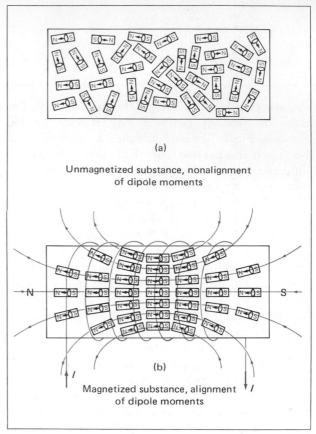

(a)

Unmagnetized substance, nonalignment of dipole moments

N   S

(b)

Magnetized substance, alignment of dipole moments

Figure 6-3f *In a piece of soft iron the domains are aligned randomly (top). A current-carrying coil of wire wrapped around the iron creates a magnetic field as shown. The domains then line up (bottom).*

are not magnetic or only slightly magnetic. In some atoms, however, the magnetic effect can be strong.

Materials that have large magnetic fields associated with them are called **ferromagnetic.** These materials include the elements iron, nickel, cobalt, several rare earth elements, and certain alloys of these and other elements. In ferromagnetic materials many atoms combine their magnetic fields in the same direction. This gives rise to magnetic **domains** or local regions of alignment. A single magnetic domain is very small and acts as a tiny bar magnet.

In iron, the domains can be aligned or non-aligned. A piece of iron with the domains randomly oriented is not magnetic and is sometimes referred to as a piece of soft iron. This effect is shown in Fig. 6-3f.

When the soft iron is placed in a magnetic field such as produced by a coil of wire, the domains line up and the iron is magnetized as shown in Fig. 6-3f.

When the magnetized iron is taken out of the magnetic field, the domains will tend to return to a mostly random arrangement due to heat effects which cause disordering. The amount of alignment remaining after the iron is taken out depends on the strength of the magnetic field in which the iron was placed.

Permanent magnets are made by placing ferromagnetic materials in strong magnetic fields. Another way to make a permanent magnet is to stroke the ferromagnetic material with a magnet. This aligns the domains. When permanent magnets are heated or dropped, the domains are shaken from their aligned positions and the magnet becomes weaker. In fact, above a certain high temperature called the *Curie temperature*, the material ceases to be ferromagnetic.

At the beginning of the seventeenth century, William Gilbert believed the earth acted as a huge magnet. Today, through knowledge gained from navigation, we know that the magnetic effect does indeed exist over the entire surface of the earth. Experiments have shown a magnetic field to exist within the earth and many hundreds of miles out into space surrounding the earth. The aurora borealis, or northern lights, a common sight in the higher northern latitudes, is associated with the earth's magnetic field. This effect is discussed in Section 21-3.

The origin of the earth's magnetic field is not known, but the most acceptable theory indicates that the field is caused by internal currents of charged particles deep within the earth, rather than the notion of the existence of a huge mass of magnetized magnetite. The earth's interior is believed to be much too hot to be ferromagnetic. Also, the magnetic poles continue to slowly change their positions, which suggests currents.

Presently, the geomagnetic poles are about 20° from the corresponding geographic poles. The north magnetic pole is located at approximately 94° W and 74° N. Figure 6-3g is an illustration of the mag-

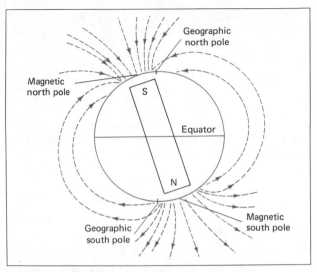

Figure 6-3g *The earth's magnetic field is probably caused by internal currents. It can be thought of as a huge bar magnet as shown. The pole of the bar magnet near geographic north is labeled S since a "north-seeking" compass needle is labeled N and opposite poles attract. Note the variation between geographic and magnetic poles.*

netic field of the earth.

In general, the magnetic meridian (an imaginary line running from the magnetic north pole to the magnetic south pole) does not coincide with the geographic meridian. The variation between the two is called the **magnetic declination,** or the variation of the compass from true, or geographic, north. The declination may vary east or west of the geographic north-south meridian. See Figs. 6-3h and 6-3i.

In addition to the variation of the compass from true north, the compass needle points down from the horizontal plane at all positions on the earth's surface except at the magnetic equator, providing the compass needle is free to move in a vertical plane. The angle the needle varies downward from the horizontal plane is called the **magnetic dip.** At the magnetic poles, the dip is 90°, and at the magnetic equator, 0°. A dip needle is illustrated in Fig. 6-3j.

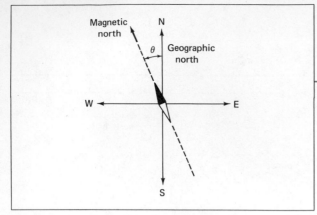

Figure 6-3h *Magnetic declination. The angle θ is measured in degrees east or west of geographic north.*

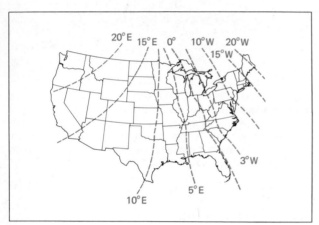

Figure 6-3i *Map showing some isogonic (same magnetic declination) lines for the United States.*

Figure 6-3j *A dip needle measures the direction of the earth's magnetic field in the vertical direction.*

6-4
ELECTROMAGNETISM

The interaction of electrical and magnetic effects is known as **electromagnetism.** Electromagnetism is one of the most important concepts in physical science, and much of our current technology is directly related to this crucial interaction. We can summarize the two basic principles of electromagnetism as follows:

1. Moving and spinning electrical charges create magnetic fields.
2. A magnetic field will deflect a moving charge.

The first idea was discussed in Section 6-3 when we discussed the source of magnetism. One of the most important applications of this principle is called an electromagnet. An **electromagnet** consists of a current-carrying coil of insulated wire wrapped around a piece of soft iron. This is illustrated in Fig. 6-4a. When the current is turned on, a magnetic field

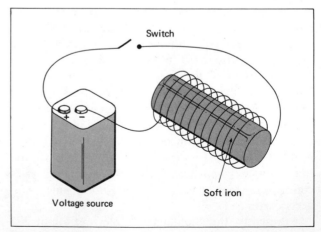

Figure 6-4a *An electromagnet consists of a coil of wire wrapped around a piece of soft iron. When the switch is closed, current flows in the wire and magnetizes the iron, creating a magnet. When the switch is open, no current flows and the iron is not magnetized.*

Figure 6-4b *A simplified diagram of a (one-directional) telephone circuit. Sound waves are converted to a varying electric current in the transmitter. The varying current travels along the phone lines to the receiver where it is converted back to sound waves.*

is created inside the coil (see Fig. 6-3e, top picture). The soft iron is magnetized by this field and makes the magnetic field about 2000 times stronger. When the current is turned off, the iron loses most of its magnetism. The advantages of the electromagnet are many. First, it can be switched on and off. Second, the strength of the electromagnet is controlled by the amount of current flowing in the wire. Third, the poles of the magnetic field can be reversed by reversing the current.

Electromagnets are found in a variety of devices and appliances such as doorbells and telephones, and in devices used for moving scrap metal. The mechanism in a telephone receiver will serve as an example of one use. A simplified diagram of a telephone circuit is shown in Fig. 6-4b.

When a telephone number is dialed, a circuit is completed with another telephone's bell. When the ringing phone is lifted, the circuit between the speakers and receivers of the two telephones is completed. A telephone conversation consists of converting the sound waves into a varying electric current. This varying current travels along the wire to the other telephone's receiver where it is converted back to sound waves.

The transmitter of a phone contains a diaphragm which vibrates due to the sound waves that are spoken. The diaphragm vibrates against a chamber which contains carbon granules. As the diaphragm vibrates, the pressure on the carbon granules varies, and this causes more or less electrical resistance in the circuit. The resistance is low when the granules are pressed together and increases as they spread apart. Because of Ohm's law, this varying resistance causes a varying electric current to be transmitted.

At the receiver end there is an electromagnet and a permanent magnet which is attached to a disk. The varying electric current gives the electromagnet a varying magnetic strength. The electromagnet attracts the permanent magnet with a variable force. The changing force on the permanent magnet and disk causes the receiver disk to vibrate. This vibration sets up sound waves which closely resemble the original sound waves, and the voice is heard at the other telephone.

The second basic principle of electromagnetism has been stated in a qualitative way: A magnetic field will deflect moving charges. Quantitatively, we want to understand how the moving charges are deflected. The formula for the force on a moving charge due to a magnetic field is given by

$$F_{\text{mag}} = qv \times B \qquad \text{Eq. 6.11}$$

where F_{mag} = force on a moving charged particle
due to the magnetic field B,
q = charge on the particle,
v = velocity of the particle,
B = magnetic field.

It should be noted that if the velocity is zero, there is no force on the particles. The force F_{mag} is a *vector quantity* as are all forces. The direction of F_{mag} takes some practice to understand. The direction of F_{mag} is perpendicular to both the velocity vector and the direction of the magnetic field. In addition, in order to have a force, the velocity vector must have some component perpendicular to the magnetic field. If the velocity vector and magnetic field vector are parallel or antiparallel, there will be no force.

The best way to understand this deflecting force is by looking at some examples. Figure 6-4c shows a beam of electrons traveling from left to right. The magnetic-field direction is vertical. The force on the

Figure 6-4c *Electrons approaching a vertical magnetic field B, as shown, experience a force F_mag that pushes them out of the page.*

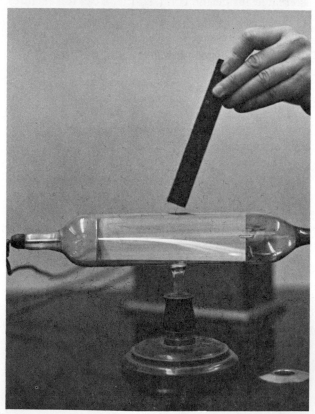

electrons is then out from the paper or towards the viewer. Two photographs of an electron beam are shown in Fig. 6-4d. In the upper photo, the beam is undeflected. In the lower photo, a vertical magnetic field is set up by the bar magnet and the beam is deflected out towards the viewer similar to Fig. 6-4c.

The electrons in conducting wires also experience effects caused by magnetic fields. In Fig. 6-4e, a wire is shown in a magnetic field. Since the electrons are not moving, there is no force on the wire. In Fig. 6-4f, the electrons in the wire are moving to the right. This is a similar situation to Fig. 6-4e except now the force causes the whole wire to be forced out of the paper toward the viewer. This demonstrates the general principle that a current-carrying wire placed in a magnetic field will experience a force.

In Fig. 6-4g, a wire is pushed into a vertical field as shown. The velocity caused by the mechanical push causes the electrons to be deflected to the right, and we say that a current has been induced; that is, an electric current can be set up by an entirely mechanical method. In Fig. 6-4g, we have created a current without any batteries, plugs, or other external voltage sources. This is the basic principle behind a

Figure 6-4d *An undeflected beam of electrons is shown at the top. A magnet produces a magnetic field which deflects the electrons out of the page similar to Fig. 6-4c.*

Figure 6-4e *A stationary wire in a magnetic field experiences no force on it because the electrons are not moving.*

generator. A **generator** converts mechanical work or energy into electrical energy. Generators are used at power plants to create electricity. Falling water and pressurized steam from burning coal or fuel oil are two of the sources of mechanical energy. The electricity is carried to homes where it is converted back into mechanical energy or heat energy.

A **motor** is a device which converts electric energy into mechanical energy. The basic principle involved has already been illustrated in Fig. 6-4f. Motors generally have many windings of wire around a piece of iron in order to enhance the basic effect shown in 6-4f. The conversion of electric energy into mechanical energy is enhanced by more loops of wire and stronger magnets. Motors are heavy because of the iron that is used to create the magnetic field.

Figure 6-4f *A current-carrying wire in which the electrons are moving to the right as shown will be deflected out of the page by the vertical magnetic field.*

Figure 6-4g *A wire is pushed into a vertical magnetic field as shown. This sets up a current in the wire.*

Figure 6-4h *A diagram of Faraday's experiment showing electromagnetic induction. The reading on the meter indicates a current flow in the circuit as the magnet is plunged downward into the coil.*

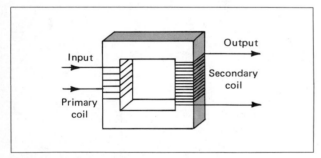

Figure 6-4i *The basic features of a transformer consist of a piece of iron, a primary coil, and a secondary coil. The alternating current in the primary coil creates a varying magnetic field in the iron which in turn induces an alternating current in the secondary coil.*

In the formula for the force due to a magnetic field, the velocity v is the relative velocity between the charged particles and the magnetic field. So far, all of our examples have been with stationary magnetic fields. Michael Faraday, in 1831, did a famous experiment in which he moved the magnet rather than the wire in order to create or induce an electric current. A diagram of this experiment is shown in Fig. 6-4h.

We discussed in Section 6-2 how transformers were used to increase or decrease voltages associated with alternating current. Transformers are a good example of the Faraday effect shown in Fig. 6-4h. The principle behind a transformer is illustrated in Fig. 6-4i. Alternating current creates a varying magnetic field in the iron core. The changing magnetic field also passes through the secondary coil where it induces a current. The voltage and current in the secondary coil are different from the primary coil. The amount of voltage is directly proportional to the number of turns in the coil. The current is inversely proportional to the number of turns in the coil. By putting more turns in the secondary coil than in the primary, we can step up the voltage. Similarly, we can decrease the voltage by putting fewer turns in the secondary than in the primary. For example, a transformer with 100 windings in the primary and 1000 in the secondary will step up the voltage by a factor of 10. A transformer with 500 windings in the primary and 25 in the secondary would step down the voltage by a factor of 20.

IMPORTANT WORDS AND TERMS

Electrons
Protons
Neutrons
Coulomb
Current
Ampere
Coulomb's law
Conductors
Insulators
Electricity
Electric potential energy

Voltage
Volt
Resistance
Ohm's law
Ohm
dc
ac
Transformers
Series circuit
Parallel circuit
Magnetic field

Electric field
Magnetic monopole
Ferromagnetic
Domains
Magnetic declination
Magnetic dip
Electromagnetism
Electromagnet
Generator
Motor

LEARNING OBJECTIVES

After reading and studying this chapter, you should be able to do the following without referring to the text:

1. Give the mks system units of electric charge, electric current, voltage, resistance, energy, and power.
2. State the basic force law for (a) electric charges (b) magnetic poles.
3. State one way in which Coulomb's law and Newton's law of universal gravitation are similar. State three important ways in which they are different.
4. Explain briefly why some materials are conductors and some are insulators.
5. Explain why electrons, rather than protons, move along wires.
6. State Ohm's law in equation form.
7. Explain the difference between ac and dc and tell why ac is used in homes.
8. Explain what transformers do and how they work.
9. Draw diagrams of a series circuit and a parallel circuit, and explain why home appliances are wired in parallel rather than series.
10. Explain why a shock from a 110-volt ac source may or may not be fatal.
11. Sketch the magnetic field around the earth.
12. Briefly give an atomic description of what causes iron to be magnetic.
13. Explain how one magnetizes a piece of iron.
14. Briefly describe how an electromagnet works and give three advantages they have over permanent magnets.
15. Draw a simple diagram to show the basic idea behind an electric generator.
16. Draw a simple diagram to show the basic idea behind an electric motor.

QUESTIONS

1. State the three basic constituents of an atom. Give their fundamental properties.
2. What is a quark?
3. Give the constants k and G in the mks system. What does this indicate about the "strength" of the electric and gravitational forces?
4. Explain how a charged rubber comb attracts bits of paper.
5. Why do conductors carry a current but insulators do not?
6. Draw a diagram of a simple electric circuit, including a switch, and tell how it works.
7. Why is ac used in homes?
8. What would happen if power were transmitted from the generating station to your home at 110 volts. What is done to prevent this from happening?
9. Why are home appliances wired in parallel rather than series?
10. What is a fuse and why is it used?
11. What causes electric shock and what precautions should be taken to avoid it?
12. What is the difference between an electric dipole and a magnetic dipole?
13. What happens if a bar magnet is cut in half?
14. Why are some materials ferromagnetic?
15. How does a permanent magnet attract a piece of soft iron?
16. What happens to a piece of iron above the Curie temperature?
17. Distinguish between magnetic dip and declination.
18. How does an electromagnet work?
19. How do telephone transmitters and receivers work?
20. What basic principle makes a motor work?
21. What basic principle makes a generator work?

PROBLEMS

1. How many electrons make a single coulomb of charge?

2. What happens to the force of attraction between the electron and proton if the distance between them is tripled?

3. The heating element of an electric iron operates on 110 volts at a current of 10 amperes.
   (a) What is the resistance of the iron?
   (b) What power is dissipated by the iron?

   Answer: (b) 1100 watts

4. In Fig. 6-2g, the voltage is 110 volts.
   (a) What will be the current flowing in a light bulb which has a resistance of 10 ohms?
   (b) What will be the current in a toaster with a resistance of 100 ohms?
   (c) What will be the total current flowing if only the light bulb and toaster are turned on?

5. (a) What is the current in a string of 10 Christmas tree lights each having a resistance of 4 ohms when plugged into a 110-volt source?

    Answer: (a) 2.75 amperes

   (b) How many watts of power are generated?

   (c) How much will it cost at $.05 per kWh to have the Christmas tree lights on for 4 hours a day for 20 days?

    Answer: (c) $1.21

6. A person with dry skin has a body resistance of 50,000 ohms.

   (a) If he or she touches the two terminals of a 2-volt dry cell, what will be the current flowing through the body?

   (b) The person's fingers are wet and the body resistance drops to 2000 ohms. What will be the current now?

7. Sketch the magnetic field (a) between a north and south pole, (b) between two north poles.

8. Use Fig. 6-3h to find the magnetic declination where you live.

9. A transformer with 1000 turns in the primary coil has to decrease the voltage from 2200 volts down to 110 volts for home use. How many turns should be in the secondary coil?

    Answer: 50

Chapter 7

HEAT

The heating effects produced by fire, the sensation received when a piece of ice is held in the hand, and the warmth produced when hands are rubbed together are well known. Explaining what takes place in each of these cases, however, is not easy. Both heat and temperature are commonly used when referring to hotness or coldness. However, heat and temperature are not the same thing. They have different and distinct meanings as we shall see.

Count Rumford (1753–1814), an American (born Benjamin Thompson), was the first to recognize the relation between mechanical work and heat. In 1789 he published the results of an experiment in which he used a blunt boring tool to drill a cannon barrel immersed in water. The temperature of the cannon and water rose to the boiling point of water in two and one-half hours. This convinced Rumford that large quantities of heat were produced by friction. He concluded from this that heat was a form of energy and appeared to be due to the motion of the drill. Later, James Prescott Joule (1818–1878), an Englishman, determined the quantitative relationship between mechanical energy and heat. He also established the important law of conservation of energy and originated many of the basic ideas for the kinetic theory of gases.

The concepts of heat and temperature play an important part in our daily lives. We like our coffee hot and our ice cream cold. The temperature of our living and working quarters must be carefully adjusted to our bodies' heat demands. The daily temperature reading is probably the most important datum given in a weather report. How cold or how

Then cold and hot and moist and dry,
In order to their stations leap.

John Dryden

hot it will be affects the clothes we wear and the plans we make. We will learn in detail in Chapter 21 how the sun provides heat to our earth. The heat balance between various parts of the earth and its atmosphere gives rise to wind, rain, and other weather phenomena. The thermal pollution of rivers caused by hot water from nuclear power plants is a recent cause for concern because of its effect on the ecology of these rivers. On a more cosmic scale, the temperature of various stars gives clues to their ages and the origin of the universe. What is temperature? What is heat? What causes heat? How is heat transferred? The answers to these questions will be examined in this chapter. Knowing these answers we can explain many phenomena occurring around us.

7-1
TEMPERATURE

Loosely, the temperature of a substance is a measure of its hotness or coldness. However, temperature is not necessarily a measure of heat. A simple experiment will illustrate this: Place ice cubes and water in a pan and heat. As heat is applied the ice melts. However, when the temperature of the ice and water mixture is checked with a thermometer, we find the temperature remains the same. Even with continued heating (and stirring) there is no change in temperature *until all the ice is melted.* Obviously, heat was added, but it didn't change the temperature. We see, then, that temperature does not necessarily measure heat. To define temperature satisfactorily, we must consider the microscopic motion of matter. This will be done in the latter portion of this chapter.

The quantitative measurement of temperature is accomplished through the use of a thermometer. A **thermometer** is a precise and reliable instrument which utilizes the physical properties of materials for the purpose of accurately determining temperature. The most common thermometer is the mercury-in-glass type. This instrument consists of a thin-walled glass bulb attached to a slender capillary glass tube, both containing mercury.* The upper end of the glass tube is sealed off; and, in most cases, the air above the mercury is removed. A scale is attached or etched on the glass for measuring the expansion of the mercury. When heat is applied, the volume of the mercury increases and more of the slender tube is filled. The glass also expands, but the mercury expands much more; thus, the instrument utilizes the difference between the two expansions to measure the temperature. After a thermometer is constructed, it is calibrated with a standard to insure proper temperature readings. Figure 7-1a is a drawing of a clinical thermometer used for measuring body temperature. In this type of thermometer a constriction is made in the capillary tube just above the glass bulb. When the thermometer is removed from the body, the mercury starts to contract but breaks at the constriction, leaving the mercury in the capillary tube. This serves the purpose of holding the reading of the thermometer at its highest value and allowing time for reading the thermometer. The mercury must be forced back down by shaking the thermometer.

The construction of a temperature scale requires two reference points and the choice of a unit. The reference points chosen are the ice point and the steam point. The **ice point** is the temperature of a mixture of ice and air-saturated water at normal atmospheric pressure. The **steam point** is the temperature at which pure water, at normal atmospheric pressure, boils. Two common temperature scales are the Fahrenheit and Celsius scales. The **Fahrenheit scale** is based on an ice point of 32° (read 32 degrees) and a steam point of 212°. The difference between the ice point and steam point is evenly divided into 180 units. Each unit is called a degree. Thus, a **degree Fahrenheit,** abbreviated °F, is $\frac{1}{180}$ of the temperature change between the ice point and the steam point. The **Celsius scale** is based on an ice point of 0° and a steam point of 100°. There are 100 equal units or

*Some home thermometers contain an alcohol, which often is colored red.

Figure 7-1a *Drawing of a clinical thermometer. The reading on the thermometer is 101°F, indicating the patient has a fever.*

Figure 7-1b *Three temperature scales. The Celsius, Kelvin, and Fahrenheit scales are shown.*

divisions between these points. So, a **degree Celsius,** abbreviated °C, is $\frac{1}{100}$ of the temperature change between the ice point and the steam point.

Have you ever wondered what doubling the temperature means? On both the Fahrenheit and Celsius scales, this is a confusing question. For instance, what is double zero degrees, or what is double −10°? These questions are meaningless unless there is an absolute limit on how cold an object can become. There is such a limit. It occurs at about −273°C, or −460°F.

The **Kelvin scale** is based on this limit, using zero as the lowest temperature possible, which is −273°C. This scale is sometimes called the absolute scale. A **degree Kelvin,** abbreviated °K, is the same as a degree Celsius, but doubling the temperature has meaning only on the Kelvin scale.

The three temperature scales are shown in Fig. 7-1b.

It is easy to convert from the Celsius scale to the Kelvin scale. Simply add 273° to the Celsius temperature and you have the Kelvin temperature. Mathematically,

$$K = C + 273$$

where K = temperature on the Kelvin scale,
C = temperature on the Celsius scale.

As examples, zero degrees Celsius equals 273°K, and a Celsius temperature of 34°C equals 307°K. It is easy to see that the ice point and the steam point are at 273°K and 373°K. Converting from Fahrenheit to Celsius or vice versa is not so easy. The equations used are

$$C = \tfrac{5}{9}(F - 32)$$

and

$$F = \tfrac{9}{5}C + 32$$

where C = temperature on the Celsius scale,
F = temperature on the Fahrenheit scale.

As examples, 59°F equals 15°C and 20°C equals 68°F. Of course, F or C or both can be negative numbers.

7-2
HEAT

Heat is a form of energy. So far we have studied two kinds of energy—kinetic energy and potential energy. In a way, heat is a third kind of energy. However, if we were to examine the microscopic sources of heat energy, we would discover that the kinetic and potential energy of very small particles, called *molecules*, are the ultimate sources of heat. Thus, heat can be viewed as either another form of energy, or as the manifestation of molecular kinetic and potential energy on a macroscopic scale. In any event, heat is energy.

The standard unit for measuring heat energy is the calorie. A **calorie** is defined as the amount of heat necessary to raise one gram of pure liquid water by one degree Celsius at normal atmospheric pressure. Water exists in the liquid phase from 0°C to 100°C, but the heat required to raise the temperature for each degree is not exactly the same. In fact, the value of the calorie varies almost one percent over the 100-degree range of temperature. Thus, the question arises as to which degree to use in defining the standard calorie. The most commonly used standard is the **15-degree calorie.** This is the amount of heat necessary to raise one gram of pure water from 14.5°C to 15.5°C at normal atmospheric pressure. The 15-degree calorie is chosen because it is approximately $\frac{1}{100}$ of the value required to raise one gram of water from 0°C to 100°C. The 15-degree calorie is equal to 4.186 joules. Remember that a joule is the unit of energy in the *mks* system of units. The **kilocalorie,** or large calorie, is equal to 1000 calories. It is the kilocalorie that is used in specifying the heat value of food. Through common usage of the term, the prefix "kilo" has been dropped. For example, if a portion of food is said to contain 45 "calories," that means it has 45,000 calories, according to our definition. In deference to the more portly of our readers, we will not dwell on this point. The unit of heat in the British system of units is the British Thermal Unit, or Btu. One **Btu** is the amount of heat required to raise one pound of water one degree Fahrenheit at normal atmospheric pressure.

Since heat is a form of energy, it can be related to kinetic energy, potential energy, and work. In fact there is a definite ratio between the heat that appears in a mechanical process and work done on it; that is, heat generated is proportional to the work done. This can be written as

$$H \propto W$$

or

$$H = JW$$

solving for J

$$J = H/W = 1 \text{ calorie}/4.185 \text{ joules}$$

J is known as the mechanical equivalent of heat.

When we do work, there is frequently a substantial amount of heating caused by friction. The conservation of energy tells us that the work done plus the heat energy produced must equal the original amount of energy available.

$$E_T = W + H$$

where E_T = total energy available initially,
W = work done,
H = heat energy produced, usually by friction.

▪ *Example 1* A sled and the child on it, having a total mass of 20 kg, slide down a hill that is 4 m high, starting from rest. Their final velocity is 8 m/s. How much heat was produced by the friction between the sled runners and the snow?

The initial energy is all potential energy. The work done is equal to the final kinetic energy. We have

$$H = E_T - W$$

$$E_T = mgh$$

$$W = mv^2/2$$

$$H = mgh - mv^2/2$$

where m = mass of sled and child,
g = acceleration of gravity.
  = 9.8 m/s²,
h = height of hill,
v = final velocity.

So,
$$H = (20 \text{ kg} \times 9.8 \text{ m/s}^2 \times 4 \text{ m})$$
$$- (20 \text{ kg} \times (8 \text{ m/s})^2/2)$$

$$= 784.0 \text{ joules} - 640 \text{ joules}$$

$$= 144 \text{ joules}$$

$$= \frac{144 \text{ joules}}{4.186 \text{ joules/calorie}}$$

$$= 34.4 \text{ calories}$$

One very well known effect of heating a material is that of expansion. Almost all matter—solids, liquids and gases—expand when heated. As a general rule, a substance expands when heated and contracts when cooled. The most important exception to this rule is water. If water is frozen, it expands. That is, ice at 0°C occupies a larger volume than the same mass of water at 0°C.

The change in length or volume of a substance due to a change in temperature is a major factor in the design and construction of items ranging from steel bridges and automobiles to watches and dental cements. The cracks in a highway are designed so that in the summer the concrete will not buckle as it expands due to the heat. Large cracks are designed into bridges for the same reason. Heat-expansion characteristics are used to control such things as the flow of water in car radiators and the flow of heat in homes through the operation of thermostats.

7-3
SPECIFIC HEAT

The calorie has been defined as the amount of heat necessary to raise the temperature of one gram of water one degree Celsius. Other substances require different amounts of heat to raise the temperature of one gram of the substance one degree. The **specific heat** of a substance is the amount of energy in calories necessary to raise the temperature of one gram of the

Table 7-3a *Specific Heat*

Substance	*Specific heat* $cal/g\,°C$
Aluminum	0.22
Copper	0.092
Glass	0.16
Ice	0.50
Iron	0.105
Silver	0.056
Soil (average)	0.25
Steam	0.50
Water	1.000

substance one degree Celsius. Obviously, the specific heat of pure water is 1.000. The specific heats of a few common substances are given in Table 7-3a.

The definition of specific heat supplies us with an equation for the amount of heat necessary to change the temperature of a given substance. It is

$$H = mc\,\Delta T$$

where H = heat gained or lost by the substance in calories,

m = mass of the substance in grams,

c = specific heat of the substance in cal/g°C,

ΔT = change in temperature of the substance in °C or °K.

■ *Example 2* A 4-g aluminum pellet at 30°C gains 88 cal of heat. What is its final temperature?

In this example $H = 88$ cal, $m = 4$ g, $c = 0.22$ cal/g°C, and ΔT is not known. We have

$$H = mc\,\Delta T$$

or
$$\Delta T = \frac{H}{mc}$$

$$= \frac{88 \text{ cal}}{4 \text{ g} \times 0.22 \text{ cal/g°C}}$$

$$= \frac{88}{0.88}°C = 100°C$$

Figure 7-3a *Graph of temperature versus heat energy for a typical substance.*

Since the original temperature was 30°C and the change in temperature is 100°C, the final temperature is 130°C.

The **heat capacity** of a substance is the amount of heat (expressed in calories) required to raise the temperature of a substance one degree Celsius. Heat capacity is calculated by finding the product of the mass and the specific heat. In the preceding example, the mass was 4 g and c was 0.22 cal/g°C, so the heat capacity of the pellet was 0.88 cal/°C.

Substances found in our environment are usually classified as being in either the solid, liquid, or gaseous phase. Referring to Fig. 7-3a, we observe how the temperature of the different phases changes as heat is applied. In the lower left-hand corner the substance is represented in the solid phase. As heat is applied the temperature rises. When point A is reached, adding more heat does not change the temperature. Instead, the heat energy is used to change the solid into a liquid. The amount of heat necessary to change one gram of a solid into a liquid at the same temperature is called the **latent heat of fusion** of the substance. In Fig. 7-3a this is simply the amount of heat necessary to go from A to B. When point B has been reached, the substance is all liquid. The temperature of the substance at which this change from solid to liquid takes place is known as the **melting point.** After point B, further heating again causes a rise in temperature. The temperature continues to rise as heat is added until point C is reached. From B to C the substance is in the liquid phase. When point C is reached, adding more heat does not change the temperature. The

added heat is now changing the liquid into a gas. The amount of heat necessary to change one gram of a liquid into a gas is called the **latent heat of vaporization** of the substance. In Fig. 7-3a this is simply the amount of heat necessary to go from C to D. When point D has been reached, the substance is all in the gas phase. The temperature of the substance at which this change from liquid to gas phase occurs is known as the **boiling point.** After point D is reached, further heating again causes a rise in the temperature.

The heat necessary to change a solid to a liquid is determined by

$$H = mH_f$$

where H = heat in calories,
  m = mass in grams,
  H_f = latent heat of fusion in cal/g.

The heat necessary to change a liquid into a gas is

$$H = mH_v$$

where H = heat in calories,
  m = mass in grams,
  H_v = latent heat of vaporization in cal/g.

For the case of water, the melting point occurs at 0°C, and the boiling point is at 100°C. The latent heat of fusion is 80 cal/g and the latent heat of vaporization is 540 cal/g. The heat necessary to convert ice to steam can be computed in five stages.

■ *Example 3* Calculate the amount of heat necessary to change 3 g of ice at −10°C into steam at +120°C.

The total heat necessary is given by

$$H = H_1 + H_2 + H_3 + H_4 + H_5$$

where H_1 = heat to raise ice 10°C,
  H_2 = heat to change ice to water at 0°C,
  H_3 = heat to raise water 100°C,
  H_4 = heat to change water to steam at 100°C,
  H_5 = heat to raise steam 20°C.

We can calculate each of these as follows:

$$H_1 = mc\,\Delta T$$
$$= 3\,\text{g} \times 0.5\,\text{cal/g}^\circ\text{C} \times 10^\circ\text{C}$$
$$= 15\,\text{cal}$$

$$H_2 = mH_f$$
$$= 3\,\text{g} \times 80\,\text{cal/g}$$
$$= 240\,\text{cal}$$

$$H_3 = mc\,\Delta T$$
$$= 3\,\text{g} \times 1.0\,\text{cal/g}^\circ\text{C} \times 100^\circ\text{C}$$
$$= 300\,\text{cal}$$

$$H_4 = mH_v$$
$$= 3\,\text{g} \times 540\,\text{cal/g}$$
$$= 1620\,\text{cal}$$

$$H_5 = mc\,\Delta T$$
$$= 3\,\text{g} \times 0.5\,\text{cal/g}^\circ\text{C} \times 20^\circ\text{C}$$
$$= 30\,\text{cal}$$

$$H = 15\,\text{cal} + 240\,\text{cal} + 300\,\text{cal} + 1620\,\text{cal} + 30\,\text{cal}$$
$$= 2205\,\text{cal}$$

From Fig. 7-3a, we get an idea of the differences between temperature and heat. We see that the temperature of a body rises as heat is added only if the substance is not undergoing a change in phase.

The important and general law of conservation of energy is a useful concept when dealing with heat energy. The law of conservation of energy gives us the general principle that the heat lost in a process must equal the heat gained.

An example of the use of this law is the determination of the specific heat of an unknown substance. If a hot metal pellet is dropped into cool water in a glass we can say, assuming that no heat is lost to the surrounding air,

Heat lost by metal = Heat gained by water
+ Heat gained by glass

If the masses and the initial and final temperatures are all known, the specific heat of the metal can be found.

■ *Example 4* A 45-g aluminum pellet is heated to 98.5°C and dropped into an 80-g glass beaker ($c = 0.20\,\text{cal/g}^\circ\text{C}$) containing 200 g of water at 20°C. The final maximum equilibrium temperature is 23.5°C. Determine the specific heat of aluminum.

$$H_{\text{aluminum}} = H_{\text{water}} + H_{\text{glass}}$$

or
$$(mc\,\Delta T)_{\text{aluminum}} = (mc\,\Delta T)_{\text{water}}$$
$$+ (mc\,\Delta T)_{\text{glass}}$$

$$45\,\text{g} \times c \times (98.5^\circ\text{C} - 23.5^\circ\text{C})$$
$$= 200\,\text{g} \times 1\,\text{cal/g}^\circ\text{C} \times (23.5^\circ\text{C} - 20.0^\circ\text{C})$$
$$+ 80\,\text{g} \times 0.20\,\text{cal/g}^\circ\text{C} \times (23.5^\circ\text{C} - 20.0^\circ\text{C})$$

Since everything is in the proper units, we can now solve for c, the specific heat of aluminum. Its units will be cal/g°C.

$$45 \times c \times 75 = 200 \times 3.5 + 80 \times 0.2 \times 3.5$$
$$3375\,c = 700 + 56$$
$$c = 756/3375$$
$$= 0.221\,\text{cal/g}^\circ\text{C}$$

7-4
THERMODYNAMICS

Thermodynamics means the dynamics of heat, that is, the flow, production, and conversion of heat for use by people. We use heat energy for most of the work done in our environment. The operation of refrigerators and other heat engines is based upon the laws of thermodynamics. These laws are important because they give relations between heat energy and energy associated with mechanical work.

The **first law of thermodynamics** is based on the law of conservation of energy and can be written as

$$\Delta H = \Delta E_i + W$$

where ΔH = the heat added to (or taken from) system,

ΔE_i = the change in internal energy of system,

W = work done by (or on) system.

All three quantities in this equation must be expressed in energy units, such as joules.

The law states that the heat energy added to a system must go into increasing the internal energy of that system, or any work done by the system, or both. The law also states that heat energy removed from a system must produce a decrease in the internal energy of the system, or any work done on the system, or both. A process in which ΔH remains constant, that is no heat is added or removed from the system, is known as an **adiabatic process.** In the above equation, if $\Delta H = 0$, then $\Delta E_i = -W$. This means that a system which is doing work will have a decrease in its internal energy and the temperature of the system decreases.

A device that uses heat energy to perform useful work is called a **heat engine.** The gasoline engine used in our automobiles is an example. Other examples are the steam and diesel engines.

The **second law of thermodynamics** can be stated in several ways. One of these is the following:

It is impossible for heat to flow spontaneously from an object having a lower temperature to one having a higher temperature.

This law is well known to all of us. If a cold object and a hot object are placed in contact, the hot object cools down and the cold object heats up (Fig. 7-4a). The reverse never happens, even though there is nothing in the first law of thermodynamics to indicate that it could not happen. For example, a cold drink always warms up to room temperature. It doesn't get colder while the room gets hotter! In all of our experience the second law seems to hold.

Figure 7-4a *Heat always flows from a substance with a higher temperature to one with a lower temperature. A cold object never gets colder when placed in contact with a warmer object.*

A stone on a hillside will on its own accord roll down the hill to a level of lower potential energy. If we want the stone to roll up hill, we must apply a force and do work on the stone to raise it to a higher potential level. The stone will not, on its own accord, roll up hill.

Heat will flow on its own accord from a higher temperature to a lower temperature, but work must be done to transfer heat energy from a low temperature reservoir to a high temperature reservoir. A device used to perform this task is called a **heat pump.** (See Fig. 7-4b.) Common heat pumps are the household refrigerator and air conditoner. Heat is transferred from inside the volume of the refrigerator to the outside by the use of electrical energy. The heat transferred, however, is much greater than the electrical energy needed to perform the work. A heat pump for the home is used to extract heat from the ground or from the air outside in order to heat the air inside the home. The heat pump is used more extensively in the south where the climate is milder. When installed in a home where the winter months are very cold, an auxiliary heating unit (usually an electric heater) must be used in conjunction with the heat pump to supply extra heat energy when needed. Heat pump installation in a home is usually more expensive than a regular furnace.

Another way of stating the second law is to say: Heat energy cannot be transformed completely into mechanical work. By use of the word "completely,"

Figure 7-4b *Schematic diagram of a heat pump. The substance used for the heat exchange is usually Freon which is pumped through the cold and hot reservoirs.*

this law implies that there will always be some heat wasted when heat energy is changed to mechanical work. Mathematically, the second law says that ΔE_i, the change in the internal energy of the system, is always greater than zero.

In 1854, Rudolf Clausius (1822–1888), a German physicist formulated the second law as follows: Natural changes tend to take place in an isolated system with continually increasing entropy. **Entropy** is defined as a measure of the disorder of a system. As a natural process takes place, the disorder increases. For example, when a solid melts, the molecules are more free to move in a random motion in the liquid phase than in the solid phase. Likewise, when evaporation takes place, there is greater disorder and an increase in entropy. If we let the change in entropy be represented by ΔS and the change in the heat energy by ΔH, then

$$\Delta S = \Delta H/T$$

where H is measured in joules and T in degrees Kelvin. S has the units of joules per degree Kelvin in the mks system of units.

Processes that are left to themselves tend to become more and more disordered, never the reverse. A heat engine is able to change some disorder (heat) into order but the engine cannot transform all of it. A student's dormitory room left to itself becomes disordered . . . never the reverse. True, the room can be cleaned and items put in order and the entropy seems to decrease. But to put things in order, energy from the food we eat must be changed into other forms of motion so the total entropy of the universe has decreased.

The second law of thermodynamics has important implications. From the first formulation we see that if the law is universally valid for all time, then the universe—the stars and galaxies—will eventually cool down to a final common temperature, at which it will remain forever. This possible fate for the universe is sometimes referred to as its "heat death."

Another less cosmic implication of the second law is that perpetual motion machines are impossible. The fault with many perpetual motion machines is that their "inventors" assume that heat can be transformed completely into work with no increase in the machine's internal energy. This is in violation of the second law.

There is also a **third law of thermodynamics.** It states that a temperature of absolute zero can never be attained. However, temperatures as low as 0.000001°K (one-millionth degree) have been reached.

7-5
PHASES OF MATTER

The three phases of matter are the solid phase, the liquid phase, and the gaseous phase. All substances exist in all three phases of matter at some temperature and pressure. At normal room temperatures a substance will be in one of the three phases. For instance, at room temperature oxygen is a gas, water a liquid, and copper a solid.

The principal distinguishing features of solids, liquids, and gases can be understood if we look at the various phases from a molecular point of view. Most substances are made up of very small particles called molecules. You may know that a **molecule** is the smallest division of a substance which still retains the properties of the substance. In a pure **solid,** the molecules are usually arranged in a specified three dimensional manner. This orderly arrangement of molecules is called a **lattice.** Figure 7-5a shows an example of a lattice in only two dimensions. The molecules (represented in the figure by small circles) are bound to each other by electrical forces. Upon heating, the molecules gain kinetic energy and vibrate about their position in the lattice. The more heat that is added, the stronger the vibrations become, and the molecules move further apart, and as shown diagramatically by Fig. 7-5a, the solid expands.

When the melting point of a solid is reached, the heat energy is great enough to break apart the bonds which hold a molecule in place. When this happens, a hole is produced in the lattice, and a nearby molecule can move to that hole. As more and more holes are produced, the lattice becomes significantly distorted. Figure 7-5b shows an arrangement of the molecules in a liquid. There are many holes in the liquid, which has only a slight lattice structure. A molecule can easily move to a new spot in the lattice, since there are so many holes in the lattice. A **liquid** is an arrangement of molecules with many holes, so that the molecules are free to move and assume the shape of the container. Upon heating a liquid, the individual molecules gain kinetic energy, and even more holes are produced. This results in the expansion of the liquid.

When the boiling point is reached, the heat energy is sufficient to break the molecules completely apart from each other. The heat of vaporization is the heat necessary to free the molecules completely from

Figure 7-5a *Schematic diagram of a typical crystal lattice in two dimensions. Heating causes the molecules to vibrate with greater amplitude about their position in the lattice, thus increasing the volume of the solid.*

Figure 7-5b *Diagram showing arrangement of molecules in a liquid. The small circles indicate molecules. They are packed closely together and form only a slight lattice structure.*

each other. Because the electrical forces holding the different molecules together are quite strong, the heat of vaporization is fairly large. When the molecules are completely free from each other, the gaseous stage is reached. A **gas** is made up of molecules which exert no forces on one another except when they collide. The distance between molecules in a gas is quite large compared to the size of the molecules. The molecules in a gas are moving rapidly.

What happens if we continue to add heat to the gas? Large amounts of added heat energy will cause the atoms of the gas to collide with sufficient energy to knock electrons free from the atoms and create a plasma. A **plasma** is referred to as a fourth phase of matter. This phase of matter is a disordered population of ionized atoms where the ion cores and the free electrons are not in thermal equilibrium. The sun, for example, is in the plasma phase. The energy released by the sun comes from the fusion of hydrogen to form helium. (See Section 9-6.) The temperature of the interior of the sun is over 15 million degrees Celsius. This provides sufficient energy to strip the atoms of all their electrons and create completely ionized matter, or what is known as a *perfect* plasma. Thus we have huge amounts of plasma in our universe, since the universe is composed of billions upon billions of stars.

Partial ionization of a gas occurs in the fluorescent lights that we use to light our classrooms and to advertise commercial establishments. The sodium vapor and mercury vapor lamps used in the laboratory are also examples of a plasma. The energy needed for ionization in these devices is supplied by high electrical voltages. Plasmas are presently being studied by large numbers of research groups with the hope of finding ways to control the fusion process and help meet the increasing demands for more energy.

We have seen that when a substance is in a single phase, heating increases the kinetic energy of the molecules of which it is composed. When a substance is changing phases, heating supplies the energy to overcome the attractive forces holding the different molecules together. Since the temperature rises only when the substance is not changing phase, we conclude that temperature can be defined as a measure of the average kinetic energy of the molecules.

7-6
THE GAS LAWS

The **pressure** of a gas is the force which the moving gas molecules exert upon a unit area. A common unit of pressure is pounds per square inch, or lb/in^2. The normal atmospheric pressure is about $15 \, lb/in^2$. An important relationship exists between the pressure, volume, and temperature of a gas at any two times,

which we shall call 1 and 2. This relation is known as the **perfect gas law** and can be stated as

$$\frac{P_1 V_1}{T_1} = \frac{P_2 V_2}{T_2}$$

where P_1, V_1, T_1 are pressure, volume, and absolute temperature at time 1, and P_2, V_2, T_2 are pressure, volume, and absolute temperature at time 2. A gas is called perfect, or ideal, when it obeys this law. A gas ceases to be ideal and the law breaks down either for high pressures or for temperatures near the liquefying point of the gas. For the range of temperatures and pressures which we shall be considering, the perfect gas law is obeyed to a high degree of accuracy for any gas. It doesn't matter if the gas is hydrogen, oxygen, nitrogen, or a combination of gases, such as air.

If the volume of a gas is held constant and we plot the pressure against the temperature of a typical gas such as helium, we get a plot similar to that of Fig. 7-6a. From the perfect gas law we see that if $V_1 = V_2$, then

$$\frac{P_1}{T_1} = \frac{P_2}{T_2}$$

Figure 7-6a *Graph of pressure versus temperature for a typical gas. The volume is constant.*

Figure 7-6b *The relationship between volume and temperature at constant pressure. As the temperature increases, the volume increases in a direct proportion. This is known as Charles' law.*

Figure 7-6c *The relationship between pressure and volume at constant temperature. The volume is inversely proportional to the pressure. This is known as Boyle's law.*

or rearranged,
$$\frac{P_1}{P_2} = \frac{T_1}{T_2}$$

This equation means that an increase in temperature results in a proportional increase in pressure. Thus, a plot of pressure and temperature data produces the straight-line graph shown in Fig. 7-6a. If we extrapolate the straight line to a pressure of zero (broken line on the graph), we see that the temperature is $0°K$, or $-273°C$. This can be seen from the last equation. If $P_1 = 0$, then T_1 must also equal zero.

A similar relation exists between the volume and temperature of a gas. Consider a gas at $0°C$ ($273°$ Kelvin, or absolute) confined in a tube with a movable piston as in Fig. 7-6b. The piston serves to keep the gas at a constant pressure. As the gas is heated to $100°C$ ($373°K$), the piston is forced upwards. If the gas occupied 546 cm^3 initially, it will expand to 746 cm^3, forcing the piston upward (Fig. 7-6b). Thus, we see that the volume of a gas at a constant pressure is in direct proportion to the absolute temperature. This proportionality is known as **Charles' law,** after the French physicist who discovered it in the eighteenth century. From the perfect gas law when $P_1 = P_2$, we get Charles' law, which can be written as

$$\frac{V_1}{T_1} = \frac{V_2}{T_2}$$

where V_1 = volume of gas at absolute temperature T_1,

V_2 = volume of gas at absolute temperature T_2.

The relationship between pressure and volume, when the temperature is kept constant, was discovered by the British physicist Robert Boyle over three centuries ago. A process in which the temperature remains constant is called an **isothermal** process. Suppose a column of a gas occupies 400 cm^3 when a 50-g mass is placed on top of it, as in Fig. 7-6c (left). If we now keep the temperature constant and double the pressure by placing another 50 g on top of the column, the volume of the gas is cut in half, as shown in Fig. 7-6c (right). This relationship is known as **Boyle's law** and can be written from the perfect gas law with $T_1 = T_2$ as

$$P_1 V_1 = P_2 V_2$$

where P_1 = pressure of gas at volume V_1,
P_2 = pressure of gas at volume V_2.

The perfect gas law is a powerful tool for computing unknown pressures, volumes, or temperatures. Some examples of its use follow:

■ *Example 5* A steel oxygen tank on a space ship has a volume of 0.50 m^3. It is filled with oxygen at a pressure of 600 lb/in². How big a volume would the oxygen occupy at a normal atmospheric pressure of 15 lb/in² at the same temperature?

We use the perfect gas law, with $T_1 = T_2$

$$\frac{P_1 V_1}{T_1} = \frac{P_2 V_2}{T_1}$$

Since the temperatures are the same they cancel, leaving Boyle's law.

$$P_1V_1 = P_2V_2$$

From the conditions stated in the problem we know that

$$P_1 = 600 \text{ lb/in}^2$$
$$V_1 = 0.50 \text{ m}^3$$
$$P_2 = 15 \text{ lb/in}^2$$
$$V_2 = ?$$

so

$$600 \text{ lb/in}^2 \times 0.50 \text{ m}^3 = 15 \text{ lb/in}^2 \times V_2$$

$$\frac{600 \text{ lb/in}^2 \times 0.50 \text{ m}^3}{15 \text{ lb/in}^2} = V_2$$

$$40 \times 0.50 \text{ m}^3 = V_2$$

$$20 \text{ m}^3 = V_2$$

Thus, we see that by storing the oxygen at great pressure, we can greatly increase the supply of oxygen available to the astronauts.

■ *Example 6* A stratosphere balloon is launched with a volume of 4000 m³ and a pressure and temperature of 15 lb/in² and 27°C. What was the balloon's volume when it reached a much higher altitude where the pressure and temperature were 3 lb/in² and −33°C?

We use the perfect gas law

$$\frac{P_1V_1}{T_1} = \frac{P_2V_2}{T_2}$$

with
$$P_1 = 15 \text{ lb/in}^2$$
$$V_1 = 4000 \text{ m}^3$$
$$T_1 = 273° + 27° = 300°\text{K}$$
$$P_2 = 3 \text{ lb/in}^2$$
$$V_2 = ?$$
$$T_2 = 273° - 33° = 240°\text{K}$$

Solving, we get

$$\frac{15 \text{ lb/in}^2 \times 4000 \text{ m}^3}{300°} = \frac{3 \text{ lb/in}^2 \, V_2}{240°}$$

$$\frac{15 \text{ lb/in}^2 \times 240° \times 4000 \text{ m}^3}{3 \text{ lb/in}^2 \times 300°} = V_2$$

$$5 \times \frac{4}{5} \times 4000 \text{ m}^3 = V_2$$

$$16,000 \text{ m}^3 = V_2$$

Thus, we see that the balloon will expand to four times its original volume.

7-7
KINETIC THEORY

So far, we have concerned ourselves mainly with the macroscopic properties of substances such as heat, temperature, and pressure. These things can be perceived by our senses quite easily. When substances are considered from a microscopic point of view as a collection of rapidly moving molecules, we reach the domain of what is known as *kinetic theory*.

According to kinetic theory, a gas is made up of a fantastically large number of molecules, moving at very high speeds (typically 1000 mi/h) and continually colliding with and bouncing off each other (about one billion collisions each second!), as well as bombarding other nearby objects. For instance, the molecules in the air are continually colliding with themselves and everything around them—tables, chairs, walls, people, etc. When they bounce off any surface, they exert a force per unit area. This is what we perceive as air pressure. (See Fig. 7-7a).

The absolute temperature of a substance is directly related to the average kinetic energy of the molecules in the substance. The faster the molecular movement, the higher the absolute temperature. Conversely, as the motion of the molecules slows

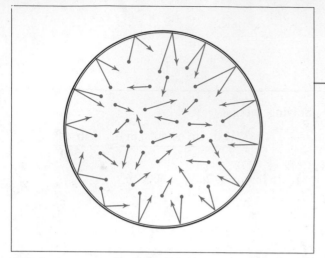

Figure 7-7a *The air molecules inside a basketball are continually hitting and bouncing off the walls, exerting pressure and forcing the walls outward. As long as the pressure is greater on the inside than on the outside, the ball remains inflated.*

down, the absolute temperature decreases. In a simple model of solids, the temperature of absolute zero would then be explained as the point at which molecules cease to move. In the modern theory of matter, however, it is known that there is a slight amount of energy and motion left, even at absolute zero. This is called the *zero point energy*.

The molecules in a gas move in a random fashion, and, because of their high speeds and very small mass, the effect of gravity is negligible on a small scale. For example, inside any building there are practically as many molecules near the ceiling as there are near the floor. On the scale of the earth's atmosphere however, gravity has a noticeable effect, and this results in more molecules being held nearer the earth's surface. Thus, as we go up in altitude, the number of molecules decreases, and the air pressure undergoes a corresponding decrease.

With the kinetic theory we can now better explain the gas laws of Section 7-6. First, let us consider the relationship between the pressure and temperature of a gas. If we heat a rigid container full of gas, the molecules gain kinetic energy; that is, their velocities increase. This increase in velocity enables them to hit the walls of the container with a greater force, and the evidence of this is an increase in pressure. So, the higher the temperature the higher the pressure. This corresponds to Fig. 7-6a.

Charles' law was illustrated by Fig. 7-6b. As the temperature was increased, the gas molecules gained kinetic energy, and this increase in kinetic energy caused work to be done on the piston. The piston was raised, and this changed the volume of the gas, although the pressure remained the same. So at constant pressure an increase in temperature increased the volume of the gas.

Boyle's law can also be understood by using the kinetic theory. If we compress a gas, the molecules are confined to a smaller volume. They then bounce off the wall more often, and this is detected as an increase in the pressure. Making the volume smaller results in a pressure increase.

We noted in Section 7-6 that the perfect gas law was no longer valid at either very high pressures or temperatures near the liquefying point of a gas. This can now be explained from a kinetic theory viewpoint. At very high pressures, the gas molecules are squeezed together so that they no longer occupy a negligible volume compared to the distance between them. The same effect occurs for gases at temperatures near the liquefying point. Under these conditions the intermolecular forces and the sizes of the molecules themselves become important. At this point, the perfect gas law must be modified to take account of these factors.

The kinetic theory also helps us to understand the difference between heat and temperature. Heat is a form of energy. It is related to both the kinetic and potential energy of molecules. If a substance is being heated, and the velocity and kinetic energy of the molecules is increasing, then the temperature will be rising. If, however, the heat energy goes into changing the potential energy of the substance, then the substance is changing phase and no temperature rise will be detected. Thus, we see that absolute temperature is a measure of the average kinetic energy. Temperature

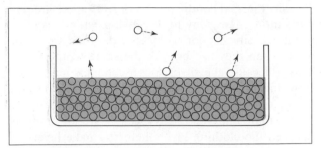

Figure 7-7b *Evaporation of water. Water molecules are held together by intermolecular forces that are electrical in nature. When a molecule on the surface has enough kinetic energy to break free, it becomes water in the vapor stage. The latent heat of vaporization supplies the energy necessary for evaporation to take place.*

7-8
HEAT TRANSFER

The transfer of heat can be accomplished by one or more of the following three methods: convection, conduction, and radiation.

The transfer of heat by **convection** requires the movement of a substance, or mass, from one position to another. The movement of air or water is an example of heat transfer by convection. Most homes are heated by convection (movement of hot air). The air is heated at the furnace, then circulated throughout the house by way of metal ducts. When the air has "lost its heat," it passes through a cold air return on its way back to the furnace to be reheated and recirculated. See Fig. 7-8a. Circulation of air in a room occurs naturally

is not related to the potential energy of the molecules.

Not all of the molecules in a substance are moving at the same speed. That is why we defined absolute temperature as a measure of the *average* kinetic energy of a molecule. Some of the molecules are moving faster and some are moving slower than average. When water (or any liquid) evaporates, it is due to the escape of the more rapidly moving molecules. The slower moving molecules cannot escape from the liquid's surface, but the faster ones can. We can now easily see why stepping out of a shower produces a cooling sensation. The cooling occurs because of the evaporation of the water droplets on the skin. The heat from your body is used to supply the water molecules with enough energy to break away from the skin's surface and go into the air. Because heat is being drawn out of your body, a cooling sensation is experienced. (See Fig. 7-7b.)

Our bodies are subjected to the bombardment of air molecules all the time. The faster the motion of the molecules, the higher the temperature. When the temperature changes and the molecules impinging upon us move faster or slower, our body perceives a change. We call that change hotter or colder.

Figure 7-8a *Heat transfer by convection. The air heated by the furnace rises and circulates throughout the house to return to the furnace when it has "lost its heat."*

because cold air is heavier than warm air: therefore, it will sink to the floor and displace the warm air, which is forced to rise. The cold air duct is located at floor level to provide a path for the air to return to the furnace to be reheated. The transfer of heat energy by means of convection currents, including both air and water currents, is essential in distributing heat energy over the surface of the earth.

Conduction is the transfer of heat energy by molecular activity. The kinetic energy of the molecules is transferred from one molecule to another through collisions. The heat flows from the higher temperature to the lower temperature, with the rate of flow being directly proportional to the temperature difference. Since the energy transfer is brought about by collisions of molecular or submolecular particles, a substance must be present to act as a medium for transfer. Some substances are very good conductors of heat, others are not. The **thermal conductivity** of a substance is a measure of its ability to conduct heat. Metals such as silver, copper, and aluminum are good conductors of heat; thus, they have high thermal conductivities. Other materials such as wood, paper, and air are poor conductors of heat; their thermal conductivities are low. Substances which are poor conductors of heat are known as **thermal insulators.** Table 7-8a gives the relative thermal conductivities of several substances.

The thermal conductivities of substances are important in many day-to-day living situations. For instance, a silver spoon will feel colder than a piece of wood when they are both at the same temperature. This is because the silver spoon conducts heat away from the body at a much higher rate. As another example, it is warmer to wear layers of clothing in winter than one very heavy coat. The air trapped between the clothing layers helps to make layers of clothing a better insulator than one heavy coat.

The conduction of heat plays an important role in making ice. In order to make ice, heat energy must be removed from the water. Since metal trays conduct heat quite well, they are frequently used. The rate at which ice is made depends on how fast the heat can be conducted out of the water. In some refrigerators, hot water freezes faster than cold. This is true even though hot water has more heat which must be taken out than cold water. Figure 7-8b shows ice trays in a typical freezing compartment. The residue of ice on the surface makes for an uneven contact between the cool ice tray and the freezing surface. The metal touches the ice in only a few places. For this reason the rate of cooling is quite slow. When hot water is put in, the hot metal tray melts part of the ice residue on the surface, so that the tray sits evenly on the ice. Because the entire metal bottom is touching, the cooling rate is very high, and the hot water will

Table 7-8a *Relative Thermal Conductivities*

Silver	0.99
Copper	0.91
Aluminum	0.49
Iron	0.16
Lead	0.084
Glass	0.002
Water	0.0015
Wood	0.0005
Paper	0.0003
Air	0.000054

Figure 7-8b *Diagram to explain how hot water can freeze faster than cold. The hot water melts the ice beneath it, making a smooth contact. Smooth contact means faster loss of heat and thus quicker freezing.*

quickly freeze. The hot water freezes faster than the cold water because the cooling rate is much higher for a smooth contact between ice tray and freezing compartment.

The process of transferring heat energy through space by means of electromagnetic waves is known as **radiation.** Electromagnetic waves carry energy and can travel through a vacuum. The heat which we get from the sun is transmitted through the vacuum of space by radiation. Heat energy in the form of electromagnetic waves is both absorbed and reflected when it hits a surface. Black objects completely absorb the heat energy, while objects that are white are good reflectors and very poor absorbers. For this reason, light-colored clothing is cooler in summer than dark-colored clothing. Similarly, dark clothing is warmer in winter.

A good example of the methods used to impede heat transfer is shown in Fig. 7-8c. The vacuum bottle is designed to prevent heat from escaping or entering the glass bottle which holds the hot or cold liquid. Heat transfer by convection and conduction is prevented due to the evacuated space. By removing most of the mass (air molecules) in this volume of space,

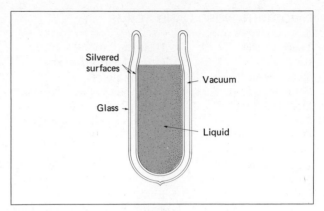

Figure 7-8c *A vacuum bottle.*

the means for convection and conduction are removed. To prevent transfer by radiation, the inner surfaces of the glass bottle are silvered. These silvered surfaces reverse the direction of the heat radiation. Thus wanted heat energy is kept in for hot coffee or kept out for cold tea.

IMPORTANT WORDS AND TERMS

Thermometer
Ice point
Steam point
Fahrenheit scale
Degree Fahrenheit
Celsius scale
Degree Celsius
Kelvin scale
Degree Kelvin
Heat
Calorie
15-degree calorie
Kilocalorie
Btu
Specific heat

Heat capacity
Latent heat of fusion
Melting point
Latent heat of vaporization
Boiling point
First law of thermodynamics
Adiabatic process
Heat engine
Second law of thermodynamics
Heat pump
Entropy
Third law of thermodynamics
Molecule
Solid
Lattice

Liquid
Gas
Plasma
Pressure
Perfect gas law
Charles' law
Isothermal
Boyle's law
Temperature
Convection
Conduction
Thermal conductivity
Thermal insulators
Radiation

LEARNING OBJECTIVES

After reading and studying this chapter, you should be able to do the following without referring to the text:

1. Define, explain, and give a physical interpretation (an example) of the important terms in the following list:
   Heat
   Temperature
   Specific heat
   Heat capacity
   Entropy
   Plasma
   Latent heat of fusion
   Latent heat of vaporization
2. State the differences between the Celsius, Fahrenheit, and Kelvin temperature scales.
3. Define and give examples for the metric and British units of heat.
4. State, explain, and give an example of each of the following:
   Boyle's law
   Charles' law
   Perfect gas law
   First law of thermodynamics
   Second law of thermodynamics
5. Distinguish between an isothermal and an adiabatic process.

6. Calculate the amount of heat necessary to change a known mass of ice at a temperature below 0°C to steam at a temperature above 100°C.
7. Determine the value of P (pressure), V (volume), or T (temperature) when all of these quantities except one are given.
8. State, explain, and give examples of the four phases of matter.
9. State, explain, and give examples of the three methods of heat transfer.
10. State the differences between thermal conductors and thermal insulators.

QUESTIONS

1. In terms of kinetic and potential energy of molecules, distinguish between heat and temperature.
2. Is there such a thing as a temperature of absolute zero? What is its meaning?
3. What are the three common temperature scales and how do they differ?
4. Which common substance has the highest specific heat?
5. A temperature of $-1°K$ was reported by an observer working in cryogenics. Does the reported temperature have a meaning? Explain.
6. Why does it take a long time for lakes and rivers to freeze when the air temperature is below freezing?
7. Which requires more heat to raise its temperature from 25°C to 30°C—water or copper?
8. When does an increase in heat not result in an increase in temperature?
9. Does heat flow from hot substances to cold substances or from cold substances to hot ones?
10. A pendulum clock is calibrated to run accurately at 21°C. When the air temperature in the room where the clock is located increases to 22°C, will the clock run faster or slower? Explain your answer.
11. A round metal plate has a large hole at its center. The temperature of the metal plate is increased by 10 degrees. Will the area of the hole increase, decrease, or remain the same? Explain.
12. Which can give the more severe skin burn, water at 100°C or steam at 100°C? Explain your answer.
13. Our grandparents placed a large tub of water in their root cellars to prevent canned fruit from freezing during the winter months when the outside temperature was very low. Explain how the tub of water prevented the fruit from freezing.
14. Pike's Peak in Colorado is over 14,000 ft above sea level. Why is it possible to place one's hand in a container of boiling water at this altitude?
15. The atmosphere provided a pressure on the surface of the oceans of 76 cm of mercury. What would be the results on the oceans, if the atmosphere were removed? Explain your answer.
16. Why does a person in a closed room held at a constant temperature feel cooler if a fan is running?

17. What does the second law of thermodynamics say about a "heat death" for the universe?
18. From a kinetic theory viewpoint, why does an increase in temperature lead to an increase in pressure?
19. Why does the pressure inside an auto tire increase during long continuous driving?
20. Which gas law studied in this chapter represents an isothermal process? Explain.
21. If the temperature of a gas increases and the volume is held constant, what happens to the pressure? Explain your answer.
22. If the temperature of a gas is held constant and the volume is decreased, what happens to the pressure? Explain your answer.
23. How can tight metal lids on glass jars be removed easily?
24. What causes water pipes to break sometimes during cold weather?
25. Why does a plastic tile floor feel colder than a rug even though both are at the same temperature.

PROBLEMS

1. Normal body temperature is 98.6°F. What is the equivalent temperature on the Celsius scale?

2. Normal room temperature is about 68°F. What is the equivalent temperature on the Celsius scale? On the Kelvin scale?

3. Determine (from an outside source) the temperature of the vegetable compartment of a household refrigerator. Express the value in Fahrenheit, Celsius, and Kelvin degrees.

4. A college student when resting produces about 60 kilocalories of heat per hour. Determine the equivalent in watts.

5. Determine the change in entropy when 1 kg of ice at 0°C melts.

6. To make iced tea, 1 kg of ice at 0°C is added to 1.6 kg of hot tea at a temperature of 98°C. What is the final temperature of the mixture. Assume the specific heat of tea is the same as water.

7. Twenty joules of work are expended in stirring 400 g of water. If the initial temperature of the water was 25°C, what will be the final temperature of the water? Assume all the work went to heat the water.

8. (a) How much heat is necessary to raise the temperature of 20 g of aluminum from 25°C to 100°C? (b) What is the heat capacity of 20 g of aluminum?

9. A 100-g copper cup holding 500 g of water is at 25°C. If one kilocalorie of heat is added, what will be the final temperature of the cup and the water? Assume that no heat is lost to the surrounding air.

10. A skier comes down a 10-m-high slope, reaching the bottom at a speed of 10 m/s. If the skier's mass is 60 kg, how much heat did he produce during his run?

11. A 20-g aluminum pellet at 100°C is dropped into 600 g of water in an 80-g copper cup, both at a temperature of 20°C. Using the specific heats in Table 7-3a, calculate the final temperature of the mixture.

12. Calculate the heat necessary to change 30 g of ice at −30°C into steam at 120°C.

13. A balloon is blown up to a pressure of 29 lb/in² so that it occupies a volume of 900 cm³ at room temperature of 17°C. The next day the temperature of the room is up to 27°C, and the balloon has expanded to 1000 cm³. What is the new pressure in the balloon.

14. A container is filled with helium gas at 22°C. What is this in °K? If heat is applied so that the average kinetic energy of the gas molecules doubles, what is the new temperature of the gas in degrees Celsius and degrees Kelvin?

Chapter 8

RELATIVITY

Many statements in our conversations involve the "relativeness" of things. For example, we say: Beauty is in the eye of the beholder. A boring class seems to run on and on, while time spent with someone we love flies by. The earth goes around the sun or the sun goes around the earth, depending on how you look at it. We will learn in Chapter 17 that it is much simpler to describe the solar system by saying that the earth and other planets revolve around the sun and spin as they go. The spinning of the earth causes day and night. However, to us who live on the earth, it certainly does appear that the sun goes around the earth. In fact, we still speak of sunrise and sunset. We see, then, that when we talk about things that occur in our daily life, it is important to make clear what our frame of reference is. The things that we observe and the values we hold depend on a frame of reference.

When we discussed Newton's laws of motion in Chapter 3, we based our thinking on our common experience of space and time. For example, an astronaut can set his watch with a clock in Houston and fly to the moon and back at high speeds. When he returns, his watch and the clock in Houston are still synchronized. This illustrates that time doesn't change even when we travel at speeds as high as 25,000 miles per hour. Our ideas of space are based on similar experiences at low velocities. In 1905 Albert Einstein (Fig. 8-1a) published a landmark paper in which he put forth a theory that revolutionized man's concept of space and time. For instance, he showed that if the astronaut were to travel at speeds approaching the speed of light, his watch would no

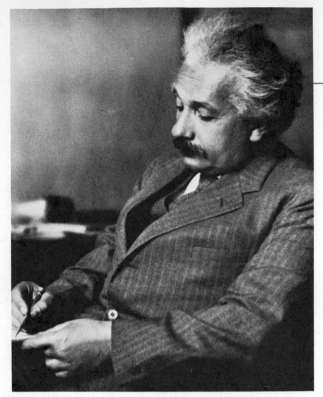

Figure 8-1a *Albert Einstein (1879–1955). Possibly the most brilliant scientist of all time, Einstein shook the world of science when he presented his theory of relativity in 1905. At the time he was working as a clerk in a Swiss patent office studying for his doctorate at night. In 1921 he received the Nobel prize in physics but, curiously, not for his theory of relativity. (Photo courtesy American Institute of Physics Niels Bohr Library.)*

8-1
MOVING SYSTEMS OF REFERENCE

Whenever a scientist makes a measurement in a laboratory, he employs a system of reference. A **system of reference** is something to which a measurement is referred. Another scientist, making the same measurement but doing so when he is moving with a constant velocity, would get different results because he uses a different system of reference. The Doppler effect is one example of this phenomenon. The frequency of a sound which a person hears from a moving source is different from the actual frequency of the sound at the source. Thus, when a speeding ambulance approaches a person waiting on a street corner, that person hears a sound from the siren with a higher frequency than does the driver of the ambulance.

As another example, let us assume that Dr. Albert is standing in a train station as a train whizzes past at 60 miles per hour. Dr. Baker is on the train and walking at a velocity of 5 miles per hour toward the front of the train. Dr. Baker would comment that he was traveling at a velocity of 5 miles per hour. However, Dr. Albert would contend that Dr. Baker was traveling at a velocity of 65 miles per hour relative to him, since the train was already running at 60 miles per hour and Dr. Baker's walking speed* should be added to that. Who is right? Of course, both are correct. We see, then, that before we make comments about Dr. Baker's velocity, we must specify whether we are talking about velocity relative to the train or relative to the train station.

In our study of motion, thus far, we have been concerned with objects going at velocities which are relatively very low compared to the speed of light. Just about everything which we have experienced is moving slowly relative to the speed of light (186,000

*We will be using "speed" and "velocity" interchangeably throughout this chapter.

longer keep the same time as the stationary clock. In other words, time is relative. It is affected by the velocity of the timepiece. (Of course, contrary to the limerick at the start of the chapter, it is impossible for time to go backwards under any circumstances.) Einstein's theory deals with space and distances in a similar fashion. They too are relative. They depend on the speed of the observer and are not absolute. Einstein's theory is quite mathematical, but the results which arise are so fascinating as to be well worth the reader's effort to understand them.

Figure 8-1b *Distance relationships in moving reference systems, O and O'. $\Delta s = v_R \Delta t + \Delta s'$. This equation is true only if v_R is much lower than the velocity of light.*

Figure 8-1c *Velocity relationships in moving reference systems. $v = v_R + v'$. This equation is true only if v_R is much lower than the velocity of light.*

miles per second!). At these relatively low velocities, it is not hard to change from one system of reference to another. Suppose we are in a system of reference which is moving with velocity v_R with respect to a fixed system, say the earth. Now suppose that during the time change, Δt, we change our position in our moving system by $\Delta s'$. What will the total change in our position be? Let us refer to Fig. 8-1b. Because we were on a moving system, we have changed our position by at least $v_R \Delta t$. If we had remained motionless at O', our displacement relative to O would have been exactly $v_R \Delta t$. But since we changed our position in the moving system from O' to A', there is an additional change of $\Delta s'$. The total change in our position in the fixed system is given by

$$\Delta s = v_R \Delta t + \Delta s'$$

We can also write this as

$$\Delta s' = \Delta s - v_R \Delta t \qquad \text{Eq. 8.1}$$

where $\Delta s'$ = change in position in the moving system,
Δs = change in position in the fixed system,
Δt = change in time,
v_R = velocity of moving system relative to fixed system.

In most of our experience, time seems to be absolute; that is, it does not depend on the coordinate system. (Shortly, we will see that this is not always true.) If for the moment, we assume that time is absolute, clocks in both the moving and fixed systems will run at the same speed. So we have

$$\Delta t' = \Delta t \qquad \text{Eq. 8.2}$$

where $\Delta t'$ = change in time in the moving system,
Δt = change in time in the fixed system.

For low speeds we know that Eq. 8.2 is adequate, since a commercial jet pilot's watch keeps the same time as a clock on the ground.

The velocities in the two reference systems of Fig. 8-1b can now be related by dividing Eq. 8.1 by Eq. 8.2

$$\frac{\Delta s'}{\Delta t'} = \frac{\Delta s}{\Delta t} - v_R \frac{\Delta t}{\Delta t}$$

or

$$v' = v - v_R \qquad \text{Eq. 8.3}$$

where v' = velocity in the moving system,
v = velocity in the fixed system,
v_R = velocity of the moving system with respect to the fixed system.

This concept of addition of velocities is illustrated in Fig. 8-1c.

In our train example, Dr. Baker's velocity (v') in the moving system (the train) was 5 mi/h. The train was running at a velocity of $v_R = 60$ mi/h. So the velocity according to Dr. Albert in the fixed system (the train station) was $v = 65$ mi/h. Thus, in Eq. 8.3

$$5 \text{ mi/h} = 65 \text{ mi/h} - 60 \text{ mi/h}$$

Equations 8.1, 8.2, and 8.3 are the foundations of relativity at low velocities. We will next examine what happens as we consider velocities which are comparable in magnitude to the speed of light.

8-2
EINSTEIN'S POSTULATES

Figure 8-2a *The non-Einsteinian way of adding high veloc- ities. This method works for low velocities but not for high velocities.*

Between 1850 and 1900 several experiments were performed* which gave very strange results if inter- preted in terms of Eqs. 8.1, 8.2, and 8.3. But in 1905, Albert Einstein explained these experiments by postu- alting two things:

Einstein's Postulates:
1. The physical laws of nature (such as Coulomb's law) are the same in all reference systems which are moving with constant velocity with respect to each other.
2. The velocity of light is independent of the motion of the source.

As an example of the first postulate, an experi- ment on a speeding spaceship concerning Coulomb's law of electric charges will yield the same results as an identical experiment carried out in a laboratory on the earth. The second postulate seems strange at first. It is in direct contradiction with Eq. 8.3. According to Eq. 8.3, if a star is moving away from the earth at three-fourths the velocity of light (as in Fig. 8-2a),† the speed of light from the star when it reaches the earth should be only one-fourth the velocity of light. If worked out, using Eq. 8.3, we have

$$v' = v - v_R$$

or $$v = v_R + v'$$

Given $$v_R = \tfrac{3}{4}c$$

and $$v' = -c$$

Then, $$v = \tfrac{3}{4}c - c$$

$$v = -\tfrac{1}{4}c \qquad \text{(WRONG)}$$

*Some of the experiments which baffled description were Fizeau's experiments on the aberration of starlight, and the Michelson- Morley experiment which measured the speed of light in different directions.

†We let c equal the speed of light in empty space.

(The minus signs mean that the velocities in Fig. 8-2a are in the opposite direction to those shown in Fig. 8-1c.)

According to Eq. 8.3, the speed of light reaching the earth from distant stars will be less than c. Ein- stein's second postulate, on the other hand, says that the speed of light reaching the earth will always be equal to c, no matter how fast the stars move away from us. Something is wrong, and it is Eq. 8.3. This means that Eqs. 8.1 and 8.2 are also inadequate at very high speeds, since we derived Eq. 8.3 from them. If Einstein's postulates are correct, we must therefore replace Eqs. 8.1, 8.2, and 8.3 with more general ones.

8-3
THE LORENTZ TRANSFORMATION

Einstein, using his second postulate, showed that Eqs. 8.1 and 8.2 should be rewritten as

$$\Delta s' = \frac{\Delta s - v_R \,\Delta t}{\sqrt{1 - (v_R^2/c^2)}} \qquad \text{Eq. 8.4}$$

$$\Delta t' = \frac{\Delta t - \Delta s v_R/c^2}{\sqrt{1 - (v_R^2/c^2)}} \qquad \text{Eq. 8.5}$$

Equations 8.4 and 8.5 are known as the **Lorentz transformation.** They tell how distances and times in systems moving relative to each other are related.

Note that distances depend on times and vice versa. Also, note that if the ratio of v_R to c is very small (i.e., $v_R/c \approx 0$), we obtain Eqs. 8.1 and 8.2.

An important consequence of Eqs. 8.4 and 8.5 is that distances and times in the two coordinate systems are no longer the same. The relationship between distances in the two systems, measured at the same time ($\Delta t = 0$), is

$$\Delta s' = \frac{\Delta s}{\sqrt{1 - (v_R^2/c^2)}}$$

or $\qquad \Delta s = \Delta s' \sqrt{1 - (v_R^2/c^2)}$ $\qquad$ Eq. 8.6

Equation 8.6 means that a moving meter stick appears to be shorter to an observer who is standing still than to an observer moving with the meter stick. This phenomenon is known as the **Lorentz-Fitzgerald contraction.**

- *Example 1* Dr. Baker has a meter stick in his rocket. The rocket is traveling at a speed of $\frac{9}{10}$ the speed of light. As he whizzes past Dr. Albert's space station, how long does the meter stick look to Dr. Albert?

This is solved by recognizing that in the moving system, $\Delta s' = 1$ m. We want to solve for Δs. The speed $v_R = 0.9c$.

$$
\begin{aligned}
\Delta s &= \sqrt{1 - (v_R^2/c^2)} \times \Delta s' \\
&= \sqrt{1 - ((.9c)^2/c^2)} \times 1 \text{ m} \\
&= \sqrt{1 - .81} \times 1 \text{ m} \\
&= \sqrt{.19} \times 1 \text{ m} \\
&= 0.4359 \text{ m}
\end{aligned}
$$

So, the meter stick appears to be only .4359 m to Dr. Albert. Dr. Baker would say that it was 1 m. Of course, when Dr. Baker's rocketship stops ($v_R = 0$), Dr. Albert will then say that it is 1 m long.

The relationship between times in the two systems is not easy to derive, but it is given by

$$\Delta t = \frac{\Delta t'}{\sqrt{1 - (v_R^2/c^2)}} \qquad \text{Eq. 8.7}$$

According to Eq. 8.7, clocks in the two systems would not keep the same time! This phenomenon is called **time dilation.**

A good example of the veracity of time dilation is provided by the lifetime of small particles called **muons,** or mu mesons. When muons are created in laboratory experiments, they have an average lifetime of $T_o = 2.2 \times 10^{-6}$ s. Within this lifetime they would travel a distance of $\Delta s = vT_o$ before disappearing by decaying into other particles. Even at the speed of light ($v = c$), the distance traveled (Δs) during T_o would be only about $\frac{4}{10}$ mi. However, muons created by cosmic ray photons at altitudes of about 15 mi reach the earth's surface. The reason that this is possible is that the muon's "clock" goes slower than an earth clock. In the muon's reference frame, the lifetime is still $T_o = 2.2 \times 10^{-6}$ s. But in the earth's frame of reference, the lifetime T_e is given by Eq. 8.7 as

$$T_e = \frac{T_o}{\sqrt{1 - (v_R^2/c^2)}}$$

where T_e = muon's lifetime according to an earth clock,

$\qquad T_o$ = muon's lifetime,

$\qquad v_R$ = velocity of muons relative to the earth.

If $v_R = 0.9998\,c$ (almost the speed of light), we find that $T_e = 50\,T_o$, and the muons will not decay until they have traveled 20 mi. Thus, they will reach the earth before decaying. Figure 8-3a shows the difference between the two "clocks" in a schematic fashion.

Equation 8.7 can be very confusing if we are not extremely careful in interpreting Δt and $\Delta t'$. Consider the following "mind-bending" situation. Assume that Dr. Baker and Dr. Albert are twins. Dr. Baker gets in a spaceship and leaves the earth, traveling at a speed .866 times the velocity of light. When 1 hour has

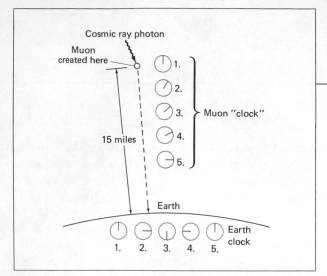

Figure 8-3a *Diagram of time differences at high velocities. The muon "clock" runs slower than an earth clock because the muons have very high velocities.*

elapsed on Dr. Baker's ship, the amount of time elapsed on the earth is

$$\Delta t = \frac{1 \text{ hour}}{\sqrt{1 - \frac{(.866 \, c)^2}{c^2}}}$$

$$= \frac{1 \text{ hour}}{\sqrt{\frac{1}{4}}} = 2 \text{ hours} \qquad \text{Eq. 8.7a}$$

Thus, Dr. Baker's clock runs twice as slow as Dr. Albert's clock on earth. But now if we consider Dr. Baker's spaceship as fixed and consider that the earth is moving away from it, the roles of Δt and $\Delta t'$ are reversed. The Δt would now refer to Dr. Baker's clock on the spaceship and $\Delta t'$ to the earth clock of Dr. Albert. Using Eq. 8.7, we get

$$1 \text{ h} = \frac{\Delta t'}{\sqrt{1 - \frac{(.866 \, c)^2}{c^2}}} = \frac{\Delta t'}{\sqrt{\frac{1}{4}}} = 2 \, \Delta t'$$

or $$\Delta t' = \tfrac{1}{2} \text{ h} \qquad \text{Eq. 8.7b}$$

According to this result, when Dr. Baker's clock has run for 1 hour, the earth clock will have run for $\frac{1}{2}$ hour, and so Dr. Baker's clock now runs twice as fast

as Dr. Albert's. But we just concluded that Dr. Baker's clock ran twice as slow as Dr. Albert's. This dilemma is sometimes called the **twins paradox.** If Dr. Baker's spaceship traveled for 20 earth years, Dr. Albert, using the reasoning of Eq. 8.7a, would say that Dr. Baker had aged only 10 years. But if Dr. Baker used the reasoning of Eq. 8.7b, he would calculate that 40 rather than 10 of his spaceship years corresponded to 20 earth years. If they were twins, which twin would really be older? This question has plagued many a student of relativity.

The explanation of the twins paradox is quite subtle—and quite profound, for it calls into question the very basis of science. This is the question of how a measurement is made. When Dr. Baker leaves the earth, he can set his clock to coincide with Dr. Albert's clock, but how do the two ever get back together to compare their clocks again or to see who is older? The only way is for Dr. Baker's spaceship to slow down, stop, and come back. But to do this, he must decelerate and then accelerate his spaceship. Eqs. 8.4, 8.5, and 8.7 hold only for systems which move at constant velocities relative to one another. They do not hold for systems which accelerate. Thus, when Dr. Baker's spaceship decelerates, new equations must be used. The fact that Dr. Baker and Dr. Albert can never compare their clocks or ages more than once without acceleration is the solution to the twins paradox. In Section 8-6 this problem will be considered again.

8-4
ADDITION OF VELOCITIES

Equations 8.4 and 8.5 yield an interesting result when applied to the relativistic addition of velocities. If we divide Eq. 8.4 by Eq. 8.5 we obtain

$$v' = \frac{v - v_R}{1 - \frac{v v_R}{c^2}}$$

Figure 8-4a *Illustrating relative velocity. The relative velocity of two spaceships, each moving at 0.75c, is not 1.5c. It is 0.96c.*

If we let $v = v_1$ and $-v_R = v_2$, this can be written as

$$v_3 = \frac{v_1 + v_2}{1 + \dfrac{v_1 v_2}{c^2}} \qquad \text{Eq. 8.8}$$

where v_3 = relative velocity,

v_1, v_2 = two velocities to be added.

Equation 8.8 can now be used to show that no velocity can ever exceed c, the speed of light in empty space. As an example, we consider the problem of two spaceships approaching each other, each ship traveling with a velocity of three-fourths the speed of light (Fig. 8-4a).

According to our classical equation (Eq. 8.3), occupants in each spaceship would claim that the other ship is approaching it at a velocity of 1.5 c. However, Eq. 8.8 yields a different result. Using this equation, we obtain

$$v^3 = \frac{\frac{3}{4}c + \frac{3}{4}c}{1 + \dfrac{\frac{3}{4}c\,\frac{3}{4}c}{c^2}} = \frac{1.5c}{1 + \frac{9}{16}}$$

$$v_3 = .96c$$

So, occupants in each ship now claim that the other is approaching it at a speed of 96 percent of the speed of light in empty space. No matter how fast two objects approach each other, Eq. 8.8 will yield a relative velocity of c or less. An interesting result of

Einstein's postulates is that nothing can exceed the speed of light in empty space.

The speed of light in empty space has been measured to be the following:

mks system, $c = 3.0 \times 10^8$ m/s

cgs system, $c = 3.0 \times 10^{10}$ cm/s

British system, $c = 186,000$ mi/s

This, of course, is much faster than any speed with which we are acquainted. However, it is not particularly fast when we consider interstellar distances. Light from the sun takes about eight minutes to reach the earth, and light from the other stars takes years to reach us.

To say that nothing can go faster than the speed of light is an imprecise statement. We should say that nothing can go faster than c, the speed of light in empty space. When light traverses matter, its velocity is c/n where n is the index of refraction, which is always greater than one. In substances such as water, where the speed of light is $\frac{3}{4}c$, it is possible for subatomic particles such as electrons to go faster than light through water. That is, the electrons go slower than c, but faster than $\frac{3}{4}c$. This effect—of particles going faster than light in a medium—has been experimentally verified.

8-5
MASS-ENERGY RELATIONSHIP

Another important relationship which follows from Einstein's postulates is the famous mass-energy relationship. It is given by

$$E = mc^2 \qquad \text{Eq. 8.9}$$

where E = total energy residing in a particle at rest of mass m,

m = mass of particle,

c = speed of light.

What this equation says is that a very small mass can produce a fantastically large quantity of energy. For instance, the energy in 1 g of matter at rest is given by

$$E = mc^2$$
$$= 1 \text{ g} \times (3 \times 10^{10} \text{ cm/s})^2$$
$$= 9 \times 10^{20} \text{ ergs}$$

This is equivalent to 25 million kilowatt hours of energy, or enough energy to drive an automobile around the earth 400 times. It is now easy to see how a lump of uranium can supply so much nuclear energy. Incidentally, only a small fraction of the uranium's mass is converted into energy.

The mass-energy relation tells us that mass is a form of energy. The energy which a particle has because it has a mass is called its **rest energy.** If a particle is moving with a velocity v, its total energy is given by

$$E = \frac{mc^2}{\sqrt{1 - (v^2/c^2)}} \qquad \text{Eq. 8.10}$$

For low velocities, this is the same as

$$E = mc^2 + \tfrac{1}{2}mv^2$$
(for v, small compared to c)

which is just the rest energy plus kinetic energy. Often the equation for the total energy of a particle is written

$$E = m'c^2$$

where

$$m' = \frac{m}{\sqrt{1 - (v^2/c^2)}} \qquad \text{Eq. 8.11}$$

Writing the equation like this puts it on a similar footing to Einstein's equation $E = mc^2$, where m is the mass of the particle at rest, or the **rest mass.** When the particle is in motion, we can still write $E = m'c^2$, but now if we say m' is the mass of the particle, then m' is bigger than the rest mass. As v increases, m' also increases.

8-6
GENERAL THEORY OF RELATIVITY

So far, we have been considering what is sometimes called Einstein's **special theory of relativity.** The word "special" refers to the fact that it holds only for systems which are moving at a constant velocity with respect to one another. The more general case would involve systems which are accelerating with respect to one another. The extremely complicated theory which describes such systems is called the **general theory of relativity.** An important aspect of this theory is that the force of gravity produces an acceleration which is equivalent to any other acceleration. So, any effect which occurs on earth with its acceleration due to gravity of 9.8 m/s² will also occur in a spaceship traveling with an acceleration of 9.8 m/s², and vice versa. An important corollary to this result is that light is bent by gravitational forces. This can be seen if we consider the light traversing a spaceship, which is accelerating as shown in Fig. 8-6a. If a beam of starlight passes through point A, it would normally pass through point B, which is almost directly across from point A. However, because the rocket ship is accelerating, by the time the light has reached the other side of the ship, the ship has accelerated upwards as shown, and the light passes through point C. An occupant of the spaceship would observe that the light was bent as shown in Fig. 8-6b. Now, since a gravitational acceleration should have the same effect as a rocket ship acceleration, we conclude that light is bent in a gravitational field.

Einstein's prediction that light is bent in a gravitational field was verified in 1919 during an eclipse of the sun. A star was known from astronomical predictions to be at a spot in the sky so as to appear to be at the edge of the sun during the eclipse. (If the sun had not been eclipsed, the star, of course, could not have been seen.) Figure 8-6c shows a schematic diagram of the situation. From its known course in the sky, the

Figure 8-6a *Diagram showing the path of light through an accelerating rocketship. For an observer, a light beam entering the rocket at point* A *should hit point* B. *However, because of the acceleration of the rocket, the beam hits point* C.

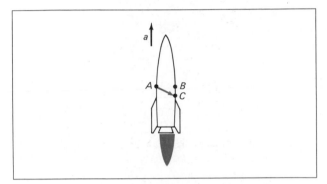

Figure 8-6b *Illustrating the bending of light. To an observer on the rocket the light appears to bend from A to C. At the same time the observer feels a downward pull caused by the acceleration of the rocket. This downward pull being similar to gravitational acceleration leads the observer to conclude that gravity caused the light to bend.*

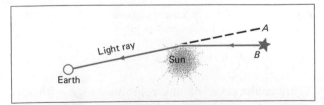

Figure 8-6c *Starlight passing near the sun is bent by its gravitational field. The star is actually at B, but its light is bent by the sun's gravity so that it appears to be at A. The drawing is not to scale, by any means.*

star was predicted to be at *B*. However, during the eclipse it was observed as apparently being at point *A*. The only acceptable explanation assumed that as the star's light passed by the sun, it was bent by the gravitational pull of the sun, so that it took the path designated by the colored line. The difference in the apparent position and the actual position was predicted by Einstein to be 1.75 seconds of arc, or an angle of about 0.0005°. The difference was later found to be 1.61 ± .30 seconds of arc, thus verifying Einstein's theory.

The very improbable phenomenon of twins aging at separate rates can also be explained using the general theory of relativity. Assume Dr. Baker and Dr. Albert are 30-year-old twins. Dr. Baker then takes off in his spaceship and accelerates at *g*, the acceleration of gravity, for one year according to his clock. He then decelerates at a rate *g* for one year. He turns his spaceship around, accelerates for a year, decelerates for a year and lands back on earth. According to the theory of general relativity, he is now 34 years old, but while he was gone his twin brother has grown to the ripe old age of 97 years. Thus, it is evident that times in different systems are truly different. Space travelers will find older people back at home! The philosophical, scientific, and sociological implications of this fact should prove to be interesting. Of course, our example is somewhat impractical at the present stage of space.

8-7
GRAVITATIONAL COLLAPSE

The possibility of a rather remarkable phenomenon has recently been proposed to explain some strange astronomical findings of the past 10 years. This phenomenon may or may not explain the new discoveries, but it is interesting in its own right. The phenomenon is called **gravitational collapse** and is the collapse of a very massive body because of its attraction for itself. Recall that Newton's law of universal gravitation was given as

$$F = G\frac{m_1 m_2}{r^2} \qquad \text{Eq. 8.12}$$

where F = force of gravity between m_1 and m_2,
 m_1 = a mass,
 m_2 = a second mass,
 G = a universal constant,
 r = distance between the centers of m_1 and m_2.

From Eq. 8.12 it is evident that if the distance r is made smaller, the force of gravitation increases.

In a large mass such as the earth or the sun, parts are always attracting each other. In Fig. 8-7a, parts A and B are attracting each other. All other parts of the mass are also attracting one another, so that the net result is that all parts are attracted toward the center. Normally, in matter such as this, electromagnetic forces tend to keep the various particles (e.g., atoms) apart. What would happen if the body were so massive that the gravitational forces of attraction were stronger than the electromagnetic forces of repulsion? Parts A and B and all other parts would be drawn closer together, as in Fig. 8-7b. But according to Eq. 8.12, the forces F_1 and F_2 increase as the distance r decreases, and so A and B would move closer and closer together. But the closer they are, the greater the gravitational force becomes. The whole mass would continue to contract until nothing was left but a fantastically massive point, as in Fig. 8-7c. This very massive point would not only bend light rays, but it would also absorb those rays that came sufficiently close. In fact, no signals such as light rays could ever be emitted from such a gravitationally collapsed point, because they would be attracted so strongly that there is no way that they could get free of the point. This absorption of light is shown in Fig. 8-7d. The normal bending of light is shown in Fig. 8-7e.

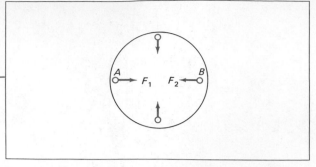

Figure 8-7a *The beginning of gravitational collapse. The force of gravitation attracts all particles toward the center of the object.*

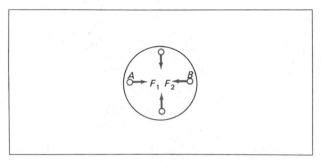

Figure 8-7b *Gravitational collapse. The smaller the distances between the particles, the greater the force of gravity becomes.*

Figure 8-7c *A gravitationally collapsed point. All the mass is concentrated at a very small point in space.*

There could never be any communication with such points. The philosophical implications of such processes are fascinating.

In 1968 a strange phenomenon was observed for the first time by astronomers. Pulses of electromag-

Figure 8-7d *The effect of a gravitationally collapsed point on light. Light rays passing close to the star are captured by the star; others farther away are only bent.*

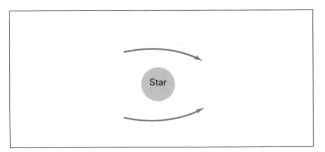

Figure 8-7e *Diagram of how the gravitational force of a massive star causes passing light to bend.*

netic radiation were detected coming from a few stars in the sky. The pulses were very regular and lasted anywhere from $\frac{1}{30}$ second to a few seconds, depending on the star. Stars which emit this type of pulsating radiation are called **pulsars.** (For more on pulsars see Section 20-3.) They are believed to be extremely dense objects, which have undergone a partial gravitational collapse. Thus, they are very dense—about one trillion times as dense as lead. In fact, a baseball-sized object consisting of this super-dense material would weigh about 10 billion tons! An example of such a star is shown in the photograph at the beginning of this chapter. The pulsar blinks on and off 15 times a second. In the top photo it is on; in the bottom photo it is off. In order to understand these fascinating objects in the sky, astronomers must use Einstein's general theory of relativity.

IMPORTANT WORDS AND TERMS

System of reference	Muons	Special theory of relativity
Lorentz transformation	Twins paradox	General theory of relativity
Lorentz-Fitzgerald contraction	Rest energy	Gravitational collapse
Time dilation	Rest mass	Pulsars

LEARNING OBJECTIVES

After reading and studying this chapter you should be able to do the following without referring to the text:

1. Explain what is meant by a system of reference.
2. State the two postulates upon which Einstein's special theory of relativity is based.

3. State the differences between Einstein's special theory of relativity and his general theory of relativity.
4. Describe some consequences of the special theory of relativity, such as time dilation, length contraction, and mass increase.
5. Describe mass-energy equivalence and state whether the equivalence is a consequence of the special or general theory of relativity.
6. Describe the phenomenon called gravitational collapse.
7. Define or explain the important words and terms listed at the end of the chapter.

QUESTIONS

1. Why is Einstein's theory called the theory of relativity?
2. Approximately, what was the fastest speed attained by a person at the time Einstein published his theory of relativity (1905)? Approximately, what is the fastest speed a person can now attain? Are either of these speeds such that relativistic effects could be observed?
3. Distinguish between *rest mass* and *relativistic mass* increase.
4. Explain the meaning of *rest energy*.
5. Does a length contraction depend on whether a system is moving away or moving toward an observer at rest?
6. Suppose there are two spaceships in outer space, one traveling at 5,000,000 mi/h and one traveling at 10,000,000 mi/h. How will experiments on board the two spaceships differ in their results?
7. Do you know of any phenomena which are due to the Lorentz-Fitzgerald contraction? If not, why haven't you experienced these phenomena?
8. Have you ever experienced any phenomena which are due to time dilation? If not, why?
9. An article appears in the newspaper stating that electrons can go faster than light in water. How do you explain this to a 13-year-old who insists that nothing can go faster than light?
10. Give two very important concepts contained in Einstein's formula $E = mc^2$.
11. What do the words "special" and "general" mean in Einstein's theory of relativity?
12. Is every light beam bent by every mass? Explain.
13. Why don't people collapse to a point due to the attractive gravitational forces within their bodies, all forces pulling toward a common center?
14. It is the year 2100, and you have lived to be over 140 years old. You see an article in the paper about a teenage astronaut who claims to be the mother of a 46-year-old man. In spite of your senility, how would you explain this to your great-great-grandchildren?
15. An electron is traveling with a velocity of $0.99c$ in an accelerator. Is it possible to double the kinetic energy of the electron by increasing its velocity? Explain.
16. Are mass and energy equivalent? Explain.

PROBLEMS

1. If Dr. Baker throws a baseball forward at 40 mi/h while in a train car that is moving at 60 mi/h, how fast will the baseball appear to be moving to Dr. Albert who is on the station platform?

2. When Dr. Baker turns on a flashlight pointed forward, the light beam travels at the speed of light. If Dr. Baker is on a train traveling at 60 mi/h, how fast is the light beam traveling, as seen by Dr. Albert, who is on the station platform?

   Answer: Also the speed of light

3. As a spectator at the Universal Relativistic Olympics, you are watching a pole vaulter sprint down the runway at $\frac{3}{5}$ times the speed of light. When he started from rest, his pole was 20 ft long. How long does his pole appear to you while he is running past?

   Answer: 6 ft

4. A dispute arises at the Universal Relativistic Olympics. A runner is timed by a judge standing at the finish line as using 10 s for the 1,000,000-mi dash. The runner insists that his watch reads 8.2 s and that he has broken the galactic record. Who is correct?

5. How long does it take for light to reach us from (a) the moon, 240,000 mi away (b) Jupiter, when it is 558,000,000 mi away (c) a star, which is 12 light years away?

   Answer: (b) 3000 s, or 50 min

6. Each second the sun gives off 4×10^{33} ergs of energy. (a) How many grams of mass are being lost by the sun each second? (b) The sun's mass is about 2.000×10^{33} g. At the rate calculated in (a) about how many years will it be until the sun's mass is reduced to 1.999×10^{33} g?

   Answer: (b) About 7 billion years

7. One reason that nothing can exceed c is that as the velocity of a particle increases, its mass also increases. Thus, since $a = F/m'$, it takes more and more force to maintain an acceleration. What is the mass of a particle moving at $\frac{4}{5}$ the speed of light if its rest mass is 1 g?

8. If a meter stick passes Sarah, who is at rest, with a velocity of 0.6c in a direction parallel to its length, what would be its length as observed by Sarah?

9. How fast is an atomic particle moving when its relativistic mass is double its rest mass?

10. An electron has a rest mass of 9.11×10^{-31} kg. What is the electron's relativistic mass when moving with a velocity of 0.8c?

Chapter 9

NUCLEAR PHYSICS

Children of yesterday, heirs of tomorrow
What are you weaving? Labor and Sorrow?
Look to your looms again.
  Faster and faster
Fly the great shuttles prepared by the
  Master.

Mary A. Lathbury

Einstein's equation ($E = mc^2$), which relates mass to energy, has been known since 1905. But its importance to the human race was not realized until 1939. In December of 1938 the German scientists Otto Hahn and Fritz Strassman showed that, under appropriate conditions, uranium atoms would split into two approximately equal parts with the accompanying release of large amounts of energy. Hahn conveyed the startling discovery to Lise Meitner, his co-worker of 25 years. Miss Meitner, forced to flee Germany in 1938, and her physicist nephew, Otto Frisch, then worked out the theory of the splitting of the atom during Christmas vacation in Sweden. Frisch communicated their results to the great Danish physicist Neils Bohr, whose atomic theory will be covered in the next chapter. Bohr was just setting out for a scientific meeting in Washington, D.C., and so he conveyed to America the startling news Frisch had told him. In 1939 the Italian physicist Enrico Fermi (Fig. 9-1a), first proposed that through the splitting of the atom large amounts of mass energy could be released for useful work. Soon after, because of the world political climate, the first censorship of scientific information took place. In 1945 the awesome results of atom splitting became known, when the first atomic bomb was exploded above a New Mexico desert. The world has never been the same since.

Figure 9-1a *Enrico Fermi (1901–1954). Fermi was one of the few scientists capable of both great theoretical and experimental work. After winning the 1938 Nobel Prize in physics, he persuaded the Italian dictator Mussolini to let him take his family to Stockholm for the presentation ceremony. He never returned to Italy but sailed instead to America. Thus it was that an "enemy alien" was in charge of the research that brought about the first controlled nuclear chain reaction in 1942, and later the atomic bomb. (Courtesy American Institute of Physics, Meggers Gallery of Nobel Laureates.)*

Figure 9-1b *Schematic diagram of a helium atom (not to scale). The nucleus is made up of two protons and two neutrons (designated p and n, respectively). The electrons orbit the nucleus.*

The nucleus forms a very small core of the atom, having a diameter of about 10^{-12} cm. In contrast to this, the electrons may be thought of as traveling in orbits* which are about 10^{-8} cm from the nucleus. A schematic diagram of a helium atom is shown in Fig. 9-1b. It is not drawn to scale because, if the nucleus were drawn an inch in diameter, the electrons would be about one-fifth of a mile away. The cost of paper prohibits page sizes of these dimensions.

9-1
BUILDING BLOCKS OF MATTER

All matter encountered in day-to-day living is made up of atoms. An **atom** is the basic building block of matter. It is composed of negatively charged particles, called **electrons,** going in orbits around a positively charged nucleus. The **nucleus** is the central core of the atom. It consists of **protons,** which are positively charged, and **neutrons,** which have no electrical charge. The basic properties of these three constituents of an atom are given in Table 9-1a.

9-2
FORCES IN NATURE

There are now thought to be four forces in nature. One of these, called the weak force, will not be discussed. Gravitational forces and electromagnetic forces have already been discussed. The other force is called the nuclear force. It came into public promi-

---

*We will see in Chapter 11 that the idea of electrons traveling in orbits is not precisely correct. However, it is a convenient model to use.

Table 9-1a *Constituents of an Atom*

Particle	Mass	Charge	Comments
Electron	0.91×10^{-27} g	-1.6×10^{-19} coulomb	Goes in orbit around nucleus.
Proton	1674×10^{-27} g	$+1.6 \times 10^{-19}$ coulomb	1836 times heavier than electron.
Neutron	1677×10^{-27} g	None	Slightly heavier than proton.

nence in the early 1940s with the development of the atomic and hydrogen bombs. In order to appreciate the relevance of the gravitational, electromagnetic, and nuclear forces, a brief description of each will be given.

The **gravitational force** is an attractive force, the strength of which depends, in part, on the masses of the objects involved. Although gravitational force was described in Chapters 3 and 8, it will be reviewed here.

The formula for finding the gravitational force between two objects is

$$F_{\text{grav}} = G \frac{m_1 m_2}{r^2} \qquad \text{Eq. 9.1}$$

where G = Newton's constant of universal gravitation,
m_1 = mass of object 1,
m_2 = mass of object 2,
r = distance between centers of m_1 and m_2.

Gravitational forces become significant only when at least one of the masses is very large. Because the earth is very massive, the force of gravity causes objects to stay on its surface and to fall toward its center. The force of gravity is also important when dealing with large objects, such as the sun, the moon, and the stars. As an example, the motion of the planets in our solar system is determined by the effects of the gravitational force. There is gravita-

tional attraction between any two objects, such as two people. However, for two small masses such as these, the force is extremely weak and can be neglected.

The **electromagnetic force** can be either an attractive or a repulsive force. Its formula is very similar to the one for the gravitational force

$$F_{\text{elec}} = k \frac{q_1 q_2}{r^2} \qquad \text{Eq. 9.2}$$

where k = a constant,
q_1 = charge of particle 1,
q_2 = charge of particle 2,
r = distance between the particles.

If q_1 and q_2 are like charges, F_{elec} is repulsive. If they are unlike charges, F_{elec} is attractive. Whenever charged particles are moving, there is also a magnetic force on the charged particle. This magnetic force is related to the fact that there is an electric charge, which explains why the force between charged particles is called the electromagnetic force. The electromagnetic force is the only important force on the electrons in an atom. The electromagnetic force between a proton and an electron in an atom ($r = 10^{-8}$ cm) is 10^{39} times greater than the corresponding gravitational force. It is the electromagnetic force which is responsible for the structure of atoms, molecules, and hence, matter in general. The fact that paper tears, rocks are hard, grass is green, and pencils write is due to the action of the electromagnetic force. In fact, about 95 percent of life's experiences are governed by this one force.

When we consider forces between massive objects, the gravitational force becomes important. When we consider very small distances of 10^{-12} cm or less, the nuclear force is important. One question which should have occurred to the reader at this point is why the protons in the nucleus do not repel each other (they have like charges), causing the nucleus to fly apart? We can see from the formula for the electromagnetic force that if the distance between like-charged particles is small, the repulsive force becomes very large. Because the nucleus does not fly apart, a more strongly attractive force must exist. This strongly attractive force, which acts only over very short distances, is called the **nuclear force.** The exact formula for the nuclear force is not known. It is too complicated to have a simple formula. However, two things about the nuclear force are well known. For distances between nuclear particles much greater than the diameter of a nucleus ($r = 10^{-12}$ cm), the force is zero. For the very short nuclear distances, it is very strongly attractive. So we can write its *approximate* formula

$$F_{nucl} = \text{Very strongly attractive,}$$
$$\text{if } r \text{ is less than } 10^{-12} \text{ cm} \qquad \text{Eq. 9.3}$$
$$F_{nucl} = 0 \text{ if } r \text{ is greater than } 10^{-12} \text{ cm}$$

where r = distance between two particles inside the nucleus.

The nuclear force between either two protons, two neutrons, or one proton and one neutron is approximately the same. It should be noted that an electrical repulsion still exists between the protons in the nucleus. However, if there are not too many protons (83 turns out to be the actual maximum number), the nuclear force attraction can overcome the electrical repulsion. With more than 83 protons, electrical repulsion within is stronger than the nuclear attraction, and the nucleus splits apart (usually into two unequal parts) of its own accord.

9-3
THE ATOMIC NUCLEUS

As we have said, the nucleus of the atom consists of protons and neutrons. These two particles have almost the same mass (see Table 9-1a), but the proton has a positive charge, while the neutron has no charge. In order for an atom to be neutral (have a total charge of zero), the number of electrons and protons must be the same. Since it is the electromagnetic force which is important in the structure of the atom, atoms are distinguished primarily by the number of charged particles (protons) in the nucleus. An **element** is a substance which has the same number of protons in all of its atoms. As an example, each atom of the element oxygen has eight protons in its nucleus, and each neutral oxygen atom has eight electrons in motion about its nucleus. The atomic properties are thus distinguished primarily by the number of protons in the nucleus.

The **atomic number** of an atom is the number of protons in the nucleus of the atom. The **neutron number** is, as the term indicates, the number of neutrons in the nucleus. The **mass number** is the number of protons plus neutrons in the nucleus. The properties of the nucleus itself depend not only on the number of protons it has but also on the number of neutrons in it. Collectively the protons and neutrons are referred to as **nucleons. Isotopes** are atoms whose nuclei have the same number of protons but different numbers of neutrons. As an example, an oxygen atom may have seven, eight, nine, or ten neutrons. Each of these possibilities, in combination with eight protons and eight electrons, results in an isotope of oxygen.

The nomenclature used to distinguish isotopes in nuclear and atomic physics is best described by an example. Let us consider the oxygen isotope $^{16}_{8}O$. The letter O is the chemical symbol for oxygen.* The

---

*The chemical symbols for all the elements are given in the periodic table in Chapter 12 and the atomic weight table, front endleaf.

Mass number ——→ 16

Atomic number ——→ 8 O 8 ←—— Neutron number

Figure 9-3a *Atomic nomenclature. The symbol for the chemical element is shown in the center.*

left superscript designates the mass number (many textbooks use a right superscript instead). The left subscript designates the atomic number and the right subscript the neutron number. (The right superscript position is left vacant because often there is a charge on the atom and the charge is indicated at this position. For instance, an oxygen atom with only seven electrons would have a total positive charge of 1 because it would have one more proton than electrons.) The atomic number plus the neutron number must equal the mass number by definition. Usually, the neutron number is omitted from an isotopic symbol, as it can be easily determined by subtracting the atomic number from the mass number. The atomic number and the chemical symbol specify the same thing since all atoms of the same element have the same number of protons. However, only a few tenacious souls have memorized the number of protons in every element, so it is usually specifically designated. Other possible isotopes of oxygen are $^{15}_{8}O_7$, $^{17}_{8}O_9$, and $^{18}_{8}O_{10}$. Figure 9-3a summarizes the nomenclature just described.

9-4
NUCLEAR STABILITY

It is an experimental fact that when only two neutrons are present in a nucleus, the nuclear force is not attractive. Of course, there is always a gravitational attraction, but it is negligible compared to the nuclear force. The nuclear force is generally attractive when there are several neutrons and protons present, but in the case of two neutrons it is weakly repulsive. Two protons will not attract each other, either. In this case, there is an electrical repulsion also present. However, for a combination of a neutron and a proton, commonly called a **deuteron,**[*] the force is attractive and the nucleus is stable. Figure 9-4a summarizes these facts. (A **stable** nucleus is one that will not separate of its own accord into two or more groups of nuclear particles.)

When more neutrons and protons are grouped together, certain combinations will be stable while others will be unstable. For low mass numbers of about 40 or less, most combinations are **unstable** unless the number of neutrons is about the same as the number of protons. Examples of stable nuclei are $^4_2He_2$, $^{12}_6C_6$, $^{23}_{11}Na_{12}$, $^{27}_{13}Al_{14}$, $^{40}_{18}Ar_{22}$. For higher mass numbers, the number of neutrons must be greater than the number of protons to have a stable nucleus. This is because when many protons are present, the repulsive electrical forces are very large, and so many neutrons are necessary to make the attractive nuclear force greater than the repulsive electrical force. Ex-

Figure 9-4a *Drawing illustrating the force between two nuclear particles. Two neutrons or two protons will not attract each other, but a neutron and a proton will attract. In heavier nuclei the converse effect appears. Two protons or two neutrons inside a large nucleus seem to be a more stable combination than one neutron and one proton.*

*Hydrogen with deuterons as the nuclei is called *deuterium*.

Figure 9-4b *A plot of the combinations of proton and neutron numbers resulting in stable nuclei. Each dot represents a stable nucleus. The dashed line indicates where the proton number equals the neutron number. The plot stops at the proton number 83. There are no stable nuclei with more than 83 protons.*

amples of elements with stable nuclei in this category are $^{44}_{20}Ca_{24}$, $^{62}_{28}Ni_{34}$, $^{89}_{39}Y_{50}$, $^{114}_{50}Sn_{64}$, $^{165}_{67}Ho_{98}$, $^{208}_{82}Pb_{126}$, proton number and mass number of any stable nucleus. All nuclei with more than 83 protons are unstable, because of the large repulsive electrical forces. A chart of all the combinations of protons and neutrons forming stable nuclei is given in Fig. 9-4b.

An important fact that is not too evident in Fig. 9-4b is that there are many stable isotopes when both the neutron number and the proton number are even numbers, and very few stable nuclei when they are both odd numbers. The fact that a nucleus with an even number of protons and an even number of neutrons is more stable than one with uneven numbers is a result of what is called the **pairing effect.** In Table 9-4a, the number of stable isotopes is divided

Table 9-4a *The Pairing Effect in Stabilizing Nuclei*

Proton Number	Neutron Number	Number of Stable Nuclei
1. Even	Even	165
2. { Even / Odd	Odd / Even }	105
3. Odd	Odd	4

into three categories, depending on whether the number of protons and neutrons is even or odd. There are 165 stable isotopes in which both of these numbers are even, 105 in which one is even and one odd, and only four in which both are odd. The four odd-odd stable isotopes are all low in mass number. They are $^{2}_{1}H_{1}$, $^{6}_{3}Li_{3}$, $^{10}_{5}B_{5}$, and $^{14}_{7}N_{7}$. What the pairing effect says, then, is that in a nucleus consisting of many protons and neutrons, the protons and neutrons tend to pair up among themselves. That is, two protons pair up and two neutrons pair up, but one proton will not pair up with one neutron. An example of the pairing effect is that $^{23}_{11}Na_{12}$ is stable, but $^{22}_{11}Na_{11}$ is not stable.

So far, we have considered only the qualitative aspects of nuclear stability. In order to describe quantitatively this phenomenon, it is necessary to introduce the concept of energy into the discussion.

Albert Einstein introduced his special theory of relativity in 1905. This was discussed in the previous chapter. One of the results of Einstein's assumptions was that mass is a form of energy. If it were possible to convert a very small amount of mass completely into energy, the total energy produced would be tremendous. Einstein's famous equation relating mass and energy is

$$E = mc^2 \qquad \text{Eq. 9.4}$$

where E = energy,
m = mass,
c = velocity of light in a vacuum.

Since $c = 3 \times 10^{10}$ cm/s, c^2 is 9×10^{20} cm^2/s^2 and 1 gram of mass is equivalent to 9×10^{20} ergs of energy. This is a fantastically large amount of energy. It can be readily perceived that if we could convert even a minute amount of mass into energy, we would get a lot of energy out of the process.

The units of mass an energy commonly used in nuclear physics are quite different from those used in preceding chapters. Because protons and neutrons are so small they are usually measured in very small units

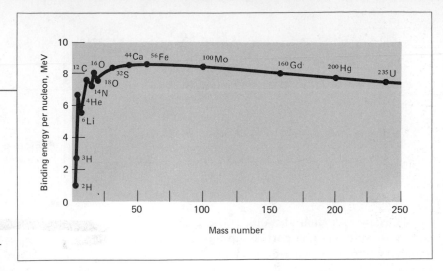

Figure 9-4c *Binding energy per nucleon versus mass number.*

of mass called atomic mass units (amu). An **amu** is defined as $\frac{1}{12}$ the mass of the ^{12}C isotope. That is

$$\text{Mass of } ^{12}\text{C} = 12.0000 \text{ amu} \qquad \text{Eq. 9.5}$$

One amu is equal to 1.66×10^{-24} g. The masses of nuclear particles in terms of these units are

$$\text{Mass of proton} = 1.00783 \text{ amu}$$

$$\text{Mass of neutron} = 1.00867 \text{ amu}$$

The unit of energy commonly used in nuclear physics is the **MeV.** This stands for one million electron volts. **An electron volt** is the amount of kinetic energy an electron (or proton) acquires when it is accelerated through an electrical potential of 1 volt. One electron volt (eV) is equal to 1.6×10^{-19} joule. If one amu of mass could be converted into energy, 931 MeV would be obtained. So we have the relation

$$(1 \text{ amu}) c^2 = 931 \text{ MeV} \qquad \text{Eq. 9.6}$$

The concept of *binding energy* plays an important role in the discussion of nuclear stability. The **total binding energy** of a nucleus is the energy necessary to separate a nucleus into protons and neutrons that are free to move independently of one another. The **binding energy per nucleon** is the total binding energy divided by the number of nucleons in the nucleus. Figure 9-4c illustrates how the binding energy per nucleon varies with the total number of nucleons in the nucleus. Note the steep rise at the beginning of

Figure 9-4d *The binding energy of the deuteron. It takes 2.22 MeV to split the deuteron into a proton and a neutron.*

the curve then a general leveling off at about 8 MeV per nucleon for mass number 20 and above. The **neutron separation energy** and the **proton separation energy** are defined as the amounts of energy necessary to separate one neutron or one proton from the rest of the nucleus. As an example of these concepts, the amount of energy necessary to separate a deuteron is 2.22 MeV. See Fig. 9-4d. Thus, the total binding energy is 2.22 MeV; the neutron separation energy and proton separation energy are the same, in this very simple case, and the binding energy per particle is 2.22 MeV $\div$ 2, or 1.11 MeV.

A more instructive example of binding energy is provided by the alpha particle, or $^4_2$He$_2$. This nucleus consists of two protons and two neutrons and is very stable. In Fig. 9-4e, the various binding and separation energies are shown. The total binding energy of the alpha particle is 28.3 MeV. Its neutron and proton separation energies are 20.6 MeV and 19.8 MeV, respectively, and its binding energy per particle is 28.3 MeV $\div$ 4, or 7.08 MeV.

(i) $\binom{pp}{nn}$ + 28.3 MeV $\longrightarrow$ (p) + (p) + (n) + (n)   Total binding energy of an alpha particle

(ii) $\binom{pp}{nn}$ + 20.6 MeV $\longrightarrow$ $\binom{pp}{n}$ + (n)   Neutron separation energy

(iii) $\binom{pp}{nn}$ + 19.8 MeV $\longrightarrow$ $\binom{p}{nn}$ + (p)   Proton separation energy

Figure 9-4e *The binding energy of an alpha particle.*

Thus, we see that it would take 28.3 MeV of energy to split an alpha particle up into two protons and two neutrons, which are sufficiently far apart that they are no longer attracted to each other by nuclear forces. One way in which we can convert mass into energy is, in effect, to turn the arrow around in Fig. 9-4e. For instance, if we could put two protons and two neutrons together to form $^4_2\text{He}_2$, we could get 28.3 MeV of energy out of the process. Processes of this sort, in which light nuclei come together to form more stable heavier nuclei and thus releasing energy in the process, are called **fusion** reactions. The energy which we get from the sun results from a fusion process. In the sun protons are being fused into $^4_2\text{He}_2$ nuclei plus some electrons to balance the charge. Neutrons are also created in the process. It is this reaction which accounts for the light and heat from the sun. As more and more protons fuse together in the sun, the sun's supply of protons will diminish, and the sun's energy output will begin to change. When this change becomes significant, life as we know it will cease to exist. However, take heart; astrophysicists think the sun will continue to shine and give off energy at its present rate for another 5 to 10 billion years.

From the definition of total binding energy in Fig. 9-4e, we can find a relation for the mass of any nucleus. It is

Nuclear mass = Mass of all protons
+ Mass of all neutrons − Total binding
energy in equivalent mass units   Eq. 9.7

Thus, we see that any stable nucleus weighs less than the total weight of the protons and neutrons contained in it. When a nucleus possesses negative binding energy, it is unstable and will fly apart spontaneously.

Each isotope has a specified mass. In nature, a given element usually occurs as several different isotopes. Hence, the **atomic weight** of an element is defined as the *average* mass (weight) of a naturally occurring element. The atomic weight is computed by multiplying the mass of each isotope by its fractional abundance in nature and adding the products. To give an example, oxygen naturally occurs 99.759 percent of the time as ^{16}O, 0.037 percent as ^{17}O, and 0.204 percent as ^{18}O. The masses of these isotopes are 15.99491 amu, 16.99914 amu, and 17.99916 amu. The atomic weight of oxygen is then

Atomic wt of
oxygen = 0.99759 × (15.99491)
+0.00037 × (16.99914)
+ 0.00204 × (17.99916)
= 15.9994

The atomic weight of each element is given in the periodic table of Chapter 12.

9-5
RADIOACTIVE NUCLEI

So far we have limited our discussion to stable nuclei. What about nuclei that are unstable, that is, those which tend to fly apart of their own accord? These nuclei which disintegrate of their own accord are called **radioactive**. There are four common ways by which a radioactive nucleus can disintegrate. They are referred to as alpha decay, beta decay, gamma decay, and fission.

Alpha decay is the disintegration of a nucleus into an alpha particle ($_2^4$He$_2$ nucleus) and the nucleus of another element. An example of this is

$$_{90}^{232}\text{Th}_{142} \longrightarrow {}_{88}^{228}\text{Ra}_{140} + {}_2^4\text{He}_2$$

Note that in this decay, the sums of the atomic numbers and the mass numbers on both sides of the arrow are equal. This is a general law which holds for all nuclear processes. For the atomic numbers, $90 = 88 + 2$, and for the mass numbers, $232 = 228 + 4$. The atomic numbers must add up to be the same because of a law which states that we cannot gain or lose electric charge. The mass numbers must add up to be the same because of a similar conservation law which states that we cannot gain or lose mass number. In general, the neutron numbers do not have to be equal on both sides of the arrow although they frequently are. Alpha decay is a very common type of decay for elements with atomic number greater than 82. The heavy nucleus which results from the decay (^{228}Ra in our example), decays again. A chain of decays results until a stable nucleus such as $_{82}^{208}\text{Pb}_{126}$ is formed.

Beta decay is the disintegration of a nucleus into a beta particle (electron) and the nucleus of another element. An example is

$$_6^{14}\text{C}_8 \longrightarrow {}_7^{14}\text{N}_7 + {}_{-1}^0\text{e}_0$$

In this decay we have represented and electron as $_{-1}^0\text{e}_0$. Since its electric charge is opposite to that of a proton, it has an atomic number of -1. Also, since its mass is negligible compared to that of a proton (see Table 9-1a), its mass number is zero. The total atomic number and total mass number on both sides of the arrow are seen to be equal in our example of beta decay. However, the neutron number in this beta decay example has gone from 8 to 7. What happens in beta decay is that a neutron is changed into a proton, an electron, and a strange particle (called a **neutrino**), which has no rest mass or electric charge but does possess energy and momentum. The proton remains in the nucleus and the electron and neutrino are emitted.

Gamma decay is slightly different from either alpha or beta decay. In **gamma decay** the atomic number, mass number, and neutron number remain the same, but the nucleus decreases its energy by emitting a "particle" of electromagnetic energy called a **gamma ray.** The gamma ray has energy but no mass or charge. A typical example of gamma decay is

$$_{82}^{204}\text{Pb}_{122}^* \longrightarrow {}_{82}^{204}\text{Pb}_{122} + \gamma$$

In this type of decay the symbol γ (Greek letter gamma) represents the particle of electromagnetic radiation that is emitted by the nucleus. The asterisk implies that the ^{204}Pb* nucleus is at a higher energy than normal ^{204}Pb. Because energy must be conserved, we get

Energy of ^{204}Pb*

$\quad$ = Energy of ^{204}Pb + Energy of gamma ray

Or, rearranged,

Energy of gamma ray

$\quad$ = Energy of ^{204}Pb* − Energy of ^{204}Pb

The energy of the gamma ray is thus seen to be equal to the difference in energy between ^{204}Pb* and ^{204}Pb. Gamma rays are different from alpha and beta rays in that gamma rays have no charge or mass associated with them.

A fourth important decay mode which a nucleus can undergo is called fission. **Fission** is the splitting of the nucleus into two nuclei of approximately equal size. It is accompanied by the emission of neutrons and the release of large amounts of energy. As an example, consider a typical fission decay of $_{92}^{236}\text{U}_{144}$.

$$_{92}^{236}\text{U}_{144} \longrightarrow {}_{54}^{140}\text{Xe} + {}_{38}^{94}\text{Sr} + 2\,({}_0^1\text{n}_1)$$

(Note that the symbol $_0^1\text{n}_1$ designates a neutron.) This is just one of the many possible fission decay reactions of ^{236}U. Many others may occur, such as

$$_{92}^{236}\text{U}_{144} \longrightarrow {}_{50}^{132}\text{Sn}_{82} + {}_{41}^{101}\text{Mo}_{59} + 3\,({}_0^1\text{n}_1)$$

$$_{92}^{236}\text{U}_{144} \longrightarrow {}_{60}^{150}\text{Nd}_{90} + {}_{32}^{81}\text{Ge}_{49} + 5\,({}_0^1\text{n}_1)$$

In all radioactive processes energy is released. This energy can be destructive if it impinges on living

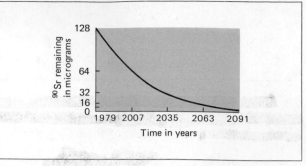

Figure 9-5a *Graph of the radioactive decay of strontium-90. This graph assumes an initial sample of 128 micrograms.*

tissue. In alpha and beta decay the energy is in the form of kinetic energy of either alpha or beta particles. In gamma radiation it is in the form of high energy electromagnetic radiation. In fission its most common form is in the kinetic energy of neutrons. Alpha and beta particles ($_2^4$He, $_{-1}^0$e) are charged particles and can be easily stopped by materials such as paper and wood. The alpha and beta particles interact with charged particles within the materials, following Coulomb's law of electrical attraction and repulsion. Gamma rays and neutrons are not charged particles and are difficult to stop. Thick shielding is needed to protect living things from these kinds of radiation. This shielding is usually in the form of either lead or concrete.

Some unstable nuclei take a long time to decay; others decay very rapidly. The relevant quantity which describes the rate of decay is called the half-life. The **half-life** of a given radioactive substance is the time it takes for half of the material to decay. The half-lives of typical radioactive isotopes range from millionths of a second to billions of years. The concept of a half-life is often misunderstood. An example best illustrates this concept. Consider the reaction:

$$_{38}^{90}\text{Sr} \longrightarrow {}_{39}^{90}\text{Y} + {}_{-1}^{0}\text{e}$$

This is a typical beta decay, and the half-life is 28 years. This means that if in 1979 you had 128 micrograms (0.000128 gram) of ^{90}Sr, then 28 years later (2007), you would have half as much, that is, 64 micrograms of ^{90}Sr. In 2007, you would have 64 micrograms so 28 years later, in 2035, you would have half again as much, or 32 micrograms of ^{90}Sr. This process continues as follows:

Year	Amount of ^{90}Sr
1979	128 micrograms
2007	64 micrograms
2035	32 micrograms
2063	16 micrograms
2091	8 micrograms
⋮	⋮

If we plot the amount of the sample of ^{90}Sr remaining versus the time, we get Fig. 9-5a.

A radioactive series is shown in Fig. 9-5b. The nuclide $_{92}^{238}$U (parent nuclei) shown at the top of the graph emits an alpha particle and decays to $_{90}^{234}$Th (daughter nuclei) which in turn emits a beta particle and decays to $_{92}^{233}$U. The half-life of each radioactive atom is given on the graph. Note that alpha decay lowers the numerical value of the atomic mass by four and the atomic number by two whereas in beta decay the numerical value of the atomic mass remains the same but the atomic number increases by one for each beta emitted.

Some isotopes have half-lives on the order of millions or even billions of years. Consequently, they can still be found in nature. Some of the naturally occurring radioactive isotopes are shown in Table 9-5a. Isotopes with short half-lives have all decayed away, and these are not found in nature.

Radioactive isotopes have many uses. One important use is found in the dating of very old artifacts. All organic matter (all matter which was once alive, such as wood, coal, bones, and parchment) contains carbon. Natural carbon occurs in three isotopes, ^{12}C, ^{13}C, and ^{14}C. The first two of these are stable, while ^{14}C decays by beta emission into ^{14}N with a half-life of 5730 years. In the atmosphere, ^{13}C is converted into ^{14}C due to bombardment by cosmic rays from outer space. The ^{14}C decays at about the same rate at which it was created; so the percentage of carbon which is ^{14}C remains relatively constant. Organic materials acquire the ^{14}C in the form of carbon dioxide, and in living things there is enough ^{14}C to cause

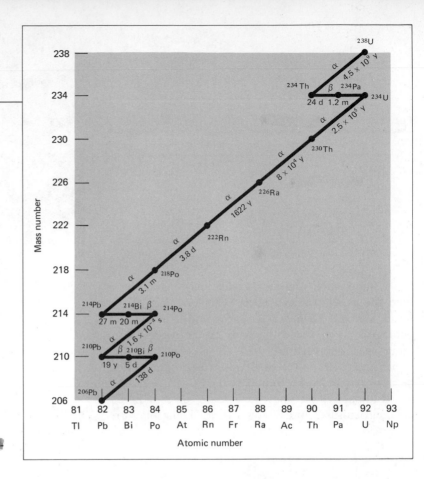

Figure 9-5b *The radioactive uranium series.*

Table 9-5a *Half-Life of Some Naturally Occurring Radioactive Nuclei*

Isotope	Half-Life
^{8}Be	2×10^{-16} second
^{214}Po	1.6×10^{-4} second
^{26}Al	7 seconds
^{34}Cl	33 minutes
^{90}Th	26 hours
^{22}Na	2.6 years
^{3}H	12.5 years
^{14}C	5.7×10^3 years
^{235}U	7.1×10^8 years
^{40}K	1.4×10^9 years
^{87}Rb	6×10^{10} years
^{115}In	6×10^{14} years
^{124}Sn	1.7×10^{17} years

about 16 beta emissions per minute per gram of carbon. Once the organic matter dies, it ceases to take in the ^{14}C from the atmosphere, and the amount of ^{14}C which it contains slowly decays. By measuring the beta decay rate in old materials, the age of the material can be determined, since the ^{14}C half-life is known.

■ *Example* What is the age of an old bone which emits 4 beta rays per minute per gram of carbon?

Living material would give a rate of 16 beta rays per minute per gram. After one half-life, or 5730 years, this would be decreased to 8 and after two half-lives, or 11,460 years, this would be cut to 4 beta rays per minute per gram of carbon. Thus, the bone is about 11,460 years old. (See Table 9-5b.)

Table 9-5b Decay of ^{14}C

Time since Death	Beta rays/min/g of C
0	16
5,730 years	8
11,460 years	4
17,190 years	2
22,920 years	1

Figure 9-5d *Radioactive cancer therapy. A new technique for cancer therapy, using small pieces of radioactive chromium-51 wire, has been developed at Argonne Cancer Research Hospital, Argonne, Illinois. The photo shows an X-ray of the pelvis of a patient with implanted "seeds" of radioactive chromium wire that can be left in the body permanently. (Photo courtesy Argonne Cancer Research Hospital.)*

Using similar methods, archeologists are able to date things as old as 40,000 years (see Fig. 9-5c). Using isotopes with much longer half-lives enables geologists to measure the ages of meteorites from outer space or very old rock. In this manner, the approximate age of the earth has been determined to be approximately 4.5 billion years.

Radioactive isotopes have many uses in medicine, agriculture, and biology. In medicine they are often used as tracers. The movement of isotopes can easily be traced in plant and animal systems, using radiation detecting instruments called geiger counters. Tracers therefore can be used to monitor the flow of blood in biological systems, to study the metabolism of plants and animals, and to determine the structure of complex molecules. An example of tracers used in cancer research is shown in Fig. 9-5d.

Radioactive materials are quite harmful to living cells. They can cause mutations in genes and are particularly harmful to cells of the fetus. Radioactive material from the nuclear bomb tests of the 1950s is believed to be the cause of the unusual number of birth defects and the higher than expected infant mortality rate in the United States for that decade. But radioisotopes can also be used to treat diseases, since the diseased cells are often more easily damaged by radioactive substances than normal cells, and therefore can be destroyed. As an example, the radioactive iodine isotope ^{131}I can be used in measurement connected with the thyroid gland. The patient ingests a prescribed amount of the isotope and it is absorbed by the thyroid gland. The radioactive ^{131}I then can be traced as it is released into the bloodstream in the form of protein-bound iodine.

Figure 9-5c *Bones, 14,000 years old, that were uncovered at an ancient burial site in the Aswan area in Sudan. To determine the age of the site, the age of the bones was determined using radioactive dating techniques with ^{14}C. (Courtesy Anthropology Department, Southern Methodist University.)*

9-6
NUCLEAR REACTIONS

The goal of the alchemists was to convert base metals, such as mercury and lead, into gold. We now know that this cannot be accomplished by chemical means, but it can be done by means of nuclear reactions. To initiate a nuclear reaction a nucleus is bombarded by another nucleus. The initial target nucleus and the projectile nucleus are called the **reactants.** The nuclei which are formed are called the **products.** The products of the reaction are determined by the energy of the bombarding nuclei and by the conservation laws of atomic number and mass number.

One of the first artificial nuclear reactions was performed in about 1912 by Ernest Rutherford. This reaction was

$$^{14}_{7}N + ^{4}_{2}He \longrightarrow ^{17}_{8}O + ^{1}_{1}H$$

The alpha particles ($^{4}_{2}He$) came from a radioactive source. Note that the total atomic and mass numbers on both sides of the arrow are the same. As mentioned before, this is the way it must be because of the laws of conservation of charge and mass. Another example is the bombarding of the base metal mercury with protons. Gold can be made in this manner as follows:

$$^{200}_{80}Hg + ^{1}_{1}H \longrightarrow ^{197}_{79}Au + ^{4}_{2}He$$

This way of making gold does not fulfill the alchemists' dream, however, because it is far too expensive ever to be economical.

One way in which nuclear reactions are carried out in a laboratory is by bombarding a nucleus with accelerated charged nuclei such as protons or alpha particles. This can be done by means of a potential difference, or voltage, employing millions of volts. Figure 9-6a is a schematic diagram of how nuclear reactions are performed in modern laboratories on a machine called a particle accelerator. See Fig. 9-6b. A proton or other charged nucleus is repelled by positive charges at one end of the accelerator and at-

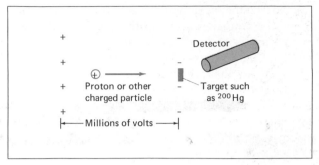

Figure 9-6a *Schematic diagram of a particle accelerator.*

Figure 9-6b *A nuclear particle accelerator. A voltage of several million volts is applied on the two ends of the tank in the upper center. The projectile beam, consisting of protons, for example, comes out through the beam pipe. The charged nuclear particles are bent by the magnet at center and then go out the pipe into a room on the right where the target is located. (Courtesy Yale University.)*

tracted by the negative charges at the other end. Since the voltages can be in the million-volt range, the proton is said to have so many MeV of kinetic energy. The products of the reactions, such as alpha particles, are then measured by a detector.

In every nuclear reaction total energy must be conserved. Energy may be either released or absorbed. A reaction in which energy is absorbed is called **endothermic.** A reaction is **exothermic** if energy is released. The energy of a reaction comes from converting mass into energy. It is easy to tell if energy is absorbed or released by examining the masses of the isotopes which are involved. Let us examine again Rutherford's first nuclear reaction,

$$^{14}_{4}\text{N} \quad + \quad ^{4}_{2}\text{He} \quad \longrightarrow \quad ^{17}_{8}\text{O} \quad + \quad ^{1}_{1}\text{H}$$
$$14.0033 \quad 4.0026 \quad\quad 16.9991 \quad 1.0078$$

with the masses (in amu) of the isotopes given under each one. The masses on the left total 18.0059 amu while those on the right total 18.0069 amu. Thus, the reaction absorbs energy. The amount of energy absorbed is

Energy absorbed = (18.0069 − 18.0059) amu
$$\times \ 931 \text{ MeV/amu}$$
$$= 0.931 \text{ MeV}$$

This is an endothermic reaction. An example of an exothermic reaction is the fusion reaction

$$^{2}_{1}\text{H} \quad + \quad ^{2}_{1}\text{H} \quad \longrightarrow \quad ^{3}_{2}\text{He} \quad + \quad ^{1}_{0}\text{n} \quad + \text{ Energy}$$
$$2.01410 \quad 2.01410 \quad\quad 3.01603 \quad 1.00867$$

In this reaction the total mass on the left is greater than the mass on the right. The energy released is

Energy released = [(2.01410 + 2.01410)
$$- (3.01603 + 1.00867)] \text{ amu} \times 931 \text{ MeV/amu}$$
$$= 0.00350 \text{ amu} \times 931 \text{ meV/amu}$$
$$= 3.26 \text{ MeV}$$

When energy is released, mass is converted into kinetic energy. When energy is absorbed in a reaction, kinetic energy gets converted into mass.

In the process of fission, mass is converted into kinetic energy. If ^{235}U is bombarded with low energy neutrons, ^{236}U is formed. This ^{236}U then fissions; that is, it splits into smaller nuclei. These two reactions are

$$^{235}_{92}\text{U} + ^{1}_{0}\text{n} \longrightarrow ^{236}_{92}\text{U}$$

$$^{236}_{92}\text{U} \xrightarrow{\text{Fission}} \text{Two or more lightweight nuclei}$$
$$+ \text{ Neutrons} + \text{Energy}$$

The three important features of fission are given by the three quantities on the right side of the fission reaction. First, the lightweight nuclei emitted are almost always readioactive, and their disposal is a troublesome problem. Second, large amounts of energy, about 200 MeV, are produced in a typical fission reaction. The third and most important feature of fission reactions is that neutrons are released. This is of importance because it was neutron bombardment of ^{235}U that started the process. There is, then, the possibility of a chain reaction. In a **chain reaction** one initial reaction triggers a series of subsequent reactions. In the case of fission, one neutron hitting a nucleus of ^{235}U forms ^{236}U, which fissions to give off more neutrons. These neutrons then hit other ^{235}U nuclei and fission occurs again and again, each time with the release of great quantities of energy and more neutrons. The reaction continues until all of the available ^{235}U is used up. A schematic diagram of the process is shown in Fig. 9-6c.

Of course the process just described is not as simple as it appears here. In order for a chain reaction to occur, at least one neutron from each fission event must strike another ^{235}U nucleus and cause another fission event. If more than one neutron from each fission process hits more ^{235}U, then the chain reaction is uncontrolled and will proceed very quickly. The result is that an enormous amount of energy is released in a very short time. This concept led to the development of the atomic bomb. However, not all the neutrons hit ^{235}U. Some may pass through the empty space between nuclei and strike nothing; others may hit other nuclei such as ^{238}U or impurities

= n

= ^{235}U

Figure 9-6c *Diagram illustrating a chain reaction. A neutron hits a ^{235}U nucleus; fission then occurs and neutrons are given off. Some of these neutrons hit other ^{235}U nuclei, and the process continues until all the ^{235}U is used up.*

and be captured. For this reason, it is important to have enriched ^{235}U. This is not easy, since natural uranium is 99.3 percent ^{238}U and only 0.7 percent ^{235}U. During World War II obtaining enriched ^{235}U from natural uranium was a top priority project. The ^{235}U was to be used in early versions of the atomic bomb, in particular, the one dropped on Hiroshima, Japan. The industrial complex and city of Oak Ridge, Tennessee, were built to manufacture enriched ^{235}U.

If the neutrons resulting from fission are absorbed by materials other than the fissionable material, the chain reaction can be controlled. A nuclear reactor operates on the principle of a controlled chain reaction. A simplified diagram of a nuclear reactor is shown in Fig. 9-6d. After the reaction has been initiated, neutron-absorbing rods are inserted so that the neutrons often hit them instead of the fissionable material. The rods are adjusted so that for every fission reaction, exactly one neutron hits more fissionable material. In this way, the chain reaction proceeds at a certain fixed rate. The energy released is maintained at a constant level. In this manner, nuclear power plants generate electricity for use by industry and in homes. Figure 9-6e is a photograph of one of the first electric generating plants in the United States powered by a nuclear reactor.

Because small amounts of fissionable material can be transformed into vast quantities of energy, man has another important supply of energy at his disposal. Unfortunately, the disadvantages of nuclear reactors that have appeared with their increased use

are thought by many people to outweigh their advantages.

In addition to ^{235}U, the other fissionable material of importance is $^{239}_{94}$Pu. This isotope is produced by the decay of ^{239}U. The ^{239}U is produced when neutrons are captured by ^{238}U. The important reactions involved in the production of ^{239}Pu are

$$^{238}_{92}U + ^{1}_{0}n \longrightarrow ^{239}_{92}U$$

$$^{239}_{92}U \longrightarrow ^{239}_{93}Np + ^{0}_{-1}e \qquad \text{Half-life: 24 minutes}$$

$$^{239}_{93}Np \longrightarrow ^{239}_{94}Pu + ^{0}_{-1}e \qquad \text{Half-life: 2.35 days}$$

The isotope ^{239}Pu has a half-life of 24,400 years. Because natural uranium is 99.3% ^{238}U, the conversion of ^{238}U into ^{239}Pu can greatly increase the amount of fissionable material available as an energy source. This is the purpose of the breeder reactor. In a

Figure 9-6d *The principal components of a nuclear reactor. The neutron-asborbing control rods are adjusted so that the chain reaction proceeds at a constant rate. The lead shielding is necessary for protection from harmful radioactive materials produced in the fission process. A liquid flows through the reactor and becomes heated.This heat energy is then converted to electricity.*

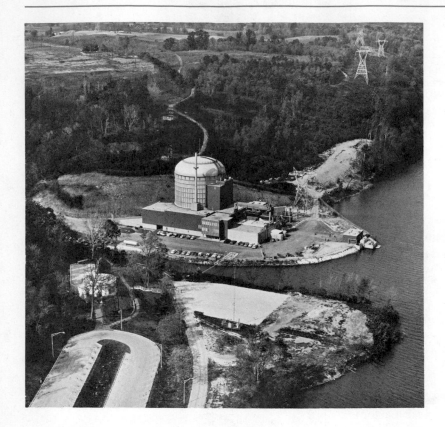

Figure 9-6e *A nuclear reactor and electric generating power plant located at Peach Bottom, Pennsylvania. (Courtesy Philadelphia Electric Company.)*

breeder reactor ^{238}U is bombarded with neutrons to create ^{239}Pu. The ^{239}Pu can be chemically separated from the uranium and fission by-products and used as the fuel in an ordinary nuclear reactor. However, even with breeder reactors the world supply of uranium is limited and will not supply the world's energy needs for more than 50 years.

The most important problem which arises in using breeder reactors is the toxicity as well as the radioactivity of ^{239}Pu. One major accident spreading this poisonous plutonium throughout the atmosphere could destroy much of the human race. As more and more breeder reactors become operational and plutonium becomes more prevalent, the chances of such an accident become more and more likely. For this and

other reasons, many environmentalists oppose the utilization of breeder reactors.

Another hazard which occurs in nuclear test explosions and one of the drawbacks of nuclear power plants is the lighter radioactive nuclei which result from the fission process. Radioactive substances from the fission process are very dangerous to human beings and must be handled with extreme caution. Radioactive nuclei with short half-lives are not too troublesome, because they decay away rather quickly. However, nuclei such as ^{14}C and ^{90}Sr with half-lives of 5730 and 28 years, respectively, are very hazardous and must be stored or disposed of with the utmost care. The isotope ^{90}Sr is particularly harmful, because its chemical properties are similar to those of calcium.

Thus, it combines chemically in the bones and through subsequent radioactivity greatly increases the likelihood of bone cancer. As a result of nuclear tests in the atmosphere by the United States and Russia in the 1940s and 1950s, there is now a significant amount of radioactive material in ordinary milk. It is not believed to be harmful in the present amount, but if further extensive tests are carried out, it would attain alarming proportions. This is one of the main reasons for the nuclear test ban treaty now in effect between the United States and Russia.

Another important way in which mass energy can be transformed into kinetic energy is through fusion reactions. In the **fusion** process, lightweight elements are fused together to form more stable heavy nuclei, with the release of energy. Some typical fusion reactions are

$$^2_1H + {}^2_1H \longrightarrow {}^3_1H + {}^1_1H + \text{Energy}$$

$$^3_1H + {}^2_1H \longrightarrow {}^4_2He + {}^1_0n + \text{Energy}$$

However, the isotopes on the left side of the equations do not readily fuse. In fact, as can be seen from Fig. 9-6f they repel each other, because both isotopes are positively charged. In order to fuse they must be traveling extremely fast when they come together. This means that very high temperatures—on the order of a million degrees—are needed. These high temperatures exist in very hot stars, such as our sun, but they are very difficult to achieve in the laboratory.

The hydrogen bomb is based on the fusion process. The high temperatures are created by first exploding a conventional fission-type atomic bomb. The fusion processes then take place, supplying many times more energy than the fission reaction. Figure 9-6g shows a possible arrangement of the components of a hydrogen bomb.

Because of the necessity of sustaining the extremely high temperatures of a fusion reaction, it is not yet known how to make a controlled fusion reactor. Such a reactor would produce large amounts of energy, and fusion could become an important source of energy in supplying electricity for the entire world.

Figure 9-6f *Two ^2_1H nuclei repel each other because they are both positively charged. In order for them to collide they must be moving towards each other at an extremely high rate of speed.*

Figure 9-6g *Diagram of a hydrogen bomb. The TNT explodes, pushing the ^{235}U together. This starts an uncontrolled chain reaction that heats the lithium hydride to over $180,000,000°F$ and creates ^3_1H and ^2_1H atoms that fuse together. The resulting explosion is followed by a fission reaction in the ^{235}U outer lining that tops off the awesome blast.*

There are two important advantages which fusion reactors would have over fission reactors:

1. The source of fuel is hydrogen instead of uranium. Hydrogen is much more prevalent—it occurs in water—than uranium. The fuel supply would last thousands of years rather than less than a hundred.
2. There are many fewer radioactive by-products in fusion reactors and none of them have long half-lives. Thus, the radioactive waste disposal problem is avoided.

Figure 9-6h *Design for a fusion reactor. Many laser beams are concentrated on a dropped pellet of fusionable material. The temperature of the pellet increases to the point where a fusion reaction occurs.*

As mentioned previously, a working fusion reactor has not been developed. One possible design (Fig. 9-6h) involves the use of powerful beams of light produced by lasers. The laser beams would pulsate, striking dropped pellets of fusionable material, and cause fusion. The energy of the resulting mini-explosions could then be used in the production of electricity.

IMPORTANT WORDS AND TERMS

Atom	Deuteron *neut prot stables*	Alpha decay $H_e^4 + X$
Electrons	Stable	Beta decay $X + {}_{-1}^{0}e$
Nucleus	Unstable	Neutrino
Protons	Pairing effect	Gamma decay — *electromagnetic energy*
Neutrons	amu	Gamma ray
Gravitational force	MeV	Fission — *emits neutrons, nucleus splits*
Electromagnetic force	Electron volt — *amt. of kinetic energy acquired when accelerated thru*	Half-life
Nuclear force	Total binding energy	Reactants
Element	Binding energy per nucleon	Products
Atomic number *# of protons*	Neutron separation energy	Endothermic *absorbed*
Neutron number *# of neutrons*	Proton separation energy	Exothermic *released*
Mass number *# of protons+neutrons*	Fusion	Chain reaction
Nucleons *pro+neutron*	Atomic weight	Nuclear reactor *controls chain reaction*
Isotopes *different # of neutrons*	Radioactive	Breeder reactor

LEARNING OBJECTIVES

After reading and studying this chapter, you should be able to do the following:

1. Describe the general structure of an atom in respect to nuclear and atomic size plus the location of the electrons, protons, and neutrons.
2. State the number of protons and neutrons in the nucleus of any atom listed in the periodic table. (Refer to the periodic table.)
3. Distinguish between electromagnetic, gravitational, and nuclear forces and state which forces are important in holding the atom together.

4. Write the symbol notation plus the correct numerical value for the mass, atomic, and neutron number for any atom listed in the periodic table. (Refer to the periodic table.)

5. State the general criteria for nuclear stability.

6. Define and explain *atomic mass unit* and *electron volt*.

7. Explain the concept of binding energy and give some examples.

8. Explain radioactivity and state the physical properties of alpha, beta, and gamma radiation.

9. Calculate the mass remaining in a sample of a radioactive element after a given time period when the elements half-life is known.

10. Distinguish between fission and fusion and state the physical characteristics of their by-products.

11. Describe the general operation of a nuclear reactor.

12. Write a balanced reaction for the nuclear decay of a given atom.

13. Calculate the amount of energy released in a nuclear reaction.

14. Define or explain the important words and terms listed at the end of the chapter.

QUESTIONS

1. Compare and contrast the distinctions between electromagnetic, gravitational, and nuclear forces.

2. What are the basic differences between a proton and a neutron?

3. Distinguish between atomic number, mass number, and neutron number.

4. What are nucleons?

5. Give two examples of how mass can be converted into energy.

6. Explain total binding energy and binding energy per nucleon.

7. Distinguish between one atomic mass unit and one electron volt.

8. Why don't the protons in a stable nucleus separate due to the repulsive electrical forces?

9. Is the nuclear force stronger at 10^{-11} cm or at 10^{-14} cm?

10. Is the electrical force stronger at 10^{-11} cm or at 10^{-14} cm?

11. What happens to an unstable nucleus?

12. Does a stable nucleus weigh more or less than the free protons and neutrons that compose it?

13. Explain the pairing effect.

14. Compare the physical properties of alpha, beta, and gamma radiation in respect to mass, charge, and penetrating ability.

15. Compare the mass number and the atomic number of the parent and daughter nuclei in (a) alpha decay, (b) beta decay, (c) gamma decay.

16. In what way can the half-life of a radioactive atom be altered?

17. Explain in terms of half-life why radioactive radon introduced into a patient can be left in the patient permanently while radioactive radium must be removed as soon as treatment is completed.

18. How long will it take for the ^{90}Sr released in the 1954 H-bomb explosion to decay away? (Half-life, 28 years)

19. In a nuclear reaction, is mass converted into energy or is energy converted into mass?

20. State the importance of the three products occurring when ^{236}U fissions.

21. Describe how a nuclear reactor works.

22. What are the drawbacks of nuclear reactors?

23. Why would fusion reactors be better than fission reactors, if they could be developed?

24. Why is it possible to make a fusion bomb but not a controlled fusion reactor at this time?

PROBLEMS

1. Write the names for each of the following isotopes. Then give the atomic number, proton number, neutron number, electron number, and the mass number for each.
$^{6}_{3}$X, $^{10}_{5}$X, $^{12}_{6}$X, $^{34}_{16}$X, $^{90}_{38}$X, $^{235}_{92}$X, $^{239}_{93}$X, $^{240}_{94}$X

2. Give three criteria for determining when an isotope will be unstable. Apply these and other reasons to determine if the following isotopes are unstable or stable:
(a) ^{12}C   (b) ^{14}N   (c) ^{16}O
(d) ^{18}F   (e) ^{90}Y   (f) ^{96}Sn
(g) ^{22}N   (h) ^{249}Cf   (i) ^{226}Ra

3. Only two isotopes of $_{19}$K (potassium) are stable. Pick which two they are from the following list:
^{38}K, ^{39}K, ^{40}K, ^{41}K, ^{42}K

4. The half-life of ^{131}I is 8 days. If a thyroid cancer patient is given 16 micrograms of ^{131}I, how many micrograms will still be in his thyroid after 24 days?

5. An old skeleton of a prehistoric man is found to yield 2 beta rays per minute per gram of carbon. About how old is the skeleton?

Answer: 17,190 years

6. Bismuth-214 may decay by emitting an alpha or a beta particle. Write the symbol notation for the daughter nucleus in each case.

7. What period of time is required for the radioactive hydrogen isotope tritium $^{3}_{1}$H to lose $\frac{3}{4}$ of its activity? Tritium has a half-life of 12.3 years.

8. Complete the following nuclear decay schemes and state whether the process is alpha decay, beta decay, gamma decay, or fission:
(a) $^{47}_{21}$Sc $\longrightarrow$ $^{47}_{22}$Ti + _____   (b) $^{210}_{84}$Po* $\longrightarrow$ $^{210}_{84}$Po + _____
(c) $^{210}_{84}$Po $\longrightarrow$ $^{206}_{82}$Pb + _____   (d) $^{240}_{94}$Pu $\longrightarrow$ $^{97}_{38}$Sr + $^{139}_{56}$Ba + _____
(e) $^{229}_{90}$Th $\longrightarrow$ $^{4}_{2}$He + _____   (f) $^{47}_{21}$Sc* $\longrightarrow$ $^{47}_{21}$Sc + _____
(g) $^{237}_{93}$Np $\longrightarrow$ $^{0}_{-1}$e + _____

Answer: (c) $^{4}_{2}$He, alpha decay

9. Complete the following nuclear reaction equations:

(a) $^{56}_{26}Fe + ^{4}_{2}He \longrightarrow ^{56}_{27}Co +$ ___ $= 3^{1}_{0}n + ^{1}_{1}H$

(b) $^{48}_{20}Ca + ^{1}_{1}H \longrightarrow ^{2}_{1}H +$ ___ $^{47}_{20}$ calcium 47

(c) $^{23}_{11}Na + \gamma \longrightarrow$ ___ $^{23}_{11}Na$

(d) $^{236}_{92}U \longrightarrow ^{131}_{53}I + 3^{1}_{0}n +$ ___ $^{102}_{39}Y$ *Answer:* (d) $^{102}_{39}Y$

(e) $^{20}_{10}Ne + ^{16}_{8}O \longrightarrow ^{12}_{6}C +$ ___ $^{24}_{12}$

(f) $^{6}_{3}Li + ^{6}_{3}Li \longrightarrow ^{7}_{3}Li +$ ___ $^{5}_{3}Li$

10. Given the following masses, calculate the amount of energy released in the fusion reactions below:

$^{1}_{0}n$ = 1.00867 amu $^{3}_{1}H$ = 3.01605 amu

$^{1}_{1}H$ = 1.00783 amu $^{3}_{2}He$ = 3.01603 amu

$^{2}_{1}H$ = 2.01410 amu $^{4}_{2}He$ = 4.00260 amu

(a) $^{2}_{1}H + ^{2}_{1}H \longrightarrow ^{3}_{1}H + ^{1}_{1}H +$ energy $2(2.0141) \rightarrow 3.01605 + 1.00783 + E$ *Answer:* (a) 4.02 MeV

(b) $^{3}_{1}H + ^{2}_{1}H \longrightarrow ^{4}_{2}He + ^{1}_{0}n +$ energy $3.01605 + 2.01410 \rightarrow 4.00260 + 1.00867 + E$

(Einstein theory $E = mc^2$)

1 amu = 931 MeV

.0043 2 × 931 MeV

.01888

(b) 17.58 MeV

11. Use the masses given in problem 10 for ^{3}H and ^{3}He.
   (a) Which of these two is stable and which is radioactive?
   (b) How much energy is given off when one of these decays into the other?
   (c) Write the decay reaction.

12. Natural chlorine is 75.8% ^{35}Cl and 24.2% ^{37}Cl. The mass of ^{35}Cl is 34.97 amu and the mass of ^{37}Cl is 36.97 amu. Compute the atomic weight of chlorine.   *Answer:* 35.45

Cl .758 × 35 = 26.53

.242 × 37 = 8.95

35.48

Chapter 10

THE BOHR THEORY OF THE ATOM

*And who loves Nature more
Than he, whose painful art
Has taught and skilled his heart
To read her skill and lore?*

Robert Bridges

In the previous chapter, the atom was described by using a model. According to this model, the nucleus, containing the protons and neutrons, is a small core, and the negatively charged electrons travel in almost circular orbits about the nucleus, much as the planets travel around the sun. In early 1900s, this was but one of several models of the atom. Before the theory of Niels Bohr, this model was not accepted as correct for the following very good reason. In the late 1800s, much elegant theoretical work had been done in the field of electromagnetism. One well-known result of this work was the discovery that any charged particle which is accelerating radiates (gives off) energy. Since the electron is a charged particle and is accelerating (anything traveling in a circle is accelerating), it should radiate energy. If the electron is giving off energy, then it must itself lose kinetic energy, since the law of conservation of energy will be obeyed. If it loses kinetic energy, it will also lose velocity and soon spiral into the nucleus. The time it would take for this to happen would be less than one second. Thus, according to classical electromagnetic theory, all the atoms of the universe should have collapsed in less than one second. Obviously, this had not happened. Was something wrong with electromagnetic theory which was known to work so well, or was something wrong with the physicists' basic understanding of nature? It will be seen in this chapter that it was the physicists' concept of nature which had to be changed.

10-1

THE DUAL NATURE OF LIGHT

In 1900 Max Planck (see Fig. 10-1a) had shaken the foundations of classical physics with his quantum hypothesis.* Planck's quantum hypothesis states that, in most situations, energy is radiated or absorbed by atoms only in discrete amounts. This was in contrast to the idea that any amount of energy can be radiated or absorbed. In 1905, Albert Einstein used Planck's hypothesis to show that the energy of a photon of light can be found by

$$E = hf$$ Eq. 10.1

where E = energy of the photon,
f = frequency of the electromagnetic wave,
h = **Planck's constant**
$= 6.625 \times 10^{-27}$ ergs · s.

In Einstein's description of light or any electromagnetic radiation, a **photon** is a "particle" of electromagnetic radiation. Each photon carries an amount of energy which is proportional to its frequency. Remembering that

$$c = \lambda f$$

where c = velocity of electromagnetic radiation,
λ = wavelength of electromagnetic radiation,
f = frequency of electromagnetic radiation,

we find that Eq. 10.1 can be written as

$$E = hf = \frac{hc}{\lambda}$$

According to this equation, the shorter the wavelength (higher the frequency), the greater the energy. As an example, a photon of red light with a wavelength of 6500 angstroms will have less energy than a photon of blue light with a wavelength of 4400 ang-

*A **quantum** is a discrete amount.

Figure 10-1a *Max Planck (1858–1947). Planck was a professor of physics at the University of Berlin. On December 14, 1900, he founded what is now known as modern physics (as opposed to classical physics) by proposing that the energy of electromagnetic radiation exists only in discrete amounts called quanta. The important small constant that appears over and over again in quantum physics is known as Planck's constant. In 1918 Planck was awarded the Nobel Prize in physics for his contributions to quantum theory. (Photo courtesy American Institute of Physics, Meggers Gallery of Nobel Laureates.)*

stroms. (Remember that an angstrom is a unit of length and is 10^{-8} cm.)

This concept of light waves carrying energy in the form of discrete packets is confusing to many people. How can a wavelike entity be composed of particles? Large particles with which we have had experience, (such as billiard balls) do not show wavelike effects such as diffraction, interference, and polarization. Light must be a wave because it shows these effects. But we have just commented on Einstein's description of light as particles, and this theory has been shown experimentally to be correct. We have a confused situation. Is light a wave or a particle? The solution to this dilemma is couched in the phrase "the **dual nature of light.**" This phrase simply means that light acts sometimes like a wave and sometimes like a particle. Many world-renowned scientists think this is not really an explanation of the nature of light, but a description of our confusion. However, the statement has generally been accepted as a good summary of our present state of knowledge about light.

10-2

Figure 10-2a *Relations between velocity and acceleration of an electron traveling in a circular path about a proton. The direction of the velocity vector is constantly changing; hence, the electron is being accelerated. The acceleration vector (a) always points towards the proton at the center of the circle.*

In 1913, Niels Bohr (see photo page 172) incorporated Planck's quantum hypothesis into a theory regarding the simplest of atoms—hydrogen. His consequent theoretical results were in startlingly good agreement with experiment. His theory is summarized in the following paragraphs.

The hydrogen atom is the simplest atom. It normally consists of one proton, no neutrons, and one electron. The force between the proton and electron is due entirely to the electrical attraction of their opposite charges. The gravitational and nuclear forces present are negligible and play no role in the physics of the atom. The electrical attractive force which exists between the proton and the electron is given as

$$F = k\frac{q^2}{r^2}$$

where k = a constant,
q = charge on the proton and electron,
r = distance between proton and electron.

Any object moving in a circle at constant speed is being accelerated. This fact follows from Newton's first law, which states that an object which has no unbalanced force acting on it, and hence no acceleration, moves in a straight line. The acceleration comes about because the direction of the velocity is changing, even though the speed is not. The acceleration for an object moving in a circle at constant speed v is called **centripetal acceleration** and is given by $a = v^2/r$. (See Section 3.5.) The direction of the acceleration is towards the center of the circle. All of these relations for an electron traveling around a proton are shown in Fig. 10-2a. Applying Newton's second law of motion to the electron, we have

$$F = ma$$

where F = force on the electron,
m = mass of the electron,
a = acceleration of the electron.

Since we know both the force on the electron and the acceleration of the electron, we write

$$F = ma$$

$$k\frac{q^2}{r^2} = m\frac{v^2}{r}$$

This was Bohr's first equation for the hydrogen atom.

Bohr's second equation for the hydrogen atom was based on the concept of energy. We can derive the equation by considering the total energy of the electron, which is

$$E = E_p + E_k$$

where E = total energy of the electron,
E_p = potential energy of the electron,
E_k = kinetic energy of the electron.

Before proceeding, however, it is necessary to discuss the fact that the potential energy of an object can be considered either positive or negative. Consider, for example, a ball on a stand at a height h

Figure 10-2b *Diagram illustrating that potential energy can be either positive or negative. A ball above ground has a positive potential energy. A ball below ground has a negative potential energy. (See also Fig. 4-4b.)*

above the ground. Its potential energy with respect to ground level is

$$E_p = mgh$$

where m = mass of the ball,
g = acceleration of gravity,
h = height above the ground.

Consider also this same ball lying in a well at a distance h below the ground. (Fig. 10-2b.) Now it has a negative potential energy. When the ball is above ground, it does work by falling to the ground. When the ball is below the ground, work must be done on the ball—it must be raised—to bring it to ground level. We express this latter situation by saying the ball does negative work. Thus, it has a negative potential energy. It is common physics terminology to say that a particle with a negative potential energy is in a **potential well.** The reason for this terminology is obvious from Fig. 10-2b.

For an electron traveling around a proton, the potential energy is negative. This can be seen by the analogy illustrated in Fig. 10-2b. For the ball in the well to be able to move about freely at the ground level, it must overcome the gravitational force which holds it in the well. Similarly, for the electron to be able to move about freely, it must overcome the attractive electrical force which holds it close to the proton. An object's potential energy in a well is negative and is given by

$$E_p = -Fs$$

where F = force on the body,
s = displacement.

For an electron circling a proton, the displacement s is the radius and F is the coulomb force. Hence,

$$E_p = \frac{-kq^2}{r^2} r$$

or

$$E_p = \frac{-kq^2}{r}$$

The electron's kinetic energy is $E_k = mv^2/2$. From Newton's second law, we found that Bohr's first equation was

$$\frac{kq^2}{r^2} = \frac{mv^2}{r}$$

or

$$\frac{kq^2}{2r} = \frac{mv^2}{2}$$

The total energy for the electron is then

$$E = E_p + E_k$$

$$E = -k\frac{q^2}{r} + \frac{1}{2}k\frac{q^2}{r}$$

If we factor out kq^2/r on the right side, we get

$$E = +k\frac{q^2}{r}\left(-1 + \frac{1}{2}\right)$$

or

$$E = -\frac{kq^2}{2r}$$

This is an equation for the total energy of the electron. Notice that the total energy is negative. The larger the radius r becomes, the more positive the total energy becomes. For an infinite radius, $E = 0$. This means, of course, that the electron is no longer bound to the proton. So far, we have derived two equations from classical physics, which had become completely established by 1913. We now proceed to the crucial step which Bohr took.

The quantity called *angular momentum* is defined for a particle going in a circle as the linear momentum (mass $\times$ velocity) times the radius of the circle. The all important step which Bohr took was to **quantize the angular momentum.** This means that he allowed only certain discrete values for the angular momentum of the electron. Specifically, he wrote

$$mvr = nh/2\pi$$

where m = mass of the electron moving in a circle,
   v = velocity of the electron moving in a circle,
   r = radius of the circle,
   h = Planck's constant (6.625×10^{-27} erg · s),
   n = any positive integer such as 1 or 2 or 3 . . .

This was quite a bold step. It says that not just any value of angular momentum can exist. Only integer numbers times $h/2\pi$ are allowed! The values $h/2\pi$, $2h/2\pi$, $3h/2\pi$, $4h/2\pi$, and so on, are possible, but no values in between these such as $1.3h/2\pi$ and $2.731h/2\pi$ are allowed.

Let us now see where this audacious assumption led Bohr. We now have three equations. They are

$$k\frac{q^2}{r^2} = \frac{mv^2}{r}$$

$$E = -\frac{kq^2}{2r}$$

$$mvr = \frac{nh}{2\pi}$$

From these three equations it is possible to show (see Appendix VI) that the radius of the electron's orbit and the energy of the electron can be expressed in terms of constants and a positive integer, n. The important results are

$$r = 0.529n^2 \text{ angstroms} \qquad \text{Eq. 10.2}$$

$$E = \frac{-13.6}{n^2}\text{eV} \qquad \text{Eq. 10.3}$$

Remember that an angstrom, Å, is 10^{-8} cm and eV stands for electron volts.

10-3
ENERGY LEVELS FOR THE HYDROGEN ATOM

Let us now examine the results of Bohr's work. As a consequence of quantizing the angular momentum, we found that both the radius of the electron's path and the energy of the electron assume only certain discrete values. Thus, the allowed values of the radius are given by

$$r_n = 0.529 \, n^2 \text{ Å}$$

where n = 1, 2, 3, etc.

For $n = 1$, then $r_1 = 0.529$ Å; for $n = 2$, then $r_2 = 0.529 \times 4 = 2.116$ Å, etc.

Table 10-3a shows the allowed values of r_n for the electron. From this we can deduce a picture of the hydrogen atom as an electron going in various possible orbits around a nucleus. The larger the value of n, the greater the distance between the nucleus and the electron. This picture of the atom is sketched in Fig. 10-3a.

The allowed values of the energy of the electron in the hydrogen atom are given by

$$E_n = \frac{-13.60}{n^2}\text{eV}$$

where E_n = total energy of the electron for a given value of n,
   n = 1, 2, 3, etc.

Table 10-3a *Allowed Values of the Electron's Radius and Energy for Low Values of n*

n	r_n	E_n
1	0.529 Å	-13.60 eV
2	2.116 Å	-3.40 eV
3	4.761 Å	-1.51 eV
⋮	⋮	⋮ .85 .544
∞	∞	0 .397

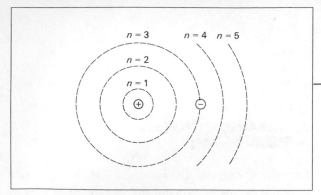

Figure 10-3a *A schematic diagram of an atom according to the Bohr theory. The possible orbits are shown as dashed lines. In this diagram, the value of n is three.*

This equation states that the energy of the electron in the hydrogen atom has only certain discrete values. As examples,

$$E_1 = \frac{-13.60}{1^2}\,\text{eV} = -13.60\,\text{eV}$$

$$E_3 = \frac{-13.6}{3^2}\,\text{eV} = \frac{-13.6}{9}\,\text{eV} = -1.51\,\text{eV}$$

Table 10-3a shows the allowed values of the energy of the electron for low values of n.

A plot of allowed electron energies, drawn to scale, is called an **energy level diagram**. The energy level diagram for the hydrogen atom is shown in Fig. 10-3b. As n increases, the energy increases. The removal of the electron from an atom is called *ionization*, and the amount of energy required to accomplish the removal is called the **ionization potential.** It is measured in electron volts. The ionization potential of the hydrogen atom is simply the energy necessary to completely separate the electron and the proton. This value varies and depends on the level (location) of the electron. From Table 10-3a it is evident that the electron will be completely separated from the proton when $n = \infty$. The energy necessary to achieve complete separation is the energy necessary to raise the electron's energy from E_n to zero. So we obtain

$$\text{Ionization energy} = -E_n$$

Thus, the ionization energy for $n = 1$ is $+13.6$ eV. To get the ionization energy for any value of n, simply change the sign of the value for E_n.

From Fig. 10-3a it is easy to see why the ionization energy decreases as n increases. As n increases, the electron gets progressively farther away from the proton, resulting in less and less electrical attraction. This results in a decrease in the ionization energy—the energy necessary to completely separate the proton and the electron.

When we speak of the energy of an atom, we are speaking of the system as a whole. For the hydrogen atom the energy can be thought of as the energy of its single electron. With more complicated atoms, we need to consider all of the electrons. The energy levels are then the energy levels for the whole atom.

In the Bohr model of the hydrogen atom, the electron circles the proton with energy values corresponding to certain values of the radius. The quantity n is called the **principal quantum number.** A **quantum number** is a number which specifies a property of the system. For the hydrogen atom, the principal quantum number, n, specifies the energy of the system. In Chapter 11, we will discuss several other quantum numbers. The description of an atomic system in-

Figure 10-3b *Energy level diagram for the hydrogen atom. Only certain energies are possible in the quantum theory.*

cludes its energy, angular momentum, and other properties. The lowest energy level of an atom is called the **ground state.** The **excited states** are the states of the atom with energies above the ground state. In the hydrogen atom the state with $n = 1$ is the ground state; $n = 3$ is called the second excited state, and so forth.

Figure 10-4a *A schematic drawing of the de-excitation of an electron in an excited hydrogen atom. A photon is emitted when the electron loses energy.*

10-4
PHOTON EMISSION AND ABSORPTION

Let us now consider in more detail the dilemma previously described in the introduction to this chapter. According to the classical theory of electromagnetism, the electron should gradually lose energy as it circles the proton and spiral in to the proton. Bohr's theory shows that our preconceived ideas about slowly and continuously losing energy do not apply to a single electron circling a proton. By quantizing the angular momentum, Bohr showed that the electron can occupy only certain discrete energy levels. It does not radiate energy as it circles the proton because it cannot lose energy in a continuous manner. The only way in which it can lose energy is to change energy levels. When this happens, the energy lost by the electron is gained by a photon. A photon is emitted every time an electron loses energy by changing energy levels. An electron must be in an excited state to lose energy. Once it is in its ground state, it cannot lose any energy. This, then, is Bohr's explanation of why the classical electromagnetic theory did not apply and why atoms do not collapse.

When an electron changes energy levels, the total energy of the system must be conserved. If the electron is initially in an excited state, it may lose energy by changing to a less excited state. When this happens the electron's energy is carried away by a photon—a light quantum. The total energy *before* must equal the total energy *after* the process, so we write

$$E_{ni} = E_{nf} + E_{photon}$$

where E_{ni} = initial energy of electron,
E_{nf} = final energy of electron,
E_{photon} = energy of photon, which is given off.

Thus, $$E_{photon} = E_{ni} - E_{nf}$$

A schematic diagram of the process of photon emission is shown in Fig. 10-4a.

Photon energies can be visualized in terms of the energy level diagram for the hydrogen atom. This is shown in Fig. 10-4b. The electron's loss in energy—which is equivalent to the photon energy—is designated by an arrow between energy levels. The photon energy is indicated by the length of the arrow. Some of the possible ways in which an electron can lose energy are shown in Fig. 10-4b.

The inverse reaction can also occur. A photon can be absorbed by an atom. The result is a change in an electron's energy level. A schematic diagram of this process, known as **photon absorption,** is shown in Fig. 10-4c. The change of energy levels of the electron are just the opposite of those shown in Fig. 10-4b. For example, in photon emission an electron might go from $n = 3$ to $n = 2$, thus losing a fixed amount of energy in the form of a photon. In photon absorption a photon with the same fixed amount of energy could be absorbed, causing the electron to go from $n = 2$ to $n = 3$.

Visible light is electromagnetic radiation with wavelengths between 4000 and 7800 Å. The color we see depends on the wavelength. Since photons are

Figure 10-4b *Energy level diagram for the hydrogen atom used to show some of the de-excitation energies of the electron. The lengths of the arrows indicate some of the possible energies of the photon.*

Figure 10-4c *A schematic drawing of the absorption of a photon by an atom. The electron gains the photon's energy and goes to a higher energy level.*

particles of electromagnetic radiation, they can be perceived as visible light only when they have wavelengths between 4000 and 7800 Å. The frequency and wavelength of the electromagnetic radiation, which is emitted when an electron changes energy levels, are determined by the energy of the emitted photon. An equation for the wavelength can be derived as follows: From Eq. 10.1

$$hf = E_{\text{photon}}$$

or

$$hf = E_{ni} - E_{nf}$$

But since

$$f = c/\lambda$$

we have

$$\frac{hc}{\lambda} = E_{ni} - E_{nf}$$

or

$$\lambda = \frac{hc}{E_{ni} - E_{nf}}$$

If we express E_{ni} and E_{nf} in eV, then Planck's constant h and the velocity of light c can be expressed in appropriate units to give the simple formula

$$\lambda = \frac{12{,}400}{E_{ni} - E_{nf}} \text{Å} \qquad \text{Eq. 10.4}$$

To find the wavelength of light in angstroms, simply divide 12,400 by the difference in energy in eV between the initial and final energies of the electron.

■ *Example 1* What is the wavelength emitted when an excited electron goes from the $n = 3$ to the $n = 2$ level?

$$E_3 - E_2 = -1.51 \text{ eV} - (-3.40 \text{ eV})$$

$$= -1.51 \text{ eV} + 3.40 \text{ eV}$$

$$= 1.89 \text{ eV}$$

$$\lambda = \frac{12{,}400}{1.89} \text{Å}$$

$$\lambda = 6{,}560 \text{ Å} \qquad \text{(red)}$$

■ *Example 2* A blue-green color (about 4860 Å) is seen to be emitted by excited hydrogen atoms. What are the quantum numbers of the initial and final levels of the electron?

$$\lambda = 4860 \text{ Å}$$

$$= \frac{12,400}{E_{ni} - E_{nf} (\text{eV})}$$

Solving for the difference in energy

$$E_{ni} - E_{nf} = \frac{12,400}{4860}$$

$$E_{ni} - E_{nf} = 2.55 \text{ eV}$$

The problem now is to find a transition of 2.55 eV in the hydrogen atom, that is, we need to find two energies which differ by 2.55 eV. Referring to Fig. 10-4b, the final quantum number (nf) cannot be $nf = 1$

because all transitions to this level have an energy greater than 10.2 eV. The final quantum number cannot be 3 or higher because all transitions to it will have an energy lower than 1.89 eV. So, the final quantum number must be $nf = 2$. The initial quantum number is $ni = 4$. Then, $E_{ni} = -0.85$ eV, $E_{nf} = -3.40$ eV, and $E_{ni} - E_{nf} = -0.85$ eV $- (-3.40$ eV$) = +2.55$ eV.

Figure 10-4d illustrates the quantum transitions of the electron from excited states in a hydrogen atom. A transition of the **Lyman series** corresponds to a quantum jump of the electron from an initial orbit of quantum number greater than one to the final orbit

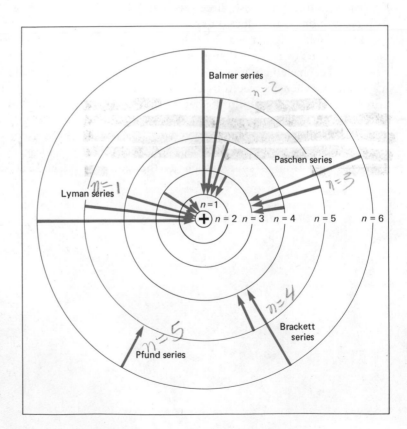

Figure 10-4d *Model of the hydrogen atom illustrating quantum jumps that give rise to the spectral lines for the different spectral series. An electron emits electromagnetic radiation of a definite frequency when it falls from an orbit of higher energy to one of lower energy.*

for which $n = 1$. The spectral lines of the Lyman series are all in the ultraviolet region of the electromagnetic spectrum. A transition of the **Balmer series** corresponds to a quantum jump of the electron from an initial orbit of quantum number greater than two to the final orbit for which $n = 2$. The spectral lines of the Balmer series are all in the visible region of the electromagnetic spectrum. The hydrogen alpha line is the red line (wavelength = 6563 Å), the hydrogen beta line is the green line (wavelength = 4862 Å), the hydrogen gamma line is the blue line (wavelength = 4341 Å), and the hydrogen delta line is the violet line (wavelength = 4102 Å). The Paschen, Brackett, and Pfund series have final orbits of $n = 3$, $n = 4$, $n = 5$, respectively, as shown in Fig. 10-4d. The spectral lines for these three series are all in the infrared region of the electromagnetic spectrum.

In the next chapter we shall discover that each element has its own particular energy level diagram. Therefore, each element gives off photons of different wavelengths when its electrons change to lower energy levels. A **spectrum** (plural: spectra) is an ordered display of the various frequencies or wavelengths produced by an electromagnetic source such as an atom, nucleus, or molecule. By examining the spectrum of light from a source such as the sun, the various elements present in the source can be identified. This identification procedure is an extremely important and useful tool to chemists, astronomers, and physicists. This specialized branch of physical science is called **spectroscopy.**

The instrument used to separate light into its component wavelengths is called a **spectroscope.** A simple spectroscope is shown in the upper photograph of color plate I. White light coming in from the left is separated into various wavelengths by a glass prism. You can see that the spectrum of white light consists of the various colors of the rainbow. If, instead of white light, the light from a pure element is observed, various kinds of spectra are seen. The spectra of five elements are shown in the lower photograph of Plate I. The spectra are labeled with the chemical symbols of the elements.

The crucial test of the Bohr theory was whether it could explain the hydrogen atom spectrum, which was well known in 1913. Bohr's theoretical values for the photon wavelengths were in almost perfect agreement with experiment. This was the justification of his theory, and it went a long way towards convincing the scientists of his day that Planck's quantum theory had far-reaching ramifications. In the next chapter we shall see that it gave birth to a whole new physics.

IMPORTANT WORDS AND TERMS

Quantum
Planck's constant
Photon
Dual nature of light
Centripetal
   acceleration
Potential well

Quantization of the angular
   momentum
Energy level diagram
Ionization potential
Principal quantum number
Quantum number
Ground state

Excited states
Photon absorption
Lyman series
Balmer series
Spectrum
Spectroscopy
Spectroscope

LEARNING OBJECTIVES

After reading and studying this chapter, you should be able to do the following without referring to the text:

1. Explain what is meant by the dual nature of light.
2. State Planck's quantum hypothesis.
3. Describe the Bohr model of the hydrogen atom.
4. State how the radius of an electron's orbit of the hydrogen atom varies as a function of the principal quantum number.
5. State how the energy of an electron of the hydrogen atom varies as a function of the principal quantum number.
6. Distinguish between quantum, quantum number, and principal quantum number.
7. State the two important postulates Bohr made concerning radiation from the atom.
8. Describe and explain an energy level diagram.
9. Distinguish between a ground state and an excited state of an electron.
10. Contrast and explain how an atom radiates or absorbs a photon.
11. Describe a spectroscope and explain the spectrum obtained with a spectroscope.
12. Distinguish between the Lyman, Balmer, Paschen, Brackett, and Pfund spectral series.
13. Define or explain the important words and terms at the end of the chapter.

QUESTIONS

1. Is light a particle or a wave? Explain.
2. Einstein said in 1905 that light consists of "particles" called photons. What proof can you give that light consists of waves?
3. Why are nuclear forces negligible in the theory of the hydrogen atom? Why are gravitational forces negligible?

4. Some hydrogen atoms have one proton and one neutron in the nucleus. How will this affect Bohr's theory?
5. Why are the energies of the hydrogen atom all negative?
6. What is the length of an angstrom in the mks system of units?
7. How does the radius of the electron's orbit of the hydrogen atom vary with the principal quantum number?
8. How does the energy of the electron of the hydrogen atom vary with the principal quantum number?
9. Give two examples of values for the energy of a hydrogen atom which are not permitted by Bohr's theory.
10. Are the photon wavelengths absorbed by hydrogen atoms the same as those emitted?
11. Explain the difference between a ground state and an excited state for an electron.
12. Explain a spectral series for the hydrogen atom. Give an example.
13. Explain the difference between a continuous and a line spectrum.
14. What information must be known to draw an energy level diagram?
15. The hydrogen atom has only one electron; however, the hydrogen spectrum consists of many lines. How do you account for so many spectrum lines?
16. What is the range of electromagnetic radiation (expressed as wavelengths in angstroms) to which the human eye is sensitive?

PROBLEMS

1. The human eye is most sensitive to light of approximately 5500 Å.
   (a) Determine the frequency of this electromagnetic radiation.
   (b) What is the energy in ergs of such a photon?

2. Which has more energy, a photon of red light, or a photon of blue light? Give proof for your answer.

3. What is the radius of a hydrogen atom for each of the following?
   (a) $n = 2$,   (b) $n = 6$,   (c) $n = 10$.

4. What is the energy in electron volts of a hydrogen atom for each of the following?
   (a) $n = 2$,   (b) $n = 6$,   (c) $n = 10$.

5. What is the ionization energy in electron volts of a hydrogen atom for each of the following?
   (a) $n = 2$,   (b) $n = 6$,   (c) $n = 10$.

6. What is the energy in eV emitted in the following transitions of an electron in the hydrogen atom?
   (a) $n = 4$ to $n = 2$,   (b) $n = 6$ to $n = 2$,   (c) $n = 3$ to $n = 1$,   (d) $n = 4$ to $n = 3$.

Answer: (b) 3.02 eV

7. (a) What are the wavelengths corresponding to the transitions in problem 6?
   (b) Which are visible? (c) Which are in the ultraviolet region and which are in the infrared?

Answer: (a) $n = 6$ to $n = 2$, $\lambda = 4106$ Å

184 / CHAPTER 10 THE BOHR THEORY OF THE ATOM

8. A hydrogen atom absorbs a photon of wavelength $\lambda = 4341$ Å. (a) How much energy did the atom absorb? (b) What were the initial and final states of the electron in the hydrogen atom?

Answer: (b) $n = 2$ to $n = 5$

$E = h\frac{v}{\lambda}$ 4.6×10^{-12} ergs $n \, 2.8 \, eV$

9. Calculate the wavelength of the hydrogen beta line of the Balmer series. Hydrogen beta is from $n = 4$. Show your work.

10. What is the shortest wavelength present in the Brackett series?

$\lambda = \frac{hc}{E_i - E_f}$ $E_i - E_f = \frac{hc}{\lambda}$

$\lambda = \frac{hc}{E_i - E_f}$ Ev

units of energy is m

$\frac{136}{42}$

diff =

$\frac{136}{2^m}$

$E = hf$

$= h\frac{v}{\lambda}$

$h = 6.63 \times 10^{-27}$ erg·sec

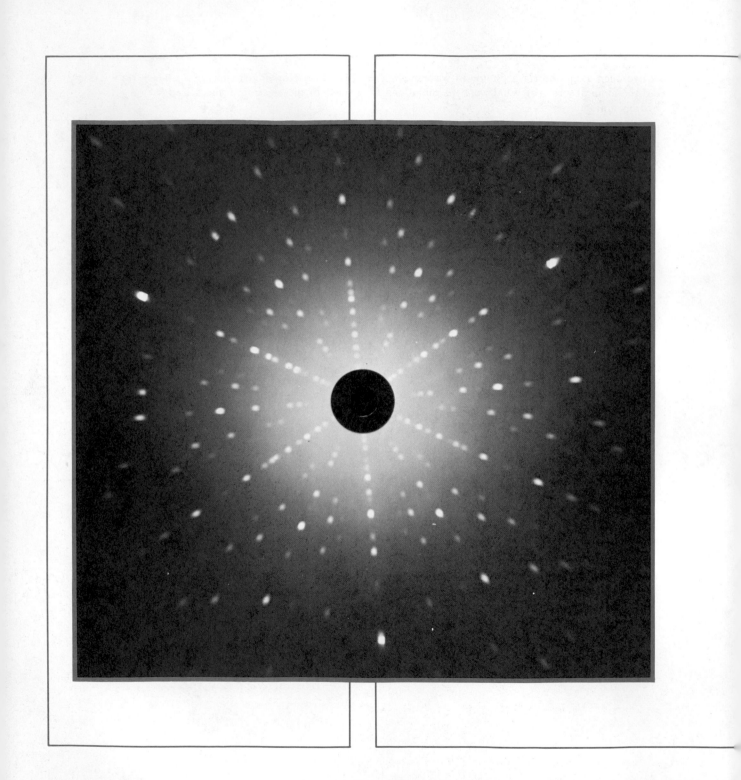

Chapter 11

QUANTUM MECHANICS AND ATOMIC PHYSICS

The beautiful agreement between Bohr's theory and the experimental hydrogen spectrum indicated that Planck's quantum hypothesis was correct. However, Bohr's theory could not adequately explain any of the elements with two or more electrons. The great Russian chemist, Dmitri Mendeleev, had established an orderly array of the various known elements in 1868, but the theoretical basis of this arrangement was not known as late as 1925. In 1925 a whole new kind of physics, based on the quantum idea, was born. This new physics, called **quantum mechanics,** revolutionized man's understanding of the submicroscopic aspects of nature. Using this new physics, the theoretical basis of Mendeleev's periodic table of the elements was discovered, and the stage was set for our modern understanding of all the physical sciences.

Quantum mechanics has applications in astronomy, nuclear physics, chemistry, meteorology, geology, and other fields of science. But the quantum idea has a much wider scope than that of science. It presented us with a precise formulation of the uncertainty involved in making any measurement. The impossibility of ever knowing anything exactly was the dawn of a new era in philosophy and theology. The old idea of a mechanistic universe, in which all things moved according to nature's laws, was replaced by a new concept of probability, with everything being, at least to some degree, uncertain. In this chapter some of the basic ideas of quantum mechanics, such as the uncertainty principle, will be discussed, and the modern understanding of atoms will be described.

Large elements in order brought,
  and tracts of calm from tempest made,
  and world wide fluctuation sway'd,
In vassal tides that follow'd thought.

Tennyson

11-1

MATTER WAVES

The concept of light acting both as a wave and as a particle was disturbing to physicists in the 1920s. This problem spurred the thinking of Louis de Broglie (Fig. 11-1a), a French physicist, who in 1925 postulated that matter, as well as light, has properties of both waves and particles. According to the hypothesis of de Broglie any moving particle has a wavelength, which is given by the formula

$$\lambda = \frac{h}{mv} \qquad \text{Eq. 11.1}$$

where λ = wavelength of moving particle,
  m = mass of moving particle,
  v = velocity of moving particle,
  h = Planck's constant.

This is truly a revolutionary equation! It states that everything that moves has a wavelength associated with it. The waves produced by moving particles are called **matter waves.**

■ *Example 1* What is the wavelength of an electron traveling at 3×10^8 cm/s?

$$\lambda = \frac{h}{mv}$$

$$= \frac{6.625 \times 10^{-27} \text{ erg} \cdot \text{s}}{9.1 \times 10^{-28} \text{ gm} \times 3 \times 10^8 \text{ cm/s}}$$

$$= 2.42 \times 10^{-8} \text{ cm}$$

$$= 2.42 \text{ Å}$$

De Broglie's equation was met with skepticism at first, but it was experimentally verified in 1927 by G. Davisson and L. H. Germer in the United States. They showed that a beam of electrons exhibits a diffraction pattern. Since diffraction is a wave phenomenon, a beam of electrons must have wavelike

Figure 11-1a *Louis de Broglie (1892–   ). Prince Louis Victor de Broglie, a French nobleman, studied medieval history at the Sorbonne. He enlisted in the French army and was a radioman in World War I. This experience created an interest in physics, and he presented his concept of matter waves in 1925 in a doctoral thesis. It was met with much skepticism at first; one cynic dubbed his theory "la comédie française." When it was shown that electrons can produce a diffraction pattern, however, de Broglie had the last laugh. He was awarded the Nobel Prize in physics in 1929. (Photo courtesy American Institute of Physics, Niels Bohr Library.)*

properties, such as frequency and wavelength. To show that something diffracts, it is necessary to have the wave pass through a slit whose width is approximately the size of the wavelength. Visible light has wavelengths of about 6000 Å, and slits with widths of this size can be made quite easily. Electrons, as was seen in the example, have wavelengths of about 2 Å. Slits of this width cannot be made. However, nature has provided us with suitable slits in the form of crystal lattices. The atoms in these crystals are arranged in rows (or other orderly arrangements), and

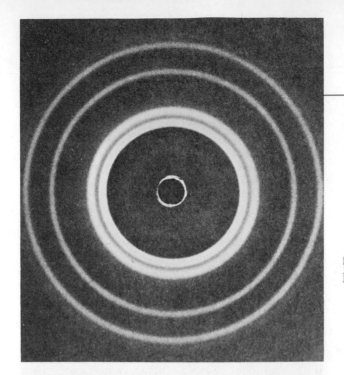

the rows of atoms make natural "slits." Davisson and Germer bombarded nickel crystals with electrons and obtained a diffraction pattern on a photographic plate. Diffraction patterns made by shooting X-rays (electromagnetic radiation of very short wavelength) and an electron beam at thin aluminum foil are shown in Fig. 11-1b. (The pattern at the top was made by X-rays.) The similarity in the diffraction patterns from the X-rays and the electrons is evident. The fact that electrons diffract demonstrates that there is a dual nature of matter. Electrons and other particles such as protons and neutrons have wave properties similar to those of light. The next big step towards understanding the nature of atoms and nuclei was taken in 1926. It was based on de Broglie's concept of matter waves.

11-2
THE SCHRÖDINGER EQUATION

In 1926 Erwin Schrödinger (Fig. 11-2a), an Austrian physicist, presented a widely applicable mathematical equation which described the matter waves of de Broglie. This equation is known as the **Schrödinger equation.** Understanding this formula takes a considerable background in the mathematics of calculus. However, the basic structure of the equation is interesting, and for curiosity's sake, let us see what it looks like. The Schrödinger equation is

$$\left(-\frac{h^2}{8\pi^2 m}\nabla^2 + E_p\right)\psi = E\psi$$

where m = mass of particle,
  h = Planck's constant,
  E_p = potential energy,
  E = total energy,
  ∇^2 (pronounced "del squared") is a calculus quantity,
  ψ (pronounced "sigh") is wavefunction (to be discussed later).

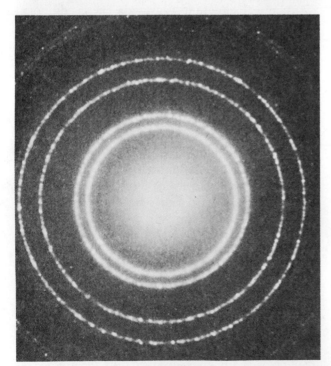

Figure 11-1b *Diffraction patterns produced by X-rays and electrons. The pattern at the top was made by X-rays. The bottom pattern was made by electrons.*

Figure 11-2b *Graph of the probability of finding an electron at a given radius in the hydrogen atom in the ground state. There is a probability that the electron will be at any radius, but the greater probability occurs at the Bohr radius of 0.529 Å.*

Figure 11-2a *Erwin Schrödinger (1887–1961), along with Heisenberg, was one of the founders of quantum mechanics. Born in Austria, he served in World War I, afterwards becoming a physics professor in Germany. In 1925 he wrote his paper which mathematically treated moving particles as waves. In 1933 he left Germany when the Nazis came to power. Later, he was a professor of physics at Dublin. (Photo courtesy Ullstein.)*

The first quantity in Schrödinger's equation is related to the kinetic energy.

$$E_k = -\frac{h^2}{8\pi^2 m}\nabla^2$$

where E_k = kinetic energy. Thus, Schrödinger's equation is simply

$$(E_k + E_p)\psi = E\psi$$

So, we see that the equation can be thought of as a statement of the well-known principle that kinetic energy plus potential energy equals the total energy.

Schrödinger's equation is the basis of practically all our present knowledge of atomic and nuclear physics. It is so basic that it would be difficult to overestimate its importance. While it may look simple, using it, except in a very few cases, is either impossible or requires a high-speed computer.

One simple case which was solved almost immediately using Schrödinger's equation was that of the hydrogen atom. The possible energies were found to be exactly the same as Bohr had obtained in 1913. But additional results also came out when Schrödinger's equation was used. These additional results concerned the quantity ψ, called the **wavefunction.** At first, it was not clear how ψ should be interpreted. After much thought and debate it was decided that ψ^2 (the wavefunction squared) represented the *probability* that the electron would be a certain distance from the nucleus. In Bohr's theory, the electron was a particle and could exist only at certain discrete distances from the proton. These distances are known as the Bohr radii and were given in Chapter 10 as $r = 0.529n^2$ Å. The Schrödinger wave equation altered the idea that the electron was a particle existing at a discrete distance from the proton. This idea was replaced by one which pictured an electron at any distance from the proton but it was most likely to be at the radius given by Bohr. Fig. 11-2b shows the probability, given by ψ^2, that an electron in the ground state will be at a given radius. The maximum probability occurs at the Bohr radius of 0.529 Å, but we see now that it is possible for the electron to be at various distances from the proton.

The idea that an electron has a certain probability of being at any given distance from the proton was quite new in 1926. To illustrate it, let us again refer to Fig. 11-2b. If you look near zero radius (near the proton), the curve shows that there is not much chance that you will find the electron. The greater the radius, the greater the probability, until you reach 0.529 Å. Here, at the Bohr radius, is the maximum probability. The probability then decreases as you continue to go out from the proton, until past about 2 Å, there is very little probability that you will find it. However, at any distance, no matter how near or far from the proton, there is a chance that the electron will be there.

11-3
THE HEISENBERG UNCERTAINTY PRINCIPLE

Shortly after Schrödinger's wave equation was published, Werner Heisenberg (Fig. 11-3a), a German physicist, proposed what is now known as the Heisenberg uncertainty principle. Its importance is better understood with a brief "flashback." After Newton's equations of motion were published in the seventeenth century, there was considerable speculation as to whether or not the future was automatically determined by the present. Using Newton's equations of motion, the resulting motions of objects could be determined, if it were known how fast and in what direction the objects were traveling and if the forces acting on them were known. Therefore, if the velocities and positions of all the electrons, protons, and neutrons in the universe were known at some present instant of time, their future velocities and positions could be determined completely. This led some philosophers to envision a deterministic universe, that is, one whose future could never be changed. The Heisenberg uncertainty principle states that it is impossible to know *simultaneously* the exact velocity and position of even a single particle. This principle puts a definite limit on our ability to predict the future of that particle. On a more cosmic scale, the Heisenberg uncertainty principle says that the future can never be determined by our knowledge of the present.

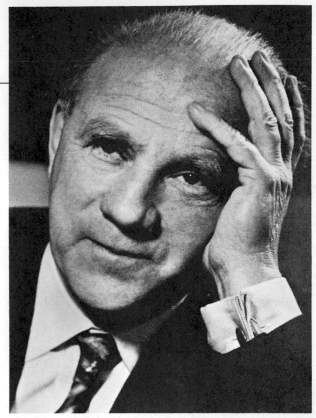

Figure 11-3a *Werner Heisenberg (1901–1976). At the age of 23, Heisenberg wrote a paper explaining quantum mechanics. His theory was entirely different from Schrödinger's, yet it produced the same results. Schrödinger later wrote a paper showing that his and Heisenberg's theories were mathematically identical. Heisenberg's name is attached to the uncertainty principle which he conceived as a result of his quantum theory. During World War II, Heisenberg was in charge of the German nuclear energy program. (Photo courtesy American Institute of Physics, Niels Bohr Library.)*

The Heisenberg uncertainty principle is based on the process of making a measurement. Let us now examine this process. In order to see an object, photons of light reflected from the object must enter our eye. Let us consider the example of recording, visually, the position of an electron, illustrated in Fig. 11-3b. In order to see the electron, a photon must come from the light source and bounce off the electron into an eye or other recording device. But photons carry energy and momentum. This latter quantity must be conserved when the photon collides with the electron. Since the direction and magnitude of the

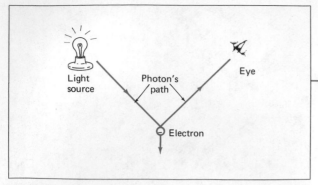

Figure 11-3b *Recording the position of an electron as an illustration of the process of measurement. The photon of light bounces off the electron before it hits the eye. In so doing, it changes the velocity of the electron.*

photon's momentum have changed as a result of hitting the electron, the electron must have gained some momentum, thereby altering its velocity. Since the electron's velocity has changed slightly, we don't know where it will be in the future, because we don't know its new velocity. We can record only where the electron was when the photon hit it.

Heisenberg formulated his uncertainty principle mathematically. In order to understand the mathematical formulation, let us recall how we record measurements in the laboratory. If we record the length of an object measured with a meter stick, we must tell what the possible limits on our measurement are. As an example, let us say we obtain a length of 10.32 ± 0.03 m. That is, the length is somewhere between 10.29 and 10.35 m, or we have an uncertainty of 0.06 m in our measurement. Heisenberg's uncertainty principle is

$$m \, \Delta v \, \Delta x = h/2\pi \qquad \text{Eq. 11.2}$$

where m = mass of object being measured,
Δv = minimum uncertainty in velocity,
Δx = minimum uncertainty in position,
h = Planck's constant
$(6.625 \times 10^{-27} \text{ erg} \cdot \text{s})$.

We see, then, that the Heisenberg uncertainty principle incorporates the minimum uncertainties of velocity and position as limitations. This is the best that we can ever hope to do. Most measurements are much more uncertain than these limits. This principle says that even if we have the most precise instruments possible, these minimum uncertainties would always exist because of the process of measurement. From the equation it is evident that the minimum uncertainties are smaller for large masses than for small masses. Minimum uncertainties become significant only when we are dealing with very small masses, such as electrons or protons. This is because h is such a small quantity. The following example illustrates the importance of the Heisenberg uncertainty principle when dealing with electrons.

■ *Example 2* If the uncertainty in the velocity of an electron is 3×10^8 cm/s, what is the minimum uncertainty in its position?

$$m \, \Delta v \, \Delta x = h/2\pi$$

$$(9.1 \times 10^{-28} \text{ g})(3 \times 10^8 \text{ cm/s}) \, \Delta x$$
$$= \frac{6.625 \times 10^{-27}}{2(3.14)} \text{ erg} \cdot \text{s}$$

Since all the units are in the cgs system, we know that the answer will be in centimeters. Solving for Δx, the minimum uncertainty in position, we get

$$\Delta x = \frac{6.625 \times 10^{-27}}{6.28 \times 3 \times 10^8 \times 9.1 \times 10^{-28}}$$
$$= 0.386 \times 10^{-8} \text{ cm}$$
or $\quad = 0.386 \text{ Å}$

This uncertainty is rather large, when we consider that the Bohr radius is 0.529 Å.

As stated above, the uncertainty principle is a precisely formulated physical law. It can also be thought of as a statement of the fact that, on a submicroscopic scale, the very process of measurement alters what is being measured. In the social sciences this same phenomenon is met on a macroscopic scale. It is very difficult, if not impossible, to measure human behavior without affecting the behavior. For instance, the presence of news reporters at the scene of a disturbance frequently changes the very nature of the disturbance.

11-4
ATOMIC QUANTUM NUMBERS

When the Schrödinger equation is solved for the hydrogen atom, several quantum numbers result. One of these is **n**, the principal quantum number, which occurred in the Bohr theory. However, several other such numbers also arise, each with its limitations. The quantum number ℓ is associated with the value of the angular momentum of the electron. (Recall that the angular momentum of a particle going in a circle is given by its mass times its velocity times the radius of the circle.) The quantum number ℓ can be 0, 1, 2, 3, . . . , but it must always be less than the value for n, the principal quantum number.

■ *Example 3* If the principal quantum number, n, is 1, what are the possible values of ℓ?

If $n = 1$, ℓ must $= 0$, since ℓ must be less than n.

■ *Example 4* If $n = 3$, what are the possible values of ℓ?

If $n = 3$, ℓ can be 0, 1, or 2.

The value of the quantum number ℓ is usually designated by a letter for notational convenience. Historically, these letters are s, p, d, f, g, etc. for $\ell = 0, 1, 2, 3, 4$, etc., respectively. (The first four letters originally stood for the words sharp, principal, diffuse, and fundamental. These words were used by early spectroscopists.) That is,

$\ell = 0$ is designated by s
$\ell = 1$ is designated by p
$\ell = 2$ is designated by d
$\ell = 3$ is designated by f
$\ell = 4$ is designated by g

After f and g, the letters continue in alphabetical order for higher values of ℓ. However, in this book we shall have no occasion to use values of ℓ above 3.

A third quantum number is the magnetic quantum number which is designated by the symbol m. The magnetic quantum number governs the orientation of the angular momentum and structures the shape of the region which the electron occupies. This number can be any positive or negative integer from $-\ell$ to $+\ell$, including zero. Therefore, there can be $2\ell + 1$ values of m. As an example, if $\ell = 2$, there are five possible values of m.

■ *Example 5* If $\ell = 0$, what are the possible values of m?

If $\ell = 0$, $m = 0$.

■ *Example 6* If $\ell = 2$, what are the possible values of m?

If $\ell = 2$, m can $= -2, -1, 0, +1,$ or $+2$.

The fourth and last quantum number is the spin magnetic quantum number which is designated by the symbol m_s. As the electron orbits around the proton, it also spins on its axis. This is analogous to the earth's spinning as it revolves around the sun. The electron can spin in one of two directions, just as a top can spin clockwise or counterclockwise. The quantum number m_s is arbitrarily designated either $+\frac{1}{2}$ or $-\frac{1}{2}$, depending on how the electron is spinning. No matter what the other three quantum numbers (n, ℓ, and m) are, the quantum number m_s is always either $+\frac{1}{2}$ or $-\frac{1}{2}$, for each value of m.

The four quantum numbers are summarized in Table 11-4a.

For the hydrogen atom, only the principal quantum number, n, determines the energy of the electron. However, for all other elements, those with two or more electrons, the quantum number ℓ also is necessary to designate the value of the energy of each electron. The quantum numbers m and m_s have nothing to do with the energy of the orbiting electron. Since only n and ℓ affect the value of the energy of the electron, each energy level is labeled by these two quantities. The notation used is to write a number which stands for the value of n followed by a letter which stands for the value of ℓ. As examples, $1s$ means the energy level when $n = 1$ and $\ell = 0$; the $3d$ level means $n = 3$ and $\ell = 2$; $4p$ means $n = 4$ and $\ell = 1$. For each energy level there are various sets of all four quantum numbers possible, because m and m_s can have different values.

		Values for m and m_s	Number of electrons
$n = 1$	$\ell = 0$	$\underset{+\frac{1}{2}\quad-\frac{1}{2}}{\overset{0}{\wedge}}$	2
$n = 2$	$\ell = 0$	$\overset{0}{\wedge}$	2
	$\ell = 1$	$\overset{-1,\ 0,\ +1}{\wedge\ \wedge\ \wedge}$	6
$n = 3$	$\ell = 0$	$\overset{0}{\wedge}$	2
	$\ell = 1$	$\overset{-1,\ 0,\ +1}{\wedge\ \wedge\ \wedge}$	6
	$\ell = 2$	$\overset{-2,\ -1,\ 0,\ +1,\ +2}{\wedge\ \wedge\ \wedge\ \wedge\ \wedge}$	10
$n = 4$	$\ell = 0$	$\overset{0}{\wedge}$	2
	$\ell = 1$	$\overset{-1,\ 0,\ +1}{\wedge\ \wedge\ \wedge}$	6
	$\ell = 2$	$\overset{-2,\ -1,\ 0,\ +1,\ +2}{\wedge\ \wedge\ \wedge\ \wedge\ \wedge}$	10
	$\ell = 3$	$\overset{-3,\ -2,\ -1,\ 0,\ +1,\ +2,\ +3}{\wedge\ \wedge\ \wedge\ \wedge\ \wedge\ \wedge\ \wedge}$	14

Figure 11-4a *Diagram illustrating the quantum numbers possible for $n = 1$ through $n = 4$. Note that the number of electrons increases by four for each additional value of ℓ. The m_s values of $+\frac{1}{2}$ or $-\frac{1}{2}$ are shown for $n = 1$ only. All others are represented by the symbol $_\wedge$.*

■ *Example 7* What are the possible sets of quantum numbers which an electron can have in the 3*p* energy level?

The 3 indicates the principal quantum number ($n = 3$); the *p* means $\ell = 1$, so *m* can be $+1, 0, -1$; and m_s can be $+\frac{1}{2}$ or $-\frac{1}{2}$ for each value of *m*. So there are six sets of quantum numbers. See Fig. 11-4a.

Table 11-4b shows various sets of quantum numbers that are possible for some energy levels of interest. From this table we see also that the number of different sets of quantum numbers depends only on the value of ℓ. When $\ell = 0$ (*s* levels), there are 2 different sets of quantum numbers possible. When $\ell = 1, 2,$ or 3 (*p*, *d*, *f*), there are 6, 10, or 14 sets possible. The significance

Table 11-4a *Quantum Numbers for an Atomic Electron*

Designation	Meaning	Possible Values
n	Principal quantum number designates in general the effective volume and determines energy	Any positive integer: $1, 2, 3, \ldots$
ℓ	Orbital quantum number governs magnitude of angular momentum also helps to determine energy	Any nonnegative integer less than n: $0, 1, 2, \ldots, n - 1$
m	Magnetic quantum number governs the orientation of angular momentum	$-\ell$ to $+\ell$ in integer steps (including zero)
m_s	Spin magnetic quantum number either counterclockwise or clockwise	$+\frac{1}{2}$ or $-\frac{1}{2}$

Table 11-4b *Possible Quantum Numbers for Some Energy Levels in Atoms* (Note that after the 4p level the different sets are only indicated.)

Energy Level	n	ℓ	m	m_s	Number of Different Sets
1s	1	0	0	$+\frac{1}{2}$	
	1	0	0	$-\frac{1}{2}$	2
2s	2	0	0	$+\frac{1}{2}$	
	2	0	0	$-\frac{1}{2}$	2
2p	2	1	$+1$	$+\frac{1}{2}$	
	2	1	$+1$	$-\frac{1}{2}$	
	2	1	0	$+\frac{1}{2}$	
	2	1	0	$-\frac{1}{2}$	
	2	1	-1	$+\frac{1}{2}$	
	2	1	-1	$-\frac{1}{2}$	6
3s	3	0	0	$+\frac{1}{2}$	
	3	0	0	$-\frac{1}{2}$	2
3p	3	1	$+1$	$+\frac{1}{2}$	
	3	1	$+1$	$-\frac{1}{2}$	
	3	1	0	$+\frac{1}{2}$	
	3	1	0	$-\frac{1}{2}$	
	3	1	-1	$+\frac{1}{2}$	
	3	1	-1	$-\frac{1}{2}$	6
4s	4	0	0	$+\frac{1}{2}$	
	4	0	0	$-\frac{1}{2}$	2
3d	3	2	$+2$	$+\frac{1}{2}$	
	3	2	$+2$	$-\frac{1}{2}$	
	3	2	$+1$	$+\frac{1}{2}$	
	3	2	$+1$	$-\frac{1}{2}$	
	3	2	0	$+\frac{1}{2}$	
	3	2	0	$-\frac{1}{2}$	
	3	2	-1	$+\frac{1}{2}$	
	3	2	-1	$-\frac{1}{2}$	
	3	2	-2	$+\frac{1}{2}$	
	3	2	-2	$-\frac{1}{2}$	10
4p	4	1	$+1$	$+\frac{1}{2}$	
	4	1	$+1$	$-\frac{1}{2}$	
	4	1	0	$+\frac{1}{2}$	
	4	1	0	$-\frac{1}{2}$	
	4	1	-1	$+\frac{1}{2}$	
	4	1	-1	$-\frac{1}{2}$	6
5s	5	0	0	$+\frac{1}{2}, -\frac{1}{2}$	2

Table 11-4b (Continued)

Energy Level	n	ℓ	m	m_s	Number of Different Sets
$4d$	4	2	$+2, +1, 0, -1, -2$	$+\frac{1}{2}, -\frac{1}{2}$	10
$5p$	5	1	$+1, 0, -1$	$+\frac{1}{2}, -\frac{1}{2}$	6
$6s$	6	0	0	$+\frac{1}{2}, -\frac{1}{2}$	2
$4f$	4	3	$+3, +2, +1, 0, -1, -2, -3$	$+\frac{1}{2}, -\frac{1}{2}$	14
$5d$	5	2	$+2, +1, 0, -1, -2$	$+\frac{1}{2}, -\frac{1}{2}$	10
$6p$	6	1	$+1, 0, -1$	$+\frac{1}{2}, -\frac{1}{2}$	6
$7s$	7	0	0	$+\frac{1}{2}, -\frac{1}{2}$	2
$5f$	5	3	$+3, +2, +1, 0, -1, -2, -3$	$+\frac{1}{2}, -\frac{1}{2}$	14
$6d$	6	2	$+2, +1, 0, -1, -2$	$+\frac{1}{2}, -\frac{1}{2}$	10

Figure 11-4b *Energy level diagram for the hydrogen atom. The number of different sets of quantum numbers is shown in parentheses above each level. Note that for the hydrogen atom, the value of ℓ does not affect the energy level. The figure is not drawn to scale and only some of the levels are shown.*

of the number of possible sets will be discussed in Section 11-5.

Let us now consider the energy level diagram for the hydrogen atom. This is shown in Fig. 11-4b. This diagram is similar to Fig. 10-3b of Chapter 10 except that both n, and ℓ quantum numbers, rather than only n, are indicated. Also, in Fig. 11-4b the number of different sets of quantum numbers is shown in parentheses on the line designating the level. Figure 11-4b is not drawn to scale; this is done so that the higher levels can be identified.

The energy situation for atoms other than hydrogen is not so simple. The energies of these atoms depend on both n and ℓ. The energy levels in order of increasing energy are as follows: $1s$, $2s$, $2p$, $3s$, $3p$, $4s$, $3d$, $4p$, $5s$, $4d$, $5p$, $6s$, $4f$, $5d$, $6p$, $7s$, $5f$, $6d$, and so on. The reason for this ordering is not simple. It is difficult for the following reasons: In the hydrogen atom, the only force is the electrical attraction between the nucleus and the single electron. However, when more than one electron orbits the nucleus, there are additional electrical forces between the electrons, which push the electrons away from each other. Solving this problem is extremely complicated, and a simple explanation of why the levels are ordered as they are cannot be given. However, there is a simple way to remember the order: Write down the levels in a tri-

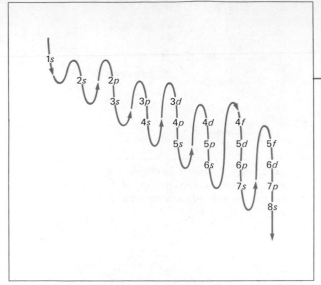

Figure 11-4c *Mnemonic device for ordering the energy levels in many-electron atoms. Order is from upper right to lower left.*

7s (2) 7p (6) 6d (10) 5f (14)
6s (2) 6p (6) 5d (10) 4f (14)
5s (2) 5p (6) 4d (10)
4s (2) 4p (6) 3d (10)
3s (2) 3p (6)
2s (2) 2p (6)

Energy →

1s (2)

Figure 11-4d *A typical energy level diagram for a many-electron atom. The number of different sets of quantum numbers is shown in parentheses above each level. (The diagram is now drawn to scale.)*

angle as shown in Fig. 11-4c. Then make diagonal lines going from the upper right-hand to the lower left-hand portion as shown. The correct ordering is achieved by starting with the top diagonal line, going from the upper right to the lower left, and then con-

tinuing to the next diagonal line down and repeating the procedure.

We can now draw an energy level diagram for a typical many-electron atom. This is shown in Fig. 11-4d. The various levels are now shifted to get the desired energy ordering.

11-5
MANY-ELECTRON ATOMS AND THE PAULI EXCLUSION PRINCIPLE

Solving the Schrödinger equation for the hydrogen atom is not particularly difficult. There is only one force involved—the electrostatic force between the proton and the electron. Quantum numbers arise naturally from the possible solutions to the equation, and the result is that the electron can occupy any of the various energy levels shown in Fig. 11-4a. The ground state of an atom is the combination of energy levels with the lowest total energy. For the hydrogen atom, the ground state is the 1s level. Almost all naturally occurring atoms are in the ground state. If an atom is not in the ground state, its electrons will give off energy in the form of photons and fall into the ground state. To understand more complicated atoms, then, it is necessary to know how the electrons distribute themselves in the ground state.

The **Pauli exclusion principle** set forth by Wolfgang Pauli, a German physicist, in 1928 brought order to the seeming chaos of many-electron atoms. The Pauli exclusion principle states that no two electrons can have the same set of quantum numbers. The underlying physical reason why this should be so is still not understood, but application of the Pauli principle makes it fairly simple to distribute the electrons among the energy levels. To illustrate its application, let us consider the case of aluminum, which has 13 electrons. How are we to distribute these 13 electrons to produce the ground state, that is, the level of lowest energy? We do so by putting the electrons into

Figure 11-5a *Energy level diagram for the ground state of the aluminum atom. Electrons are represented by colored circles. For the ground state, the lowest levels are filled.*

Figure 11-5b *Energy level diagram for an excited state of the sodium atom. When the electron loses energy and changes to a lower level as indicated by the arrow, a photon of yellow light is emitted.*

the lowest energy levels possible. There are only two different sets of quantum numbers in the $1s$ energy level, so only two electrons can go there. No more than two can go into this level, because if there were three or more, at least two would then have the same set of quantum numbers, in violation of the Pauli exclusion principle. Next, we put two into the $2s$ level. Then six go into the $2p$ level and two more go into the $3s$ level. The last electron then goes into the $3p$ level. We can represent the electrons as small circles and make an energy level diagram for the ground state of aluminum. This is shown in Fig. 11-5a. Because of the Pauli principle, the numbers which we put in parentheses above the various energy levels in Figs. 11-4b and 11-4d now take on a new significance. They are the maximum numbers of electrons which can go into the various energy levels.

The electrons in many-electron atoms are usually in the ground state which is made up of the lowest energy levels possible. However, one or more electrons is sometimes in a higher level. As an example, the energy level diagram for an excited state of the sodium atom is shown in Fig. 11-5b. When the last electron is in the $4p$ state, it can lose energy and fall to the $3s$ energy level, giving off a photon of light that looks yellow to us. The spectrum of the sodium atom and several other atoms is given in color Plate I. Because each atom has a unique set of energy levels, each atom has a unique spectrum. The sixth spectrum

from the bottom of this color plate is the spectrum of sunlight. Here, one can see that several wavelengths are absorbed. By comparing the sun's spectrum to known spectra, the elements in the sun's outer atmosphere can be determined. This is a good example of the usefulness of spectroscopy. It is the best way we have of identifying the elements in stars or in substances of unknown composition.

Rather than drawing energy level diagrams like Figs. 11-5a and 11-5b each time we want to represent the electron arrangement in an atom, we use a shorthand notation called the electron configuration. When writing an **electron configuration,** the levels are written in order of increasing energy and the number of electrons in each level is designated by a superscript. For example, the electron configuration of the aluminum atom in the ground state (Fig. 11-5a) is

$$1s^2 2s^2 2p^6 3s^2 3p^1$$

The electron configuration for the excited state of sodium in Fig. 11-5b is

$$1s^2 2s^2 sp^6 4p^1$$

A more complicated electron configuration is the one for radium, which has 88 electrons. Radium's configuration in the ground state is

$$1s^2 2s^2 2p^6 3s^2 3p^6 4s^2 3d^{10} 4p^6 5s^2 4d^{10} 5p^6$$
$$6s^2 4f^{14} 5d^{10} 6p^6 7s^2$$

11-6
ELECTRON PERIODS

If we look closely at Fig. 11-4d, we see that spaces between the energy levels are not equal. For instance, there is a big gap between the $1s$ and $2s$ levels. In general, there are big energy gaps between the s levels and the levels below them, with smaller gaps between the other levels, such as between $4s$-$3d$ and $3d$-$4p$. Energy levels (such as $4s$, $3d$, and $4p$) between the big gaps all have approximately the same energy. An **electron period** consists of a set of energy levels, all of which have about the same energy. We can indicate the energy gaps between periods by vertical lines drawn between energy levels in an electron configuration as follows:

$$1s^2 \mid 2s^2 2p^6 \mid 3s^2 3p^6 \mid 4s^2 3d^{10} 4p^6 \mid 5s^2 4d^{10} 5p^6 \mid$$
$$6s^2 4f^{14} 5d^{10} 6p^6 \mid \text{etc.}$$

Each group of energy levels between the lines forms a period. The different periods are listed in Table 11-6a.

A many-electron atom can be thought of in terms of the Bohr model of electrons orbiting the nucleus. As in the hydrogen atom, the lower the value of n, the closer the orbit to the nucleus. An **electron shell** consists of all the orbits with the same value of n. In the Bohr model all the electrons in the same shell are represented as being in the same orbit. The first shell, or orbit, can hold two electrons; the next shells have maxima of 8, 18, 32, 50, 72, . . . , but except for the first four shells, the maxima are never reached. Examples of the Bohr model, as applied to carbon and sodium, are shown in Fig. 11-6a. Since carbon has only six electrons, two go into the first shell and four into the second. Sodium has eleven electrons: two go into the first orbit, or ring, around the nucleus, eight go into the second, and one goes into the third. The electron configurations for carbon and sodium in the ground state are

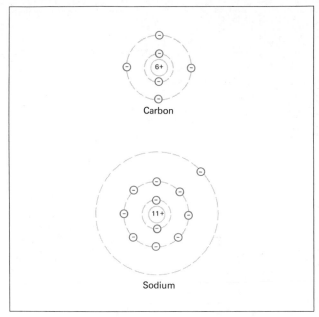

Figure 11-6a *Schematic diagrams of carbon and sodium atoms. The solid ring in the center represents the nucleus. Each shell is drawn as a broken ring around the nucleus, with the appropriate number of electrons shown for each shell of the particular atom.*

Carbon: $1s^2 2s^2 2p^2$

Sodium: $1s^2 2s^2 2p^6 3s^1$

These configurations should be compared with Fig. 11-6a.

The Bohr model of electrons orbiting the nucleus is a simple model and easy to visualize. However, it cannot be exact, because the electrons do not travel in definite orbits. Instead, as was illustrated in Fig. 11-2b for a $1s$ electron, each electron has a probability of being at any particular distance from the nucleus. The dashed circles in Fig. 11-6a and in similar figures in the following chapters simply give a rough *s block*

Z	Symbol	Config
1	H	$1s$
3	Li	$2s$
4	Be	$2s^2$
11	Na	$3s$
12	Mg	$3s^2$
19	K	$4s$
20	Ca	$4s^2$
37	Rb	$5s$
38	Sr	$5s^2$
55	Cs	$6s$
56	Ba	$6s^2$
87	Fr	$7s$
88	Ra	$7s^2$

d block

21 Sc $3d\,4s^2$	22 Ti $3d^2\,4s^2$	23 V $3d^3\,4s^2$	24 Cr $3d^5\,4s^1$	25 Mn $3d^5\,4s^2$	26 Fe $3d^6\,4s^2$	27 Co $3d^7\,4s^2$	28 Ni $3d^8\,4s^2$	29 Cu $3d^{10}\,4s^1$	30 Zn $3d^{10}\,4s^2$
39 Y $4d\,5s^2$	40 Zr $4d^2\,5s^2$	41 Nb $4d^4\,5s^2$	42 Mo $4d^5\,5s^1$	43 Tc $4d^6\,5s^1$	44 Ru $4d^7\,5s^1$	45 Rh $4d^8\,5s^1$	46 Pd $4d^{10}$	47 Ag $4d^{10}\,5s^1$	48 Cd $4d^{10}\,5s^2$
57 La* $5d\,6s^2$	72 Hf $5d^2\,6s^2$	73 Ta $5d^3\,6s^2$	74 W $5d^4\,6s^2$	75 Re $5d^5\,6s^2$	76 Os $5d^6\,6s^2$	77 Ir $5d^7\,6s^2$	78 Pt $5d^9\,6s^1$	79 Au $5d^{10}\,6s^1$	80 Hg $5d^{10}\,6s^2$
89 Ac† $6d\,7s^2$									

p block

						2 He $1s^2$
5 B $2p$	6 C $2p^2$	7 N $2p^3$	8 O $2p^4$	9 F $2p^5$	10 Ne $2p^6$	
13 Al $3p$	14 Si $3p^2$	15 P $3p^3$	16 S $3p^4$	17 Cl $3p^5$	18 Ar $3p^6$	
31 Ga $4p$	32 Ge $4p^2$	33 As $4p^3$	34 Se $4p^4$	35 Br $4p^5$	36 Kr $4p^6$	
49 In $5p$	50 Sn $5p^2$	51 Sb $5p^3$	52 Te $5p^4$	53 I $5p^5$	54 Xe $5p^6$	
81 Tl $6p$	82 Pb $6p^2$	83 Bi $6p^3$	84 Po $6p^4$	85 At $6p^5$	86 Rn $6p^6$	

f block

*Lanthanides

58 Ce $4f\,5d\,6s^2$	59 Pr $4f^3\,6s^2$	60 Nd $4f^4\,6s^2$	61 Pm $4f^5\,6s^2$	62 Sm $4f^6\,6s^2$	63 Eu $4f^7\,6s^2$	64 Gd $4f^7\,5d^1\,6s^2$	65 Tb $4f^9\,6s^2$	66 Dy $4f^{10}\,6s^2$	67 Ho $4f^{11}\,6s^2$	68 Er $4f^{12}\,6s^2$	69 Tm $4f^{13}\,6s^2$	70 Yb $4f^{14}\,6s^2$	71 Lu $4f^{14}\,5d^1\,6s^2$

†Actinides

90 Th $6d^2\,7s^2$	91 Pa $5f^2\,6d^1\,7s^2$	92 U $5f^3\,6d^1\,7s^2$	93 Np $5f^4\,6d^1\,7s^2$	94 Pu $5f^6\,7s^2$	95 Am $5f^7\,7s^2$	96 Cm $5f^7\,6d^1\,7s^2$	97 Bk $5f^9\,7s^2$	98 Cf $5f^{10}\,7s^2$	99 Es $5f^{11}\,7s^2$	100 Fm $5f^{12}\,7s^2$	101 Md $5f^{13}\,7s^2$	102 No $5f^{14}\,7s^2$	103 Lw $5f^{14}\,6d^1\,7s^2$

Figure 11-6b *A chart of the elements in terms of electron periods. Each row is a separate period. The chart shows the order of entry of electrons with increasing atomic number. Only those subshells are given that electrons have just entered.*

Table 11-6a *Electron Periods*

Period Number	Energy Levels in Period	Number of Electrons in Period
1	1s	2
2	2s, 2p	8
3	3s, 3p	8
4	4s, 3d, 4p	18
5	5s, 4d, 5p	18
6	6s, 4f, 5d, 6p	32
7	7s, 5f, 6d, 7p	32
⋮	⋮	⋮

(A vertical arrow labeled "Energy" points downward alongside the Period Number column.)

idea of where the electron is most likely to be found, as it goes around the nucleus.

If we examine Table 11-6a, we see that after the first three periods, electrons from different shells can go into the same period. This is because, although the radius of an electron's path depends mostly on the value of n, the electron's energy depends on both n and ℓ. Electrons with the same value of n are in the same shell. Electrons with the same value of n and ℓ are said to be in the same **electron subshell.** A subshell is similar to an energy level, since all electrons in the same subshell have the same energy. Several subshells make up a period, as can be seen from Table 11-6a.

A chart of the elements in terms of electron periods is shown in Fig. 11-6b. Each row is a separate period. The chart shows the order of entry of elec-

trons with increasing atomic number. Only those subshells are given in which electrons have just entered. Four separate groups are shown in the chart. The letters s, p, d, and f above each group indicates that these orbitals are being filled in the group. There are only two electrons in the first period, which is 1s. However, we have put the number 2 on the far right of Fig. 11-6b because it is the last number in the period. All numbers on the far right represent the last number in the period. The second and third rows represent the second and third periods. They contain only s and p energy levels. The next four periods have d levels as well. The f levels of the sixth and seventh periods are listed below the chart. Note that after numbers 56 and 88, the next number has been put into the d level instead of the f level. These discrepancies in the chart are too complicated for us to fret over. After numbers 57 and 89, the f levels begin being filled. The chart ends after seven periods, since there are only 105 known elements. We can represent the various elements by designating how many electrons there are in the last subshell being filled in the ground state of the element. For instance, in the electron configuration for carbon, the sixth and last electron is the second electron in the 2p subshell. Note that all elements listed in the vertical column with carbon have the p^2 orbital being filled but at a different energy level. The chart in Fig. 11-6b is a representation of the periodic chart of elements that will be discussed completely and with great detail in the next chapter.

IMPORTANT WORDS AND TERMS

Quantum mechanics	Heisenberg uncertainty principle	Electron period
Matter waves	n, ℓ, m, and m_s quantum numbers	Electron shell
Schrödinger equation	Pauli exclusion principle	Electron subshell
Wavefunction	Electron configuration	

LEARNING OBJECTIVES

After reading and studying this chapter, you should be able to do the following without referring to the text:

1. Write the electronic configuration for any atom listed in the periodic table.
2. State de Broglie's hypothesis concerning matter waves and give an example calculation.
3. Describe each term of the Schrödinger equation and state the results the equation gives in relation to de Broglie's matter waves.
4. State and explain Heisenberg's uncertainty principle.
5. List the four quantum numbers generated by the Schrödinger wave equation, state their meaning, and give their possible values.
6. State Pauli's exclusion principle.
7. Determine the value of the wavelength for two different masses moving at different velocities.
8. Calculate uncertainties using Heisenberg's principle.
9. Determine the value of the quantum number l when given the value for n.
10. Determine the value of the quantum number m when given the value for l.
11. Draw ground state energy level diagrams for a given element.
12. Draw a schematic diagram illustrating the electron structure of a given atom.

QUESTIONS

1. What are matter waves?
2. What does the Schrödinger equation describe?
3. How should the quantity ψ in the Schrödinger equation be interpreted?
4. Is it possible for the electron to be (a) inside the nucleus of an atom? (b) 100 Å away from the nucleus?
5. Compare the number of electrons of a neutral atom to its atomic number.
6. How accurately can the position and velocity of a particle be known simultaneously?
7. State the four quantum numbers and explain the general meaning of each.
8. What do the letters s, p, d, and f mean?
9. How many possible values of the quantum number l are there for each value of the quantum number n?
10. How many possible values of the quantum number m are there for each value of the quantum number l?
11. (a) How many additional and different values of the l quantum number are there for each additional consecutive value of the quantum number n?
    (b) How many electrons are there for these additional and different l values?
12. State Pauli's exclusion principle.
13. Distinguish between electron period and electron shell.

14. What shorthand notation is used to represent the electron arrangement in an atom? Give an example.
15. What is the basis of the periodic chart of the elements? (Hint: Compare Fig. 11-6b with the periodic table, Table 12-2a, in the next chapter.)

PROBLEMS

1. The Bohr radius for $n = 1$ is 0.529 Å.
   (a) What is the circumference of an atom with the Bohr radius?
   (b) Set this circumference equal to a wavelength of the electron. Now use de Broglie's equation to find how fast the electron is moving. Compare your answer to the speed of an electron according to the Bohr theory, that is, $v = 2.19 \times 10^8$ cm/s.

   Answer: (b) The same as in the Bohr theory.

2. Calculate the de Broglie wavelength of a 0.5-kg baseball moving with a velocity of 2600 cm/s (approximately 60 mi/h).

3. Calculate the de Broglie wavelength of the earth. $(m_E = 6 \times 10^{24}$ kg; $v_E = 3 \times 10^4$ m/s).

4. What are the possible values of ℓ when
   (a) $n = 3$, (b) $n = 5$?

5. What are the possible values of quantum number m when quantum number
   (a) ℓ is 2, (b) ℓ is 3?

6. (a) List the possible sets of quantum numbers for an electron in the $4d$ subshell.
   (b) How many electrons are there?

7. Write the electron configurations for each of the following elements:
   (a) $_{17}$Cl, (b) $_{19}$K, (c) $_{53}$I.

8. Draw the ground state energy level diagram similar to Fig. 11-5a for each of the following elements:
   (a) $_{17}$Cl, (b) $_{19}$K.

9. Draw schematic diagrams of electrons orbiting the nucleus similar to Fig. 11-6a for each of the following elements:
   (a) $_{17}$Cl, (b) $_{19}$K.

Chapter 12

THE PERIODIC TABLE

The objects in our environment are composed of matter. **Matter** may be defined as anything that exists in time, occupies space, and has mass. The elements are the different kinds of atoms which make up the matter in our everyday environment. The physical, chemical, and combining properties of the various elements affect us constantly. To cite two examples, the element oxygen is necessary to sustain human life, while the diamond in an engagement ring is pure carbon. With only a few exceptions, the 105 known elements, either singularly or in combination, are the components of all matter. The science of chemistry deals with the transformations which this matter undergoes.

The modern understanding of matter, based on the electron structure of atoms, is relatively recent, but chemistry had its beginnings early in recorded history. Egyptian hieroglyphs which date back to 3400 B.C. show wine presses; wine making, of course, required a chemical fermentation process. By 2000 B.C. the Egyptians and Mesopotamians possessed the knowledge required to produce and work the metals gold, silver, lead, copper, iron, and bronze—a mixture of copper and tin. Dyes were discovered during this period, and methods of fixing them were known.

In the fourth century B.C., the Greek philosopher Aristotle developed the false idea that all matter was composed of four elements: earth, air, fire, and water. As the Greek culture faded, the pseudoscience of alchemy was born. Its main goals were to change base metals into gold and to find the "elixir of life," which could restore an aging human body to its youthful vigor. Alchemy was a mixture of magic and experimentation, and it was practiced in China as well as the West. Although it was practiced by many charlatans, its finest practitioners were dedicated experimentalists, who contributed to the real science of chemistry. The goals of the alchemists were never achieved, but by the time the scientific method was born in the sixteenth century, much had been discovered about the properties of matter.

In 1869, Mendeleev in Russia and Julius Meyer in Germany, working independently, published a classification of the known elements showing a relationship between the magnitude of the atomic weights and the properties of elements. This relationship led to the conclusion that the properties of elements are periodic functions of their atomic weights. In 1914, Henry Moseley, an English physicist, showed that atomic numbers rather than atomic weights are fundamental in determining the properties of elements, thus laying the foundation for the modern periodic table. In this chapter we shall learn the properties of elements and how to use the periodic table to determine many of these properties.

Figure 12-1a *Robert Boyle (1627–1691). Born in Ireland, the fourteenth child of the wealthy Earl of Cork, young Robert proved to be a child prodigy. After graduation from Oxford, Boyle used his wealth to finance his research in many facets of science. He is probably best known for Boyle's law (see Chapter 7). Considered as one of the founders of chemistry, he was the first person to define an element and the first to suggest that heat was due to molecular motion. He also proved that air transmitted sound. His work did much to advance the cause of experimentation in chemistry. He was sick much of his life, but managed to write more than 40 books on language, religion, and science. (Photo courtesy Culver Pictures)*

12-1
ELEMENTS

In the very earliest of civilizations, nine elements were isolated: gold, silver, lead, copper, tin, iron, carbon, sulfur, and mercury. The alchemists added five more: antimony, arsenic, bismuth, phosphorus, and zinc. As the list grew, a precise definition of an element was needed. The concept of an element as we know it today was formulated by Robert Boyle (Fig. 12-1a), an English chemist, in 1661. In his book *The Skeptical Chemist,* Boyle proposed that the name "element" be applied to those simple bodies that

could not be separated into components by chemical methods. In 1789, the French scientist, Antoine Lavoisier, expanded Boyle's definition and applied the term element to the matter which remained after complete chemical decomposition had occurred.

Lavoisier, who is sometimes called the father of chemistry, introduced quantitative methods into chemistry and isolated the element oxygen. In 1789, his list of elements included 23 which we recognize today as elements. The modern definition of the word element is "a substance in which all the atoms have the same atomic number." Our modern listing of the elements has grown to 105, with more expected to be added.

In order to designate the different elements, a symbol notation is used. The notation used today was first conceived by the Swedish chemist Jons Jacob Berzelius. He used the first one or two letters of the Latin name for each element. Thus, sodium is designated Na for "natrium," silver is Ag for "argentum," tin is Sn for "stannum," and so forth. Since Berzelius' time, most elements have been symbolized by the first one or two letters of the English name. Examples are C for carbon, O for oxygen, H for hydrogen, and Ca for calcium. Note that the first letter is always capitalized and the second is lower case. This symbol notation of the elements has proven to be quite useful.

When the atoms of two or more elements react chemically, they often join together to form what chemists call **molecules** of a compound. The **formula** for the compound is written by putting the element symbols adjacent to each other and using subscripts to designate the number of atoms of each element in the compound. For instance, water is written H_2O. This means that in a single molecule of the compound water there are two atoms of hydrogen and one atom of oxygen. If more than one molecule is to be designated, a number is placed in front of the molecule. For instance, $3 H_2O$ means there are three molecules of water, each having two atoms of hydrogen and one atom of oxygen.

In Table 12-1a, each element is listed in alphabetical order, along with its symbol, atomic number, and atomic weight. There are 105 elements listed in this table and number 106 has recently been created. Efforts are constantly being made to create the heavier elements 107, 108, etc.

You will recall (Section 9-3) that the atomic number of an element refers to the number of protons in the nucleus of the atom. The atomic number also is equal to the number of electrons in the neutral atom.

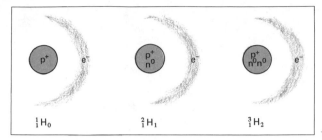

Figure 12-1b *The three isotopes of hydrogen. Protium, deuterium, and tritium.*

We will see shortly, however, that sometimes electrons are gained or lost, so that the atom is no longer neutral. It is the number of protons, not electrons, which determines what element you have. Each element can have several isotopes. Isotopes are forms of the same element whose nuclei have the same number of protons but different numbers of neutrons.

The element hydrogen is found in nature as three isotopes. See Fig. 12-1b. The three isotopes have similar chemical properties since they all have the same electron structure. They differ in physical properties and can be separated in a mass spectrograph because they have different masses. Ordinary hydrogen ($_1^1H$) is called protium, hydrogen two ($_1^2H$) is called deuterium, and hydrogen three ($_1^3H$) is called tritium. In a given sample of hydrogen, about one atom in 5000 is deuterium and about one atom in 10,000,000 is tritium. Heavy water (D_2O) is composed of two atoms of deuterium and one atom of oxygen. Protium and deuterium are stable atoms, whereas tritium is radioactive with a half-life of 12.4 years.

The atomic weight of each element is the average weight of the isotopes contained in the naturally occurring element. Atomic weights today are based on the ^{12}C isotope, which is given a *relative* atomic weight of exactly 12.0000. The reason naturally occurring carbon has an atomic weight slightly greater

Table 12-1a *Alphabetical Listing of the Elements*
The atomic weights are based on $^{12}C = 12.0000$. If the element does not occur naturally, the mass number of the most stable isotope is given in parentheses.

	Symbol	Atomic Number	Atomic Weight		Symbol	Atomic Number	Atomic Weight
Actinium	Ac	89	(227)	Europium	Eu	63	151.96
Aluminum	Al	13	26.9815	Fermium	Fm	100	(253)
Americium	Am	95	(243)	Fluorine	F	9	18.9984
Antimony	Sb	51	121.75	Francium	Fr	87	(223)
Argon	Ar	18	39.948	Gadolinium	Gd	64	157.25
Arsenic	As	33	74.9216	Gallium	Ga	31	69.72
Astatine	At	85	(210)	Germanium	Ge	32	72.59
Barium	Ba	56	137.34	Gold	Au	79	196.967
Berkelium	Bk	97	(247)	Hafnium	Hf	72	178.49
Beryllium	Be	4	9.01218	Hahnium	Ha	105	(260)
Bismuth	Bi	83	208.9806	Helium	He	2	4.00260
Boron	B	5	10.81	Holmium	Ho	67	164.9303
Bromine	Br	35	79.90	Hydrogen	H	1	1.0080
Cadmium	Cd	48	112.40	Indium	In	49	114.82
Calcium	Ca	20	40.08	Iodine	I	53	126.9045
Californium	Cf	98	(251)	Iridium	Ir	77	192.22
Carbon	C	6	12.011	Iron	Fe	26	55.847
Cerium	Ce	58	140.12	Krypton	Kr	36	83.80
Cesium	Cs	55	132.9055	Lanthanum	La	57	138.9055
Chlorine	Cl	17	35.453	Lawrencium	Lr	103	(257)
Chromium	Cr	24	51.996	Lead	Pb	82	207.12
Cobalt	Co	27	58.9332	Lithium	Li	3	6.941
Copper	Cu	29	63.545	Lutetium	Lu	71	174.97
Curium	Cm	96	(247)	Magnesium	Mg	12	24.305
Dysprosium	Dy	66	162.50	Manganese	Mn	25	54.9380
Einsteinium	Es	99	(254)	Mendelevium	Md	101	(256)
Erbium	Er	68	167.26	Mercury	Hg	80	200.59

Table 12-1b *Abundant Elements as Percentages of the Earth's Crust*

Element	Percentage (by weight)
Oxygen	47
Silicon	28
Aluminum	8
Iron	5
Calcium	3.6
Sodium	2.8
Potassium	2.6
Magnesium	2
All others	1

than 12.0000 is that there is some ^{13}C and ^{14}C in naturally occurring carbon as well as ^{12}C.

Only a few of the elements exist in large amounts in the earth's crust. About 75 percent of the weight of the earth's crust is composed of only two elements—oxygen and silicon. Aluminum and iron, two important metals, are quite abundant also, but many elements exist naturally in only minute quantities. Several, of course, do not occur at all in nature and can only be made artificially. Table 12-1b gives the approximate percentage by weight of the more abundant elements in the earth's crust.

Table 12-1a *(Continued)*

	Symbol	Atomic Number	Atomic Weight		Symbol	Atomic Number	Atomic Weight
Molybdenum	Mo	42	95.94	Scandium	Sc	21	44.9559
Neodymium	Nd	60	144.24	Selenium	Se	34	78.96
Neon	Ne	10	20.179	Silicon	Si	14	28.086
Neptunium	Np	93	(237)	Silver	Ag	47	107.868
Nickel	Ni	28	58.71	Sodium	Na	11	22.9898
Niobium	Nb	41	92.9064	Strontium	Sr	38	87.62
Nitrogen	N	7	14.0067	Sulfur	S	16	32.06
Nobelium	No	102	(253)	Tantalum	Ta	73	180.9479
Osmium	Os	76	190.2	Technetium	Tc	43	(99)
Oxygen	O	8	15.9994	Tellurium	Te	52	127.60
Palladium	Pd	46	106.4	Terbium	Tb	65	158.9254
Phosphorus	P	15	30.9738	Thallium	Tl	81	204.37
Platinum	Pt	78	195.09	Thorium	Th	90	232.0381
Plutonium	Pu	94	(244)	Thulium	Tm	69	168.9342
Polonium	Po	84	(209)	Tin	Sn	50	118.69
Potassium	K	19	39.102	Titanium	Ti	22	47.90
Praseodymium	Pr	59	140.9077	Tungsten	W	74	183.85
Promethium	Pm	61	(145)	Uranium	U	92	238.029
Protactinium	Pa	91	(231)	Vanadium	V	23	50.9414
Radium	Ra	88	(226)	Xenon	Xe	54	131.30
Radon	Rn	86	(222)	Ytterbium	Yb	70	173.04
Rhenium	Re	75	186.2	Yttrium	Y	39	88.9059
Rhodium	Rh	45	102.9055	Zinc	Zn	30	65.37
Rubidium	Rb	37	85.4678	Zirconium	Zr	40	91.22
Ruthenium	Ru	44	101.07				
Rutherfordium	Rf	104	(257)				
Samarium	Sm	62	150.4				

12-2
THE PERIODIC TABLE

By 1870, a total of 65 elements had been discovered. Using the methods of Lavoisier and other dedicated chemists of that time, the atomic weight and other properties of many elements were determined. However, except for a few sketchy attempts, a system of classifying the elements had not been found. It remained for the great Russian chemist Dmitri Mendeleev (1834–1907) to formulate a satisfactory classification scheme. Mendeleev surveyed all data on the known elements and was able to arrange them in a table. A modern version of his table, called the periodic table of the elements, is shown in Table 12-2a and on the front inside cover of this book. The **periodic table** arranges the elements in rows according to their electron periods with all the elements in the same column having similar properties. Mendeleev's original table was based on atomic weights, but H. G. J. Moseley, an English physicist, showed in 1913 that properties of the elements are determined by their atomic numbers. The modern statement of the **periodic law** is as follows:

Table 12-2a Modern Periodic Table

The number of electrons in filled shells is shown in the column at the extreme left; the remaining electrons for each element are shown below the symbol and atomic number for each element. The atomic weights shown above the symbols are based on Carbon-12.

REPRESENTATIVE ELEMENTS — TRANSITION ELEMENTS — REPRESENTATIVE ELEMENTS — NOBLE GASES

PERIODS	I A	II A	III B	IV B	V B	VI B	VII B	VIII	VIII	VIII	I B	II B	III A	IV A	V A	VI A	VII A	0
1 — 0	1.0080 H[1] 0																1.0080 H[1] 1	4.00260 He[2] 2
2 — 2	6.941 Li[3] 1	9.01218 Be[4] 2											10.81 B[5] 3	12.011 C[6] 4	14.0067 N[7] 5	15.9994 O[8] 6	18.9984 F[9] 7	20.179 Ne[10] 8
3 — 2,8	22.9898 Na[11] 1	24.305 Mg[12] 2											26.9815 Al[13] 3	28.086 Si[14] 4	30.9738 P[15] 5	32.06 S[16] 6	35.453 Cl[17] 7	39.948 Ar[18] 8
4 — 2,8	39.102 K[19] 8,1	40.08 Ca[20] 8,2	44.9559 Sc[21] 9,2	47.90 Ti[22] 10,2	50.9414 V[23] 11,2	51.996 Cr[24] 13,1	54.9380 Mn[25] 13,2	55.847 Fe[26] 14,2	58.9332 Co[27] 15,2	58.71 Ni[28] 16,2	63.545 Cu[29] 18,1	65.37 Zn[30] 18,2	69.72 Ga[31] 18,3	72.59 Ge[32] 18,4	74.9216 As[33] 18,5	78.96 Se[34] 18,6	79.90 Br[35] 18,7	83.80 Kr[36] 18,8
5 — 2,8,18	85.4678 Rb[37] 8,1	87.62 Sr[38] 8,2	88.9059 Y[39] 9,2	91.22 Zr[40] 10,2	92.9064 Nb[41] 12,1	95.94 Mo[42] 13,1	(99) Tc[43] 14,1	101.07 Ru[44] 15,1	102.906 Rh[45] 16,1	106.4 Pd[46] 18	107.868 Ag[47] 18,1	112.40 Cd[48] 18,2	114.82 In[49] 18,3	118.69 Sn[50] 18,4	121.75 Sb[51] 18,5	127.60 Te[52] 18,6	126.905 I[53] 18,7	131.30 Xe[54] 18,8
6 — 2,8,18	132.906 Cs[55] 18,8,1	137.34 Ba[56] 18,8,2	[57-71] *	178.49 Hf[72] 32,10,2	180.948 Ta[73] 32,11,2	183.85 W[74] 32,12,2	186.2 Re[75] 32,13,2	190.2 Os[76] 32,14,2	192.22 Ir[77] 32,15,2	195.09 Pt[78] 32,17,1	196.967 Au[79] 32,18,1	200.59 Hg[80] 32,18,2	204.37 Tl[81] 32,18,3	207.12 Pb[82] 32,18,4	208.981 Bi[83] 32,18,5	(209) Po[84] 32,18,6	(210) At[85] 32,18,7	(222) Rn[86] 32,18,8
7 — 2,8,18,32	(223) Fr[87] 18,8,1	(226) Ra[88] 18,8,2	[89-103] †	(257) Rf[104] 32,10,2	(260) Ha[105] 32,11,2	[106]	[107]	[108]										

INNER TRANSITION ELEMENTS

* LANTHANIDE SERIES

138.906 La[57] 18,9,2	140.12 Ce[58] 20,8,2	140.908 Pr[59] 21,8,2	144.24 Nd[60] 22,8,2	(145) Pm[61] 23,8,2	150.4 Sm[62] 24,8,2	151.96 Eu[63] 25,8,2	157.25 Gd[64] 25,9,2	158.925 Tb[65] 27,8,2	162.50 Dy[66] 28,8,2	164.930 Ho[67] 29,8,2	167.26 Er[68] 30,8,2	168.934 Tm[69] 31,8,2	173.04 Yb[70] 32,8,2	174.97 Lu[71] 32,9,2

† ACTINIDE SERIES

(227) Ac[89] 18,9,2	232.038 Th[90] 18,10,2	(231) Pa[91] 20,9,2	238.029 U[92] 21,9,2	(237) Np[93] 23,8,2	(244) Pu[94] 24,8,2	(243) Am[95] 25,8,2	(247) Cm[96] 25,9,2	(247) Bk[97] 26,9,2	(251) Cf[98] 28,8,2	(254) Es[99] 29,8,2	(253) Fm[100] 30,8,2	(256) Md[101] 31,8,2	(253) No[102] 32,8,2	(257) Lw[103] 32,9,2

The properties of elements are periodic functions of the atomic number.

The periodic table is most easily understood in terms of the electron configurations of the elements. The atomic number of each element corresponds to the number of electrons in a neutral atom of that element. The element's position in the table depends on the energy level of its last several electrons. For example, the electron configuration of sulfur, atomic number 16, is

$$1s^2 2s^2 2p^6 3s^2 3p^4$$

The last 4 electrons are in the $3p$ level in the third period. Thus, sulfur occupies the fourth position in the $3p$ level in the third row of the periodic table. In each box of the table the element's atomic number, symbol, atomic weight (or most stable isotope, if the element is not found in nature), and number of electrons in the outermost shells are given. Thus, for sulfur, there are six electrons in the outer, unfilled shell; two are in the $3s$ subshell, and four are in the $3p$ subshell. As another example, the electron configuration of $_{76}$Os is

$$1s^2 2s^2 2p^6 3s^2 3p^6 4s^2 3d^{10} 4p^6 5s^2 4d^{10} 5p^6 6s^2 4f^{14} 5d^6$$

It has 2 electrons in its outer, or sixth shell, 14 electrons in its fifth shell, and 32 electrons in its fourth shell. So it has the numbers 32, 14, 2 under its symbol in the periodic table.

The rows of the periodic table are called **periods**, and they correspond to the electron periods given in Table 11-6a. The obvious grouping of columns into the first 2 on the left side, the middle 10, and the last 6 on the right side, correspond to the s, d, and p energy levels, which can contain a maximum of 2, 10, and 6 electrons, respectively. Two groups of 14 elements each corresponding to the f levels, which hold a maximum of 14 electrons, are placed separately at the bottom of the table.

12-3
CLASSIFICATION OF ELEMENTS

By examining the periodic table, we can classify the elements into four broad categories:

1. *Elements in which the s and p subshells are completely filled with electrons.* Each of these elements has electrons completely filling up all the subshells in a period. They are helium, with 2 electrons in the $1s$ level, and neon, argon, krypton, xenon, and radon, each of which has 8 electrons in the outer shell—2 in the s subshell and 6 in the p subshell. These elements are in the column at the far right labeled 0 and normally occur as gases. Only rarely do they react with other elements. They are called the **noble gases.**

2. *Elements in which an s or p subshell is incomplete.* The s and p subshells together contain a maximum of 8 electrons. Groups of elements which have their last electrons in one of these subshells are called **representative elements.** They are labeled according to the number of electrons in their s and p subshells. Elements in columns labeled IA, IB, IIA, IIB, IIIA, IVA, VA, VIA, and VIIA are included in this category. The Roman numeral indicates the number of s and p electrons in the outer shell. Columns IB and IIB are distinguished from IA and IIA as follows: Elements in columns IB and IIB have filled d subshells in the last period while elements in columns IA and IIA have empty d subshells in the last period.

3. *Elements in which the last electrons fill the d subshell in the period.* These elements are placed in the middle columns of the periodic table. The columns are labeled IIIB, IVB, VB, VIB, VIIB, and VIII. The Roman numeral designates the number of s and d electrons in the period. (The exceptions to this are the three columns labeled VIII. These have from 8 to 10 electrons in the s and d levels.) The elements in this category are

called **transition elements.** Sometimes column IB or both columns IB and IIB are included in this category.

4. *Elements in which the last electrons are in the f subshell.* These elements are placed in two rows below the main table. They are in the **lanthanide, or rare earth, series:** cerium (Ce) through lutetium (Lu), and the **actinide series:** thorium (Th) to lawrencium (Lw). These two groups of elements make up the **inner transition elements.**

Many of the properties of elements are determined by the electrons in the outermost shell. These electrons are called **valence electrons.** In any period only the *s* and *p* electrons are in the outermost shell. The *d* and *f* electrons always have a lower value for the principal quantum number, so they are in inner shells compared to the *s* and *p* electrons in the same period. As an example, the sixth period contains the levels 6*s*, 4*f*, 5*d*, 6*p*. Only the *s* and *p* electrons are in the outer, or sixth, shell. The *d* and *f* electrons are in the inner shells 5 and 4. Since only *s* and *p* electrons occupy the outermost shell, there can only be between 0 and 8 valence electrons.

As mentioned previously, it is the valence electrons which determine many of the properties of the elements. Elements tend to gain or lose valence electrons so that they will have a filled outer shell of electrons. Except for helium, elements which have one or two electrons in the *s* level tend to lose these electrons easily, so that there will be no electrons in the outer shell. Elements which have six or seven *s* and *p* level electrons tend to acquire electrons to fill up the *s* and *p* subshells. Elements with 3, 4, or 5 valence electrons either gain or lose electrons to get 0 or 8 in the outer shell.

One of the earliest classifications of elements was into metals and nonmetals. This classification was originally due to certain distinctive properties. Our modern definition is that a **metal** is an element which tends to lose its valence electrons, while a **nonmetal** is an element which tends to gain electrons to complete an outer shell. The large majority of elements are metals. The reason for so many metals can be readily seen in terms of electron structure. When the last electrons in the electron configuration occupy *d* and *f* levels, the electrons go into inner shells. The elements in which this happens (those in the transition series and inner transition series) thus have only one or two *s* electrons in their outermost shells. Therefore, all of these elements tend to lose their valence electrons, so they are metals. Only the elements on the far right of the periodic table can be nonmetals. The actual dividing line between metals and nonmetals cuts through the periodic table like a staircase. It is shown as a heavy line on the right side of Table 12-2a.

The physical properties of metals and nonmetals can be summarized as follows:

Metals

1. Good conductors of heat and electricity
2. Malleable—capable of being beaten into thin sheets
3. Ductile—capable of being stretched into wire
4. Of metallic luster
5. Opaque as a thin sheet (translucent if extremely thin)
6. Solids at room temperature. The one exception is mercury, Hg.

Nonmetals

1. Poor conductors of heat and electricity
2. Brittle—if a solid
3. Nonductile
4. Not of a metallic luster
5. Transparent as a thin sheet
6. Solids, liquids, or gases at room temperature.

The chemical properties of metals and nonmetals can be summarized as follows:

Figure 12-3a *Periodic arrangement of the metals, metalloids, and nonmetals.*

Metals

1. Have one to four electrons in their outer energy shell
2. Lose their outer electrons easily
3. Form hydroxides that are basic
4. Are good reducing agents
5. Are electropositive—they tend to become positive charges in compounds.

Nonmetals

1. Have four to eight electrons in their outer energy shell
2. Gain electrons readily
3. Form hydroxides that are acidic
4. Are good oxidizing agents
5. Are electronegative—they tend to become negative charges in compounds.

One of the first ways in which metals and nonmetals were distinguished was by their chemical properties. When an element combines with oxygen, the resulting compound is called an **oxide.** It was discovered early that metal oxides react with water to form acids. (Acids will be defined and discussed in Chapter 15.) In general, the metallic character of the elements increases as you go down and to the left of the periodic chart. The nonmetallic properties increase upward and to the right. Thus, francium is the most metallic element and fluorine the most nonmetallic. Some elements display properties of both metals and nonmetals. Elements which exhibit this hybrid behavior are called **metalloids.** Five of these elements are found on the border between the metals and nonmetals. They are boron, silicon, germanium, arsenic, and tellurium. The periodic arrangement of the metals, nonmetals, and metalloids is shown by contrasting shades of color in Fig. 12-3a.

The elements can also be classified according to whether they are solids, liquids, or gases at room temperature and pressure. Only two, bromine and mercury, occur as liquids. Eleven occur as gases. They are hydrogen, nitrogen, oxygen, fluorine, chlorine, and the six noble gases. All the rest are solids.

12-4
FAMILIES OF ELEMENTS

A row of the periodic table is called a period. The elements in a column of the table are said to be in the same group, or family, of elements. All the elements in a family have similar electron configurations; therefore, they have similar properties. In particular, if one element in a family reacts with a given substance, then all the elements in the family will react in an analogous manner with that substance. The molecular formulas of the compounds produced will also be similar. In this section we will discuss several of these families of elements.

The Noble Gases

The noble gases and their electron configurations are

$_2$He:$1s^2$

$_{10}$Ne:$1s^22s^22p^6$

$_{18}$Ar:$1s^22s^22p^63s^23p^6$

$_{36}$Kr:$1s^22s^22p^63s^23p^64s^23d^{10}4p^6$

$_{54}$Xe:$1s^22s^22p^63s^23p^64s^23d^{10}4p^65s^24d^{10}5p^6$

$_{86}$Rn:$1s^22s^22p^63s^23p^64s^23d^{10}4p^65s^24d^{10}5p^6$

$$6s^24f^{14}5d^{10}6p^6$$

Each of the noble gases contains complete periods of electrons; and except for helium, they all contain eight electrons in their outermost shells. They are monatomic; that is, they exist as single atoms in nature. This group of elements almost never reacts with other elements to form compounds. We can conclude, then, that electron configurations with completed periods, or eight electrons in the outer shell, are quite stable. This conclusion is of great importance when considering the stability of other atoms. All atoms, except those of hydrogen and helium, tend to accept, donate, or share electrons with other atoms so as to have zero or eight electrons in their outer shell.

The noble gases were once called inert gases because they did not react. However, in recent years chemists have caused them to react in certain circumstances. Hence, the terminology inert gases no longer applies. The noble gases are sometimes called the rare gases because of their scarcity. Helium can be obtained from natural gas, and radon is a radioactive byproduct of radium's decay; but the other noble gases are found only in the air. Their chief use is in "neon signs." These are made by putting minute amounts of various rare gases into a sealed glass vacuum tube. When an electric current is passed through the tube, the gases glow. The particular color of the glow depends on the mixture of rare gases used.

Because the noble gases have similar electron configurations, many of their physical and chemical properties are similar. In Table 12-4a some of their

Table 12-4a Physical Properties of the Noble Gases

Element	Symbol	Atomic Number	Atomic Weight	Melting Point (°C)	Boiling Point (°C)	Atomic Radius (Å)
Helium	He	2	4.00260	−272	−268.9	0.93
Neon	Ne	10	20.179	−248.6	−246.1	1.12
Argon	Ar	18	39.948	−189.3	−186	1.54
Krypton	Kr	36	83.80	−157.2	−153.4	1.69
Xenon	Xe	54	131.30	−111.0	−108.1	1.90
Radon	Rn	86	(222)	−71	−62	2.2

physical properties are summarized. From this table we see that the melting points, boiling points, and atomic radii increase with increasing atomic number. However, the important fact here is that they have similar melting points and boiling points, particularly when compared to other atoms in the periodic table. The atomic radii will be discussed in Section 12-5.

The Alkali Metals

The elements in the far left column of the periodic table, except for hydrogen, are called the **alkali metals.** Hydrogen is omitted because of its unique spot in the table. It is the only element which can achieve a closed outer shell by either gaining or losing an electron. The electron configurations of the alkali metals all end with one s electron in the outer shell. Thus, they have one valence electron. The alkali metals tend to lose this outer electron quite easily so that a stable, closed-shell electron configuration will be reached. Thus, they react readily with other elements. The elements sodium and potassium are abundant in the earth's crust, but lithium, rubidium, and cesium are rare. Francium which is radioactive is very rare. The most common compound containing an alkali metal is table salt, NaCl. Other alkali metal compounds known to the ancients were potash (potassium carbonate, K_2CO_3) and washing soda (sodium carbonate, Na_2CO_3).

The formulas and names of some common stable compounds of sodium are

NaCl	Sodium chloride (common table salt)
NaOH	Sodium hydroxide (caustic soda)
Na_2CO_3	Sodium carbonate (washing soda)
$NaHCO_3$	Sodium hydrogen carbonate (baking soda; bicarbonate of soda)
Na_2O	Sodium oxide

By knowing these compounds of sodium and the fact that all the elements in a family produce similar compounds, we automatically know a great deal about the compounds of the other alkali metals. Thus, we expect potassium oxide to have the molecular formula K_2O. We do not expect it to be KO, KO_2, KO_3, K_2O_5, and so on. Similarly, lithium carbonate should be, and is, Li_2CO_3.

Some of the physical properties of the alkali metals are given in Table 12-4b. The concept of families can be used to estimate the melting and boiling points (at present unknown) of the radioactive element francium. We expect all the alkali metals to have similar properties, so we would guess from Table 12-4b that francium's melting and boiling points should be about 20°C and 680°C, respectively. These guesses, of course, are not exact but should be close.

Table 12-4b *Physical Properties of the Alkali Metals*

Element	Symbol	Atomic Number	Atomic Weight	Melting Point (°C)	Boiling Point (°C)	Atomic Radius (Å)
Lithium	Li	3	6.941	180	1326	1.23
Sodium	Na	11	22.9898	97.5	889	1.57
Potassium	K	19	39.102	63.4	757	2.03
Rubidium	Rb	37	85.4678	38.8	679	2.16
Cesium	Cs	55	132.9055	28.7	690	2.35
Francium	Fr	87	(223)	—	—	2.7

Table 12-4c *Physical Properties of the Halogens*

Element	Symbol	Atomic Number	Atomic Weight	Melting Point (°C)	Boiling Point (°C)	Atomic Radius (Å)	Physical State
Fluorine	F	9	18.9984	−218	−188	0.72	Gas
Chlorine	Cl	17	35.453	−101	−34	0.99	Gas
Bromine	Br	35	79.90	−7.3	58.8	1.14	Liquid
Iodine	I	53	126.9045	114	184	1.33	Solid
Astatine	At	85	(210)	—	—	1.40	Solid

The Halogens

The elements of the periodic table in the column next to the far right are called the **halogens.** They have seven electrons in the outer s and p subshells and need only one more electron to achieve a closed-shell electron configuration. Thus, they tend to react quite strongly with other elements and so are not found free in nature. The reactivity of the halogens decreases as you go down the periodic table. Generally, the halogens are all classified as nonmetals. Fluorine and chlorine occur at room temperature as gases, bromine as a liquid, and iodine and astatine as solids. The type of iodine, found in a medicine cabinet is not the pure solid element. It is a tincture—solid iodine dissolved in a denatured alcohol. Some basic properties of the halogens are given in Table 12-4c.

Fluorine is the most reactive of all the elements. It corrodes even platinum, a metal that withstands most other chemicals. In a stream of fluorine gas, wood, rubber, and even water burst into flame. Fluorine is a pale-yellow, highly poisonous gas. it was responsible for the deaths of several very able chemists before it was finally isolated. Chlorine is also a poisonous gas. It is used as a disinfectant in swimming pools and as a purifying agent in public water supplies. About one part chlorine is used per one million parts water. Bromine is a foul-smelling poisonous liquid. It is also used as a disinfectant. Iodine is a blue-black solid. Most table salt is now "iodized" to supplement the human diet; an iodine deficiency causes thyroid trouble. Some formulas and compound names for the halogens are

NaI	Sodium iodide
$AlCl_3$	Aluminum chloride
SnF_4	Stannic fluoride
PCl_5	Phosphorus pentachloride
HBr	Hydrogen bromide
$CaCl_2$	Calcium chloride

By analogy with these formulas, we can write the correct formulas of many other stable compounds, for example, NaBr, AlF_3, and $SnCl_4$.

The Alkaline Earths

The **alkaline earths** are found in column IIA of the periodic table. They contain two valence electrons; hence, they are metals. They are quite active chemically, but they are not as reactive as the alkali metals. They also have higher melting points and are harder and stronger than their neighbors in column IA. Some of the properties of the alkaline earths are given in Table 12-4d. Calcium is used by the body of

Table 12-4d *Physical Properties of the Alkaline Earths*

Element	Symbol	Atomic Number	Atomic Weight	Melting Point (°C)	Boiling Point (°C)	Atomic Radius (Å)
Beryllium	Be	4	9.01218	1,283	1,500	0.89
Magnesium	Mg	12	24.305	650	1,120	1.36
Calcium	Ca	20	40.08	850	1,490	1.74
Strontium	Sr	38	87.62	790	1,348	1.91
Barium	Ba	56	137.34	704	1,638	1.98
Radium	Ra	88	(226)	700	1,500	—

some mammals in the formation of bones and teeth. When radioactive strontium-90 is ingested, it goes into bone marrow because of its similarity to calcium. If enough strontium-90 is ingested it can destroy the bone marrow or perhaps cause cancer. Magnesium is used in making lightweight metal alloys. Beryllium has a similar use. Some typical alkaline earth compounds are

$CaCO_3$	Calcium carbonate
SrO	Strontium oxide
$Mg(OH)_2$	Magnesium hydroxide
$BaSO_4$	Barium sulfate
$BeCl_2$	Beryllium chloride

There are other groups of elements which, because of similar group characteristics, can be classified as families. We have considered only four as examples. As mentioned previously, one element, hydrogen, is unique. It has one valence electron and lacks one valence electron. It acts sometimes like an alkali metal, forming HCl, H_2S, etc. (similar to NaCl, Na_2S) and sometimes like a halogen, forming NaH, CaH_2, etc. (compare with NaCl, $CaCl_2$). It is the lightest of the elements. At room temperature, it is a colorless, odorless, highly flammable gas. It has a melting point of $-259°C$ and a boiling point of $-252.7°C$.

12-5
PERIODIC CHARACTERISTICS

In this section the periodic characteristics of several properties of the elements will be discussed. One of these is the size of the atoms that compose the elements. Figure 12-5a is a plot of the atomic radius *versus* atomic number of elements in the periodic table. Generally speaking the atomic radii decrease from left to right in a period. Notice the large radius of each alkali metal, with respect to the other elements of that period. This is because the outer electron of an atom of an alkali metal is loosely bound to the nucleus. In a family of elements, the atomic radii increase from top to bottom. Each successive element of the family or group has an additional energy level containing electrons. As the charge on the nucleus increases (more protons), the outer electrons are bound more firmly, thus decreasing the atomic radius. This continues until the outer shell of an element is filled. The radius then jumps to a much higher value, since the next element has an electron in a new outer shell.

When an atom gains or loses electrons it acquires a net electric charge. The net electric charge is the number of protons minus the number of electrons. For example, if there are 8 protons and 10 electrons, the net electric charge is minus 2, written 2−. An

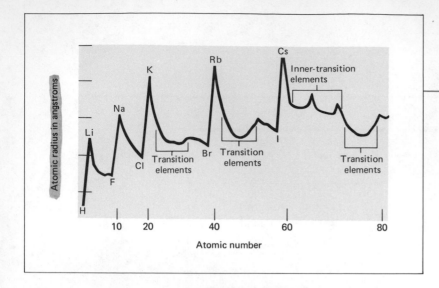

Figure 12-5a *Plot of the relationship between the atomic radius and the atomic number of elements in the periodic table. At the beginning of each period the radius is greatly increased. This accounts for the relatively large radii of the alkali metals.*

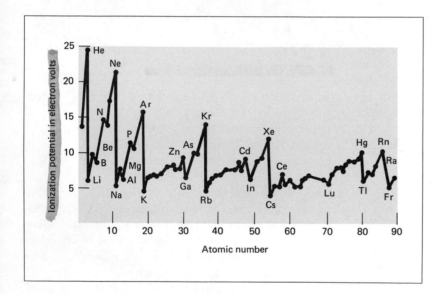

Figure 12-5b *Plot showing the relationship between the first ionization potential and the atomic number for the first 90 elements in the periodic table.*

atom or group of atoms with a net electric charge is called an **ion.** Consider magnesium, atomic number 12. It normally has 12 protons and 12 electrons for a net electric charge of zero. When it loses its two outer (valence) electrons, it has 12 protons and 10 electrons for a net charge of 2+. The net charge on the atom is sometimes called the **valence.** The valence also can be thought of as the number of electrons an atom can give up (or acquire) to achieve a filled outer shell. Magnesium has a valence of 2+. Recall that magne-

sium has two valence electrons also. The valence of positive ions is frequently equal to the number of valence electrons. The negative charge of an ion can also be an indication of valence. Consider oxygen with 8 protons and 10 electrons, for a net charge of 2−. Thus, oxygen has a valence of 2−. Note that when an element gains or loses electrons it does not lose its chemical identity. This is because it is the atomic number, not the number of electrons, which determines the identity of the element.

The amount of energy necessary to remove one electron from an atom is called its **ionization potential.** The first ionization potential is the energy required to remove the most loosely held electron. The second ionization potential is the energy to remove a second electron, and so on. The periodic nature of the first ionization potential is shown for the first 90 elements in Fig. 12-5b. Notice that the alkali metals have the lowest first ionization potentials. This is because their last electron is in an outer shell, and it does not take much energy to remove the electron. Elements located to the right of the alkali metals in the table have more electrons added in the same shell when protons are added to the nucleus. The added protons bind the electrons more and more strongly until the shell is completely filled. In general, the ionization potential increases from left to right in the periodic table. There are a few exceptions. The noble gases have the highest ionization potentials.

IMPORTANT WORDS AND TERMS

Matter	Lanthanide series	Family
Element	Rare earth series	Monatomic
Molecules	Actinide series	Rare gases
Formula	Inner transition elements	Alkali metals
Periodic table	Valence electrons	Halogens
Periodic law	Metal	Alkaline earths
Periods	Nonmetal	Ion
Noble gases	Oxide	Valence
Representative elements	Metalloid	Ionization potential
Transition elements	Group	

LEARNING OBJECTIVES

After reading and studying this chapter, you should be able to do the following with the use of the periodic table:

1. Name the symbol for each element given in the periodic table and specify the element's atomic number, atomic weight, atomic mass, proton number, neutron number, electron number, and valence.
2. Name the elements in any family or group of elements.
3. Name the elements in any period.
4. Name the four main classifications of elements and specify the elements in each classification.
5. State the number of electrons located in the main shell of each atom.
6. Write the electron configuration for each element listed in the table.

7. State the phase of each element when the element is at normal atmospheric pressure and temperature.
8. Name the elements that are metals, nonmetals, or metalloids.
9. State the relative size of one atom in respect to another atom located in the same group or family.
10. Arrange a given number of atoms in order of increasing ionization potential.
11. Write the electronic configuration for any atom.
12. Define or explain the important words and terms at the end of the chapter without referring to the text.

QUESTIONS

1. Distinguish between an atom and an element.
2. Explain why there are many more different kinds of atoms when we know of only 105 different kinds of elements.
3. What was the view held by the Greek philosopher Aristotle concerning the composition of matter?
4. State the differences between deuterium and ordinary hydrogen.
5. Distinguish between atomic number and atomic weight.
6. What is the modern basis for the periodic law?
7. Looking at the periodic table, we notice that there are four major groupings of columns. There are 2 columns on the left, 10 in the middle, 6 on the right, and 14 below. What causes these groupings of 2, 10, 6, and 14 columns?
8. Define the terms metal and nonmetal.
9. What are metalloids? Give three examples.
10. State some physical properties that distinguish metals from nonmetals.
11. State some chemical properties that distinguish metals from nonmetals.
12. What are the two most abundant metals?
13. On which isotope are all atomic weights based?
14. What determines the properties of elements?
15. List the elements that are gases and liquids at room temperature and atmospheric pressure.
16. List several elements that do not react readily. List several elements that are highly reactive.
17. A search is on for the element with atomic number 114. Which element will it most resemble?
18. Explain why the alkali metals have larger radii than the elements with one or two, more or less, protons.
19. How would you expect the ionization potential and atomic radius of francium to compare with the other alkali metals?
20. State the group number of each of the following elements: (a) scandium, (b) lead, (c) niobium, (d) hahnium, (e) uranium.
21. State the period of each of the following elements: (a) helium, (b) zinc, (c) radium, (d) europium, (e) plutonium.

PROBLEMS

1. Using the periodic table, determine the atomic number, proton number, atomic weight, atomic mass, neutron number, and number of valence electrons of each of the following elements and state whether each is a metal, or nonmetal, or an inert gas: (a) lithium, (b) neon, (c) calcium, (d) chlorine, (e) niobium, (f) gold, (g) holmium, (h) uranium.

2. Arrange the elements cesium, lithium, potassium, rubidium, and sodium according to increasing atomic radii.

3. Arrange the elements argon, helium, krypton, and neon according to decreasing ionization potential.

4. Which atom has the largest radius in each of the following pairs of elements: sodium or potassium, lead or nickel, polonium or selenium, argon or xenon?

5. Classify the following elements as noble gas, representative element, transition element, or inner transition element:
(a) krypton, (b) iron, (c) iodine, (d) strontium, (e) hydrogen, (f) helium, (g) nitrogen, (h) mercury, (i) uranium.

6. Determine the most common valence of each of the elements in problem 5.

7. Arrange the atoms, cobalt, helium, cesium, tin, and chlorine in order of increasing atomic radius.

8. Arrange the atoms helium, magnesium, rubidium, zinc, and xenon in order of increasing ionization potential.

9. Write the electron configuration for
(a) The first transition element scandium (Sc).
(b) The second transition element yttrium (Y).

10. Write electron configurations for the coinage metals—copper, silver, and gold. Consult the periodic table for help. (Note: These configurations are exceptions to the rules of Chapter 11 for writing configurations. The three coinage metals only have one valence electron.

Chapter 13

COMPOUNDS, MOLECULES, AND IONS

*Happy the man who, studying nature's
  laws,
Through known effects can trace the secret
  cause.*

Virgil

There are 105 known elements. In the previous chapter, they were arranged in order of increasing atomic number to form the periodic table. With the use of the table, we learned many of the properties of these known elements.

Matter found in our environment is usually in the form of mixtures. A **mixture** is a nonchemical combination of two or more substances of varying proportions. When we look around, we see examples of mixtures, such as ocean water, soil, concrete, plants, and animals. Our environment is composed of millions of compounds and mixtures. In this chapter we lay the foundations for compound formation and study the properties of some of the compounds that make up our world. A **compound** is a substance composed of two or more elements chemically combined in a definite proportion. As you now know, an element is made up of only one kind of atom. Some compounds are made up of **ions**, others of molecules. We stated in Chapter 12 that an ion is an atom or a chemical combination of atoms with a net electric charge. A **molecule** is an uncharged particle of an element or a compound. The properties of many ions and molecules play an important role in our environment. Some molecules, such as the nitrogen oxides, sulfur dioxide, carbon dioxide, and carbon monoxide pollute the air we breathe. Others play a vital role in our being alive. For instance, we will see in Chapter 21 that the ozone, carbon dioxide, and water molecules in the air absorb much of the sun's energy, thus keeping the earth at a liveable temperature. Similarly, the various minerals and rocks that we find in our earth obtain their properties from those of the various molecules and ions that constitute them.

The formation of ions and molecules is due to the electromagnetic forces between all the various electrons and the nuclei of the atoms involved. It is a very complicated problem, because there are so many forces to consider. Modern chemists use large computers and the techniques of quantum mechanics to try to understand how molecules and ions form. Although there is still a lot to be understood, chemists are now able to produce new compounds having almost any desired property. We begin our study with some of the basic principles of compound formation and the properties of some of the different types of compounds.

13-1
PRINCIPLES OF COMPOUND FORMATION

The noble gases, with their outer shells filled with electrons, are said to have closed shell electron configurations. The noble gases are monatomic, and it is extremely difficult to make them combine with any of the other elements. From these facts we can conclude that closed shell configurations result in extremely stable substances. Another important fact to remember in compound formation (see Section 12-4) is that elements in the same family form similar compounds. Since all the elements in the same family have an equal number of valence electrons, we conclude that it is the outer electrons which are important in compound formation. We now have a justification for the two basic assumptions which are made to explain compound formation.

1. The only electrons which take part in compound formation are those in the outer shell.
2. Compounds are formed when each atom in the molecule or ion achieves a closed shell electron configuration.

Both of these assumptions, however, are violated at various times in compound formation. In the transition and inner transition elements, some of the inner d or f shell electrons sometimes take part in compound formation. This violates the first assumption. There are also exceptions to the second assumption. Occasionally, a compound is formed in which one or more of the atoms does not have a closed shell configuration. However, since most compounds do obey our basic assumptions we will use them and neglect the few exceptions which occur.

Except for elements in the first period (hydrogen and helium), there can be a maximum of eight electrons in the outer shell of elements in every period. In order for an atom to achieve a noble gas (closed shell) configuration, it must have either eight or zero valence electrons. Atoms achieve eight or zero valence electrons by gaining, losing, or sharing electrons with other atoms. The concept that atoms tend to have eight electrons in their outer shell when forming molecules or ions is called the octet rule. Hydrogen, of course is the most important exception to this rule. It must have two electrons in its outer shell when it combines with other atoms. Other exceptions to the octet rule occur with elements in the two transition groups. However, in the remaining chapters we shall be dealing almost exclusively with the representative elements, for which the octet rule holds, generally.

There are two ways in which individual atoms can achieve a closed shell electron configuration. They can transfer electrons or they can share electrons. In the transfer of electrons one or more elements lose some or all of their outer electrons, and another one or more more elements gain these same electrons to achieve closed shell configurations. Compounds which are formed by this electron transfer process are called **ionic compounds.** Because these compounds in the liquid phase conduct electricity, they are sometimes referred to as **electrovalent compounds.** The majority of compounds are formed when electrons are shared between different atoms. Compounds formed by the electron sharing process are called **covalent compounds.** Now let us examine these two kinds of compounds.

13-2
IONIC COMPOUNDS

Compounds which are formed when electrons are lost or gained are referred to as ionic compounds. This is because when the electrons are lost or gained by the various atoms, ions are formed. The energy released when an atom gains an electron, forming a negative ion, is called **electron affinity.** The nonmetals have high electron affinities and the metals have low electron affinities. Thus only the nonmetals tend to form negative ions by gaining electrons. Atoms that need only one or two electrons to fill their outer shell can easily acquire the electrons from atoms that have low ionization energy, i.e. atoms with only one or two electrons in the outer shell such as those in groups IA and IIA of the periodic table. To illustrate this, let us consider how common table salt, NaCl, is formed. The sodium atom can be drawn schematically with two electrons in the first shell, eight in the second, and one in the third. The chlorine atom has two in the first, eight in the second, and seven in the third. Diagrams for these two atoms are shown in Fig. 13-2a. The sodium atom tends to lose its outer electron so that it will have zero electrons in its third shell. The chlorine atom tends to gain an electron so that it will have eight electrons in its outer shell. When the sodium atom loses its electron, it becomes a sodium ion with 11 positive charges and 10 negative charges, for a net electric charge of $1+$. The chlorine atom gains the electron lost by the sodium atom and

Figure 13-2a *Schematic diagrams of the two atoms that comprise common table salt. The sodium atom has 11 positive charges in its nucleus, the chlorine atom 17. Each small circle represents a negatively charged electron orbiting in the shells. Charge on the neutral atoms is zero.*

becomes a chlorine ion with 17 positive charges and 18 negative charges. It has a net charge of $1-$. The positively charged sodium ion is then attracted to the negatively charged chlorine ion. This situation is illustrated in Fig. 13-2b.

Figures 13-2a and 13-2b can be represented very easily, using what is known as the electron-dot notation. In this notation the nucleus and inner electrons of an element are represented by the element symbol and the valence electrons by dots. The sodium and chlorine atoms of Fig. 13-2a would be represented in the electron dot notation as

$$\text{Na} \cdot \qquad : \overset{..}{\underset{.}{\text{Cl}}} :$$

The sodium and chlorine ions of Fig. 13-2b would be represented as

$$\text{Na}^+ \qquad : \overset{..}{\underset{..}{\text{Cl}}} : ^-$$

The charge of the ion is represented by the number of plus or minus signs. An ion with two negative charges

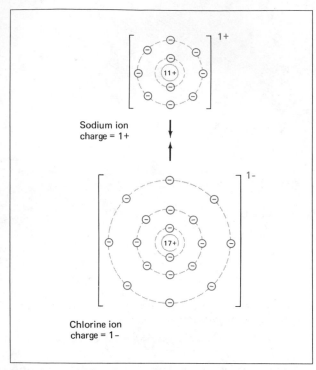

Sodium ion
charge = 1+

Chlorine ion
charge = 1−

Figure 13-2b *Diagram illustrating the formation of table salt, NaCl. The oppositely charged ions are attracted to each other.*

is sometimes designated by a double minus sign ($=$) but most often by $^{2-}$. Consider the following examples:

$$Mg^{++} \text{ or } Mg^{2+}$$

$$:\!\overset{..}{O}\!:= \quad \text{or} \quad :\!\overset{..}{O}\!:^{2-}$$

The charge on most ions of the representative elements can easily be determined as follows: The positive charge will be equal to the number of valence electrons. The negative charge will be eight minus the number of valence electrons. For example, aluminum, in column IIIA of the periodic table, has three valence electrons. The aluminum ion thus will have a charge of $3+$. Sulfur, with six valence electrons, tends

to fill its outer shell. Its ion will have a negative charge of $8 - 6$, or a charge of $2-$. Some common examples of atoms and their ions in the electron dot notation follow:

Atom	Ion	Atom	Ion
H·	H⁺	Na·	Na⁺
Li·	Li⁺	Mg:	Mg²⁺
Be:	Be²⁺	Al:	Al³⁺
B:	B³⁺	·P:	P⁵⁺
·N:	N⁵⁺ or :N:³⁻	:S:	:S:²⁻
:O:	:O:²⁻	:Cl:	:Cl:⁻
·F:	:F:⁻		

We have not included elements in column IVA of the periodic table because these elements share electrons and thus form covalent rather than ionic compounds.

In forming simple ionic compounds one element loses its electrons and the other element gains them, resulting in ions. The ions are then held together by the electrical attraction between them. The formation of NaCl can be represented as

$$Na· \; + \; :\!\overset{..}{\underset{..}{Cl}}\!: \; \longrightarrow \; Na^+ :\!\overset{..}{\underset{..}{Cl}}\!:^{-}$$

In this example, the sodium atom lost one electron and the chlorine atom gained one. Thus, we see that for every ion of sodium there is one ion of chlorine. This one-to-one correspondence is necessary to balance the charges on the ions. Another example of an ionic compound is calcium oxide, CaO. Its formation is represented by

$$Ca: \; + \; :\!\overset{..}{O}\!: \; \longrightarrow \; Ca^{2+} :\!\overset{..}{\underset{..}{O}}\!:^{2-}$$

In forming this compound the calcium atom has lost two electrons and the oxygen atom has gained two. Thus, to get a neutral compound, we need one ion of each element. Now, consider what happens with calcium and chlorine. Calcium has two electrons to lose, but each atom of chlorine can gain only one. There must be two atoms of chlorine to accept both of the electrons and still maintain a neutral charge on the compound. We have

$$Ca: + :\overset{..}{\underset{..}{Cl}}: + :\overset{..}{\underset{..}{Cl}}: \longrightarrow :\overset{..}{\underset{..}{Cl}}:^- Ca^{2+} :\overset{..}{\underset{..}{Cl}}:^-$$

All of the atoms in the compound have closed shells and the total charge is zero. The molecular formula of calcium chloride is $CaCl_2$. We are now beginning to see how molecular formulas arise. The numbers of atoms of the various elements involved in the compound are determined by the requirements that the total charge must be zero and that all the atoms are to have closed shell electron configurations. As another example, consider sodium and sulfur. Each sodium atom loses one electron and each sulfur gains two, so there must be two sodium atoms for every sulfur atom.

$$Na\cdot + Na\cdot + :\overset{..}{\underset{.}{S}}: \longrightarrow Na^+ :\overset{..}{\underset{..}{S}}:^{2-} Na^+$$

The formula for the compound formed, sodium sulfide, is thus seen to be Na_2S.

13-3
PROPERTIES OF IONIC COMPOUNDS

Ionic compounds are formed by the electrical attraction of the ions. These compounds generally have higher melting and boiling points than covalent compounds. Ionic compounds occur in the solid phase as crystals. A **crystal** is an orderly arrangement of atoms. A crystal of NaCl consists of one ion of sodium for every one of chlorine. Part of a sodium chloride, or table salt, crystal is shown in Fig. 13-3a. It can be seen from the figure that each ion has six ions of the opposite charge surrounding it in a three-dimensional array. Of course, Fig. 13-3a is simply a sketch of part of the crystal. An individual crystal has a large number of Na^+ and Cl^- ions. The NaCl crystal shown in

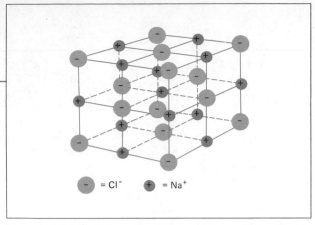

Figure 13-3a *Schematic diagram of the sodium chloride crystal. The sodium ions are the smaller spheres with the plus charge indicated. The chloride ions have one more shell so they are drawn as bigger spheres with the minus charge indicated.*

Fig. 13-3a is an example of a cubic crystal; there are many other possible crystal arrangements. The cubic nature of the NaCl crystal can be seen by looking closely at a grain of ordinary table salt.

When we write the formula NaCl, it means that in the compound sodium chloride, there is one ion of Na^+ to one ion of Cl^-. On a microscopic scale it is impossible to associate any one ion of Na^+ with a particular ion of Cl^-. However, there will always be a one-to-one correspondence between the Na^+ and Cl^- ions. Similarly, the formula $CaCl_2$ means that, in the compound calcium chloride, there are two chlorine ions for every calcium ion.

One of the more important properties of ionic compounds is their behavior when an electric current is passed through the compound in its molten phase. This behavior is illustrated as follows: If a light bulb is connected in series with a battery, it lights up. This is due to electrons flowing from the negative terminal of the battery, through the light bulb, and back to the positive terminal. If the wires are cut, the electrons cannot flow (no electric current), and the light bulb will not light. Now, if the cut wires are connected to graphite (pure carbon) rods, which are inserted into a melted ionic compound, such as NaCl, the bulb lights. That is, ionic compounds in the liquid form

Figure 13-3b *Diagram illustrating how melted NaCl conducts electricity. The lighted bulb indicates the flow of an electric current. The electric circuit is achieved because of the reactions of the ions at the electrodes.*

conduct electricity. Figure 13-3b is a diagram of this situation.

The explanation of how melted ionic compounds conduct electricity is not simple. *The electrons do not flow through the liquid.* The process is more complicated. The graphite rods which were inserted into the melt are charged positively and negatively, because they are connected to the positive and negative battery terminals. These charged rods are called *electrodes.* The positively charged rod is called the *anode;* the negatively charged rod is called the *cathode.* The positively charged sodium ions are attracted to the cathode while the negatively charge chlorine ions are drawn towards the anode. For this reason a positively charged ion is called a **cation,** and a negatively charged ion is called an **anion.** When a sodium ion reaches the cathode, an electron is taken from the cathode, and the sodium ion becomes a sodium atom. The reaction which takes place at an electrode is called a **half-reaction.** The half-reaction at the cathode is represented by

$$Na^+ + e^- \longrightarrow Na$$

The net effect of this process is to take electrons off the cathode. At the anode, the negatively charged chlorine ions give up an electron to the positively charged anode. Then, two chlorine atoms form a

covalent chlorine gas molecule. These appear as bubbles on the anode. The half-reaction is

$$2\,Cl^- \longrightarrow Cl_2 + 2\,e^-$$

This half-reaction donates electrons to the anode. The electric current is thus achieved because of the reaction of the ions at the electrodes, and the light bulb lights up. To summarize, at the cathode electrons are removed; at the anode they are added. The net result is an electron flow, or current. Eventually, all the sodium and chlorine ions react. At this point the current ceases to flow, and the light goes out. An electric current will also be conducted if a solution of table salt in water is used instead of molten NaCl. Most ionic compounds dissolve in water, and the solutions will conduct an electric current in much the same way that molten NaCl does.

The half-reaction which occurs at the anode is an example of oxidation. **Oxidation** is the process in which electrons are lost. At the cathode, reduction takes place. **Reduction** is the process in which electrons are gained. The word oxidation comes from the word oxygen, since it is the material which frequently causes oxidation. For example, when pure iron (Fe) is exposed to the air, the oxygen atoms combine with the iron to form rust, which for our purpose can be designated by Fe_2O_3. The oxygen atoms gain electrons, so they will have a filled shell, and are said to be reduced. The iron atoms lose electrons. They are oxidized.

The total reaction in which electrons are gained and lost is called an **oxidation-reduction reaction.** If we add the two half-reactions in our example, we get

$$
\begin{aligned}
2\,Na^+ + 2\,e^- &\longrightarrow 2\,Na \\
2\,Cl^- &\longrightarrow Cl_2 + 2\,e^- \\
\hline
2\,Na^+ + 2\,Cl^- + 2\,e^- &\longrightarrow 2\,Na + Cl_2 + 2\,e^-
\end{aligned}
$$

We have multiplied the half-reaction at the cathode by two before comparing it to the anode half-reaction. This is done to balance the number of electrons

Figure 13-4a *Illustrating the forces between two hydrogen atoms. Particles with unlike charges attract each other; those with like charges repel each other. Each particle has three forces on it due to charges on the other three particles. Arrows indicate the forces and the directions in which they are acting.*

gained and lost. The two electrons on both sides of our oxidation-reduction reaction cancel, and we get

$$2\,NaCl \longrightarrow 2\,Na + Cl_2$$

as our final reaction. By sending an electric current into melted NaCl, sodium and chlorine ions can be converted into pure sodium metal and chlorine gas. In general, the production of an oxidation-reduction reaction by means of an electric current is called **electrolysis.** The electrolysis of sodium and many other metals is an important commercial process which produces the metals in a pure form. Electrolysis is also the means by which electroplating is accomplished. The material to be plated is used as the cathode in electrolysis and a melted ionic compound or a solution of the ionic compound in water is the source of the metal to be plated.

So far, we have discussed several properties of ionic compounds. Their high binding energies, produced by the strong electrostatic attractions between ions, causes them to have high melting points. They do not exist as single molecules, but the individual ions occur in a ratio given by the molecular formula. They do not conduct electricity in the solid form but do in the liquid form. Finally, most ionic compounds dissolve in water, and in solution they conduct electricity. The mechanism by which this dissolving process occurs will be discussed after we have learned something about the structure of the water molecule.

13-4
COVALENT COMPOUNDS

In the formation of ionic compounds electrons are transferred; in covalent compounds they are shared. Hydrogen gas, H_2, is the simplest example of a covalent compound. It contains two atoms in each molecule. Molecules containing two atoms are called *diatomic.* When two hydrogen atoms are brought together, there is an attraction between each electron and proton and repulsion between the two electrons and between the two protons. This is shown schematically in Fig. 13-4a. When this phenomenon is investigated using quantum mechanics and the Schrödinger equation, it is found that there is more attraction than repulsion up to a certain distance between atoms. When the positively charged nuclei get about an angstrom apart, they begin to repel each other. The hydrogen molecule consists of a total system of two nuclei and two electrons. Once the atoms are close together, the two electrons no longer orbit around individual nuclei. They both orbit around both of the protons, tracing out a circular path about one or the other. Of course, as was previously mentioned, the concept of electrons in orbits is somewhat naive. The quantum mechanical description is based on the probability that an electron will occupy any given spot around the nuclei. In the hydrogen molecule both electrons are shared equally by both nuclei, and it is this sharing of electrons which tends to hold the atoms together. There is said to be a **covalent bond** between the two atoms. Contrast this to an **ionic bond,** which is formed when two oppositely charged ions attract each other. In the electron-dot notation the separated hydrogen atoms are written

$$H \cdot \qquad H \cdot$$

and the H_2 molecule is written

$$H : H$$

The two dots between the hydrogen atoms indicate that these electrons are being shared.

Not all atoms will share electrons. For instance the helium atom, with a closed $1s$ shell, will not form a stable molecule with another helium atom. Two helium atoms repel each other, but two hydrogen atoms attract. The molecule H_2 is stable, but He_2 is not. It takes complicated quantum mechanical calculations to determine which molecules will be stable and which will not. However, our basic assumption serves as a good guide. Stable covalent molecules are formed when the atoms share electrons in such a way as to give all atoms a closed outer shell configuration.

Let us consider more examples of covalent bonding. Two chlorine atoms in the electron dot notation are given by

$$\ddot{\underset{..}{\text{Cl}}}\cdot \qquad \cdot\ddot{\underset{..}{\text{Cl}}}:$$

Each chlorine atom needs one electron to have a closed shell. If each shares its unpaired electron, we get the following Cl_2 molecule.

$$:\ddot{\underset{..}{\text{Cl}}}:\ddot{\underset{..}{\text{Cl}}}:$$

In the Cl_2 molecule each chlorine atom has six electrons plus two shared electrons giving it a closed outer shell.

Covalent bonds can also be formed between unlike atoms. Consider the molecule HCl. The hydrogen and chlorine atoms have the dot structures

$$\text{H} \qquad \cdot\ddot{\underset{..}{\text{Cl}}}:$$

If each shares one electron, they both have closed outer shells.

$$\text{H}:\ddot{\underset{..}{\text{Cl}}}:$$

Sometimes more than one electron is shared by each atom. Consider how carbon dioxide, CO_2, is formed. In the dot notation two oxygen atoms and one carbon atom look like

$$:\ddot{\underset{..}{\text{O}}}: \qquad \cdot\dot{\text{C}}\cdot \qquad :\ddot{\underset{..}{\text{O}}}:$$

In order for us to get a stable molecule, we need to have eight electrons around each atom. If only two electrons are shared between each oxygen atom and the carbon atom, we get the following structure:

$$:\ddot{\underset{..}{\text{O}}}:\dot{\underset{..}{\text{C}}}:\ddot{\underset{..}{\text{O}}}: \qquad \text{(Unstable)}$$

This structure is unstable because, although the carbon atom has eight electrons around it, each oxygen atom has only six electrons around it. In order to get a stable molecular structure, there must be four electrons shared between the carbon atom and each oxygen atom. The correct dot structure of the CO_2 molecule is

$$:\ddot{\text{O}}::\text{C}::\ddot{\text{O}}:$$

An ordinary single covalent bond consists of two shared electrons. A sharing of four electrons produces a double bond, and a sharing of six electrons produces a triple bond. Nitrogen gas, N_2, is an example of a triple bond. Two nitrogen atoms may be represented as

$$\cdot\dot{\text{N}}\cdot \qquad :\dot{\text{N}}\cdot$$

To satisfy the octet rule each nitrogen atom must share three electrons. The molecule N_2 looks like

$$:\text{N}:::\text{N}:$$

Another example of the triple bond is carbon monoxide, CO. Carbon and oxygen in dot notation look like

$$\cdot\dot{\text{C}}\cdot \qquad :\ddot{\text{O}}:$$

The carbon molecule needs four more electrons and the oxygen molecule needs two to satisfy the octet rule. The only way this can be accomplished is for the oxygen atom to share four of its electrons with the carbon atom and for the carbon atom to share two of its electrons with the oxygen atom. Altogether, six electrons are shared.

$$:\text{C}:::\text{O}:$$

One of the most common molecules containing three atoms is water, H_2O. The individual atoms are, of course,

$$H \cdot \quad : \ddot{O} : \quad \cdot H$$

When they combine, each hydrogen atom shares its electron with the oxygen atom, which shares two of its electrons.

$$H : \ddot{O} : H$$

The electron-dot notation is used merely to indicate how the electrons are shared among atoms, not to indicate the spatial arrangement of the atoms. We could also have written H_2O as

$$H : \ddot{O} : \quad \text{or} \quad : \ddot{O} : H \quad \text{or} \quad \begin{matrix} & \ddot{O} \\ H & H \end{matrix}$$
$$\quad \; H \qquad\qquad\qquad H$$

The actual molecule looks more like these last three. The two $O:H$ bonds form an angle of about $105°$ with each other.

The compound ammonia, NH_3, is a covalent combination of one nitrogen atom and three hydrogen atoms. Each hydrogen atom shares its electron with the nitrogen atom as follows:

$$\begin{matrix} & \ddot{} \\ H : \ddot{N} : H \\ H \end{matrix}$$

Methane gas, CH_4, is an example of a simple molecule with five atoms. its structure can be represented by

$$\begin{matrix} H \\ H : \ddot{C} : H \\ H \end{matrix}$$

More complicated electron-dot structures arise when elements other than hydrogen are involved. The halogens combine analogously to hydrogen. Consider

CH_3Cl. Its structure is just like CH_4 with a chlorine atom replacing a hydrogen atom.

$$\begin{matrix} H \\ H : \ddot{C} : \ddot{Cl} : \\ H \end{matrix}$$

Other compounds with halogens are NH_2Cl, NI_3, $CHBrCl_2$. Their dot structures are

$$\begin{matrix} & & & & : \ddot{Br} : \\ H : \ddot{N} : H & & : \ddot{I} : \ddot{N} : \ddot{I} : & & : \ddot{Cl} : \ddot{C} : H \\ : \ddot{Cl} : & & : \ddot{I} : & & : \ddot{Cl} : \end{matrix}$$

Note that in each of the preceding structures the total number of electrons for the molecule is equal to the total number of valence electrons in the atoms making up the molecule. Also note that each atom has a closed shell of electrons around it.

It is sometimes necessary for a molecule to have single, double, and even triple bonds simultaneously in order to satisfy the octet rule. The molecules C_2H_3Br and C_2HI are examples.

$$\begin{matrix} H & & \ddot{Br} \\ \ddot{C} : : \ddot{C} & & H : C : : : C : \ddot{I} \\ H & & H \end{matrix}$$

In these structures the carbon atoms are bonded with double and triple bonds, while the other elements are held to the carbon atoms by single bonds.

Frequently, atoms combine together with a net charge. Such combinations are called *complex ions*, or polyatomic ions. They are also known as radicals. The atoms within the complex ion are covalently bonded. The whole aggregation then behaves like an ion in forming compounds. There is usually a strong bonding between atoms in complex ions, so that it is difficult to break them up. In chemical reactions they frequently act as a single unit. Some important examples of complex ions are given in Table 13-4a.

The structure of the polyatomic ions of known charge can be determined by writing down the free

Table 13-4a *Some Important Complex Ions*

Ammonium	NH_4^+	Carbonate	CO_3^{2-}
Hydroxide	OH^-	Sulfate	SO_4^{2-}
Nitrate	NO_3^-	Sulfite	SO_3^{2-}
Cyanide	CN^-	Phosphate	PO_4^{3-}
Nitrite	NO_2^-		

atoms in their dot notation, adding or subtracting electrons depending on the charge on the ion, and then following the rules of compound formation. The structures of NO_3^- and PO_4^{3-} are examples.

$$\left[\begin{array}{c} \ddot{\text{O}} : : \text{N} : \ddot{\text{O}} : \\ \ddot{\text{O}} : \end{array} \right]^- \quad \left[\begin{array}{c} : \ddot{\text{O}} : \\ : \ddot{\text{O}} : \text{P} : \ddot{\text{O}} : \\ : \ddot{\text{O}} : \end{array} \right]^{3-}$$

Some compounds contain both ionic and covalent bonds. Sodium hydroxide, NaOH, is an example. Its structure is

$$\text{Na}^+ \left[: \ddot{\text{O}} : \text{H} \right]^-$$

The hydroxide ion, $[OH]^-$, is covalently bonded together as a unit. In sodium hydroxide the hydroxide unit and sodium are bound by an ionic bond. Like most ionic compounds, sodium hydroxide dissolves in water. Other examples of ionic compounds formed by complex ions are silver nitrate, $Ag^+[NO_3]^-$, and barium sulfate, $Ba^{2+}[SO_4]^{2-}$.

So far, we have classified compounds as being either ionic or covalent. In ionic bonding the electrons are transferred from one element or group of elements to another. In covalent bonding the electrons are shared. Actually, most bonds are intermediate between these two extremes. For most bonds, the electrons are unequally shared between two atoms, because the atoms are unlike. Whenever two unlike atoms share electrons, one will have a greater attraction for the shared electrons than the other. The

tendency to attract shared electrons is called **electronegativity,** the measure of the ability of an atom to attract electrons in the presence of another atom. Consider the polar molecule HCl. The chlorine atom needs an electron to fill its outer shell. The hydrogen atom also needs an electron to fill its outer shell, but if it loses its electron it will have a closed shell. The chlorine atom has a higher electro-negativity than the hydrogen. Electronegativity increases from left to right within a period. The result is that although the electrons are shared between the two atoms, they tend to spend more time at the chlorine end than at the hydrogen end of the molecule. We get a molecule which looks like

$$\overset{+}{\text{H}} : \ddot{\ddot{\text{Cl}}} : ^-$$

The + and − indicate that the chlorine end of the molecule tends to be negatively charged and the hydrogen end tends to be positively charged. A bond such as this, in which the electrons are unequally shared, is called a **polar bond.** All bonds between unlike atoms are polar bonds. Bonds between like atoms have a perfect sharing and are called **nonpolar bonds.**

Sometimes a molecule has several polar bonds arranged symmetrically. An example of this is methane, CH_4. Its structure is

$$\begin{array}{c} \overset{+}{\text{H}} \\ \overset{..}{} \\ \text{H} : \underset{..}{\overset{..}{\text{C}}} = : \overset{+}{\text{H}} \\ \overset{+}{\text{H}} \overset{..}{} \\ \text{H}+ \end{array}$$

Methane has four polar bonds, but they are arranged so that no one side or end of the molecule has a more positive or negative charge than any other side or end. Because of this symmetry it is a **nonpolar molecule.** Most molecules have one end or side positively charged and one end or side negatively charged. Such molecules are called **polar molecules.** The HCl molecule is an example of a polar molecule.

Another example of a polar molecule is water. Its structure is

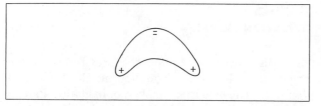

Figure 13-4b *Schematic diagram of a water molecule showing its polar character.*

In the water molecule we see that the oxygen end is negatively charged, and the hydrogen ends are positively charged. To emphasize the polar property of the water molecule it is sometimes represented as in Fig. 13-4b. In ionic solutions the negative side of the water molecule attracts positive ions, while the positive ends attract negative ions.

13-5
PROPERTIES OF COVALENT COMPOUNDS

Because of the nature of the bonding involved, covalent compounds have quite different properties from those of ionic compounds. Although the covalent bond is strong within a molecule, the various molecules in the compound only weakly attract each other. Their binding energies in the solid and liquid form are therefore not very high, and their melting points and boiling points are low compared to ionic compounds. Many covalent compounds occur as liquids or gases at room temperature. For instance, carbon tetrachloride, CCl_4, melts at $-23°C$ compared to $800°C$ for the ionic compound NaCl.

Many molecules are bonded by polar bonds. The more polar the bond, the more unequal the electron sharing. An ionic bond is actually a very polar bond. Polar molecules are partially covalent and partially ionic. In general, with the exception of the hydrogen atom, a combination of atoms from the far left and far right (ignoring the noble gases) of the periodic table form ionic compounds, for example, NaCl. Also, compounds formed from complex ions tend to be ionic. Compounds tend to be covalent if they are formed from elements from the same or adjacent columns of the periodic table, or if they are formed from any two nonmetallic atoms, such as hydrogen and nitrogen. In addition, all compounds formed with carbon are covalent.

Covalent compounds occur in the gas phase as individual molecules with a specific molecular formula. A molecule of gaseous CCl_4 consists of one carbon atom and four chlorine atoms. Ionic compounds have electric charges on the ions, and an ionic gas molecule might have any number of atoms in the ratio given by the molecular formula.

In liquid form, covalent compounds do not conduct electricity well. Benzene, a nonpolar covalent compound, does not conduct electricity at all. Pure water, a polar covalent compound, conducts electricity only very slightly. Ordinary tap water is far from pure. It usually has many ions dissolved in it, and therefore, tap water is an excellent conductor of electricity. This is why if a lightning bolt were to hit a lake or swimming pool, the occupants would be killed by the electric shock.

Water is a polar compound. When a solid ionic compound, such as table salt, is put into water, the ionic compound usually dissolves. What occurs when NaCl dissolves in water can be explained with the aid of Figs. 13-5a and 13-5b. In Fig. 13-5a, four ions of a salt crystal are shown surrounded by water molecules. In Fig. 13-5b, the positively charged ends of the water molecules are attracted to the Cl^- ions, and the negatively charged ends of the water molecules are attracted to the Na^+ ions. The water molecules surrounding the ions tend to decrease the attraction of the Na^+ and Cl^- ions for one another. This dissolves

Figure 13-5a *Schematic diagram of a NaCl crystal in water before the crystal is dissolved. A portion of the salt crystal is shown along with a few water molecules.*

Figure 13-5b *Schematic diagram of NaCl dissolved in water. The positive ends of the water molecule attract the negative chloride ions and the negative end the sodium ions. These forces act to break up the sodium chloride crystal.*

the salt crystal. Ions surrounded by water molecules are called *solvated ions.* The dissolving process, illustrated by Figs. 13-5a and 13-5b, is called *solvation.*

13-6
OXIDATION NUMBER

A useful concept in writing chemical formulas and equations is that of the oxidation number, or oxidation state. The **oxidation number,** or **oxidation state,** is essentially a measure of the electric charge which an atom in a molecule would have if it were ionically bound. When assigning oxidation numbers, the atom with the greater electron affinity is given a negative oxidation number. The atom with the lesser electron affinity is given a positive oxidation number.

The following are some general rules for assigning oxidation numbers:

1. With only a few exceptions (the peroxides: H_2O_2, Na_2O_2, etc.), oxygen always has an oxidation number of -2.
2. With the exception of the metal hydrides, hydrogen always has an oxidation number of $+1$.
3. The oxidation number of an ion is equal to the charge on the ion.
4. The oxidation number of a free element or an element combined only with itself is zero.
5. The oxidation number of the halogens is usually -1.
6. The oxidation number of the alkali metals is usually $+1$.
7. The oxidation number of the alkaline earth is usually $+2$.

Using these rules, the oxidation numbers of other elements in a neutral molecule or charged ion can be determined. The sum of the oxidation numbers of all the atoms involved in a molecule or ion must be equal to the charge on the molecule or ion. In the most basic sense, the concept of oxidation number is just a means of keeping track of the various charges involved in molecules and ions.

Some examples follow:

■ *Example 1* What is the oxidation number of Cl in HCl?

Since H has an oxidation number of $+1$, Cl has an oxidation number of -1.

■ *Example 2* What is the oxidation number of iron in Fe_2O_3?

Oxygen has an oxidation number of -2. The total negative charge on the molecule is $3 \times (-2) = -6$. Since the molecule is neutral, there must be six positive charges. There are two iron atoms so each must have $+3$. For the total charge on the molecule we get $2 \times (+3) + 3 \times (-2) = 0$.

■ *Example 3* What is the oxidation number of phosphorus in the $(PO_4)^{3-}$ ion?

The oxidation number of oxygen is -2. There are four oxygen atoms so there are $4 \times (-2) = -8$ negative charges. The total charge is -3, so phosphorus must have an oxidation number of $+5$. We get $+5 + [4 \times (-2)] = -3$.

■ *Example 4* What is the correct formula for calcium phosphate?

The phosphate ion, $(PO_4)^{3-}$, has an oxidation number of -3; the calcium ion has an oxidation number of $+2$. The total charge on the molecule must be zero. To have an equal number of positive and negative charges we need three calcium ions and two phosphate ions. Then we have $3 \times (+2) + 2 \times (-3) = 0$. The formula is then $Ca_3(PO_4)_2$.

Many elements have several possible oxidation states. In $FeCl_3$ iron has an oxidation state of $+3$; in $FeCl_2$, $+2$. Tin has an oxidation number of $+2$ and $+4$ in $SnCl_2$ and $SnCl_4$, respectively. Chlorine usually has an oxidation number of -1, as in the preceding examples. However, it has an oxidation number of $+3$ in $NaClO_2$, $+5$ in $NaClO_3$, and $+7$ in $NaClO_4$.

13-7
NAMING COMPOUNDS

The naming of simple compounds is straightforward. Compounds with only two elements in them are called binary compounds.

Binary compounds with one oxidation state are named by giving the name of the more metallic element, followed by the name of the more nonmetallic element but with the suffix "ide." Examples are as follows:

NaCl	Sodium chloride
NaH	Sodium hydride
CaI_2	Calcium iodide
HCl	Hydrogen chloride
KBr	Potassium bromide
Ba_3P_2	Barium phosphide

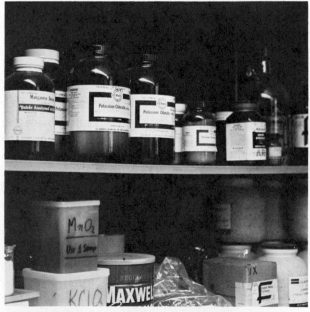

Figure 13-7a *Some of the compounds used in experimental work in beginning chemistry. (Photo courtesy James Crouse.)*

If there is a complex ion involved, it is given its usual name without a new ending. Some examples follow:

NH_4Cl	Ammonium chloride
$(NH_4)_2SO_4$	Ammonium sulfate
H_3PO_4	Hydrogen phosphate
$CaCO_3$	Calcium carbonate

When elements can exist in more than one oxidation state in a binary compound with another element, the compounds are distinguished by using the prefixes "mono" (one), "di" (two), "tri" (three), "tetra" (four), "penta" (five), "hexa" (six), "hepta" (seven), and "octa" (eight). The prefixes designate the number of atoms of the element which occur in the molecule. The prefix "mono" is sometimes omitted.

CO	Carbon monoxide
CO_2	Carbon dioxide
NO_2	Nitrogen dioxide
N_2O_4	Dinitrogen tetraoxide
N_2O_5	Dinitrogen pentaoxide
SO_2	Sulfur dioxide
SO_3	Sulfur trioxide

If the metallic element has two possible oxidation numbers, these are sometimes distinguished by Roman numerals indicating the oxidation number.

$FeCl_2$	Iron(II) chloride
$FeCl_3$	Iron(III) chloride
Hg_2O	Mercury(I) oxide
HgO	Mercury(II) oxide

Under an older system of nomenclature the lower oxidation state has the ending "ous," and the higher oxidation state has the ending "ic." Under this older system, the four compounds above have the names ferrous chloride, ferric chloride, mercurous oxide, and mercuric oxide.

13-8

SOME SIMPLE ORGANIC COMPOUNDS

One of the most important classes of compounds are the **organic compounds.** These are covalent compounds which contain the element carbon. It was originally believed that organic compounds (compounds derived from living matter) were exclusively of plant or animal origin, hence the name; but this has proven to be false, since many of the so-called organic compounds can be made in the laboratory from minerals (inorganic compounds). The study of carbon compounds is known as **organic chemistry.** The carbon atom has four electrons in its outer shell. This gives it the ability to form long-chain and ring-shaped molecules. There are over a million carbon compounds, and others are being added to the list constantly. Many of these carbon compounds are found in things we use daily, such as food, fuels, drugs, detergents, perfumes, and synthetic fibers. Carbon is the essential element of life, because it is the basic constituent of the complex molecules of proteins, fats, and carbohydrates, plus the nucleic acids. In this section we will examine the element carbon and some of its simpler compounds.

The element carbon exists in two crystalline structures, graphite and diamond. A third form, called amorphous carbon, has the same crystalline structure as graphite. Coke and charcoal are examples of amorphous carbon. Graphite is soft, black, slippery, and is a good conductor of electricity. Diamond is very hard (the hardest substance known, with perhaps the exception of a substance called boron carbide), colorless, transparent, and it will not conduct an electric current. Since both diamond and graphite are composed of carbon atoms which are identical in properties, the difference between graphite and diamond must be in the bonding or crystalline structure.

Figure 13-8a *Crystalline structure of graphite. Each carbon atom is bonded to three other carbon atoms in the same plane. Atoms in one plane are weakly bonded to atoms in adjoining planes.*

Figure 13-8b *The crystalline structure of diamond. Note that each carbon atom is bonded to four others, spatially arranged at the corners of a regular tetrahedron.*

The carbon atoms in graphite are arranged in hexagonal rings. Each carbon atom is bonded to three other carbon atoms which lie in the same plane, and thus each carbon atom forms a part of three hexagons (Fig. 13-8a). The hexagons extend in all directions within the formed plane. The remaining electron of the carbon atom is relatively free and shifts from one bond to another in an endless fashion. Thus, there is a free valence electron which gives graphite its ability to conduct an electric current. The attractive forces between planes are very weak and allow the planes to slide easily over one another. This accounts for the slippery feeling of graphite. Graphite is used in many lubricants because of this property, but it is probably the best known as the black substance used in making "lead" pencils.

The crystalline structure of diamond is quite different from that of graphite. The four valence electrons of carbon atoms in diamond are arranged with those of four neighboring carbon atoms to form a tetrahedral structure as shown in Fig. 13-8b. The carbon atoms at the vertices are in turn bonded to four other carbon atoms. Thus, the entire structure is one big crystal, which is extremely strong in all directions. Besides being very hard, diamond has a high melting point. When heated to 1000°C in the absence of air, diamond will change to graphite. (See Fig. 13-8c.)

Compounds containing only carbon and hydrogen are known as **hydrocarbons.** The **alkanes** are

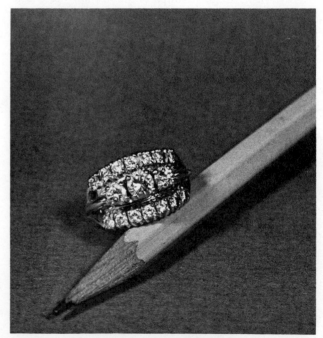

Figure 13-8c *Two forms of carbon: graphite and diamond. (Photo courtesy James Crouse.)*

hydrocarbons which have a composition that satisfies the general formula

$$C_nH_{2n+2}$$

where n = the number of carbon atoms,
$2n + 2$ = the number of hydrogen atoms,
$n = 1, 2, 3, \ldots$

The formula says that the number of hydrogen atoms present in the compound is twice the number of carbon atoms plus two.

Methane (CH_4) is the first member of the alkane series. Ethane (C_2H_6) is the second member, propane (C_3H_8) the third, and butane (C_4H_{10}) the fourth. Table 13-8a lists some other members of the series. The prefix "n" in front of the name, in Table 13-8a, stands for the word *normal*. Normal hydrocarbons are those whose molecules are all straight chains, with no carbon-hydrogen groups branching off the chain.

The hydrocarbon's structure is easy to visualize when we write the **structural formula** (a picture representation of the way the atoms are connected to one another) instead of its molecular formula. For example, the structural formula for methane (CH_4) is written

$$H-\underset{\displaystyle H}{\overset{\displaystyle H}{C}}-H$$

The structural formula is similar to the electron-dot notation. Each dash corresponds to two shared electrons and so represents a covalent bond. Ethane and propane structural formulas are

Ethane   Propane

The alkanes make up many well-known petroleum products. Methane is the principal constituent in natural gas used for cooking stoves. Gasoline contains many of the alkanes from pentane to decane

Table 13-8a *Some Compounds of the Alkane Series of Hydrocarbons*

CH_4	Methane
C_2H_6	Ethane
C_3H_8	Propane
C_4H_{10}	n-Butane
C_5H_{12}	n-Pentane
C_6H_{14}	n-Hexane
C_7H_{16}	n-Heptane
C_8H_{18}	n-Octane
C_9H_{20}	n-Nonane
$C_{10}H_{22}$	n-Decane
$C_{11}H_{24}$	n-Undecane
$C_{12}H_{26}$	n-Dodecane
$C_{15}H_{32}$	n-Pentadecane
$C_{20}H_{42}$	n-Eicosane

($n = 5$ to $n = 10$). Kerosene contains the alkanes with $n = 10$ to 16. The alkanes with higher values of n make up other products such as diesel fuel ($n = 12$ to 20), fuel oil ($n = 14$ to 22), lubricating oil ($n = 20$ to 30), petroleum jelly ($n = 22$ to 40), and paraffin wax ($n = 25$ to 50). The alkanes are also used as starting materials for many other products such as paints, plastics, drugs, detergents, insecticides, and cosmetics.

The alkanes are highly combustible. They react with the oxygen in air to form heat, carbon dioxide, and steam. If the combustion is not complete, carbon in the form of black soot and carbon monoxide are formed. The alkanes are colorless. Recall that paraffin wax, kerosene, and "white" gasoline used in camp stoves all lack color. Most gasolines used in automobiles, however, are colored, because a little dye has been added to indicate that it contains tetraethyl lead, a deadly poison. The alkanes are nonpolar compounds. The alkane molecules thus are not attracted to the polar molecules of water and so solvation does not occur to any appreciable extent.

When two hydrogen atoms are removed from a molecule of an alkane, two electrons of the carbon atoms become unpaired. In order to satisfy the octet rule, a double bond must be formed. Hydrocarbons with a double bond are called **alkenes.** For example,

Ethane   Ethene
(or Ethylene)

The general formula for the alkene series is C_nH_{2n}. The series begins with C_2H_4 (ethene, or ethylene), shown by the structural formula above. Some of the normal alkenes are listed in Table 13-8b. A number, instead of n (normal), preceding the name indicates the presence of a double bond and its position. The carbons in the chain are numbered starting at the end—which end doesn't matter. For example, 1-pentene and 2-pentene have the following structural formulas:

1-pentene

2-pentene

The alkanes are all saturated hydrocarbons. A **saturated hydrocarbon** contains only single bonds, and its hydrogen content is at a maximum. The alkenes are known as **unsaturated hydrocarbons,** because they can take on hydrogen atoms to form saturated hydrocarbons or atoms of substances other than hydrogen to form derivatives.

When a second pair of hydrogen atoms is removed from the carbon atoms that have a double bond, a triple bond is formed. The general formula for this series, **alkyne,** is C_nH_{2n-2}, and the simplest

Table 13-8b *Some Compounds of the Alkene Series of Hydrocarbons*

CH_2CH_2	Ethene (ethylene)
CH_3CHCH_2	Propene (propylene)
$C_2H_5CHCH_2$	1-Butene
$C_3H_7CHCH_2$	1-Pentene
$C_3H_7CHCHCH_3$	2-Hexene
$C_5H_{11}CHCH_2$	1-Heptene
$C_6H_{13}CHCH_2$	1-Octene
$C_7H_{15}CHCH_2$	1-Nonene
$C_{10}H_{20}$	1-Decene
$C_{15}H_{30}$	1-Pentadecene
$C_{20}H_{40}$	1-Eicosene

member is ethyne (or acetylene), $HC{\equiv}CH$. The members of this series are more reactive than those of the alkene series. Some members of the alkyne series are listed in Table 13-8c.

The general formula C_nH_{2n}, which gives the structure of the alkenes, also gives the structure for a saturated hydrocarbon series called the cycloalkanes. The prefix "cyclo-" indicates the ring structure, and the ending "-ane" the saturated condition of the chain. The carbon atoms of this series are arranged in a ring with two hydrogen atoms connected to each carbon atom in the ring. These are called the **cyclic**

Table 13-8c *Some Compounds of the Alkyne Series of Hydrocarbons*

$HC{\equiv}CH$	Ethyne (acetylene)
$CH_3C{\equiv}CH$	Propyne (methylacetylene)
$C_2H_5C{\equiv}CH$	1-Butyne
$C_3H_7C{\equiv}CH$	1-Pentyne
$C_3H_7C{\equiv}CCH_3$	2-Hexyne
$C_5H_{11}C{\equiv}CH$	1-Heptyne
$C_6H_{13}C{\equiv}CH$	1-Octyne
$C_7H_{15}C{\equiv}CH$	1-Nonyne
$C_{10}H_{18}$	1-Decyne
$C_{15}H_{28}$	1-Pentadecyne
$C_{20}H_{38}$	1-Eicosyne

hydrocarbons. The first member of the series is cyclopropane, which has the structural formula

Cyclopropane

Various numbers of double bonds can also be introduced into cyclic structures. One of the most important cyclic hydrocarbons is benzene (C_6H_6). It has the structural formula

H—C benzene ring structures

Benzene ring

Since the molecule takes the structure of a ring, it is known as the **benzene ring.** The molecule has the ability to switch back and forth between the two structures shown. This property is called resonance and makes the benzene molecule more stable than the regular hydrocarbons. Sometimes the benzene structure is written

Benzene ring

Benzene is obtained from coal tar, a by-product of soft coal. When other atoms or groups of atoms are substituted for the hydrogen atoms in the benzene ring, a vast number of different compounds can be produced. These compounds include such things as perfumes, explosives, drugs, solvents, insecticides, lacquers, and a host of others. (See Fig. 13-8d.)

So far we have been naively drawing structural formulas as if molecules had only two dimensions.

Figure 13-8d *The aromatic character of the compounds of benzene gives rise to many exotic fragrances. (Photo courtesy James Crouse.)*

The seemingly obvious principle that molecules exist in three dimensions was not realized until 1874, some 10 to 20 years after the concepts of ring and chain molecules were discovered. In 1874 Jacobus van't Hoff in Holland and Joseph Le Bel in France independently reached the conclusion that many unexplained chemical mysteries could be explained if molecules were viewed in three dimensions.

The molecule CH_4 is actually a tetrahedron with a carbon atom in the center. Its three-dimensional structure looks like

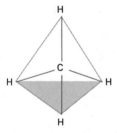

This tetrahedral arrangement is the basic geometric shape of organic molecules. The way chains are built up in three dimensions is illustrated by *n*-butane.

n-butane

From this model of *n*-butane we can see that the following structural formulas are all the same.

```
        H
        |
    H—C—H
        |
H  H  H  H        H
|  |  |  |        |
H—C—C—C—C—H   H—C—C—H
|  |  |  |        |   |
H  H  H  H        H   H
                  H—C—H
                      |
                      H
   n-butane        n-butane
```

```
    H  H  H
    |  |  |
H—C—C—C—H
    |  |  |
    H  H
H—C—H
    |
    H
   n-butane
```

All of these represent the *n*-butane molecule. They all have the carbon and hydrogen atoms attached in the same way: There are three hydrogen atoms on the two end carbon atoms and two hydrogen atoms attached to the middle two carbon atoms. However, another arrangement of the atoms of the butane molecule, is possible. It is an isomer of *n*-butane. **Isomers** are molecules with the same molecular formula but a different structure or arrangement of atoms. Isobutane also has 4 carbon atoms and 10 hydrogen atoms, but its structural formula is

```
    H  H  H
    |  |  |
H—C—C—C—H
    |  |  |
    H     H
    H—C—H
        |
        H
   Isobutane
```

Isobutane's structural formula is different from the three structural formulas given for *n*-butane, because it has three hydrogen atoms attached to three of the carbon atoms, while the fourth carbon atom has only one hydrogen atom bonded to it.

IMPORTANT WORDS AND TERMS

Mixture	Oxidation	Oxidation state
Compound	Reduction	Organic compounds
Molecule	Oxidation-reduction reaction	Hydrocarbons
Octet rule	Electrolysis	Alkanes
Ionic compounds	Covalent bond	Structural formula
Electrovalent compounds	Ionic bond	Alkenes
Covalent compounds	Electronegativity	Saturated hydrocarbon
Electron affinity	Polar bond	Unsaturated hydrocarbon
Crystal	Nonpolar bond	Alkynes
Cation	Nonpolar molecule	Cyclic hydrocarbons
Anion	Polar molecule	Benzene ring
Half-reaction	Oxidation number	Isomers

LEARNING OBJECTIVES

After reading and studying this chapter you should be able to do the following without referring to the text:

1. Define, explain, and give an example of the important terms in the following list: (a) compound, (b) mixture, (c) molecule, (d) ion.
2. State the two basic assumptions made to explain compound formation.
3. Describe ionic bonding, state the properties of ionic compounds, and give a few examples of ionic compounds.
4. Describe covalent bonding, state the properties of covalent compounds, and give a few examples of covalent compounds.
5. Distinguish between an oxidation and a reduction process and give examples of each.
6. Distinguish between polar and nonpolar molecules and give some examples of each.
7. Define *oxidation number* and state the general rules for assigning oxidation numbers.
8. State the general rules for naming compounds.
9. Describe organic compounds and state the general formula for some compounds known as hydrocarbons.
10. Draw the structural formula for benzene, state where benzene is obtained, and name some of the compounds produced by substitution.
11. Define or explain the important words and terms listed at the end of the chapter.

QUESTIONS

1. Distinguish between a compound and a mixture. Give an example of each.
2. Distinguish between a compound and a molecule. Give an example of each.
3. Which electrons of an atom take part in the formation of compounds?
4. What is the electron configuration for the most stable state of an atom or ion that take part in the formation of a compound?
5. What is a chemical bond?
6. What force is responsible for molecular formation?
7. During compound formation, how is the number of valence electrons in an atom related to its tendency to lose or gain electrons?
8. Distinguish between ionic and covalent bonding.
9. Give an example of a compound not composed of molecules.
10. Why is hydrogen a diatomic gas while helium is monatomic?
11. Compare the physical properties of ionic compounds with those of covalent compounds.
12. Compare oxidation number to the number of electrons in the outermost shell of a neutral atom.
13. Where on the periodic chart do you find elements with (a) high electronegativity, (b) low electronegativity?
14. Why are single bonds sometimes formed while at other times double or triple bonds are formed?

15. A compound has a boiling point of $-10°C$. Is it ionic or covalent? Why?
16. A compound melts at a very high temperature of $1000°C$. Would you expect it to be ionic or covalent? Why?
17. How does water dissolve ionic compounds?
18. List some of the important alkanes and their uses.
19. What is the structure of benzene? List some of the products in which benzene occurs.
20. Explain how an ionic liquid conducts electricity.
21. Write the general formula for (a) the alkanes, (b) the alkynes.

PROBLEMS

1. Draw the electron dot notation for the following atoms and their ions:
   (a) Mg, (b) S, (c) Br, (d) K, (e) Hg, (f) Ni, (g) Al.

   Answer: (a) Mg: and Mg^{2+}

2. (a) Specify the following as either ionic or covalent substances:
   CO_2, NaCl, CaF_2, NH_4^+, $(NO_2)^-$, N_2H_3Cl
   (b) Draw the electron dot formulas for the above compounds and ions.

3. Specify the following molecules as either polar or nonpolar:
   H_2O, CO, HCl, CH_4, O_2, NH_3

   Answer: NH_3, polar

4. Find the oxidation numbers of all elements in the following formulas:
   (a) H_2SO_4 (e) SO_3^{2-}
   (b) N_2 (f) SO_4^{2-}
   (c) CO   (g) CrO_4^{2-}
   (d) CO_2 (h) $Cr_2O_7^{2-}$

   Answer: (g) O, 2−; Cr, 6+

5. Write formulas for the following compounds:
   (a) Barium nitrate   (e) Calcium phosphate
   (b) Aluminum sulfate   (f) Ammonium nitrite
   (c) Vanadium(II) oxide   (g) Calcium iodide
   (d) Vanadium(III) oxide   (h) Sodium sulfite

   Answer: (e) $Ca_3(PO_4)_2$

6. What are the names of the following compounds:
   (a) CuO   (c) $Na_2(SO_4)$ (e) C_8H_{18}
   (b) Cu_2O (d) C_6H_6

7. Draw structural formulas for the following organic compounds:
   (a) benzene, (b) propane, (c) propene, (d) propyne.

8. Draw structural formulas for the isomers glucose and fructose. (Hint: see Chapter 16.)

outline properties of ionic & covalent compounds

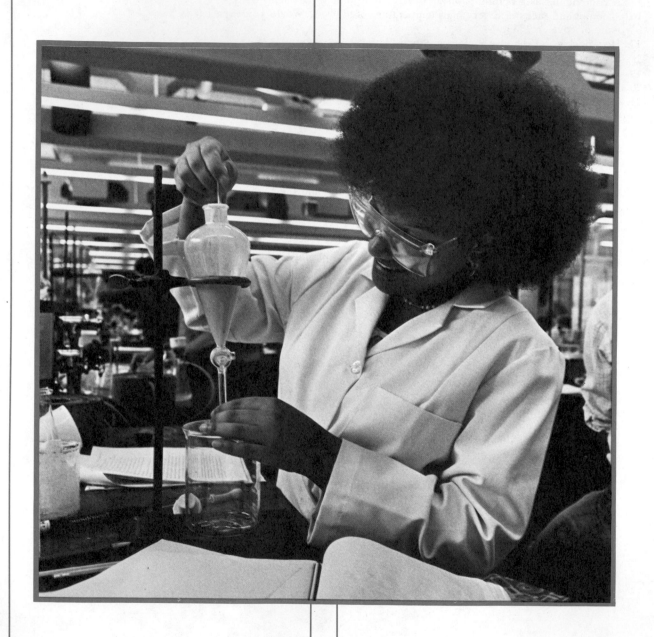

Chapter 14

SOME CHEMICAL PRINCIPLES

The chemists are a strange class of mortals, impelled by an almost insane impulse to seek their pleasure among smoke and vapor, soot and flame, poisons and poverty; yet among all these evils I seem to live so sweetly that I may die if I would change places with the Persian King.

Johann Joachim Becher

In this chapter some of the basic principles of chemistry will be discussed. Most of these were discovered before 1850 by chemists, such as Becher (1635–1682), who earnestly pursued their research in small labs which they themselves financed. Historically, these laws and principles were known before Mendeleev constructed his periodic table in 1868. It was long after 1868, of course, that the structure of atoms and molecules was understood. In fact, the principles we shall study in this chapter helped to establish our modern understanding of molecular structure. When these principles were first discovered, the reasons behind them were unknown. The explanation for them lies in our present understanding of atoms and molecules discussed in the previous three chapters.

Many of these very old laws concerning proportions and mixtures are important to us today. For instance, if the exhaust of a car is emitting black smoke we know that an incorrect mixture of air and gasoline vapor is being burned in the engine. Such exhaust gases are one of the chief pollutants of our atmosphere. Instances of where mixtures are important are also well known to all cooks. If a recipe is not followed closely the wrong proportions of ingredients will yield an unpalatable dish. The chemists of Becher's day have left us a legacy which we would be wise to guard carefully.

14-1
TYPES OF MATTER

Matter is classified in several different ways. One of the ways, which was mentioned previously, is the classification of matter into four phases: solid, liquid, gas, and plasma. In the solid phase, matter is rigid and has a definite shape. It expands or contracts comparatively little with pressure and temperature changes. In the liquid phase, matter assumes the shape of the container and maintains a horizontal surface because of the effect of gravity. Similar to solids, liquids do not expand or contract much when pressure and temperature are varied. In the gas phase, matter fills the container completely and uniformly. Gravitational effects are small since the individual gas molecules have such a small mass. Gases expand and contract quite easily when pressure and temperature are varied. In fact, the gas laws tell us that the volume of gas is inversely proportional to its pressure and directly proportional to its temperature.

Matter which is of uniform composition throughout is described as **homogeneous;** that is, any sample will be like any other sample. Matter which is of nonuniform composition is described as **heterogeneous.** As examples, a well-stirred cup of coffee with cream and sugar is a homogeneous mixture, while a pizza is a heterogeneous mixture. A **substance** is a homogeneous sample of matter all specimens of which have identical properties and composition. A compound is a substance composed of two or more elements chemically combined in a definite proportion. That is, a compound is composed of only one kind of molecule. All compounds are substances, but not all substances are compounds. Pure elements are substances, but they are not compounds. For example, pure hydrogen gas is a substance, but it is not a compound, while steam is both a substance and a compound. A sample of matter composed of two or more substances in varying amounts that are not chemically combined is called a mixture. Mixtures can be homogeneous or heterogeneous. A **solution** is

Figure 14-1a *The types of matter.*

a homogeneous mixture. From this definition a solution need not be a liquid; it may also be a solid or a gas. Salt water is a liquid solution. Metal alloys are examples of solid solutions. Air is an example of a gaseous solution. In Fig. 14-1a, these different types of matter are summarized.

The characteristics of a substance are known as its properties. There are two basic types of properties, physical and chemical. **Physical properties** are those which do not involve a change in the chemical composition of the substance. Among these properties are density, hardness, color, melting point, boiling point, electrical conductivity, thermal conductivity, and specific heat. These properties may change with a change in pressure or temperature. Those changes, which do not alter the chemical composition of the substance, are called **physical changes.** The processes of freezing or boiling water are good examples of physical changes brought about by changing the temperature. Other examples of physical changes are dissolving table salt in water, heating a piece of metal, and evaporating water.

The properties involved in the transformation of one substance into another are known as **chemical**

properties. When wood burns, oxygen in the air unites with the different substances in the wood to form new substances. When iron corrodes, it combines with oxygen and water to form a new substance commonly known as rust. These are examples of chemical properties of wood and iron, respectively. Changes which result in the formation of new substances are known as **chemical changes.** The fermenting of wine, burning of gasoline, souring of milk, discharging of a battery, and the exploding of gun powder are all examples of chemical changes. All chemical changes involve a production or absorption of energy.

14-2
EARLY CHEMICAL LAWS

We have seen in previous chapters that mass is a form of energy. Mass and energy are related by Einstein's formula, $E = mc^2$. Total energy, including mass energy, is always conserved in any process. In chemical reactions energy is produced or absorbed in the form of heat, kinetic, or potential energy. There is, then, a corresponding gain or loss in mass energy, but the total energy does not change. However, the change in mass in a chemical reaction is so infinitesimal that it cannot be detected. If the total mass involved in chemical reaction is precisely weighed before and after the reaction takes place, no change can be detected. This law is known as the **law of conservation of mass** and is stated as follows:

> **There is no detectable change in the total mass during a chemical process.**

This law was formulated in 1774 by Antoine Lavoisier. Lavoisier had developed balances which could measure $\frac{1}{100}$ the weight of a drop of water. He put tin in a vessel, sealed it, and carefully weighed it. The vessel was then heated so that the tin reacted with the air in the vessel to form a new compound. La-

voisier next reweighed the sealed container and found the weight to be identical with the original weight. Because the container had been sealed, no air could go in or out. When Lavoisier opened the vessel, air rushed in. This showed that the tin had reacted with the air inside, or something in the air, to form the new compound. This experiment of Lavoisier's was one of the first great quantitative chemical experiments. Now, two centuries later, Lavoisier's law of conservation of mass is still almost exactly true.

One of the greatest theoretical chemists of the early 1800s was John Dalton (Fig. 14-2a). He was the first man of his time to use the old Greek idea of atoms. He also proposed the notion of a molecule. Along with many other contributions to chemistry, he proposed what is now known as the **law of definite proportions,** sometimes called the law of definite composition:

> **Different samples of a pure compound always contain the same elements in the same proportion by weight.**

For instance,

$$9 \text{ g H}_2\text{O} = 8 \text{ g oxygen and } 1 \text{ g hydrogen}$$

$$36 \text{ g H}_2\text{O} = 32 \text{ g oxygen and } 4 \text{ g hydrogen}$$

In both cases the ratio by weight of oxygen to hydrogen is 8 to 1. This is true, because each molecule of H_2O is composed of one atom of oxygen (atomic weight 16) and two atoms of hydrogen (atomic weight 1, total weight of hydrogen $= 2 \times 1$ or 2). Thus, each molecule of H_2O is composed of 16 parts oxygen by weight and 2 parts hydrogen by weight or a ratio of 8 to 1. Since a pure covalent compound is simply a very large collection of identical molecules, the proportion by weight of any element in a covalent compound will be the proportion by weight that it occupies in an individual molecule of that covalent compound.

■ *Example 1* If 11 g of carbon dioxide, CO_2, are composed of 3 g of carbon and 8 g of oxygen, how

Figure 14-2a *John Dalton (1766–1844). Dalton was born in the village of Eaglesfield, England, the son of a poor weaver. He was a child prodigy, opening his own school at the age of 12. In his mid-twenties he became a teacher at the University of Manchester. His greatest contribution to science was his atomic theory of matter. He took the word "atom" from the Greek word* atomos, *meaning indivisible. Dalton was a bachelor all his life—because he "never had time to marry." One thing that occupied much of his time was making daily weather observations. He made thousands of these during his lifetime. (Photo courtesy Culver Pictures.)*

many grams of carbon and oxygen will be in 55 g of carbon dioxide?

Since $\frac{3}{11}$ of every sample of CO_2 must be carbon and $\frac{8}{11}$ of every sample of CO_2 must be oxygen, the amount of carbon is $\frac{3}{11} \times 55$ g $= 15$ g and the amount of oxygen is $\frac{8}{11} \times 55$ g $= 40$ g. In both the 11-g and 55-g sample of CO_2, the ratio is 3 parts carbon to 8 parts oxygen. That is, 3:8 or 15:40.

For ionic compounds, a similar submicroscopic description is easily given. Since the ions occur in a definite proportion, the weight of the elements in the compound are in the same proportion by weight as they are in the smallest combination of ions which gives the formula of the compound.

■ *Example 2* If 25 g of calcium carbonate ($CaCO_3$) consist of 10 g calcium, 3 g carbon, and 12 g oxygen, how many grams of each element are contained in 100 g of $CaCO_3$?

Calcium carbonate is an ionic compound with one ion of calcium for every carbonate ion (CO_3^{2-}). Using the law of definite proportions, we know that $\frac{10}{25}$ of every sample must be calcium, $\frac{3}{25}$ carbon, and $\frac{12}{25}$ oxygen. Thus in 100 g of $CaCO_3$:

$$\text{Amount of calcium} = (\tfrac{10}{25}) \times 100 = 40 \text{ g}$$

$$\text{Amount of carbon} = (\tfrac{3}{25}) \times 100 = 12 \text{ g}$$

$$\text{Amount of oxygen} = (\tfrac{12}{25}) \times 100 = 48 \text{ g}$$

Sometimes the elements which are to be combined to form a compound are not in the correct proportions. In this situation one of the elements will be completely used and the one in excess will be partially used in the correct proportion with some of it left over.

■ *Example 3* How many grams of CO_2 can be made from 36 g of carbon and 40 g of oxygen?

The ratio of carbon to oxygen in CO_2 is 3:8. So 8 g of oxygen are needed for every 3 g of carbon. If we divide 8 g into the 40 g of oxygen available, we get 5. If we divide 3 g of carbon into the 36 g available, we get 12. Since 12 is bigger than 5, the carbon is in excess. We have 5×8 g ($= 40$ g) of oxygen available. This will combine with 5×3 g ($= 15$ g) of the carbon to make 55 g of CO_2. Then 21 g of carbon are left over. In summary,

Compound	Amount of Nitrogen (g)	Amount of Oxygen (g)	Ratio
NO	100	114.2	
N_2O	100	57.1	57.1 : 114.2 or 1 : 2
N_2O_3	100	171.3	171.3 : 114.2 or 3 : 2
NO_2	100	228.4	228.4 : 114.2 or 2 : 1
N_2O_5	100	285.5	285.5 : 114.2 or 5 : 2
NO_3	100	342.6	342.6 : 114.2 or 3 : 1

Amount of oxygen available $= 40$ g

Amount of carbon available $= 36$ g

Amount of carbon needed
(to combine with 40 g
of oxygen to produce CO_2) $= 15$ g

Amount of CO_2 produced $= 40$ g $+ 15$ g
$= 55$ g

Amount of carbon left over $= 21$ g

In the preceding example we also see that the law of conservation of mass is obeyed. We began with a total of 76 g (40 g + 36 g) and we ended with a total of 76 g (55 g + 21 g).

A third law, which was discovered in the early 1800s, is called the law of multiple proportions. This law describes a relationship that exists for different compounds that are formed from the same elements. Carbon monoxide, CO, and carbon dioxide, CO_2, are examples. The **law of multiple proportions** is stated as follows:

> **Whenever two elements, call them E_1 and E_2, combine to form different compounds, the various amounts of E_2 combined with the same amount of E_1 are in a ratio of small whole numbers.**

Some examples will clarify the statement of this law. First, consider CO and CO_2. These two compounds are both composed of the same two elements, carbon

and oxygen. The two amounts by weight of oxygen which combine with the same weight of carbon are in a ratio of small whole numbers and so conform to the law. In CO, 16 g of oxygen will combine with 12 g of carbon; in CO_2, 32 g of oxygen will combine with 12 g of carbon. The numbers 16 and 32 are in the ratio 1:2, which is a ratio of small whole numbers as predicted by the law of multiple proportions. It is easy to see from our understanding of molecules why this law must be so. In a single molecule of CO there is one oxygen atom; in a single molecule of CO_2 there are two oxygen atoms. This ratio of 1:2 is then reflected in the amounts of oxygen that will combine with equal amounts of carbon to form the two different compounds. A good example of the multiple proportion law is found in the compounds of nitrogen and oxygen. There are six of these compounds. The amounts of oxygen which unite with 100 g of nitrogen are shown in the table at the top of the page.

All the ratios of the amounts of oxygen that combine with 100 g of nitrogen are in small whole numbers. That is, ratios such as 1.738:1 or 1:1.23 do not occur. Again, the molecular understanding of compounds readily explains the ratios. The compounds have the formulas NO, N_2O, N_2O_3, NO_2, N_2O_5, and NO_3, and the ratios are easily related to these molecular formulas. In the early 1800s, discovery of the law of multiple proportions was an important step towards the molecular understanding of chemical compounds. A modern chemical laboratory balance is shown in Figure 14-2b.

Figure 14-2b *In a modern chemical laboratory balances of the type shown here are used for accurate measurement of a substance's mass. (Photo courtesy James Crouse.)*

14-3
MOLECULAR WEIGHTS AND VOLUMES

John Dalton was the first to study systematically the weight relationships among the various elements. These weight relationships are given in the periodic table as atomic weights. In the present day table, the weight relationships are based on an arbitrary scale that assigns the ^{12}C isotope the value 12.0000. All other naturally occurring elements are given a weight value relative to that of the ^{12}C isotope (see Table 12-1a). The atomic weights for most elements have been determined to several decimal places. For purposes of simplicity, however, we shall round these values off to the nearest integer or half integer. For example, we shall consider the atomic weights of hydrogen, carbon, oxygen, and chlorine to be approximately 1, 12, 16, and 35.5, respectively. Note that since these are only relative weights, they have no units. In order to give units to atomic weights, the concept of gram atomic weight is necessary. A **gram atomic weight** is defined as the weight of an atom of an element relative to ^{12}C, expressed in grams. The gram atomic weights of hydrogen and carbon are then 1 g and 12 g, respectively. The number of gram atomic weights of a given sample of an element is found by dividing the mass of the sample by one gram atomic weight. As an example, in 20 g of carbon there are $\frac{20}{12} = 1.67$ gram atomic weights.

The **formula weight** of a compound is the sum of the atomic weights given in the formula of the compound. The formula weight of $CaCO_3$ is thus $40 + 12 + (3 \times 16) = 100$. The formula weight is a relative quantity like the atomic weight. The **gram formula weight** is the formula weight of the compound expressed in grams. The gram formula weights of $CaCO_3$ and H_2O are 100 g and 18 g, respectively. A gram formula weight is also referred to as a **mole.** In 200 g of $CaCO_3$ there are 2 moles.

■ *Example 4* How many moles are there in (a) 9 g of H_2O, (b) 65 g of NaCl?

(a) The gram formula weight of H_2O is 18 g, so in 9 g there are 9 g/18 g = 0.5 mole.
(b) The gram formula weight of NaCl is 23 g + 35.5 g = 58.5 g. Thus, in 65 g there are 65 g/58.5 g = 1.11 moles.

In covalent compounds, the formula given represents a discrete molecule of the compound. Therefore, the terms **molecular weight** and **gram molecular weight** are often used for covalent compounds instead of formula weight and gram formula weight. However, in ionic compounds, distinct molecules do not exist in the crystal or in solution. Thus, the terms molecular weight and gram molecular weight do not apply to ionic compounds and so formula weight and gram formula weight are the terms used.

The **percentage composition** of an element in a compound is the percent by weight of each element in the compound. It can be found by computing the total weight of an element in a compound by dividing it by the formula weight of the compound. To get a percentage, the resulting answer must then be multiplied by 100.

■ *Example 5* What is the percentage composition of the elements in table salt, NaCl? In sulfuric acid, H_2SO_4?

First find the formula weight. The formula weight of NaCl is $23 + 35.5 = 58.5$. The percent composition is then

$$Na = \frac{23}{58.5} \times 100 = 39.3\%$$

$$Cl = \frac{35.5}{58.5} \times 100 = 60.7\%$$

The formula weight of sulfuric acid is $2 \times 1 + 1 \times 32 + 4 \times 16 = 2 + 32 + 64 = 98$

The percentage composition is then

$$H = \frac{2}{98} \times 100 = 2.04\%$$

$$S = \frac{32}{98} \times 100 = 32.65\%$$

$$O = \frac{64}{98} \times 100 = 65.31\%$$

In 1808, the French chemist Joseph Louis Gay-Lussac performed some experiments on gases that had a profound effect on the modern understanding of atoms and molecules. Gay-Lussac mixed different gases together to form new gases. As he did this, he took pains to measure the volumes of the initial and final gases. He found that when he mixed 2 volumes of hydrogen gas with 1 volume of oxygen gas, he obtained 2 volumes of steam. Similarly, when he mixed 1 volume of nitrogen gas with 3 volumes of hydrogen gas, he got 2 volumes of ammonia. In all the experiments, he measured the gas volumes at the same temperature and pressure. From these and many other experiments, he developed what is called **Gay-Lussac's law of combining volumes:**

When gases combine to form new gaseous compounds, all at the same temperature and pressure, the volumes of the initial and final gases are in the ratio of small whole numbers.

The explanation of Gay-Lussac's law of combining volumes was first proposed by the Italian chemist Amadeo Avogadro in 1811. Avogadro explained Gay-Lussac's experiments by making two assumptions. His first assumption, which has since been confirmed many times, is now known as **Avogadro's law:**

Equal volumes of gases at the same temperature and pressure contain equal numbers of molecules.

Avogadro's second assumption (which is also correct) was that sometimes the molecules of a gaseous element consist of two atoms. That is, hydrogen gas, nitrogen gas, oxygen gas, and many others all have two atoms in a molecule. They are diatomic gases. Of course, the noble gases are all monatomic. We have seen previously that molecules are formed in this way, because each atom needs to have eight electrons to fill its outer shell.

Avogadro's two (correct) assumptions form the basis for the modern understanding of Gay-Lussac's experiments. Figure 14-3a shows schematically what takes place when 2 volumes of hydrogen and 1 volume of oxygen unite to form 2 volumes of steam.

Figure 14-3a *A graphic representation of Avogadro's explanation of Gay-Lussac's experiment. There are four molecules in each volume. This satisfies Avogadro's law.*

Figure 14-3b *A graphic representation of Avogadro's explanation of the formation of ammonia. There are two molecules in each volume. This satisfies Avogadro's law.*

According to Avogadro's law, equal numbers of molecules are shown in each volume of gas. Two of the hydrogen atoms combine with one oxygen atom to form one steam molecule.

Figure 14-3b is an explanation of a similar situation in which nitrogen and hydrogen combine to form ammonia gas. Since experimentally we find that 3 volumes of hydrogen combine with 1 volume of nitrogen to form 2 volumes of ammonia gas, the molecular formula of ammonia can be deduced. To do this, Avogadro's law is used, along with his assumption that H_2 and N_2 are the correct molecular formulas for hydrogen and nitrogen gas. If we then put 2 molecules of each gas into each volume, we find that the molecular formula of ammonia can only be NH_3.

Equal numbers of gas molecules occupy equal volumes at the same temperature and pressure. However, the weights of different molecules are different, and equal volumes do not, of course, weigh the same. For instance, an oxygen molecule is about 16 times as heavy as a hydrogen molecule. Thus, one volume of oxygen gas will weigh about 16 times the weight of an equal volume of hydrogen gas. The volume unit used in chemistry is a liter. A **liter** is equal to 1000 cm³, or roughly equivalent in volume to a quart.

One liter of oxygen gas weighs 1.43 g at standard conditions. **Standard conditions** are 0°C and sea level atmospheric pressure. Throughout the rest of the book, when a volume of gas is used the temperature and pressure will be assumed to be at the standard conditions unless otherwise specified. One liter of hydrogen gas weighs about $\frac{1}{16}$ of the weight of a liter of oxygen, or about 0.09 g. The weights of other gases can be found using the ratio of their molecular weights to that of oxygen.

▪ *Example 6* What is the weight of one liter of nitrogen gas?

The molecular weight of N_2 is 28. The molecular weight of O_2 is 32. The weight of a volume of any gas is proportional to its molecular weight.

$$\frac{\text{Wt. of one liter } N_2}{\text{Mol. weight } N_2} = \frac{\text{Wt. of one liter } O_2}{\text{Mol. weight } O_2}$$

$$\frac{\text{Wt. of one liter } N_2}{28} = \frac{1.43 \text{ g}}{32}$$

or

$$\text{Wt. of one liter } N_2 = \frac{28}{32} \times 1.43 \text{ g} = 1.25 \text{ g}$$

One mole of oxygen gas weighs 32 g. Since one liter of oxygen gas weighs 1.43 g, the volume occupied by 32 g is given by a simple ratio

$$\frac{\text{Vol. of 32 g}}{32 \text{ g}} = \frac{\text{Vol. of 1.43 g}}{1.43 \text{ g}}$$

or

$$\frac{\text{Vol. of one mole } O_2}{32 \text{ g}} = \frac{1 \text{ liter}}{1.43 \text{ g}}$$

or

$$\text{Vol. of one mole } O_2 = \frac{32 \text{ g}}{1.43 \text{ g}} \times 1 \text{ liter}$$

$$= 22.4 \text{ liters}$$

Similarly, we find that one mole of H_2 equals about $\frac{2}{0.09} \times 1$ liter = 22.2 liters. If we had used a more exact figure for the molecular weight of hydrogen (2.016 instead of 2), we would have obtained 22.4 liters again. In fact, one mole of any gas occupies 22.4 liters. The quantity 22.4 liters is known as the **gram molecular volume.** It is the volume of one mole of any gas at standard conditions.

▪ *Example 7* What volume will 27 g of steam occupy?

Molecular formula of steam is H_2O. The gram molecular weight is $1 \times 2 + 16 \times 1 = 18$ g. There are, therefore, $\frac{27}{18} = 1.5$ moles of steam in 27 g of steam. Now 1.5 moles of any gas occupies 1.5×22.4 liters = 33.6 liters.

■ *Example 8* What is the weight of 1 liter of nitrogen gas?

(This problem has been worked in a previous example but now we will do it in a simpler fashion.) First we find how many moles in 1 liter of nitrogen gas (or any gas).

$$\text{Number of moles} = \frac{1 \text{ liter}}{22.4 \text{ liters}} = 0.0446 \text{ mole}$$

The weight of 0.0446 mole of nitrogen is $0.0446 \times$ the gram molecular weight of nitrogen gas or $0.0446 \times 28 = 1.25$ g.

We see from the examples that it is important to find the number of moles of the gas. The volume can then be found by multiplying the number of moles by 22.4 liters.

Volume of gas = No. of moles × 22.4 liters

If the weight of the gas is desired, it can be found by multiplying the number of moles by the gram molecular weight of the substance. This works for liquids and solids as well as gases.

Weight = Number of moles × Gram formula
weight of substance

14-4
AVOGADRO'S NUMBER

Avogadro's law states that equal volumes of gases at the same temperature and pressure contain equal numbers of molecules. **The number of molecules in one mole of a gas or liquid or solid is called Avogadro's number.** Avogadro's number can be found by

many different methods to be 6.02×10^{23} molecules per mole. Thus, in 22.4 liters of any gas, there will be 6.02×10^{23} molecules of that gas. The number of molecules in any covalent compound can be easily found, once the number of moles of the covalent compound is known. The formula is

No. of molecules = No. of moles ×
Avogadro's number

■ *Example 9* How many molecules are there in 6 g of water?

In 6 g of water there are $\frac{6}{18} = \frac{1}{3}$ mole. The number of molecules is then

$$\frac{1}{3} \times 6.02 \times 10^{23} = 2.0 \times 10^{23} \text{ molecules}$$

in 6 g of water.

Again, we see that if we can find the number of moles, we can find the quantity we are looking for, in this case the number of molecules.

In order to work problems concerning volumes of gases, weights of substances, or the number of molecules in a covalent compound, we first need to find the number of moles from information given in the problem. The number of moles can be found from any of three different equations. They are

$$\text{Number of moles} = \frac{\text{Weight in grams}}{\text{Gram formula weight}}$$
(or gram atomic weight if a substance is monatomic)

$$\text{Number of moles} = \frac{\text{Volume of gas in liters}}{22.4 \text{ liters}}$$

$$\text{Number of moles} = \frac{\text{Number of molecules}}{6.02 \times 10^{23}}$$

The correct equation to use depends on the information given in the problem.

■ *Example 10* How many molecules are there in 11.2 liters of nitrogen gas?

First find the number of moles.

$$\text{Number of moles} = \frac{11.2 \text{ liters}}{22.4 \text{ liters}} = 0.5$$

$$\text{Number of molecules} = 0.5 \times 6.02 \times 10^{23}$$

$$= 3.01 \times 10^{23}$$

Avogadro's number also enables us to get an approximate idea of the size of molecules. For instance, water has a density of 1 g/cm³. Using this information and Avogadro's number, the approximate size of a water molecule can be found. To do this, the number of molecules in 1 cm³ of water must be found. There will be $\frac{1}{18}$ mole in 1 g of water and $(\frac{1}{18}) \times 6.02 \times 10^{23}$ molecules $= 0.33 \times 10^{23}$ molecules in 1 g, or 1 cm³, of water. Next, we assume that the molecules are packed closely together in the liquid. The volume occupied by one molecule is then the total volume (1 cm³) divided by the number of molecules in that volume.

$$\text{Vol. of one water molecule} = \frac{1 \text{ cm}^3}{0.33 \times 10^{23}}$$

$$= 3 \times 10^{-23} \text{ cm}^3$$

If we then assume that each H_2O molecule is a sphere (which we know it is not, it looks more like a triangle), we can find the radius of that sphere.

$$\frac{4}{3}\pi r^3 = 3 \times 10^{-23} \text{ cm}^3$$

We then estimate the radius of the sphere

$$r^3 = \frac{3}{4\pi} \times 3 \times 10^{-23} \text{ cm}^3$$

$$= .72 \times 10^{-23} \text{ cm}^3$$

$$= 7.2 \times 10^{-24} \text{ cm}^3$$

Take the cube root to find r.

$$r = \sqrt[3]{7.2} \times 10^{-8} \text{ cm}$$

The cube root of 8 is 2 so the cube root of 7.2 is about 2 and the radius of a water molecule is about 2×10^{-8} cm or about 2 angstroms. This seems a reasonable value since we already know that the Bohr

radius of the hydrogen atom is 0.529 angstrom. We would expect the water molecule to be a few times larger than a single hydrogen atom.

14-5
SOLUTIONS

Solutions are homogeneous mixtures. They can be solid, liquid, or gaseous solutions. Metal alloys and air have previously been cited as examples of solid and gaseous solutions. The most common solutions, however, are liquids. The substance in excess in a solution is called the **solvent.** The substance dissolved is the **solute.** Solutions in which water is the solvent are called **aqueous solutions.** A solution in which the solute is present in only a small amount is called a **dilute solution.** If the solute is present in large amounts, the solution is **concentrated solution.** When the maximum amount of solute possible is dissolved in the solvent, the solution is called a **saturated solution.**

The concentration, or amount of solute dissolved, is frequently expressed in terms of the molar concentration. The **molar concentration,** or **molarity,** is the number of moles of solute per liter of solution. Thus, a one molar solution, written 1.0*M*, has one gram formula weight of solute dissolved in one liter of solution. In general

$$\text{Molarity} = \frac{\text{Number of moles of solute}}{\text{Number of liters of solution}}$$

Note that it is the *number of liters of solution,* not the number of liters of solvent, which is used. For example, a one-molar solution of sodium chloride (NaCl), or table salt, in water is made by dissolving one mole (58.5 g) of NaCl in water so that the final solution is one liter. One liter of water is not added to the mole of solute; only enough water is added to make the volume of the solution one liter.

The concentration of a solution can be any value up to the saturation limit. The molarity desired can be obtained in any number of ways, depending on the amount of solution desired. For example, a 2.0M solution can be made by dissolving 2 moles of solute in water to make a liter of solution, or, if we wish to make a smaller quantity, by dissolving 1 mole of solute in 0.5 liter of solution. The concentration is the same either way.

■ **Example 11** What is the molarity of 40 g of salt dissolved in water to give 0.80 liter of solution?
First find the number of moles of salt.

$$\text{Number of moles} = \frac{40 \text{ g}}{58.5 \text{ g}} = 0.684$$

$$\text{Molarity} = \frac{0.684}{0.80} = 0.855M$$

■ **Example 12** How many grams of H_2SO_4 are present in 0.6 liter of a 3.0M solution of sulfuric acid?
First find the number of moles, then convert to number of grams.

$$3.0M = \frac{\text{Number of moles}}{0.6 \text{ liter}}$$

$$\text{Number of moles} = 3.0 \times 0.6 = 1.8$$

Now find the gram formula weight of H_2SO_4.

$$\text{Grams } H_2SO_4 \text{ in 1 mole} = (2 \times 1)$$
$$+ (1 \times 32) + (4 \times 16)$$
$$= 98 \text{ g}$$

$$\text{Grams in 1.8 moles} = 1.8 \times 98 \text{ g}$$
$$= 176.4 \text{ g}$$

Remembering that Avogadro's number (6.02×10^{23}) is the number of molecules in a mole of any substance, we can find the number of molecules or ions in any concentration of solution. For example, a single liter of a 2.0M concentration of table sugar, a covalent compound, contains $2 \times 6.02 \times 10^{23} = 12.04 \times 10^{23}$ sugar molecules. If we have an ionic compound such as NaCl, we can compute the number of ions in the solution. A 0.5 liter quantity of a

3.0M NaCl solution would contain $2 \times 3.0 \times 0.5 \times 6.02 \times 10^{23} = 18.06 \times 10^{23}$ ions. The factor of two comes about because NaCl contains two ions, Na$^+$ and Cl$^-$.

When a solute is added to a solvent, it dissolves and, if thoroughly stirred, distributes itself uniformly throughout the solvent. The distribution of molecules or ions is the same at the top and bottom of the solution. As more solute is added, the solution becomes more and more concentrated; more and more solute at the top and at the bottom. Finally, after a given quantity has been added, no more solute dissolves and we say the solution is saturated. The **solubility** of a given solute is the amount of solute that will dissolve in a specified volume of solvent (at a given temperature) to produce a saturated solution. The solubility depends on the temperature of the solution. If the temperature is raised, the solubility of solids almost always increases. For this reason, hot water dissolves more of the solute than cold water.

If a gas such as CO_2 is being dissolved in water, the solubility increase is directly proportional to the pressure. This principle is used in the manufacture of soft drinks. The carbon dioxide is first forced into the beverage at high pressure. Then the beverage is bottled and tightly capped to maintain the pressure on the CO_2. Once the bottle is opened, the pressure inside the bottle is reduced to normal atmospheric pressure, and the CO_2 starts escaping from the liquid as shown in Fig. 14-5a. If the bottle is allowed to remain open for some time, most of the CO_2 escapes, and the drink loses its zingy taste. The solubility of gases in most liquids decreases with increasing temperature. This effect is also demonstrated with soft drinks. If the beverage is allowed to warm before opening the bottle, the solubility of the CO_2 decreases, and when the bottle is opened, the CO_2 may escape so fast that the beverage shoots out of the bottle.

When saturated solutions of solids are prepared at high temperatures and then cooled, the solubility drops and the excess solid crystallizes (forms crystals) and separates from the solution. However, if there are

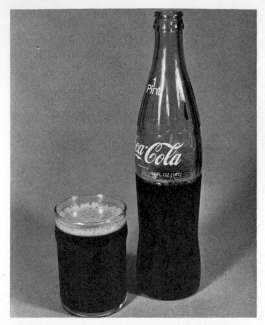

Figure 14-5a *Dissolved carbon dioxide escapes from a carbonated beverage when pressure on the liquid surface is lowered. (Photo courtesy James Crouse.)*

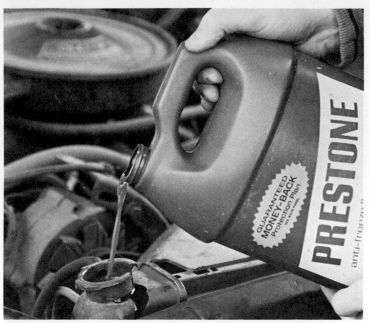

Figure 14-5b *In sufficient amounts, an antifreeze added to water in an automobile radiator produces a solution with a freezing point low enough to prevent freezing during cold weather. (Photo courtesy James Crouse.)*

no crystals of the solid already present in the saturated solution, crystallization may not take place if the saturated solution is carefully cooled. The solution will then contain a larger amount of solute than the solubility of the solute dictates. Solutions which contain more than the normal maximum amount of dissolved solute are said to be **supersaturated solutions.** Such solutions are unstable. The introduction of a "seed" crystal will cause the excess solute to crystallize. As discussed later in Chapter 23, this is the basic principle behind cloud seeding to cause rain. If the air is supersaturated with water vapor, the introduction of certain types of crystals into the clouds greatly increases the probability that the water vapor will form raindrops. Under certain conditions the carbon dioxide solution comprising a soft drink also may be supersaturated. Vigorously shaking the solution causes the excess gas to be liberated.

Whenever a solid is dissolved in water, the freezing point of the water solution is lowered and the boiling point is raised. These changes are related to the number of molecules or ions of solute present in the solution. Ordinary tap water boils at a slightly higher temperature and freezes at a slightly lower temperature than pure water, because of the presence

of impurities. Similarly, the salt in ocean water causes the freezing point of salt water to be lower than fresh water. A well-known application of this principle of lowering the freezing point is the use of "antifreeze" in car radiators (see Fig. 14-5b.) The antifreeze in use today is either an alcohol or a glycol. The more antifreeze that is present, the more the freezing point is lowered.

As stated previously, water is a polar molecule. Water dissolves ionic compounds or polar covalent compounds, because of its electrical attraction for such substances. However, when a nonpolar covalent compound such as carbon tetrachloride or gasoline is combined with water, it is not attracted by the water molecules. The water molecules attract only each other, and the nonpolar molecules are forced away from the water. For this reason, nonpolar compounds will not dissolve in water. Nonpolar compounds will dissolve each other, however, but they will not dissolve ionic or polar covalent compounds. As a general rule, "like dissolves like." If you wish to dissolve stains made by a nonpolar compound such as grease, use nonpolar solvents such as gasoline and benzene. For a polar compound, such as sugar, water should be used.

IMPORTANT WORDS AND TERMS

Homogeneous
Heterogeneous
Substance
Solution
Physical properties
Physical changes
Chemical properties
Chemical changes
Law of conservation of mass
Law of definite proportions
Law of multiple proportions
Gram atomic weight

Formula weight
Gram formula weight
Mole
Molecular weight
Gram molecular weight
Percentage composition
Gay-Lussac's law of
  combining volumes
Avogadro's law
Liter
Standard conditions
Gram molecular volume

Avogadro's number
Solutions
Solvent
Solute
Aqueous solution
Dilute solution
Concentrated solution
Saturated solution
Molar concentration
Molarity
Solubility
Supersaturated solution

LEARNING OBJECTIVES

After reading and studying this chapter, you should be able to do the following without referring to the text:

1. Describe the various types of matter and give examples of each.
2. Distinguish between physical and chemical properties of matter.
3. State and explain the following laws:
    (a) Conservation of mass
    (b) Definite proportions
    (c) Multiple proportions
    (d) Gay-Lussac's law of combining volumes
    (e) Avogadro's law
4. Distinguish between gram atomic weight and gram formula weight.
5. Distinguish between gram molecular weight and molecular weight.
6. State the physical characteristics required for standard conditions.
7. State the numerical value of Avogadro's number.
8. Distinguish between solvent and solute.
9. Describe the principle characteristics of solutions.
10. Distinguish between molar and molar solution.
11. Calculate the formula weight of a given compound.
12. Calculate the percentage composition of a compound.
13. Calculate the number of grams of a compound that can be obtained from a given amount of known elements.
14. Calculate the number of grams of an element that can be obtained from a given amount of a known compound.
15. Calculate the number of moles present in a given amount of a compound.
16. Calculate the number of molecules in a given amount of a compound.
17. Calculate the volume at standard conditions that will be occupied by a given amount of a known gas.

18. Calculate the molarity of a solution when the volume and the amount of the substance in the solution are given.
19. Define or explain the important words and terms listed at the end of the chapter.

QUESTIONS

1. Describe the physical properties that distinguish solids, liquids, and gases.
2. Identify each of the following as a physical or chemical change:
   (a) Melting of ice   (b) Burning of a match
   (c) Fermenting of wine   (d) Dissolving sugar in water
   (e) Rusting of steel   (f) Magnetizing a sewing needle
3. Distinguish between homogeneous and heterogeneous matter.
4. Identify three pure substances and list some of the properties used to identify them as pure substances.
5. Give some examples of solutions that are not in the liquid phase.
6. State the law of conservation of mass.
7. Is the law of conservation of mass exactly true or only approximately true?
8. What is the molecular explanation of the law of definite proportions?
9. What is the molecular explanation of the law of multiple proportions?
10. State Gay-Lussac's law of combining volumes.
11. What were Avogadro's assumptions that explained Gay-Lussac's law of combining volumes?
12. Distinguish between gram atomic weight and gram molecular weight. Give an example of each.
13. How is it possible to mix one volume of a gas with three volumes of another gas and end up with only two volumes?
14. Why is the volume of a gas occupied by one mole always the same, no matter what the gas (assuming standard conditions)?
15. What is the meaning of the term "standard conditons"?
16. What is Avogadro's number? State its numerical value.
17. Describe the following terms as applied to solutions: solute, solvent, aqueous, dilute, concentrated, saturated, supersaturated.
18. Define a one-molar solution. Give an example.
19. Why do soft drinks contain bubbles? What are the bubbles?
20. Why does an opened soft drink go flat after a period of time?
21. Why is antifreeze added to the water in a car's radiator during the winter?
22. On what type of compounds is carbon tetrachloride a useful cleaning agent? Explain your answer.

PROBLEMS

1. Calculate the formula weight of the following compounds:
   (a) Water (H_2O)   (b) Carbon dioxide (CO_2)
   (c) Methane (CH_4)   (d) Benzene (C_6H_6)

2. Calculate the formula weight of the following compounds:
   (a) Sodium chloride (NaCl)
   (b) Calcium carbonate ($CaCO_3$)
   (c) Hydrochloric acid (HCl)
   (d) Sodium bicarbonate ($NaHCO_3$)

3. Determine the percentage composition of the following compounds:
   (a) Water (H_2O)
   (b) Sugar ($C_{12}H_{22}O_{11}$)
   (c) Table salt (NaCl)
   (d) Aspirin ($CH_3COOC_6H_4COOH$)

4. Determine the percentage composition of the following compounds:
   (a) Carbon monoxide (CO)
   (b) Potassium chlorate ($KClO_3$)
   (c) Manganese dioxide (MnO_2)
   (d) Iron rust (Fe_2O_3)

5. Calculate the number of grams of sodium chloride (NaCl) that can be formed from 100 g of sodium and 105 g of chlorine. What element is not completely used? How much is not used?

6. Calculate the number of grams of carbon dioxide (CO_2) that can be formed from 36 g of carbon and 100 g of oxygen. What element is not completely used? How much is not used?

7. Calculate the number of grams of sulfur that can be formed from 128 g of sulfur dioxide (SO_2).

8. Calculate the number of grams of lime (CaO) and carbon dioxide (CO_2) that can be produced from 200 g of calcium carbonate ($CaCO_3$).

9. Find the number of moles present in each of the following:
   (a) 56 g of carbon monoxide
   (b) 30 liters of carbon monoxide gas at standard conditions
   (c) 3.01×10^{23} molecules of carbon monoxide

10. Determine the number of molecules contained in the following quantities:
    (a) 1 g of water
    (b) 44.8 liters of carbon monoxide gas
    (c) 4.0 moles of sugar

    Answer: (a) 0.33×10^{23}

11. Compute the volume at standard conditions which will be occupied by the following gases:
    (a) 40 g of argon
    (b) 150 g of xenon
    (c) 48 g of oxygen
    (d) 14 g of carbon dioxide

    Answer: (c) 33.6 liters

12. Find the molarity of the following solutions:
    (a) 2.5 moles of sugar in 1 liter of solution
    (b) 49 g of H_2SO_4 in 2 liters of solution
    (c) 12 g of HCl in 0.7 liter of solution

    Answer: (c) $0.47M$

Chapter 15

CHEMICAL REACTIONS

There's nothing constant in the universe,
All ebb and flow, and every shape that's
     born
Bears in its womb the seed of change.

Ovid, Metamorphoses

Chemical change in nature takes place continuously. Green plants absorb carbon dioxide from the atmosphere and with energy from the sun and chlorophyll as a catalyst react with water from the soil to form sugar and free oxygen. The complex chemical reaction is called photosynthesis. The animal life on our planet ingest the sugar and inhale the oxygen to obtain energy. Chemical change in the body then returns to the environment water and carbon dioxide. This interaction between plants and animals and their environment constitutes an ecosystem.

Our environment is made up of atoms that combine with one another to form molecules, which in turn react with other atoms or molecules to produce the many substances we need and use. In the production of new products, energy is released or absorbed, depending on the nature of the chemical reaction. There are countless chemical reactions, but only a few will be studied in this chapter.

Something of the nature of chemical reactions can be learned by studying examples of two specific types of reactions: oxidation-reduction and acid-base reactions. Everyone is acquainted with a match burning in air and the rusting of iron. These are examples of a chemical reaction known as an oxidation-reduction reaction. Most of us are also acquainted with a sour or acid stomach and the substances used to neutralize it. Neutralization is an example of an acid-base reaction. We will see in Chapters 16, 21, and 31 that chemical reactions are important in the chemistry of large molecules, in atmospheric processes, and in geology. The present chapter discusses the basic chemistry involved in chemical reactions, in particular, oxidation-reduction and acid-base reactions.

15-1
BASIC CONCEPTS

When two or more substances are brought together, they may react to form new substances. Consider an example in which substances A and B react to form substances C and D. This is written

$$A + B \longrightarrow C + D$$

The expression is called a chemical equation. It is read: Substances A and B react to yield the substances C and D. The initial substances A and B are called the **reactants,** and the final substances C and D are called the **products.**

In any chemical reaction three things take place.

1. The reactants disappear or are diminished.
2. Different substances appear as products. These products have chemical and physical properties which are significantly different from the original reactants.
3. Energy is either released or absorbed. This energy may take the form of heat, light, electricity, etc.

The energy characteristics of chemical reactions will be discussed more fully in Section 15-2.

Now let us suppose that we turn our original reaction around, so that when C and D are combined, A and B are formed as the products. We get

$$C + D \longrightarrow A + B$$

If we start with A and B, we get C and D; if we start with C and D, we get A and B. A reaction of this type

is called a **reversible reaction.** The reversible reaction can be written

$$A + B \Longrightarrow C + D$$

The double arrows mean that the reaction proceeds in both directions. An example of a reversible reaction involves iron, steam, and hydrogen gas. The equation is

$$3\,Fe + 4\,H_2O \Longrightarrow Fe_3O_4 + 4\,H_2$$

If steam is passed over hot iron, the reaction proceeds from left to right. If hydrogen gas is passed over heated Fe_3O_4, the reaction goes from right to left.

In a reversible reaction such as

$$A + B \Longrightarrow C + D$$

all four substances (A, B, C, and D) are present in the vessel containing the substances. On a molecular scale, molecules of A and B are combining to form molecules of C and D; and at the same time, molecules of C and D are combining to form molecules of A and B. When the two competing reactions are occurring at the same rate, we say that the reactions have reached a state of equilibrium. **Equilibrium** is a dynamic process in which there is a fixed number of molecules of all reactants and products present. However, individual molecules are continuously changing back and forth, due to interactions with each other.

The amounts of A, B, C, and D present at equilibrium depend on many things. Among these are temperature, pressure, and the excess amounts of reactants and products present. As an example, consider again the reaction

$$3\,Fe + 4\,H_2O \longrightarrow Fe_3O_4 + 4\,H_2$$

and the reverse reaction

$$Fe_3O_4 + 4\,H_2 \longrightarrow 3\,Fe + 4\,H_2O$$

In both of these reactions a stream of gas is passed over a hot solid. If steam is passed over hot iron, the hydrogen is swept out by the steam, and the first reaction dominates. If hydrogen gas is passed over

Fe_3O_4, the steam is swept out by the excess hydrogen, and the second reaction occurs. This example demonstrates just one of many methods by which a reversible reaction may be shifted from right to left or left to right.

Every reaction is to some extent reversible. That is, there will always be some of every kind of atom or molecule present which can possibly occur in a reaction. For example, consider the reaction

$$2\,HI \longrightarrow H_2 + I_2$$

This reaction says that hydrogen iodide decomposes into hydrogen and iodine. However, in any sample of hydrogen and iodine there will be some hydrogen iodide, even though it will be a very small amount. To show explicitly that this is the case, we would write the reaction with a small arrow going from right to left.

$$2\,HI \rightleftharpoons H_2 + I_2$$

Writing the reaction this way emphasizes the reversibility, however slight, of the reaction. However, in the discussions that follow, we will omit the small arrow from right to left. It will be understood that at equilibrium there are always some molecules of all the various reactants and products in the reaction.

A chemical reaction is sometimes referred to as a chemical equation. The reason for calling a reaction an equation is that the law of conservation of mass means that in chemical reactions, atoms cannot be created or destroyed. Thus, there must be the same number of atoms on each side of the arrow in any equation for a chemical reaction. As an example, consider the "roasting" of zinc ore (ZnS). In this process sulfide ore is heated in air in order to convert the zinc sulfide to an oxide from which pure zinc can be extracted. The chemical equation for the roasting of the ore zinc sulfide is

$$2\,ZnS + 3\,O_2 \longrightarrow 2\,ZnO + 2\,SO_2$$

The numbers in front of the chemical formulas in this equation are called coefficients. They indicate the relative amounts of each substance which participates in the reaction. For instance, in the above reaction, for every 3 moles of oxygen gas reacting with every 2 moles of zinc sulfide, the result will be 2 moles of zinc oxide and 2 moles of sulfur dioxide. An inspection of the equation will show that it is indeed balanced. A tabulation of the number of atoms on both sides of the equation shows

$$\left.\begin{array}{l} \text{2 atoms of zinc} \\ \text{2 atoms of sulfur} \\ \text{6 atoms of oxygen} \end{array}\right\} \longrightarrow \left\{\begin{array}{l} \text{2 atoms of zinc} \\ \text{2 atoms of sulfur} \\ \text{6 atoms of oxygen} \end{array}\right.$$

This equality of atoms is because atoms cannot be created or destroyed in chemical reactions.

In balancing chemical equations the subscripts on the various elements represent the number of atoms in each molecule, and these numbers cannot be changed. The coefficients are to be found such that there is an equal number of atoms on both sides of the arrow. As an example, consider balancing the equation for the decomposition of potassium chlorate ($KClO_3$) when heated. (This is a common method of producing oxygen in the laboratory.)

- *Example 1* Balance the following:

$$KClO_3 \longrightarrow O_2 + KCl$$

To balance the equation, inspect it to see which elements have a different number of atoms on the two sides of the equation. We see that the potassium and chlorine atoms are balanced already. Only the oxygen is not. Of course, once we adjust the equation to balance the oxygen atoms, we may unbalance the potassium and chlorine atoms.

To get the oxygen atoms equal on both sides we must put a 2 in front of the $KClO_3$ (potassium chlorate) and a 3 in front of the O_2. This gives us

$$2\,KClO_3 \longrightarrow 3\,O_2 + KCl$$

The oxygen is now balanced, but the potassium and chlorine are unbalanced. However, this is easily rectified by putting a 2 in front of the KCl. Finally we have the balanced equation

$$2 \, KClO_3 \longrightarrow 3 \, O_2 + 2 \, KCl$$

In addition to balancing the number of atoms on both sides of an equation, the electric charge must also be balanced. This is because electric charge cannot be created or destroyed. Consider the equation for the ionization of sulfuric acid.

$$H_2SO_4 \longrightarrow 2 \, H^+ + SO_4^{2-}$$

In this equation the charges on each side of the arrow add up to zero. The number of atoms on both sides of this equation is also the same, as it must be.

Now let us consider balancing a more difficult equation.

■ *Example 2* Balance the following:

$$Pb + PbO_2 + H^+ + SO_4^{2-} \longrightarrow PbSO_4 + H_2O$$

This is a reaction which occurs in a lead storage battery when it is being used. To balance the equation we first inspect it to see what is not balanced. The lead is not balanced so we put a 2 in front of the $PbSO_4$ on the right. To balance the sulfur, we put a 2 in front of the SO_4^{2-} on the left. We then have

$$Pb + PbO_2 + H^+ + 2 \, SO_4^{2-} \longrightarrow 2 \, PbSO_4 + H_2O$$

Now we need a 4 in front of the H^+ to get a total charge of zero on the left (4 + charges and 4 − charges) and a 2 in front of the H_2O on the right to balance the equation. The equation is correctly written as

$$Pb + PbO_2 + 4 \, H^+ + 2 \, SO_4^{2-} \longrightarrow$$
$$2 \, PbSO_4 + 2 \, H_2O$$

15-2
ENERGY AND RATE OF REACTION

All chemical reactions involve a change in energy. The energy is either released or absorbed in the form of heat, light, electrical energy, sound, etc. If energy is

Figure 15-2a *A gas flame. Visible evidence of an exothermic chemical reaction. (Photo courtesy James Crouse.)*

released in a chemical reaction the reaction is called an **exothermic reaction;** if energy is absorbed the reaction is called an **endothermic reaction.** An example of a common exothermic reaction is the burning of natural gas, which is composed primarily of methane.

$$CH_4 + 2 \, O_2 \longrightarrow CO_2 + 2 \, H_2O + Energy$$

This reaction occurs when a gas stove is lighted. (See Fig. 15-2a.) An example of an endothermic reaction is the production of ozone, the triatomic molecule of oxygen.

$$3 \, O_2 + Energy \longrightarrow 2 \, O_3$$

This reaction occurs in the upper atmosphere, where the energy is provided by the ultraviolet radiation from the sun. It also occurs near electric discharges. Ozone has a pungent odor, which is readily detected even in very small amounts. It is the unusual odor one notices when near a toy electric train that is operating and its motor is producing sparks.

In the two preceding equations we have explicitly included the word "energy" to show the exothermic or endothermic nature of the reactions. This is usually not done, however.

The energy associated with a chemical reaction is related to the bonding energies between the atoms that form the molecules. The process is complex, because during a chemical reaction some chemical bonds are broken while others are formed, and it is difficult to determine the energies associated with each. In general, energy is absorbed when bonds are broken and given off when they are formed. In the ozone reaction the energy balance can be summarized as follows:

$$O_2 \text{ bonds broken} \longrightarrow O_3 \text{ bonds formed}$$
(more   (less energy re-
energy absorbed)   leased)
Net result: Energy is absorbed

When methane burns in air the energy balance is

$$\left.\begin{array}{l} CH \text{ bonds broken} \\ O_2 \text{ bonds broken} \end{array}\right\} \quad \left\{\begin{array}{l} CO \text{ bonds formed} \\ OH \text{ bonds formed} \end{array}\right.$$
(less   (more energy re-
energy absorbed)   leased)
Net result: energy is released

A chemical reaction may occur when the reactants are brought together; that is, when two or more molecules or ions collide with one another, they may react chemically, but *only if the kinetic energy of the molecules is above a minimum value.* In other words, a certain amount of energy is needed to start a chemical reaction. As an example, we are all aware of the necessity of striking a match before it ignites. In this example, we must contribute some energy—through friction—to initiate the chemical reaction of a match burning. Another example is the burning of methane gas. A flame is needed to ignite the methane, because the CH and O_2 bonds must be broken initially. Once the stove is lighted the net energy released suffices to break the bonds of still more CH_4 and O_2 molecules, and the reaction proceeds continuously, giving off energy in the form of heat and light.

The energy necessary to get a chemical reaction started is called the **activation energy.** This is the kinetic energy colliding molecules must possess in order to react chemically. However, once the activation energy is supplied, an exothermic reaction can release more kinetic energy than was supplied. This is

Figure 15-2b *A graphic illustration of the energetics of an exothermic chemical reaction. If enough energy is supplied to activate the reaction (lifting the boulder), the total energy released (the falling boulder) will equal the activation energy plus the net energy released.*

illustrated on a different scale in Fig. 15-2b. In the figure a boulder is on top of a cliff. If enough energy is supplied to raise the boulder to the edge of the cliff, the boulder will roll down the cliff, releasing great amounts of energy. A similar situation exists for an exothermic reaction. Once the activation energy is supplied, the formation of new molecular bonds can release more kinetic energy than was absorbed in breaking the original bonds. In endothermic reactions, more kinetic energy is absorbed to break molecular bonds than is generated when new bonds are formed. In this case there is a net loss of kinetic energy. This is usually perceived on a large scale as a "loss of heat."

When an exothermic chemical reaction takes place so very rapidly that there is a large change in the volume of the gases and an almost instant liberation of the energy of the reaction, an explosion occurs. Some of the energy in an explosion is given off in the form of sound energy. A reaction which proceeds more slowly than an explosion and yet is still quite rapid is termed combustion. In combustion, heat and light are produced in the form of fire. Common examples of combustion are the burning of natural gas, wood, paper, or coal in air. When gasoline in the form of heptane is burned, the combustion reaction is

$$C_7H_{16} + 11\,O_2 \longrightarrow 7\,CO_2 + 8\,H_2O$$

In the automobile engine there is insufficient oxygen to completely burn all of the carbon, and the reaction is more commonly

$$C_7H_{16} + 9\,O_2 \longrightarrow 4\,CO_2 + 2\,CO + C + 8\,H_2O$$

In this reaction the poisonous gas carbon monoxide, CO, is formed, along with the element carbon, which is black. The gases coming out of an automobile tailpipe consist of significant amounts of poisonous carbon monoxide. For this reason, running an automobile engine in a closed garage is very dangerous. The carbon monoxide accumulating from the exhaust can be fatal. The black color of some exhaust gases indicates the presence of large amounts of carbon, and this indicates that incomplete oxidation is occurring.

The rate of a reaction is dependent on several factors. One of the most important factors affecting the rate of reaction is the temperature. This is due to the fact that the average kinetic energy of the molecules is directly proportional to the temperature. In many cases the rate of reaction is significantly altered by even a slight change in temperature. In fact, as a general rule of thumb, a 10°C increase in temperature leads to a doubling of the reaction rate. This is sometimes referred to as the **ten-degree rule.**

The boundary of an atom or molecule is defined by its outermost orbiting electrons. These outer electrons are repelled by the electrons of neighboring atoms or molecules and thus, the particles tend not to react unless they possess enough energy to overcome this tendency. The kinetic energy possessed by any one atom or molecule may or may not be great enough for reaction. Molecules and atoms which have low energies of activation readily react because a greater number of the collisions are effective. If the temperature of the reactants is raised, the kinetic energy of the molecules will be increased, and the reaction rate will increase.

In some reactions the energy of activation of the molecules or atoms can be partially altered by adding

Figure 15-2c *As molecules increase their speed, the probability of a collision effectively leading to a reaction increases. The top portion of the drawing shows a hydrogen iodide molecule that has insufficient energy to cause a reaction. The HI molecules merely bounce off each other. In the lower collision the molecules are moving fast enough to break their internal bonds and form H_2 and I_2 molecules, according to the equation $2\,HI \rightarrow H_2 + I_2$.*

a substance called a catalyst. A **catalyst** is a substance which speeds or slows the rate of reaction but is not itself consumed in the reaction. Catalysts have no effect on the equilibrium mixture of a reaction. They only lessen the amount of time required to reach equilibrium conditions. Some catalysts act by providing a surface on which the reactants are adsorbed and concentrated. Others act as "carriers"; that is, they unite with a reactant to form an intermediate substance which takes part in the chemical process and then decomposes to release the catalyst. The use of nitrogen oxide (NO) to produce SO_3 used in the manufacture of sulfuric acid is an example of a catalyst which acts as a carrier.

$$2\,SO_2 + O_2 \longrightarrow 2\,SO_3 \text{ (slow reaction)}$$

$$2\,NO + O_2 \longrightarrow 2\,NO_2 \ \Big\}$$
$$SO_2 + NO_2 \longrightarrow SO_3 + NO \ \Big\} \quad \text{(fast reactions)}$$

The NO combines with oxygen to form NO_2. The NO_2 then reacts with SO_2 to form SO_3, releasing NO, which can be used again.

Another example of using a catalyst occurs in the decomposition of potassium chlorate ($KClO_3$). A common method for preparing oxygen in the laboratory is the decomposition of potassium chlorate by heating. When heated to 400°C the white crystalline solid breaks down, producing potassium chloride and releasing oxygen. The reaction can be written as

$$2 KClO_3 + 400°C \longrightarrow 2 KCl + 3 O_2$$

The process, however, is slow and heat requirements are high. If a small amount of manganese dioxide (MnO_2) is mixed with the $KClO_3$, the reaction takes place rapidly at 250°C. The manganese dioxide does not enter into the reaction but acts only as a catalyst.

The way in which many catalysts work is not known. A substance that acts as a catalyst for one reaction may or may not act as a catalyst for another. Catalysts are used extensively in manufacturing processes. They also play an important part in biochemical processes. The human body has many thousands of catalysts called **enzymes,** which act to control various physiological reactions. It is also believed that vitamins and hormones are catalysts.

15-3
OXIDATION-REDUCTION REACTIONS

The element oxygen was discovered in 1774 by Joseph Priestley (1733–1804), an English clergyman, who referred to the element as "air." Priestley made the discovery by heating red rust of mercury (mercuric oxide) in a closed glass flask with a burning glass (lens). The reaction which Priestley observed was

$$2 HgO \longrightarrow 2 Hg + O_2$$

In this reaction, the oxide has decomposed, but it is much more common for oxygen to combine with another substance to form the oxide. We have seen that combustion is a reaction that frequently involves oxygen. When oxygen combines with another substance we call the reaction oxidation. For instance, the equation for the oxidation of carbon to form carbon dioxide is

$$C + O_2 \longrightarrow CO_2$$

Another example of oxidation is the souring of wines, that is, the changing of alcohol to acetic acid. The reaction can be written as

$$\underset{\text{Ethyl alcohol}}{CH_3CH_2OH} + O_2 \longrightarrow \underset{\text{Acetic acid}}{CH_3COOH} + H_2O$$

Acetic acid is the principal component of vinegar. Obviously, you should keep your wine sealed airtight!

When oxygen is removed from a compound, the reaction is called reduction. For example, when the ore hematite, iron(III) oxide, is heated with coke (carbon) in a blast furnace reduction occurs, the iron oxide being reduced to iron. This reaction is extremely important in the steel-making industry. It can be written as

$$2 Fe_2O_3 + 3 C \longrightarrow 4 Fe + 3 CO_2$$

We observe from the equation that oxygen is removed from the iron oxide. We say that the iron has been reduced. At the same time, we see that oxygen has reacted with the carbon to form carbon dioxide, that is, the carbon has been oxidized. Thus, two reactions, oxidation and reduction, occur together in this example. A reaction in which both oxidation and reduction occur is called an **oxidation-reduction reaction.**

Oxidation-reduction reactions do not always involve oxygen. In order to apply to a larger number of reactions, oxidation-reduction reactions are also stated in terms of electrons gained or lost by atoms. By definition the loss of electrons by an atom or ion is called **oxidation.** The gain of electrons by an atom or ion is called **reduction.** When an atom or ion gives up electrons, it is **oxidized,** and when an atom or ion gains electrons, it is **reduced.** For example, when copper (Cu) combines with oxygen to form copper(II) oxide, the copper atoms lose electrons and are thus said to be oxidized. The oxygen atoms gain electrons and are reduced.

$$2\,Cu + O_2 \longrightarrow 2\,(Cu^{2+}O^{2-})$$

Since all the electrons lost by atoms or ions must be gained by other atoms or ions, it is evident that oxidation and reduction occur at the same time and at the same rate. The atom that is oxidized is the **reducing agent;** the atom that is reduced is called the **oxidizing agent.** In the example above, the metal (copper) reduced the oxygen; therefore, it was the reducing agent. Likewise, the nonmetal (oxygen) caused the oxidation to occur, so it was the oxidizing agent. Generally speaking, metals are reducing agents and nonmetals are oxidizing agents.

As just mentioned, according to the electron concept the element oxygen does not have to take part in an oxidation-reduction reaction. If we place a shiny iron nail in a copper sulfate solution, the iron nail becomes coated with a film of copper, and we find iron ions in the solution. This is an oxidation-reduction reaction. The copper has replaced the iron on the nail. The iron atoms lose electrons and are oxidized into ferrous (2+) ions. The copper ions in solution gain electrons and are reduced to copper atoms. The reaction is shown in the following equation:

$$Fe + Cu^{2+}SO_4^{2-} \longrightarrow Fe^{2+}SO_4^{2-} + Cu$$

or $\qquad Fe + Cu^{2+} \longrightarrow Fe^{2+} + Cu$

In the reaction of copper sulfate with iron,

$$CuSO_4 + Fe \longrightarrow Cu + Fe^{2+}SO_4^{2-}$$

the iron displaces the copper. Such reactions are called **displacement reactions.** Thus, we see that some metals give up their valence electrons while others gain electrons. The activity of a metal, that is, its propensity to lose electrons, is due to its electron configuration. When metals compete for electrons,

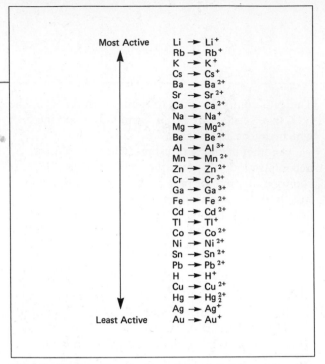

Figure 15-3a *The electromotive, or activity, series. The higher the element is in the series, the more active it is. For example, Fe is more active than Cu, and Zn is more active than Fe.*

the more active metal yields to the less active. Thus, iron atoms, being more active than copper atoms, lose electrons to the copper ions and become iron ions. The copper ions, on the other hand, gain electrons and are converted to neutral copper atoms.

The relative activity of any metal can be discovered by merely placing the metal in an ionic solution containing another metal and seeing if the metal replaces the one in solution. If it does, it is more active, because it has given electrons to the metal in solution. If this process is carried out for all metals, an order-of-activity series can be obtained. Such a series is called the **electromotive series.** It is shown in Fig. 15-3a.

There are thousands of important oxidation-reduction reactions. We will mention only a few more examples. In the commercial production of metals the metals must be separated from other elements present in the metals' ores. Frequently, the ores are sulfides or carbonates. The first step is to "roast" the ores. This consists of heating them to very high temperatures in the presence of air. Some typical reactions are

$$2\,ZnS + 3\,O_2 \longrightarrow 2\,ZnO + 2\,SO_2$$

$$PbCO_3 \longrightarrow PbO + CO_2$$

The roasting process converts the metal to its oxide. The next step is to reduce the oxide to the pure metal. Carbon and carbon monoxide are frequently used in this step. For example,

$$ZnO + C \longrightarrow Zn + CO$$

$$Fe_2O_3 + 3\,CO \longrightarrow 2\,Fe + 3\,CO_2$$

These reactions are extremely important in the commercial manufacturing of metals.

Glass consists of molten material which has cooled very rapidly. Although glass seems to be a solid, it does not have the crystalline structure typical of solids, and for this reason it is usually thought of as a liquid which flows very, very slowly. The common varieties of glass (such as that used in windows and clear bottles) are made by heating mixtures of Na_2CO_3, $CaCO_3$, and SiO_2 (sand). The reactions which occur are

$$Na_2CO_3 + SiO_2 \longrightarrow Na_2SiO_3 + CO_2$$

and

$$CaCO_3 + SiO_2 \longrightarrow CaSiO_3 + CO_2$$

A mixture of Na_2SiO_3 and $CaSiO_3$ are the basic substances in ordinary glass. Pyrex glass is ordinary glass with boron oxide, B_2O_3, added to it. The addition of B_2O_3 gives the glass a low coefficient of expansion, which enables it to withstand changes in temperatures without breaking. If a colored glass is desired, an appropriate substance is added during the manufacture. A few of the compounds used in coloring glass are listed in Table 15-3a.

Table 15-3a *Compounds Used in Glassmaking to Color Glass*

Compound	Color
Iron(II) compounds	Green
Iron(III) compounds	Yellow, brown
Uranium compounds	Yellow, green, fluorescence
CoO	Blue
MnO_2	Violet
CaF_2	Milky
SnO_2	Opaque
Cu_2O	Red, blue, green

15-4
ELECTROCHEMICAL REACTIONS

As we have seen in previous chapters, the properties of elements and compounds can be explained in terms of their electron structures. Microscopically, chemistry is intimately intertwined with electric charges. In this section we concern ourselves with several important chemical reactions which generate or consume electric charges.

Substances that dissolve in water to give solutions which will conduct an electric current are called **electrolytes.** In Section 13-3 we discussed how molten table salt, NaCl, conducts an electric current. All ionic compounds are electrolytes. Acids and bases are also electrolytes. They also ionize in water and conduct an electric current. Also in Section 13-3, we discussed electrolysis and identified the cathode and the anode. In electrolysis the **cathode** is the electrode at which electrons enter the electrolytic solution. The **anode** is the electrode where electrons are withdrawn from the electrolytic solution. In electrochemistry the important reactions occur at the anode and the cathode.

Figure 15-4a *Cross section of a dry cell, showing the important parts.*

Electrochemical cells are extremely important. The dry cell is used in flashlights, transistor radios, etc. to provide an electrical potential. The common dry cell is shown schematically in Fig. 15-4a, and some common examples are shown in Fig. 15-4b. The dry cell is basically a zinc can with a carbon rod imbedded in a paste of ammonium chloride, NH_4Cl. A porous separator is inserted between the zinc can and the NH_4Cl. Some MnO_2 surrounds the carbon rod also. When the cell is connected, electrons flow externally from the zinc can to the carbon rod. The chemical reactions occurring at the cathode and the anode depend on the size of the current required from the cell. When small currents are drawn, the following reactions probably occur:

$$Zn + 4\,NH_3 \longrightarrow Zn(NH_3)_4^{2+} + 2\,e^- \quad \text{(Anode)}$$

$$2\,MnO_2 + 2\,NH_4^+ + 2\,e^- \longrightarrow$$
$$Mn_2O_3 + 2\,NH_3 + H_2O \quad \text{(Cathode)}$$

The dry cell is called an irreversible cell, because the application of an external potential will not recharge it. An example of a reversible cell is the lead storage battery of the type used in automobiles.

The lead storage battery consists of a set of lead grids in a dilute sulfuric acid solution (Fig. 15-4c).

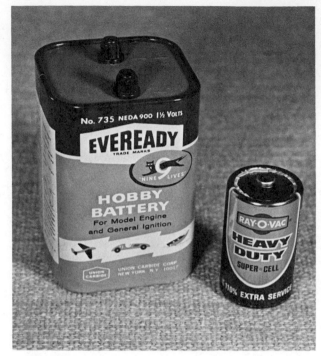

Figure 15-4b *Some typical dry cell batteries. (Photo courtesy James Crouse.)*

Figure 15-4c *Schematic diagram of a lead storage battery.*

One grid, or plate, contains spongy lead, the other contains lead dioxide, PbO_2. These are mounted in the solution of sulfuric acid, H_2SO_4, which acts as an electrolyte by ionizing in water to yield H^+ and SO_4^{2-} ions.

The lead grid acts as the anode, and the reaction during discharge is

$$Pb + SO_4^{2-} \longrightarrow PbSO_4 + 2\,e^- \quad \text{(Anode)}$$

The PbO_2 acts as the cathode. The reaction during discharge is

$$PbO_2 + 4\,H^+ + SO_4^{2-} + 2\,e^- \longrightarrow PbSO_4 + 2\,H_2O \quad \text{(Cathode)}$$

If an external electrical potential is applied to a discharged storage battery it can be recharged. The reactions which occur during recharge are just the reverse of those which occur during discharge. During discharge the sulfuric acid is consumed. Conversely, it is regenerated by the charging process. Since the presence of H_2SO_4 indicates a charged state, the condition of a lead storage battery may be determined by measuring the density of the sulfuric acid solution. The higher the density, the more the battery is charged. When the auto engine is running, the battery is being charged (provided the alternator is working properly). The battery is discharging when the car is being started. A single set of lead plates, such as shown in Fig. 15-4c, generates a potential of 2.05 volts. Thus, a 12-volt battery contains 6 sets of plates, while a 6-volt battery contains 3 sets of plates.

The new maintenance-free battery is similar to the lead storage battery in operation but requires no maintenance over its full life. There are three major factors that qualify a battery as maintenance-free: (a) terminal corrosion has been reduced to a point where it is no longer a factor in limiting battery life, (b) no additional water is required during the life of the battery, and (c) the battery should be able to stand on the shelf for a reasonable period (18 months) without charging and still start an automo-

Figure 15-4d *A maintenance-free battery. No filler caps required because no additional water is ever needed, minimal terminal corrosion and low stand loss. (Photo courtesy J. C. Penny Company.)*

bile or other vehicle. These characteristics have been achieved with the new J.C. Penney battery shown in Fig. 15-4d. A maintenance-free battery has lead alloyed grids with approximately 1.5 percent cadmium and 1.5 percent antimony. The positive and negative grids can be of this construction but the J.C. Penney battery also uses a calcium-lead alloy in the negative grid and the lead alloyed cadmium-antimony for the positive grid. The new battery is called calcium-plus.

The single most destructive chemical reaction which occurs in the United States is the rusting, or corrosion, of iron (see Fig. 15-4e). In fact, about one out of every four men in the steel industry is working to replace iron products which have been ruined by rusting. The corrosion of iron is brought about by the presence of both oxygen and water. Iron will not rust in dry air. The explanation of corrosion involves electrochemistry, as we shall see.

When iron is in contact with water containing an electrolyte, the iron, being very active, tends to give up its electrons according to the reaction

$$Fe \longrightarrow Fe^{2+} + 2\,e^- \quad \text{(Anode)}$$

This reaction can be considered a reaction at an anode. Usually, one part of the iron will be more active than other parts. At the less active part of the

Figure 15-4e *The process of corrosion has seriously damaged this automobile exterior. (Photo courtesy James Crouse.)*

Figure 15-4f *The burning of coal produces many pollutants. One of these is sulfur dioxide, SO_2, which forms a weakly acidic solution with water. (Photo courtesy James Crouse.)*

iron, H^+ ions react to form hydrogen gas, which collects on the iron surface.

$$2\,H^+ + 2\,e^- \longrightarrow H_2 \quad \text{(Cathode)}$$

The accumulation of hydrogen gas on the iron surface tends to inhibit the reaction. However, dissolved oxygen in the water reacts with the H_2 to form water.

$$2\,H_2 + O_2 \longrightarrow 2\,H_2O$$

The iron ions react with the OH^- ions present in the water and then with oxygen to form $Fe_2O_3 \cdot 2\,H_2O$, which is iron rust.

$$Fe^{2+} + 2\,OH^- \longrightarrow Fe(OH)_2$$

$$4\,Fe(OH)_2 + O_2 \longrightarrow 2(Fe_2O_3 \cdot 2\,H_2O)$$
$$\text{Iron rust}$$

The pollutant sulfur dioxide, SO_2, which is emitted during the burning of coal (Fig. 15-4f), will combine with water to form dilute H_2SO_3. This, then, supplies H^+ ions for the cathode reaction in the rusting process. Thus, SO_2 pollution is partially responsible for the destruction of billions of dollars worth of structural materials each year.

Rusting can be prevented by coating the iron surface with a material such as paint or a ceramic enamel, like that used in sinks, stoves, or refrigerators. Some alloys of iron, such as stainless steel [a mixture of iron, chromium (15%) and nickel (8%)], are corrosion resistant.

15-5
ACIDS AND BASES

The acid-base concept originated with the beginning of chemistry and is an important classification of chemical compounds. The word acid is derived from the Latin *acidus*, which means sour. The sour taste is one of the properties of acids, but the student should never actually taste an acid in the laboratory. An **acid** is a substance which, when dissolved in water, has the following properties:

1. Conducts electricity.
2. Changes the color of litmus dye from blue to red.
3. Tastes sour. (The student should *never* taste any acid.)
4. Reacts with a base to neutralize its properties.
5. Reacts with metals (e.g., Zn, Mg) to liberate hydrogen gas.

A **base** is a substance which, when dissolved in water, has the following properties:

1. Conducts electricity.
2. Changes the color of litmus dye from red to blue.
3. Tastes bitter and feels slippery (as with acids, the student should *never* taste a base).
4. Reacts with an acid to neutralize the acid's properties.

One of the first theories formulated to explain the acid-base concept was put forth by Svante Arrhenius (1859–1927), a Swedish chemist who proposed in 1887 that the characteristic properties of aqueous solutions of acids were due to the hydrogen ion (H^+), and the characteristic properties of aqueous solutions of bases were due to the hydroxide ion (OH^-).

According to the **Arrhenius acid-base concept,** when a substance like hydrogen chloride (HCl) is added to water, some HCl molecules dissociate into H^+ ions and Cl^- ions which exist in the solution in equilibrium with the HCl molecules which are not dissociated. The acidic properties of HCl are due to the H^+ ions. Similarly, when a substance such as sodium hydroxide (NaOH) is added to water, partial dissociation also occurs and the Na^+ ions and OH^- ions exist in equilibrium with NaOH molecules which are not dissociated. The basic properties of NaOH are due to the OH^- ions.

However, it is not necessary for a substance to have H^+ or OH^- ions in order to have acid and base properties. For example, ammonia, NH_3, and sodium carbonate, Na_2CO_3, contain no OH^- ions but in water solutions these compounds are bases. In water the reactions are

$$NH_3 + H_2O \longrightarrow NH_4^+ + OH^-$$

$$Na_2CO_3 + H_2O \longrightarrow HCO_3^- + OH^- + 2\,Na^+$$

The OH^- ions come from the water, so aqueous solutions of NH_3 and Na_2CO_3 have the same properties as other bases.

A base which is common to many households is sodium bicarbonate, $NaHCO_3$, known as baking soda. Its solution is weakly basic, the reaction being

$$NaHCO_3 + H_2O \longrightarrow H_2CO_3 + OH^- + Na^+$$

Acids act upon the carbonate ion of baking soda to give off carbon dioxide, a gas.

$$NaHCO_3 + H^+(\text{from an acid}) \longrightarrow$$
$$Na^+ + H_2O + CO_2$$

This reaction is involved in the leavening process in baking. Baking powders contain baking soda plus an acidic substance such as $KHC_4H_4O_6$ (cream of tartar). When this combination is dry, no reaction occurs, but when water is added, CO_2 is given off. Some common bases are shown in Figure 15-5a.

The Arrhenius theory, as we noted, is restricted to aqueous solutions containing H^+ or OH^- ions. Another theory of acids and bases was conceived independently in 1923 by J. N. Brönsted (1879–1947), a Danish physical chemist, and T. M. Lowry (1874–1936), an English chemist, and developed by others since then. The Brönsted-Lowry theory defines an acid as a substance which acts as a proton donor and a base as a substance which acts as a proton acceptor. According to this theory reactions involving the transfer of a proton are known as **acid-base reactions.**

Since a hydrogen ion (H^+) is in reality a proton, a proton donor must be a substance which has hydro-

Figure 15-5a *Some common bases found around the home are ammonia, NH_3, and milk of magnesia, $Mg(OH)_2$. Baking soda, $NaHCO_3$, in water is slightly basic, and baking powder, since it contains baking soda, has basic characteristics in water solution. (Photo courtesy James Crouse.)*

gen atoms. For example, when hydrogen chloride (HCl) is placed in water, the hydrogen ion (H^+) is transferred from the HCl to the water (H_2O) molecule, as shown by the following equation:

$$HCl + H_2O \longrightarrow H_3O^+ + Cl^-$$

This reaction proceeds almost entirely from left to right. The H_3O^+ ion, called the **hydronium ion,** and the Cl^- ion are formed in this reaction. In solution the H^+ ions of acids are joined to water molecules to give hydronium ions, H_3O^+.

Sulfuric acid (H_2SO_4) has two hydrogen atoms per molecule. When this acid molecule is placed in a water solution, dissociation occurs as shown.

$$H_2SO_4 + H_2O \longrightarrow \underset{\substack{\text{Hydronium} \\ \text{ion}}}{H_3O^+} + \underset{\substack{\text{Hydrosulfate} \\ \text{ion}}}{HSO_4^-}$$

The hydrosulfate ion, in turn, can act like an acid, as its proton is transferable to a water molecule.

$$HSO_4^- + H_2O \longrightarrow H_3O^+ + \underset{\text{Sulfate ion}}{SO_4^{2-}}$$

Acids are classified as strong or weak. A strong acid is one that ionizes almost completely in solution. The high propensity of an acid to ionize, and thus be classified as strong, is due to a weak binding energy of the H^+ ion. The acids HCl, HNO_3, and H_2SO_4 have weakly bound H^+ ions and therefore are strong acids. A weak acid is one that does not ionize to any great extent; that is, only a small number of the acid's protons unite with H_2O to form the H_3O^+ ion. Weak acids are formed from compounds which strongly bind the H^+ ion.

The Brönsted-Lowry concept holds that there must be a base that corresponds to every acid; that is, every acid when it donates or gives up a proton (H^+ ion) becomes a base and every base which accepts a proton becomes an acid. This can be expressed as

$$\text{Acid} + \text{Base} \longrightarrow \text{Base} + \text{Acid}$$

or, using HCl as an example, as

$$HCl + H_2O \longrightarrow Cl^- + H_3O^+$$

and

$$H_3O^+ + Cl^- \longrightarrow H_2O + HCl$$

According to the Brönsted-Lowry concept, acids and bases are classified as strong or weak according to their ability to donate or accept protons. Hydrochloric acid (HCl) is a strong acid, that is, the HCl molecule gives up its proton very readily. Accordingly, the chloride ion is a weak base, that is, it does not accept a proton readily to form a HCl molecule. Table 15-5a lists a few common acids and the base produced when the acids donate their protons. The base produced from an acid is called the **conjugate base,** and an acid produced from a base by the addition of a proton is called the **conjugate acid.** It follows from what has been said that the stronger the acid the weaker its conjugate base, and the stronger the base the weaker its conjugate acid.

Table 15-5a *List of Common Acids and Their Conjugate Bases*

	Acid Name	Reaction	Conjugate Base	
Strong	Hydrochloric	$HCl \longrightarrow H^+ + Cl^-$	Chloride ion	Weak
↑	Nitric	$HNO_3 \longrightarrow H^+ + NO_3^-$	Nitrate ion	↑
	Sulfuric	$H_2SO_4 \longrightarrow H^+ + HSO_4^-$	Bisulfate ion	
	Phosphoric	$H_3PO_4 \longrightarrow H^+ + H_2PO_4^-$	Dihydrogen phosphate ion	Bases
Acids	Lactic	$HC_3H_5O_3 \longrightarrow H^+ + C_3H_5O_3^-$	Lactate ion	
↓	Acetic	$HC_2H_3O_2 \longrightarrow H^+ + C_2H_3O_2^-$	Acetate ion	
	Carbonic	$H_2CO_3 \longrightarrow H^+ + HCO_3^-$	Bicarbonate ion	
	Ammonium ion	$NH_4^+ \longrightarrow H^+ + NH_3$	Ammonia	↓
Weak	Water	$H_2O \longrightarrow H^+ + OH^-$	Hydroxide ion	Strong

Acids are very important industrially. Sulfuric acid is one of the most important chemical products in the United States. It is used in the refining of petroleum, the "pickling" of steel, and in the manufacturing of fertilizers and numerous other products. A dilute solution of hydrochloric acid is present in the human stomach and is used in the digestion of our foods. There are persons who must take a daily supplement of this acid because it is lacking in their bodies. Many acids can be found in our foods. A few examples are given in Table 15-5b and Figure 15-5b.

An important property of an acid is the disappearance of its characteristic properties when brought in contact with a base and vice versa. This mixing of

acid and base has to be in the proper proportions, however, for the properties of each to disappear. When the proper proportions are mixed and the characteristics of both the acid and the base disappear, we say that neutralization has taken place. **Neutralization** is the mutual disappearance of the H^+ ions which are responsible for the characteristic properties of the acid and the OH^- ions which are responsible for the characteristic properties of the base. The other ions associated with the acid and the base undergo no chemical change and remain as ions in the aqueous solution or appear as a crystalline solid if the water is evaporated. For example, if hydrochloric acid (HCl) is added to potassium hydroxide (KOH) in the proper proportions, neutralization takes place as shown by the following equation:

$$H^+ + Cl^- + K^+ + OH^- \longrightarrow K^+ + Cl^- + HOH$$

This equation represents the typical reaction of an acid with a hydroxide base. Note that the metal ion (K^+) and the nonmetal ion (Cl^-) appear as ions on both sides of the equation, indicating that they do not take part in any chemical reaction, whereas the H^+ ion combines with the OH^- ion to form a molecule of water. The water molecule is usually written H_2O, but

Table 15-5b *Some Common Acids Found in Foods*

Acid	Formula	Food
Acetic	$HC_2H_3O_2$	Vinegar
Citric	$H_3C_6H_5O_7$	Citrus fruits
Carbonic	H_2CO_3	Soft drinks
Lactic	$HC_3H_5O_3$	Sour milk
Oxalic	$H_2C_2O_4$	Rhubarb
Tartaric	$H_2C_4H_4O_6$	Grapes

Figure 15-5b *Some common acids found around the home are carbonic acid, H_2CO_3, in soft drinks; acetic acid, $HC_2H_3O_2$, in vinegar, citric acid, $H_3C_6H_5O_7$, in citrus fruits; and tartaric acid, $H_2C_4H_4O_6$, in grape juice. (Photo courtesy James Crouse.)*

it was written HOH in this equation to show that the water molecule is formed by the proton donated by the acid and the hydroxide ion donated by the base. If equal numbers of H^+ ions and OH^- ions are mixed together, they will combine to form water molecules and the acid and base properties of the original substances will disappear.

The following examples illustrate the principles involved for neutralization to occur:

■ *Example 1* How many cubic centimeters of $1.0M$ HCl are needed to neutralize 10 cm^3 of $1.0M$ KOH? (Refer to Section 14-5 for molarity.)

The number of H^+ ions must equal the number of OH^- ions for neutralization to take place. Since the acid and the base are both $1.0M$ and the ion concen-

tration is the same, that is, there is one H^+ ion for each HCl molecule and one OH^- ion for each KOH molecule, then the volumes must be equal. Thus, 10 cm^3 of HCl are needed to neutralize 10 cm^3 of NaOH. They combine equally to form water (HOH) and the acid and base characteristics disappear.

■ *Example 2* How many cubic centimeters of $1.0M$ H_2SO_4 are needed to neutralize 10 cm^3 of $2.0M$ NaOH? Again the number of H^+ ions must equal the number of OH^- ions for neutralization to take place. Since the acid is $1.0M$ and the base $2.0M$, the acid is twice as strong as the base. But the H^+ ion concentration for the acid is two and the OH^- ion concentration for the base is one. Thus, 10 cm^3 of acid are needed to neutralize 10 cm^3 of the base.

More involved neutralization problems with odd values of molarity can be solved using the following:

$$M_a V_a C_a = M_b V_b C_b$$

where M_a is the molarity of the acid,
M_b is the molarity of the base,
V_a is the volume of the acid,
V_b is the volume of the base,
C_a is the concentration of H^+ ions per molecule of acid,
C_b is the concentration of OH^- ions per molecule of base.

■ *Example 3* How many cubic centimeters of $4.8M$ phosphoric acid H_3PO_4 are needed to neutralize 10 cm^3 of $1.6M$ calcium hydroxide $Ca[OH]_2$?

$$4.8M \times 10 \text{ cm}^3 \times 3 = 1.6M \times V_b \times 2$$

$$V_b = \frac{4.8 \times 10 \times 3}{1.6 \times 2} = \frac{144}{3.2} = 45 \text{ cm}^3$$

The positive and negative ions that remain after the H^+ ions have combined with the OH^- ions constitute a **salt.** If the water of the solution is removed by evaporation, the remaining solid is the salt in crystalline form. All salts are ionic compounds consisting, usually, of a positive metallic ion and a negative nonmetallic ion. To summarize, when an acid and a base are mixed together they produce water plus a salt. This is shown by the following equations:

$$H^+Cl^- + Na^+OH^- \longrightarrow HOH + Na^+Cl^-$$
$$\text{Acid} \qquad \text{Base} \qquad \text{Water} \qquad \text{Salt}$$

$$H^+NO_3^- + K^+OH^- \longrightarrow HOH + K^+NO_3^-$$
$$\text{Acid} \qquad \text{Base} \qquad \text{Water} \qquad \text{Salt}$$

The most common salt is table salt (NaCl). It is frequently called simply salt, but there are many other salts according to the strict chemical definition.

All salts are ionic compounds; most salts are crystalline solids at ordinary temperatures. Some salts have a salty taste, while others taste sweet, bitter, sour, or have no taste. Aqueous salt solutions may have either acidic, basic, or neutral characteristics. In addition to being formed when neutralization occurs, salts are produced by the direct union of a metal and a nonmetal by electron transfer, and also by reactions between metallic and nonmetallic oxides.

Some salts occur in our environment in the **hydrated** form, that is, they contain one or more molecules of water bonded with the salt. The salt and water are combined according to the law of definite proportions. As examples, sodium thiosulfate, $Na_2S_2O_3 \cdot H_2O$, (hypo), magnesium sulfate, $MgSO_4 \cdot 7\ H_2O$, (epsom salt), and sodium carbonate, $Na_2CO_3 \cdot 10\ H_2O$ (washing soda), always have 1, 7, and 10 molecules, respectively, of water bonded to the molecule of salt. **Anhydrous salts** are those which do not have water bonded to the molecule of salt. Sodium chloride, NaCl (table salt), is an example.

The partial neutralization of an acid or a base produces what are known as acid salts or basic salts. A common acid salt is potassium acid tartrate, $KHC_4H_4O_6$ (cream of tartar); a common basic salt is lead carbonate, $Pb_3CO_3(OH)_4$ (white lead).

15-6
ACIDS AND BASES IN SOLUTION

Pure water has the unusual property of acting as either an acid or a base because it dissociates to a small extent, producing both hydrogen and hydroxide ions. That it does slightly dissociate is shown by the fact that pure water conducts a very feeble electric current. The ionization of pure water is shown in the following equation:

$$H_2O + H_2O \longrightarrow H_3O^+ + OH^-$$

Thus, we see that water can act as an acid or a base, according to the definitions given previously. We have the formation of H^+ (or H_3O^+) and OH^- ions, and one molecule acts as a proton donor and the other as a proton acceptor, as stated by the Brönsted-Lowry theory.

As was mentioned, the dissociation of water to produce H^+ and OH^- ions is very slight. The concentration is only one ten-millionth of a mole of H^+ and OH^- ions in one liter of water at 25°C, or expressed in powers of 10, the concentration is 1×10^{-7} mole/liter. This concentration of H^+ and OH^- ions is used to define a neutral solution; that is, any solution containing 1×10^{-7} mole/liter of both H^+ ions and OH^- ions at 25°C is defined as a **neutral solution.** An **acid solution** is defined as one in which the H^+ ion concentration is *greater* than 1×10^{-7} mole/liter, and an **alkaline solution** is one in which the OH^- ion concentration is greater than 1×10^{-7} mole/liter. We can also define an alkaline solution as one in which the H^+ ion concentration is *less* than 1×10^{-7} mole/liter. In all cases, the temperature is assumed to be 25°C.

The concentration of the H^+ ions is a measure of the acidity or basicity of a solution. The concentration can be expressed in powers of 10, as stated above, but this has proven to be rather awkward, and a method known as the *p*H method has been devised for ex-

Table 15-6a *The pH Concept*

pH	Concentration of H^+ Ions	Acidic or Basic Property
−2	1×10^2 mole/liter	Very acidic
1	1×10^{-1} mole/liter	↑
4	1×10^{-4} mole/liter	
6	1×10^{-6} mole/liter	Acidic
7	1×10^{-7} mole/liter	Neutral
8	1×10^{-8} mole/liter	Basic
10	1×10^{-10} mole/liter	↓
15	1×10^{-15} mole/liter	Very basic

pressing the H^+ ion concentration. Stated in simple terms, the **pH** is the exponent of the negative power to which 10 is raised when used to express the concentration. For example, a solution containing H^+ ions with a concentration of 0.0000001 (1×10^{-7}) mole/liter has a *p*H of 7. A concentration of 1×10^{-11} has a *p*H of 11. A neutral solution has a *p*H of 7. A *p*H less than 7 indicates an acid solution and a *p*H greater than 7 indicates a basic solution. A solution with a *p*H of 5 is said to be slightly acidic and one with a *p*H of 9 is slightly basic. Soda water (carbonic acid) has a *p*H of 4.5, blood 7.4, sea water 8.2, and milk of magnesia 10.5. Solutions in which the *p*H is 7 or more points from the neutral position of 7 are strongly acidic or basic. The concept of *p*H is illustrated in Table 15-6a.

IMPORTANT WORDS AND TERMS

Reactants	Oxidized	Acid-base reaction
Products	Reduced	Hydronium ion
Reversible reaction	Reducing agent	Conjugate base
Equilibrium	Oxidizing agent	Conjugate acid
Exothermic reaction	Displacement reaction	Neutralization
Endothermic reaction	Electromotive series	Salt
Activation energy	Electrolytes	Hydrated
Ten-degree rule	Cathode	Anhydrous salts
Catalyst	Anode	Neutral solution
Enzymes	Acid	Acid solution
Oxidation-reduction reaction	Base	Alkaline solution
Oxidation	Arrhenius acid-base concept	*p*H
Reduction	Brönsted-Lowry theory	

LEARNING OBJECTIVES

After reading and studying this chapter, you should be able to do the following without referring to the text:

1. Describe a chemical reaction and distinguish between the reactants and the products.
2. Explain chemical equilibrium and give an example.
3. Distinguish between exothermic and endothermic reactions. Give an example of each.

4. Explain activation energy.
5. Explain how temperature effects the rate of a chemical reaction.
6. Describe chemical reaction rates in the pressure of a catalyst.
7. Distinguish between oxidation and reduction.
8. Explain oxidation-reduction reactions and give an example.
9. Explain the electromotive series.
10. Define and give an example of (a) an acid, (b) a base, (c) a salt.
11. Explain neutralization.
12. Describe the Arrhenius acid-base concept.
13. Describe the Brönsted-Lowry theory of acids and bases.
14. Explain the meaning of the term pH.
15. Define or explain the important words and terms at the end of the chapter.

QUESTIONS

1. What is a chemical reaction?
2. What three things occur in every chemical reaction?
3. List the factors that influence the rate of a chemical reaction.
4. Give an example of (a) a fast chemical reaction, (b) a slow reaction.
5. Explain the concept of dynamic equilibrium.
6. Explain why chemical reactions proceed faster as the temperature is increased.
7. Why do chemical reactions proceed at a faster rate when the concentration of the reactants are increased?
8. What is the origin of the energy released in a chemical reaction?
9. Distinguish between exothermic and endothermic reactions.
10. How does a catalyst affect a reaction's equilibrium mixture?
11. How is the activation energy related to the total energy released by a chemical reaction?
12. Describe the role of a catalyst in a chemical reaction.
13. Why is it necessary to keep wines sealed or covered?
14. The human body converts sugar into carbon dioxide and water at body temperature, 98.6°F or 37°C. Why are much higher temperatures required for the same conversion in the laboratory?
15. Which would be easiest to oxidize, sodium or chlorine? Explain.
16. Which would be easiest to reduce, sodium or chlorine? Explain.
17. Which is the better oxidizing agent, nitrogen or fluorine? Why?
18. Why are the precious metals gold and silver found free in nature while the metals iron and magnesium are found in ores?
19. Identify some uses of the electromotive series.
20. Describe two theories that define an acid and a base.
21. In respect to the Arrhenius concept of acids, bases, and salts, what is meant by neutralization?
22. Distinguish between a hydrogen ion and a hydronium ion.
23. How is glass made? What causes its colors, if any?

24. Explain the difference between a reversible and an irreversible electrochemical cell.
25. Why is an automobile battery so heavy?
26. What is the most destructive chemical reaction in terms of dollars?
27. What two substances are necessary for iron to rust?
28. Explain how water can be both an acid and a base.
29. How is the strength of an acid or base expressed?

PROBLEMS

1. Balance the following chemical equations:
   (a) When an acid reacts with a metal, H_2 gas is given off.
   $$2HCl + Fe \longrightarrow FeCl_2 + H_2$$
   (b) Nitric acid may be prepared by the following reaction:
   $$H_2O + NO_2 \longrightarrow 2HNO_3 + NO$$

   Answer:
   (b) $H_2O + 3 NO_2 \longrightarrow$
   $2 HNO_3 + NO$

2. Balance the following chemical equations:
   (a) When octane gasoline is burned completely, the reaction is
   $$C_8H_{18} + O_2 \longrightarrow CO_2 + H_2O$$
   (b) Borax is a good water-softening agent. One of the important reactions of the borate ion in water is
   $$B_4O_7^{2-} + H_2O \longrightarrow H_3BO_3 + OH^-$$

3. State which element or ion is oxidized and which reduced in the following reactions:
   (a) The Haber process for manufacturing ammonia is
   $$N_2 + 3 H_2 \longrightarrow 2 NH_3$$
   (b) An important reaction in steel production is
   $$Fe_2O_3 + 3 CO \longrightarrow 2 Fe + 3 CO_2$$

   Answer: (b) Iron is reduced, carbon is oxidized.

4. State which element or ion is oxidized and which reduced in the following reactions:
   (a) Sulfur in coal burns to form the pollutant sulfur dioxide.
   $$S + O_2 \longrightarrow SO_2$$
   (b) Hydrofluoric acid is so strong it can etch glass. It is formed as follows:
   $$H_2 + F_2 \longrightarrow 2 HF$$

5. Using the electromotive series, use an arrow to indicate the direction of the following displacement reactions:
   (a) $Na + K^+ \quad Na^+ + K$
   (b) $Ni + Cu^{2+} \quad Ni^{2+} + Cu$

   Answer:
   (a) $Na + K^+ \longleftarrow$
   $Na^+ + K$

6. Using the electromotive series, use an arrow to indicate the direction of the following displacement reactions:
   (a) $Sn + Fe^{2+} \quad Sn^{2+} + Fe$
   (b) $Pb + Fe^{2+} \quad Pb^{2+} + Fe$

7. State whether the following solutions are acidic or basic:
   (a) 10 g of H_2SO_4 dissolved in 1 liter of water
   (b) Blood

(c) A solution with a pH of 10

(d) A solution with 1×10^{-9} mole/liter of OH^- ion concentration

8. State whether the following solutions are acidic or basic:
   (a) 10 g of $Mg(OH)_2$ dissolved in 1 liter of water _base_
   (b) Pure water _neutral_
   (c) A solution with a pH of 7 _neutral_
   (d) A solution with 1×10^{-5} mole/liter of H^+ ion concentration _acidic_

9. One molar sulfuric acid (H_2SO_4) is added to $2.0M$ potassium hydroxide (KOH) to form a salt plus water, as shown by the following equation:

$$H_2SO_4 + Ca(OH)_2 \longrightarrow \underset{\text{salt}}{\underline{\hspace{2cm}}} + HOH$$

   (a) Write the formula for the salt.
   (b) Balance the equation.
   (c) Calculate the volume of H_2SO_4, in cubic centimeters, needed to neutralize 10 cm^3 of the KOH.

10. One molar hydrochloric acid (HCl) is added to $1.0M$ calcium hydroxide [$Ca(OH)_2$] to form a salt plus water, as shown by the following equation:

$$HCl + Ca(OH)_2 \longrightarrow \underset{\text{salt}}{\underline{\hspace{2cm}}} + HOH$$

   (a) Write the formula for the salt.
   (b) Balance the equation.
   (c) Calculate the volume of HCl, in cubic centimeters, needed to neutralize 10 cm^3 of the $Ca(OH)_2$.

Chapter 16

COMPLEX MOLECULES

I am part of all you see
In Nature: part of all you feel:
I am the impact of the bee
Upon the blossom; in the tree
I am the sap—that shall reveal
The leaf the bloom—that flows and flutes
Up from the darkness through its roots.

Madison Cawein

Previously, we learned that atoms of the chemical elements are nature's building blocks. In Chapter 13, we discovered that compounds are formed by combining the atoms of various elements. Many naturally occurring compounds are made up of very complex molecules. Since the fundamental elements are known, it would be hoped that these complex molecules could be analyzed and understood. It is true that a great degree of success has been achieved in analyzing those compounds whose molecules contain only a few atoms. However, these small molecules comprise only the inanimate portion of nature such as gases, water, and minerals. The chemical structure of the living portion of nature is grossly complex. **Complex molecules** may be either an aggregate of atoms or repeating units of smaller molecules chemically bonded together, thus giveing rise to great molecular weights and chain-like structures.

A simple example of building molecules from repeating units is the hydrocarbon series discussed in Chapter 13. Specifically, the fundamental repeating unit, or **monomer,** of the alkane series could be structurally written

$$\left[\begin{array}{c} H \\ | \\ -C- \\ | \\ H \end{array} \right]_n$$

where it is understood that other fundamental units are attached to the open bonds and the series is terminated by hydrogen atoms. The subscript n indicates the fundamental unit may be repeated n times and the resulting chainlike structure is called a **polymer.** For example, if $n = 3$, the resulting compound is propane (see Section 13-8). This simple example is meant to introduce the concept of repeating units, and it should be kept in mind that the polymers occurring in nature, or prepared synthetically, may have repetition numbers (n) which may be in the thousands or hundreds of thousands. In addition, the fundamental unit may itself be quite complex and the chain configurations and variations may differ greatly.

Nature has been in the business of making large complex molecules for a long time. All living and growing matter is composed of these molecules, some being so complex that they are still not completely understood. The complexity of nature's molecules should come as no surprise since they compose the structure of life itself. The newest frontier of science is concerned with determining the structures and interactions of the complex molecules of which living things are made. Society someday may be forced to make some fundamental decisions regarding how far it will allow science to go in solving the mysteries of life.

16-1
COMMON ORGANIC COMPOUNDS

The number and complexity of the organic compounds in nature can, in large part, be attributed to the ability of the carbon atom to form single, double, and even triple bonds with itself and other elements. There are over a million organic compounds, and their identification, classification, and structural determination has been—and still is— the task of workers in the field of organic chemistry. Only a few of the more common compounds will be presented here, most of which should be familiar to the student because of their usage and occurrence in everyday life.

First, consider the following list of classes of compounds. A specific example from each class is given to show how the complexity of molecules can develop, starting with just a simple hydrocarbon.

Hydrocarbon (Ethane)

1. *Alkyl halides*

(Ethyl chloride)

2. *Alcohols*

(Ethyl alcohol)

3. *Aldehydes*

(Acetaldehyde)

4. *Acids* (*organic*)

(Acetic acid)

5. *Esters*

(Ethyl acetate)

The colored portions of the formulas indicate the organic functional group that characterizes the general physical and chemical properties of these compounds.

Alkyl Halides

An **alkyl halide** is formed by replacing the hydrogen atom(s) in hydrocarbons with halogens—chlorine, bromine, fluorine, and iodine. Ethyl chloride, the example given in the list, is used as a local anesthetic and is the product of the reaction of ethane with chlorine.

$$C_2H_6 + Cl_2 \longrightarrow \left[H-\underset{\underset{H}{|}}{\overset{\overset{H}{|}}{C}}-\underset{\underset{H}{|}}{\overset{\overset{H}{|}}{C}}-Cl \right] + HCl$$

Ethane   Chlorine   Ethyl chloride

The halogen may replace more than one of the hydrogen atoms; in fact, they may all be replaced as the following series shows:

$$H-\underset{\underset{H}{|}}{\overset{\overset{H}{|}}{C}}-Cl \qquad H-\underset{\underset{H}{|}}{\overset{\overset{Cl}{|}}{C}}-Cl$$

Methyl   Dichloro-
chloride   methane

$$Cl-\underset{\underset{H}{|}}{\overset{\overset{Cl}{|}}{C}}-Cl \qquad Cl-\underset{\underset{Cl}{|}}{\overset{\overset{Cl}{|}}{C}}-Cl$$

Chloroform   Carbon
tetrachloride

The latter two compounds are common as an anesthetic and a cleaning fluid, respectively. However, the toxicity of carbon tetrachloride is a serious hazard, and caution should be exercised when using it.

Alcohols

Alcohols are organic compounds that contain one or more OH groups that have been substituted for one or more hydrogen atoms. Examples of some common alcohols are shown in Fig. 16-1a. Ethyl

Figure 16-1a *Commercial products that contain a common alcohol. (Photo courtesy James Crouse.)*

alcohol (C_2H_5OH) is probably the most important alcohol known. It is made from sugars by the action of yeast in the process of fermentation.

$$C_6H_{12}O_6 \xrightarrow{\text{Yeast}} 2 \left[H-\underset{\underset{H}{|}}{\overset{\overset{H}{|}}{C}}-\underset{\underset{H}{|}}{\overset{\overset{H}{|}}{C}}-OH \right] + 2\,CO_2$$

Ethyl alcohol

or synthetically from ethylene and water

$$H-\underset{\underset{H}{|}}{\overset{\overset{H}{|}}{C}}=\underset{}{\overset{\overset{H}{|}}{C}}-H + H_2O \xrightarrow{H_2SO_4} H-\underset{\underset{H}{|}}{\overset{\overset{H}{|}}{C}}-\underset{\underset{OH}{|}}{\overset{\overset{H}{|}}{C}}-H$$

Ethylene   Ethyl alcohol

Ethyl alcohol is a colorless liquid that mixes with water in all proportions. It is the least toxic of all alcohols and is used in alcoholic beverages. Ethyl alcohol is also used as a solvent and in the production of many substances including perfumes, dyes, varnishes, antifreeze, and ethyl ether.

Alcohols are characterized by the —OH, or hydroxyl group; hence they are organic equivalent to the inorganic bases. Many alcohols exist, some with one (—OH) group, others with two or more (—OH) groups. Ethylene glycol is an example of an alcohol with two hydroxyl groups.

$$
\begin{array}{c}
H \\
| \\
H-C-OH \\
| \\
H-C-OH \\
| \\
H
\end{array}
$$

Ethylene glycol

Ethylene glycol is widely used an an antifreeze in automobiles.

Aldehydes

Aldehydes are characterized by the $-C{\overset{O}{\underset{H}{}}}$ group and are formed when alcohols react with oxygen (are oxidized). When ethyl alcohol is oxidized, acetaldehyde is formed.

$$
2\left[\begin{array}{c} H\ \ H \\ |\ \ \ | \\ H-C-C-H \\ |\ \ \ | \\ H\ \ OH \end{array}\right] + O_2 \longrightarrow 2\left[\begin{array}{c} H \\ | \\ H-C-C{\overset{O}{\diagdown}}_{H} \\ | \\ H \end{array}\right] + 2\ H_2O
$$

Ethyl alcohol   Acetaldehyde

A more common aldehyde, formaldehyde is prepared similarly from methyl alcohol (CH_3OH). Formaldehyde is used as a disinfectant and tissue preservative.

Organic Acids

The further action of oxygen on aldehydes produces a group of compounds known as **organic acids,**

which are characterized by the molecular arrangement $-C{\overset{O}{\diagdown}}_{OH}$ called the **carboxyl** group. There are many of these carboxylic acids. Acetic acid, whose dilute natural form is vinegar, is formed by the following reaction:

$$
2\left[\begin{array}{c} H \\ | \\ H-C-C{\overset{O}{\diagdown}}_{OH} \\ | \\ H \end{array}\right] + O_2 \longrightarrow 2\left[\begin{array}{c} H \\ | \\ H-C-C{\overset{O}{\diagdown}}_{OH} \\ | \\ H \end{array}\right]
$$

Acetaldehyde   Acetic acid

The simplest carboxyl acid, formic acid, is common in insects and is the cause of painful discomfort from insect bites. It is prepared by the oxidation of formaldehyde.

Esters

When a carboxylic acid reacts with an alcohol, an ester is formed. For example,

$$
\begin{array}{cc}
\begin{array}{c} H \\ | \\ H-C-C{\overset{O}{\diagdown}}_{OH} \\ | \\ H \end{array} &
\begin{array}{c} H\ \ H \\ |\ \ \ | \\ + H-C-C-H \\ |\ \ \ | \\ H\ \ OH \end{array}
\end{array} \longrightarrow
$$

Acetic acid   Ethyl alcohol

$$
\begin{array}{c}
H \\ | \\ H-C-C{\overset{O}{\diagdown}}\ \ \ \begin{array}{c} H\ \ H \\ |\ \ \ | \\ O-C-C-H \\ |\ \ \ | \\ H\ \ H \end{array} \\ | \\ H
\end{array} + H_2O
$$

Ethyl acetate
(an ester)

An **ester** is a compound which conforms to the general formula

$$
R-\overset{O}{\overset{\|}{C}}-O-R'
$$

where R and R' are any alkyl groups. **Alkyl** groups have the general formula C_nH_{2n+1}; that is, they are alkanes less one hydrogen atom. R and R' may be the same group, but they are usually different. Esters are analogous to salts. Inorganic acids and bases react to form salts. Organic acids and alcohols react to form esters.

Esters possess very distinct and pleasant odors. The fragrance of many flowers and the pleasant taste of ripe fruits are due to esters. For example, bananas contain the ester amyl acetate,

$$CH_3\overset{\displaystyle O}{\overset{\|}{C}}-O(CH_2)_4CH_3$$
Amyl acetate

and oranges contain octyl acetate,

$$CH_3\overset{\displaystyle O}{\overset{\|}{C}}-O(CH_2)_7CH_3$$
Octyl acetate

The above groups of compounds may not seem to have very complex molecules, however, the examples given were selected for their simplicity. For the opposite extreme, we need look only as far as ourselves or other living matter. The molecules of living matter are extremely large and complex. These molecules contain tens of thousands of atoms (molecular weights of 1 billion) connected in three-dimensional patterns which are very complex in structure and very difficult to analyze. Common organic compounds in living matter are carbohydrates, proteins, fats (lipids), nucleic acids, and vitamins.

Carbohydrates

Carbohydrates are compounds composed of carbon, hydrogen, and oxygen with the hydrogen-to-oxygen ratio usually 2 to 1, the same as in the water molecules. The general formula for the carbohydrates is $C_n(H_2O)_x$. Thus, the name carbohydrates was given to those substances which were known as hydrates of

carbon, although the nature of these compounds is not what the name implies. The most important carbohydrates are sugars, starches, and cellulose. The simplest sugars, glucose ($C_6H_{12}O_6$) and fructose ($C_6H_{12}O_6$), are isomers. Their structural formulas are

Glucose   Fructose

Glucose, also known as dextrose and grape sugar, is found in sweet fruits, such as grapes and figs, and in flowers and honey. It is present in the blood to the extent of 0.1 percent, but is present in much greater amounts in persons suffering from diabetes. Glucose is formed in plants by the action of sunlight and chlorophyll on carbon dioxide (CO_2) from the air, and water (H_2O) from the soil. The chemical reaction in simple terms is

$$6\ CO_2 + 6\ H_2O + energy \longrightarrow \underset{Glucose}{C_6H_{12}O_6} + 6\ O_2$$

with energy being absorbed in the process; that is, the energy coming from the sunlight is stored as chemical energy in the plant.

Fructose, also called fruit sugar, the sweetest of all sugars, is found in fruits and honey. Combined with glucose, it forms the sugar sucrose known as common sugar or cane sugar. The reaction is written as follows:

$$\underset{Fructose}{C_6H_{12}O_6} + \underset{Glucose}{C_6H_{12}O_6} \longrightarrow \underset{Sucrose}{C_{12}H_{22}O_{11}} + H_2O$$

Sucrose has two important isomers, maltose or malt sugar and lactose or milk sugar, which are sweet, crystalline solids soluble in water. Lactose is present in milk (about 4 percent) and is an essential ingredient in the food for infants.

Starch has the general formula $(C_6H_{10}O_5)_n$ where n may take on values up to 3000. Thus, starch is a polymer (many monomers associated together) which consists of long chains of glucose units. Starch, a noncrystalline substance, is formed by plants in seeds, tubers, and fruits. Starch as a food is present in potatoes, grains (corn and rice), and vegetables. In the digestion process, the starches are converted to glucose. Glycogen (animal starch), a smaller and more highly branched polymer of glucose, is stored in the liver and muscles of animals as a reserve food supply which is easily converted to energy.

The energy in foods is given in terms of the energy they can deliver when oxidized to their final products. These final products are carbon dioxide, water, and nitrogenous compounds which are given off as waste products. For example, when glucose is oxidized, 674 kilocalories of energy per mole are provided. Since there are 180 grams in one mole of glucose, 1 gram of glucose provides 3.74 kilocalories of energy. This energy is used to do work and provide body heat in living organisms.

Cellulose has the same general formula $(C_6H_{10}O_5)_n$ as starch, but its structure is different; thus, it has different properties. Cellulose is the main component of the cell walls of plants and receives its name for this reason. It is a carbohydrate which cannot be digested by humans, but is used for making paper, guncotton (which is used in smokeless gun powder), rayon, and cellophane.

Proteins

Proteins are one of the essential components of living cells. Proteins are composed of carbon, oxygen, nitrogen (approximately 16 percent in all proteins), and hydrogen. Some also contain small amounts of sulfur, phosphorus, iron, and copper. The molecular structure of proteins is complex, but the basic chemical units are amino acids. An amino acid is an organic compound that contains an amine group and an acid group. The **amino** functional group has the structural formula

When an amino group is combined with a hydrocarbon group, an amine is formed. Similarly, when the carboxyl functional group

is combined with a hydrocarbon group, an acid is formed. Thus, an amino acid contains both of these groups. There are over twenty known amino acids; eight of these are essential in the human diet. The simplest amino acid is glycine, whose structural formula is

When two molecules of glycine combine, a molecule of water is eliminated. Two molecules of glycine can be shown as

Since a water molecule is given off, a bond (called a peptide bond) is formed and the resulting protein is

Peptide bond

This process can be repeated by linking more glycine molecules to the protein whose structural formula appears above. Repeated again and again, long-chain proteins of very high molecular weight are formed. The molecular weight of proteins ranges from 1×10^4 up to 1×10^7 for the most complex. The human body contains more than 1×10^5 different proteins.

Lipids

Lipid is a general term that includes such substances as fats, oils, and waxes.

Fats are used in the diets of humans. In the digestive process, the fats are broken down into glycerin and acids, which are absorbed into the blood stream and oxidized to produce energy that may be used immediately or stored for future use. Fats are also used by the body as an insulator to prevent loss of heat.

Fats are esters composed of glycerol ($C_3H_5(OH)_3$, an alcohol) and organic acids known as fatty acids. Fatty acids are acids that are present in any organic acid. A typical fatty acid is stearic acid, $C_{17}H_{35}COOH$ (a component of beef fat). When stearic acid is combined with glycerol, glyceryl stearate (a fat) is obtained. This is shown in the following reaction:

$$3\left[\begin{array}{c} O \\ \| \\ C_{17}H_{35}-C-OH \end{array}\right] + \begin{array}{c} H \\ | \\ H-C-OH \\ | \\ H-C-OH \\ | \\ H-C-OH \\ | \\ H \end{array} \longrightarrow$$

Stearic acid   Glycerol

$$\begin{array}{c} H \\ | \\ C_{17}H_{35}COO-C-H \\ | \\ C_{17}H_{35}COO-C-H + 3 H_2O \\ | \\ C_{17}H_{35}COO-C-H \\ | \\ H \end{array}$$

Glyceryl stearate (a fat)

The stearate structure is a large chain hydrocarbon containing 18 carbon atoms, but written in short form here to save space.

Liquid fats, such as vegetable oils, are composed of hydrocarbon chains with double bonds between some of the carbon atoms; that is, they are unsaturated. These oils can be changed to solid fats by a process called **hydrogenation.** This is done simply by adding hydrogen to the carbon atoms which have the double bond; thus, the hydrocarbon chains become saturated. For example, when cottonseed oil (a liquid) is hydrogenated, margarine (a solid) is obtained. This is shown by the following equation:

$$\left[\begin{array}{c} CH_3(CH_2)_7CH{=}CH(CH_2)_7COOCH_2 \\ | \\ CH_3(CH_2)_7CH{=}CH(CH_2)_7COOCH \\ | \\ CH_3(CH_2)_7CH{=}CH(CH_2)_7COOCH_2 \end{array}\right] + 3\ H_2 \longrightarrow$$

Cottonseed oil (a liquid)

$$\left[\begin{array}{c} CH_3(CH_2)_{16}COOCH_2 \\ | \\ CH_3(CH_2)_{16}COOCH \\ | \\ CH_3(CH_2)_{16}COOCH_2 \end{array}\right]$$

Margarine (a solid)

Thus, liquid fats (oils) are esters of unsaturated acids, and solid fats are esters of saturated acids.

Waxes are esters derived from long-chain monohydric alcohols (an alcohol with only one OH group per molecule) and fatty acids. Beeswax from the honeycomb of the bee contains myricyl palmitate ($C_{15}H_{31}COOC_{31}H_{63}$) and is used in pharmaceutical preparations. Carnauba wax, which comes from the Brazilian wax palm, is used in commercial waxes for wood and metal surfaces. Chinese wax, which is secreted by an insect, is used in medicine.

The term **detergent** means cleansing agent and commonly refers to a soap substitute, but ordinary soap, which is a cleansing agent, is also a detergent (see Fig. 16-1b). Ordinary soap is made by combining fats or oils with alkalies such as sodium or potassium

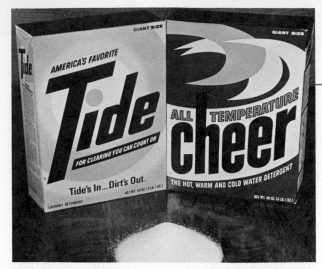

Figure 16-1b *Two laundry detergents and ordinary soap powder. The soap is also a detergent. (Photo courtesy James Crouse.)*

hydroxide. When fats are treated with sodium hydroxide, glycerol, and sodium salts of the fatty acid, sodium palmitate, sodium stearate, and sodium oleate are formed. A typical soap is sodium stearate ($C_{17}H_{35}COO^-Na^+$) whose structural formula is

$$H-\underset{\underset{H}{|}}{\overset{\overset{H}{|}}{C}}---(C_{15}H_{30})---\underset{\underset{H}{|}}{\overset{\overset{H}{|}}{C}}-\overset{\overset{O^-}{\nearrow}}{\underset{\underset{O}{\searrow}}{C}} \quad Na^+$$

Soaps have a cleansing action because the carboxyl end of the soap ion is water-soluble and the hydrocarbon end is oil-soluble. Thus, the hydrocarbon end can dissolve fats and the carboxyl or ion end surrounds the dirt particles and forms a film around them which allows them to be washed away. Thus, through a combination of emulsifying action and particle suspension, soaps serve as detergents.

Although soaps have a cleansing action, they have the disadvantage of forming precipitates when used in acidic solutions or with hard water, which contains calcium, magnesium, and iron ions.

The modern detergents, which are soap substitutes, contain a long hydrocarbon chain which is nonpolar (for example, $C_{12}H_{25}—$), and a polar group such as sodium sulfate ($—OSO_3Na$). This detergent has the structural formula

$$H-\underset{\underset{H}{|}}{\overset{\overset{H}{|}}{C}}---(CH_2)_{10}---\underset{\underset{H}{|}}{\overset{\overset{H}{|}}{C}}-O-\underset{\underset{O}{\overset{O}{|}}}{\overset{}{S}}-O^- \quad Na^+$$

Sodium lauryl sulfate

Synthetic detergents such as sodium lauryl sulfate do not form precipitates with the calcium, magnesium, and iron ions. Therefore, they are effective cleansing agents in hard water.

Nucleic Acids

Nucleic acids are very high molecular weight polymers, present in living cells. It is generally believed that they control the processes of heredity and take an active role in building proteins from amino acids. There are two types of nucleic acids; *deoxyribonucleic acid*, **DNA**, which is located primarily in the nucleus of the cell and *ribonucleic acid*, **RNA**, which is located in the cytoplasm or outer structure of the cell. Figure 16-1c is a diagram of a living cell with some of its components. DNA has the molecular structure of a double-stranded helix (see Fig. 16-1d) composed of simple sugars (deoxyribose), phosphates, and nitrogen compounds of which there are four: adenine, thymine, cytosine, and guanine, commonly labeled A, T, C, and G. The sugars and phosphates form the spiral frames of the double-stranded helix and the nitrogen bases form the connecting links, with adenine always joined with thymine and cytosine always joined with guanine to form links between the spiral frames. The molecules vary in length, but they may be several inches long and contain billions of cross linkages. The DNA molecular weight is greater than 1 million. RNA has a similar helical structure except the linkage is different, and the molecular weight is smaller, ranging from 20,000 to one million.

Figure 16-1c *Diagram of a typical living cell. Every living cell contains a nucleus with cytoplasm surrounding it. The cell membrane encloses everything. The whole mass of living matter is called protoplasm.*

Figure 16-1d *Schematic diagram of the DNA molecule. The DNA molecule has the structure of a double-stranded helix. The spiral frames are formed from sugars and phosphates. The connecting links are composed of nitrogen compounds.*

The biological significance of DNA and how it affects heredity is opening up an exciting new field of study because DNA is found in every living cell (plant and animal) in our environment. This will be discussed in Section 16-2.

Vitamins

Vitamins are organic substances which are needed in minute amounts to perform specific functions for normal growth and nutritional needs of the human body. The original name vitamine originated in 1912 by researchers who thought the substances were amines which were vital for good health. Since then the "e" has been dropped because all vitamins are not amines.

Vitamin A (Retinol) was the first vitamin to be discovered and identified. Elmer V. McCollum, an American scientist, identified Vitamin A in 1912 through an experiment in nutrition with rats. He noted that rats grew and remained healthy when fed a fat-source diet of butter and egg yolk, but their health declined and they eventually died when lard or olive oil was substituted for the butter. This led McCollum to believe some mysterious substance was present in the butter or egg yolk or both. Through a series of experiments he extracted this substance, a yellow-colored oil, from the butter. He called it "fat-soluble A," which today is known as Vitamin A. Vitamin A is an alcohol with the empirical formula $C_{20}H_{29}OH$ and the structural formula

It is found in eggs, milk, butter, and green and yellow vegetables. It increases the resistance to infections and prevents night blindness in humans.

Table 16-1a *Chemical Formulas of Common Vitamins*

Vitamin	Formula
A_1	$C_{20}H_{29}OH$
A_2	$C_{22}H_{31}OH$
B_1	$C_{12}H_{17}ON_4SClHCl$
B_2	$C_{17}H_{20}N_4O_6$
B_6	$C_8H_{11}NO_3HCl$
B_{12}	$C_{63}H_{90}N_{14}O_{14}PCo$
C	$C_6H_8O_6$
D	$C_{28}H_{44}O$
E	$C_{20}H_{50}O_2$
K_1	$C_{24}H_{35}O_2$

There are many other important vitamins, some whose functions are well known, and others whose importance is not yet understood. Table 16-1a lists some of the important vitamins with their empirical formulas.

16-2
THE INGREDIENTS OF LIFE AND THE GENETIC CODE

As we have noted, the molecular structures of living matter are quite complex. However, it is generally agreed by scientists that when the earth was formed only simple compounds such as water vapor (H_2O), hydrogen (H_2), ammonia (NH_3), and methane (CH_4) were present. How, then, did the complex molecules of life form from these simple elements? The first experimental insight to this question was gained in 1952 when Stanley Miller, a graduate student in chemistry, performed his now famous **Miller experiment.** In studying the process of forming complex molecules, Miller circulated through a glass container a mixture of water vapor, hydrogen, ammonia, and methane which he thought might be similar to our primordial atmosphere. Some form of energy must be present to initiate a chemical reaction in the mixture, so a sparking device was placed in the container. This

could conceivably be likened to lightning flashes in the early atmosphere. As the gases continuously passed through the spark, drops of liquid began to collect on the bottom of the container. When this liquid was analyzed, it was found to contain several organic compounds which are found in living matter, the most significant of which were four amino acids, the building blocks of proteins. Thus, Miller's simple, but very significant, experiment demonstrated a possible way in which the compounds of living matter might have been formed originally. A schematic diagram of the apparatus used in Miller's experiment is shown in Fig. 16-2a.

Subsequent experiments yielded as many as 10 amino acids from similar reactions. This synthesis of the amino acids is extremely significant because of their relation to proteins. As discussed previously, proteins are one of the essential components of living matter. They are produced by living cells and make up approximately 75 percent of the dry weight of our bodies. Proteins are also the main component of plant cell interiors. Proteins are quite complex in molecular structure, being composed of chemical combinations of about 20 amino acids. The Miller experiment demonstrated how nature might have prepared the amino acids. The next logical question, ignoring that of the beginning of life itself, might be, How do living cells form the correct protein from the various combinations of amino acids? That is, how does an animal cell generate proteins characteristic of the animal, and a plant cell, proteins characteristic of the plant? In short, How are the characteristics of life reproduced? The answer to this, of course, involves heredity.

Why offspring resemble their parents has long fascinated us. It was thought for a long time that the characteristic traits were carried in the blood and were passed to a child by mixing of the parental blood. This misconception gave rise to the idea of blood lines such as royal blood, having bad blood, or being full-blooded for some race. Science, however, has traced the trait-carrying agent to structures, called **chromosomes,** within the nucleus of the living cell.

Figure 16-2a *Apparatus used in Miller's experiment. Steam, ammonia, methane, and hydrogen were introduced into the reaction chamber, and the products cooled and collected in the trap. It was found that 10 amino acids could be made in this way.*

In the figure the following labels appear: Electrodes, Valve, Gases → NH_3, CH_4, and H_2, Gas mixture, Sparks, Condenser, Water out, Water vapor, Droplets of amino acids, Water in, Boiling water, Heat source, Trap, Bunsen burner.

Chromosomes are thread-like structures that split in such a way that when a body cell divides to produce new cells, the two resulting daughter cells have the same number of identical chromosomes as the parent cell. This process of cell division is called **mitosis.** Because of mitosis, cells are reproduced with the same characteristics. All mature body cells in the same animal or plant have the same number of chromosomes.

In another type of cell division, called **meiosis,** cells result with just half the number of chromosomes as the original cell. Meiosis occurs in the reproductive organs of sexually reproducing plants and animals and involves the sex cells—sperm in the male and egg in the female. Since the sperm and the egg cells unite in the fertilization process, it is evident that for the cell of the offspring to have the proper number of chromosomes and combination of traits, the new egg and sperm cells must each have half as many chromosomes as the original cell. Thus, as a result of meiosis, each child gets a set of chromosomes from each parent and passes half of these to its own offspring. The difference between mitosis and meiosis is illustrated in Fig. 16-2b.

Human body and sex cells have 46 chromosomes, but humans have thousands of inherited traits and characteristics. Therefore, the chromosomes must have a multiple structure to carry all these traits. On a closer look at the threadlike structure of the chromosome, it was found that what appeared to be a thread was actually made up of small beads hooked together in a necklacelike fashion. Each human chromosome might contain as many as a thousand of the beads, or **genes,** which are the carriers of inherited traits. Thus, a cell may inherit two genes for the same characteristic in meiosis, only one of which is **dominant** or influences the trait. The other is **recessive** and does not influence the characteristics of the offspring. The recessive gene is, however, a part of the cell and may be transferred again in meiosis and become dominant in some later generation. The appearance of a child with red hair in a family whose generations have had all black hair is such an example. Some ancestor of the child must have added a gene for red hair which remained recessive for several generations of the family.

Science has focused much attention on the gene. It has been found to consist of the nucleic acid DNA. As noted previously, DNA is composed of sugars, phosphates, and the nitrogen compounds adenine (A), thymine (T), cytosine (C), and guanine (G). The molecule is a spiral ladder-like structure with the

Mitosis

Chromosome

Cell

sugars and phosphates forming the sides and the nitrogen compounds the rungs (see Fig. 16-1d). The nitrogen compounds can form in only certain combinations, A with T and C with G. However, the combined nitrogen compounds may form the rungs of the molecular DNA ladder in any order, such as AT, CG, TA, AT, GC, CG, AT, and so on. It is now believed that the arrangement of the nitrogen compound sequences, or **genetic code,** gives each gene its special characteristic. A simple gene might be a small portion of the DNA molecule a few hundred or thousand rungs long. In a single human cell, the DNA molecule is thought to be about 3 feet long when uncoiled and contain some 6 billion rungs—a truly complex molecule. J. Watson and F. Crick were awarded the Nobel prize in 1962 for their work in elucidating the helical structure of the DNA molecule.

According to current theory, it is believed that when a chromosome of the living cell divides in mitosis, the DNA ladder literally unzips and the two individual parts pick up more raw materials from the nucleus around them. Since an A must combine with a T and a C with a G, the old substances pick up new partners such that two duplicate DNA molecules are formed with the same code, and hence the same genes. This unzipping and duplicating action is illustrated in Fig. 16-2c. Should an "accident" occur in the reconstruction of the molecule which affects the order of the DNA nitrogen compound pairs, the characteristic of the gene is altered and a **mutation,** or genetic change may occur. Mutations may arise in a number of ways not fully understood. They may be caused by photons (gamma rays, X-rays, light, and so on) which hit the molecule, or mutations may be caused by chance. According to quantum mechanics, events have only a probability of occurring. That is, there is a high probability the gene will exactly reproduce. Sometimes, by chance, it does not.

To complete the story and return to the original question of how the proteins are reproduced, one must consider another nucleic acid, RNA, which acts as the messenger of the DNA in forming proteins. RNA is a single-strand molecule which is similar to

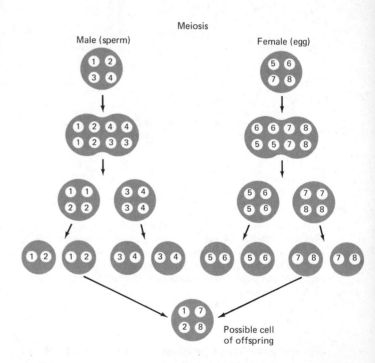

Meiosis

Male (sperm)

Female (egg)

Possible cell of offspring

Figure 16-2b *Schematic illustration of the processes of mitosis and meiosis. In mitosis, the chromosomes divide in such a way before the cell divides that each daughter cell has the original number of chromosomes. In meiosis, the immature reproductive cells of both male and female divide so that one-half of the original number of chromosomes is present in the new cell. Thus, the correct number of chromosomes are present in the body cells of the offspring when a male and female cell unite.*

Figure 16-2d *Diagram illustrating the process of building proteins. The long strand messenger RNA and the short strand transfer RNA are formed from DNA. The transfer RNA molecules are coded to link with a given amino acid. The transfer RNA then hooks up with the messenger RNA, and the amino acids join as prescribed to build a protein.*

Figure 16-2c *The duplication of a DNA molecule. The ladder structure "unzips" in the middle. Then the appropriate nitrogen compounds join each half of the ladder, and two new identical ladders are formed. In reality the ladders are spirals as shown in Fig. 16-1d.*

the DNA molecule. The DNA molecule produces RNA, in two varieties: a long strand "messenger" RNA and a short strand "transfer" RNA. It is these molecules which go into the **cytoplasm,** or outer portion of the cell, to produce protein from amino acids. The messenger RNA carries a fixed nitrogen code from the DNA molecule for the production of a certain type protein. The transfer RNA is coded to seek out a particular amino acid and then to recombine with the messenger RNA according to the given code. Thus, the amino acids are assembled in a predesignated way and the proteins and cells grow as dictated by the individual's inherited traits contained in his genetic DNA code. This process is illustrated in Fig. 16-2d.

16-3
ARTIFICIAL MOLECULES—PLASTICS

Chemists have for a long time tried to duplicate the compounds of nature. Alchemy, or the attempt to transform chemically base metals into gold, was an early example of one such effort. As the science of chemistry progressed and the formulas and basic components of some of the natural compounds became known, chemists were able to synthesize some of these natural compounds by causing the appropriate combinations of elements or compounds to react. During this early trial and error development, there were, no doubt, as many accidental as deliberate discoveries.

From the attempts to synthesize nature's compounds, synthetics were discovered. A **synthetic** is a material whose molecule has no duplicate in nature. The first synthetic was prepared by Leo Baekeland in 1907. By mixing phenol (C_6H_5OH) and formaldehyde (HCHO), a polymeric combination of these two compounds, $HOCH_2C_6H_4OH$, was formed. The polymer is commercially known as Bakelite, a common electrical insulator. This discovery set off a series of investigations for synthetic materials. Necessity also helped catalyze these endeavors. During World War II many natural supplies, such as rubber, were cut off, thus providing impetus to the search for synthetic products which would take the place of the natural materials in short supply. Chemists became aware that substituting different elements or compounds in a complex molecule would change its physical properties. For example, the substitution of a chlorine atom for a hydrogen atom in ethane produces ethyl chloride, which has vastly different properties from ethane. By knowing the general properties of the substituted groups, a chemist can tailor his chemical molecule to satisfy a given requirement. As a result of this scientific approach, a host of man-made molecules have been constructed. Probably the best known of these is the group of synthetic polymers commonly called **plastics.** Plastics have become an integral part of our modern life, being used in clothing, shoes, buildings, autos, sports, art, electrical appliances, toothbrushes, toys, and a myriad of other things.

Not only are the properties of plastics affected by their chemical composition, but the alignment of their long molecular chains can also cause differences in their qualities. For example, if a thin sheet of polymer is heated to its softening point then stretched and cooled while held fixed, the molecular chains are aligned along the direction of stretch and the material is said to be oriented, being less flexible than before. Should the material be reheated without being held, the sheet will shrink back to its original unstretched unaligned state. This property is of great use in the packaging industry where meats or other items are wrapped in stretched material which is then heated. The result is a skin-tight polymer wrapping. The next time you have occasion to have a piece of plastic wrapping, stretch it to align the molecules. If it will not stretch the material, then it is probably already oriented. Heating will unalign the molecular chains if it has been stretched while heated. Other chemical "tricks" such as putting short side-chains or branches on the main molecular chains can affect the physical properties of a complex molecule. The long polymeric chains may also be chemically crosslinked, or tied together, to obtain desirable properties.

A few of the more common plastics along with their monomer unit will be considered to show how man has constructed these molecules. Three polymers of similar structure are listed below, with the monomer of each.

Monomers of Three Common Polymers

Polyethylene
$$\left[\begin{array}{c} \begin{array}{cc} H & H \\ | & | \\ -C-C- \\ | & | \\ H & H \end{array} \end{array} \right]_n$$

Polypropylene

$$\left[\begin{array}{c} \overset{\displaystyle H}{\underset{\displaystyle H}{\overset{|}{\underset{|}{C}}}}-\overset{\displaystyle H}{\underset{\displaystyle CH_3^*}{\overset{|}{\underset{|}{C}}}} \end{array}\right]_n$$

Polystyrene

$$\left[\begin{array}{c} \overset{\displaystyle H}{\underset{\displaystyle H}{\overset{|}{\underset{|}{C}}}}-\overset{\displaystyle H}{\underset{\displaystyle \bigcirc^{**}}{\overset{|}{C}}} \end{array}\right]_n$$

*—CH_3 is referred to as the methyl group.

**⬡ is the symbol for benzene with one hydrogen missing ($-C_6H_5$), which is the phenyl group.

Polyethylene is the simplest of the artificial molecules and may be thought of as a very long alkane; nature's alkanes do not have the thousands of repeating units as in man's molecules. By replacing a hydrogen atom in the polyethylene monomer with a methyl group or a phenyl group, one get polypropylene and polystyrene respectively. By thus changing the molecular structure, the physical properties of the polymers are changed. For example, the melting points of the three polymers are 135°, 175°, and 220°C, respectively.

These polymers have many applications. Because of their chemical inertness, they are used for chemical storage containers as well as many other packaging applications.

Another common polymer of simple structure that has found many uses, especially in coating cooking utensils, is Teflon, a fluorocarbon resin. Its structure is

$$\left[\begin{array}{c} \overset{\displaystyle F}{\underset{\displaystyle F}{\overset{|}{\underset{|}{C}}}}-\overset{\displaystyle F}{\underset{\displaystyle F}{\overset{|}{\underset{|}{C}}}} \end{array}\right]_n$$

Teflon

By replacing all the hydrogen atoms in the polyethylene monomer with fluorine atoms, along a long carbon chain, Teflon, a material of high melting point and low surface friction, is obtained. When a frying pan is coated with Teflon, foods will not "stick." It should also be mentioned that the majority of the ploymers are good electrical insulators and are used to great advantage in this application.

Some polymers may be made in fiber or thread form, hence clothing, rugs, and even artificial grass for supporting events are made from plastics. One such polymer is nylon which has found wide usage in hosiery. Nylon has a somewhat complicated monomer

Nylon

$$\left[\begin{array}{c} \underset{\displaystyle O}{\overset{|}{C}}(CH_2)_4\underset{\displaystyle O}{\overset{|}{C}}-\overset{\displaystyle H}{\overset{|}{N}}(CH_2)_6\overset{\displaystyle H}{\overset{|}{N}} \end{array}\right]_n$$

Other polymers such as rayon and dacron are well-known substitutes for natural fibers.

Technology and the development of polymers have gone hand in hand, each aiding the other. As mentioned, early wartime necessity spurred the development of synthetics. When the supply of natural rubber was cut off, a synthetic rubber called neoprene was developed. The different structures of natural rubber and neoprene are

Rubber
(polymer of
isoprene)

$$\left[\begin{array}{c} \overset{\displaystyle H}{\underset{\displaystyle H}{\overset{|}{\underset{|}{C}}}}-\overset{\displaystyle CH_3}{C}=\overset{\displaystyle H}{C}-\overset{\displaystyle H}{\underset{\displaystyle H}{\overset{|}{\underset{|}{C}}}} \end{array}\right]_n$$

Neoprene

$$\left[\begin{array}{c} \overset{\displaystyle H}{\underset{\displaystyle H}{\overset{|}{\underset{|}{C}}}}-\overset{\displaystyle Cl}{C}=\overset{\displaystyle H}{C}-\overset{\displaystyle H}{\underset{\displaystyle H}{\overset{|}{\underset{|}{C}}}} \end{array}\right]_n$$

However, need did not precede discovery in all cases. Our entry into the space age has created a big demand for different and unusual materials. However, some materials which could stand the rigorous space environment were already available. Two examples are polyethylene terephthalate and polycarbonate, known commercially as Mylar polyester film and Lexan polycarbonate resin, respectively. They are extremely tough materials—the Echo satellites were constructed of thin sheets of Mylar, while the durable Lexan serves as astronauts' helmet visors. They also have everyday applications. Another group of polymers with applications from bathtub sealer to space ships is the silicones. These compounds are rubbery and flexible. One such polymer, RTV 615, is elasticlike rubber, transparent, and has a melting point four times that of steel. It has been used as heat shields for space vehicle reentry where the temperatures are in excess of 8000°C. This material transmits virtually no heat, due to a slow layer by layer decomposition of its molecules which uses excess heat as the latent heat of vaporization.

The list of polymers and their uses could go on and on. However, it should be evident from the few mentioned that artificial molecules are a tribute to our ingenuity.

16-4
DRUGS

A **drug** is a compound that may produce a physiological change in human beings or other animals and may become a poison when used in excessive amounts. The molecular composition of drugs is usually quite complex, as the majority of drugs are extracted from natural compounds. Because of the human involvement and the complexity of the human body, the development of new or synthetic drugs is relatively slow compared to other materials such as plastics. A great degree of caution must be exercised in administering new drugs, and many experiments must be run on laboratory animals in an

effort to find any side effects that may be harmful to humans. Occasionally these effects are latent and emerge only after widespread usage, such as in the case of thalidomide, the sleeping compound believed to be the cause of deformities in the human fetus. Often there is no conclusive evidence of a compound being harmful to humans, but preventive action is sometimes taken, as in the case of the banning of cyclamate sweeteners when it was observed that large doses caused cancer in rats.

From our definition of a drug, it is evident that drugs are useful for purposes other than medicinal, although they are in large part used to correct some body malfunction. One of the oldest drugs known to man is alcohol. As a by-product of fermented natural beverages, ethyl alcohol has been consumed by humans probably since the beginning of civilization. Some alcohols have severe physiological effects; methyl alcohol, for example, may cause blindness if taken internally. In small amounts, ethyl alcohol acts as a depressant although it is commonly, and falsely, considered to be a stimulant. Acting on the brain's control mechanisms, it can cause loss of muscular control, such as slurred speech, mental disorganization, sleep, and nausea when taken in moderate amounts. In addition, alcohol acts on the parts of the brain that control blood circulation and the kidneys. When alcohol is taken into the body the capillaries dilate, thus bringing more blood nearer the surface of the body. This causes a warm feeling but also a more rapid loss of body heat. Its effect on the kidneys is to depress the control center of the brain which ordinarily slows the kidney's excretion. Addiction to alcohol is also possible after prolonged usage. Figure 16-4a shows some common substances containing drugs.

Another very common drug is aspirin. **Aspirin** is the compound acetylsalicylic acid.

Aspirin (acetylsalicylic acid)

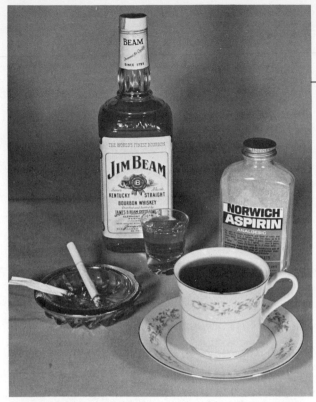

Figure 16-4a *Five common substances containing a drug: marijuana, nicotine, alcohol, caffeine, and aspirin.* (*Photo courtesy James Crouse.*)

Aspirin is used primarily as an analgesic. An **analgesic** is a drug that relieves pain without dulling consciousness. Aspirin also has the ability to reduce the body temperature, and it is thus effective for reducing fevers. It is estimated that enough acetylsalicylic acid is produced in the United States to supply every person in the United States with 400 aspirin tablets each year.

Relatively recently developed drugs are the antibiotics, or "wonder" drugs. An **antibiotic** is an antibacterial substance which is produced by a living organism and is administered to fight bacterial infections in the body. Before the development of the wonder drugs, death was quite common from serious infections, and their discovery was one of the major milestones in medicine. Probably the best known of these drugs are the penicillins, which are derived from molds of a certain type. There are now many other wonder drugs such as the mycins, which have proven to be more effective than penicillin in many cases.

Another group of drugs called **alkaloids** are complex compounds found in plants. These include morphine ($C_{17}H_{19}NO_3$), quinine ($C_{20}H_{24}N_2O_2$), nicotine ($C_{10}H_{14}N_2$), and strychnine ($C_{21}H_{22}N_2O_2$). Strychnine is best known as a heart stimulant although in large amount it is a poison, as all drugs are in overdoses. Tobacco is a source of nicotine and contains from approximately 0.5 to 5 percent nicotine content in its leaves. Because it is a dangerous poison, nicotine finds little use as a drug, but it is an effective insecticide. Quinine was used as an early treatment for malaria, but has been replaced by more effective drugs. Morphine is a narcotic found in crude opium which is extracted from the oriental poppy. It is used to relieve pain and induce sleep. Addiction to this drug may occur with prolonged usage.

There has been increasing concern over the misuse of drugs, especially the narcotics. **Heroin,** an artificial derivative of morphine, is a habit-forming drug. Once the habit is formed or the person is "hooked," the craving for drugs is insatiable, and the user will go to any extreme, including violence, to obtain a certain drug or the money to buy it, thus creating a serious social problem. Controversy is still waging as to the overall effect of the use of marijuana. **Marijuana** is the leaves and flowering tops of the Indian hemp plant. Smoking or ingestion of marijuana causes dizziness, hilarity, delusions of grandeur, heightened mental awareness, and mental confusion, depending on the user. It is generally agreed that marijuana is not habit-forming and its effects are similar to that of alcohol. However, the concern is that a person may become accustomed to the effects of using marijuana (which diminish with prolonged usage) and might try a more potent, habit-forming drug such as heroin.

Another group of drugs are the psychedelics, or hallucinogenic, drugs, one of which is **LSD** (lysergic acid diethyl amide). This dangerous drug causes prolonged stimulation, which may be followed by insanity. Sensations are amplified, and exaggerations and hallucinations are experienced. The LSD molecule is quite complex.

LSD

The lower colored portion of the formula shows an *indole* ring; such a ring is found in a similar structure called serotonin. **Serotonin** is a compound found in the brain and is believed to transfer signals to the nerve cells. It is theorized that because of a similar structure to serotonin, LSD interferes with this transfer, and the signals are altered. Thus the drug could change the sensory signals, producing hallucinations and possibly almost any other reaction. Because of the risk of lasting brain disorders, LSD is a drug taken at risk to the user.

Of a somewhat different social impact are the drugs contained in birth control pills, another success with complex molecules. The drugs used in these pills are synthesized and have proved to be more effective than similar natural compounds. The first one of these drugs to be synthesized was **norethynodrel.** It was known for some time that the hormone progesterone would prevent ovulation in the human female about 85 percent of the time; this, however, is hardly an effective contraceptive. By synthesizing compounds similar in structure to progesterone a compound was found, norethynodrel, which turned out to be virtually 100 percent effective at one-third the dosage of progesterone when given with a small amount of another hormone, estrogen.

IMPORTANT WORDS AND TERMS

Complex molecules
Monomer
Polymer
Alkyl halides
Alcohols
Aldehydes
Organic acids
Carboxyl
Ester
Alkyl
Carbohydrates
Starch
Cellulose
Proteins
Amino

Lipids
Hydrogenation
Waxes
Detergents
Nucleic acids
DNA
RNA
Vitamins
Miller experiment
Chromosomes
Mitosis
Meiosis
Genes
Dominant
Recessive

Genetic code
Mutation
Cytoplasm
Synthetic
Plastics
Drug
Aspirin
Analgesic
Antibiotic
Alkaloid
Heroin
Marijuana
LSD
Serotonin
Norethynodrel

LEARNING OBJECTIVES

After reading and studying this chapter you should be able to do the following without referring to the text:

1. Write the organic functional group for a few types of organic compounds.
2. List one or two uses for each of the organic compounds formed using the functional groups referred to in Learning Objective 1.
3. Distinguish between a monomer and a polymer.
4. State the product formed when an organic acid reacts with an alcohol.
5. State and explain the general formula for carbohydrates.
6. Describe the chemical structure of proteins.
7. State the similarities and differences of fats, oils, and waxes.
8. Distinguish between a detergent and a soap.
9. Describe the general structure of the DNA and the RNA molecules.
10. Explain the function of DNA and RNA in cell reproduction and the genetic code.
11. Describe the Miller experiment and state its significance.
12. Distinguish between mitosis and meiosis.
13. Describe the group of synthetic polymers known as plastics.
14. Define a drug, list some examples, and give the beneficial and/or harmful effect the drug has on the human body.
15. Define or explain the important words and terms at the end of the chapter.

QUESTIONS

1. Describe the structure of a polymer.
2. What functional group characterizes an alkyl halide? an alcohol? an aldehyde? an acid? an ester?
3. How are aldehydes formed chemically?
4. By what general chemical reaction are alcohols formed?
5. Name the alcohol used in alcoholic beverages. Draw its structural formula.
6. What is the result of reacting oxygen with an aldehyde?
7. Write in words the general equation for the reaction of an organic acid with an alcohol.
8. What is the meaning of the word carbohydrate? Is the nature of carbohydrate compounds what the name implies? Explain.
9. What is the chemical composition of proteins?
10. What is hydrogenation?
11. Distinguish between mitosis and meiosis.
12. How is the DNA molecule constructed?
13. How does the DNA molecule reproduce itself?
14. Explain the function of DNA and RNA in cell reproduction and the genetic code.
15. What is the acetylsalicylic acid content of regular aspirins?
16. Describe the effects of LSD use by humans and give the probable cause.

PROBLEMS

1. Considering the monomer unit CH_2, what is the repetition number of each of the following:
   (a) *n*-octane (b) *n*-eicosane (c) polyethylene?

2. Complete and balance the following equation. Name the compound formed.
$$CH_3COOH + C_2H_5OH \longrightarrow \underline{\hspace{1.5cm}} + H_2O$$
   Acetic   Ethyl
     acid   alcohol

3. Complete and balance the following equation. Name the compound formed.
$$CH_3OH + O_2 \longrightarrow \underline{\hspace{1.5cm}} + H_2O$$
   Methyl
   alcohol

4. Complete and balance the following equation:
$$CH_2O + O_2 \longrightarrow \underline{\hspace{1.5cm}}.$$
                Formic acid

5. Complete and balance the following equation:

$$CO_2 + H_2O + Energy \longrightarrow \underset{\text{Glucose}}{\underline{\hspace{3cm}}} + O_2$$

6. From Table 16-1a prepare a list of the food sources of the important vitamins. Explain the effects of a deficiency of these important vitamins in a person's diet. (Outside source work will be necessary.)

7. Refer to an outside source and give the reason why termites can digest cellulose, the most abundant carbohydrate, but humans cannot.

8. Given the general formulas below, write molecular and structural formulas for the first member of each group. Then name the compound for each.

$$R-OH \qquad R-\overset{\displaystyle O}{\overset{\|}{C}}-H \qquad R-\overset{\displaystyle O}{\overset{\|}{C}}-OH \qquad R-\overset{\displaystyle O}{\overset{\|}{C}}-O-R$$

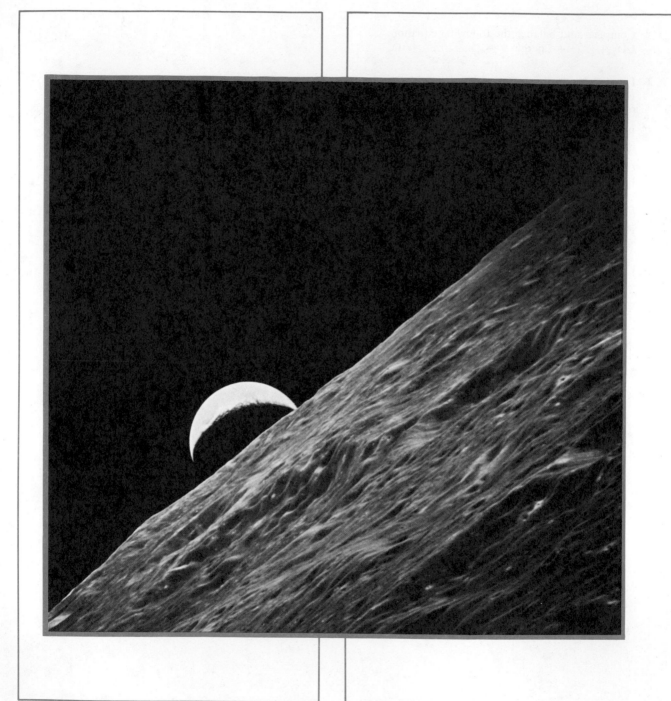

Chapter 17
THE SOLAR SYSTEM

The acquisition of knowledge begins with simple fundamental ideas which expand to broader and more complex concepts.

What is the universe? Did the universe have a beginning? If so, when in the past did the beginning take place? If there was a beginning, can we expect an end in the future? What is the size, shape, form, and substance of the universe? What are we? When and where did we originate? What is our future? Is it possible for us to know all things and control the forces of the universe?

The universe as it now exists for us is changing. We observe very few changes taking place in respect to our own brief lifetimes, but information collected over hundreds of generations is convincing proof that it is changing.

One field of science concerned with these questions is astronomy, one of the oldest sciences known. The ancient Babylonians and Greeks were watchers of the sky and originated many of the words and terms used today in describing our universe.

The basic tool of astronomy is measurement. In the beginning there were limits on the accuracy of measurements of celestial objects because of the limitations of the naked eye. With the invention of the optical telescope and the discovery and development of photography, more accurate measurements of celestial objects were made possible. Recently the invention of the radio telescope and electronic instrumentation broadened particle detection methods and increased tremendously the depth of penetration into space, which expanded the known size of the universe to billions of light years. At first thought, this seemed to be an extremely great distance, but it still is not great enough to provide information on the curvature of space for determining the shape and size of the universe.

Present knowledge indicates a known universe consisting of particles with certain characteristics and relationships. Beyond this known universe (defined as that which human beings are in contact with directly or indirectly), there is believed to exist an unknown universe, which may or may not be limitless. We do know that, as we develop new instruments and better techniques for taking better measurements, we observe more and more matter.

We are just beginning to acquire knowledge concerning the great mysteries of the universe. As we journey through space and time, we observe from our space platform (Earth) the many stars, star clusters, and interstellar matter that make up our home galaxy. Beyond the home galaxy, known as the Milky Way, we observe countless other galaxies which seem to extend into the limitless depths of space.

In the twentieth century, we are no longer bound to our space platform. Space travel has become a reality. We are lifting instruments and ourselves above the dense, hazy atmosphere which blurs and blanks out many celestial objects. Astronauts, having taken their first giant step to the moon, have their sights set on the planets and the more distant stars. As the view becomes clearer and our instruments become more sophisticated, we shall gain deeper insight into the many mysteries that have puzzled us since ancient times.

17-1

A LOOK AT THE EARTH

A description of the physical universe as it exists here and now provides a starting point for obtaining answers to questions raised in the introduction. The description should include the size, shape, and form of all bodies observed, with the characteristics and relationships of the various bodies. Any description of the present state must necessarily be incomplete and have some degree of error. All objects within range of our instruments have not been detected, and corrections must be continually made as better methods and techniques are developed.

The earth is one of nine planets that revolve around the sun. These nine planets and their 35 known satellites, plus thousands of asteroids and countless comets and meteors, make up the local **solar system.** Other similar planetary systems may exist in the universe, but present instruments are unable to detect them.

A description of the earth depends on the observer's position. If the observer is located on the earth's surface, he sees a planet that is part sky and part earth. The upper visible part, called the sky, may be clear blue, some blue with white, or dark clouds. The lower part, from the horizon down, may be all water, if the observer is at sea, or all grassland if he is located on the prairie. It may also appear to be concrete roads and tall buildings, or to be made up of many combinations of views.

Viewed from an airplane at several thousand feet above the surface, the earth appears as patches of green and brown over ordinary land masses, white in snow-covered areas, or shades of blue or green over large bodies of water. As the observer moves outward, away from the surface by hundreds of miles, the geometric patterns change to outlines of large land forms and water areas the size of large lakes and seas. Viewed from thousands of miles away, the curvature of the earth's sphere stands out dramatically and large land and water masses are clearly outlined where the absence of cloud cover permits a good view.

Our earth viewed from the moon would pass through phase changes similar to those in our view of the moon, but the earth would appear much larger. The full earth appears four times as large in diameter and more than 60 times as bright as the full moon when viewed from a distance of 240,000 miles. The earth appears much brighter because the clouds and water areas are so much better reflecting surfaces than the dull, dark surface of the moon.

From the sun, earth would always appear as a full disk if observed through a telescope. If viewed with the naked eye, the earth would appear as we see other planets of our solar system, as "stars."

If the observer traveled to Alpha Centauri, the star nearest our sun, the earth could no longer be seen, not even with our best telescope. The light of the sun reflected from the earth would be too faint to detect.

17-2
SHAPE AND SIZE OF THE EARTH

The earth is an oblate spheroid—flattened at the poles and bulging at the equator. This is due primarily to the rotation of the earth on its axis. Although the difference in the diameter at the poles and at the equator is about 27 miles, the difference is very small when compared with the total diameter of the earth, which is about 8000 miles. The ratio of 27 to 8000 is approximately $\frac{1}{300}$, which is a rather small fraction. If the earth were represented by a basketball, which is approximately 10 inches in diameter, the eye would not detect a difference of $\frac{1}{30}$ of an inch in the diameter. The earth is a more perfect sphere than the average basketball.

The Greek scientist Eratosthenes calculated the circumference of the earth in 250 B.C. He knew that the noon sun on the first day of summer was directly overhead (on the zenith) at Syene (now Aswan), Egypt. Eratosthenes lived at Alexandria, which was located 500 miles due north of Syene. He had received reports that deep wells at Syene were lighted all the way to the bottom on the first day of summer, which meant that the sun was directly overhead at

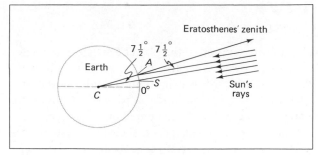

Figure 17-2a *Diagram showing the method Eratosthenes used to calculate the size of the earth. Point A represents Alexandria and point S represents Syene.*

Syene. At Alexandria, Eratosthenes discovered that a vertical stick cast a shadow on this same date, which positioned the sun $7\frac{1}{2}°$ south of his zenith. Assuming the earth to be a sphere and the sun to be such a great distance away that the rays of light would come in parallel to one another, he calculated the diameter of the earth to be 7850 miles, which is very close to the value of 7918 that we use today.

Figure 17-2a illustrates the principles involved in Eratosthenes' calculations. Since there are 360° in a circle, and the distance between Syene and Alexandria is known, the *circumference* can be calculated by using the following relationship:

$$\frac{\text{Circumference}}{360°} = \frac{\text{Dist. Syene to Alexandria}}{7.5°}$$

$$\text{Circumference} = \frac{360° \times 500 \text{ mi}}{7.5°}$$

$$\text{Circumference} = 48 \times 500 \text{ mi}$$

$$\text{Circumference} = 24,000 \text{ mi}$$

As the earth goes around the sun, it sweeps out a plane called the **ecliptic plane.** The earth rotates on an axis that is tilted by 23.5° from the ecliptic plane. The earth has the largest density of any planet. The earth's density is 5.5 g/cm^3 or 5.5 times as dense as water.

17-3

PLANETARY MOTION

The rotating motion of the planet Earth is not an easy concept to prove. In early times, most people were convinced the earth was motionless and that the sun, moon, planets, and stars revolved around the earth, which was considered the center of the universe.

This incorrect theory of the solar system was called the earth-centered or **geocentric theory.** The geocentric theory was supported and extended by Aristotle (384–322 B.C.) who was considered a great philosopher.

There were several important problems with the geocentric theory of Aristotle. Two outstanding faults were (1) a failure to explain the variation of brightness of the planets at different times of the year, and (2) a failure to explain the occasional backward or retrograde motion of some planets.

The retrograde motion of a planet is seen by observing the planet night after night relative to the fixed stars. Observed in this way, the planets move, or wander, among the fixed stars. Normally, a planet such as Mars moves eastward relative to the fixed stars. However, it sometimes moves westward relative to the fixed stars and describes a loop as shown in Fig. 17-3a.

In order to solve the difficulties of varying brightness and retrograde motion, Claudius Ptolemy of Alexandria (second century A.D.) modified and extended the geocentric theory. In the Ptolemaic model, the planets moved in circles within circles, called epicycles and deferents. The Ptolemaic theory was extremely complicated, but it was accepted with few questions for over 1200 years.

The first philosopher to give serious thought to the fact that the earth rotates about an axis was Ponticus Heracleides (375–310 B.C.), a member of the Greek Academy. The Greek astronomer, Aristarchus of Samos (about 310–230 B.C.), is credited with extending the idea of daily rotation of the earth and originating the sun-centered or **heliocentric theory,**

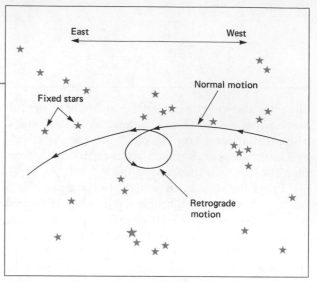

Figure 17-3a *Retrograde motion is a backward motion of a planet relative to the fixed stars. It is seen by observing a planet's position each night for several months. It easily explained by the heliocentric theory of the solar system.*

which placed the sun at the center of the universe with the earth, moon, and planets revolving around the sun in different orbits and at different speeds. These ideas were not popular and were forgotten, probably because it was difficult for people to believe the ground upon which they lived could be in rapid motion. The theory also disagreed with the philosophic doctrine that the earth was the center of the universe. Aristarchus also failed to support his theory with mathematical details for use in predicting future planetary positions. Lastly, he failed to satisfy his critics concerning the apparent motion stars would have if the earth's position would change through large distances, as he proclaimed, in revolving around the sun.

The heliocentric theory remained buried for 18 centuries. The rebirth of the theory and the development of the solar system as we think of it today began with Nicholas Copernicus (1473–1543), a Polish astronomer (Fig. 17-3b), who viewed the heliocentric theory of Aristarchus as absurd when he first studied it. But much dissatisfaction was building up against the Ptolemaic, or geocentric, system. This theory, which had been useful for more than a dozen centu-

Figure 17-3b *Nicholas Copernicus (1473–1543). Polish astronomer who developed mathematical proof of the heliocentric theory.*

ries for predicting positions of heavenly bodies, was now becoming unreliable. There was doubt about the exact length of the year, and the calendar needed to be revised. Copernicus was invited by the church to express his opinion on the revision, but he refused, saying that the position of the sun and moon were not known with sufficient accuracy to make a reliable calendar.

Copernicus observed and studied the problem for many years and finally became convinced of the validity of the heliocentric theory. He published his work under the title of *Revolutions* in 1543, the year of his death. Copernicus had succeeded where Aristarchus had failed, because of mathematical proofs which could be used to predict future positions of the planets. Although the results were no more accurate than those attained with the Ptolemaic method, they were much simpler to use. Since Copernicus held to the idea of uniformly moving concentric spheres, he had to retain the epicycle motions, but they were greatly reduced in number.

There was general public oppositon to Copernicus' sun-centered theory. The same arguments that were used against Aristarchus were again applied to Copernicus, but the scientists who were calculating the positions of the planets in most cases used the heliocentric theory because of its simplicity. The full acceptance of the theory that the earth rotates about an axis and is one of the planets that revolve around a fixed sun had to wait another century, until more accurate predictions could be made of planetary motion and the reason why the planets move as they do.

After the death of Copernicus, the study of astronomy was continued and developed by several men, three of whom made their appearance in the last half of the sixteenth century. Notable among these men was Tycho Brahe (1546–1601), a Danish astronomer who built an observatory on the island of Hven near Copenhagen and spent most of his life observing and studying the stars and planets. He is considered the greatest practical astronomer since the Greeks. His measurements of the planets and stars, all made with the naked eye (the telescope had not been invented), proved to be more accurate than any previously made. Brahe's data, published in 1603, was edited by his colleague Johannes Kepler (1571–1630), a German mathematician and astronomer who had joined Brahe during the last year of his life. After Brahe's death, his lifetime of observations were at Kepler's disposal, and they proved very useful in providing the data necessary for the formulation of the laws we know today as Kepler's laws of planetary motion.

Kepler was very interested in the irregular motion of the planet Mars. He spent considerable time and energy before he came to the conclusion that the uniform circular orbit proposed by Copernicus was not a true representation of the observed facts. Perhaps because he was a mathematician, he saw a simple type of geometric figure, which would give the correct solution to his calculations while removing all the epicycles used by Copernicus. Kepler's first law, known as the **law of elliptical paths,** states that all planets move in elliptical paths around the sun with the sun at one focus of the ellipse. Note that there is nothing at the other focus of the ellipse.

Figure 17-3c *An ellipse can be drawn by using two thumb tacks, a closed piece of string, a pencil, and a sheet of paper.*

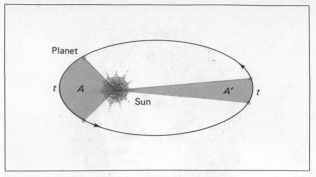

Figure 17-3d *Kepler's law of equal areas. As the earth revolves about the sun, equal areas are swept in equal periods of time. Area A = area A'.*

An ellipse is a figure that is symmetrical about two unequal diameters. See Fig. 17-3c. An ellipse can be drawn by using two thumb tacks, a closed piece of string, paper, and pencil. The points where the two tacks are positioned are called the foci of the ellipse.

Kepler's first law gives the shape of the orbit, but fails to predict when the planet will be at any position in the orbit. Kepler, aware of this, set about to find a solution from the mountain of data he had at his disposal. After a tremendous amount of work, which was carried out with no indication that a solution was possible, he discovered what is now known as Kepler's second law of planetary motion, or **law of equal areas,** which states that an imaginary line (radius vector) joining a planet to the sun sweeps out equal areas in equal periods of time (Fig. 17-3d).

From the illustration, it can be seen that the speed of the revolving body will be greatest when the radius vector is the least (that is, when the planet is closest to the sun), and its speed the slowest when the planet is farthest away. Thus, a method was provided for determining the speed, which allows the position of the planet to be predicted at some future time.

After the publication of his first two laws in 1609, Kepler began a search for a relationship between the motions of the different planets and an explanation to account for these motions. Ten years later he published *De Harmonica Mundi* (*Harmony of the World*), in which he stated his third law of planetary motion. Kepler's third law, known as the **harmonic law,** states that the ratio of the square of the period to the cube of the semimajor axis (one-half the larger axis of an ellipse), is the same for all planets. This can be written as

$$\frac{\text{Period}^2}{\text{Radius}^3} = \text{Constant}$$

or algebraically as

$$\frac{T^2}{R^3} = k$$

where T = the period,
R = the length of the semimajor axis,
k = a constant that has the same value for all planets.

Planets go around the sun in elliptical paths, but these ellipses are very close to being circles (except for Pluto). So R in Kepler's third law is approximately the distance between the planet and the sun. For the earth, T = 365.25 days (one year) and R = 93 million miles.

Kepler's third law allows a simple explanation of retrograde motion. As shown in Fig. 17-3e, retrograde motion is due to the fact that as the earth and another planet such as Mars orbit the sun, the earth moves faster than Mars. A person watching Mars from the earth sees retrograde motion as a natural consequence of the movement of the earth and Mars.

Galileo Galilei (1564–1642), Italian astronomer, mathematician, and physicist who is usually called

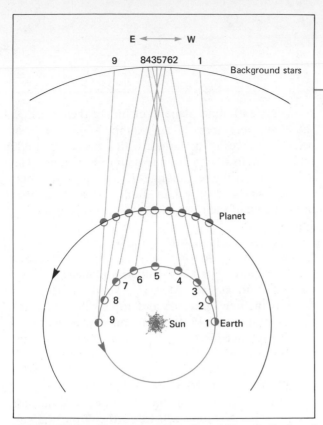

Figure 17-3e *The retrograde motion of Mars. As Earth and Mars revolve eastward around the sun, Mars appears to stop and move westward, or backward, then later stops and moves eastward again. This motion is easily explained as shown in the heliocentric theory.*

Galileo, was one of the greatest scientists of all times. The most important of his many contributions to science were in the field of mechanics. He originated the basic ideas for the formulation of Newton's first two laws of motion, and founded the modern experimental approach to scientific knowledge. The motion of bodies, especially the planets, was of prime interest to Galileo. His concepts of motion and the forces which produce motion opened up an entirely new approach to astronomy. In this field he is noted for his contribution to the heliocentric theory of the solar system.

In 1609 Galileo was the first person to observe the moon and planets through a telescope. With the telescope he discovered four of Jupiter's 14 moons, thus proving that Earth was not the only center of motion in the universe. Equally important was his discovery that the planet Venus went through a change of phase similar to that of the moon as called for by the heliocentric theory, but contrary to the Ptolemaic theory which called for a new or crescent phase of Venus at all times.

Galileo also observed the craters of the moon, dark spots on the sun, star clusters, and individual stars in the Milky Way. He published *Sidereus Nuntius* (*Starry Messenger*) in 1610, in which he reported his observations with the telescope. He published his major work, *Dialogue of the Two Chief World Systems, Ptolemaic and Copernican*, in 1632; it was received enthusiastically, but the authority of the church banned the book and Galileo, a Catholic, was forced to renounce his belief in a moving earth. He was placed under house arrest and forbidden for the rest of his life from any work on astronomy. Thus, he had failed to convince his opponents to believe the sun-centered theory, even though the telescope had presented scientific facts for its correctness. Kepler did not have such conflicts with the church because he was a German Protestant.

Only the passing of time was to remove the prejudice against scientific knowledge and allow the scientist freedom to express the facts of nature as he observed them. The works of Copernicus, Kepler, and Galileo were integrated by Newton in 1687 with the publication of the *Principia*.

Sir Isaac Newton (1642–1727), English physicist regarded by many as the greatest physical scientist the world has known, formulated the principles of gravitational attraction between bodies and established physical laws determining the magnitude and direction of the forces which cause the planets to move in elliptical orbits in accordance with Kepler's laws (cf. Introduction and Chapter 3). Newton invented calculus and used it to help explain Kepler's first law. Newton also used the law of conservation of angular momentum to explain Kepler's second law.

To explain Kepler's third law, Newton showed that the constant in Kepler's equation was

$$\frac{T^2}{R^3} = \frac{4\pi^2}{Gm_{sun}}$$

where T = period of a planet,
R = distance between the planet and the sun,
G = gravitational constant,
m_{sun} = mass of the sun.

Newton's explanations of Kepler's laws unified the heliocentric theory of the solar system and brought an end to the confusion of the past. He gave us an ordered system of the sun and planets satisfactory for the present time.

17-4
THE SOLAR SYSTEM

Today our solar system is known to consist of one star (the sun, which contains 99.87 percent of the material of the system), nine planets (including the earth), 35 satellites (our moon is an example), thousands of asteroids (Ceres is the largest with a diameter of 600 miles), billions of comets, and countless meteors. The distribution of the remaining 0.13 percent of the solar system's mass is shown in Table 17-4a. Note that more than half the remaining mass is concentrated in Jupiter.

Planets that have orbits smaller than the Earth's are classified as inferior planets and those with orbits greater than the Earth's, as superior planets. Another method is to classify Mercury, Venus, Earth, and Mars as the **inner planets;** and Jupiter, Saturn, Uranus, Neptune, and Pluto as the **outer planets.**

The relative distances of the planets from the sun are shown in Fig. 17-4a. The orbits are all elliptical, but nearly circular except for Pluto. Pluto's orbit actually goes inside Neptune's orbit. The actual position of Pluto is shown in 1981. It can be seen that for several years Neptune will be further from the sun than Pluto! Study Fig. 17-4a for a minute and notice how far from the sun Jupiter is compared to Mars.

Table 17-4a *The Solar System*

Name	Mean Distance from Sun			Diameter (miles)	Mass with Respect to the Earth	Density g/cm³ (water = 1)	Period of Revolution	Period of Rotation	Inclination of Axis with the Vertical	Inclination of Orbit with the Ecliptic	Magnetic Field	Satellites
	Million Miles	Astron. Units	Bode's Law									
Sun				864,000	332,000.	1.4		25 days			Yes	9
Mercury	36	0.39	0.4	3,026	0.055	5.4	88 days	59 days	less than 28°	7°	Weak	None
Venus	67	0.72	0.7	7,575	0.81	5.2	225 days	243 days (retrograde)	3°	3.4°	No	None
Earth	93	1.00	1.0	7,918	1.00	5.5	365.25 days	24 h	23.5°	0°	Yes	1
Mars	142	1.52	1.6	4,216	0.11	3.9	687 days	24.6 h	24°	2°	No	2
Asteroids	257	2.77	2.8				5 yr typical			10° average	No	
Jupiter	483	5.20	5.2	86,000	318.	1.3	12 yr	10 h	3.1°	1.3°	Yes	14
Saturn	886	9.54	10.0	72,000	95.	0.7	29.5 yr	10.2 h	26.7°	2.5°	Yes	10
Uranus	1783	19.19	19.6	33,000	14.6	1.2	84 yr	11 h (retrograde)	82°	0.8°	?	5
Neptune	2793	30.07	38.8	31,000	17.3	1.7	165 yr	16 h	29°	1.8°	?	2
Pluto	3666	39.46	77.2	1,600(?)	0.001(?)	1(?)	248 yr	6.39 days	?	17°	?	1

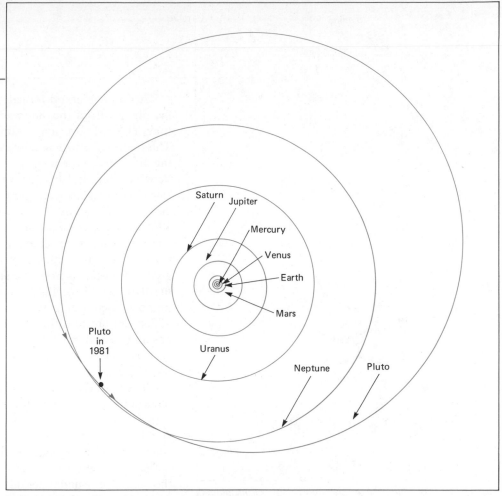

Figure 17-4a *The orbits of the planets to scale are shown. The actual position of Pluto in 1981 is also shown. The orbits of the planets are all counterclockwise.*

Note that the distance from Saturn to Neptune is further than from the sun to Saturn. The distinction between the four inner planets and the five outer planets can also be seen in this illustration.

When viewed from above the solar system (i.e., looking down on the north pole of the earth), the planets all move counterclockwise around the sun. The planets also spin with a counterclockwise rotation when viewed from above. The exceptions are Venus and Uranus which have a retrograde rotation.

The relative sizes of the planets are shown in Fig. 17-4b. Note the huge size of the outer planets compared to the inner planets. Also note how small Pluto is.

The inclinations of the orbits of the planets relative to the earth's orbit are shown in Fig. 17-4c. Note that the solar system is contained within a disk shape rather than a spherical shape. Also, note the large inclination of Pluto's orbit.

A simple way to remember the order of some of the planets is to realize that the first letter of the words *S*aturn, *U*ranus, and *N*eptune spell SUN.

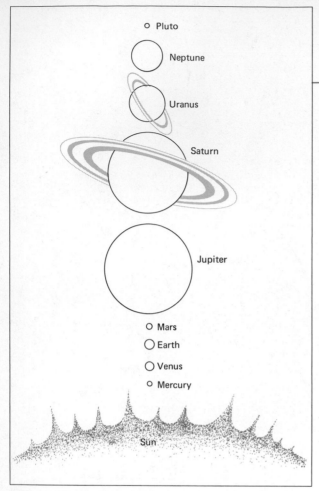

Figure 17-4b *The relative sizes and order of the planets from the sun. Distances are not drawn to scale.*

A simple method for calculating the distances of the planets from the sun was discovered by Johann Bode (1747–1826), a German astronomer, in 1772. This method, known as **Bode's law,** was very useful in the discovery of the asteroids (starlike bodies) located between Mars and Jupiter. Values for the distances in astronomical units from the sun to the different planets, as obtained by Bode's law, are given in Table 17-4a. One **astronomical unit,** the abbreviation of which is "a.u.," is equal to the mean distance from the earth to the sun which is 9.3×10^7 miles. The law gives the distance from the sun to the planets when the figures 0, 3, 6, 12, 24, and so on (doubling the number each time, except for the zero) are each added to four, and the sum divided by ten. The law has no physical interpretation, but it does provide an easy method for remembering the distance to most of the planets.

The period of time required for a planet to travel one complete orbital path is referred to as either the sidereal or synodic period. The **sidereal period** is defined as the time interval between two successive conjunctions of the planet with a star (planet and star are together on the same meridian) as observed from the sun. The **synodic period** is the time interval between two successive **conjunctions** (either inferior or superior) of the planet with the sun (planet and sun

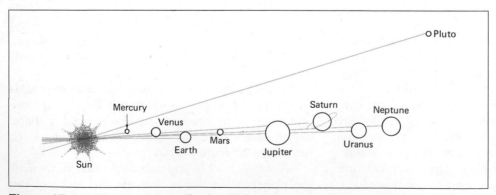

Figure 17-4c *Diagram of the inclination of planetary orbits with the orbital plane of the earth.*

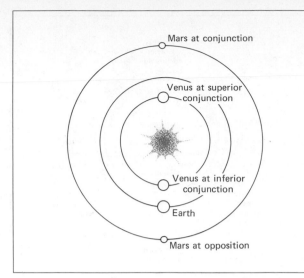

Figure 17-4d *The positions of Venus and Mars at conjunction and at opposition. Venus, an inferior planet, is shown at inferior and superior conjunction with Earth. Mars, a superior planet, is shown at conjunction and opposition with Earth.*

on same meridian) as observed from the earth. The true period of revolution is the sidereal period. **Opposition** is the term used to describe the position of a planet when the planet is 180 degrees from the sun; that is, the planet is on the opposite side of the earth from the sun. See Fig. 17-4d for an illustration of conjunction and opposition.

Each of the planets in the solar system has interesting characteristics. We will discuss each planet in order of increasing distance from the sun. We will also include in our discussion the characteristics of asteroids, comets, and meteors.

Mercury

The planet Mercury is closest to the sun and has the shortest period of revolution. The early Greeks named it after the speedy messenger of the gods, and it is the fastest moving of the planets due to its position closest to the sun.

The appearance of Mercury is similar to that of the moon as can be seen from Fig. 17-4e. However, it has a very high density, almost as high as the earth's.

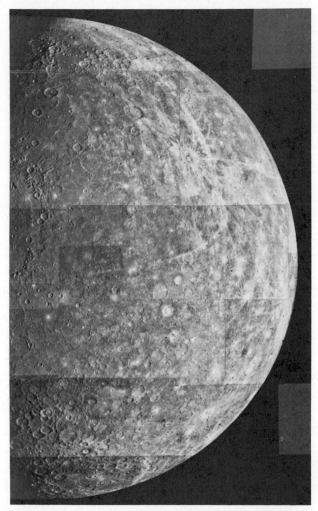

Figure 17-4e *Photomosaic of Mercury. This mosaic was constructed of 18 photographs taken at 42-second intervals by Mariner 10 from a distance from Mercury of about 210,000 kilometers (130,000 miles), six hours after the spacecraft flew past the planet on March 29, 1974. The north pole is at the top, and the equator extends from left to right about two-thirds down from the top. A large circular basin, about 1300 kilometers (800 miles) in diameter, is emerging from the day-night terminator at the left center. Bright-rayed craters are prominent in this view of Mercury. One such ray seems to join both east-west and north-south directions. (Courtesy NASA.)*

This indicates that it probably has an inner core of iron as does the earth.

Mercury's rotation period is exactly two-thirds as great as its period of revolution. Thus, it rotates exactly three times while circling the sun twice. This probably results from tidal gravitational effects from the sun. As it rotates, the side facing the sun has temperatures of approximately 700°K while the dark side is at about 100°K.

Like the moon, Mercury has no atmosphere. This is expected since Mercury's small mass would have little gravitational force of attraction for gas molecules. Mercury also receives about seven times as much solar energy per unit area as the earth. This would give the molecules of any past atmosphere sufficient energy to escape the gravitational pull of the planet.

The most perplexing property of Mercury is its weak magnetic field which was first detected in 1974 and confirmed in 1975. Mercury's magnetic field is about 1% as strong as the earth's. We believe the earth's magnetic field is caused by the earth's rapid rotation, but with Mercury's slow rotation no magnetic field was expected. We will have to wait and see what can explain Mercury's weak magnetic field.

Venus

Venus is our closest planetary neighbor, approaching Earth at a distance of 26 million miles at inferior conjunction. It is the third brightest object in the sky, exceeded only by the sun and our moon. Because of its brightness, it was named in honor of the Greek goddess of beauty.

The relative position of Venus in respect to the sun and Earth and the important positions in respect to the sun as observed from Earth are shown in Fig. 17-4f. When Venus is at superior conjunction, it is in full phase (full illumination of the side facing the earth) for an observer located on the earth, but is not visible to the observer at this time because of the brightness of the sun. As Venus moves eastward, it appears to the earth observer as the evening star until

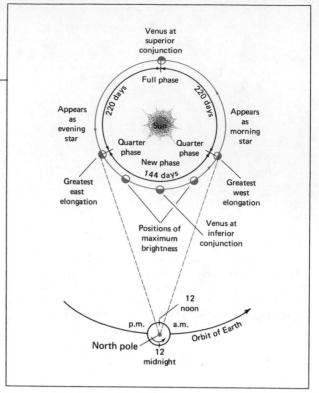

Figure 17-4f *Drawing showing the positions Venus must have in order to appear as either our evening or morning star.*

it reaches inferior conjunction. The greatest eastern elongation (greatest angular distance from the sun) occurs 220 days after superior conjunction, but maximum brightness does not occur until Venus has about 39° elongation from the sun. This occurs about 36 days before and after inferior conjunction.

The period of rotation of Venus is 243 earth days retrograde, that is, the motion is clockwise, or westward, the opposite of its revolving motion about the sun. Thick, opaque, and slightly yellowish clouds prevent us from seeing the surface of Venus, but recently (1973) radar astronomers have produced a map (Fig. 17-4g) of cloud-shrouded Venus, showing huge shallow craters in the equatorial zone up to 160 kilometers (100 miles) across. The thick atmosphere of Venus gives rise to an atmospheric pressure at the surface 90 times greater than that on Earth. The atmosphere of Venus contains an abundance of carbon dioxide (90–95%), some nitrogen (less than 5%), less than 1% oxygen, and 0.01% water. The surface temperature is around 750°K or about 900°F, due mainly to the "greenhouse effect," so life as we know it cannot exist on the planet. An additional effect which presents a hazard to spacecraft is that the clouds of Venus may contain droplets of sulfuric acid which may rain down upon the surface.

The synodic period of Venus is 584 earth days. Viewed from the earth with a telescope, Venus shows phases, an example of which is shown in Fig. 17-4h. An examination of Fig. 17-4f explains the phases and why the planet appears as the familiar morning and evening star.

Mars

From Earth, Mars has a reddish color and was named after the bloody Greek god of war. Mars is about one and one-half times as far from the sun as the earth. It is tilted on its axis at an angle of 24°

Figure 17-4h *When viewed with a telescope, Venus as the morning star shows a phase like that of the last quarter moon. (Courtesy Mount Wilson and Palomar Observatories.)*

Figure 17-4g *Radar maps of Venus. Radar astronomers have produced a dramatic map (left) of cloud-shrouded Venus showing a 1500-kilometer (910-mile) wide area that revealed huge shallow craters in the equatorial zone up to 160 kilometers (100 miles) across. The black area through the center depicts a belt that cannot be accurately tested from the radar-tracking station located on California's Mojave Desert. The map on the right shows, encircled, the position of the heavily cratered area on the planet's face. (Courtesy NASA.)*

Figure 17-4i *A topographic chart of Mars. Note the many craters, basins, mountains, and canyons. (Courtesy NASA.)*

which is very close to the earth's 23.5° angle of tilt. It rotates once every 24 hours and 37.4 minutes which is very close to a single earth day. It takes 687 days or about 23 earth months to go around the sun.

Mars has two small satellites, or moons, named Phobos and Deimos, which are about 20 and 10 miles across, respectively. Both are irregularly shaped and extensively cratered. Like our moon, they keep one side always facing Mars.

The mass of Mars is about one-tenth as large as that of the earth. Its density is also much less (3.9 g/cm³) than that of Mercury, Venus, and Earth (5.5 g/cm³). This indicates that Mars probably does not have a large iron core in the center like the Earth. Probably because of this, Mars has no magnetic field associated with it.

Many of the surface features of Mars can be seen in the topographic map of Fig. 17-4i. From this map, one can observe that Mars has many craters similar to the moon. The largest of the so-called basins is located in the southern hemisphere from 50 to 90 degrees east longitude. It is about 1000 miles across. Its smoothness is attributed to dust which has settled out from dust storms. The large volcanic mountain in the upper left of the figure is called *Olympus Mons*. It is very large, towering 16 miles high. Three smaller volcanic mountains can be seen to the southeast. The large canyon further southeast is called *Valles Marineris* and stretches for about 3000 miles.

The topographic map shown in Fig. 17-4i does not cover the polar region. Figure 17-4j shows the northern polar cap of Mars. The polar caps are composed

Figure 17-4j *The northern hemisphere of Mars from the polar cap to a few degrees south of the equator is seen in this mosaic of three photos. The north polar ice cap (upper left) is shrinking during the late Martian spring. At the bottom, the huge Martian volcanoes are visible. The west end of the great equatorial canyon can be seen at the right. (Courtesy NASA.)*

of dry ice (frozen CO_2) and water ice. The water ice is believed to change directly to water vapor in the summer and freeze again in winter.

The atmosphere of Mars is much more tenuous than that of the earth. The atmospheric pressure is slightly less than 1% that on the earth. The atmosphere is composed of 95% carbon dioxide, 2.5% nitrogen, 1.5% argon and traces of oxygen, carbon monoxide, neon, krypton, and xenon. Because of the low atmospheric pressure, water cannot exist as a liquid on the Martian surface. Traces of water vapor have been measured in the atmosphere.

Although Mars is exceedingly dry now, there is evidence for water in the past. This is shown in Figs. 17-4k and 17-4l. Figure 17-4k shows a channel and Fig. 17-4l shows teardrop features. Both of these effects are believed to indicate that water once flowed on the surface of Mars.

Our best photographs of the surface of Mars have come from the Viking landers which have carried out experiments to search for life. Figure 17-4m shows the first close-up photograph ever taken of the surface of Mars. These Viking missions have carried out numerous experiments searching for evidence of life on Mars. So far, evidence of life has not been found.

Figure 17-4k *Mars channel. Though not unique on the Martian surface, this meandering "river" is the most convincing piece of evidence that a fluid once flowed over the surface of Mars, draining an extended area, and eroding a deep channel. The feature is some 575 kilometers (355 miles) long and 5–6 kilometers (3–$3\frac{1}{2}$ miles) wide. (Courtesy NASA.)*

Figure 17-4l *Tear-drop shaped features indicate that water flowed on the surface of Mars in the past. (Courtesy NASA.)*

Figure 17-4m *The first photograph ever taken on the surface of Mars was taken by the Viking I lander on July 20, 1976. The large rock in the center is about 4 inches across. One footpad of the lander is visible on the right. (Courtesy NASA.)*

Asteroids

Bode's law calls for the next planet beyond Mars to be 2.8 astronomical units from the sun, but no large planetary body is found at this distance. On January 1, 1801, the first of the many planetary bodies was discovered by Giuseppi Piazzi, an Italian astronomer. This small body, named Ceres, is slightly less than 600 miles in diameter and is the largest of more than 1600 bodies which have since been discovered orbiting the sun between Mars and Jupiter. Because the objects appear starlike when viewed through a telescope (only one other, Vesta, 240 miles in diameter, can be seen with the naked eye), they were named **asteroids** (Greek *asteroeides,* starlike).

The diameters of the known asteroids range from that of Ceres down to only a mile. There are a dozen or more in the 100-to-200-mile range, perhaps 100 in the 50-to-100-mile range, less than 500 in the 25-to-50-mile range, and perhaps billions the size of boulders, marbles, and grains of sand. Not all asteroids are spherical in shape. Eros, which can approach the earth to within 14 million miles, is roughly rod-shaped, being 14 miles long and 5 miles thick. Many others are odd and irregular in shape. They all revolve counterclockwise around the sun, as viewed from above the orbits, or eastward, with an average inclination (degree of incline) to the ecliptic plane of 10°. Although most move in an orbit between Mars and Jupiter, some have orbits that range beyond Saturn or inside the orbit of Mercury.

The asteroids are presently believed to be early solar system material that never collected into a single planet. One piece of evidence supporting this view is that there seem to be several different kinds of asteroids. Asteroids at the inner edge of the belt seem to be stony in nature, while the ones further out are darker, indicating more carbon content. A third group may be mostly composed of iron and nickel.

In 1977, the first asteroid was found that has an elliptical orbit lying mostly between Saturn and Uranus. It has been named Object Kowal after its discoverer, Charles Kowal of the Hale Observatories.

Jupiter

Jupiter, named after the god of the skies because of its brightness and giant size, is the largest planet of the solar system, both in volume and mass. The motion about its axis is faster than that of any other planet, since it takes only ten hours to make one rotation. Jupiter possesses more than half of the angular momentum of the total solar system.

Jupiter is presently known to have 14 moons and even more may be discovered. The four moons closest to Jupiter were first seen by Galileo. By 1951 twelve moons were known, and in 1974 and 1975 Charles Kowal discovered two more very small moons. This has led to speculation that there may be a number of large chunks of rock orbiting Jupiter and how many are to be labeled as moons may become a question of definition.

Jupiter's diameter is 11 times as large as that of the earth and it has 318 times as much mass. However, its density is only 1.3 g/cm^3. Jupiter is thought to be mostly molecular hydrogen compressed into a solid state by extremely high pressure. A tentative model of Jupiter's interior is shown in Fig. 17-4n. Jupiter's atmosphere is composed of about 85% hydrogen and about 14% helium with a remaining 1% or so of methane, ammonia, and several other molecules.

Jupiter's temperature ranges from a very cold −150°F in the upper atmosphere to above the boiling point further down in the atmosphere. Jupiter's atmosphere gets more and more dense until finally a liquid state is reached as shown in Fig. 17-4n.

Jupiter has the interesting property of actually giving off twice as much heat as it receives from the sun. This makes it, in a sense, a star as well as a planet. If Jupiter had about 75 times more mass, nuclear reactions could have started in its interior similar to those in the center of the sun. Jupiter might be thought of as the smallest star ever observed, but it is usually considered our solar system's largest planet.

In the color photograph of Jupiter on color Plate IV, the surface features are easily seen. The clouds of Jupiter show a band structure and the great red spot stands out. The red spot has an erratic movement and changes color and shape. It sometimes completely disappears. The most recent theory of the great red spot states that it is a huge cyclonic storm similar to a hurricane on earth but much longer lasting.

Jupiter possesses a tremendous magnetic field. At the top of the atmosphere it is 10 times as strong as the Earth's field. Jupiter's magnetic poles are reversed from those on the earth, and Jupiter's magnetic poles are 10° from its geographic poles which are defined by its rotation axis. Its rotation axis is inclined by only a few degrees, so Jupiter does not experience seasonal effects like the earth and Mars.

CLOUD TOPS

AMMONIA CRYSTALS
AMMONIA DROPLETS
AMMONIA VAPOR
ICE CRYSTALS
WATER DROPLETS
WATER VAPOR

UPPER ATMOSPHERE

LIQUID AND/OR SOLID HYDROGEN

METALLIC HYDROGEN

INTERNAL ENERGY SOURCE GRAVITATIONAL OR RADIOACTIVE

ROCKY SILICATES METALLIC ELEMENTS

Figure 17-4n *A tentative model of Jupiter. Jupiter has its own internal energy source, and it is so massive that it is almost a star. Its atmosphere is mostly hydrogen and helium which gets more and more dense until it becomes liquid and finally solid. (Courtesy NASA.)*

Saturn

The most distinctive feature of Saturn is its system of three prominent rings, as seen through a telescope. See Fig. 17-4o. To the naked eye, Saturn appears as a fairly bright yellow star. The rings, believed to be composed of chunks of rock or ice from 1 inch to several feet across, are known as the outer, middle, and inner rings. Since they are at different distances from the planet, the rings revolve at different speeds, according to Kepler's third law. The middle ring is also known as the bright ring because it appears to be the brightest. The inner ring, known as the "crepe" ring, is the faintest, and is not visible in Fig. 17-4o. The inner ring begins about 7000 miles above Saturn's atmosphere and is about 11,000 miles in width. The bright ring begins at this altitude and extends outward an additional 16,000 miles. At this point begins Cassini's division, a gap of about 3000 miles discovered by G. D. Cassini (1625–1712) in 1675, while he was director of the Paris Observatory. Beyond the Cassini gap, the outer ring begins and extends outward 10,500 miles. The total diameter of the ring system is approximately 170,000 miles. The average thickness of the rings is believed to be less than 10 miles.

Outside the rings lie Saturn's 10 moons. The largest moon, named Titan, is about 3600 miles in diameter, larger than Mercury and slightly smaller than Mars. Titan is very cold (175°K or −144°F),

but it is still warm enough for some life. Some scientists speculate that there is more likely to be life on Titan than on Mars.

Saturn's density is 0.7 g/cm³, much lower than any other planet. Its structure is believed to consist of a small solid core surrounded by a layer of ice, a layer of metallic hydrogen and an atmosphere of molecular hydrogen. Helium is also believed present in the atmosphere. Saturn's upper atmospheric temperature is about −240°F.

Saturn is similar to Jupiter in that it rotates about once every 10 hours. It has a weak magnetic field, and like Jupiter it emits more energy than it receives from the sun.

Uranus

The planet Uranus was discovered in 1781 by William Herschel (1738–1822), an English astronomer who named the planet after the grandfather of Jupiter, thus retaining the naming pattern set by the ancient astronomers. The planet is just visible to the naked eye and, seen through a telescope, appears a greenish color.

In March of 1977, data were taken that showed for the first time that Uranus, like Saturn, has rings around it. An artist's conception of Uranus shows the rings in Fig. 17-4p. The discovery was made by a team of scientists led by Dr. James Elliott of Cornell University.

Figure 17-4o *Saturn. This photograph, showing the ring system of Saturn, was made in red light. Only the middle, or bright, ring and the outer ring are shown in the photograph. The inner, or crepe, ring is too faint to be seen. (Courtesy Mount Wilson and Palomar Observatories.)*

Figure 17-4p *An artist's rendering of the rings of Uranus, discovered in 1977. The sun is on the right, and the rings are almost perpendicular to Uranus' orbital path around the sun. (Courtesy NASA.)*

Figure 17-4q *Neptune. One of the planet's two known satellites can be seen also. (Lick Observatory Photographs.)*

Uranus has several other interesting features besides its rings. Its rotation axis is inclined 82° with respect to its orbital motion, which positions the sun nearly overhead at the north and south poles as Uranus revolves around the sun. Its rotation is retrograde with a period of 20 hours.

Uranus' physical makeup is similar to Saturn's except that the rocky core of Uranus is relatively larger compared to the planet's total size. Because it is so distant from the sun it is very cold, probably less than −300°F. Uranus is known to have five moons, but little is known about them.

Neptune

Neptune was discovered in 1846 by John G. Galle (1812–1910), a German astronomer at the Berlin Observatory. Partial credit is also shared by Englishman John Couch Adams and Frenchman U. J. J. Leverrier, two mathematicians. Using Newton's law of gravitation, Adams and Leverrier made calculations which produced information on where to look for a supposed planet that was disturbing the motion of the planet Uranus. The name Neptune was proposed by D. F. Arogo, French physicist who had suggested that Leverrier begin the critical investigation of the planet. He first suggested Leverrier as the name, but later withdrew this suggestion because of little acceptance of it outside of France.

The planet cannot be observed with the naked eye and appears to have a greenish hue when viewed through a telescope. See Fig. 17-4q. The surface temperature is less than −333°F. The physical makeup of the planet is believed to be similar to that of Uranus.

Pluto

The planet Pluto, named for the god of outer darkness, is the most distant planet from the sun. It was discovered by C. W. Tombaugh in 1930 at the Lowell Observatory in Arizona, after a thorough search near the position predicted by theoretical calculations. The planet had been predicted because

discrepancies appeared in the orbital motion of Uranus and of Neptune. General information concerning Pluto is given in Table 17-4a. Because of the planet's small size and great distance from the sun, no surface details have been discovered (Fig. 17-4r).

In 1978 a satellite of Pluto was discovered by James W. Christy. Named Charon, Pluto's moon is about 500 to 600 miles in diameter, making it the largest satellite in relation to its parent planet. Simultaneously, Pluto was found to be much smaller than previously believed. Its diameter is now thought to be about 1600 miles, making it the smallest planet.

The planet is so far away that its temperature is difficult to measure. Also the diameter, density, and period of rotation as given in Table 17-4a are estimated values and may change when more precise data are obtained.

Some astronomers believe that Pluto was once a moon of Neptune. There are three reasons for this:

1. Pluto is much smaller than the other four outer planets (see Fig. 17-4b).
2. Pluto does not lie along the planetary disk (see Fig. 17-4c).
3. Pluto's orbit is highly elliptical. It actually goes inside the orbit of Neptune (see Fig. 17-4a).

Pluto went inside Neptune's orbit in late 1978 and will exit in the year 2000. Thus, for over 20 years Neptune will be the most distant planet from the sun.

There is one very strong reason to doubt that Pluto was once a moon of Neptune. For Pluto to have been a moon of Neptune it once must have been very close, and the present orbits suggest that this did not

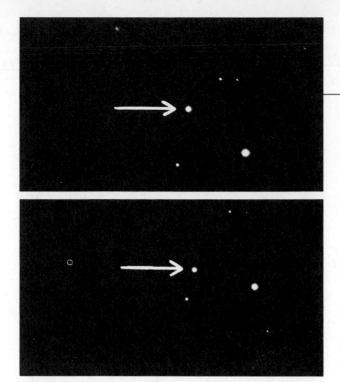

Figure 17-4r *Pluto. These two photographs show the planet's change of position in 24 hours. (Courtesy Mount Wilson and Palomar Observatories.)*

occur in the past. Detailed studies show that in their present orbits, Pluto and Neptune can never come closer than 18 astronomical units. Whether or not Pluto was once a moon of Neptune is thus not at all certain.

Is Pluto the outermost planet from the sun? The answer is yes, if we are referring to known planets. But there may be others beyond Pluto which have not been detected because of their small sizes and great distances.

Comets

Comets are named from the Latin word *cometes*, which means "longhaired." They are the solar system members that periodically appear in the sky for a few weeks or months, then disappear. A comet occupies a very large volume but has a very small mass; therefore, it must be composed of very fine particles. A **comet** consists of four parts: (1) the **nucleus,** typically a few miles in diameter and composed of rocky or metallic material, solid water, ammonia, methane, and carbon dioxide; (2) the **head,** or **coma,** which surrounds the nucleus and which can be as much as several hundred miles in diameter, and is formed from the nucleus as it approaches the sun; (3) a long, voluminous, and magnificent **tail,** also formed from the coma by solar winds and radiation, and which can be millions of miles in length; and (4) a spherical cloud of hydrogen, believed to be formed from the dissociation of water molecules in the nucleus, surrounding the coma. The sphere of hydrogen in some comets may have a radius exceeding that of the sun (Fig. 17-4s).

Comets are seen by reflected light from the sun and the fluorescence of some of the molecules comprising the comet. As comets approach and move around the sun, the material in the coma and tail gets larger (Fig. 17-4s). This increase in size is evidently caused by the sun, perhaps by (1) the solar wind which consists of streams of particles (electrons, protons, and the nuclei of light elements) given off by the sun and driven outward at speeds of hundreds of miles per second; and (2) radiation pressure generated by the radiant energy given off by the sun. It is believed that only a thin outer shell of the comet's nucleus is heated. As it moves toward its closest approach (perihelion) to the sun, the increasing solar radiation causes the surface materials to melt and evaporate. The evaporating surface particles form the coma and the long tail, which is driven away from the sun by the solar wind and radiation pressure. Each time the comet passes near the sun on its long journey through space, it loses part of its mass. Eventually the comet loses all its mass and disappears as a comet, but the rocky or metallic material continues to move around the sun.

Halley's comet, named after Edmond Halley (1656–1742), a British astronomer, is the brightest and best known comet. Halley was the first to suggest and predict the periodic appearance of comets. He observed the comet which bears his name in 1682 and, using Newton's laws of motion, correctly pre-

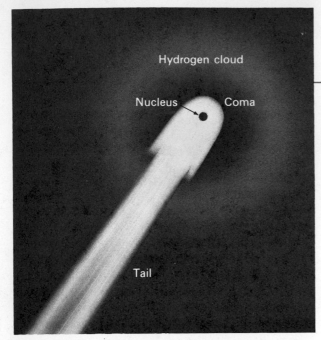

Figure 17-4s *The principal parts of a comet.*

dicted it would return in 76 years. Halley's comet last appeared in 1910 (see Fig. 17-4t), and is scheduled for its next appearance in 1986.

A **meteor,** or shooting star, is a metallic or stony object that becomes luminous as it enters and burns in the earth's atmosphere because of the tremendous heat generated by friction with the air. Millions of objects enter the earth's atmosphere every day. Most of these are completely burned in the atmosphere, but many strike the earth and are known as **meteorites.** The meteors are members of the solar system that probably come from the remains of comets and fragments of shattered asteroids. The largest known meteorite weighs more than 120,000 pounds and fell in southwest Africa. The largest known meteorite in the United States, weighing about 79,000 pounds, was

Figure 17-4t *Halley's comet. Fourteen views of Halley's comet taken in 1910 between April 26 and June 11. Note the change in size of the coma and tail. (Courtesy Mount Wilson and Palomar Observatories.)*

Figure 17-4u *Ahnighto–Cape York meteorite. Iron meteorite weighing about 79,000 pounds. (Courtesy American Museum of Natural History.)*

Figure 17-4v *Meteor crater. This crater, located near Winslow, Arizona, is 600 feet deep and 4000 feet in diameter. (Courtesy U.S. Air Force.)*

found near Cape York, Greenland, in 1895, and is on display at the Hayden Planetarium, New York (Fig. 17-4u).

Meteorites can be of any size or shape, and are classified as metallic or stony. The metallic ones have about 91 percent iron, 8 percent nickel, small quantities of cobalt, and phosphorus, and traces of many other elements. The stony ones have about 36 percent oxygen, 26 percent iron, 18 percent silicon, and 14 percent magnesium, with smaller quantities of several other elements. These are average compositions; individual meteorites can vary greatly. See Fig. 17-4v.

17-5
THE ORIGIN OF THE SOLAR SYSTEM

Any theory proposed to explain the origin and development of the solar system must account for the system as it exists now in respect to size, shape, form, substance, and change. The preceding section has been a general description of the system in its present state, which, according to our best measurements, has lasted for at least five billion years.

What are some of the major items which must be accounted for if an acceptable theory of origin is to be obtained? First, we must account for the origin of the material used to create the system. Second, we must account for the forces which were needed to form the system. Third, we must account for the size, shape, substance, positions, and motion of the sun, planets, satellites, asteroids, comets, and meteors.

Presently, there are two types of theories that attempt to explain the origin of the solar system. One group of theories are called **catastrophe theories.** In these theories our solar system arises from a catastrophe such as the near collision of a star with our sun. This group of theories has several drawbacks. Among these drawbacks are (1) hot gases drawn from the sun would not collect to form planets, but would be dispersed, and (2) the angular momentum of the present planetary system requires an approaching star to be at a minimum distance to account for the angular momentum; but in order for the material to be withdrawn from the sun, the star would have to approach closer than this minmum distance.

A second group of theories are called **evolutionary theories.** In these theories the solar system forms out of a single rotating cloud of gases due to the mutual gravitational attraction of all the particles.

The main drawback of these theories is that the sun is now spinning much more slowly than seems reasonable using these theories. Recent advances indicate that there may be a way to explain this apparent discrepancy.

The theory currently in fashion is called the **protoplanet* hypothesis.** This hypothesis proposes that initially there was a large swirling volume of cold or rarefied gases and dust positioned in space among the stars of the Milky Way. This is supported by the fact that today we observe many such nebulae throughout the universe. For some reason, such as a supernova shock wave passing through, the gases began to condense and a larger mass began to form, which produced greater force, causing the entire cloud of gases and dust to shrink (Fig. 17-5a).

The collection of particles was slow at first but became increasingly faster as the central mass became larger. As the particles moved inward, the rotation of the mass had to increase to conserve angular momentum. Because of the rapid turning, the cloud began to flatten and spread out in the equatorial plane. Kepler's third law is a statement that the central part must move faster than the outer parts. This set up shearing forces which, coupled with variations in density, produced the formation of other masses which moved around the large central portion, the protosun, sweeping up more material and forming the protoplanets.

During the early stages of development, the space between the protosun and the protoplanets was filled with large amounts of gas and dust, shielding the protoplanets from the protosun, which was beginning to burn and radiate as a star. With the passing of time, the space between the protosun and protoplanets became transparent and the protoplanets' atmospheres were heated and driven off by the pressure of solar radiation. The protoearth was perhaps 1000 times more massive than the planet Earth. The inner protoplanets, receiving more heat than the outer protoplanets, were greatly affected by this heating and lost most of their atmospheres. The outer

*"Proto" is Greek for *primitive.*

protoplanets, being at great distances from the solar radiation, were less affected and appear today in a similar protoplanet stage. Planetary satellites are believed to have formed from similar condensations, surrounding their rotating protoplanets, although our moon may have been a smaller protoplanet.

The asteroids, which have irregular size and shape, are believed to be the remnants of a planet that never formed. Meteorites are thought to be small asteroids, and by studying the composition of meteorites we have important clues to the origin of our solar system.

The comets were formed in the outer parts of the nebula, beyond the orbit of Neptune. Since very little material was available at this distance, the comets were very small in size, about one mile in diameter, composed of icy ammonia, methane, water, and some silicates. Some of the formed comets have been scattered from their birthplace by disturbances from the planets. Countless others remain in the vast space of their birthplace.

If this improved theory or hypothesis is the true picture concerning the origin of our solar system, then it is very reasonable to believe there are many, many other systems located throughout the universe, perhaps some with life similar to our own.

17-6
THE PLANET EARTH

Although we are unable to sense directly the motion of our home planet, it is undergoing several motions simultaneously. Two which have major influences on our daily lives will be explained in this section: (1) the daily rotation of the earth on its axis, and (2) the annual revolution of the earth around the sun. A third motion, precession, will be discussed in Section 18-5.

It is important to know the difference between rotating and revolving. A mass rotates, or is in **rotation,** when it turns around an axis within the mass.

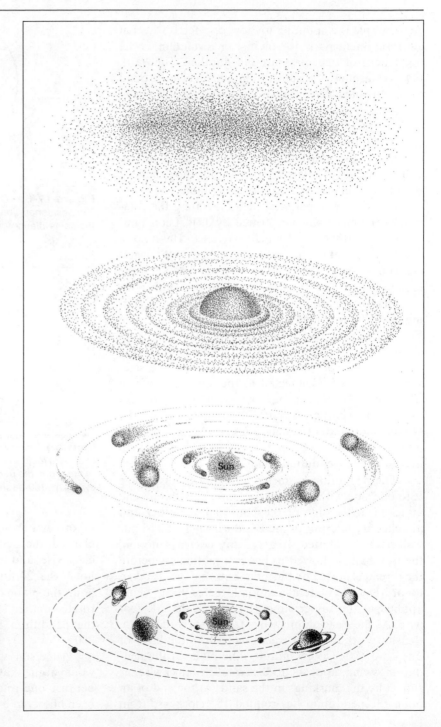

Figure 17-5a *Graphic representation of the origin of the solar system according to the protoplanet hypothesis. Interstellar gases and dust particles, with some net angular momentum, collected because of gravitational forces. Most of this collapsed mass formed the sun. The remainder of the mass formed the planets, which now revolve around the sun according to Kepler's laws.*

An example is a spinning toy top, or a ferris wheel at an amusement park. Revolving, or **revolution,** is the movement of one mass around another. The earth revolves around the sun, the moon revolves around the earth, and electrons revolve around the nucleus of an atom.

The fact that the earth rotates on its axis was not accepted until the eighteenth century. A few men had considered the possibility of a rotating earth, but no definite proof was given to support their beliefs; therefore, their ideas were not acceptable.

In 1851, an experiment demonstrating the rotation of the earth was performed by J. B. Leon Foucault (1819–1868), a French physicist. He used a 200-foot pendulum that today is called a **"Foucault pendulum."** More noticeable results can be seen if the experiment is performed at the north or south pole.

Picture a large single-room building with a ceiling over 200 feet high located at the north pole. See Figs. 17-6a and 17-6b. Fastened to the ceiling, precisely above the north pole, is a support hook having very little friction, from which a 200-foot, fine steel wire is attached. Connected to the lower end of the wire is a massive iron ball with a short sharp steel needle attached permanently to its underneath side. On the floor, under the pendulum, is placed a layer of fine sand which can be slightly furrowed by the sharp needle as the pendulum swings back and forth.

The pendulum is started swinging by displacing it to one side with a strong fine thread, with one end attached to the side of the ball, and the other end attached to one wall of the building where a 24-hour wall clock is mounted. To avoid any sideways motion, the iron ball is allowed to become motionless before the thread is parted by burning. Extreme care is taken to prevent any lateral external forces from being applied to the upper support point of the 200-foot wire. As the pendulum swings freely back and forth, the sharp needle point traces its path in the layer of fine sand. After a few minutes, the plane of the pendulum swing appears to be rotating clockwise as shown by the markings in the sand. At the end of an hour, the pendulum has rotated 15° clockwise from

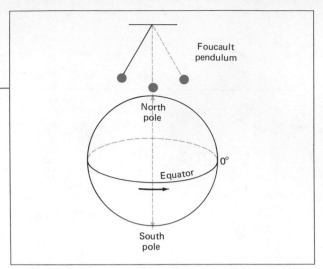

Figure 17-6a *Diagram of the earth with a Foucault pendulum positioned above the north pole. The pendulum swings in the same plane while the earth rotates about its axis.*

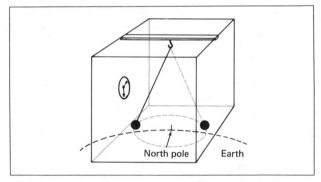

Figure 17-6b *Diagram of a Foucault pendulum positioned in a room located at the north pole of the earth. To an observer in the room, the pendulum would appear to change its plane of swing by 360 degrees every 24 hours.*

its original swinging position. When six hours have elapsed, the pendulum appears to have rotated 90° clockwise, and is swinging parallel to the wall which holds the 24-hour clock. With the passing of each hour, the plane of the swinging pendulum appears to rotate another 15° clockwise. At the end of 24 hours, the pendulum has made an apparent rotation of 360°.

A person who believes in a motionless earth would argue that the pendulum actually rotated 360°, because one rotation of the swinging pendulum has been observed by anyone stationed in the large room.

A different view of the experiment can be obtained if we make the walls of the building of a transparent material such as glass, and perform the experiment sometime during the winter months for the northern hemisphere. The north pole is having 24 hours of darkness during these months and the stars are always visible. When starting the pendulum this time, we take care to place the iron ball in direct line with the stars Alpha and Beta, the pointers in the cup of the Big Dipper. As the minutes pass, we observe as before the apparent rotation of the plane of the swinging pendulum in a clockwise direction in reference to the large room and the clock on the wall. We also observe through the transparent walls of the room that the pendulum still swings in the same direct line with the stars Alpha and Beta; that is, the pendulum, Alpha, and Beta are in the same plane. There has been no rotation of the pendulum in reference to the fixed stars. No forces have been acting on the pendulum to change its plane of swing. Only the force of gravity has been acting vertically downward to keep it swinging. Therefore the pendulum does not rotate, but the building and the earth rotate eastward or turn counterclockwise once during the 24-hour period. See Fig. 17-6c for a photograph of a Foucault pendulum. A rotating earth has a major influence on our weather, deflecting the winds from their paths and generating cyclones and other cyclic storms.

The Foucault pendulum is an experimental proof of rotation. What kind of an effect can we expect due to the earth's motion around the sun?

As the earth orbits the sun once each year, the relative positions of nearby stars change with respect to faraway stars. This effect is called parallax. In general, **parallax** is the apparent motion, or shift, that occurs between two fixed objects when the observer changes position. To see parallax for yourself, hold your finger at a fixed position in front of you. Close one eye, move your head from side to side, and notice the apparent motion between your finger and some object out the window. Note also that the apparent motion becomes less as the finger is moved farther away. Figure 17-6d is an illustration of the parallax of

Figure 17-6c *A Foucault pendulum. Note the small wooden sticks that have been knocked over by the swinging pendulum. Is the building turning clockwise or counterclockwise? (Courtesy Center of Science and Industry, Columbus, Ohio.)*

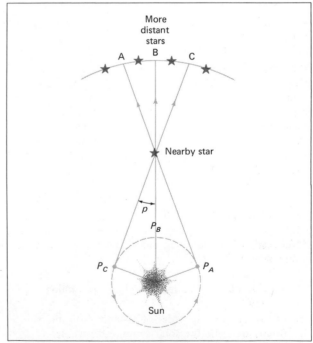

Figure 17-6d *Parallax of a star is the apparent displacement of a star that is located fairly close to the earth, in respect to more distant stars. When the observer is at P_A the star appears in the direction A. As the earth revolves counterclockwise, the star appears to be displaced and appears in the direction indicated for different positions of the earth. Positions P_A and P_C are a few months apart. The angle of parallax p is also shown.*

a nearby star as measured from the earth in relation to stars that are more distant.

As earth revolves around the sun, there is an apparent shift in the positions of the nearby stars in respect to the stars which are more distant, because of the motion of the earth. This apparent annual shift is called parallax. Since the stars are at very great distances from the earth, the parallax angle is very small, and distances greater than 20 million astronomical units cannot be measured with precision.

Because the stars are so far away, the parallax of the stars illustrated in Fig. 17-6d cannot be seen with the naked eye. It was first observed with a telescope in 1838 by Friedrich W. Bessel (1784–1846), a German astronomer and mathematician. The observation of parallax was undisputed proof that the earth really does go around the sun. Today, the measurement of the parallax angle is the best method we have of determining the distances to nearby stars.

IMPORTANT WORDS AND TERMS

Solar system	Astronomical unit	Meteor
Ecliptic plane	Sidereal period	Meteorites
Geocentric theory	Synodic period	Catastrophe theories
Heliocentric theory	Conjunction	Evolutionary theories
Kepler's law of elliptical paths	Opposition	Protoplanet hypothesis
Kepler's law of equal areas	Asteroids	Rotation
Kepler's harmonic law	Comet	Revolution
Inner planets	Nucleus	Foucault pendulum
Outer planets	Coma	Parallax
Bode's law	Tail	

LEARNING OBJECTIVES

After reading and studying this chapter, you should be able to do the following without referring to the text:

1. State Kepler's three laws of planetary motion, and draw figures illustrating the first two laws.
2. List the planets in order of distance from the sun.
3. Briefly describe the solar system as a whole.
4. Give one or more distinguishing feature of each planet.
5. List the inner planets and the outer planets and give several ways in which they are different.
6. State three reasons why Pluto is thought to be an escaped moon of Neptune.
7. Describe and differentiate between comets, meteors, and asteroids.
8. State briefly the protoplanet hypothesis for the origin of the solar system.
9. Define *parallax* and explain why it cannot be seen with the naked eye.

QUESTIONS

1. What objects comprise the solar system?
2. Briefly describe the solar system as a whole.
3. What is the explanation of Kepler's second law?
4. What distinguishes the inner planets from the outer planets?
5. Which planet is furthest from the sun? Explain.
6. Which planet has the most elliptical-shaped orbit?
7. Why do all of the planets revolve in the same direction around the sun?
8. What is the retrograde motion of Mars and how is it explained?
9. What is retrograde rotation?
10. Which planet is the largest? Which has the most moons?
11. Which planet moves fastest? Which one moves slowest?
12. Which planet is hottest? Which planet is coldest?
13. How do the atmospheric pressures on Venus, Earth, and Mars differ?
14. Which planets are known to have rings?
15. Is there intelligent life on Mars? on Earth?
16. How do asteroids, comets, and meteors differ?
17. Which planet is most dense? Which is least dense?
18. Why is Pluto thought to have once been a moon of Neptune?
19. Before 1900, what proof was there that the earth rotated and revolved?
20. Why cannot parallax be seen with the naked eye?

PROBLEMS

1. Determine the distance from the sun to the first five planets using Bode's law. Compare your answers to the actual distances.

2. Using Newton's formula $T^2/R^3 = 4\pi^2/Gm_{sun}$, show how the mass of the sun could be calculated.

3. Show that Kepler's third law holds approximately for Earth and Mars. That is, show that

$$\frac{(365.25)^2}{(93 \times 10^6)^3} = \frac{(687)^2}{(142 \times 10^6)^3}$$

4. Use Kepler's third law to show that the closer a planet is to the sun, the faster it revolves around the sun.

5. Use Kepler's second law to show that comets spend most of their time far from the sun.

6. List the following distances in order of increasing length: (a) Sun to Earth, (b) Mars to Jupiter, (c) Jupiter to Saturn, (d) Saturn to Uranus.

7. List the planets in order of decreasing size (largest first).

8. List the planets in order of decreasing density (largest first).

Chapter 18
PLACE AND TIME

What place have you?
What place have I?

What time have you?
What time have I?

Only here and now
Have you and I.

In physical science we observe and examine events that take place in our environment. These events occur at different places and at different times. Some occur immediately and are observed, while others occur at great distances and are not observed until a later time. For example, you see a flash of lightning, but do not hear the sound of thunder until later. A star explodes at some great distance from the planet Earth and years later the event is observed as the light radiation reaches the earth. Thus the events or happenings taking place in our environment are separated in space and time. Albert Einstein was the first to point out that space and time are related and exist as a single entity. Einstein associated gravitational fields with space and developed the concept of four-dimensional space, which has given us an entirely different concept concerning the reality of our environment. In this chapter we shall study the basic concepts in physical science which refer to our location and the objects of our environment in space and time. The questions of where and when are experiences we have in our daily lives: "The meeting of the next physical science class will be in Heath Hall at 10:00 A.M. on Tuesday." "Sally will spend the Christmas holidays in Florida." "Viking I made a soft landing on Mars on July 20, 1976." All of the familiar phrases include place and time. Thus, each day of our lives we experience the march of time and the change of position.

18-1

CARTESIAN COORDINATES

The location of an object in our environment requires a reference system which has one or more dimensions. A one-dimensional system is pictured by the number line shown in Fig. 18-1a, which illustrates two fundamental features of every coordinate system. A straight line is drawn, which may extend to plus infinity in one direction, and minus infinity in the opposite direction. For the line to represent a coordinate system, an origin must be indicated and unit length along the line must be expressed. Temperature scales, left-right, above ground–below ground, time past–time future, profit-loss are all examples of one-dimensional coordinate systems.

A two-dimensional system is shown in Fig. 18-1b, in which two number lines are drawn perpendicular to each other and the origin assigned at the point of intersection. The two-dimensional system is called a **cartesian coordinate system,** in honor of the French philosopher and mathematician Rene Descartes (1596–1650), the inventor of coordinate geometry. It is also referred to as a rectangular coordinate system. The horizontal line is normally designated the x-axis, and the vertical line the y-axis. Every position or point in the plane is assigned a pair of coordinates (x and y), which gives the distance from the two lines, or axes. The x number gives the distance from the y-axis, and the y number gives the distance from the x-axis. Many of the cities in the United States are laid out in the cartesian coordinate system. Usually one street runs east and west, corresponding to the x-axis, as mentioned above, while another street runs north and south, which corresponds to the y-axis.

Our interest in the cartesian coordinate system is for determining the location of any position on the surface of the spherical earth, and the location of any objects on the celestial sphere.

A spherical surface is a curved surface on which all points are equidistant from a point called the

Figure 18-1a *A one-dimensional reference system.*

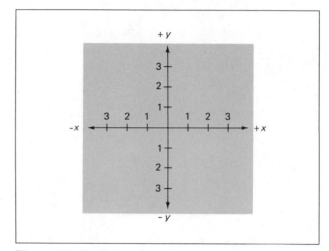

Figure 18-1b *A two-dimensional reference system.*

center. The location of any position on a spherical surface can be made using two reference circles similar to the coordinate system mentioned above.

18-2

LATITUDE AND LONGITUDE

The location of an object on the surface of the earth is accomplished by means of a coordinate system known as latitude and longitude. Since the earth is turning about an axis, we can use the *geographic poles,* which are defined as those points on the surface of the earth where the axis projects from the sphere, as reference points. The *equator*, defined in respect to the poles, is an imaginary line circling the earth at the surface, halfway between the north and south geographic poles. The equator is a great circle; that is, a circle on the surface of the earth located in a plane that passes through the center of the earth. Any such plane would divide the earth into two equal halves.

The **latitude** of a surface position can be defined as the angular measurement, in degrees, north or south of the equator. The latitude lines are circles drawn parallel to the equator; that is, they are circles drawn around the surface of the sphere parallel to the equator. Any number can be drawn. The circles become smaller as the distance from the equator becomes greater (Fig. 18-2a). These circles are called **parallels,** and when a person travels due east or west, he follows a parallel. Latitude has a minimum value of zero degrees at the equator and a maximum value of 90° north or 90° south at the poles (Fig. 18-2a).

Imaginary lines drawn along the surface of the earth running from the north geographic pole, perpendicular to the equator, to the south geographic pole are known as **meridians.** Meridians are half circles, which are portions of a great circle, since the circle is located in the same plane as the center of the earth. An infinite number of lines can be drawn as meridians. **Longitude** can be defined as the angular measurement, in degrees, east or west of the reference meridian, which is called the prime, or Greenwich, meridian. Longitude has a minimum value of zero at the prime meridian and a maximum vlaue of 180° east and west (Fig. 18-2b).

The latitude and longitude of one point (Washington, D.C., 39° N, 77° W) are shown in Figs. 18-2a and 18-2b.

The latitude and longitude shown in Figs. 18-2a and 18-2b are combined in Fig. 18-2c and shown in a cutaway of the earth. The **Greenwich meridian** was chosen for the prime, or zero, meridian because a large optical telescope was located at Greenwich,

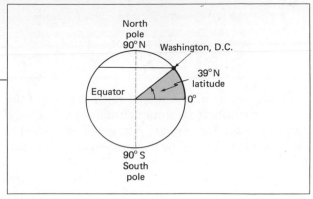

Figure 18-2a *Diagram showing the latitude of Washington, D.C.*

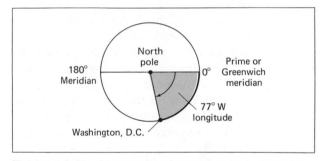

Figure 18-2b *Diagram showing the longitude of Washington, D.C.*

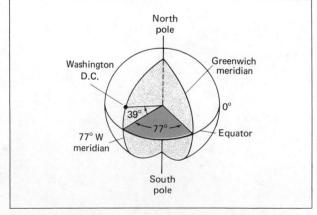

Figure 18-2c *Diagram showing both the latitude and longitude of Washington, D.C.*

England, and because England ruled the seas at the time the coordinate system of latitude and longitude was originated. The primary purpose of the system at the time was to determine the location of ships at sea.

18-3
TIME

The continuous measurement of time requires the periodic movement of some object as a reference. On October 8, 1964, the 12th General Conference of Weights and Measures, meeting in Paris, adopted an atomic definition of the *second* as the international unit of time. The definition is based on a change in energy levels of the cesium-133 atom. The exact wording of the definition for the atomic second is as follows: "The standard to be employed is the transition between the two hyperfine levels $F = 4$, $M_F = 0$, and $F = 3$, and $M_F = 0$ of the fundamental state $2s_{1/2}$ of the atom of cesium-133 undisturbed by external fields and the value 9,192,631,770 hertz is assigned." When a transition from one energy state to another occurs, the atom emits or absorbs radiation, the frequency of which is proportional to the energy difference in the two states. The cesium-133 atom provides a highly accurate and stable reference frequency of 9,192,631,770 cycles per second, which can be referred to by well-known electronic techniques. The National Bureau of Standards Frequency Standard, NBS-III, a cesium beam with a 3.66-meter interaction region is shown in Fig. 18-3a.

For everyday purposes, we are interested in the earth as a time reference, since our daily lives are influenced by the day and its subdivisions of hours, minutes, and seconds. The day has been defined in two basic ways: (1) The solar day has been defined as the elapsed time between two successive crossings of the same meridian by the sun. This is also known as the **apparent solar day,** since this is what appears to happen. Since the earth travels in an ellipse, the orbital velocity of the earth is not constant; therefore, the apparent days are not the same in duration. To remedy this situation, the mean solar day is computed from all the apparent days during a one-year period. The variation of the apparent solar day from the **mean solar day** is as much as 16 minutes. (2) The

Figure 18-3a *The National Bureau of Standard's NBS-111 cesium beam frequency generator for establishing the time standard. (Courtesy the National Bureau of Standards.)*

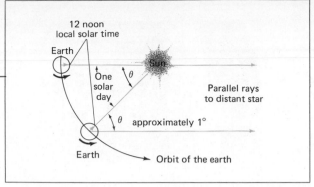

Figure 18-3b *The difference between the solar day and the sidereal day. One rotation of the earth on its axis in respect to the sun is known as one solar day. One rotation of the earth on its axis in respect to any other star is known as one sidereal day. Note that the earth turns through an angle of 360° for one sidereal day and approximately 361° for one solar day.*

sidereal day has been defined as the elapsed time between two successive crossings of the same meridian by a star other than the sun (cf. Section 1-4). Figure 18-3b illustrates the difference between the solar and sidereal days.

Since the earth rotates 365.25 times during one revolution, the magnitude of the angle through which the earth revolves in one day is

$$360°/365.25 \text{ days} = 0.985°$$

or slightly less than 1° per day. The earth must rotate through an angle of this same magnitude for the completion of one rotation in respect to the sun; therefore, the solar day is longer than the sidereal day by approximately 4 minutes, since the earth rotates 360 degrees/24 hours or 15 degrees/hour or 1 degree/ 4 minutes.

The earliest measurement of solar time was accomplished with a simple device known as a gnomon (Greek *gnomon*, a way of knowing), which is a vertical rod erected on level ground that casts a shadow when the sun is shining (Fig. 18-3c). The vertical pointer of the sundial is called a gnomon. By the third century B.C., the water clock had been invented. In hot and dry regions of the earth, sand was used to count the hours. The burning of candles was also used, since they burn at a fairly constant rate. It was not until the fourteenth century that the first mechanical clock was invented and it was not accurate. In the sixteenth century, Galileo originated the plans for the construction of a pendulum clock. The story is written that while attending church, Galileo noticed a large hanging lamp swinging back and forth at regular intervals of time as measured by his pulse rate. This provided him with the incentive to experiment with the pendulum and thus pave the way for the building of the first pendulum clock by Christian Huygens after Galileo's death.

Today, we have developed the electric clock, which is controlled by the frequency of the alternating current delivered by the power company; and the quartz-crystal clocks, which are precise to one part in

Figure 18-3c *A gnomon. A vertical rod positioned so as to cast a shadow to indicate the time of day.*

10^9. Still greater precision is obtained by the atomic clocks mentioned earlier in this chapter.

The 24-hour day, as we know it, begins at midnight and ends 24 hours later at midnight. By definition, when the sun is on the meridian, it is 12 noon *local solar time* at this meridian. The hours before noon are designated A.M. (**ante meridiem,** before noon) and P.M. (**post meridiem,** after noon). The time of 12 o'clock should be stated as 12 o'clock noon or 12 o'clock midnight, with the dates. For example, we write 12 o'clock midnight, December 3–4.

Figure 18-3d *Time zones of the continental United States. Theoretical time zones are shown. Actual boundaries are irregular.*

Our modern civilization runs efficiently because of our ability to keep accurate time. Since the late nineteenth century, most of the countries of the world have adopted the system of **standard time zones.** This scheme theoretically divides the earth into 24 time zones, each containing 15° of longitude or 1 hour. The first zone begins at the prime meridian which runs through Greenwich, England, and extends $7\frac{1}{2}$° each side of the prime meridian. The zones continue east and west from the Greenwich meridian with the centers of the zones being multiples of 15°. The actual widths of the zones vary because of local conditions, but all places within a zone have the same time, which is the time of the central meridian of that zone. For example, Washington, D.C. is located at 77° W longitude and is within $7\frac{1}{2}$° of the 75° meridian. See Fig. 18-3d.

Figure 18-3e shows the time and date on the earth for any Tuesday at 7 A.M. (PST), 8 A.M. (MST), 9 A.M. (CST), and 10 A.M. (EST). As the earth turns eastward the sun appears to move westward taking 12 noon with it. Twelve midnight is 180° or 12 hours eastward of the sun, and as 12 noon moves westward, 12 midnight follows bringing the new day.

When you travel westward into a different time zone, the time kept by your watch will be one hour fast or ahead of the standard time of the westward zone; therefore, the hour hand must be moved back one hour if the watch is to have the correct time. This process would be necessary as you continue westward

through additional time zones. A trip all the way around the earth in a westward direction would mean the loss of 24 hours or one complete day. When traveling eastward the opposite is true, that is, your watch will be one hour slow for each zone, and the hour hand is set ahead one hour.

A better understanding of why a day is lost in traveling around the earth in a westward direction can be obtained if we take a make-believe trip. Suppose we leave the Dulles International Airport in Washington, D.C. by jet plane at exactly 12 noon local mean solar time on Tuesday, December 5, and fly westward at a speed equal to the apparent westward speed of the sun. Since Washington is located at 39° N latitude, the plane must travel the 39° N parallel westward at about 800 miles per hour. As we leave the airport, we observe the sun out the left window of the plane about 32° above the southern horizon. One hour after leaving the airport, we notice the sun can still be seen out the left window at the same altitude. Six hours later, our watches indicating 7 P.M., the sun still has the same apparent position as observed from the left window of the plane. Since the plane is flying at the same apparent speed as the sun, the sun will continue to be observed out the left window of the plane. Twenty-four hours later we arrive back in Washington with the sun in the same apparent position. During the 24-hour trip, our time remained at 12 noon local mean solar time. If we had been observing the time with a sundial, it would have remained at 12 noon. We are aware of the passing of the 24 hours, because we kept track of the times with our watches, but we did not see the sun set or rise, and we did not pass through 12 midnight; therefore, the time to us is still 12 noon Tuesday. To friends meeting us at the airport, it is 12 noon Wednesday.

To remedy situations such as this, the **International Date Line** (IDL) was established at the 180° meridian. When crossing the IDL traveling westward, the date is advanced into the next day and when crossing the IDL traveling eastward, one day is subtracted from the present date.

Figure 18-3e *Diagram showing the times and dates on the earth for any Tuesday at 10 A.M. EST. As time passes, the sun moves westward; thus, 12 noon moves westward and midnight follows (180° or 12 hours) behind, bringing the new day, Wednesday, with it.*

If the local solar time is known at one longitude, the local time at another longitude can be determined by remembering there are 15° for each hour of time, or four minutes for each degree. Should the calculation extend through midnight or if the International Date Line is crossed, the date would change.

■ *Example* What is the corresponding standard time and date in Tokyo (36° N, 140° E) when the time and date in Los Angeles (34° N, 118° W) is 10 A.M. PST, November 24?

First: Determine the time zone centers of Tokyo (135°E) and Los Angeles (120°W) and draw a diagram illustrating the given date and what is to be determined.

Second: Add to the diagram the 180° meridian and the sun's meridian. The sun's meridian is not needed in the solution to the problem but helps to visualize the times at different meridians.

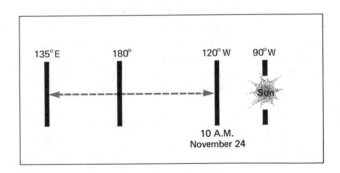

Third: Determine the minimum number of degrees difference between the two longitudes. (If the two times were given instead of the two longitudes, then the difference in time would be determined.)

180° − 120° = 60° West to the 180° meridian
180° − 135° = 45° West to the 135° E meridian

therefore 60° + 45° = 105° between the two longitudes.

Fourth: Knowing the degree difference is 105°, we calculate the time difference by using the fact that there are 15 degrees of longitude in each hour of time.

$$\frac{105°}{15°/\text{hour}} = 7 \text{ hours}$$

Fifth: Calculate the time at Tokyo by subtracting 7 hours from the 10 A.M. time at Los Angeles.

$$10 \text{ A.M.} - 7 \text{ hours} = 3 \text{ A.M.}$$

(Note: 12 noon is at 90° W since the sun is on this meridian.)

Since the 180° meridian is passed in going westward from 120° W to 135° E, we automatically go into the next day. Therefore, 3 A.M. November 25 is the answer.

During World War I, the clocks of many countries were set ahead one hour during the summer months to give more daylight hours in the evening, thus conserving fuel used for lighting. This has now become standard practice for all but three or four of the 50 states of the United States. The reason now is still to conserve fuel, as during the war. During the summer months in the United States, time known as **Daylight Saving Time** begins on the last Sunday of April and ends on the last Sunday of October.

18-4
THE SEASONS

The spinning earth is revolving around the sun in an orbit which is elliptical, yet nearly circular. If we consider the orbit to be circular, the average orbital speed may be calculated simply by dividing the distance traveled by the time, or

$$\text{Speed} = \frac{\text{Circumference of orbit}}{\text{Time}}$$

Since the circumference of a circle is $2\pi r$, then

$$\text{Speed} = \frac{2\pi r}{t}$$

$$\text{Speed} = \frac{2 \times 3.14 \times 9.3 \times 10^7 \text{ mi}}{8.64 \times 10^4 \times 3.65 \times 10^2 \text{ s}}$$

or $\text{Speed} = 18.5 \text{ mi/s}$

Remember, the actual speed varies slightly, since the earth travels in an ellipse rather than a perfect circle.

When the earth makes one complete orbit around the sun, the elapsed time is known as one *year*. We are concerned with two different definitions of the year in this text. The **tropical year,** or the year of the seasons, is the time interval from one vernal equinox to the next vernal equinox; that is, the elapsed time between one northward crossing of the sun above the equator and the next northward crossing of the sun above the equator. In respect to the rotation period of the earth, the tropical year is 365.24220 mean solar days. The sidereal year is the time interval for the earth to make one complete revolution around the sun in respect to any particular star other than the sun. The sidereal year has a period equal to 365.2536 mean solar days. This is approximately 20 minutes longer than the tropical year. The reason for the difference will be explained later in Section 18-5.

The axis of the spinning earth is not perpendicular to the plane swept out by the earth as it revolves around the sun, but is tilted $23\frac{1}{2}°$ from the vertical as illustrated in Fig. 18-4a. This position of the axis in respect to the orbital plane produces a change in the sun's overhead position throughout the year, and causes our changing seasons. Figures 18-4b and 18-4c illustrate the apparent positions of the sun over a period of one year.

The overhead position of the sun or any celestial object is given in degrees latitude and is known as the **declination** of the object. For example, the declination of the sun on June 21 is $23\frac{1}{2}°$ north latitude.

Figure 18-4a *Inclination of the earth's axis. The earth's axis is inclined 23½° from the vertical. The vertical is defined as any line perpendicular to the orbital plane of the earth.*

Later, in Section 20-2, the word declination will be used in defining the position of an object located on the celestial sphere.

The sun's overhead position is never greater than 23½° latitude, and the sun always appears due south at 12 noon local solar time for an observer located in the continental United States. When the sun is at 23½° north or south, it is at its farthest point from the equator. This farthest point of the sun from the equator is known as the solstice (sun to stand still). The most northern point is called the **summer solstice** and the most southern position is known as the **winter solstice.** This applies to the northern hemisphere. In the southern hemisphere, the dates for the summer and winter solstices are reversed from what is shown in Fig. 18-4b.

As the sun goes around the earth, its position overhead varies from 23½° north to 23½° south of the equator. When it is directly over the equator, the days and nights have 12 hours each around the world. These dates are called the equinoxes. The **vernal equinox** occurs on or about March 21 and the **autumnal equinox** occurs on or about September 21 each year. These dates are labeled in Fig. 18-4b.

When the sun is observed at 12 noon local solar time, it is on the observer's meridian and appears at its maximum above the southern horizon on that day for all observers north of the sun. The angle measured from the horizon to the line of sight to the sun at

Figure 18-4b *Earth's positions, relative to the sun, and the four seasons. As the earth revolves around the sun, its north-south axis remains pointing in the same direction. On March 21 and September 21, the sun is directly above the equator and all places on the earth are having 12 hours of daylight and 12 hours of darkness. On June 21 the sun's declination is 23½° N, and the northern hemisphere is having more daylight hours than dark hours. It is summer in the northern hemisphere and winter in the southern hemisphere. On December 21 the sun's declination is 23½° S, and the southern hemisphere is having more daylight hours than dark hours. It is winter in the northern hemisphere and summer in the southern hemisphere. The sun is drawn slightly off center to indicate the earth is slightly closer to the sun during winter in the northern hemisphere than during summer.*

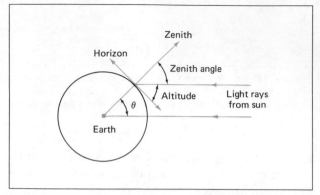

Figure 18-4e *The relationship between zenith angle, altitude, horizon, and the angle θ. Since the incoming rays of light from the sun are parallel, the angle θ and the zenith angle are equal in magnitude.*

Figure 18-4c *Diagrams of the declination of the sun at different times of the year. The upper drawing is an illustration of a greatly magnified earth in respect to the sun showing the sun's declination on the dates indicated. The lower graph plots the sun's declination (degrees latitude) versus time (months).*

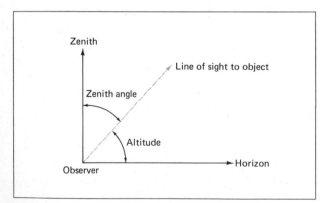

Figure 18-4d *Zenith angle and altitude. Since the zenith is perpendicular to the horizon, the zenith angle plus the altitude equals 90°.*

noon is called its **altitude,** and the angle from the **zenith** (position directly overhead) to the line of sight to the sun at noon is called its **zenith angle.** See Fig. 18-4d. The zenith is 90° from the **horizon** (the dividing line where the earth and sky appear to meet), therefore the sum of the zenith angle and the altitude is 90°. The altitude of the sun can easily be determined by measurement with a *sextant.* If the declination of the sun is known, the observer's latitude can be determined. The relation between the above terms is illustrated in Fig. 18-4e.

The maximum and minimum altitudes of the sun for an observer at Washington, D.C. (39° N) can be determined by using data from Fig. 18-4c. The solutions are as shown in Figs. 18-4f and 18-4g. The relationship between the two solutions is illustrated in Fig. 18-4h. The altitude of the sun for all other days of the year, as observed from Washington, would be between these two values. A similar solution would give the sun's altitude from any latitude.

The seasons have a tremendous effect on the lives of everyone. Our yearly life cycles are ordered by the season's progressions. Many of our holidays were

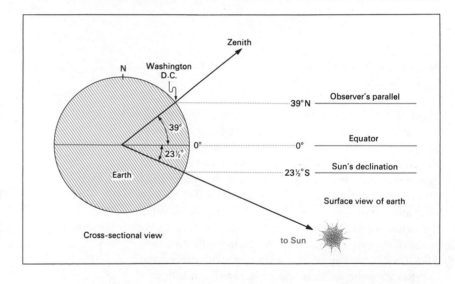

Figure 18-4f *Two methods of finding the approximate altitude of the sun as observed from Washington, D.C. on June 21. Since the angle between the sun and the observer is $39° - 23\frac{1}{2}° = 15\frac{1}{2}°$, the altitude of the sun is $90° - 15\frac{1}{2}° = 74\frac{1}{2}°$.*

Figure 18-4g *Two methods of finding the approximate altitude of the sun as observed from Washington, D.C., on December 21. Since the angle between sun and observer is $39° + 23\frac{1}{2}° = 62\frac{1}{2}°$, the altitude of the sun is $90° - 62\frac{1}{2}° = 27\frac{1}{2}°$.*

originally celebrated as commemorating a certain season of the year. The celebration that has evolved into our Easter holiday was originally a celebration of the coming of spring and a renewal of nature's life. Halloween originally commemorated the beginning of the winter season, while Thanksgiving commemorated the end of the harvest. The ancient festival of the winter solstice (on December 21 or December 22) and the beginning of the northward movement of the sun has evolved into our Christmas holiday on December 25. The reasons for celebrating our various holidays have changed over the course of time, but the original dates were set by nature's annual timepiece—the movement of the earth around the sun.

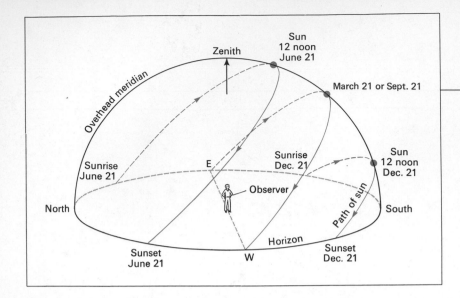

Figure 18-4h *Diagram showing the apparent path of the sun across the sky on June 21 and December 21, as observed from Washington, D.C. (39° N latitude).*

18-5
PRECESSION OF THE EARTH'S AXIS

Most of us are acquainted with the action of a toy top which has been placed in rapid motion and allowed to spin about its axis. After spinning a few seconds, the top begins to wobble or do what physicists call "precess." See Fig. 18-5a. The top, a symmetrical object, will continue to spin about an axis if the center of gravity remains above the point of support. When the center of gravity is not in a vertical line with the point of support, the axis slowly changes its direction. This slow rotation of the axis is called **precession.**

Because the earth is spinning rapidly, it bulges at the equator and cannot be considered a perfect sphere. The moon and sun also apply a gravitational torque to the earth; this tends to bring the earth's equatorial plane into its orbital plane. Because of this torque, the axis of the earth slowly rotates clockwise or westward about the vertical or the north ecliptic pole (Fig. 18-5b). The period of the precession is 25,800 years, that is, it takes 25,800 years for the axis to precess through 360°.

Since the precession is clockwise and the earth is revolving counterclockwise around the sun, the tropi-

Figure 18-5a *Precession of a top.*

cal year is approximately 20 minutes shorter than the sidereal year. As the axis precesses, Polaris will no longer be the north star. The star Vega in the constellation Lyra will be the north star some 12,000 years from now. The Southern Cross, a constellation of stars located within 27° of the present south celestial pole, will then be visible from Washington, D.C.

Our calendar is based on the seasons of the year. We want our summers to be warm and our winters to be cold. As the earth precesses, we define June 21 to be when the northern hemisphere tilts towards the

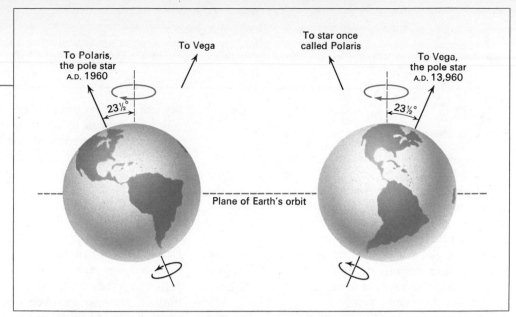

Figure 18-5b *Precession of the earth's axis clockwise about a line perpendicular to the earth's orbit. The earth's axis is presently pointing toward the star Polaris, which we call the pole, or north, star. In about 12,000 years the axis will be pointing toward the star Vega in the constellation Lyra.*

sun at a maximum angle. This means that as the earth precesses, the stars seen at various seasons change. This is shown in Fig. 18-5c. The stars we see on summer nights will slowly change with a period of 25,800 years. In the year 14,900 A.D. stars seen on summer nights will be our present winter night stars. Due to precession, the 12 constellations in the zodiac will slowly cycle through different months with a one-month change occurring every 2150 years (25,800/12). The stars seen overhead on June 21 some 2150 years ago are seen overhead on July 21 today, and will be seen overhead on August 21 in another 2150 years.

In the early 1970s, a popular song called *The Age of Aquarius* took its title from the fact that the constellation Aquarius was moving from its January 21 to February 21 time slot of thousands of years ago towards the March 21 and April 21 time slot. Since older cultures started the year on March 21, the "age" or 2150-year period took on aspects of the March 21 to April 21 constellation. The song spoke of "the dawning of the age of Aquarius" which we see would slowly occur as the earth precessed.

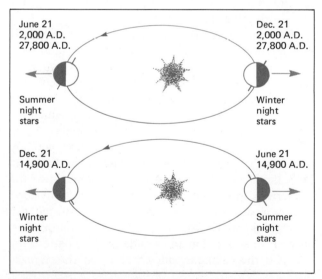

Figure 18-5c *The precession of the earth's axis will cause future generations to see different stars in the summer than are seen today. After 12,900 years our winter night stars will be their summer night stars and vice versa. After another 12,900 years, the constellations will be similar to what we see now.*

18-6

THE CALENDAR

The continuous measurement of time requires the periodic movement of some object as a reference. Various lengths of time had a direct influence on the life of ancient men and women. Because of these influences, they had more than one reference for measuring the events of their lives. It is reasonable to believe that the first unit for the measurement of time was the day; because of the daily need for food people spent the daylight hours hunting. During the dark hours they slept.

A longer period of reference was based on the periodic movement of the moon. Some societies probably used the moon as a basic division of time, since they were unable to count the days over a long period. The first appearance of the crescent moon and the time of the full moon was a time of worship for many primitive tribes.

Our month of today originated from the periodic movement of the moon, which requires $29\frac{1}{2}$ solar days to orbit the earth. The plans for the first calendar seem to have originated before 3000 B.C. with the Sumerians, who ruled Mesopotamia. Their calendar was based upon the motion of the moon, which divided the year into 12 lunar months of 30 days each. Since $30 \times 12 = 360$ days, and the year actually contains $365\frac{1}{4}$ days, corrections had to be made to keep the calendar adjusted to the seasons. How the Sumerians made the corrections remains a mystery, but their successors, the Babylonians, adjusted the length of the months and added an extra month when needed. This Babylonian calendar set the pattern for many of the calendars adopted by ancient civilizations.

The calendar we use today originated with the Romans. The early Roman calendar contained only 10 months, and the year began with the coming of spring. The months were named March, April, May, June, Quintilis, Sextilis, September, October, November, and December. The winter months of January and February did not exist. This was the period of waiting for spring to arrive. It was about 700 B.C. that Numa, who reigned in Rome, added the month of January at the beginning of the year, before March; and February at the end of the year, after December. It was about 275 years later, in 425 B.C., that the two months were changed to the present order.

The so-called Julian calendar was adopted in 45 B.C., during the reign of Julius Caesar. Augustus Caesar, who ruled the Roman Empire after Julius, renamed the month Quintilis as July in honor of Julius, and the month Sextilis as August in honor of himself. He also removed one day from February, which he added to August to make it as long as the other months.

The Julian calendar had 365 days in a year. In every year divisible by four, an extra day was added to make up for the fact that it takes approximately $365\frac{1}{4}$ days for the earth to orbit the sun. Thus, in 1981, 1982, and 1983 there are 365 days. In 1980 and 1984 there are 366 days. The Julian calendar was fairly accurate and was used for over 1600 years.

In 1582, Pope Gregory realized that the Julian calendar was slightly inaccurate. The vernal equinox was not falling on March 21, and religious holidays were coming at the wrong time. A discrepancy was found, and it was decreed that 10 days would be skipped to correct it. The discrepancy was due to the fact that there are 365.2422 days in a year and not 365.25 as the Julian calendar used.

Pope Gregory set the calendar straight and devised a method to keep it correct. He decreed that every 400 years, 3 leap years would be skipped. The leap years to be skipped were the century years not divisible by 400. Table 18-6a shows examples of the number of days in some century years. Remember, however, that all other years divisible by four are leap years and have 366 days. Our present day calendar with these leap year designations is called the **Gregorian calendar.**

Pope Gregory's reform of 1582 was not universally accepted. Most Protestant countries did not immediately go along. England and the American colonies finally changed their calendars in 1752 when there was an 11-day discrepancy. To correct it, September 2 was followed by September 14 and some people thought they were losing 11 days of their life. Other problems also arose when landlords asked for a full month's rent and banks sought to collect a full month's interest. In fact, riots were touched off in London over the change. When Russia and China accepted the Gregorian calendar in the early 1900s, it marked the first time that the whole world used the same calendar. There are other calendars still in use for religious purposes, but for civil events the Gregorian calendar is used around the world.

In our calendar we have seven days in each week. The origin of the seven-day week is not definitely known, and not all cultures have had a seven-day week. One possible origin of the seven-day week is that it takes approximately seven days for the moon

Table 18-6a *Number of Days in Century Years*

Year	Number of Days
1600	366
1700	365
1800	365
1900	365
2000	366
2100	365
2200	365
2300	365
2400	366
2500	365

Table 18-6b *The Days of the Week*

Heavenly Object	English Name	French Name	Saxon Name
Sun	Sunday	dimanche	Sun's day
Moon	Monday	lundi	Moon's day
Mars	Tuesday	mardi	Tiw's day
Mercury	Wednesday	mercredi	Woden's day
Jupiter	Thursday	jeudi	Thor's day
Venus	Friday	vendredi	Fria's day
Saturn	Saturday	samedi	Saturn's day

to go from one phase to the next (e.g., new to first quarter phase).

A more likely explanation concerns the fact that as the ancients watched the sky night after night, there were exactly seven celestial bodies that moved relative to the fixed stars. These seven objects visible to the naked eye are the sun, moon, and five visible planets: Mars, Mercury, Jupiter, Venus, and Saturn.

Our present days of the week can still be connected by their names to the sun, moon, and five visible planets. Table 18-6b shows the heavenly object, English name, French name, and the Saxon name for the seven days of the week. Note that either the English or French word is similar to the name of a heavenly object.

Our English days, Tuesday, Wednesday, Thursday, and Friday, come from the words Tiw, Woden, Thor, and Fria, who were Nordic gods. Woden was the principal Nordic god; Tiw and Thor were the gods of law and war; and Fria was the goddess of love. Sunday, Monday, and Saturday retain their connection to the sun, moon, and Saturn.

IMPORTANT WORDS AND TERMS

Cartesian coordinate system
Latitude
Parallels
Meridians
Longitude
Greenwich meridian
Apparent solar day
Mean solar day
Sidereal day

Ante meridiem
Post meridiem
Standard time zones
International Date Line
Daylight Saving Time
Tropical year
Declination
Summer solstice
Winter solstice

Vernal equinox
Autumnal equinox
Altitude
Zenith
Zenith angle
Horizon
Precession
Gregorian calendar

LEARNING OBJECTIVES

After reading and studying this chapter, you should be able to do the following without referring to the text:

1. Define latitude and longitude and give their ranges.
2. State the latitude of the equator and north and south poles.
3. State how the prime meridian and International Date Line are determined.
4. Calculate the date and time at any latitude and longitude given the date and time at another latitude and longitude.
5. Explain, with the aid of a diagram, why we have different seasons of the year.
6. Tell when we have the vernal equinox, autumnal equinox, summer solstice, and winter solstice each year.
7. Compute the altitude of the sun at noon for 40° N latitude on March 21, June 21, September 21, and December 21.
8. Explain the origin of the dates of several holidays (e.g., Christmas, Easter).
9. Draw a diagram and use it to explain the precession of the earth's axis.
10. Explain why we have 12 months in a year and seven days in a week in our present calendar.
11. Explain when we have leap years in the Gregorian calendar.

QUESTIONS

1. On the earth's surface, can you go north indefinitely?
2. On the earth's surface, can you go west indefinitely?
3. How is zero degrees defined for latitude? for longitude?
4. What is the possible range of latitude? of longitude?
5. What do A.M. and P.M. mean?
6. How are standard time zones approximately determined?
7. How many time zones are there in the 48 contiguous states? in the world?
8. What is the purpose of Daylight Saving Time?

9. Determine when the following occur in the northern hemisphere: autumnal equinox, spring equinox, winter solstice, summer solstice.
10. Exactly how many days are in the tropical year?
11. What causes the seasons?
12. How would the seasons be modified if the earth's axis tilted at 10° instead of 23.5°?
13. What are the origins of the dates for Halloween (October 31) and Christmas (December 25)?
14. How long does it take the earth to precess one time?
15. What is the origin of the month?
16. What is the origin of the seven-day week?
17. How often was there a leap year in the Julian calendar?
18. How often is there a leap year in the Gregorian calendar?

PROBLEMS

1. How far away is the point at 90° N, 130° E from 90° N, 150° E?

2. Draw a diagram and explain why the points 60° N, 130° E and 60° N, 150° E are closer together than the points 30° N, 130° E and 30° N, 150° E.

3. What are the latitude and longitude of the point on the earth that is opposite Washington D.C. (39° N, 77° W)?   *Answer: 39° S, 103° E*

4. A professional basketball game is to be played in Portland, Oregon. It is televised in New York beginning at 9 P.M. What time must the game begin in Portland?   *Answer: 6 P.M.*

5. If an Olympic event begins at 10 A.M. on August 14 in Moscow (56° N, 38° E), what date and time will it begin in Los angeles (34° N, 118° W)?

6. What is the latitude of someone in the United States who sees the sun at an altitude angle of 71.5° on June 21?   *Answer: 42° N*

7. What is the latitude angle of the sun on March 21 for someone at the north pole?

8. What is the latitude angle of someone in the United States who sees the sun at an altitude angle of $24\frac{1}{2}$° on December 21?   *Answer: 42° N*

9. How many days are in each of the following years: 1983, 1984, 1985, 1986, 1987, 1988, 2000, 2001, 2004, 2100, 2200, 2300, 2400?

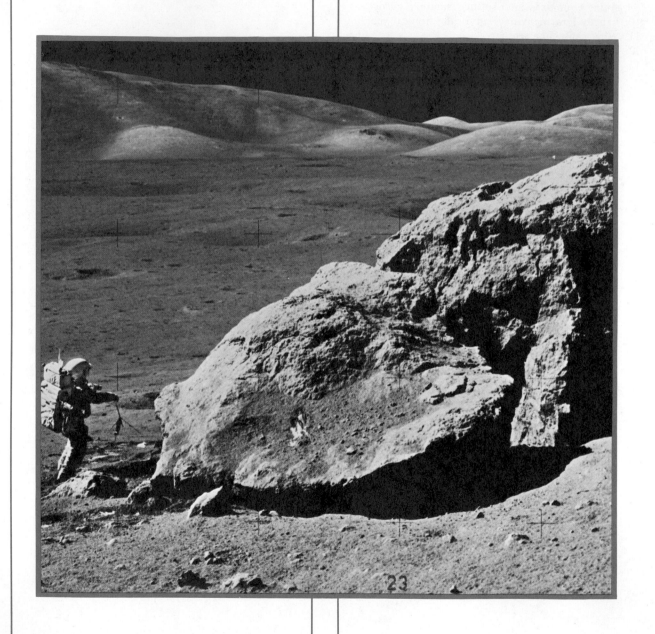

Chapter 19

THE MOON

Mysterious wanderer of the night
Many have been captured by your light,
So now try on the other shoe
We have finally captured you.

The exact origin of the word "moon" seems to be unknown, but many writers believe the original meaning related to the measurement of time. We do know that the length of our present month is based upon the motion and the phases of the moon, that primitive people worshipped the moon, and that many societies today base their religious ceremonies on the new and full phases of the moon. We also know the human reproduction cycle is synchronized to a lunar cycle with the ovaries producing ova about every 28 days.

On July 20, 1969 human beings first landed on the moon, and Apollo 11 astronauts placed a retro-reflector (an optical reflector designed to return the reflected ray of a laser beam exactly parallel to the incident ray) array on the moon's surface. The retro-reflector is part of a lunar-ranging experiment which measures the distance to the moon with an accuracy of 15 centimeters. Measurements with this high accuracy can be taken over long periods of time and will show the variation in the orbital distance of the moon in great detail. Such measurements can be used to (1) determine the rate of continental drift on the earth (latitude and longitude of a place can be determined with great accuracy); (2) detect any change in the location of the north pole; (3) determine the orbit of the moon with greater accuracy; and (4) determine if the gravitational constant (G) is decreasing very slowly with time, because the universe is believed by some scientists to be expanding.

The moon is a rather insignificant body if viewed from outside our solar system, but it is our largest

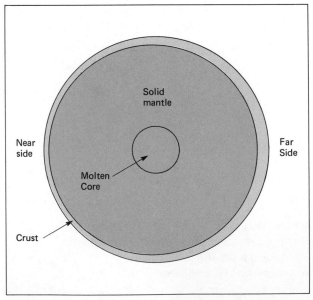

Figure 19-1a *The first landing on the moon, July 20, 1969. (Courtesy NASA.)*

natural satellite and appears as the second brightest object in the sky to the earth observer because it is very close to us. Its average distance from the earth is about 240,000 miles. Because of the moon's nearness and its influence on our lives, this chapter is devoted to its study.

19-1
GENERAL FEATURES

The earth's moon at its brightest is a wondrous sight as it reflects the sun's light back to earth. The moon appears quite large to earth observers. In fact, our moon is the largest moon of any inner planet. Mercury and Venus have no moons and the moons of Mars are quite small. Our moon is the fifth largest in the whole solar system.

The moon revolves around the earth once every $29\frac{1}{2}$ days, and it rotates at the same rate as it revolves. For this reason, we only see one side of the moon. An observer on the side of the moon facing the earth

Figure 19-1b *The interior of the moon is believed to consist of a molten or plastic core, a solid mantle rich in silicate materials, and a crust that varies in depth from about 40 miles on the near side to 80 miles on the far side.*

would always be able to see the earth, but the sun would appear to rise and set and rise again once every 29½ days. Thus, all sides of the moon are heated by the sun's rays.

The moon is spherical in shape, with a diameter of 2160 miles, a value slightly greater than one-fourth the earth's diameter. The slow rotation of the moon, coupled with the tidal bulge caused by the earth's gravitational pull on the solid material, produces an oblateness which is very small. The best measurements indicate a difference of less than a mile between the polar and equatorial diameters.

The moon has mass of $\frac{1}{81}$ that of the earth, and an average density of 3.3 g/cm³. The earth's average density is 5.5 g/cm³. The surface gravity of the moon is only $\frac{1}{6}$ that of the earth. This means your weight on the surface of the moon would be about $\frac{1}{6}$ of your weight at the earth's surface. The moon's interior is thought to be made up of a molten metallic core at the center, a solid mantle, and a crust that is about 40 miles thick on the near side and 80 miles thick on the far side (Fig. 19-1b). In spite of the molten core, the moon has only a very weak magnetic field. This may be partially due to the moon's very slow rotation.

Except for the change in the phases of the moon, its most predominant feature is the appearance of the surface, which is marked with craters, plains, rays, rills, mountain ranges, and faults (Fig. 19-1c). These features vary in size, shape, and structure. The most

Figure 19-1c *Lunar earthside hemisphere. USAF Lunar Reference Mosaic showing the features of the moon. Numbers 1, 2, 6, and 7 are Mare Imbrium, Mare Serenitatis, Mare Tranquillitatis, and Mare Crisium, respectively; number 3 the Apennines Mountains; number 8 the Great Straight Wall; and numbers 4, 5, 9, and 10 are the craters Kepler, Copernicus, Tycho, and Clavius, respectively. The numbers R7, R8, and R9 represent the landing positions of the United States lunar probes of the Ranger (hard-landing spacecraft) series. The first six flights were unsuccessful. Numbers S1 and S3 represent the landing positions of the U.S. lunar probes of the Surveyor (soft-landing spacecraft) series. Numbers A11, A12, A14, A15, A16, and A17 represent the landing positions of Apollo 11, 12, 14, 15, 16, and 17, respectively. The numbers L9 and L13 represent the landing positions of Russia's soft-landing probes Luna 9 and Luna 13. (Courtesy U.S. Air Force.)*

Figure 19-1d *Photograph of the lunar surface taken from Apollo 11 spacecraft in lunar orbit. The large central crater is International Astronomical Union No. 308, which has a diameter of about 50 miles. (Courtesy NASA.)*

outstanding are the craters which are clearly visible to the earth observer with low-power binoculars or telescope.

Craters

The word **crater** (Greek, *kratēr*) means bowl-shaped, and the lunar craters are great pits believed to be caused by the impact and explosion of small and large objects which have come from space. The craters are rather shallow (their depths are small in comparison to their diameters) and their floors are located below the lunar surface. See Fig. 19-1d. The slope of a rim with a width of about $\frac{1}{5}$ the width of the crater from crest to crest is greater on the inside than on the outside. When the measured volume of the material in the rim is compared to the volume of the crater hole, we find they are approximately the same. This tends to support the impact hypothesis.

Plains

There are thousands of craters on the lunar surface, ranging in diameter from a few feet to the 150-mile Clavius. The large flat areas called *maria* (Italian, seas), named by Galileo, are believed to be craters formed by the impact of huge objects from space and which later were filled with lava. These areas, which are now called **plains,** are very dark in appearance because the moon's surface is a poor reflector. The surface reflects only 7 percent of the light received from the sun. The plains, which are similar to black asphalt, reflect very little light. There are 14 major plains on the front side which cover over 50 percent of the visible lunar surface. Most of the plains are located in the northern hemisphere and can be plainly seen with the naked eye during the full phase of the moon.

The surface of the moon is a terrain of rolling rounded knolls composed of a layer of loose debris or soil called regolith which has a depth less than 10 meters on the flat lunar plains (maria). The lunar highlands, since they are older, have a thicker layer of regolith. The rock samples brought back by Apollo astronauts are similar to the volcanic rock found on the earth.

Rays

Some craters are surrounded by streaks, or **rays,** that extend outward over the surface. These are believed to be pulverized rock which was thrown out when the crater was formed. The rays appear much brighter than the crater, and it is known that powdered rock reflects light better than regular-sized rock. The rays also become darker with age. Photographs show that in cases where rays from one crater overlap the rays of another, the rays on top appear to be brighter. The ray system of a crater has an average diameter of about 12 times the diameter of the crater. The lunar photographs also show that the ray systems are marked with small craters called secondary craters, which are believed to have been formed by debris thrown out from the primary crater during the explosion caused by an impinging object from space.

Rills

Another feature of the lunar surface is the existence of long narrow trenches, or valleys, called **rills.** They vary from a few feet to about three miles in width and extend hundreds of miles in length with little or no variation of width. Some rills are rather straight, while others follow a circular path. The rills have very steep walls and fairly flat bottoms that are as much as one-half mile below the lunar surface. Origin of the rills is believed to be from moonquakes. A similar separation of the earth's surface would be produced by an earthquake.

Mountain Ranges

The **mountain ranges** on the lunar surface have peaks as high as 20,000 feet, and all formations seem to be components of circular patterns bordering the great plains, or maria. This indicates they were not formed and shaped by the same processes as mountain ranges on the earth, which were formed by internal forces.

The formation of craters and other surface features of the moon took place a long time ago when the solar system was filled with large amounts of matter, and these features are much the same now as they were when formed, because of the absence of

Figure 19-1e *Moon at last quarter phase. Notice the Straight Wall of Mare Nubium in the lowest dark area. (Lick Observatory Photographs.)*

water and a lunar atmosphere. There has been some change due to impact of projectiles from space.

A **fault** is a break, or fracture, in the surface of the moon in which a vertical displacement has taken place. Several of these are observed on the lunar surface. In a photograph of the last-quarter moon, a very large fault in Mare Nubium can be seen (Fig. 19-1e). This fault (called the Straight Wall) is about 60 miles long and 1200 feet high, and its side is inclined about 40° to the horizontal.

19-2
HISTORY OF THE MOON

Before the Apollo program to land a human being on the moon was begun in the early 1960s, very little was known about the origin and history of the moon. But our exploration of the moon has changed all that.

The first landing on the moon was July 20, 1969, when the landing craft of Apollo 11 descended to the moon's surface in Mare Tranquillitatis at 0.67° North, 23.49° East. Since then, five other Apollo lunar landing missions (Apollo 12, 14, 15, 16, and 17) have been completed as shown in Fig. 19-1c. Apollo 12 landed in Oceanus Procellarum on November 19, 1969; Apollo 14 landed in Fra Mauro highlands on January 31, 1971; Apollo 15 landed in Hadley-Apennines region on July 30, 1971; Apollo 16 landed in Descartes region on April 21, 1972; and Apollo17 landed in a valley at Taurus-Littrow on December 11, 1972. The Apollo astronauts collected and brought back to Earth 379 kilograms of lunar material and erected on the lunar surface 2104 kilograms of scientific instruments that will collect data long after the astronauts have returned to Earth.

The American Apollo program has now ceased due to budgeting considerations. The Soviet Union, however, continues to send unmanned missions to the moon which have brought back additional rock samples. The most recent mission, Luna 24, brought back a sample from Mare Crisium on the far-east edge of the moon's visible surface.

The rock samples brought back from the moon, such as shown in Fig. 19-2a, have enabled us to have a much better understanding of the moon's origin and history. Samples from the maria or lowlands have yielded ages considerably younger than samples from the highlands. Rocks from the highlands were formed between 3.9 and 4.4 billion years ago, whereas the rocks from the maria or lowlands have ages between 3.8 and 3.1 billion years. No rocks older than 4.4 billion years, or younger than 3.1 billion years have been found.

Almost all of the craters on the moon are now

Figure 19-2a *Lunar sample. A close-up view of a 2-centimeter chip of breccia collected by the Apollo 11 astronauts during the landing in July 1969. (Courtesy NASA.)*

known to be due to the bombardment of meteorites of various sizes. Because the moon has no atmosphere or water on its surface, very little erosion takes place on the moon. Once formed, a crater remains for billions of years or until a meteorite hits to form a crater on top of it. Compare this to the earth, which has only a few remaining meteorite craters. The craters left by meteorites striking the earth have been eroded away for the most part.

The plains and a few (about 1%) of the craters were produced by volcanic eruptions on the moon. The plains are composed of black volcanic lava that covered over many craters. Most of the plains are on the near side of the moon. This can be seen by studying Fig. 19-2b. The lower right one-third of the moon shown in this photograph is the far side of the moon. The fact that there were fewer volcanic eruptions on the moon's far side is probably correlated with the fact that the moon's crust is thicker there (see Fig. 19-1b).

From the ages of moon rocks and other data, we now have a fairly solid understanding of the moon's origin and history. However, as more samples are

Figure 19-2b *The full moon showing the far side of the moon in the lower right one-third of the photo. Note the scarcity of maria on the far side compared to the near side. (Courtesy NASA.)*

During this period, meteorite bombardment was getting less intense as fewer and fewer rock fragments were left near the earth-moon system.

After 3.1 billion years, the moon's mantle had become so thick that it could no longer be penetrated by molten rock, and the moon has been geologically quiet since that time. Meteorites have continued to bombard the surface and have formed a thin veneer of dust several feet thick on the moon's surface.

Much of our knowledge of the moon has been gained by the analysis of rock samples. Another important avenue of investigation has been the study of moonquakes. The seismological data gathered has enabled us to come up with Fig. 19-1b. Other studies have shown that there is practically no water at all in any form on the moon. In addition, no biological life has been found.

brought back from the moon and more studies are done, changes may occur in these theories.

The earth and moon are both believed to have originated in the same part of the solar system at about the same time 4.6 billion years ago. At this time, the earth and moon formed from the agglomeration of many smaller pieces of rock and other matter. The earth was a much larger body and it was formed from higher density materials than the moon. The moon's surface was hot and molten until about 4.4 billion years ago.

The oldest rocks on the moon were formed about 4.4 billions years ago when the moon's crust was cool enough to solidify. From 4.4 to 3.9 billion years ago the moon was intensely bombarded by many meteorites that were still present near the earth-moon system. This was the period when most of the moon's craters were formed.

During the period 3.9 to 3.1 billion years ago the moon's interior had heated up enough from radioactive effects to cause volcanic eruptions to occur that formed the many plains. The lava flows from these eruptions covered much of the moon's lowlands.

19-3
LUNAR MOTION

The earth's moon revolves eastward around the earth in an elliptical orbit once every $29\frac{1}{2}$ solar days or $27\frac{1}{3}$ sidereal days. The orbital plane of the moon does not coincide with the orbital plane of the earth, but is tilted at an angle of approximately 5° with respect to the earth's orbital plane. The moon rotates eastward as it revolves, making one rotation during one revolution. Figure 19-3a is an illustration of the eastward motion of the moon and the inclination of the orbital plane to the ecliptic.

Since the moon revolves in an elliptical orbit, the distance from the earth to the moon varies as the moon revolves. At the closest point (called **perigee**), the moon is 221,463 miles from the earth, and at the farthest point (called **apogee**), the moon is 252,710 miles from the earth. This change in distance of over 31,000 miles produces only a slight change in the apparent size of the moon as viewed from the earth. The mean apparent diameter of the moon is slightly over one-half degree. The mean distance to the moon

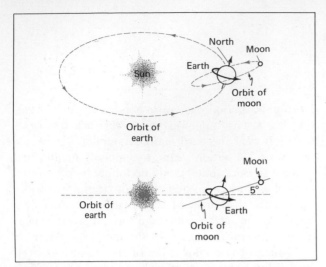

Figure 19-3a *The relative motions of the moon and earth. The top diagram is that of a view at an angle above the ecliptic plane; the lower diagram is a view parallel to the ecliptic plane.*

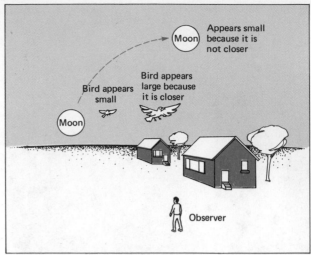

Figure 19-3c *The moon illusion. The moon appears larger on the horizon than in the sky because it is the same distance away (240,000 miles) at each point. Our eye and brain are used to objects, such as birds, that appear small on the horizon and then appear to get larger because they are approaching us. As the moon goes up in the sky, it appears to get smaller because it is not getting closer, and our brain reflexively acts as if it is getting closer.*

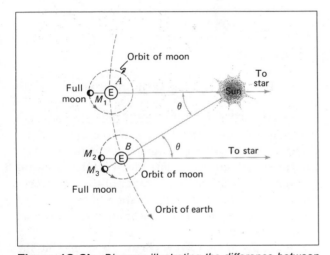

Figure 19-3b *Diagram illustrating the difference between the sidereal and synodic months. With the earth at position A and the moon at position M_1, the moon, earth, and sun are in the same plane. As the earth revolves through the angle θ, arriving at position B, the moon revolves through 360° to position M_2, and one sidereal month—one revolution in respect to a star—has elapsed. The moon must move through 360° plus an angle equal to θ before arriving at position M_3, at which time the moon, earth, and sun will again be in the same plane, and one synodic month will have taken place.*

is 238,856 miles, but 240,000 miles will be used as the mean for solving problems in this text.

As stated above, there are two different lunar months. The period of the moon in respect to a star other than the sun is $27\frac{1}{3}$ days; this is called the sidereal period, or **sidereal month,** and is the actual time taken for the moon to revolve 360°. The period of the moon in respect to the sun is $29\frac{1}{2}$ days; this is called the **synodic month,** or the month of the phases. The moon revolves more than 360° during the synodic period. See Fig. 19-3b.

To an observer on earth, the moon appears to rise in the east and set in the west each day. This apparent motion of the moon is due to the spinning of the earth on its axis once each day.

When the full moon rises at approximately 6 P.M. it appears to be very large on the horizon. Later, as it gets higher in the sky, it appears to get smaller. This optical illusion is called the **moon illusion.**

The reason for the moon illusion is illustrated in Fig. 19-3c. Normally, when something such as a bird

PLATE I

The spectrum.
(Figs. 3.2, 3.3 "Color Spectrum" in General College Chemistry, 5th edition (1976), by Charles W. Keenen, Jesse H. Wood, and Donald C. Kleinfelter. Copyright © 1957 by Harper & Row, Publishers, Inc. By permission of the publisher.)

PLATE II

SOME COMMON MINERALS

1. Quartz

Quartz is one of the most common minerals in igneous, sedimentary, and metamorphic rocks. Under favorable conditions, it occurs in hexagonal crystals. It is harder than glass.

4. Amphibole
(Var. Hornblende)

Amphibole is very abundant in igneous and metamorphic rocks. Hornblende, the most common variety, is usually dark green to black. Two cleavage planes intersect at 55° and 125° to form cleavage fragments which are commonly wedge-shaped.

2. Plagioclase

Plagioclase, a very abundant mineral, occurs in igneous, sedimentary, and metamorphic rocks. Varieties range in color from white to dark gray. It has two cleavage faces which meet at nearly right angles. Twinning lines, seen as straight, parallel striations, are helpful in identification.

5. Olivine

Olivine is abundant in dark igneous rocks low in SiO_2. It is usually olive green. Large crystals are rare. Only the green mineral in the photograph is olivine. Like quartz, olivine is harder than glass.

3. Pyroxene
(Var. Diopside)

Pyroxene is very abundant in igneous and metamorphic rocks. Varieties range in color from light green to dark green and black. Crystals are blocky. Two directions of cleavage meet at nearly right angles.

6. Mica

Mica is found in igneous, sedimentary, and especially metamorphic rocks. Biotite mica is dark brown to black. Muscovite mica is colorless to pale yellow or brown. Mica's most distinctive characteristic is perfect cleavage in one direction, which causes it to split into very thin sheets.

PLATE III

Mineral photographs by Katherine Jensen, Pittsford, N.Y.

7. Calcite

Calcite is abundant in sedimentary and metamorphic rocks and is the principal constituent of limestone and marble. It is usually colorless or white but may be light shades of green, blue, yellow, or other colors. Calcite commonly occurs in crystals. It has three well-developed directions of cleavage inclined to one another to form rhombohedral cleavage fragments.

10. Pyrite

Pyrite is scattered through igneous, sedimentary, and metamorphic rocks. Because of its brassy yellow color, it has been confused with gold and for this reason is called "Fool's Gold." It commonly occurs in crystals.

8. Dolomite

Dolomite is abundant in sedimentary and igneous rocks. Its color is commonly white to pink. Crystals as well-developed as those on the photograph are rare. Dolomite effervesces slowly in dilute hydrochloric acid.

11. Hematite

Hematite occurs in igneous and metamorphic rocks but is most common in sedimentary rocks. Its crystalline varieties are black with a metallic luster. The sedimentary varieties are usually an earthy red brown. Hematite is the most important ore of iron.

9. Halite

Halite occurs in sedimentary rocks. It is deposited by extreme evaporation of ocean water. Halite is usually colorless, but may be shades of blue, green, yellow, brown, or other colors. It has three very prominent directions of cleavage which meet at right angles to form cubic cleavage fragments. Its crystal form is also cubic.

12. Fluorite

Fluorite occurs in igneous and sedimentary rocks. It has an almost endless variety of colors. Fluorite commonly occurs in crystals and has several well-developed directions of cleavage.

PLATE IV

The planet Jupiter as photographed by Pioneer 10 spacecraft.
(Courtesy NASA)

The surface of the planet Mars as photographed by Viking I at approximately noon Mars time.
(Courtesy NASA)

is seen on the horizon, it is far away. As it approaches, it gets larger because it is getting closer. When we see the bird overhead, we expect it to be larger because we know it is closer. The moon, however, acts differently. As it goes up in the sky, it is not getting any closer. It is approximately 240,000 miles away when on the horizon and when overhead. Our eye and brain interpret events so that we assume that the moon is closer when it is higher in the sky when in fact, it is not. Thus, the moon is interpreted as being smaller when overhead. If you measure the angle subtended by the moon's diameter with a precise instrument, you will find the angle does not change significantly as the moon rises. The moon's apparent change in size is only an illusion.

19-4
PHASES

The most outstanding feature presented by the moon to an earth observer is the periodic change in its appearance. One-half of the moon's surface is always reflecting light from the sun, but only once during the lunar month does the observer on the earth see the lighted half of the moon. Throughout most of the moon's period of revolution, only a portion of its lighted side is presented to us.

The starting point for the periodic, or cyclic, motion of the moon is arbitrarily taken at the new phase position. New phase of the moon occurs when the earth, sun, and moon are in the same plane, with the moon positioned between the sun and earth. They are not necessarily in a straight line. This corresponds to the position at far left in Fig. 19-4a. At this position, the dark side of the moon is toward the earth and the moon cannot be seen from this planet. Since the sun is on the observer's meridian with the moon, the new moon occurs at 12 noon local solar time.

The **new moon,** like all the phases of the moon, occurs just for an instant—the instant it is on the

same meridian as the sun. We often speak, however, of a phase of the moon lasting for a full day of 24 hours.

The moon revolves eastward from the new phase position and for the next $7\frac{3}{8}$ solar days (one-fourth of $29\frac{1}{2}$ days) it is seen as a waxing crescent moon. The term **waxing** means that the bright portion of the moon is getting larger while **waning** means that the bright portion is getting smaller. A **crescent** moon is a moon in which less than one-quarter of the moon's surface is bright, as we observe it from the earth. A **gibbous** moon occurs when more than one-quarter of the moon's surface appears bright, when seen by an earth-based observer. Figures 19-4b and 19-4c illustrate how the phases occur and appear to look to an observer on the earth.

The moon is in the *waxing crescent phase* and appears as a crescent moon to an earth-bound observer when the moon is less than 90 degrees east of the sun. The moon is in **first quarter phase** when the

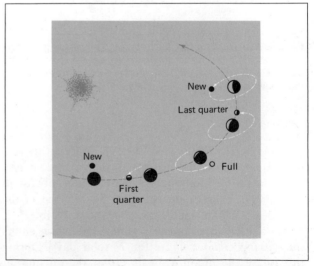

Figure 19-4a *Diagram illustrating the eastward (counterclockwise, looking from above) motion of the earth and moon around the sun.*

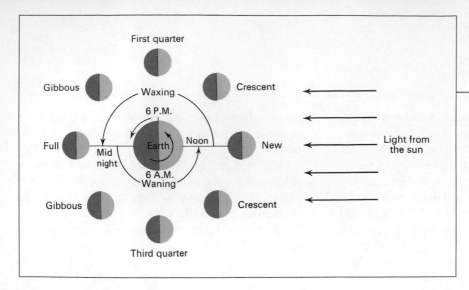

Figure 19-4b *Phases of the moon as observed from a position in space above the north pole of the earth.*

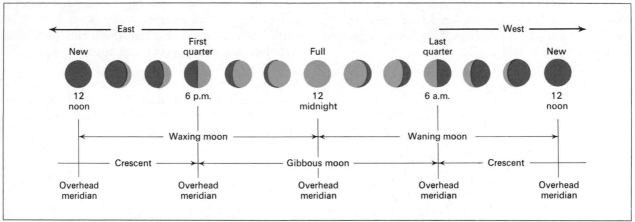

Figure 19-4c *Drawing illustrating the phases of the moon as observed from the United States. The observer is looking south; therefore, east is on the left. The sun's position can be determined by noting the time (local solar) the moon is on the overhead meridian. The time period represented in the drawing is 29½ days, or one synodic month.*

moon is exactly 90° east of the sun and appears as a quarter moon on the observer's meridian at 6 P.M. local solar time. First quarter phase occurs only for an instant, since the moon can only be 90° east of the sun for an instant. See Figs. 19-4a and 19-4b.

From the first quarter position, the moon enters the *waxing gibbous phase* for 7⅜ solar days. During this phase the moon appears larger than a quarter moon but less than a full moon. When the moon is exactly 180° east of the sun, the moon will be in full phase and will appear as a **full moon** to the earth-

bound observer. The full moon appears on the observer's meridian at 12 midnight local solar time.

From the full phase position, the moon enters the *waning gibbous phase* and remains in the waning gibbous phase for 7⅜ solar days. The appearance of the moon during this phase is the same as the waxing gibbous phase, except that the outer curved edge of the moon is toward the east and the moon will be seen in the sky at a different time. When the moon is exactly 270 degrees east of the sun, the moon will be in the third, or last, quarter phase. The **last quarter**

phase appears on the observer's meridian at 6 A.M. local solar time, with the curved edge of the moon toward the east.

From the last quarter position, the moon enters the *waning crescent phase* and remains in this phase for $7\frac{3}{8}$ solar days. During this phase the moon appears smaller than a quarter moon. Its appearance is the same as the waxing crescent moon except the outer curved edge of the moon is toward the east, and the moon appears in the sky at a different time.

Figure 19-4c illustrates the moon's appearance and position above the southern horizon as observed from a northern latitude greater than $28\frac{1}{2}$ degrees north. The moon is shown in the first drawing on the left in the new phase position. The moon is shown on the observer's meridian at 12 noon local solar time and shaded black, illustrating that it cannot be seen at this time since it is on the meridian with the sun. The next three positions illustrate the waxing crescent phase. Note the face of the moon is appearing larger in size as the moon approaches the first quarter phase, and that the outer curved edge is toward the west where the sun is located.

When the moon is in first quarter phase, the moon is on the observer's meridian at 6 P.M. local solar time. The sun will be at or near the western horizon at this time. The setting time of the sun depends upon the time of year. The moon revolves eastward entering waxing gibbous phase as shown by the next three positions. Note the moon is larger than a quarter moon and still increasing in size of face. The outer curved edge is still toward the west.

The next position shows the moon at full phase and on the observer's meridian at 12 midnight. If the date is March 21 or September 21 the moon will rise on the eastern horizon at 6 P.M., when the sun is setting in the west, and the moon will set at 6 A.M., as the sun is seen rising on the eastern horizon.

The moon revolves on eastward entering waning gibbous phase. Note that the size of the face is decreasing, and that the outer curved edge of the moon is toward the east—just the opposite from the waxing gibbous moon. When the moon is 90° west of the sun (same as 270° east of sun), it will be on the observer's meridian at 6 A.M. local solar time as shown in the next position. The moon appears as a quarter moon, but note that the curved edge is toward the east. The sun will be rising at or near this time.

After the third quarter phase, the moon enters the waning crescent phase and continues to decrease in size of face. Note the outer curved edge of the waning crescent moon is toward the east. The last position shows the moon back to the new phase position.

Table 19-4a summarizes the times for the various phases of the moon to rise, be overhead, and set. An example of what an observer in the United States sees when looking at the first quarter phase is shown in Fig. 19-4d. Figures similar to 19-4d for the other phases can be drawn using Table 19-4a.

Since the moon revolves around the earth every $29\frac{1}{2}$ solar days, it gains 360° in $29\frac{1}{2}$ days, or 12.2° per day on the sun. This means the moon is on the observer's meridian about 50 minutes later each day, because the earth must rotate through 360° plus 12.2° before the moon appears on the overhead me-

Table 19-4a *Times for the Various Phases of the Moon to Rise, Be Overhead, and Set*

Phase	Approximate Rising Time	Approximate Time Overhead	Approximate Setting Time
New moon	6 A.M.	Noon	6 P.M.
First quarter moon	Noon	6 P.M.	Midnight
Full moon	6 P.M.	Midnight	6 A.M.
Last quarter moon	Midnight	6 A.M.	Noon

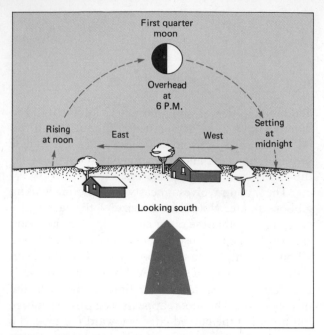

Figure 19-4d *An observer in the United States sees the first quarter moon rise in the east at noon. It is overhead at 6 P.M. and sets in the west at midnight. The right portion is bright.*

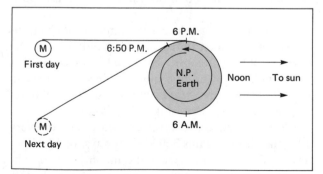

Figure 19-4e *The moon rises 50 minutes later each day because as the earth rotates, the moon is revolving around the sun. For instance, the full moon rises in the east at about 6 P.M., but the next day it would be rising at about 6:50 P.M.*

ridian. (See Fig. 19-4e.) The average time of moonrise is thus delayed about 50 minutes each day. The actual time depends on the latitude of the observer, with greater variation noted by an observer in the higher latitudes. The variation depends upon the angle between the moon's path and the horizon.

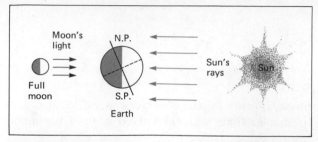

Figure 19-4f *The full moon is on the opposite side of the earth from the sun. A side view of the earth, full moon, and sun in winter is shown here. The sun's rays hit the southern hemisphere most directly. The moon's reflected light hits most directly on the northern hemisphere.*

The approximate altitude of the full moon can be found by recognizing that the full moon will be on the opposite side of the earth from the sun (Fig. 19-4f). Thus, when the sun is low in the sky in the winter, the full moon will be high in the sky. In summer, the sun is high in the sky and the full moon is low in the sky.

19-5
ECLIPSES

The word **eclipse** means the darkening of the light of one celestial body by another. The sun provides the light by which we see objects in our solar system; that is, nonluminous objects in our solar system are observed by reflected light from the sun. Since the light from the sun is falling upon objects in the solar system, objects cast shadows which extend away from the sun. The size and shape of the shadow depend on the size and shape of the object and its distance from the sun. The earth and moon, being spherical bodies, cast conical shadows, as viewed from space.

If we examine the shadow cast by the earth or moon, we discover two volumes of different degrees of darkness. The darkest and smallest region is known as the **umbra.** An observer located within this region is completely blocked from the sun during a solar eclipse. The semidark region is called the **penumbra.** An observer positioned in this region can see only a portion of the sun during a solar eclipse. See Fig. 19-5a.

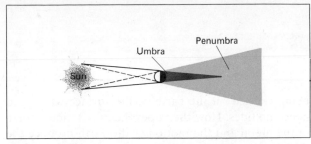

Figure 19-5a *The umbra and penumbra are, respectively, the dark and semidark shadows cast by an object exposed to light rays. Here the light rays are from the sun.*

A **solar eclipse** occurs when the moon is at or near new phase and also in the ecliptic plane. If these two events occur together, the sun, moon, and earth are all nearly in a straight line. The moon's shadow will then fall upon the earth and the sun's rays will be hidden from those observers in the shadow zone. A total eclipse occurs in the umbra region and a partial eclipse in the penumbra region.

The length of the moon's shadow varies as the moon's distance to the sun varies. The average length of the moon's umbra is 233,000 miles, which is slightly less than the mean distance between the earth and moon. Since the umbra is shorter in length than the mean distance from the earth to the moon, an eclipse of the sun can occur in which the umbra fails to touch the earth. An observer positioned on the earth's surface directly in line with the moon and sun would see the moon's disc projected against the sun, and a bright ring, or annulus, would appear outside the dark moon. This is called an **annular eclipse.** Around the zone of the total (very dark) annular eclipse would appear the larger semidark region of the penumbra. The penumbra region may be as large as 6000 miles in diameter at the surface of the earth. The maximum diameter of the umbra at the earth's surface is about 170 miles. This maximum value can exist only when the sun is farthest from the earth, which is early July, and the moon is at perigee, or at its closest distance to the earth.

The motion of the moon and earth are such that the shadow of the moon moves generally eastward during the time of the eclipse with a speed of about 1000 miles/hour. Thus, the region of total eclipse does not remain long at any one place. The greatest possible value is about $7\frac{1}{2}$ minutes, and the average is about three or four minutes.

A **lunar eclipse** occurs when the moon is at or near full phase and also in the ecliptic plane. The sun, earth, and moon must be positioned in a nearly straight line, with the earth between the sun and moon. Thus, the shadow formed by the earth blots out the face of the moon. The average length of the earth's shadow is about 860,000 miles and the diameter of the shadow at the moon's position is great enough to place the moon in total eclipse for a time slightly greater than $1\frac{1}{2}$ hours. A partial eclipse of the moon can last as long as 3 hours and 40 minutes.

The orbital plane of the moon is inclined to the ecliptic (the annual path of the sun) at an angle slightly greater than 5°; therefore, the path of the moon crosses the ecliptic at two points, as it makes its monthly journey around the earth. The points where the moon's path crosses the ecliptic are known as **nodes.** The point of crossing going northward is called the **ascending node,** and the point of crossing going southward is called the **descending node.** See Fig. 19-5b. A solar or lunar eclipse can occur only at or near the nodal points, since the earth, moon, and sun

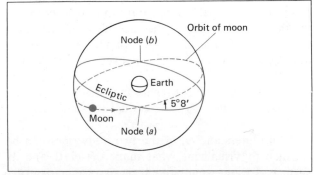

Figure 19-5b *The orbital plane of the moon in respect to the ecliptic plane. The points shown in the diagram where the two planes intersect are called the nodal points. Node (a) is called the ascending node (b) the descending node.*

must be in a nearly straight line, and this occurs only at or near the points where the moon crosses the ecliptic.

The orbital plane of the moon is precessing westward or clockwise if viewed from above the orbital plane. The precession of the moon's orbit causes the nodal points to move westward along the ecliptic, making one complete cycle in 18.6 years.

19-6
TIDES

Anyone who has been to the seashore for a day's visit is aware of the rising and falling of the surface of the ocean, which may be as great as 40 feet in some regions. The alternate rise and fall of the ocean surface is called the tides. People related tides to the passage of the moon in the first century A.D., but all efforts to explain the phenomenon failed until the seventeenth century, when Newton applied his law of universal gravitation to the problem. He related, and explained, the alternate rise and fall of the ocean's surface with the motions of the earth, moon, and sun. A few of the many factors contributing to the height that the ocean rises and falls at a particular location are (1) the rotation of the earth on its axis; (2) the position of the earth, moon, and sun in respect to one another; (3) the varying distance between the earth and moon; (4) the inclination of the moon's orbit; and (5) the varying distance between the earth and sun.

There are two high and two low tides, daily, because of the moon's gravitational attraction and the motion of the moon and earth.

An understanding of the two daily tides can be clarified by visualizing what shape the earth and its surface of water would take if there were no external gravitational forces and the earth had no motion (Fig. 19-6a). If the earth did not rotate, there would be no centripetal force. On a nonrotating earth, with no external gravitational forces, there would be no forces being exerted on the earth or the surface water and hence no tides. How then does the gravitational force of the moon and the motion of the earth produce the tides?

The answer to this question can be found and the reason for two daily tides at a given location on the earth's surface can be explained, if the moon's mass is considered to be concentrated at a point (Fig. 19-6b). The sun is also a factor in causing tides, but it is left out of this explanation to simplify the results. The magnitude and direction of the moon's gravitational forces acting on the earth at points A, B, C, D, and O are shown. The magnitude is indicated by the length of the arrow drawn to represent the gravitational force. Note that the forces at A, O, and B are all in the same direction toward the moon, but the magnitudes are not the same. The force at A is the greatest because it is closest to the moon. Remember, Newton's law of gravitational attraction says the force between two masses is inversely proportional to the square of the distance between the two masses. Thus, the force at O is less than the force at A, and the force at B is less than the force at O. So, forces at C and D have approximately the same magnitude but are slightly less than the force at O. The direction of the forces at C and D are toward the moon as shown. If all the forces shown are now added in reference to the force at O, at the center of the earth, we obtain the results as shown in Fig. 19-6c. The drawing shows the earth's crust and oceans stretched out along the line AB and brought together along line CD producing an egg-shaped body, thus giving rise to the two bulges as shown with high tides at points A and B, and low tides at points C and D.

Since the earth is rotating counterclockwise, completing one turn in 24 hours, point D will move toward point A, point A toward C, and so on for points C and B. Thus, it can be seen that two tides will occur at any of the points every 24 hours.

When the sun, earth, and moon are positioned in a nearly straight line, the gravitational force of the moon and sun combine to produce higher and lower tides than usual. That is, the variations between high

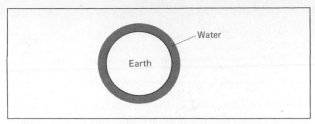

Figure 19-6a *The shape of a stationary earth with all water surfaces unaffected by any external gravitational fields.*

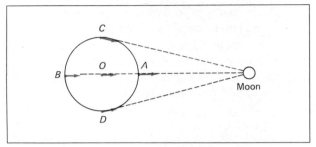

Figure 19-6b *Gravitational forces (arrows) acting on the earth by the moon, which is considered as having a mass concentrated at a point.*

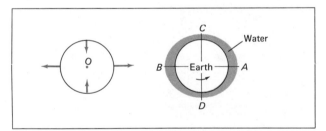

Figure 19-6c *The resultant tidal forces (arrows) acting on the earth to produce the egg-shaped contour shown at the right. View of the earth is from above the north pole.*

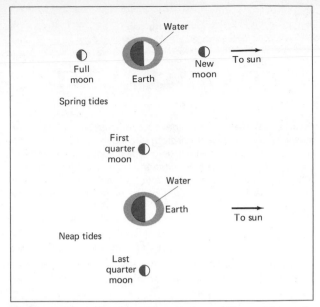

Figure 19-6d *Diagram of the relative positions of the earth, moon, and sun at the times of spring and neap tides. During spring tides the sun and moon are lined up and the tidal effects are more extreme. During neap tides, the sun and moon act against each other and the tides are more moderate.*

and low tides are greatest at this time. These tides of greatest variation are called **spring tides,** and they occur at the new and full phases of the moon. When the moon is at first or third quarter phase, the sun and moon are 90° with respect to the earth. At these times, the tidal forces of the moon and sun tend to cancel one another, and there is a minimum difference in the height of the surface of the ocean. In this case the tides are known as **neap tides.** It should be observed that two spring tides and two neap tides take place each lunar month, since the moon passes

through each of its phases once each month. There is a spring tide at new moon, a neap tide at first quarter, another spring tide at full moon, and a second neap tide at last quarter. This is shown in Fig. 19-6d.

The height of the tide also varies with latitude. See Fig. 19-6e. The tide is highest at the moon's declination and at the position opposite the moon's declination. The time of high tide does not correspond to the time of the meridian crossing of the moon. Actually, the bulge is always a little ahead (eastward) of the moon because of the earth's rotation. Since the earth rotates faster than the moon revolves, the earth carries the tidal bulge forward in the direction it is rotating, which is eastward.

The action of the tides produces a retarding motion on the earth's rotation, slowing it and lengthening the solar day about $\frac{1}{1000}$ second per century. Since the law of conservation of angular momentum must be observed, the decrease of the earth's angular momentum must appear as an increase in the moon's angular momentum. A measurement of the moon's orbit shows that the semimajor axis is increasing about one-half inch per year. If calculations were

Figure 19-6e *Spring tide occurring at time of summer solstice (sun at greatest northern declination) when the moon is in new phase. Maximum height of bulge is at 23½° N and 23½° S, since the sun's declination is 23½° N. If a spring tide were to occur three months later, when the sun is on the equator, the maximum height of the bulge would be in the equatorial region.*

made of the results of forces acting on the earth and moon over a long period of time (50 billion years), one would find that the solar day and the lunar month had increased in length until each was equal in length to 47 of our present days. The moon would then be about 340,000 miles from the earth with the earth always presenting the same face to the moon. Tidal friction would have ceased, the moon would be gradually approaching the earth, and the earth's rotational speed would be increased.

An interpretation of a calculation of this kind should be made with caution because of the extremely long period of time involved. It is possible that many factors which are not apparent at the moment should be considered in this calculation.

IMPORTANT WORDS AND TERMS

Craters	Perigee	Waxing	Last quarter moon	Lunar eclipse
Plains	Apogee	Waning	Eclipse	Nodes
Rays	Sidereal month	Crescent	Umbra	Ascending node
Rills	Synodic month	Gibbous	Penumbra	Descending node
Mountain ranges	Moon illusion	First quarter moon	Solar eclipse	Spring tides
Faults	New moon	Full moon	Annular eclipse	Neap tides

LEARNING OBJECTIVES

After reading and studying this chapter, you should be able to do the following without referring to the text:

1. State the approximate distance between the earth and moon, the approximate diameter of the moon, the time for the moon to go around the earth, and the time for the moon's rotation.
2. Explain why the moon has so many craters compared to the earth.
3. State what the plains are and what caused them.
4. Summarize the important periods in the moon's history.
5. Explain what causes the moon illusion.
6. Draw a diagram of the earth, moon, and sun when the moon is in the first quarter phase and tell when the moon will rise, be overhead, and set.
7. Draw a diagram to explain why the moon rises 50 minutes later each day.
8. Explain what causes a solar eclipse and why we do not usually have one when the moon is in the new phase.
9. Explain what causes a lunar eclipse and why they last longer than do solar eclipses.
10. Explain why we have tides; particularly, why we have two high tides and two low tides each day.

QUESTIONS

1. How often would someone on the moon see (a) the sun rise (b) the earth rise?
2. What caused each of the following on the moon: (a) craters, (b) plains, (c) rays, (d) rills.
3. How long ago did the earth and moon form?
4. Why does the moon appear to get smaller as it rises in the sky? Does it actually get smaller?
5. How often do we have a full moon?
6. Determine when the following occurs: (a) The new moon sets, (b) The full moon rises, (c) The last quarter moon sets.
7. Why does the moon rise 50 minutes later each day?
8. Which phase of the moon (a) is overhead at 6 P.M., (b) sets at 6 A.M., (c) rises at midnight?
9. What is the difference between a waxing and a waning moon?
10. Why is the full moon higher in the sky in winter than in summer?
11. Why doesn't an eclipse occur every time there is a new or full moon?
12. Why do lunar eclipses last so much longer than solar eclipses?
13. During which phases of the moon do spring and neap tides occur?
14. Why are there two high tides and two low tides each day?

PROBLEMS

1. If a person weighs 120 lb on the earth, what will be the same person's weight on the moon?

2. How many days are in 12 lunar months (synodic months)?

3. What would be the longitude of someone who sees a full moon when it is noon in Washington, D.C. (39° N 77° W)?

4. Referring to problem 3, what would be the longitude of someone who sees the full moon setting at the same time?   *Answer:* 167° W

5. Consider a person in the United States who sees the last quarter phase.
   (a) Which side of the moon would be bright?
   (b) What phase would an observer see in Australia and which side would be bright?

6. A book states that the moon rises about 50 min later each day. What is a more accurate figure?   *Answer:* 48.8 min

7. A high tide is occurring at 77° W longitude.
   (a) What other longitude is also experiencing high tide?
   (b) What two longitudes are having low tide?

8. Draw a diagram illustrating a total solar eclipse. Include the orbital paths of the earth and moon, and indicate the approximate time of day the eclipse is taking place.

9. Draw a diagram illustrating a total lunar eclipse. Include the orbital paths of the earth and moon, and indicate the approximate time of day the eclipse is taking place.

Chapter 20
THE UNIVERSE

Space is where mass and time do their thing.

The study of the stars is the oldest science. Thousands of years ago men and women under clear desert skies of the Near East watched the stars in awesome wonder. The earliest scientists plotted the positions and brightnesses of the stars as the earth went through its calendar of seasons. The sun and moon were worshipped as gods and the days of the week were named after the sun, moon, and visible planets as discussed in Chapter 18. Yet, it is only recently that we have understood the most basic nature of stars.

Fifty years ago, not even Einstein himself knew what made the stars shine. Today we know that all stars go through a cycle of stages. Stars are born, they shine, and then they expand, contract, possibly explode and eventually die out. Our knowledge of how all this happens has been made possible through our study of the atomic nucleus and by applying the laws of science in many diverse fields.

Our knowledge of the universe is also growing. We are learning more and more of its secrets, and we are beginning to understand how our present civilization fits into an overall scheme in the harmony of the universe. We now have an imperfect theory of evolution—a theory concerned not so much with the origin of human beings, but with the origin of the elements of our solar system. We can now begin to understand how carbon, nitrogen, and oxygen atoms were produced in the stars. It is these atoms that were necessary to produce life itself. As these theories evolve, new mysteries appear, but the continued search for truth is invigorating.

The stars and galaxies of our universe give off many different kinds of electromagnetic radiation. These were discussed in Chapter 6 and include radio waves, microwaves, infrared, visible, ultraviolet, X-rays, and gamma rays. Up until about 40 years ago, we looked only at the visible light given off by the stars. With the advent of radio telescopes, quasars and pulsars were discovered in 1960 and 1968, respectively. Most of the other regions of the electromagnetic spectrum are absorbed by our atmosphere. Satellites and balloons going above our atmosphere have enabled us to study other forms of radiation emitted by stars, and new developments are occurring frequently. There is now the strong possibility that we have seen black holes using X-ray astronomy techniques.

Recent experiments in the microwave region of the spectrum confirm the finding that we live in an expanding universe that had its beginning 15 to 20 billion years ago. But will our universe continue to expand? What is the ultimate fate of the universe? We hope the answers to these questions will be forthcoming, but for now there is much to understand. This chapter attempts to summarize many of the basic ideas about the stars and galaxies that make up our universe.

20-1
THE SUN

The sun is a star, spherical in shape, with a diameter of 865,000 miles (over 100 times greater than the earth's diameter). Viewed from the earth, at its mean distance, the sun's angular diameter is approximately $\frac{1}{2}$ degree. This ordinary star of the Milky Way galaxy* is the most important object in the solar system to us because it supplies heat, light, and other radiation for the processes of life on the earth. The sun rotates

*For a definition of galaxy, see Section 20-4.

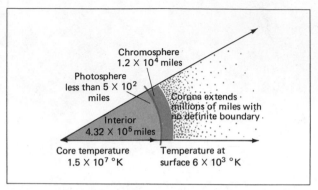

Figure 20-1a *Diagram of a radial cross section of the sun.*

about an axis every 25 earth days and moves through space with its family of planets at a speed of 12 miles per second in the direction of the constellation Hercules. The sun's equator is inclined about 7° with the orbital plane of the earth. The rotational period of the sun given above is the period at its equator; the period of rotation is longer at higher latitudes.

A cross-sectional view of the sun is shown in Fig. 20-1a. The sun's temperature is believed to be about 15,000,000°K at the center, and the temperature decreases farther out to the visible surface of the sun, called the **photosphere.** The temperature of the photosphere has been measured at about 6000°K.

The interior of the sun is so hot that individual atoms do not exist because high speed collisions continually knock the electrons loose from the atomic nuclei. The interior is composed of high speed nuclei and electrons moving about more or less independently, similar to a gas. A gas, you will recall, is composed of rapidly moving atoms or molecules. In the sun, we have very rapidly moving positively charged nuclei and negatively charged electrons. These high speed charged nuclei and electrons form a fourth state of matter called a **plasma.** The sun is a plasma with an average density of 1.4 g/cm³. Note that this density is 1.4 times as great as water! We see that the

Figure 20-1b *Solar corona photographed at total eclipse. (Courtesy Mount Wilson and Palomar Observatories.)*

sun and the stars are definitely unlike anything we normally see on earth.

The sun's photosphere is composed of about 94% hydrogen, 5.9% helium, and 0.1% of heavier elements, the most abundant being carbon, oxygen, nitrogen, and neon. We believe that the interior of the sun has a similar composition, although we have no good experimental evidence for this. Note that when we discuss elements in stars, we are really speaking about the nuclei of the elements since the electrons are stripped off the nuclei and are speeding around independently of the nuclei.

The photosphere, viewed through a telescope with appropriate filters, has a granular appearance. The granules are hot spots (about 100°K higher than the surrounding surface), a few hundred miles in diameter, that last only a few minutes. Extending more than 12,000 miles above the photosphere lies the **chromosphere** (color + sphere) which is composed mainly of hydrogen. The chromosphere can be seen as a thin red crescent for only the few seconds a total eclipse shuts out the light from the sun. At the time of a total solar eclipse the chromosphere and photosphere are hidden by the moon, and the **corona**

(outer solar atmosphere) can be seen as a white halo. See Fig. 20-1b.

A very distinct feature of the sun's surface is the appearance of sunspots. **Sunspots** are patches (some are thousands of miles in diameter) of cooler material on the surface of the sun. Each has a central darker part, called the *umbra*, and a lighter border called the *penumbra*. Figure 20-1c is a photograph of the whole disc and an enlargement of a very large sunspot. These large sunspots last for several weeks before disappearing from view.

The number of sunspots appearing on the sun varies over a 22-year period. A period begins with the appearance of a few spots or groups near 30° latitude in both hemispheres of the sun. The number of spots increases, with a maximum generally between 100 and 200 occurring about four years later near an average latitude of 15°. As time passes, the number of spots decreases until, in about seven more years, only a few spots are observed near 8° latitude.

About this same time a few spots begin to appear at 30°, and the number of sunspots begins to increase again, indicating an 11-year cycle. But there is a notable difference. The sunspots have an associated

Figure 20-1c *Whole solar disc* (top). *The great sunspot group near the center of the disc is enlarged in the lower photograph. (From a photograph taken April 7, 1947, courtesy Mount Wilson and Palomar Observatories.)*

magnetic field which is different in appearance from the previous ones. Studies indicate that if a sunspot has a north magnetic pole during the initial increase and decrease, the next 11-year cycle will show a south magnetic pole associated with the sunspot.

The 22-year or 11-year sunspot cycle has been observed since about 1715. Galileo saw sunspots through a telescope in 1610 and they were possibly observed even before that without using telescopes. There were reports of their observation after Galileo, but during the period 1645 to 1715 hardly any sunspots were reported. During this 70-year period very

few northern lights were seen in Northern Europe, and a "Little Ice Age" occurred during this time in Europe. The evidence seems to indicate that the sun's activity is not always as regular as it often appears to be.

Another distinct feature of the sun's surface is the appearance of **prominences** which seem to be connected with violent storms in the chromosphere. They are very evident to the astronomer during solar eclipses, at which time they appear as great eruptions at the edge of the sun. They are red in color, have an associated magnetic field, and take many different shapes and forms. They may appear as streamers, loops, spiral or twisted columns, fountains, curtains, or haystacks. They extend outward for thousands of miles from the surface, occasionally reaching a height of one million miles. An extraordinarily large prominence is shown in Fig. 20-1d. The elapsed time between the first and last picture in the photograph is 4 hours and 22 minutes. Since the height and times are given, the velocity of expansion can be easily calculated. The white dot represents the relative size of the earth.

The chief property of the sun, of course, is the fact that it shines, or gives off energy. However, it was not until 1938 that scientists came to understand what made the sun shine. We now know that the sun shines because of nuclear fusion reactions inside its core.

The sun's core is made up mostly of hydrogen nuclei or protons, or in nuclear notation $_1^1H$. These protons are moving at very high speeds and they occasionally fuse together as shown in Fig. 20-1e. The products of this nuclear reaction are a deuteron (proton and neutron together or $_1^2H$) and a positive electron or positron designated $_{+1}e$ and a neutrino designated ν. A **neutrino** is an elementary particle that has no charge, no mass (as far as is known), travels at the speed of light, and hardly ever interacts with other particles such as electrons or protons. This first reaction is fairly rare and for this reason the sun's hydrogen burns relatively slowly. In fact, we believe that the sun has been shining for about 5 billion years.

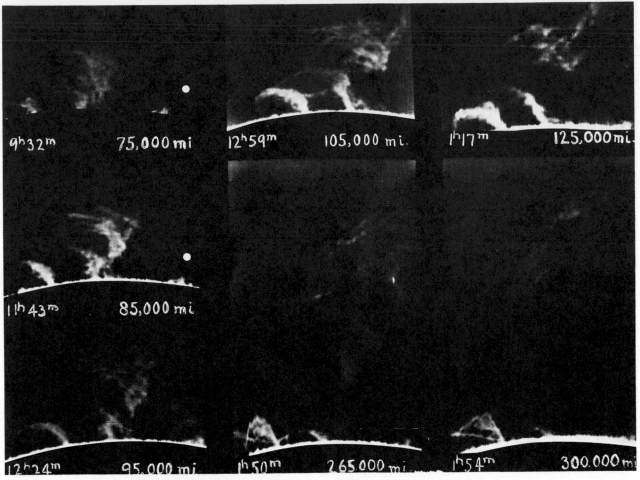

9ʰ32ᵐ   75,000 mi   12ʰ59ᵐ   105,000 mi.   1ʰ17ᵐ   125,000 mi.

1ʰ43ᵐ   85,000 mi

12ʰ24ᵐ   95,000 mi   1ʰ50ᵐ   265,000 mi   1ʰ54ᵐ   300,000 mi

Figure 20-1d *Solar prominences. (From a photograph taken October 8, 1920, courtesy Yerkes Observatory.)*

Once the deuteron is formed, it quickly reacts with a proton to form a helium-3 nucleus ($^3_2$He) and a photon or particle of light designated γ. Next, two helium-3 nuclei fuse together to form the more common helium-4 nucleus and two protons.

In each of these three fusion reactions, mass is converted into energy and energy is liberated. These three reactions are called the **proton-proton chain** and can be written as

$$^1_1\text{H} + {}^1_1\text{H} \longrightarrow {}^2_1\text{H} + {}_1\text{e} + \nu + \text{Energy} \quad \text{(Slow)}$$

$$^2_1\text{H} + {}^1_1\text{H} \longrightarrow {}^3_2\text{He} + \gamma + \text{Energy} \quad \text{(Fast)}$$

$$^3_2\text{He} + {}^3_2\text{He} \longrightarrow {}^4_2\text{He} + {}^1_1\text{H} + {}^1_1\text{H} + \text{energy} \quad \text{(Fast)}$$

If we multiply the first two reactions by two and add up both sides, we get the net reaction which is

$$^1_1\text{H} + {}^1_1\text{H} + {}^1_1\text{H} + {}^1_1\text{H} \longrightarrow$$
$$^4_2\text{He} + 2\,{}_1\text{e} + 2\gamma + 2\nu + \text{Energy}$$

In the net reaction, four protons or hydrogen nuclei react to form a helium nucleus, two positive electrons, two high-energy photons, two neutrinos and a great deal of energy. The energy factor simply means that the particles on the right possess more kinetic and radiant energy than the particles on the left. In our reaction, we have converted mass into energy via Einstein's formula $E = mc^2$.

Figure 20-1e *The reactions which make up the proton-proton chain.*

Every second in the sun's interior, 630 million tons of hydrogen are being converted into 625.4 million tons of helium with the release of the equivalent of 4.6 million tons of mass energy. Yet, at this rate, we expect the sun to shine for another 5 billion years or so.

We now have a good understanding of the basic concept of how a star shines. A star shines because of nuclear fusion reactions inside its core. But do we have any direct evidence that this is true? In an effort to get direct evidence of what goes on inside the sun's core the solar neutrino experiment is being done. The **solar neutrino experiment** is an experiment to detect neutrinos from the sun's core.

The major difference between photons and neutrinos is that photons react much more readily with matter than neutrinos. In fact, a photon produced at the center of the sun will interact countless times and take possibly millions of years before it finally reaches the sun's surface. A neutrino, on the other hand, will zip right through the sun and earth and most neutrino detection devices without interacting at all. Of course, occasionally a neutrino will interact with a detector, and we can say that a neutrino has been detected. In the solar neutrino experiment only a few neutrinos are detected in a month's time.

The results of the solar neutrino experiment, at least initially, have been perplexing. They indicate that not as many neutrinos are being produced in the sun's core as is predicted by our current best theory. Thus, we see that there are some unsolved problems with regard to our understanding of the structure and dynamics of the interior of stars.

20-2
THE CELESTIAL SPHERE

A view of the stars on a clear night makes a deep impression on an observer. As we look into the night sky, the stars appear as bright points of light on a huge dome overhead. As the time of night passes, the dome seems to turn westward as part of a great sphere, with the observer at the center. The apparent motion of the stars is due to the eastward rotation of the earth. The naked eye is unable to detect any relative motion among the stars on the apparent sphere or to perceive their relative distances from the earth. The stars all appear to be mounted on a very large sphere having a very large radius.

This huge moving (apparent motion) sphere has been named the **celestial sphere,** and the way it appears to the observer depends upon the observer's position on the earth. An observer positioned at 90° N (the north pole) would see Polaris, the north star, directly overhead. From this latitude all stars on the celestial sphere appear to move in concentric circles about the north star, never going below the horizon; that is, they never set. If the observer is located at 40° N latitude, he or she would observe the north star 40° above the northern horizon, and all stars within 40° of the north star would appear to move in concentric circles about Polaris, never going below the horizon. Detailed observations reveal the celestial sphere to rotate (apparently) about an axis that is an extension of the earth's polar axis, with the celestial equator lying in the same plane as the earth's equator. See Fig. 20-2a.

The position of a star or other object beyond our solar system is determined with the assignment of three space coordinates. The first and second are *declination* and *right ascension*, which are angular coordinates representing the direction of the star in respect to the sun. The third coordinate is *distance*, which determines the star's linear distance from the sun (Fig. 20-2b). **Declination** is the angular measure north or south of the celestial equator measured in degrees. It has a minimum value of zero at the celestial equator and increases to a maximum of 90° north, and to a maximum of 90° south. The direction in respect to the equator is indicated by a plus (+) or minus (−) sign. All angles measured north of the equator have (+) values and all angles measured south of the equator have (−) values. For example, the star in Fig. 20-2b has a declination of +20 degrees.

Right ascension is the angular measure in degrees or hours, with the hours divided into minutes and seconds. Right ascension begins with zero hours at the celestial prime meridian and continues eastward to a maximum value of 24 hours or 360°, either of which coincides with the starting point. The **celestial prime meridian** is an imaginary half-circle running from the north celestial pole to the south celestial pole and crossing perpendicular to the celestial equator at the intersection of the ecliptic at the point where the ecliptic passes northward.

The distance coordinate is usually measured in astronomical units, in parsecs, or in light years. An **astronomical unit** we have defined as the mean distance of the earth from the sun, which is 92,955,700 miles as measured by radar. The **light year** is the distance traveled by electromagnetic radiation in one year. One light year equals approximately 6×10^{12} miles or 9.5×10^{12} kilometers, calculated by multiplying the speed of light by the number of seconds in one year. One **parsec** is defined as the distance to a

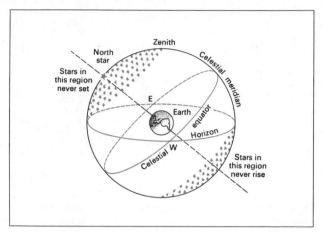

Figure 20-2a *Celestial sphere, as seen by an observer on the earth at a latitude of 40°N.*

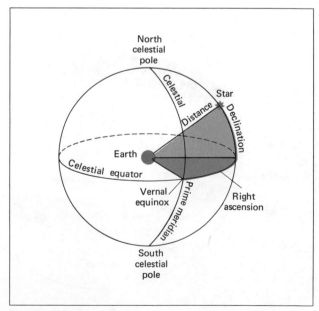

Figure 20-2b *The three space coordinates: distance, declination, and right ascension.*

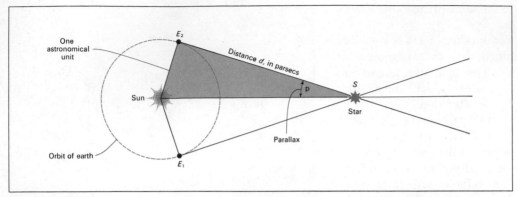

Figure 20-2c *Annual parallax of a star. The annual parallax of the star S is the angle p. By definition, when angle p is 1 second of arc, the distance d is 1 parsec. The basic relation can be written: Distance in parsecs = 1/Parallax in seconds.*

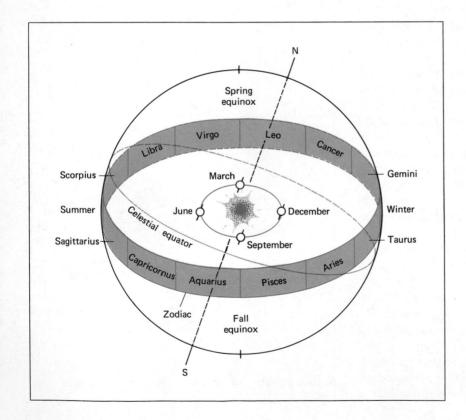

Figure 20-2d *Signs of the zodiac. The drawing illustrates the boundaries of the zodiacal constellations. Each of the 12 sections of the zodiac is 30° wide and 16° high, or 8° each side of the ecliptic.*

star when the star exhibits a parallax of one second of arc (Fig. 20-2c). One parsec equals 3.26 light years, or 206,265 astronomical units. This can be written

1 parsec = 3.26 light years
$$= 206,265 \text{ astronomical units}$$

The star S in Fig. 20-2c, observed from position E_1 and later from E_2, appears to move against the background of more distant stars. This apparent motion is, as you know, called parallax. The angle p measures the parallax in seconds of arc. Hence, the term parsec. The definition provides an easy method for determining the distance to a celestial object, since merely taking the reciprocal of the angle p, measured in seconds, gives the distance in parsecs.

■ *Example 1* What is the distance to the star Proxima Centauri if the annual parallax is 7.62×10^{-1} second?

$$d = \frac{1}{7.62 \times 10^{-1}} = 1.31 \text{ parsecs}$$

■ *Example 2* What is the distance to the star Sirius A if the annual parallax is 3.75×10^{-1} second?

$$d = \frac{1}{3.75 \times 10^{-1}} = 2.667 \text{ parsecs}$$

There are prominent groups of stars in the celestial sky that appear to the earth observer as distinct patterns. These groups called **constellations** have been given names which can be traced back to the early Babylonian and Greek civilizations. Although the constellations have no physical significance, today's astronomers find them useful in referring to certain areas of the sky and, in 1927, they set specified boundaries for the 88 constellations so as to encompass the complete celestial sphere.

We are all aware of the apparent daily motion of the sun across the sky, and when the moon is visible we are aware of its motion. The constellations also appear to move across the sky from east to west, if one observes the stars for an hour or two. Their daily motion is due to the eastward rotation of the earth on its axis. They also have an annual motion due to the earth's motion around the sun. We observe the constellations Pisces, Aquarius, and Capricornus in the autumn night sky. In the winter months Orion, the Hunter, is seen along with Gemini, Taurus, and Aries. Sagittarius, the Archer, is a summer constellation. For other summer and the spring constellations, see Fig. 20-2d.

The **zodiac** is a section (actually a volume) of the sky extending around the ecliptic 8° above and 8° below the ecliptic plane (Fig. 20-2d). The zodiac is divided into 12 equal sections, each 30° wide and 16° high. Each section has its apex at the sun and extends outward to infinity. The boundaries of the zodiac were specified such that the sun, moon, visible planets, and most of the asteroids traveled within its limits. Occasionally however, the planets Pluto and Venus are outside the boundaries of the zodiac.

20-3
STARS

Hipparchus of Nicaea, Greek astronomer and mathematician, was the greatest known observer of the stars of antiquity. He measured the celestial latitude and longitude of more than 800 stars, and compiled the first star catalog, which was completed in 129 B.C. He assigned the stars, in respect to their brightness, to six **magnitudes.** The brightest ones were listed as stars of the first magnitude; those not quite so bright as second magnitude, the next less bright as third magnitude and so on down to the sixth magnitude. A modified version of Hipparchus' scale is used today. When a comparison was made between a first magnitude star and a sixth magnitude star, it was discovered the brighter star gave off about 100 times as much radiant energy. From this, a definition of the magnitude scale was made, in which each magnitude difference is equal to the fifth root of 100. This can be written

$$\text{Magnitude difference} = \sqrt[5]{100} = 2.512$$

For example, a first magnitude star is 2.512 times as bright as a second magnitude star, and a second magnitude star is 2.512 times as bright as a third, and so on. On this scale, the brightest object in the sky, our sun, has a magnitude of -26.7; the full moon a magnitude of -12.7; the planet Venus a magnitude of -4.2; and the brightest star, Sirius, a magnitude of -1.43.

The energy output of a star is measured by its absolute magnitude. The **absolute magnitude** of a star is defined as the apparent magnitude a star would have if it were placed 10 parsecs (32.6 light years) from the earth. The absolute magnitude of the sun is about 4×10^{33} ergs/s, which is measured by determining the amount of energy falling on the earth's surface and knowing the distance from the earth to the sun. If the annual parallax (from which distance is calculated) and the apparent brightness of a star can be measured, the absolute magnitude can be calculated.

The surface temperature of a star can be inferred from its color. Cool stars are red, hotter stars are yellow, then white, and the hottest stars are blue-white. Often a star's temperature is inferred from its color index.

The color index is a number indicating the relative brightness of the blue and yellow regions of the star's spectrum. The color sensitivity of the eye is greatest in the green-yellow region of the spectrum, and by using appropriate light filters, the starlight falling upon a photoelectric cell can be adjusted to that received by the eye thus establishing a visual magnitude system. Photographic emulsions are most sensitive to the blue-violet regions of the spectrum and other filters can be used in front of the photographic plates for establishing photographic magni-

tudes. The number representing the color index of a star can be found by obtaining the photographic magnitude and subtracting the visual magnitude. This can be written as

$$\text{Color index} = m_p - m_v$$

where m_p = photographic magnitude
and m_v = visual magnitude.

A yellow star, for instance, might have $m_v = 4$ and $m_p = 5$ indicating that it is dimmer when seen through a blue filter. Its color index would be $+1$. A blue-white star would appear brighter when seen through a blue filter than through a yellow filter so m_p would be less than m_v. A blue-white star would have a negative color index. In general, stars have color indices ranging from $+2.0$ for cool red stars to -0.4 for hot blue-white stars.

When the absolute magnitudes or brightnesses of stars are plotted against their surface temperatures or colors, we get an **H-R diagram** named after Ejnar Hertzsprung, a Danish astronomer, and Henry Russell, an American astronomer. A typical H-R diagram for stars is shown in Fig. 20-3a. Note that the temperature axis is reversed. That is, the temperature increases to the left instead of to the right.

Most stars on an H-R diagram get brighter as they get hotter. These stars form the **main sequence,** a narrow band going from upper left to lower right. Stars above the main sequence that are cool and yet very bright must be unusually large to be so bright. So they are called **red giants.** Stars below the main sequence that are hot yet very dim must be very small so they are called **white dwarfs.**

Stars are plasmas having a composition by number (not weight) of about 94% hydrogen, 5.9% helium, and 0.1% other elements. Their surface temperatures range from about 2000°K to 50,000°K. They vary in mass and size. Most have a mass of between 0.1 to 5 solar masses. Others range from 0.04 solar mass to 75 solar masses. They range in size from white dwarfs, about 8000 miles in diameter, to supergiants, which

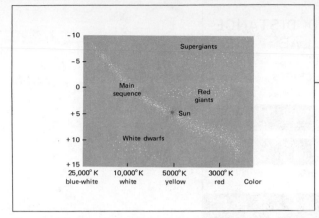

Figure 20-3a *Hertzsprung-Russell diagram. The diagram is a plot of the absolute magnitude of stars versus their surface temperature, or color.*

are over 400 million miles in diameter. Most stars appear as multiple systems. A **binary star** system consists of two stars orbiting around each other. Systems with three or more stars also occur, but they are not nearly as common as binary stars.

There are many stars in the sky that are observed to vary in brightness over a period of time. The first such star noted was Delta Cephei in the constellation Cepheus. Delta Cephei varies in brightness with a period of approximately 5.4 days. From its least bright magnitude of 5.2, it gradually becomes brighter for about two days and attains a magnitude of 4.1, after which it decreases in brightness until it reaches its minimum in another 5.4 days. The cycle begins again and goes through the same sequence. The period of other cepheid variables varies from 1 to 50 days. Presently there are over 500 known stars that vary in magnitude with a fixed period at between 1 and 50 days and they are known as **cepheid variables** after Delta Cephei.

The importance of the cepheid variables is the fact that a definite relationship exists between a star's period and its average brightness. Generally, the longer the period, the brighter the star. By measuring the period, the absolute brightness can be calculated. Then the distance can be computed from the absolute brightness. Thus, the distance to any galactic system in which cepheid variables can be detected and their periods measured can be calculated.

The distance to remote galaxies is calculated by means of the red-shift in the galaxy's spectrum. When a photograph is taken of an element in the gaseous phase located in an earth laboratory, a normal spectrum is obtained. When a photograph is taken of a galaxy containing this same element, the lines of the spectrum may show a displacement of the normal lines toward the red end of the spectrum (longer wavelengths) or toward the blue end of the spectrum (shorter wavelengths). The displacement is believed to be due to the Doppler effect and the direction of the displacement depends upon whether the galaxy is moving toward or away from the observer. Refer to Section 5-5 for a discussion of the Doppler effect.

The displacement for two lines on the calcium spectrum labeled H and K for five different galaxies is shown in Fig. 20-3b. A small shift of the lines means a low velocity of recession, a greater shift means a greater velocity, and so on. The interesting point concerning the red-shift is a correlation of velocity and the apparent brightness of the galaxy. The fainter the galaxy, the greater the velocity. Generally speaking, the farther away a galaxy is located, the less its apparent brightness. Thus, a velocity and magnitude correlation yields a velocity and distance correlation. This provides us with a means for measuring great distances which are far beyond the limits of cepheid variable observation.

When a galaxy is approaching the earth, the lines of the spectrum shift toward the blue end of the spectrum, indicating a shortening of wavelength. When the galaxy is receding from the earth, the lines shift to the red end of the spectrum, indicating a lengthening of the wavelength. The greater the velocity, the greater the shift in the spectrum lines, as shown in Fig. 20-3b.

There are many stars which appear dim and insignificant, but suddenly increase in brightness in a matter of hours by a factor of a thousand or more. A star undergoing such a drastic change in brightness is called a **nova,** or new star. The star is unstable and throwing off an expanding shell of gas. The eruption

RELATION BETWEEN RED-SHIFT AND DISTANCE FOR EXTRAGALACTIC NEBULAE

CLUSTER NEBULA IN	DISTANCE IN LIGHT-YEARS	RED-SHIFTS
VIRGO	4,500,000	750 MI/S
URSA MAJOR	600,000,000	9,300 MI/S
CORONA BOREALIS	800,000,000	13,400 MI/S
BOOTES	1,400,000,000	24,400 MI/S
HYDRA	2,200,000,000	38,000 MI/S

Figure 20-3b *Doppler shift. The velocity of a galaxy (extragalactic nebula) can be measured by the red-shift (commonly known as the Doppler shift) of its spectral lines. The relation between shifts at the red end of spectrum and distance is shown. Note the increasing length of the arrow. (Courtesy Mount Wilson and Palomar Observatories. Distance data revised.)*

may last for as many as 50 years before the star settles down to a stable state and becomes the original dim star. Several novae have been observed and studied, and it is believed the star is releasing internal energy faster than the surface can radiate. Thus, in order to remain active, the star throws off the excess energy plus a portion of its mass.

Occasionally, a star explodes and throws off large amounts of material which may be so great that the star is destroyed. These gigantic explosions are known as **supernovae,** only three of which may have been observed in our galaxy. One of the most celebrated is the Crab Nebula in the constellation Taurus. This nebula (Latin word for cloud) is expanding at the rate of approximately 70 million miles per day. Since we know the average angular radius and the expansion rate, the original time of the explosion can be calculated. The result yields about 900 years, which agrees closely with Chinese and Japanese records which report the appearance of a bright new star in the constellation Taurus in 1054.

In addition to the novae and supernovae, there are the planetary nebulae which possess a large, slowly expanding ringlike envelope, as shown in Fig.

20-3c. These nebulae, as observed with a telescope, appear greenish in color. The nebula shown in Fig. 20-3d, one of the brightest in the night sky, is known as a diffuse nebula. This type of nebula is more irregular and turbulent than the planetary nebulae. The Great Nebula in Orion is over 25 light years in diameter and contains hundreds of stars which illuminate the dark material. Some cosmologists (scientists who study the universe, that is, its history, present structure, and future evolution) believe that the dust and gaseous material found in clouds such as these provide the material needed for the birth of new stars.

Figure 20-3c *Ring nebula in the constellation Lyra. (Lick Observatory Photographs.)*

Figure 20-3d *Great Nebula in Orion. Considered the brightest of the diffuse nebulae, this one is located near the middle star in Orion's sword and is a gaseous and dusty nebula. (Lick Observatory Photographs.)*

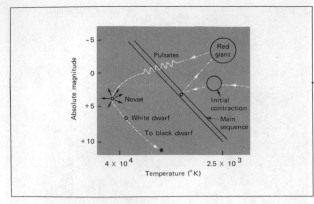

Figure 20-3e *Hertzsprung-Russell diagram used to illustrate the evolution of a star with the same mass as our sun.*

We now know that stars are born, shine, expand, possibly explode, and then die. This is the general life cycle of a star. However, the exact details depend on a star's initial composition (the percentage amounts of hydrogen, helium, and heavier elements) and on its mass. The greater the mass of a star, the faster it moves through its life cycle.

The general evolution of a star with a mass typical of our sun is shown in Fig. 20-3e. The birth of a star, according to the best accepted theory, begins with the condensation of interstellar material, mostly hydrogen, because of radiation pressure from nearby stars, supernovae and the gravitational attraction between the interstellar material. The size of the star formed depends upon the total mass available, which in turn determines the rate of contraction. As the interstellar mass condenses and loses gravitational potential energy, the temperature rises and the material gains thermal energy. As the star continues to decrease in size, the temperature continues to increase until a thermonuclear reaction begins and hydrogen is converted to helium. The star's position in the main sequence of the H-R diagram, its temperature, and the radiation given off (luminosity) are determined by the mass of the star.

On the main sequence of the H-R diagram, the star continues to convert hydrogen into helium. This process is called **hydrogen burning,** but it is a nuclear burning and not a chemical fire. The hydrogen burn-

ing stage lasts for billions of years—possibly 10 to 15 billion years for a star like our sun.

As the hydrogen in the core is converted into helium, the core begins to contract and heat up. This heats up the surrounding shell of hydrogen and causes the hydrogen burning in the shell to proceed at a more rapid rate. This rapid release of energy causes the star to expand and become a red giant.

Eventually the core gets so hot that helium can fuse into carbon, and soon other nuclear reactions occur in which all elements get created. The creation of the nuclei of elements inside stars is called **nucleosynthesis.** The physics of the problem becomes extremely complicated, but we believe that the cepheid variable stage occurs next if conditions (mass and composition) are just right. We also believe that supernovae occur sometime after the red-giant stage. The supernova stage occurs when reactions in the core proceed so rapidly that the star is blown almost completely apart, with only a remnant core remaining. The supernova stage probably does not occur for all stars.

Eventually, no more nuclear fusion reactions are possible and the star's nuclear energy has been spent. What is left of the star begins to gravitationally collapse (see Section 8-7) and go to its end stage. The end stage of a star depends upon the star's mass at the end of its active life. Average and small mass stars become white dwarfs. When a star is a white dwarf, it is very small. It has gravitationally collapsed as far down as it can go and still obey the laws of physics. It is about the size of the earth and is so dense that a single teaspoonful of matter weighs five tons. Because it is so small it is a fairly dim star.

Large-mass stars (between 1.5 and 3.0 times the sun's mass) have more gravitational attraction, and they collapse down to a size of approximately 20 miles in diameter. The electrons and protons in this superdense star combine to form neutrons and this **neutron star** is composed of about 99% neutrons. A teaspoonful of a neutron star would weigh one billion tons. Because the angular momentum of the star must

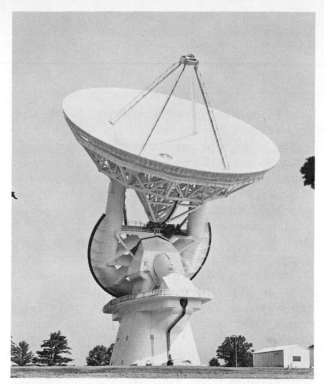

Figure 20-3f *The 140-foot radio telescope of Green Bank, West Virginia. This telescope, complete in 1965, is the largest equatorially mounted radio telescope in the world. The immense concrete base not only serves to support the telescope but also houses the controls, hydraulic system, electric equipment, transformer vault, and electronic workshop. The instrument cost about $14 million and took seven years to design and build. (Courtesy National Radio Astronomy Observatory.)*

Figure 20-3g *Crab Nebula in Taurus. Also known as NGC 1952 and Messier 1, this nebula is the remains of a supernova of 1054 A.D. At the center is a pulsar. Photographed with a 200-inch telescope. (Courtesy Mount Wilson and Palomar Observatories.)*

be conserved, the small size of the neutron star dictates that it must be spinning rapidly. Rapidly rotating neutron stars give out radio radiation in pulses and are called **pulsars.**

Pulsars were first discovered in 1968 using radio telescopes similar to the one shown in Fig. 20-3f. They pulse with a constant period that may be between 30 times per second up to four seconds. The fastest spinning pulsar is located at the center of the Crab Nebula. This pulsar is identified with the remains of a supernova that took place in 1054 A.D. This supernova was recorded by the Chinese and must have been quite bright. The period of this pulsar is slowly getting larger, indicating that the rotating

neutron star is slowing down very slowly. An optical picture of the pulsar blinking on and off is shown at the beginning of Chapter 8.

If the remaining core is greater than about three times the mass of the sun, the star is believed to end up as a gravitationally collapsed object even smaller and more dense than a neutron star. Such a star is so dense that light cannot escape from its surface. Thus is would appear black, and it is called a black hole. A **black hole** is an unbelievably dense star so massive and small that light cannot escape from its surface. Black holes can be seen by the effect they have on another star in a binary system. There is some experimental evidence for such a black hole, but the facts are not yet conclusive. Nevertheless, we think that black holes do exist.

After a star explodes in a supernova and the core goes into its end stage, what becomes of the ejected material? We believe that this material is thrown into space (see Fig. 20-3g) and eventually becomes seed material for future stars. In fact, because of its surface composition and age, we believe our own sun is a second-generation star. That is, one or more stars went through their life cycles and exploded as supernovae. The ejected material later coalesced again into our sun and solar system. Since the universe is believed to be 15 to 20 billion years old and our sun is only about 5 billion years old, this idea of the sun being a second-generation star is generally accepted.

20-4
GALAXIES

The sun and its satellites occupy a very small volume of space in a very large system of stars known as a **galaxy** (Greek *galaxis*, Milky Way). A galaxy is an extremely large collection of stars occupying an extremely large volume of space. A galactic system is classified as irregular, spiral, or elliptical, depending upon how it appears when photographed.

Edwin P. Hubble (1889–1953), U.S. astronomer at the Mount Wilson Observatory, established a system of classifying galaxies according to their appearance in photographs. The system starts with spherical ellipticals, spreads out with increasing flatness, then branches off into a normal spiral sequence, and a barred spiral sequence, with a scattering of irregulars outside the two branches. See Figs. 20-4a, 20-4b, and 20-4c.

The type Sa spirals have their spiral arms closely wound to the central region, whereas the Sb spirals have arms that spread out more from the center, and the Sc spirals have very loose spiral arms. The S0 type links the normal spirals to the smooth ellipticals.

The barred spirals, which are distinguished by a broad bar that extends outward from opposite sides of the central region, follow similar unwinding of the spiral arms.

The irregular type galaxies have no definite form but take on the appearance of shapeless clouds. Examples of this type are the two Magellanic Clouds which can be easily seen with the naked eye from the southern hemisphere. They are named in honor of Magellan, who reported seeing them on his famous voyage around the world.

The observed galaxies, which total into the billions, appear at different distances from the earth. The spiral galaxies account for about 77 percent of the brighter galaxies observed. The ellipticals account for about 20 percent of the total, and the irregulars only 3 percent. These numbers may not give a true picture of the total distribution of galaxies, because the spiral galaxies are brighter, therefore showing up more often in the photographs.

Our local galaxy is called the **Milky Way.** It is of the spiral Sb type. The Milky Way contains some 100 billion (10^{11}) stars and has an appearance similar to that of the Great Nebula in Andromeda (Fig. 20-4d). The Milky Way (Fig. 20-4e) is about 10^5 light years in diameter and has a maximum thickness in the central region of approximately 10^4 light years. It is rotating eastward or counterclockwise as viewed from above. The period of rotation, which is not the same for all regions of the galaxy, is more than 2×10^8 years for the region that contains our solar system. Our solar system is located near the plane of rotation about

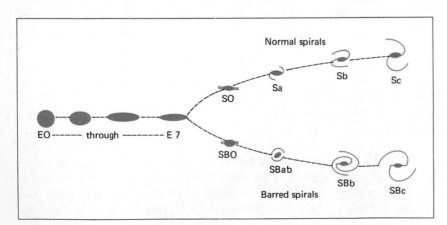

Figure 20-4a *Galactic systems. Sequence of elliptical, normal spiral, and barred spiral galaxies.*

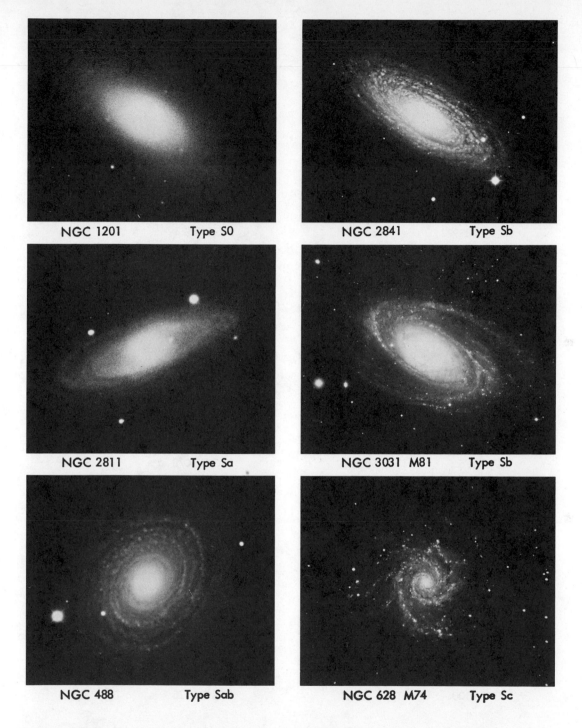

NGC 1201   Type S0

NGC 2841   Type Sb

NGC 2811   Type Sa

NGC 3031 M81   Type Sb

NGC 488   Type Sab

NGC 628 M74   Type Sc

Figure 20-4b *Normal spiral galaxies. These galaxies have a central spherical hub surrounded by spiral arms that range from the tightly wound Sa type to the loosely wound Sc type. Note that two major arms emerge from the hubs and spiral outward with decreasing brightness. (Courtesy Mount Wilson and Palomar Observatories.)*

NGC 2859   Type SB0

NGC 2523   Type SBb(r)

NGC 175   Type SBab(s)

NGC 1073   Type SBc(sr)

NGC 1300   Type SBb(s)

NGC 2525   Type SBc(s)

Figure 20-4c *Barred spiral galaxies. These spirals have a hub that is not spherical but has the appearance of a bar from which the arms wind outward, in the way they do from normal spirals. Note that the SB0 type possesses few or no spiral arms, the SBab has a slight unwinding of the arms, and the SBb and SBc types have more unwinding. (Courtesy Mount Wilson and Palomar Observatories.)*

Figure 20-4d *Great Nebula in Andromeda. This galaxy, also known as Messier 31, is a spiral-type galaxy similar to that of our Milky Way. Two ellipical galaxies are shown also. (Courtesy Mount Wilson and Palomar Observatories.)*

2.5×10^4 light years from the galactic center, and moving at a rate of more than 200 kilometers per second in the direction of the constellation Cygnus.

The galactic equator, a great circle positioned halfway between the galactic poles, is inclined about 62° to the celestial equator (Fig. 20-4f). The north galactic pole has a right ascension of 12 hours, 40 minutes, and a declination of $+28°$. The south galactic pole has a right ascension of zero hours, 40 minutes, and a declination of $-28°$.

A coordinate system of galactic latitude and longitude measures the latitude perpendicular to the galactic equator and measures the right ascension eastward from the celestial sphere's right ascension of 18 hours, 40 minutes. This is where the galactic equator intersects the celestial equator going northward.

The range of galactic latitude is from 0° to 90° north and 0° to 90° south. Right ascension begins at 0 and increases eastward 24 hours. An inspection of the celestial sphere will show the Milky Way passing through Cassiopeia for maximum plus declination, and through the Southern Cross for maximum minus declination.

A galaxy is composed of such items as individual atoms and molecules, dust, planets, stars, multiple stars, and star clusters. The closest star to our solar system is Alpha Centauri, a triple star, located about 4.2 light years away, with right ascension 14 hours, 33 minutes and $-60°$ declination. The next closest star is Barnard's star, named after Edward M. Barnard (1857–1923), an American astronomer who first observed the fast motion of the star in 1916. This star

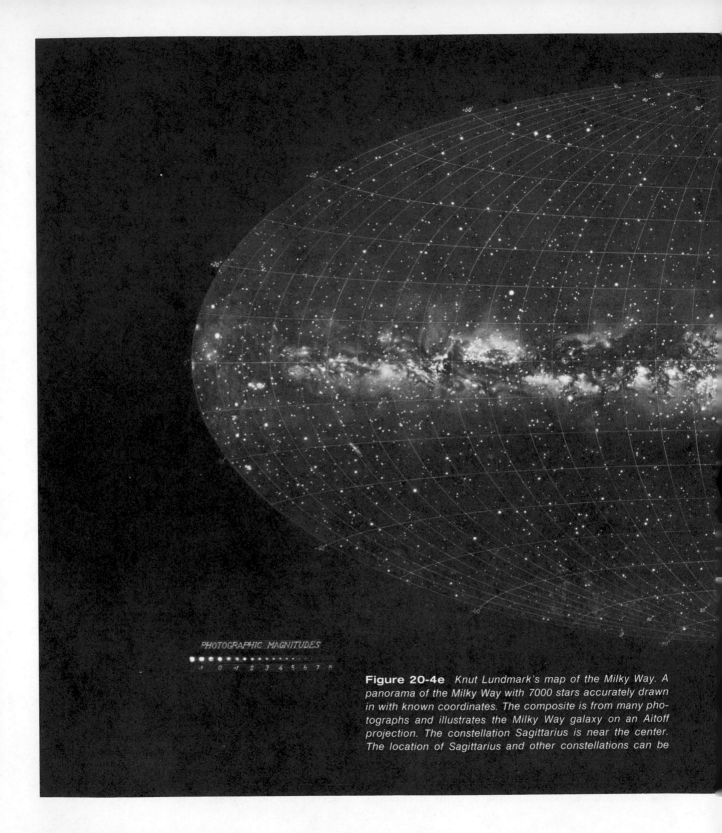

PHOTOGRAPHIC MAGNITUDES

Figure 20-4e *Knut Lundmark's map of the Milky Way. A panorama of the Milky Way with 7000 stars accurately drawn in with known coordinates. The composite is from many photographs and illustrates the Milky Way galaxy on an Aitoff projection. The constellation Sagittarius is near the center. The location of Sagittarius and other constellations can be*

LUND OBSERVATORY

MARTIN KESKÜLA
TATJANA KESKÜLA

seen by referring to Fig. 20-4g. At the lower right are the Small
and Large Magellanic Clouds. These are two irregular galaxies
located about 200,000 light years away. At the far left, at
coordinates −20° and 90°, is Messier 31, the spiral galaxy
similar to the Milky Way. (Courtesy Lund University.)

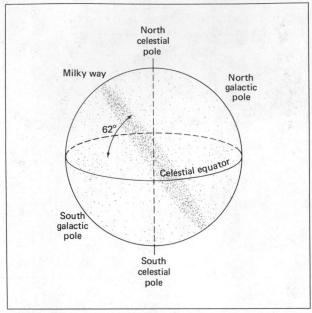

Figure 20-4f *Diagram illustrating the inclination of the Milky Way with the celestial equator.*

has a faster apparent motion than any other known star, moving at a rate of approximately 10 seconds of arc per year. The average rate for visual stars is about $\frac{1}{10}$ second per year. Barnard's star is approximately 5.9 light years from our sun and is located near 18 hours right ascension and +4° declination. Recent information concerning Barnard's star seems to indicate a large satellite associated with the star, the first evidence of another planetary system.

In the immediate neighborhood of the Milky Way and confined to an ellipsoidal volume of space—some 9×10^5 parsecs for the major axis and about 8×10^5 parsecs for the minor axis— is located a small group of galaxies known as the **Local Group.** There are 17 known members of the group, and others are believed to exist which have not been detected because of their low magnitude or brightness. Our Milky Way is a member located near one end of the major axis. Messier 31 (Great Nebula in Andromeda) is also a member and is positioned near the opposite end of the major axis. The known members along with their type, dimensions, and distance

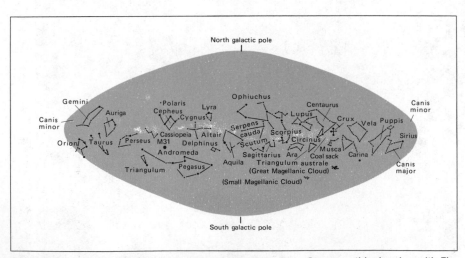

Figure 20-4g *Constellations in and near the Milky Way. Compare this drawing with Fig. 20-4e.*

Table 20-4a *The Local Group of Galaxies*

Name	Right Ascension (hours and minutes)	Declination (degrees)	Type	Approximate Distance from Earth (light years)	Apparent Diameter
Milky Way	0000	00	Sb		
Large Magellanic Cloud	0526	−69	Irr	2×10^5	9°
Small Magellanic Cloud	0056	−73	Irr	2×10^5	330′
Draco Dwarf	1756	+66	E	2.5×10^5	30′
Ursa Minor Dwarf	1600	+65	E	2.5×10^5	1°
Sculptor Dwarf	0055	−34	E	3×10^5	1°
Fornax Dwarf	0235	−34	E	6×10^5	1°
Leo II Dwarf	1045	+12	E	8×10^5	12′
NGC 6822	1943	−15	Irr	1.4×10^6	20′
Leo I Dwarf	1044	+12	E	1.6×10^6	13′
NGC 185	0042	+48	E	2×10^6	13′
NGC 147	0035	+48	E	2×10^6	13′
NGC 205	0039	+42	E5	2×10^6	10′
IC 1613	0100	+2	Irr	2×10^6	17′
Messier 31	0041	+41	Sb	2×10^6	260′
Messier 32	0041	+40	E2	2×10^6	8′
Messier 33	0132	+31	Sc	2×10^6	1°

from the earth are listed in Table 20-4a. Note that most of the galaxies are ellipticals and only three are of the spiral type.

The galaxies of the Local Group seem to be moving with random motions. The two Magellanic Clouds are approaching our galaxy and several others, including M31, are moving away.

Our Local Group is just one of thousands of clusters of galaxies. Another such cluster can be found in the constellation Hercules. This cluster is shown in Fig. 20-4h. In general, we find that galaxies tend to group themselves into clusters which group themselves into clusters of clusters (superclusters),

and so on. Thus, we see that matter in the universe does not seem to be uniformly distributed.

When Edwin Hubble began looking at Doppler shifts of galaxies, he found nothing surprising at first. In the Local Group the Doppler shifts were small: some were blue and some were red. But as he looked at galaxies farther and farther away he found only red-shifts. In fact, the farther away the galaxy, the larger the red-shift. From these measurements, the distances to remote galaxies can be determined by converting the observed red-shift of the galaxy to radial velocity and then plotting the logarithm of the velocity against the apparent magnitude. See Fig.

Figure 20-4h *Cluster of galaxies in the constellation Hercules. (Courtesy Mount Wilson and Palomar Observatories.)*

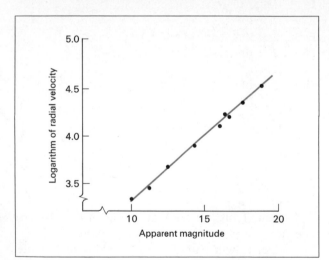

Figure 20-4i *Graph of data showing the relationship (Hubble's law) between the logarithm of the radial velocity of a galaxy and its distance from the Milky Way. Recall that apparent magnitude and distance are related.*

20-4i. Hubble's discovery, now known as **Hubble's law,** can be written as

$$v = Hd$$

where v = recessional velocity of the galaxy,
 d = distance away from the galaxy,
 H = Hubble's constant

$$= \frac{55 \text{ km/s}}{10^6 \text{ parsec}}$$

Hubble's law is extremely important because it gives us vital information about the structure of our universe.

20-5
QUASARS

About 1960, radio astronomers using their newly developed high-resolution (ability to distinguish the separation of two points) radio telescopes began detecting extremely strong radio signals from sources having small angular dimensions. The sources were named **quasars**—a shortened term for "quasistellar radio sources."

Hundreds of quasars have now been detected (see Fig. 20-5a) and they have two important characteristics. First, it was found that quasars have extremely large red-shifts. This indicates from Hubble's

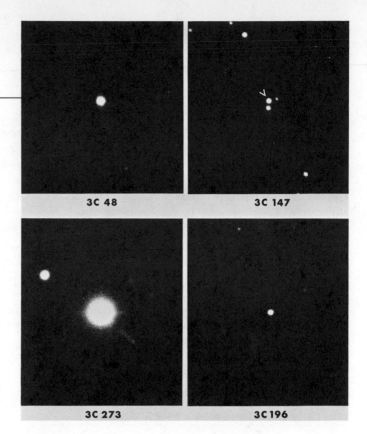

Figure 20-5a *Quasistellar radio sources. Note the jetlike prong of 3C273. This quasar has an apparent visual magnitude of 12.8 and can be seen with a good 8-inch telescope. (Courtesy Mount Wilson and Palomar Observatories.)*

3C 48

3C 147

3C 273

3C 196

law they they must be very far away. When their brightness was considered, however, it was shown that if they were indeed extremely distant, quasars must be extremely powerful sources of energy. The mystery of quasars then becomes a question of energy. Quasars have an energy output equivalent to about 10,000 times that of a typical spiral galaxy. Where could such a vast amount of energy come from?

Of course, there is no energy problem if the quasars are really closer than Hubble's law predicts. All we can measure is a quasar's apparent magnitude or brightness. If quasars are close, they would have absolute brightnesses similar to ordinary galaxies. But if they are far away, their apparent magnitude indicates an extremely large amount of energy being given off. A few astronomers claim that the large red shifts are caused by other phenomena and that Hubble's law does not hold for highly red-shifted quasars. Many astronomers disagree, however, and maintain that Hubble's law does hold.

If Hubble's law holds, the energy problem is very perplexing. An additional piece of data indicates that quasars vary in brightness over weeks or months at a time. What could cause this? Several suggestions have been given. One possibility is that quasars are somehow linked to black holes on a large scale. Possibly they are huge rotating objects collapsing due to gravity. The rotational motion might cause the variation in brightness. Another possibility is that quasars are simply another kind of galaxy—much more powerful than the galaxies near us. Quasars are billions of light years away, so we are looking at the light from quasars when they were possibly very young galaxies. These very young galaxies possibly gave off much more energy due to the burning of very large mass stars. They may also have many supernovae going off at any one time. This would give rise to variations in brightness. Unfortunately, neither of these theories has, as yet, been proven correct.

Quasars remain a mystery but possibly in the next few years the riddle of quasars will be solved.

20-6
COSMOLOGY

Cosmology is the study of the universe as a whole. We think that we know one fact concerning the basic structure of the universe. The main key to our understanding comes from Hubble's law which states that all distant galaxies are moving away from us and the farther away a galaxy is from us, the faster it is moving. This leads us to assume that the universe is expanding or exploding.

Figure 20-6a shows that in an explosion every particle moves away from every other particle. In this diagram, particle 2 is not at the center of the explo-sion; but particles 1, 3, 4, and 5 all appear to be moving away from particle 2. Particle 3 is moving faster than particle 2, that is how it got farther away from the explosion center. Particle 1 is moving slower than particle 2 and is still close to the explosion center. An observer on particle 2 would see both particles 1 and 3 moving away. Compare this to a person in an automobile going at 40 mi/h who sees a car behind him traveling at 20 mi/h and a car in front traveling at 60 mi/h—both moving away from him. To continue our analysis, it is obvious that particles 4 and 5 are also moving away from particle 2. In fact, a careful perusal of Fig. 20-6a will show that all the particles are moving away from each other.

Hubble's law leads us to conclude that our universe is expanding or exploding. The logical question

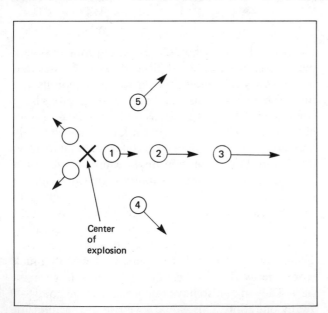

Figure 20-6a *In an explosion, all particles move away from all other particles. In this diagram, if you were sitting on particle 2, particles 1, 3, 4, and 5 would all appear to be moving away from you. Note that particle 2 is not at the center of the explosion.*

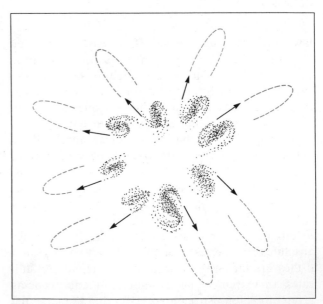

Figure 20-6b *If the mass of the universe is large enough, the receding galaxies will eventually slow down and turn around, and the universe will begin to contract.*

Figure 20-6c *The space shuttle (below) will be used to carry a space telescope (above) that measures 2.4 meters in diameter into earth orbit. This telescope will enable us to see objects seven times fainter than can now be seen.*

to ask is when this explosion occurred. Our current knowledge of the universe is contained in what is known as the Big Bang theory. The **Big Bang theory** of the universe states that 15 to 20 billion years ago the universe was concentrated in a massive glob of extremely dense material which suddenly exploded. The galaxies were hurled away from each other, and the universe has been expanding ever since.

Our next logical question is to ask about our universe's future. Will the universe continue to expand forever, or will it contract? Astronomers do not yet know the answer to this question, but the key to our understanding lies in determining the average density of the universe. If the universe is dense enough, the mass in the universe will cause the receding galaxies to slow down and fall back on each other as shown in Fig. 20-6b. This effect is similar to blasting a rocket off the earth. If the initial velocity of the rocket is not great enough, it will slow down and fall back to the earth. If the initial velocity is large enough, the rocket will continue to move away from the earth forever.

One theory of the universe says that there is not enough mass in the universe and the universe will expand forever. A second theory states that the mass is large enough and the expansion of the universe will slow down, and the universe will eventually begin to contract. This creates the possibility that eventually the universe will contract to a point and explode again. This idea is called a pulsating universe because it continuously expands, contracts, explodes, expands, contracts, explodes, etc. The choice between an ever-expanding universe and a pulsating universe cannot be made at this time because not enough data is known.

An example of the type of equipment, which may lead us to a definitive answer deciding between these theories, is the space telescope shown in Fig. 20-6c. This telescope would enable us to see seven times deeper into the outermost reaches of our universe. Past experience tells us that our knowledge continues to increase with the march of time. Perhaps in the next five or ten years we will understand the fate of our universe.

IMPORTANT WORDS AND TERMS

Photosphere
Plasma
Chromosphere
Corona
Sunspots
Prominences
Neutrino
Proton–proton chain
Solar neutrino experiment
Celestial sphere
Declination
Right ascension
Celestial prime meridian
Astronomical unit

Light year
Parsec
Constellations
Zodiac
Magnitude
Absolute magnitude
H-R diagram
Main sequence
Red giants
White dwarfs
Binary star
Cepheid variables
Nova
Supernova

Hydrogen burning
Nucleosynthesis
Neutron star
Pulsar
Black hole
Galaxy
Milky Way
Local Group
Hubble's law
Quasars
Cosmology
Big Bang theory

LEARNING OBJECTIVES

After reading and studying this chapter, you should be able to do the following without referring to the text:

1. Define the term *plasma* and describe the composition of the sun's interior.
2. Tell what sunspots are and describe their overall behavior.
3. Describe the basic process by which a star, such as the sun, shines.
4. Define the term *neutrino* and tell why the solar neutrino experiment is so important.
5. Describe the motion of the stars and planets if you look at the stars for the following amounts of time: (a) One night (b) Every Monday night at midnight for one year
6. Draw an H-R diagram and label the axes and the major features.
7. Tell why cepheid variables are important.
8. State the major events in the birth, life, and death of a star.
9. List the three possible end stages of a star and the mass a star would have to fall into each category.
10. State the sun's age and the age of the universe.
11. State Hubble's law in words.
12. List the chief properties of quasars and tell why they are puzzling.
13. State the Big Bang theory of the universe in one sentence.

QUESTIONS

1. What is a plasma?
2. What is the temperature of the sun (a) at the center, (b) at the surface?

3. Are sunspots cyclic?
4. What makes the sun shine?
5. What are neutrinos and photons and how are they different?
6. Why are the results of the solar neutrino experiment perplexing?
7. What is the zodiac?
8. Describe each of the following: (a) light year (b) parsec (c) astronomical unit
9. On an H-R diagram, give the position of each of the following: (a) the sun, (b) the main sequence, (c) red giants, (d) cepheid variables, (e) white dwarfs
10. List the following possible stages of a star in the correct order: hydrogen burning, cepheid variable, white dwarf, gravitational accretion, red giant.
11. What is a pulsar?
12. What is a black hole?
13. What is a quasar?
14. How old is (a) the sun, (b) the universe?
15. Why is the sun considered a second-generation star?
16. Describe the Milky Way.
17. Why is the universe thought to be expanding?

PROBLEMS

1. The mass of the sun is 2×10^{30} kg, or in terms of earth weight 2.2×10^{27} tons. Compute how many years it would take to consume one-tenth the sun's mass if 630 million tons are consumed every second.

   Answer: 11 billion years

2. Do the calculations mentioned in the book to derive the net reaction
   $$^1_1H + ^1_1H + ^1_1H + ^1_1H \longrightarrow ^4_2He + 2\,_{+1}e + 2\gamma + 2\nu + \text{Energy}$$
   from the reactions for the proton-proton chain.

3. Show that a light year is approximately equivalent to 6×10^{12} miles.

4. Find the distance away in parsecs for a star with a parallax of 0.2 seconds.

5. How much brighter is a star of absolute magnitude $+1$ than a star of absolute magnitude $+6$?

6. How much brighter is a cepheid variable of absolute magnitude -3 than a white dwarf of absolute magnitude $+7$?

   Answer: 10,000 times

7. The Crab Nebula is expanding at a rate of 70 million miles per day. If it is the remnants of a supernova of 1054 A.D., how many miles in diameter is it today?

   Answer: 47×10^{12} mi or 8 light years

8. It takes the sun 2×10^8 years to go around the center of the galaxy. How many times has it (and our solar system) gone around the galaxy during its life of 5 billion years?

   Answer: 25

9. How many miles away is the closest star, Alpha Centauri, which is 4.2 light years away?

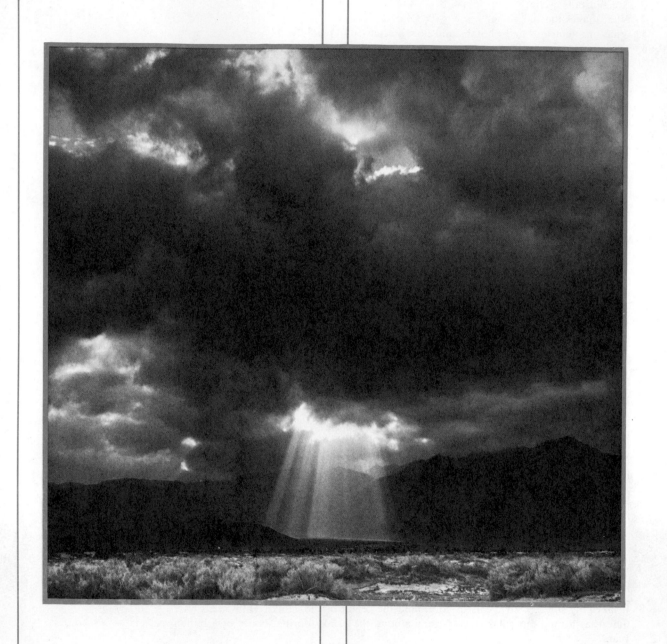

Chapter 21
THE ATMOSPHERE

Our **atmosphere** is the gaseous shell, or envelope, of air that surrounds the earth. Just as certain sea creatures live at the bottom of the ocean, humans live at the bottom of this vast atmospheric sea of gases. **Meteorology** (Greek *meteora*, the air) is the study of phenomena occurring in the atmosphere such as heat transfer, clouds, and winds. The properties and changes of the atmosphere affect our lives and our natural environment. Atmospheric variations are monitored daily and changing patterns are studied to help predict future conditions.

The phenomena of the atmosphere are important to many branches of science. They produce changes on the surface of the earth which concern the geologist, as will be discussed in later chapters. The astronomer longs for a view of the stars that is not obscured by the atmosphere and must consider the effect of atmospheric phenomena on his observations. Yet, while we attempt to learn the workings of the atmosphere and to even effect local changes for our own benefit, we indiscriminately pollute the atmospheric blanket that protects and sustains life.

The condition of our atmosphere has become a topic of major interest with respect to ecology, the study of living organisms in relationship to their environment. Since the air we breathe is such an integral part of our environment, a knowledge of the atmosphere's constituents, properties, and workings is needed to understand and appreciate current ecological problems. In other words, what are some of the normal conditions and other phenomena of the atmosphere?

". . . this most excellent canopy, the air, look you, this brave o'erhanging firmament, this majestical roof fretted with golden fire . . ."

(*Shakespeare*—Hamlet)

21-1
COMPOSITION

The atmosphere is a mixture of many gases. However, it is somewhat amazing that only two comprise about 99 percent of the volume of air near the earth. From Fig. 21-1a and Table 21-1a, it can be seen that air is primarily composed of nitrogen and oxygen, with nitrogen being about four times as abundant as oxygen. The other main constituents are argon and carbon dioxide. Many other gases of minute quantities are found in the atmosphere along with dust, pollen, salt particles, and so on. Some of these, especially water vapor and carbon monoxide, vary in concentration, depending on conditions and locality. The amount of water vapor in the air is dependent to a great extent on the temperature as will be discussed later. Carbon monoxide is a by-product of incomplete combustion. A sample of air taken near a freeway, for example, would contain a concentration of carbon monoxide considerably higher than that normally found in the atmosphere. It is estimated that over 10,000 tons of carbon monoxide are daily put into the atmosphere in and around Los Angeles, most of which results from automobile combustion. Carbon

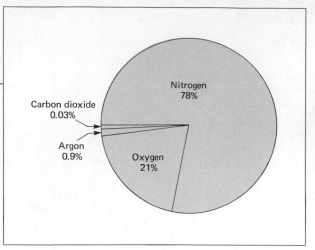

Figure 21-1a *Graphic representation of the volume composition of the major constituents of air.*

dioxide, another by-product of combustion, is also present in abnormally high concentrations in such populated areas.

In general, however, the relative amounts of the major constituents of the atmosphere remain fairly constant. Nitrogen, oxygen, and carbon dioxide are involved in the life cycles of plants and animals and are continually taken from the atmosphere and replenished as by-products of the various cycles. Animals breathe in oxygen and expel carbon dioxide, while plants use carbon dioxide and expel oxygen in their life-cycles. Nitrogen is also assimilated by some plants and released as decay. Thus, nature operates in such a way as to maintain the balance of the various gases in the atmosphere.

On the other hand, human beings may contribute to changes in the atmosphere and perhaps eventually create imbalance. Recent concern has arisen in this regard because of increased amounts of CO_2 being poured into the atmosphere. Consider the amount of burning required to heat the houses in the United States alone in the winter. Add to this the industrial burning for heat and energy. The resultant amount of gases released into the atmosphere is considerable and gives cause for concern, as will be discussed in Chapter 26.

Table 21-1a *Composition of Air*

Nitrogen	N_2	78% (by volume)
Oxygen	O_2	21%
Argon	Ar	0.9%
Carbon dioxide	CO_2	0.03%

Others (traces)		*Others (variable)*
Neon	Ne	Water vapor (H_2O) 0–4%
Helium	He	Carbon monoxide (CO)
Methane	CH_4	Ammonia (NH_3)
Nitrous oxide	N_2O	Solid particles—dust,
Hydrogen	H_2	pollen, etc.

21-2
ORIGIN

The origin of the atmosphere is generally associated with the origin of the earth which is some four to five billion years old. There are several theories of cosmogony which attempt to explain the origin of the universe. Without speculating on the validity of any one theory, it is generally agreed that at the time of the creation of our solar system, the planets were extremely hot masses, since their creation by any means would require enormous amounts of energy. In the beginning, then, the earth was a molten mass surrounded by hot gases, probably methane, ammonia, and hydrogen compounds. Because of their high temperatures and inherent kinetic energy, large amounts of these gases escaped into space. As the earth cooled, gases that were dissolved in the molten mass were released, giving rise to H_2O, N_2, and CO_2 in our atmosphere. The water vapor condensed and reevaporated, causing cooling and solidification of the earth's crust with subsequent erosion and formation of bodies of water. The gases cooled along with the planet and, being then less energetic, were retained by gravitational attraction. At this time, life began to appear.

How life occurred is still a mystery. However, scientists have shown that the fundamental building blocks of life, the amino acids, can be constructed from combinations of methane, hydrogen, ammonia, and water, all of which were available on the cooling earth. (See Chapter 16, the Miller experiment.) In any event, the appearance of plant life allowed the formation of oxygen and the evolution of animal life. Plants produce oxygen by **photosynthesis,** the process by which CO_2 and H_2O are converted into sugars (carbohydrates) and O_2, using the energy from the sun (Fig. 21-2a).

The key to photosynthesis is the ability of chlorophyll, the green pigment in plants, to convert sunlight into chemical energy. In chemical notation, the pho-

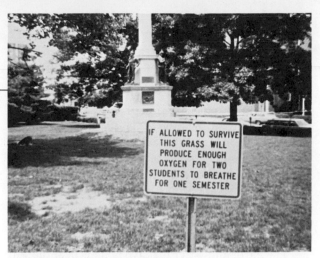

Figure 21-2a *An appeal for photosynthesis. Oxygen is a by-product of the photosynthesis process by which plants convert carbon dioxide and water to carbohydrates (sugars).*

tosynthesis process may be expressed

$$(n)\ CO_2 + (n)\ H_2O \xrightarrow[\text{chlorophyll}]{\text{sunlight}} (CH_2O)_n + (n)\ O_2$$

where the n's indicate the numbers that balance the chemical equation. Photosynthesis accounts for the liberation of approximately 130 billion tons of oxygen into the air annually, with some 2000 billion tons of CO_2 being involved in the process. Over one-half of the photosynthesis takes place in the oceans, which contain many forms of green plants. Originally, oxygen may also have come from the dissociation of water.

Over billions of years, the earth's atmosphere evolved to its present composition. The atmosphere of other planets evolved in different fashions. The atmospheres of Jupiter and Saturn contain mainly ammonia and hydrogen, while Venus has a considerable amount of CO_2. Mercury, however, lost its atmosphere, as did our moon. The retention of an atmosphere depends on a balance between gravity and the thermal kinetic energy of the atmospheric gases.

The gravitational force that attracts the gas molecules of an atmosphere to a planet is dependent on the mass of the planet. (See Chapter 3, Section 3-3.) A planetary atmosphere receives energy from the sun, and the temperature of the local atmosphere depends on the velocities of the gas molecules. If sufficient energy is acquired, some of the molecules will have enough velocity to overcome the gravitational attraction and will escape into space. As a result, a planet can gradually lose its atmosphere. Similarly, spacecraft must be given sufficient velocity (kinetic energy) to escape the earth's gravitational attraction.

The earth and other massive planets near the sun have sufficient gravitational attraction to retain their atmospheres. However, because of the relatively small masses of Mercury and the moon, their gravitational attractions were not sufficient to retain their energetic atmospheres.

21-3
VERTICAL STRUCTURE

To distinguish different regions of the atmosphere, we look for variations that occur with height or altitude. The changes in physical properties can be used to define vertical divisions of the atmosphere. Some of the atmospheric properties which show vertical variations that we will examine are (1) density and pressure, (2) temperature, (3) homogeneity and heterogeneity of gases, (4) ozone and ion concentrations.

Density and Pressure

The atmosphere extends upward with continuously decreasing density (mass per volume). There is no clearly defined upper limit of the earth's atmosphere; it simply becomes more and more tenuous. It merges without distinction into the interplanetary gases, which may be thought of as part of the extensive "atmosphere" of the sun. However, the decrease in density occurs quite rapidly in the lower atmosphere. One-half of the entire atmosphere lies

below a height of three and one-half miles, and three-fourths lies below seven miles. The greater density near the earth is due to the gravitational attraction and compression of the air.

Closely related to density is atmospheric pressure (force per area). The pressure at a particular altitude is effectively a measure of the weight (or amount) of gas above that location. Just as a sea diver experiences increased pressure when descending, due to the weight of the water above him, the atmospheric pressure is greater near the earth due to the weight of the air above. Like density, the pressure varies vertically, decreasing with increasing altitude. This occurs continuously (Fig. 21-3a), and does not make for distinct boundaries. Consequently, vertical divisions of the atmosphere are not made on the basis of pressure.

Temperature

In measuring the temperature of the atmosphere versus altitude, we find that distinct changes do occur. This leads to major divisions of the atmosphere based on temperature variations. The temperature of the atmosphere decreases with increasing altitude at an average rate of $3\frac{1}{2}°F$ per 1000 feet up to about 10 miles, as illustrated in Fig. 21-3a. This region is called the **troposphere** (Greek *tropism*, to change). The atmospheric conditions of the lower troposphere are referred to collectively as **weather.** Changes in the weather reflect the local variations of the atmosphere near the earth's surface.

The actual depth of the troposphere varies from about $10\frac{1}{2}$ miles at the equator to about five miles at the poles, with an average thickness of about seven miles. The troposphere was the only region of the atmosphere known until the twentieth century. Balloonists ascended to the upper regions of the troposphere only in the 1930's. Because the conditions of the troposphere (with the exception of those near the earth's surface) were then unknown, many unexpected problems arose. The balloonists encoun-

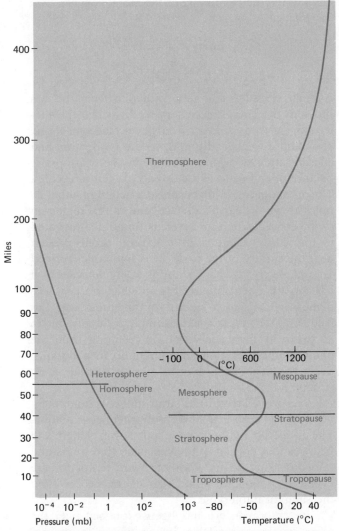

Figure 21-3a *The vertical structure of the atmosphere. Variations in the physical properties of the atmosphere, such as temperature and homogeneity of the atmospheric gases, give rise to natural divisions. The gradual decrease of density and pressure with altitude, however, offers no distinct variations upon which divisions can be based.*

tered difficulty in breathing, freezing of valves necessary for descent, and loss of the use of limbs due to frostbite. At the top of the troposphere, the temperature falls to 50° to 60°F below zero. Beyond these altitudes, sounding balloons and rockets were used for exploration. These were unmanned flights containing instruments to record atmospheric properties. More recently, satellites have been used to make atmospheric observations. Atmospheric observations are discussed in Chapter 25.

Above the troposphere, the temperature of the atmosphere increases nonuniformly over a region of about 35 miles in depth. This portion of the atmosphere, from approximately 10 to 35 miles in altitude, is called the **stratosphere** (Latin *stratum,* covering layer). The temperature then decreases rather uniformly to a height of 60 miles. This region, between 35 and 60 miles in altitude, is called the **mesosphere** (Greek *meso,* middle). Above the mesosphere, the thin atmosphere is heated intensely by the sun's rays and the temperature climbs to over 2000°F. This region extending to the outer reaches of the atmosphere—several hundred miles—is called the **thermosphere** (Greek *therme,* heat). The temperature in the thermosphere varies considerably with solar activity.

It should be realized that the boundaries of these regions are not sharply defined and extend over several miles. The intermediate regions comprising the boundaries are given the suffix "pause." The top of the troposphere is referred to as the **tropopause,** and the respective upper boundaries of the stratosphere and mesosphere as the **stratopause** and **mesopause.**

Homogeneity and Heterogeneity of Gases

The mixture of atmospheric gases provides another means of atmospheric division. In the lower portion of the atmosphere, turbulence continually mixes the gases so that the air is a homogeneous

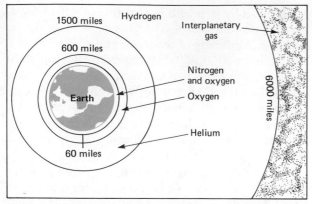

Figure 21-3b *An illustration of the heterogeneous layering of the gases in the upper atmosphere based on satellite data. (Courtesy NASA.)*

mixture. Thus, this region is called the **homosphere.**

Above the homosphere, beginning at about 55 to 60 miles from the earth's surface, is the **heterosphere.** Because no turbulence exists there to keep the gases mixed, they settle out in a layered structure, as illustrated in Fig. 21-3b. This is analogous to sand suspended in water. As long as the mixture is stirred, the sand remains in a uniform suspension. However, if the mixing is ceased, the sand settles out according to particle size and weight much the same as the gases of the heterosphere.

Ozone and Ion Concentrations

The atmosphere is also divided into the **ozonosphere** and **ionosphere.** Their boundary is approximately the same as the homosphere and heterosphere, respectively. As the names imply, there are concentrations of ozone and ions in these regions. **Ozone** (O_3) is formed by the dissociation of molecular oxygen and the combining of the atomic oxygen with the molecular oxygen.

$$O_2 + \text{Energy} \longrightarrow O + O$$

$$O + O_2 \longrightarrow O_3$$

At high altitudes, energetic radiation from the sun provides the energy necessary to dissociate the molecular oxygen. Oxygen, however, becomes less abundant at higher altitudes, so the production and concentration of ozone depend on the appropriate balance of energetic particles and oxygen molecules. The optimum condition occurs at a height of about 20 miles above the earth's surface, and in this region an ozone concentration, or layer, is found as illustrated in Fig. 21-3c. There is little natural ozone present near the surface of the earth. It is sometimes formed around electric discharges and is easily detected as it has a distinct pungent smell from which it derives its name (Greek *ozein*, to smell). In some areas, e.g., Los Angeles, ozone is classified as an air pollutant, due to its relatively high concentration resulting from the photochemical reactions of sunlight on air pollutants. Such reactions give rise to photochemical smog (Chapter 26).

The ozone layer in the stratosphere is fundamental to life, because it absorbs ultraviolet radiation. The ultraviolet radiation near 3000 Å wavelength and below is that portion of the spectrum that burns and tans our skins in the summer. Were it not for the ozone absorption, we would be badly burned and find the sunlight intolerable.

Recently, concern has been expressed over the effect on the ozone layer by gaseous chlorofluorocarbon compounds used as propellants in spray aerosol cans. Through certain chemical reactions, these chlorofluorocarbons react to change ozone to O_2. Appreciable concentrations of chlorofluorocarbons in the ozonosphere could conceivably reduce the ozone concentration and permit large amounts of harmful ultraviolet radiation to reach the earth. This

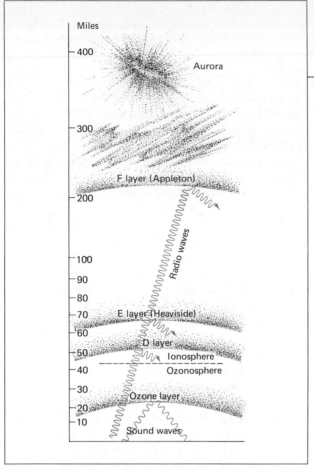

Figure 21-3c *Diagram showing the regions of the ozonosphere and ionosphere. Atmospheric divisions are based on ozone and ion concentrations. The air temperature variation, due to the ozone layer, gives rise to the reflection of sound waves, while the ion layers reflect radio waves.*

potential pollution hazard will be discussed further in Chapter 26.

Since the ozone layer absorbs the energetic ultraviolet radiation, one can expect an increased temperature in the ozonosphere. It can be seen from a comparison of Figs. 21-3a and 21-3c that the ozone layer lies in the stratosphere. Hence, the ozone absorption of ultraviolet radiation provides an explanation for the temperature increase in the stratosphere as op-

posed to the continually decreasing temperatures versus altitude experienced in the neighboring troposphere and mesosphere.

This warm region of ozone concentration was first investigated by means of sound. Samuel Pepys, the noted English writer, recorded the phenomenon of being able to hear, in London, the cannons of the fleet in the English Channel during a battle of the Dutch Trade War of 1666, while people on the coast heard nothing. Similar events were observed during the Civil War and World War I. It was reasoned that the sound of the cannon must have been reflected back to earth by something in the atmosphere. Just as light is reflected from water, perhaps the sound was reflected from the boundary of warm and cold air when incident at the proper angle. Subsequent investigations revealed the existence of the warm air layer and established the presence of ozone as the cause. In cases where sound is heard at a distance but not nearby, sound waves moving horizontally are blocked, for example by a hill, whereas those moving vertically are reflected by the layered air and heard at a distance. This explanation is illustrated in Fig. 21-3d.

In the upper atmosphere, the energetic particles from the sun cause the ionization of the gas molecules. For example,

$$N_2 + Energy \longrightarrow N_2^+ + e^-$$

The electrically charged ions and electrons are trapped in the earth's magnetic field and form ionic layers in the region of the atmosphere referred to as the ionosphere. Kennelley and Heaviside in 1902 independently discovered a layer at a height of 60 miles. Appleton found another ionic layer at an altitude of 200 miles (Fig. 21-3c). An important feature of these layers is that they reflect radio waves and hence are important in global radio communication. Radio waves, like all electromagnetic waves, travel in straight lines, making it impossible to send them directly around the curved earth. However, by re-

Figure 21-3d *An illustration of the reflection of sound in the atmosphere. Sound waves may be reflected at the boundary of warm air associated with the ozone layer and be heard many miles away from the sound source.*

Figure 21-3e *Global radio transmission. Radio waves, which normally travel in a straight line, are reflected around the curvature of the earth by ionic layers. Ionic disturbances, however, sometimes interfere with reflections from an ionic layer and disrupt transmission. Transmission via satellite is not, in general, disrupted by ionic disturbances.*

flecting or "bouncing," the waves off an ionic layer, they may be transmitted around the earth as illustrated in Fig. 21-3e. The Heaviside layer reflects longer wavelengths (10^2 to 10^3 m) than does the Appleton layer (10 to 10^2 m). These layers have been designated E and F, respectively, and are subdivided into E_1, E_2, F_1, and F_2 layers. Another layer, called D,

has also been found below the E layer. The lower layers (D and E) are prominent only during the day and the division of the F layer becomes indistinguishable at night. Radio waves beamed vertically upward are reflected back to the earth, thereby allowing calculation of the height of the reflecting layer from the knowledge of the velocity of the waves and the time elapsed.

The reflection of radio waves from the ionic layers depends on uniformity in the density of the layer. Should a solar disturbance produce a shower of energetic particles which upsets this uniformity, a communications "blackout" may occur. Radio transmission is then reduced to line-of-sight transmission as reflections from ionic layers are impaired. These disturbances caused considerable inconvenience in early worldwide communications. Cables were laid across the oceans to keep communications open, and now satellites relay radio and television communications.

Solar disturbances are also associated with beautiful displays of light in the upper atmosphere of the polar regions called northern lights, or **aurora borealis.** See Fig. 21-3f. The southern hemisphere tends to be forgotten by people living north of the equator. However, light displays of equal beauty occur in the southern polar atmosphere and are called **aurora australis.**

The location and height of the aurora depend on the interaction of the incoming charged particles and the earth's magnetic field. In general, the particles are deflected toward the earth's magnetic poles (our polar regions) over which the majority of the auroras occur. The Norsemen thought the auroras were the reflections from shields in Valhalla. Today, however, they are associated with the excitation of oxygen and nitrogen molecules in the atmosphere by charged particles (electrons and protons) and the recombination of electrons and ions. The occurrence of auroras has been correlated with solar activity, which gives rise to an influx of ionizing radiation to the earth's upper atmosphere.

Figure 21-3f *The aurora borealis. The ions of the ionosphere give rise to brilliant displays of light called auroras. (University of Alaska photograph. Courtesy U.S. Air Force.)*

21-4
ENERGY CONTENT

The sun is by far the most important source of energy for the earth and its atmosphere. At an average distance of 93 million miles from the sun, Earth intercepts only a small portion of the vast amount of the solar energy emitted. This energy traverses space in the form of radiation and the portion incident on the

earth's atmosphere is called **insolation**—*in*coming *sol*ar radi*ation*.

Because the earth is tilted on its axis $23\frac{1}{2}°$ with respect to the plane of its orbit about the sun, the insolation is not evenly distributed over the earth's surface (Fig. 21-4a). This tilt, coupled with the earth's revolution around the sun, gives rise to the seasons. Although the earth is closest to the sun in January and farthest away in July, the northern hemisphere

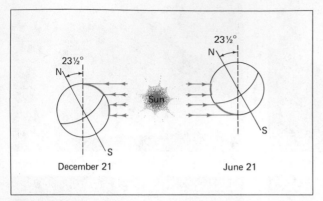

Figure 21-4a *The seasonal variation of insolation. The tilt of the earth's axis relative to the plane of its orbit causes the incoming solar radiation to be distributed differently on the earth's surface at different times of the year. In December, the southern hemisphere receives more direct rays of the sun than does the northern hemisphere. In June, the situation is reversed.*

experiences colder weather in January because this part of the earth is tilted away from the sun. As a result, the days are shorter and the radiation received north of the equator is distributed over a larger area, accounting for less warming. The southern hemisphere at this time is enjoying warm weather. Six months later, in July, the situation is reversed for the hemispheres. Thus, the tilting and resultant distribution of insolation gives rise to seasonal variations and their reversal in the northern and southern hemispheres.

The average summer temperature of the southern hemisphere is slightly higher than that of the northern hemisphere. This is, in part, due to the fact that the southern hemisphere is closer (about three percent) to the sun when tilted toward it. However, more water and less land mass in the southern hemisphere makes for less summer-winter temperature variation.

The maximum insolation for the earth is received daily around noon in both hemispheres, yearly near June 21 in the northern hemisphere, and yearly near December 21 in the southern hemisphere. The maximum daily and yearly temperatures, however, lag behind these times. That is, the maximum daily temperature occurs around 2 to 3 P.M. and August is usually the hottest month in the northern hemisphere. This is readily understood since more radiation is received after 12 noon and on those days following June 21. The additional radiation contributes to the daily and average yearly temperatures, causing their maxima to occur later than the times of maximum insolation. This is much the same as slowly turning up the gas flame under a pan of water. When the maximum flame is reached and then slowly decreased, the temperature continues to rise and reaches a maximum sometime after the maximum flame, since the water is still being heated by the decreasing flame.

The average intensity of solar radiation outside the earth's atmosphere at the mean distance of the earth from the sun is $2.0\ \text{cal/cm}^2 \cdot \text{min}$ or $3.1 \times 10^{12}\ \text{cal/mi}^2 \cdot \text{h}$. This quantity is referred to as the **solar constant** and expresses the amount of energy received at normal incidence on an area per unit time. For example, let's assume that this energy intensity is received at the earth's surface and that the area of a sunbather's back is 3000 cm². Every minute, he would receive on the average 6000 calories of energy, since

$$\frac{2.0\ \text{cal}}{\text{cm}^2 \cdot \text{min}} \times 3000\ \text{cm}^2 = 6000\ \text{cal/min}$$

If the bather remained in the sun for 30 min, he would receive a total amount of energy of

$$6000\ \frac{\text{cal}}{\text{min}} \times 30\ \text{min} = 180{,}000\ \text{cal}$$

This amount of energy could raise the temperature of 10 cups of water from room temperature (23 °C) to its boiling point. (See Section 7-3.) However, only about 50 percent or less of the insolation reaches the earth's surface, depending on atmospheric conditions.

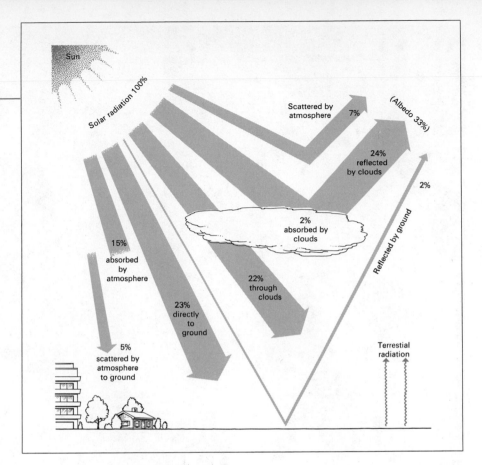

Figure 21-4b *An illustration of insolation distribution. The percentages vary somewhat, depending on atmospheric conditions.*

In considering the energy content of the atmosphere, one might think that this comes directly from insolation. However, surprisingly enough, most of the direct heating of the atmosphere comes not from the sun, but from the earth. To understand this, we need to examine the distribution and disposal of the incoming solar radiation (Fig. 21-4b).

About 33 percent of the insolation received is returned to space with no appreciable effect on the atmosphere as a result of reflection by clouds, scattering by particles in the atmosphere, and reflection from terrestrial surfaces, such as water, ice, and variable ground surfaces. The reflectivity, or the average fraction of light a body reflects, is known as its **albedo** (Latin *albus*, white). Thus, the earth has an albedo of 0.33, or reflects about one-third of the incident sunlight. The brightness of the earth as viewed from space depends on the amount of sunlight that is reflected, and clouds play an important role (Fig.

21-4c). In comparison, the moon with its dark surface and no atmosphere has an albedo of only 0.07, i.e. reflects only seven percent of the insolation. Even if the earth were as small as the moon, it would appear much brighter (about five times) than the moon to an astronaut in space.

Scattering of insolation occurs in the atmosphere from gas molecules of the air, dust particles, water droplets, etc. By scattering, we mean the absorption of incident radiation and its reradiation in all directions. As can be seen from Fig. 21-4b, some of the scattered radiation is scattered back into space, and some is scattered toward the earth. The important type of scattering for our discussion is the so-called **Rayleigh scattering,** named after Lord Rayleigh (1842–1919), the British physicist who developed the theory. Lord Rayleigh showed that the scattering for particles of molecular size was proportional to $1/\lambda^4$, where λ is the wavelength of the incident light. That is, the

Figure 21-4c *Earth from 22,300 miles in space. The cloud cover of Earth is important in reflecting sunlight and contributes significantly to the albedo. (Courtesy NASA.)*

longer the wavelength of the radiation, the less the scattering.

The gas molecules of the air account for most of the scattering in the visible region of the spectrum. In the visible spectrum, the wavelength increases from violet to red. The blue end of the spectrum is therefore scattered more than the red end. (The colors of the visible spectrum—and the rainbow—may be remembered with the help of the name of ROY G. BIV—red, orange, yellow, green, blue, indigo, and violet.) As the sunlight passes through the atmosphere, the shorter wavelengths are scattered by the gas molecules. The scattered blue end of the spectrum is rescattered in the atmosphere and eventually is scattered toward the earth. This is the light we see when looking at the sky, and hence, the sky appears blue. The eye is more sensitive to blue than to the shorter wavelength violet, and since sunlight has less violet light than blue light, the sky doesn't appear violet. If the earth had no atmosphere, the sky would appear black, other than in the vicinity of the sun. At sunset, the sunlight travels a greater distance through the atmosphere to an observer, and most of the blue end of the spectrum is scattered out of the sunlight. Consequently, sunsets appear red, and light reflected from clouds causes the clouds to appear pink.

Rayleigh scattering plays a role in the selection of amber fog lights and red tail lights for automobiles. In the case of fog lights, if the fog droplets are very small, then the red end of the visible spectrum is scattered less than the blue end, and the light from the amber fog lights penetrates the fog. Fog lights can be a definite advantage in a city where a not-too-dense fog can contain many fine smoke particles that contribute to the Rayleigh scattering. Fog lights are not too advantageous in thick fogs with large droplets. On the basis of Rayleigh scattering, you should be able to explain why automobile tail lights are red.

Larger particles of dust, smoke, haze, and those from air pollution in the atmosphere may preferentially scatter long wavelengths. These scattered wavelengths, along with the scattered blue light due to Rayleigh scattering, can cause the sky to have a milky blue appearance—white being the presence of all colors. Hence, the blueness of the sky, gives an indication of atmospheric purity. Cloud droplets and raindrops scatter even longer wavelengths. This fact is used in the principle of weather radar, which is an important means of weather monitoring as we shall learn in Chapter 25.

About 15 percent of the insolation is absorbed directly by the atmosphere. Most of this absorption is accomplished by ozone, which removes the ultraviolet radiation, and by water vapor, which absorbs strongly in the infrared region of the spectrum. The major portion of the solar spectrum lies in the narrow visible region. Nitrogen and oxygen, which compose the bulk of the atmosphere, are practically transparent to visible radiation, so the major portion of the solar radiation passes through the atmosphere without appreciable absorption.

After reflection, scattering, and direct absorption of insolation by the atmosphere, about 50 percent of the total incoming solar radiation reaches the earth. This goes into terrestrial surface heating, primarily through the absorption of visible radiation. As was noted previously, the atmosphere, in particular the troposphere, derives most of its energy content from the earth. This is accomplished in three main ways. In order of decreasing importance, these are (1) absorption of terrestrial radiation, (2) latent heat of condensation, and (3) conduction from the earth's surface.

Absorption of Terrestrial Radiation

The earth, like any warm body, radiates energy which may be absorbed by the atmosphere. The wavelength of the radiated energy is dependent on the earth's temperature. This may be seen from the equation relating the energy (E) and wavelengths of the radiation

$$E = hf = \frac{hc}{\lambda}$$

where h = Planck's constant,
c = velocity of light,
f = frequency,
λ = wavelength.

This energy is proportional to the temperature of the source producing it. Thus

$$E = \frac{hc}{\lambda} \propto T \qquad \text{Eq. 21.1}$$

Hence

$$\lambda \propto \frac{hc}{T}$$

The earth's temperature is such that it radiates energy primarily in the infrared region. That is, the earth absorbs energy in the shorter wavelength visible region of the solar spectrum, becomes warmed, and reradiates energy in the longer wavelength infrared portion of the spectrum.

Water vapor and carbon dioxide (CO_2) are the main absorbers of long wavelength radiation in the atmosphere, with water vapor being the more important because it absorbs most of the infrared spectrum (see Fig. 21-4d). We refer to the gases as being selective absorbers, since they absorb certain wavelengths and transmit others.

The absorption of long-wavelength terrestrial radiation by water vapor and CO_2 adds to the energy content of the atmosphere. This heat-retaining process of the atmospheric gases is referred to as the **greenhouse effect,** because it is analogous to a similar effect occurring in greenhouses. The absorption and transmission properties of glass are similar to those of the atmospheric gases—in general, visible radiation is transmitted, but infrared radiation is absorbed (Fig. 21-4e). We have all observed the warming effect of sunlight passing through glass, for example, in a closed car on a sunny, cold day. In a greenhouse, the objects inside become warm and reradiate long-wavelength infrared radiation, which is absorbed and reradiated by the glass. Thus, the inside of a greenhouse heats up and is quite warm on a sunny day,

Figure 21-4d *Absorption spectra for O_2 and O_3, CO_2, H_2O, and the atmosphere. CO_2 and H_2O are the chief atmospheric absorbers in the infrared region, while O_2 and O_3 absorb mainly in the ultraviolet region. Notice that the atmospheric gases are almost "transparent" or absorb very little visible radiation. (Redrawn with permission from Fleagle and Businger,* An Introduction to Atmospheric Physics, *Academic Press, New York, 1963.)*

even in the winter. Heat retention in a greenhouse is also promoted by the glass enclosure of the greenhouse which prevents loss of heat by convection.

The greenhouse effect of the atmosphere is quite noticeable in the cooling of the earth at night, particularly on cloudy nights. With a cloud and water-vapor cover to absorb the earth's radiation, the night air is comparatively warm. Without this insulating effect, the night is usually "cold and clear," as the energy from the daytime insolation is quickly lost. This effect also explains the rapid nocturnal temperature drop over desert areas because of the dryness of the air over these regions.

There are several interesting sidelights associated with the greenhouse effect and terrestrial radiation. With regard to a glass greenhouse that transmits the visible portion of insolation, the short-wavelength ultraviolet transmission depends on the iron content of the glass. Ordinary glass does not appreciably transmit wavelengths below 3100 Å (0.31μ)* or in the

---

*The wavelengths of light in the visible and near visible regions are commonly expressed in microns (μ). $1 \mu = 10^{-6}$ m $= 10^{-4}$ cm.

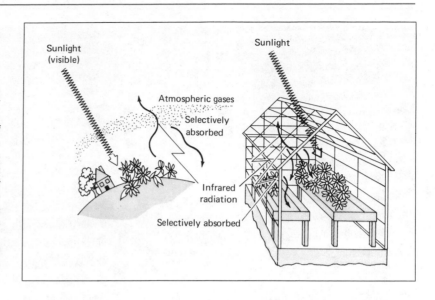

Figure 21-4e *An illustration of the greenhouse effect. The gases of the lower atmosphere transmit most of the visible portion of incident sunlight as does the glass of a greenhouse. The warmed earth emits radiation in the infrared region, which is selectively absorbed by the atmospheric gases whose absorption spectrum is similar to that of glass. This absorbed energy heats the atmosphere and helps maintain the earth's temperature like the interior of a greenhouse; hence the name greenhouse effect. (Courtesy J. D. Wilson,* An Environmental Approach to Physical Science, *Heath, Lexington, Mass., 1974.)*

ultraviolet region. Hence, although there is a considerable warming effect through glass from visible radiation, one cannot receive a suntan or sunburn through ordinary glass.

Despite the daily and seasonal gain and loss of heat, the average temperature of the earth has remained fairly constant over a long period of time. This indicates that the earth must lose or reradiate as much energy as it receives. If this were not the case, the continual gain of energy would cause the earth's average temperature to rise. The selective absorption of the atmospheric gases provides a thermostatic or heat-regulating process for the earth. For example, suppose the earth's temperature were such that it emitted radiation with wavelengths between 4 and 6μ. Referring to the absorption spectrum of the atmosphere in Fig. 21-4d, it can be seen that the absorptivity of these wavelengths is almost 1.0 (100 percent), or is almost completely absorbed by CO_2 and water vapor. As a result of this absorption, the lower atmosphere becomes warmer and effectively holds in the earth's heat or insulates it. With additional insolation, the earth becomes warmer and its

temperature rises. However, according to Eq. 21.1, the greater the temperature, the shorter the wavelength of the emitted radiation, and so the wavelength of the terrestrial radiation is shifted to shorter wavelengths. At about 3.5μ in the atmospheric absorption spectrum, there is a transmission "window," and the terrestrial radiation can pass through the atmosphere. Thus losing energy, the temperature of the earth decreases and the terrestrial radiation is shifted to a longer wavelength, which is absorbed by the atmosphere. Averaged over the total spectrum, the selective absorption of the atmospheric gases play an important role in maintaining the earth's average temperature.

Humans radiate energy in the infrared region, as does the earth. This radiation is not visible to the human eye, and requires special equipment to be detected. Warm, infrared emitting objects may thus be detected in the dark; hence, infrared detection has come to be associated with "seeing in the dark." Since the wavelength of the radiation is dependent on the temperature of the source, infrared photographs taken from satellites are currently being used to study

Figure 21-4f *Infrared photography. The upper photo shows how crop damage may be detected by use of infrared imagery. The lower photos show the infrared image of Yellowstone National Park with its geothermal areas. (Courtesy NASA.)*

temperature variations in the earth's crust arising from such things as volcanic activity and geothermal regions (Fig. 21-4f).

Latent Heat of Condensation

Approximately 70 percent of the earth's surface is covered by water. Consequently, a great deal of evaporation occurs due to the insolation reaching the earth's surface. You will recall from Chapter 7 that the latent heat of vaporization for water is 540 calories per gram. That is, 540 calories of heat energy are required to change one gram of water from a liquid to a gaseous state. Thus, a large amount of energy is transferred to the atmosphere in the form of latent heat. This energy is released in the atmosphere with the condensation of water vapor in the formation of clouds, fog, dew, etc.

Conduction from the Earth's Surface

A comparatively smaller, but significant amount of heat energy is transferred to the atmosphere by conduction from the earth's surface. Because the air is a relatively poor heat conductor, this process is restricted to the layer of air in direct contact with the earth's surface. The heated air is then transferred aloft by convection. Thus, the temperature of the air tends to be greater near the surface of the earth and decreases gradually with altitude.

The transfer of heat by conduction depends on the temperature difference between the air and the earth's surface. (Heat energy may be transferred from the air to the earth if the latter is colder.) The heat transfer to the atmosphere varies for land and water surfaces because they in general have different temperatures. The specific heat, or the amount of heat required to raise a gram of mass one degree in temperature (Section 7-3), is on the order of four times greater for water than for land. Therefore, water heats more slowly than does land given equal amounts of energy. This is augmented by water being transparent, so the sun's rays can penetrate, and by currents that effect mixing. Conversely, water cools more slowly than does land. Hence, neglecting seasonal influences, the land surface is warmer during the day, and the water is warmer at night. On a large scale, this effect also occurs seasonally, with summer and winter correspondingly to day and night.

Because of the large surface area of the earth, large amounts of heat are transferred from land and sea to the atmosphere by conduction. The redistribution of this heat energy by atmospheric convection plays an important role in weather and climate conditions.

21-5
ATMOSPHERIC MEASUREMENTS

Measurements of the atmosphere's properties and characteristics are very important in its study. Its properties are measured daily and records have been compiled over many years. Meteorologists study these records with the hope of observing cycles and trends in the atmospheric behavior so as to better understand and predict its changes. The layman listens to the daily atmospheric readings to obtain a qualitative picture of the conditions he may experience that day. Fundamental measurements made in the atmosphere include those of: (1) temperature, (2) pressure, (3) humidity, (4) wind speed and direction, and (5) precipitation.

Temperature

Temperature is a measure of heat energy. In matter it is a measure of the molecular activity or the kinetic energy of the molecules. The kinetic theory of gases shows the energy of gases to be directly proportional to the temperature (Section 7-7). The higher the temperature, the more energetic the gas molecules. When measuring the temperature of a gas,

Figure 21-5a *A diagram of Torricelli's barometer. The external air pressure on the surface of the pool of mercury supports the mercury column in the inverted tube. The height of the column therefore depends on atmospheric pressure and provides a way of measuring it.*

some of the molecules collide with the temperature measuring device and impart to it some of their energy. The more the molecular activity, the more collisions and the greater the energy transfer, which results in a higher temperature reading.

The liquid-in-glass **thermometer** is the most common device used to measure temperature. As discussed in Section 7-1, this thermometer consists of alcohol or mercury enclosed in a glass bulb which is attached to a capillary called the bore. The liquid expands linearly up the bore when heated and the temperature is read from a calibrated scale on the glass of the capillary. The common temperature scales are Fahrenheit and Celsius. A thermometer in contact with the gases of the atmosphere records the air temperature due to the energy transfer resulting from collisions of the air molecules with the glass bulb of the thermometer.

Heat transfer by radiation to a thermometer may give rise to a higher temperature reading. As a result, a thermometer in the sun may read a comfortable 70°F (21°C), but the air feels quite cool. A truer air temperature is often expressed in the summer by saying it is so many degrees "in the shade," implying the thermometer is not exposed to the direct rays of the sun.

Pressure

Pressure is defined as the force per unit area ($P = F/A$). At the bottom of the atmosphere we experience the resultant weight of the gases above us. Since we experience this weight before and after birth as a natural part of our environment, little thought is given to the face that every square inch of our bodies sustains an average weight of 14.7 lb at sea level or a pressure of 14.7 lb/in². We refer to 14.7 lb/in² as one atmosphere of pressure.

One of the first investigations of atmospheric pressure was initiated by Galileo. In attempting to pipe water to elevated heights by evacuating the air from a tube, he found that it was impossible to sus-

tain a column of water over 34 feet in height. Evangelista Torricelli (1608–1647), who was Galileo's successor as professor of mathematics in Florence, pointed out the difficulty through the invention of a device that exhibited the height of a liquid column to be dependent on atmospheric pressure. This device, a **mercury** (Hg) **barometer,** is still used to measure the atmospheric pressure. A filled tube of mercury was inverted into a pool of mercury. Although some ran out, a column of mercury 30 inches high was left in the tube as illustrated in Fig. 21-5a. A modern version of a mercury barometer is shown in Fig. 21-5b. Since the column of mercury has weight, some force must hold the column up. The only available force is that of the atmospheric pressure on the surface of the mercury pool. This force must equal the weight of the mercury column, and is given by

$$F = mg$$

where m = mass of mercury in the column,
$\quad\quad g$ = acceleration due to gravity.

Mass is related to the density of a liquid by

$$m = \rho V$$

where ρ = density of the liquid,
$\quad V$ = volume.

The volume of the mercury column is

$$V = Ah$$

where A = cross-sectional area of the tube,
$\quad h$ = height of the column.

Then, combining these relations gives

$$F = mg = \rho Vg = \rho gAh$$

Figure 21-5b *A modern mercury barometer. Modern barometers operate on the same principle as Torricelli's barometer (Fig. 21-5a) but are highly refined to give accurate measurements. (Courtesy NOAA.)*

Since pressure is $P = F/A$, the pressure needed to maintain the column, or the pressure of the atmosphere, is

$$P = \frac{F}{A} = \frac{\rho gAh}{A} = \rho gh$$

or

$$P = \rho gh \qquad \text{Eq. 21.2}$$

Hence, knowing the height of the column, the density, and the acceleration due to gravity, the pressure may be calculated. It is still common practice, however, to express the barometric reading in inches of mercury.

The density of mercury is 13.6 g/cm^3 and $g = 980 \text{ cm/s}^2$, hence

$$P = \rho gh = (13.6)(980) \times h \; (cm)$$

One atmosphere of pressure supports a column of mercury approximately 30 inches in height (actually 29.92 inches). Thirty inches is approximately 76 centimeters; thus,

$$P = 13.6 \times 980 \times 76 = 1{,}012{,}928 \text{ dynes/cm}^2$$

This is an approximate value because of the approximate values of the constants used. The pressure of one atmosphere is usually given as

$$
\begin{aligned}
1 \text{ atmosphere (atm)} &= 1{,}013{,}250 \text{ dynes/cm}^2 \\
&= 760 \text{ mm Hg} \\
&= 14.7 \text{ lb/in} \\
&= 30 \text{ in Hg}
\end{aligned}
$$

We are now in a position to understand Galileo's difficulty in not being able to sustain a column of water over 34 feet in height. Any liquid may be used in a barometer; however, the column heights will vary due to the different densities of the liquids. Water has a density of 1 g/cm^3, which is 13.6 times lighter than mercury. The atmospheric pressure then should support a column of water 13.6 times higher than the 30 inches of mercury. Thus,

$$13.6 \times 30 \text{ in} = 408 \text{ in} \times \frac{1 \text{ ft}}{12 \text{ ft}} = 34 \text{ ft}$$

Figure 21-5c *A recording aneroid barometer. The aneroid barometer uses a metal diaphragm sensitive to changes in atmospheric pressure, rather than a liquid as in the Torricellian barometer. With the automatic recording device, a continuous record of air pressure can be obtained. (Courtesy Weather Measure Corporation.)*

Hence, the atmosphere does not have sufficient pressure to support a column of water over 34 feet, just as Galileo and Torricelli found.

Meteorologists use the unit of **millibar** (mb) to record pressure. A bar is defined as 10^6 dynes/cm², and 1 bar = 1000 mb. Since one atmosphere is approximately 10^6 dynes/cm², we have

$$1 \text{ atm} \simeq 1 \text{ bar or } 1000 \text{ mb}$$

and normal atmospheric pressures are of the order of 1000 mb. Actually,

$$1 \text{ atm} = 1013.2 \text{ mb}$$

As would be expected, atmospheric pressure varies with altitude. This was suggested by the French scientist Blaise Pascal (1623–1662) and was proven by his brother-in-law who measured a reduced atmospheric pressure on top of a mountain. Daily atmospheric disturbances also cause the barometric pressure to vary. These variations are continually monitored by weather stations and are important in predicting weather changes as we shall learn.

Since mercury vapor is toxic and a column of 30 inches of mercury awkward to handle, another type of barometer in common use is the **aneroid** (without fluid) **barometer.** This is a mechanical device having a metal diaphragm which is sensitive to pressure (Fig. 21-5c). A pointer and dial arrangement is normally used to indicate the pressure changes on the diaphragm. Aneroid barometers are common around the home and are usually inlaid in wood with a dial thermometer for decorative display on the wall. The altimeters used by airplanes and skydivers are really aneroid barometers. The pressure reading decreases rather uniformly with height in the troposphere. By replacing the barometer dial face with one calibrated inversely in height, the barometer becomes an altimeter.

Atmospheric pressure quickly becomes evident to us when sudden changes are effected. A relatively small change in altitude will cause one's ears to "pop." Airplanes are equipped with pressurized cabins that maintain the normal atmospheric pressure on our bodies. The internal pressure of the body is accustomed to the external pressure of 14.7 lb/in². Should this be reduced, the excess internal pressure may be evidenced in the form of a nose bleed.

People living at high altitudes must become accustomed to the reduced pressure and "thin" air. The body's metabolism changes at higher altitudes as a result of the reduced oxygen content of the air. Criticisms of the 1968 Olympic games held in Mexico City at an elevation 7500 feet were leveled at the effect of reduced pressure and oxygen on the athletes. Another effect of high altitude or lower atmospheric pressure is the reduction of the boiling points of liquids. At 10,000 feet elevation, the boiling point of water is 10°F lower than at sea level, and the cooking time for boiling foods increases accordingly. Cake mixes must be especially prepared with a smaller proportion of "rising" ingredients. The sponginess and rising of a cake is due to the release of gas during baking. A cake which rises a normal amount at sea level may virtually "blow up" at high altitudes due to the decrease in atmospheric pressure.

A frequent use of atmospheric pressure is in drinking through a straw. It is generally believed that sucking on the straw draws the liquid up the straw. Actually, the sucking action reduces the air pressure in the straw by removing the air molecules. The atmospheric pressure on the liquid's surface, external to the straw, then pushes the liquid up the straw. It should be evident from Galileo's problem that a soda straw over 34 feet long would not work.

Humidity

Humidity is a measure of the water vapor in the air. It affects our comfort and indirectly our ambition and state of mind. In the summer many homes have the hum of a dehumidifier which removes moisture from the air, while in the winter, exposed pans of water are strategically placed to allow the water to evaporate into the air. There are several ways to express humidity. **Absolute humidity** is simply the amount of water vapor in a given volume of air and is normally measured in grains per cubic foot in the British system of units (1 lb = 7000 grains). An average value would be on the order of 4.5 gr/ft^3. Another way of expressing humidity is by **partial pressure,** or how much pressure the water vapor in the air contributes to the total atmospheric pressure. By far the most common method is **relative humidity,** which compares the amount of moisture a volume of air contains to the maximum moisture it could contain at a given temperature. Expressed in terms of a percentage, this is defined as

$$(\%)\ RH = \frac{AC}{MC}(\times\ 100\%) \qquad \text{Eq. 21.3}$$

where RH = relative humidity,
  AC = actual moisture content,
  MC = maximum moisture capacity.

The actual moisture content is the absolute humidity, or the amount of water vapor in a given volume of air. The maximum moisture capacity is the maximum amount of water vapor that the volume of air can hold at a given temperature. These quantities are commonly expressed in grains per cubic foot.

Just as a given amount of water at a certain temperature can dissolve so much table salt, a volume of air at a given temperature can hold only so much water vapor. When the maximum amount of salt is dissolved in solution, we say the solution is saturated. The addition of more salt to a saturated solution results in salt on the bottom of the container. However, more salt may be put into solution if the water is heated. Similarly, when air is heated, it can hold more water vapor. That is, warm air has a greater capacity for holding water vapor than does colder air.

Conversely, if the temperature of a nearly saturated salt solution is lowered, at a certain temperature the solution will become saturated. Additional lowering of the temperature will cause the salt to come out of solution as it will be oversaturated. In an analogous fashion, if the temperature of a sample of air is lowered, it will become saturated at some temperature. The temperature to which a sample of air must be cooled to become saturated is called the **dew point.** Cooling below this causes oversaturation and may result in condensation and loss of moisture in the form of precipitation.

Humidity may be measured by several means. **Absorption hygrometers** employ chemicals such as $CaCl_2$ or $Ca(NO_3)_2$ which absorb water vapor from the air and may be analyzed quantitatively. Other chemicals give a qualitative indication of the humidity by a change in some physical characteristic such as color. **Hair hygrometers** make use of the fact that human hair expands about 2.5 percent over the range of 0–100 percent humidity. This expansion may be measured mechanically and calibrated in terms of humidity. See Fig. 21-5d. The most common method of measuring humidity uses the **psychrometer.** This

Figure 21-5d *A hair hygrometer, recording type. The variation in the length of human hair due to absorbed moisture is used to measure humidity. The hair filaments of the hygrometer are at the tip of the thumb. The chart on the rotating drum at the left permits continuous monitoring of the humidity. (Courtesy NOAA.)*

instrument consists of two thermometers, one of which measures the air temperature while the other has its bulb surrounded by a wick which is kept wet. These thermometers are referred to as the dry bulb and wet bulb, respectively. They may be simply mounted as in Fig. 21-5e or incorporated in a **sling psychrometer.** By rotating the sling psychrometer (Fig. 21-5f) air is passed over the wet bulb and readings may be obtained more quickly than from the wall type which comes slowly to equilibrium. The dry bulb measures the air temperature while the wet bulb has a lower reading which is a function of the amount of water vapor in the air.

Consider the human body's cooling mechanism momentarily. The body perspires and the evaporation of the perspiration causes cooling due to the removal of heat from the body to change the liquid perspiration into vapor, that is, the latent heat of vaporization (Section 7-5). If the humidity is high and the air contains a lot of water vapor, it is difficult for evaporation from the body to take place. As a result, we say it is a hot and muggy day. On the other hand, if the humidity is low, evaporation can occur easily. We then feel cool and comfortable. The wet bulb of the psychrometer works in a similar fashion. If the

humidity is high, little water is evaporated and the wet bulb is only slightly cooled. Consequently, the temperature of the wet bulb is only slightly lower than the temperature of the dry bulb, or its reading is slightly depressed. If, however, the humidity is low, a great deal of evaporation will occur accompanied by considerable cooling, and the wet bulb will be considerably depressed. Hence the temperature difference of the thermometers, or the depression of the wet bulb, is a measure of the humidity.

The relative humidity, the maximum moisture capacity, and also the dew point which depends on the moisture content and hence the relative humidity, may be read directly from Tables 1 and 2, Appendix III, when the psychrometer readings are known. Consider an example in which the dry bulb has a reading of 80°F and the wet bulb a reading of 73°F. The depression of the wet bulb is then

$$80° - 73° = 7°$$

Finding the dry bulb temperature in the first column of Table 1, the maximum moisture capacity for that temperature is given in the adjacent column, in this case 10.9 gr/ft³. The relative humidity is obtained from Table 1 by finding the depression of the wet

Figure 21-5e *A psychrometer for relative humidity measurements. The dry bulb records the air temperature, while the wet bulb, kept moist by a cloth wick from the water reservoir, registers the lower temperature of the wick which is due to moisture evaporation. The evaporation depends on the humidity. Knowing the air temperature and the difference in the thermometer readings, one can obtain the relative humidity from reference tables.*

point (70°F) as can be seen from Table 1. At the dew point the air is saturated and the relative humidity is 100 percent. The actual moisture capacity is then the maximum capacity.

To visualize these concepts, consider a container of water which is partially full. The sides of this particular container are quite marvelous, inasmuch as they shrink when the temperature decreases. At 80°F the container has 7.8 g of water but is capable of holding 10.9 g. Hence, the container is $\frac{7.8}{10.9} \times 100 = 72$

Figure 21-5f *A sling psychrometer. The thermometers are whirled in a sling device to promote evaporation and thus the relative humidity can be determined more quickly. The wet bulb is covered with a tubular piece of cloth that is moistened. (Courtesy NOAA.)*

bulb in the top row of the table and coming down that column to the row which corresponds to the dry bulb reading. Doing this, the relative humidity is found to be 72 percent for the above depression. The dew point temperature is obtained by the same procedure from Table 2, Appendix III, and is 70°F. With this information, the actual moisture content may be calculated, using Eq. 21.3

$$RH = \frac{AC}{MC}$$

where the relative humidity is expressed in decimal form; or

$$AC = RH \times MC = 0.72 \times 10.9 = 7.8 \text{ gr/ft}^3$$

This is the actual moisture content and should correspond to the maximum moisture capacity at the dew

Figure 21-5g *A diagram illustrating the container analogy of relative humidity. At 80°F the beaker can hold 10.9 grains of water. However, if the sides of the container were to shrink with temperature so that at 70°F the beaker's capacity were 7.8 grains, it would be 100% full (saturated). Analogously, a cubic foot of air at 80°F is capable of "holding" 10.9 grains of water vapor; but if it holds only 7.8 grains, its relative humidity (fullness) is 72%. Should the temperature of the air be decreased to 70°F (dew point), the relative humidity is then 100%.*

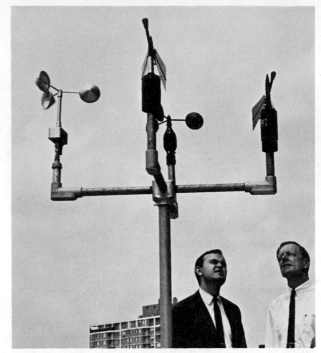

Figure 21-5h *An array of anemometers and wind vanes. The anemometers record the wind speed while the wind vanes line up with the wind to indicate the direction from which it comes. The instruments in the photo are electronically monitored. (Courtesy NOAA.)*

percent full, as shown in Fig. 21-5g. But if the temperature is lowered to 70°F, the sides of the container shrink so that it can hold only 7.8 gr. At this special temperature (dew point), the container is full (or saturated) and it is 100 percent full. Should the temperature be lowered still further and the sides shrink further, the water would overflow. In the analogous situation, the air generally loses its moisture in the form of precipitation when cooled below the dew point.

Wind Speed and Direction

Wind speed is measured by an **anemometer.** This instrument consists of three or four cups each attached to a rod that is free to rotate. The cups catch the wind, and the greater the wind speed, the faster the anemometer rotates. A **wind vane** indicates the direction from which the wind is blowing. This instrument is simply a freely rotating indicator which, because of its shape, lines up with the wind and points the wind direction. An array of anemometers and wind vanes is shown in Fig. 21-5h. These instruments are common at airports where wind speed and direction are important. Both the direction and a rough estimate of the wind's speed are indicated by the airport's "wind sock," which is a type of wind vane. A rotating canvas "sock" which catches the wind can usually be seen from small aircraft or from the ground; the way it points indicates the direction in which the wind is blowing. The sock's fullness provides a wind speed estimate. Wind vanes designed around a metal rooster were once quite popular in rural areas. Found usually atop the barn, they are commonly called weather vanes.

Precipitation

The major forms of precipitation are rain and snow. Rainfall is measured by a **rain gauge.** This may be simply a container with vertical sides and graduated in inches placed in an open area. After a rainfall, the rain gauge is read, and the amount of precipitation is reported as so many inches of rain. It is assumed that this much rainfall is distributed evenly over the surrounding area. If the precipitation is in the form of snow, the depth of snow (where not drifted) is reported in inches. The actual amount of water received depends on the density of the snow. To record the water amount, the snow is melted and the actual amount of water is reported in inches. More elaborate rain measuring instruments automatically measure and record the rain or snowfall.

IMPORTANT WORDS AND TERMS

Atmosphere	Ionosphere	Partial pressure
Meteorology	Ozone	Relative humidity
Photosynthesis	Aurora borealis	Dew point
Troposphere	Aurora australis	Absorption
Weather	Insolation	Hygrometer
Stratosphere	Solar constant	Hair hygrometer
Mesosphere	Albedo	Psychrometer
Thermosphere	Rayleigh scattering	Sling psychrometer
Tropopause	Greenhouse effect	Wind speed
Stratopause	Thermometer	Wind direction
Mesopause	Mercury barometer	Anemometer
Homosphere	Millibar	Wind vanes
Heterosphere	Aneroid barometer	Rain gauge
Ozonosphere	Absolute humidity	

LEARNING OBJECTIVES

After reading and studying this chapter, you should be able to do the following without referring to the text:

1. State the nature and general composition of air.
2. Explain why some planets have atmospheres and others do not.
3. Explain how the oxygen–carbon dioxide balance of the atmosphere is maintained.
4. Distinguish the properties that give rise to vertical divisions of the atmosphere.
5. State the divisions of the atmosphere based on temperature.
6. Explain the distribution of insolation.
7. State and explain the processes by which the lower atmosphere obtains the major portion of its energy content.
8. Explain the greenhouse effect.
9. Name the atmospheric properties that are commonly measured and describe how they are measured.

QUESTIONS

1. What is meteorology and what does the word mean literally?
2. What is the composition of the air you breathe?
3. Humans inhale oxygen and expel carbon dioxide. How is the relatively constant oxygen content of the atmosphere maintained?
4. Why do the earth and the other planets have atmospheres, while Mercury and the moon do not?
5. Describe how the temperature of the atmosphere varies in each of the following regions: (a) the troposphere, (b) the stratosphere, (c) the mesosphere, and (d) the thermosphere.
6. What is the basic distinction between the homosphere and the heterosphere?
7. How were the ozone and ion layers detected and investigated?
8. Of what importance is the atmospheric ozone layer?
9. What is believed to be the cause of the displays of light called auroras?
10. What is meant by insolation and what is the solar constant?
11. From what source does the atmosphere receive most of its *direct* heating, and how is this accomplished (three methods)?
12. Why is the sky blue?
13. Explain what is meant by the greenhouse effect.
14. How does the selective absorption of atmospheric gases provide a thermostatic action for the earth?
15. Explain why the temperature of the troposphere decreases with altitude and why the temperature of the stratosphere increases.
16. What is the principle of the liquid barometer and what is the height of the barometer column for one atmosphere of pressure?
17. At high altitudes, would boiled food have to be cooked a greater or shorter amount of time than at sea level? Why? Explain the principle of a pressure cooker.
18. Why does water condense on the outside of glass containing an iced drink?
19. How is it possible that it may be raining, yet the relative humidity is reported to be less than 100 percent? Explain the principle of the psychrometer.
20. Which way, relative to the wind direction, does a finned wind vane point?

PROBLEMS

1. On a plot of the atmosphere with altitude, locate the heights of the following:
   (a) The top of Mt. Everest—29,000 ft
   (b) Commercial airline flights—approximately 30,000 ft
   (c) Meteor trails—250,000 ft
   (d) Communication satellites—40 mi
   (e) Supersonic transports (SST)—65,000 ft
   (f) The E and F ion layers
   (g) Auroral displays—300 mi

2. A satellite in orbit around the earth near the top of the atmosphere has an area of 2 m² exposed to the sun. Assuming all of the incident radiation is absorbed, how much heat energy would it receive in one hour?

Answer: 2.4×10^6 cal

3. For a proposed solar power generating system, solar collectors would cover an area equal to a square 75 mi on a side in a southwest desert area. How much heat energy would be received by insolation on this area in one hour? (Recall only 50 percent of the total insolation reaches the earth's surface.)

4. A radio wave is directed vertically upward. The time lapse for its return is 7.0×10^{-4} s. From what ionic layer was the wave reflected? (Hint: Electromagnetic radio waves travel at the speed of light, 186,000 mi/s.)

5. A mercury barometer has a column height of 75 cm. What is the pressure in (a) dynes/cm²? (b) millibars?

Answer: (a) 999,600 dyne/cm² (b) 999.6 mb

6. What would be the height of the barometer in the preceding problem if a liquid with a density of 3.4 g/cm³ were used instead of mercury?

7. A mercury barometer has a column height of 29 in.
   (a) What is the pressure? (Hint: The pressure may be easily calculated in lb/in² by direct ratio.)
   (b) The palm of a hand is about 3 in by 4 in. How many pounds of force are there on a palm as figured from the barometric reading in part (a)?

Answer: (a) 14.2 lb/in² (b) 170 lb

8. On a day when the air temperature is 90°F, the wet bulb of a psychrometer has a reading of 86°F. Find each of the following:
   (a) Relative humidity
   (b) Dew point
   (c) Maximum moisture capacity of the air
   (d) Actual moisture content of the air

Answer: (a) 85% (b) 85°F (c) 14.8 gr/ft³ (d) 12.6 gr/ft³

9. A psychrometer has a dry-bulb reading of 75°F and a wet-bulb reading of 60°F. Find each of the quantities asked for in problem 8.

10. On a day when the air temperature is 80°F and the relative humidity is 70 percent, what is the actual moisture content of the air?

11. On a winter day, a psychrometer has a wet-bulb reading of 35°F and a dry-bulb reading of 45°F.
    (a) What is the actual moisture content of the air?
    (b) How many degrees would the air temperature have to be lowered for the actual moisture content to be equal to the maximum moisture capacity (i.e. 100 percent relative humidity)?

12. (a) How much moisture is there in one cubic mile of air (a very small volume compared to the total atmosphere), if the psychrometer reading for this volume is 75°F and 68°F for the dry and wet bulbs, respectively? (Hint: 1 mi³ = 1.5×10^{11} ft³.)
    (b) Using 7000 gr = 1 lb and the weight of one gallon of water = 8.3 lb, calculate the number of gallons of water in one cubic mile of air.

Answer: (a) 9.9×10^{11} gr (b) 1.7×10^{10} gal

Chapter 22

WINDS

Wind is the horizontal motion of air. Vertical air motions are referred to as updrafts or downdrafts, or collectively as **air currents.** Air movement mixes the gases of the troposphere and affects our weather and climate. In relation to the movement of large masses of air which will be considered in Chapter 24, winds are the agents of transportation which redistribute the energy of the lower atmosphere and bring the warming influences of spring and summer and cold chill of fall and winter. Their presence may be welcomed or abhorred. Winds also cause erosion of the earth's surface and influence ocean currents, as will be discussed in the geology part of this text.

The effects of winds on our environment are often overlooked. Winds and air currents enter into the life cycles of many plants and animals. Pollen carried by the winds is fundamental to nature. Animals sniff the wind for scents of danger, while the transportation of radioactive particles and other contaminants may result in environmental problems. Before the invention of the combustion engine, people harnessed the wind for use in transportation and developed devices that put the wind to work. Occasionally, the wind causes death and destruction. Certain winds have taken on literary personalities such as the fierce North wind, the gentle Trade wind, and the brave West wind.

In general, we are only familiar with the winds near the surface of the earth. But winds also occur in the upper troposphere and are known as jet streams. They too influence our weather. Due to the vastness of the earth and its atmosphere, the winds seem boundless and the descriptive phrase of being "as free as the wind" is often used. However, since the air is a mixture of gases, its behavior is governed by the gas laws of Chapter 7 and other physical principles. Using these, it is possible to gain some understanding of the winds and their behavior, which is essential in meteorology.

22-1
FORCES PRODUCING AIR MOTION

As in all dynamic situations, forces are necessary to produce motion and changes in motion. The gases of the atmosphere are subject to two primary forces arising from gravity and pressure differences due to temperature variations. The force due to gravity is vertically downward and acts on each gas molecule. Although this force is often overruled by forces in other directions, its downward component is ever-present, which accounts for the greater density of air near the earth.

The pressure of a gas is directly proportional to its temperature. As defined in Section 21-5, **pressure** is force per unit area ($P = F/A$). Thus, a temperature variation in air gives rise to a difference in pressure (force). It is the difference in pressure resulting from temperature differences in the atmosphere which causes air movement. A pressure difference (ΔP) corresponds to an unbalanced force; that is,

$$\Delta P = P_2 - P_1$$

but,

$$P = \frac{F}{A}$$

Thus, assuming equal areas

$$\Delta P = P_2 - P_1 = \frac{F_2}{A} - \frac{F_1}{A}$$

or

$$\Delta P = \frac{F_2 - F_1}{A}$$

From the equation it can be seen that if there is no pressure difference, $\Delta P = 0$; that is, $P_2 = P_1$, then $F_2 = F_1$, and there would be no movement, assuming the air is initially at rest. When there is a pressure difference, the air moves from a high-pressure region to a low-pressure region.

The pressure over a geographical area may be mapped by taking barometric readings at different locations. A line drawn through points (locations) of equal pressure is called an **isobar.** Since all the points on the isobar are of equal pressure, there will be no air movement along an isobar. The wind direction will be at right angles to the isobars in the direction of the lower pressure. This is illustrated in Fig. 22-1a. Topographically, the isobar is analogous to a line of equal elevation. The isobars of greater value correspond to higher elevations. Just as a ball would move down a hill to a lower elevation, the air moves down the pressure hill, or **pressure gradient,** toward a lower isobar. An isobar of minimum pressure represents the bottom of a pressure valley. This region is referred to as a **low-pressure trough.** However, it should be remembered that the isobars in Fig. 22-1a represent the pressure readings at different locations. The vertical analogy is meant only as an aid in understanding the air motion along a surface of reasonably constant elevation.

The pressure and volume of a gas are related to its absolute temperature by the ideal gas law (cf. Chapter 7)

$$PV = kT \qquad \text{Eq. 22.1}$$

where P = pressure,
 V = volume,
 k = Boltzmann's constant,
 T = temperature.

Thus, a change in temperature causes a change in pressure and volume or density ($\rho = m/V$). Regions of the atmosphere with different temperatures have different air pressures and densities. As a result, localized heating sets up air motion in a **convection cycle** and is referred to as **thermal circulation.**

Figure 22-1a *Representative isobars. Isobars are lines drawn through locations having equal atmospheric pressures. The air motion, or wind direction, is at right angles to the isobars and from a region of high pressure* (upper isobars) *to a region of low pressure* (lower isobars).

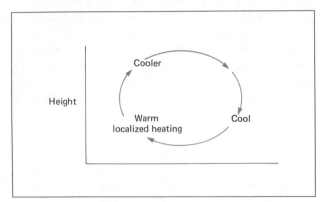

Figure 22-1b *A diagram illustrating thermal circulation of air. Localized heating, which causes the air in the region to rise, initiates the circulation. As the warm air rises and cools, cool air near the ground moves horizontally into the region vacated by the rising air. The upper, still cooler air then descends to occupy the region vacated by the cool air. These motions set up a convection cycle.*

To understand this, consider the lower left corner of the cycle in Fig. 22-1b. A warm land area may heat the immediate air by conduction. The warm air expands, becomes less dense and more buoyant, and rises, creating an updraft. Warm air can be "seen" rising from a blacktop road in the summer, for exam-

ple. The "heat waves" are actually the result of the change in the air's density and the resulting change in the index of refraction, which allows visual detection. The buoyant warm air rises and cools as it expands. What happens next can best be visualized if we think of the region which contained the warm air as a "pocket." As the warm air expands and rises from this pocket, the pressure in the pocket is lowered and cooler air moves horizontally toward this region of lower pressure. The space having been vacated by this horizontally moving cool air, then, is filled by cool air descending to occupy the void. These motions set up the convection cycle and thermal circulation of air as illustrated in Fig. 22-1b. Thermal circulation is an important atmospheric mechanism as we shall see.

Once the air has been put into motion, velocity-dependent forces act on the wind. These secondary forces are the Coriolis force and friction. The **Coriolis force,** named after the French engineer and mathematician who first described it, arises from the fact that on the earth an observer is in a rotating frame of reference. It is sometimes referred to as a pseudo-force, since it is introduced to account for the effect of the earth's rotation. Humans tend to be egocentric. We commonly consider ourselves motionless although we are on a rotating earth which has a surface velocity of 1000 mi/h near the equator. Newton's laws of motion apply to a fixed or inertial reference frame and may be used for ordinary motions on the earth without correction. However, for high velocities or huge masses such as the atmospheric gases, the correction for the earth's rotation becomes important.

To visualize this effect, consider an observer standing at the center O of a rotating turntable with an angular velocity ω, as illustrated in Fig. 22-1c. A turntable is taken for two-dimensional simplicity, although the observer at O is analogous to an observer at the north pole. A projectile is shot from O at a target X on the perimeter of the turntable. To an inertial observer not in the rotating system, during the time of projectile flight, the target would rotate from in front of the straight-path projectile while

Projectile
path

Initially   At impact

Inertial observer

Observer's line of sight

Apparent
projectile
path

Observer in the rotating frame

Figure 22-1c *An illustration of different reference frames. The upper drawing illustrates the motion of a projectile origi- nally fired at X as seen by an inertial observer—one not on the rotating disk, such as the reader of this text. The projectile path is seen by this observer as a straight line, in accordance with Newton's laws, and Y rotates into the projectile path. The lower drawing illustrates the motion of the projectile as seen by an observer on the rotating disk, at position O. The projec- tile appears to be deflected from its original path toward Y, apparently by a Coriolis force.*

the surface of the rotating earth at lower latitudes experience greater surface velocities than those at greater latitudes. In Fig. 22-1d, object *A* is at a greater latitude than object *B*. Since both objects follow their respective paths of rotation once every 24 hours, *B* must experience a greater surface velocity than *A* since it covers a greater distance in the same amount of time.

Consider now particle motion southward in the northern hemisphere as shown in the upper drawing of Fig. 22-1e. A particle initially at *A* is projected along the meridian connecting *A* and *B*. The particle has the same surface velocity as point *A* and will fall behind the moving meridian as it (the particle) ap- proaches *B*, since the earth's surface has a greater velocity in this region. To an observer at *A* who ro- tates with the meridian and always looks southward along it, the particle will appear to be deflected to the right. A similar analysis shows that a particle moving northward from *B* to *A* is also deflected to the right, as illustrated in the lower drawing of Fig. 22-1e. In

another object *Y* to the right of the target would rotate in front of the projectile and be struck. The observer at *O* moves with the turntable, so that the target *X* always is in the same position relative to his line of sight. It would appear to this observer that the projectile was deflected to the right in order to hit *Y* as shown in the lower drawing of Fig. 22-1c. He would reason, therefore, that according to Newton's laws some force must have acted to the right causing the deflection. This is the pseudo-Coriolis force, which is needed to explain the observations made in the rotating system.

To understand this effect of the earth for observ- ers not at the poles, it must be realized that objects on

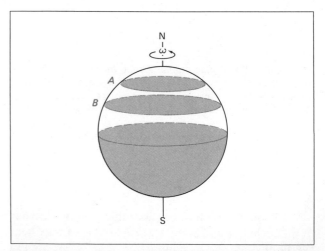

Figure 22-1d *Variations of velocity of the earth's surface with latitude. During one 24-hour rotation of the earth, B will travel a longer distance than A. Thus, B has a greater surface velocity than A.*

Figure 22-1e *The Coriolis deflection for particle motion in the northern hemisphere. The upper drawing shows that because B has a greater surface velocity than A, a particle traveling southward with the same surface velocity as A along the meridian of A and B will fall behind B. To an observer at A, the particle is apparently deflected to the right, viewed along the direction of particle motion. Similarly for particle motion northward from B to A, the apparent deflection is to the right.*

motion. The direction of the Coriolis force is always normal, or at a right angle, to the motion of the object and its magnitude varies with latitude; it is zero at the equator and becomes greater towards the poles.

Consider this effect on wind motion. Initially, air moves toward a low-pressure area (low) and away from a high-pressure area (high). Because of the Coriolis force the wind is deflected, and in the northern hemisphere the wind rotates counterclockwise around a low and clockwise around a high. These wind motions are referred to as **cyclone** and **anticyclone,** respectively. The rotations are reversed in the southern hemisphere. These circulations for the northern hemisphere are shown in Fig. 22-1f.

The wind motion around highs and lows is sometimes stated as Buys Ballot's law after the Dutch

this case, the particle has the same surface velocity as point *B* and will move ahead of the moving meridian as it approaches *A*, which has a smaller surface velocity. To an observer at *B* it would again appear that the particle was deflected to the right.

By the same reasoning, it can be shown that moving objects in the southern hemisphere are deflected to the left in the above cases. East-west motions are more difficult to analyze and require mathematics beyond the scope of this text. The results, however, are the same as above. Hence, due to the Coriolis force, moving objects are deflected to the right in the northern hemisphere and to the left in the southern hemisphere as observed along their lines of

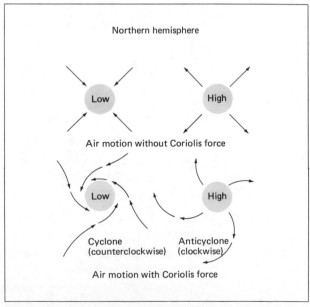

Figure 22-1f *The effect of the Coriolis force on air motion. In the northern hemisphere, the Coriolis deflection to the right produces counterclockwise motion of air around a low and clockwise rotation around a high. The situation is reversed in the southern hemisphere.*

meteorologist who first published it: "Standing in the northern hemisphere with your back to the wind, the high is on the right and the low on the left and reversed in the southern hemisphere." However, it is usually easier to remember the direction of the wind motion by understanding the effect of the Coriolis force. It should also be noted that water motion or currents in the oceans also are affected by the Coriolis force.

Friction, or drag, can also cause the retardation or deflection of air movement. Just as a liquid has a flow resistance caused by the internal friction of its molecules, moving air molecules experience frictional interaction among themselves or with terrestrial surfaces. The opposing frictional force along a surface is in the opposite direction of the air motion and its magnitude is proportional to the air speed. As a result, winds moving into a cyclonic disturbance may be deflected in the direction opposite the direction of rotation.

22-2
LOCAL WINDS

Air moves in all directions. Vertical air motion is important in cloud formation and precipitation, as will be discussed in Chapter 23. As you know, horizontal air movements near the surface of the earth produce the winds which "blow," causing their pres-

Table 22-2a *Wind Speeds*

Average Speed (mph)	Signs	Class	Meteorological Symbol*
0	Smoke rises vertically	Calm	
1–3	Smoke deflected	Light air	
4–7	Leaves rustle	Slight breeze	
8–12	Leaves and twigs move	Gentle breeze	
13–18	Small branches move, dust rises	Moderate breeze	
19–24	Small trees sway	Fresh breeze	
25–31	Large branches move, telephone lines hum	Strong breeze	
32–38	Trees in motion	High wind	
39–46	Twigs broken, walking impaired	Fresh gale	
47–54	Slight damage; e.g., to roofs and TV antennas	Strong gale	
55–63	Trees uprooted, much damage	Whole gale	
64–74	Widespread damage	Storm	
>74	Tropical storm or blizzard	Hurricane	

*Each full flag represents 5 knots (nautical miles per hour). On a weather map, the flag staff points in the direction from which the wind comes.

Table 22-2b *Determining the Wind-Chill Index*[a]

Windspeed (miles per hour)	35	30	25	20	15	10	5	0	−5	−10	−15	−20	−25	−30	−35	−40	−45
Calm	35	30	25	20	15	10	5	0	−5	−10	−15	−20	−25	−30	−35	−40	−45
5	33	27	21	16	12	7	1	−6	−11	−15	−20	−26	−31	−35	−41	−47	−54
10	21	16	9	2	−2	−9	−15	−22	−27	−31	−38	−45	−52	−58	−64	−70	−77
15	16	11	1	−6	−11	−18	−25	−33	−40	−45	−51	−60	−65	−70	−78	−85	−90
20	12	3	−4	−9	−17	−24	−32	−40	−46	−52	−60	−68	−76	−81	−88	−96	−103
25	7	0	−7	−15	−22	−29	−37	−45	−52	−58	−67	−75	−83	−89	−96	−104	−112
30	5	−2	−11	−18	−26	−33	−41	−49	−56	−63	−70	−78	−87	−94	−101	−109	−117
35	3	−4	−13	−20	−27	−35	−43	−52	−60	−67	−72	−83	−90	−98	−105	−113	−123
40	1	−4	−15	−22	−29	−36	−45	−54	−62	−69	−76	−87	−94	−101	−107	−116	−128
45	1	−6	−17	−24	−31	−38	−46	−54	−63	−70	−78	−87	−94	−101	−108	−118	−128
50	0	−7	−17	−24	−31	−38	−47	−56	−63	−70	−79	−88	−96	−103	−110	−120	−128

VERY COLD — BITTER COLD — EXTREME COLD

Equivalent Temperatures (°F)

[a] Find the temperature in the top row and the wind speed in the left column. Read the chill factor where the corresponding column and row intersect. For example, a calm air temperature of 10°F and a wind speed of 20 mi/h together are equivalent in cooling effect to −24°F.

ence to be felt and noticed. A wind is named after the direction or region *from* which it comes. For example, a wind blowing from north to south is referred to as a north wind. A wind blowing out to sea from the land is called a land breeze. The strength of the wind is determined by its speed. As previously mentioned, this is measured by an anemometer; however, an estimate of the wind speed may be obtained from Table 22-2a.

Wind also plays an important factor in how cold we feel. At moderately low temperatures in a brisk wind we may feel extremely cold. The wind promotes the loss of body heat, which adds to the chilling effect.

For example, more body heat may be lost when the temperature is 20°F and the wind speed is 18 mph than when the temperature is 0°F and there is no wind. Thus, the temperature alone does not tell us how we should dress when going out of doors. The effect of wind on how cold we feel may be expressed by a wind-chill index, commonly called the **chill factor.** This not to say that the chill factor indicates how "chilly" one would feel, because our feeling depends on other things besides wind and temperature, such as state of nourishment, individual metabolism, and protective clothing. However, the wind-chill index in Table 22-2b, which describes the cooling power of

Figure 22-2a *Daily convection cycles of air over land and water. During the day the land heats more quickly than water and a convection cycle is set up as illustrated in the upper drawing. At night the land cools more quickly than the water and the cycle is reversed (lower drawing).*

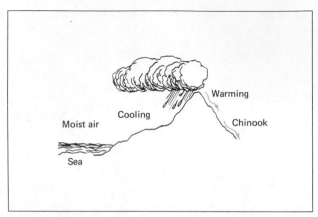

Figure 22-2b *Topography for chinook formation. After the ascending air loses moisture on the windward side of the mountain, the descending dry air is warmed on the leeward side of the mountain.*

wind on exposed flesh at various temperatures, is a good guide as to what clothing would be needed for protection on a cold windy day.

Atmospheric pressure differences involved in thermal circulations due to geographical features give rise to local winds. As discussed in Chapter 21, land areas heat up more quickly than do water areas. The heating of the land area gives rise to a convection cycle. As a result, during the day when the land is warmer than the water, a lake or **sea breeze** is experience as shown in Fig. 22-2a. At night the land loses its heat more quickly than the water and the air over the water is warmer. The convection cycle is then reversed and at night **land breezes** blow. Sea and land breezes are sometimes referred to respectively as **onshore** and **offshore winds.**

This local heating effect is also experienced in mountain valleys. During the day the air in contact with the mountain slopes heats up and rises, producing a **valley breeze.** At night the slopes cool faster than the air at the same elevation over the valley, and the cool air descends the slopes, giving rise to a **mountain breeze.**

The heating effect which produces thermal circulation also applies on a large scale. Seasonal heating and cooling of large continental land masses initiates convection cycles. In the summer, the air over the heated continents rises, creating low-pressure areas into which air flows, while in winter, high-pressure areas exist over the continents and air flows outward toward the warmer low-pressure areas over the oceans. The winds of these cycles are most pronounced on the Asian continent and are called **monsoons** (Arabic *mausin*, season). During the summer, sea air moves toward the heated continent and is called the **summer monsoon.** As the moist sea air travels inland toward the mountains, precipitation occurs. The sea wind or summer monsoon is associated with the wet season in southern and eastern Asia. In the winter the cycle is reversed, and then a cold dry wind blows from the mountains toward the sea. This prevailing land wind is called the **winter monsoon.**

Other local winds depend on the local geography and acquire names particular to that area. An example of such a wind is the **chinook,** or "snow-eater," experienced on the eastward slopes of the Rocky Mountains. Similar winds are called **föhns** in the Alps. The winds from the Pacific Ocean blow toward and up the westward slopes of the Rockies. Since the temperature decreases with altitude, much of the moisture is lost as rain or snow on the westward slopes and near the top of the mountains. As a result, dry air descends the eastern slopes becoming warm with descent. This chinook wind of warm dry air may cause a rapid rise in temperature and a quick melting of the winter snow, hence the name "snow eater" (Fig. 22-2b).

Figure 22-3a *A diagram showing global air circulation if the earth were nonrotating and had a uniform surface. The air motion would experience no Coriolis force and each hemisphere would have a single thermal cycle extending from the equator to the pole.*

22-3
WORLD CIRCULATION

Air motion changes locally with altitude, the geographical features, and the seasons. However, the earth does possess a general circulation pattern. If the earth were completely covered by water or land and did not rotate, convection cycles would circulate the surface air from the cold polar regions toward the equator. If this were the case, the prevailing winds would always assume a general north-south direction, as shown in Fig. 22-3a.

However, because of the Coriolis force, land and sea variations over the earth's surface, and other complicated reasons, the hemispheric circulation is broken up into six general convectional cycles, or pressure cells. The earth's general circulation struc-

ture is shown in Fig. 22-3b. Many local variations occur within the cells, which shift seasonally in latitude due to variations in insolation. However, the prevailing winds of this semipermanent circulation structure are important in influencing general weather movement around the world.

A low-pressure belt exists in the equatorial region due to rising warm air. Direct insolation heats this region and the air motion for the most part is rising air currents of hot, humid air. As a result, there is a vast equatorial low-pressure zone. The air cools as it rises and this zone is characterized by massive vertical clouds and frequent rain. The heavy rainfall over the land areas in this warm region gives rise to the tropical rain forests that are common near the equator.

The surface winds in the equatorial low-pressure zone are light and variable, and often calm. These latitudes are referred to as the **doldrums.** Early sailing ships were often becalmed in the doldrums and

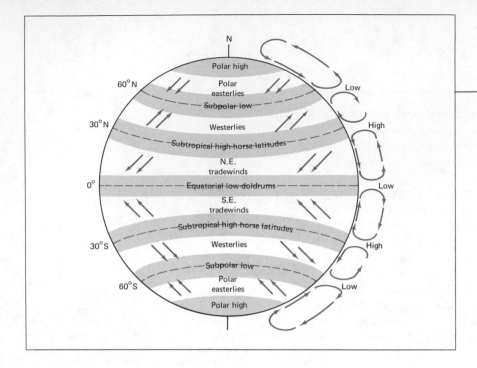

Figure 22-3b *The earth's general air circulation structure. For rather complicated reasons, the earth's general circulation structure has six convection cycles.*

floated listlessly due to lack of wind. We use the expression of "being in the doldrums" to describe a feeling or mood of bored listlessness.

The air rising from the equatorial low divides and travels horizontally north and south at high altitudes. At about 30° N and 30° S latitudes, the air loses its momentum, cools, and descends toward the surface. Hence, the subtropical regions near 30° N and 30° S are characterized by descending air, which makes them high pressure zones. There is little horizontal surface air movement. These regions became known as the **horse latitudes,** due to the fact that early Spanish sailing ships were becalmed in these latitudes and their cargoes of horses had to be eaten for food or thrown overboard to conserve fresh water. These high-pressure regions of descending, warming air are the locations of some of the world's great deserts, e.g., the Sahara in north Africa and the Kalahari in southwest Africa.

The air descending in the horse latitudes moves toward the equator, completing the thermal-convection cycle (see Fig. 22-3b). The surface winds between the horse latitudes and the doldrums are known as the **trade winds.** These winds are generally regular and steady, and were of considerable importance in establishing the early trade routes of sailing ships. Due to the Coriolis force, the trade winds blow in general from the northeast in the northern hemisphere and from the southeast in the southern hemisphere.

We might expect another single thermal circulation cell between the horse latitudes and the poles. However, for rather complicated reasons, which we will not consider, there are two pressure cells. One of these is between the regions of subtropical highs of the horse latitudes (30° N and 30° S) and subpolar lows near 60° N and 60° S. In the latitudes between 30° and 60°, the **westerlies** prevail. The general directionality of these surface winds is again due to the Coriolis force. These west winds blew the early sailing ships back across the Atlantic to Europe. The homeward wind was christened the "brave West wind." Voyages to the New World were made with the help of the gentle northeast trade winds. Care was taken to

Figure 22-3c *Early sailing routes. Exploration and trade routes were dependent on the prevailing wind zones.*

avoid the horse latitude region near 30° N, which is the latitude of the southeastern United States, by keeping within the northeast trade wind zone. It can now be understood why Columbus in sailing from Spain arrived in the West Indies and why Spanish and Portuguese explorations were done in Central and South America.

The other sailing route to America was the northern route with the aid of the **polar easterlies.** These are the surface winds of the pressure cell between the subpolar lows at 60° latitude and the polar highs. By use of this route, the Norse, French, and English explorations were made along the northeastern coast of Canada and the United States. The southeast trade winds helped promote slavery in America by providing a convenient sailing route from Africa. These early sailing routes are illustrated in Fig. 22-3c. History might have been quite different if the Coriolis effect were reversed for the hemispheres.

In the region of 40–50° latitude in the southern hemisphere, there are large stretches of ocean. With no land masses to cause variation, the westerlies gather momentum, becoming strong and blustery until they reach gale force. The winds in this region are sometimes referred to as "the Roaring Forties" and "the Howling Fifties."

The general wind circulation has a major influence on climate. Prevailing winds affect the ocean currents, which redistribute the ocean's heat energy in much the same way as do winds in the atmosphere. There is still much more to be learned about the general circulation of the air, since its movement is sparsely monitored over many regions of the earth.

The extent of the circulation patterns was widely noticed during early nuclear test periods when the prevailing winds carried radioactive fallout around the world. Food supplies began to show unnaturally high concentrations of radioactive materials. Of chief concern was milk and its strontium-90 concentration, which could potentially affect the consumer, especially infants. (See Section 9-5.) Indirectly, therefore, the earth's general air circulation was responsible for the resulting Nuclear Test-Ban Treaty which limited nuclear testing to underground explosions.

22-4
JET STREAMS

In the upper troposphere there exist fast moving "rivers" of air called **jet streams.** The boundaries of the jet streams appear to be reasonably well defined as they move in the atmosphere in a complex fashion. They were first noted in the 1930s, but did not receive much attention until a decade later, during World War II. Pilots reported that their aircraft had been held motionless by an "invisible hand" while at full throttle. The invisible hand was really a jet-stream headwind. Speed records were set with the help of jet-stream tailwinds. The Japanese also knew of the jet streams. Near the end of the war they launched incendiary balloons in the jet stream blowing toward the United States across the Pacific Ocean. The balloons were widely scattered, however, and only a small percentage reached the west coast, doing little damage.

The behavior of the jet streams is variable and not well understood. They are known to vary in altitude and latitude with the seasons. In the winter they are found at altitudes of four to six miles and in the region of 30° latitude. In the summer the altitude increases to above six miles and the general location is 50° latitude. The wind speeds of the jet streams vary. They move well over 100 mph and have been measured in excess of 300 mph. Their general direction is mainly from west to east, but shows considerable variation due to deflection by the Coriolis force, as well as by frictional forces.

Jet streams are thought to result from the general circulation structure in the regions where great high- and low-pressure areas meet. Under the proper conditions, the additive circulation of a high and low could produce winds of exceedingly high velocity. The jet streams are relatively recent to meteorological observation, and their effect on the weather is currently somewhat vague. Variations in their seasonal migrations have been associated with the severity of recent winters. It has also, been speculated that intersecting jet streams may contribute to the formation of destructive vortices of winds we call tornadoes (Section 24-3). As more is learned of their behavior, it is possible that these rivers of air in our atmosphere may become routes of air transportation much as the rivers of the earth—especially downstream.

IMPORTANT WORDS AND TERMS

Wind	Anticyclone	Summer monsoon
Air currents	Friction	Winter monsoon
Gravity	Chill factor	Chinook
Pressure differences	Sea breeze	Föhn
Isobar	Land breeze	Doldrums
Pressure gradient	Onshore wind	Horse latitudes
Low-pressure trough	Offshore wind	Trade winds
Convection cycle	Valley breeze	Westerlies
Thermal circulation	Mountain breeze	Polar easterlies
Coriolis force	Monsoon	Jet streams
Cyclone		

LEARNING OBJECTIVES

After reading and studying this chapter, you should be able to do the following without referring to the text:

1. Distinguish between winds and air currents.
2. State the forces that produce air motion.
3. Explain the Coriolis force and its effect on air motion around highs and lows.
4. Describe several local winds.
5. Explain the chill-factor index.
6. Describe the earth's general air circulation structure.
7. Discuss jet streams.

QUESTIONS

1. Are winds and air currents the same?
2. What are the primary forces producing air motion?
3. What are isobars and how do winds blow relative to isobars? What is meant by a low-pressure trough?
4. Explain how a convection cycle is set up near the earth.
5. What are the secondary forces of air motion, and what is the basic distinction between secondary and primary forces?
6. What effect does the Coriolis force have on air motion around (a) lows and (b) highs?
7. Why is the Coriolis force reversed for the northern and southern hemispheres?
8. Distinguish between cyclones and anticyclones in both hemispheres.
9. In what direction does (a) a north wind and (b) a west wind blow?
10. Determine the wind-chill factor for each of the following:
    (a) A temperature of 20°F and a wind speed of 20 mi/h?
    (b) A temperature of 30°F and a wind speed of 25 mi/h?
11. Compare land and sea breezes to valley and mountain breezes. Explain seasonal monsoon effects.
12. What are the doldrums and the horse latitudes?
13. (a) What is the general wind direction in the United States and why?
    (b) Generally speaking, on which side of town would it be best to build a home so as to avoid smoke and other air pollutants generated in the town.
    (c) Should the prevailing wind direction be of any consideration in the heating plan and insulation of a house?
14. Explain how the wind zones of the earth's general circulation pattern influenced the exploration and settlement of the Americas.
15. What are jet streams and how do they vary seasonally? Explain this variation relative to the pressure cells of the earth's circulation structure.

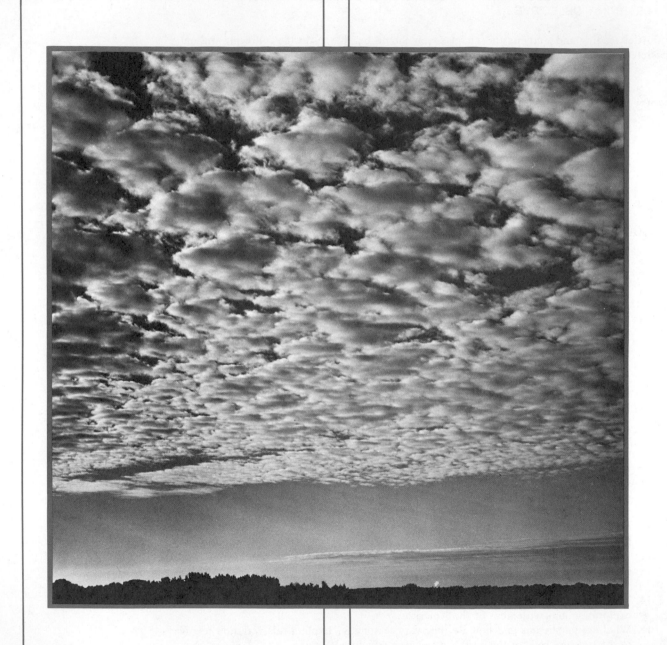

Chapter 23

CLOUDS

Clouds are buoyant masses of visible droplets of water vapor or ice crystals. Have you ever wondered what keeps clouds afloat or how they are formed? These processes are important weather phenomena which indicate atmospheric conditions. The size, shape, and behavior of clouds are the layman's key to the weather. Without the use of sophisticated meteorological instruments or a knowledge of scientific principles, one may "read" the sky and predict with surprising accuracy the forthcoming weather. Cloud signs were well known to observers of ancient times, and the empirical knowledge gained from observing clouds was probably the basis for the first attempts at weather forecasting.

clouds was probably the basis for the first attempts at troposphere, from high ice clouds to low-lying fogs. They add beauty to the sky as their billowy forms float listlessly on a summer day; or dark and foreboding, they may herald an impending storm. As stated in Chapter 21, clouds play an important role in maintaining the earth's heat balance by reflecting solar radiation and insulating against heat loss. Of equal importance is the precipitation process associated with clouds. Rain and other forms of precipitation originate in clouds. This process is fundamental to life itself as all plants and animals need water to survive.

Clouds form part of the hydrologic cycle by which moisture is distributed over the earth. An abundance of precipitation affects our environment, and the resulting water changes the features of the earth's surface, as will be discussed in Chapter 30. Also, as an integral part of the precipitation process,

Wet weather seldom hurts the most unwise
So plain the signs, such prophets the skies.

Vergil

clouds are the agents of our fledgling attempts to control the weather in the form of rainmaking.

The first step in our study of clouds begins in learning the different varieties and in identifying the various forms. The role of different clouds in the precipitation process and in giving an indication of weather phenomena will become evident as you study the next two chapters.

23-1
CLOUD CLASSIFICATION

Clouds are classified according to their *shape* and *appearance*, and *altitude*. There are four basic root names: **cirrus** (Ci) meaning *curl* and referring to wispy fibrous forms; **cumulus** (Cu) meaning *heap* and referring to billowy round forms; **stratus** (St) meaning *layer* and referring to stratified or layered forms; and **nimbus** (Nb) referring to a cloud from which precipitation is occurring or threatens to occur. These root forms are then combined to describe the types of clouds and precipitation potential.

When classified according to height, clouds are separated into four families: (1) **high clouds,** (2) **middle clouds,** (3) **low clouds,** and (4) **clouds with vertical development.** These are listed in Table 23-1a with their approximate heights and the cloud types belonging to each family. An illustration is provided for each cloud type in subsequent photographs.

High Clouds

The high clouds are visible water vapor in the form of ice crystals due to the temperature at their altitude. These ice crystals may vary from 5 to 50 microns in diameter. (One micron is one-millionth of a meter.) The cirrus member of this family is the well-known wispy mare's tail or artist's brush. They appear as though an artist had made short curling strokes with white paint on a blue background. The **cirrocumulus** clouds occur in layered patches and are referred to as mackerel scales due to their fish-scale-like structure. The cirrocumulus cloud pattern gives rise to the term mackerel sky. High winds at these altitudes often give the cirrocumulus clouds a wavy or ripple appearance. Cirrus and cirrocumulus clouds are shown in Fig. 23-1a. The **cirrostratus** clouds are in

Table 23-1a *Cloud Families and Types*

Family	Types	Illustration	Meteorological Symbols*
High clouds (above 20,000 ft)	Cirrus (Ci)	Fig. 23-1a	
	Cirrocumulus (Cc)	Fig. 23-1a	
	Cirrostratus (Cs)	Fig. 23-1b	
Middle clouds (6000–20,000 ft)	Altostratus (As)	Fig. 23-1d	
	Altocumulus (Ac)	Fig. 23-1e	
Low clouds (ground level–6000 ft)	Stratus (St)	Fig. 23-1f	
	Stratocumulus (Sc)	Fig. 23-1i	
	Nimbostratus (Ns)	Fig. 23-1j	
Clouds with vertical development (1500–60,000 ft See Section 23-2)	Cumulus (Cu)	Fig. 23-1k	
	Cumulonimbus (Cb)	Fig. 23-1l	

*Different symbols describe variations in cloud types. Consult meteorology texts for complete listings.

Figure 23-1a *Cirrus and cirrocumulus clouds. Artist's-brush cirrus are to the upper right of the photograph and the mackerel-scale cirrocumulus are to the left. (Courtesy Paul H. Taft, Cloudcroft, N.M.)*

Figure 23-1b *Cirrostratus clouds. In this photograph they completely cover the sky, as evidenced by the solar halo. (Courtesy NOAA.)*

Figure 23-1c *Contrails. The condensed water-vapor trails caused by aircraft in the upper cloud region. (Courtesy Paul H. Taft, Cloudcroft, N.M.)*

Figure 23-1d *Altostratus clouds. Thick, gray altostratus clouds hide the sun. (Courtesy NOAA.)*

Figure 23-1e *Altocumulus clouds, rolled, and arranged in flattened layers by the moving air. (Courtesy NOAA.)*

Figure 23-1f *Stratus clouds. Low-lying stratus clouds are sometimes called high fogs. (Courtesy NOAA.)*

Figure 23-1g *An advection fog. Formed when moist air moves over a cool surface. Advection fogs are sometimes considered low-lying stratus clouds. (Courtesy NOAA.)*

the form of a thin veil of ice crystals which may partially or completely cover the sky. This ice veil often gives rise to the effect of solar and lunar halos caused by the scattering of light by the ice crystals (Fig. 23-1b). This effect is analogous to a light bulb placed behind a frosted pane of glass. Occasionally, there are enough brightly reflecting faces of ice crystals to produce a white circle which passes through the sun and all the way around the sky. In some instances two bright-colored patches appear on each side of the sun just outside the regular halo. These phenomena are known as **parhelia** (side suns), or more commonly referred to as sun dogs.

Although not specifically a cloud, a common sight in the high-cloud region is the **contrail** (condensation trail) of a jet aircraft as shown in Fig. 23-1c.

Contrails result from the condensation of exhaust water vapor and are sometimes referred to as vapor trails. Like the cirrus clouds in the high-cloud region, they are composed of ice crystals. Contrails can give indications of conditions in the upper troposphere. The spreading of contrails is related to the winds aloft and the contrail length (i.e., the time required for its dissipation) depends on the humidity.

Middle Clouds

The middle clouds are distinguished by names with the prefix "alto." They vary considerably in shape and thickness. The **altostratus** member of this cloud family consists of layered forms of varying thickness (Fig. 23-1d). A thick layer may appear gray

Figure 23-1h *A radiation fog that formed overnight in an Adirondack mountain valley when radiative heat loss cooled the air sufficiently to cause condensation. (Courtesy C. McNicol, Schroon Lake, N.Y.)*

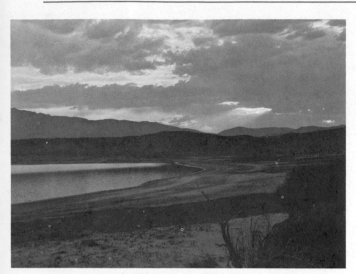

Figure 23-1i *Stratocumulus clouds. Long layers of cotton-like masses. (Courtesy NOAA.)*

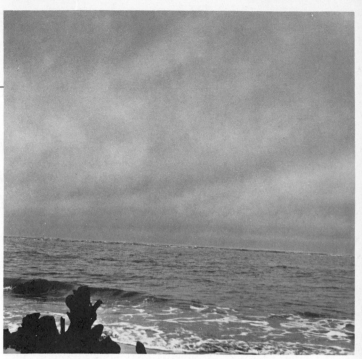

Figure 23-1j *Nimbostratus clouds. These dark stratocumulus clouds are given to precipitation. (Courtesy NOAA.)*

Figure 23-1k *Fair weather cumulus clouds. The billowy white clouds are commonly seen on a clear day. (Courtesy NOAA.)*

Figure 23-1l *A cumulonimbus cloud. A massive thunderhead with rain in progress. Note the "veil" at the top of the cloud. (Courtesy Paul H. Taft, Cloudcroft, N.M.)*

and dark. They often hide the sun or moon and cast a shadow. When their thickness is such that they are translucent to the sun or moon, an indistinctly defined halo, referred to as a corona, may be observed around the sun or moon. **Altocumulus** clouds have several varieties. They appear commonly as woolly patches or as rolled, flattened layers as shown in Fig. 23-1e. The wavy air motion resulting from air passing up and over mountains may form altocumulus clouds in a lenticular, or lens, shape on the leeward side of the mountain (see Figs. 23-2b and 23-2d). These clouds have a distinct "flying-saucer" appearance.

Low Clouds

The stratus clouds of the low-cloud family are thin layers of water vapor. They may appear dark and are common in the winter, giving rise to the sky's "hazy shade of winter." Stratus clouds are sometimes referred to as high fogs (see Fig. 23-1f). Since fog consists of visible droplets of water vapor, it falls under the definition of a cloud. In some respects ground fog may be thought of as a low-lying stratus cloud. Most fogs are either advection or radiation fogs. **Advection fog** forms when moist air moving over a colder surface is cooled below the dew point and condensation occurs. Advection fogs "roll in," as shown in Fig. 23-1g. **Radiation fog** results in station-

ary air overlying a surface which cools. This occurs typically in valleys as in Fig. 23-1h. The **stratocumulus** clouds are long layers of cotton-like masses (Fig. 23-1i). The rounded masses sometimes blend together so as to give the stratocumulus clouds a wavy appearance. When the low clouds become dark and given to precipitation, they take on the generic name of **nimbostratus.** See Fig. 23-1j.

Vertical Development

The clouds of massive grandeur are those with vertical development. Formed by rising air currents, the billowy cumulus clouds are a common sight on a typical summer day as shown in Fig. 23-1k. When the cumulus clouds turn dark and forecast an impending storm, they are referred to as **cumulonimbus** clouds. These clouds are often called thunderheads (Fig. 23-1l).

Winds and air currents cause many variations in cloud shapes. The following are some adjectives used to describe these shapes: **fracto,** meaning broken and applied to stratus and cumulus clouds; **congestus,** meaning crowded or heaped and applied to cumulus clouds; **humilis,** meaning lowly, or poorly formed, also applied to cumulus clouds; **mammatus,** referring to mammary-shaped clouds as shown in Fig. 23-1m.

Cloud height, or the cloud "ceiling," is important

Figure 23-1m *Cumulonimbus mammatus-type clouds formed on the lower surface of the anvil of a cumulonimbus cloud. (Courtesy NOAA.)*

INDICATOR located in W.B. office

90 90

a

0 0

Indicator scale

117 VAC power

1 Telephone pair

Projected light beam

a = Continuously varying angle from horizontal

a

Detected light (modulated)

Detector

117 VAC power

1 Telephone pair

Telephone pair for talk circuit

Known baseline

117 VAC power

Figure 23-1n *A diagram of a ceilometer used for measuring cloud heights. By monitoring the light reflected from clouds, their heights can be calculated. (Courtesy NOAA.)*

in meteorology and particularly in aviation. The height of the cloud coverage is determined by a **ceilometer** which operates on a reflection principle. Clouds are good reflectors of radiation. By observing the time elapsed between the emission and detection of reflected light, along with the knowledge of the speed of light, the height of a cloud may be calculated. A schematic diagram of a ceilometer is presented in Fig. 23-1n.

23-2
CLOUD FORMATION

To be visible as droplets, the water vapor in the air must condense. This requires a certain temperature, namely, the dew point. Hence, if moist air is coooled below the dew point, the water vapor contained therein will generally condense into fine droplets and form a cloud. The air is continually in motion, and when an air mass moves into a cooler region, cloud formation may take place. Since the temperature of the troposphere decreases with height, cloud formation is associated with the vertical movement of air. In general, clouds are formed by vertical air motion and are shaped and moved about by horizontal air motion, or winds.

Vertical air motion may occur by different means, such as the vertical displacement of light warm air by heavier cold air, or by the localized heating of air as in the case of a convection cycle. When a mass of air is heated locally, it rises. This occurs because the heated air is less dense than the surrounding cooler air, and it therefore is buoyant. That is, when the temperature of an air mass is increased, it expands to a larger volume. Since the density ρ is the mass per unit volume ($\rho = m/V$), the increase in volume causes a decrease in the density. Hence, a volume of warm air is lighter, or less dense, than a corresponding volume of colder air. Just as a cork held submerged in water will be buoyed up when released, the warm air rises in the surrounding colder stationary

air. As the warm air mass ascends, it becomes cooler as a result of further expansion. This is because the internal heat energy is used to do work in expanding the rising air mass against the surrounding stationary air. A region of rising air is one of low pressure.

The rate at which the temperature decreases with height is called the **lapse rate.** The normal lapse rate in stationary air is on the order of $3\frac{1}{2}°F/1000$ ft. This value may vary with latitude and changing atmospheric conditions, e.g., seasonal changes. In the case of rising air, lapse rate is referred to as dry or wet depending on whether or not condensation has taken place. The lapse rate of dry air is referred to as the **dry adiabatic lapse rate.** Should condensation occur, the lapse rate decreases due to heating from the latent heat of condensation and is called the **wet adiabatic lapse rate.** Adiabatic (Greek *adiabatos*, not passable) implies no heat loss. In the processes under consideration, the heat loss is very slow and approaches adiabatic conditions.

Because heat energy is used in the expansion of a warm air mass, a rising air column has a greater lapse rate than does the surrounding stationary air. Thus, rising air cools more quickly. When a rising air mass cools to the same temperature as the surrounding stationary air their densities become equal. The rising air mass then loses its buoyancy and is said to be in a **stable** condition. A heated air mass rises until stability is reached, and this portion of the atmosphere is referred to as a **stable layer,** that is, a layer of air of uniform temperature and density.

Clouds are formed when moisture in rising air condenses into droplets that can be seen. If the rising air reaches its dew point before becoming stable, condensation occurs at that height, and the rising air carries the condensed droplets upward, forming a cloud. Hence, when localized heating of an air mass occurs, one of the following happens:

1. The rising air reaches stability without condensation; that is, the rising air cools to the same temperature as the surrounding stationary air and becomes stable at some height without reaching

Stable layer $T_{r'} = T_s$

Condensation level $T_r = T_d$

Figure 23-2a *Vertical cloud development. The cloud begins to form at the height at which the dew point is reached and condensation occurs ($T_r = T_d$). The vertical development continues until stability is reached ($T_{r'} = T_s$). (Courtesy NOAA.)*

the dew point. No condensation takes place and no cloud forms. Or,

2. The temperature of the rising air (T_r) cools to the dew point (T_d), condensation occurs ($T_r = T_d$), and a cloud begins to form. See Fig. 23-2a. The air, still buoyant, continues to rise, but cools with a slower lapse rate (wet adiabatic) due to heating from the latent heat of condensation. It rises until its temperature ($T_{r'}$) becomes equal to the temperature of the surrounding stationary air (T_s), i.e., stability is reached ($T_{r'} = T_s$). The height at which condensation occurs is the height of the base of the cloud. The vertical distance between the level where condensation began and the level where stability is reached is the height, or thickness, of the cloud mass. It is assumed for purposes of this discussion that no precipitation occurs.

If the lapse rates are known, the air temperature may be calculated for any height by the general formula

$$T = T_0 - Rh$$

where T = temperature of the air at a height h,
T_0 = temperature of the air at the level from which height is measured, i.e., $h = 0$,
R = lapse rate,
h = any given height.

For example, if the air temperature of stationary air ($R = 3\frac{1}{2}°\text{F}/1000\text{ ft}$) at the earth's surface is 75°F, then at 4000 ft, the stationary air temperature would be

$$T = 75° - (3\tfrac{1}{2}°/1000)(4000)$$
$$= 75° - 14° = 61°\text{F}$$

The formula simply says that for every 1000 ft of altitude (height), $3\frac{1}{2}°$ is subtracted from the initial air temperature, in this case.

By inverse reasoning of the above cloudforming process, it can be easily seen that if a cloud were carried downward by air currents, it would evaporate into the unsaturated air below the height where the cloud's temperature is equal to the dew point. Hence, cloud formation indicates updrafts and dissipating

Figure 23-2b *The formation of different cloud shapes. Clouds are formed and shaped by vertical ascent and air motion.*

Stable layer $T_r = T_s$

(Altostratus)

Anvil cumulus

Lenticular cloud

Altocumulus, may cool and fall

Cumulus clouds may break up forming As and Ac clouds

Wind

Winds distort clouds

Air rising over topological barriers may form clouds

Figure 23-2c *Cumulonimbus with anvil. The vertically formed cumulonimbus cloud spreads out along a stable layer, taking the shape of an anvil. At the left of the photograph the wind has distorted the top of an anvil into cirrus clouds. (Courtesy Paul H. Taft, Cloudcroft, N.M.)*

clouds indicate downdrafts. Once formed, the wind shapes the clouds. Some examples of this action are shown in Figs. 23-2b, c, and d.

Stable layers are the limits to which air may rise. Should stability occur at a low altitude, the air is held near the surface of the earth. Under special atmos-pheric conditions, such as rapid radiative cooling near the earth, the temperature may locally increase with altitude. The lapse rate is then said to be *inverted*, giving rise to a **temperature inversion** (Fig. 23-2e). Local temperature inversions occur in several ways. Most common are radiation and subsidence inver-

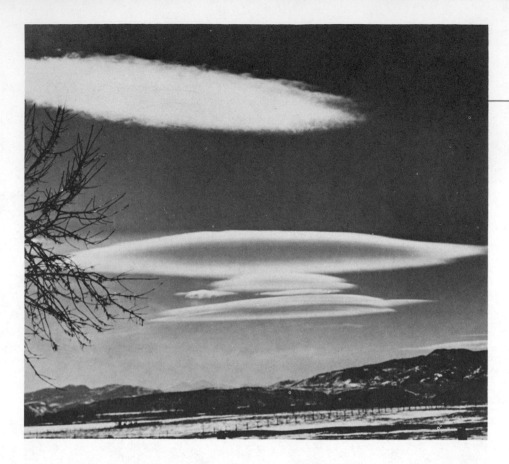

Figure 23-2d *Lenticular altocumulus clouds. The wavy air motion resulting from air rising over a topographical barrier sometimes forms lens-shaped clouds. Note the flying-saucer appearance of these clouds. (Courtesy NOAA.)*

sions. Radiation temperature inversions occur daily. They are associated with the earth's radiative heat loss. The ground is heated by insolation during the day, and at night it cools by radiating heat back into the atmosphere (cf. Section 21-4). If it is a clear night, the land surface and the air near it cool quickly. The air some distance above the surface, however, remains relatively warm, thus giving rise to a temperature inversion. Radiation fogs are common evidence of this cooling effect in valleys. Subsidence temperature inversions occur when a high-pressure air mass moves over a region and becomes stationary. As the dense air settles, it becomes compressed and heated. If the temperature of the compressed air layer exceeds the temperature of the air below it, then the lapse rate is inverted similar to that shown in Fig. 23-2e.

Inverted lapse rates prohibit locally heated air from rising to high altitudes. This condition becomes a serious problem for cities and industrial areas. The air in these areas is continually polluted (Fig. 23-2f).

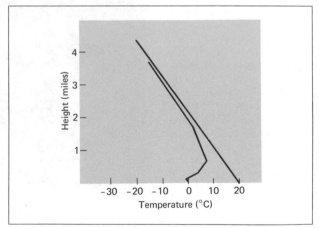

Figure 23-2e *A graph of a normal lapse rate and a temperature inversion. Ordinarily near the earth, the lapse rate decreases uniformly with altitude. However, radiative cooling of the earth may cause the lapse rate to become inverted and the temperature increases with altitude (usually below one mile). A similar condition may occur from the subsidence of a high-pressure air mass.*

Figure 23-2f *Smog arising from fog conditions along with overpollution of the atmosphere due to an inverted lapse rate.*

With the air being prevented from rising by a low, stable layer resulting from the temperature inversion, it quickly becomes overpolluted. With the inverted lapse rate, low-lying fog is likely. The condition of fog in an overpolluted atmosphere is referred to as **smog.** This can be a hazardous health problem and has occurred within recent years in Los Angeles, New York, and London, resulting in the loss of life.

Air pollutants are usually carried away from their source areas by air currents and winds, but are still normally confined to the lower troposphere by stable layers. In this region, the air is cleansed by precipitation and the pollution is "washed" back to earth. Some concern has been expressed over the possibility of commercial supersonic air transportation in the upper troposphere and lower stratosphere. With no natural cleansing mechanism in these regions, they could gradually become polluted, and the pollution could affect our environment. The sources and effects of air pollution will be considered further in Chapter 26.

23-3
CONDENSATION AND PRECIPITATION

Hath the rain a father? or who hath begotten the drops of dew? Out of whose womb came the ice? and the hoary frost of heaven, who hath gendered it? . . . Cans't thou lift up thy voice to the clouds, that abundance of water may cover thee?

Job 38

To answer these questions put to Job about precipitation, one must be familiar with the condensation process. In the previous section on cloud formation, it was simply stated that condensation occurs when the dew point is reached. It was assumed that all the essentials for condensation were present. However, it is quite possible for an air mass containing water vapor to be cooled below the dew point without condensation occurring. In this state the air mass is said to be **supersaturated,** or **supercooled.**

How, then, are the visible droplets of water vapor formed? It might seem that the collision and coalescing of the water molecules would form a droplet, but this would require the collision of millions of molecules. Moreover, only after a small droplet has reached a critical size will it have sufficient binding forces to retain additional molecules. The probability of a droplet forming by this process is quite remote. Instead, the water droplets form around microscopic foreign particles already present in the air. These particles on which the droplets form are called **hygroscopic nuclei.** They are present in the air in the form of dust, combustion residue, salt from sea water evaporation, and so forth. Since foreign particles initiate the formation of droplets which eventually fall as precipitation, the previously mentioned cleansing mechanism of the atmosphere is readily understood.

Liquid water may also be cooled slightly below the freezing point without the formation of ice, if there are no particles present or rough-container surfaces to act as nuclei or nucleation sites for the ice crystals. Similarly, bubble formation requires nucleation sites for the bubbles to grow. Bubbles rising from a carbonated beverage usually form near the container surface where nucleation sites are present. If a glass containing a beverage is extremely clean, the beverage may appear to be "flat" because of the absence of the rising bubbles. However, upon tasting, the carbonation is found to be present. This is commonly observed because of the use of modern detergents. The detergents act as "wetting agents," and wet the nuclei, allowing them to be washed away more efficiently during cleaning (Section 16-1). The classic demonstration of this "wetting" property of detergents is to apply the detergent to the bottom of a duck. Since the duck's buoyancy under ordinary circumstances is a result of the fact that its feathers repel water, or do not become wet, it sinks when placed in water after the detergent has been applied, because its feathers become wet. (Naturally, the duck is saved.)

The droplets formed by the nucleation process are very minute and have diameters on the order of 5–200 microns. (One micron is one-millionth of a meter.) Droplets of this size fall very slowly with an approximate velocity of 15 ft/min or about $\frac{1}{6}$ mi/h and form a fine drizzle. Since the condensation is formed in updrafts, the droplets are readily suspended in the air as a cloud. To have precipitation, larger droplets or drops must form. This may be brought about by two processes: (1) **coalescence** or (2) the **Bergeron process.**

Coalescence

Coalescence is the formation of drops by the collision of the droplets, with the result that the larger droplets grow at the expense of the smaller ones. The efficiency of this process depends on the variation in the size of the droplets. Raindrops vary in size, reaching a diameter of approximately 7 mm. The volume of a raindrop is calculated by

$$V = \tfrac{4}{3}\pi(d/2)^3$$

where V = volume,
  d = diameter.

To produce a drop with a diameter of 1 mm from droplets 10 microns (10^{-2} mm) in diameter, a number of droplets n with a volume V_1 would have to coalesce to form the volume V_2 of the drop. That is,

$$V_2 = nV_1$$

Then, solving for n and expressing the volumes in terms of the diameters,

$$n = \frac{V_2}{V_1} = \left(\frac{d_2}{d_1}\right)^3 = \left(\frac{1\ \text{mm}}{10^{-2}\ \text{mm}}\right)^3 = 10^6$$

Hence, it would take a million of the 10-micron diameter droplets coming together to form a drop 1 mm in diameter. However, if the droplets were 100 microns (10^{-1} mm) in diameter, a similar calculation shows that only 10^3 or 1000 droplets would be needed. From this, it is easily seen that having larger droplets greatly enhances the coalescence process.

Bergeron Process

The Bergeron process, named after the Swedish meteorologist who suggested it, is probably the more important process for the initiation of precipitation. This process involves clouds which contain ice crystals in their upper portions and have become supercooled in their lower portions. Mixing or agitation of the clouds allows the ice crystals to come into contact with the supercooled vapor. Acting as nuclei, the ice crystals grow larger from the condensing vapor. The ice crystals melt into large droplets in the lower portion of the clouds and coalesce to fall as precipitation. Air currents are the normal mixing agents. The Bergeron process is illustrated in Fig. 23-3a.

It should be noted that there are three essentials in the Bergeron process: (1) ice crystals, (2) supercooled vapor, and (3) mixing. **Rainmaking** is based on the essentials of the Bergeron process. The early rainmakers were somewhat charlatanic in nature. With much ceremony they would beat on drums, or more sophisticatedly, fire cannons and rockets into the atmosphere. In the early 1890s, the U.S. Congress appropriated almost $20,000 to finance the rainmaking efforts of General R. G. Dyrenforth. It had been observed that rain frequently occurred during a battle which involved cannon. Reasoning that the rain must in some way be associated with the cannon fire, the General designed a balloon to carry an explosive charge aloft. This was also supplemented by cannon on the ground. The experiment was carried out near San Antonio, Texas, with explosives and guns being set off simultaneously at periodic intervals. Some rain did develop after many firings, but subsequent attempts failed to produce results. General Dyrenforth called in more cannon, much to the dismay of the local residents, but this too failed to produce rain.

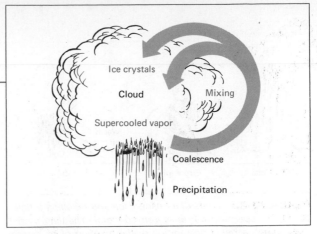

Figure 23-3a *A diagram illustrating the essentials of the Bergeron process. The mixing of ice crystals and supercooled water vapor produces larger droplets and initiates precipitation.*

The initial success may have been coincidental with natural rainfall. However, the explosives did supply the agitation or mixing which may have initiated the Bergeron process, providing the other two essentials were present.

Modern rainmakers approach the problem differently. There are usually enough air currents present for mixing, but the ice crystal nuclei may be lacking. To correct this, the cloud is "seeded" with silver iodide crystals or dry ice pellets (solid CO_2). The silver iodide crystals have a crystal structure similar to that of ice and provide a substitute for the ice crystals. The silver iodide crystals are produced by a burning process. This may be done on the ground, with the silver iodide crystals being carried aloft by the rising warm air, or the burner may be attached to an airplane and the process carried out in the cloud to be seeded. Dry ice pellets are dropped into the cloud from an airplane. The temperature of solid dry ice is $-79°C$ and it quickly sublimes. That is, it goes directly from the solid phase to gaseous phase. Rapid cooling associated with the sublimation cools the water vapor and freezes existing water droplets to provide nuclei. Precipitation then may occur if this part of the Bergeron process has been absent. Also, the released latent heat from the freezing droplets is

Figure 23-3b *Types of air ascension and resulting precipitation. Air ascension may arise from convection heating; motion over some barrier (orographic); or the meeting of air masses having different temperatures, with the warmer, less dense air being displaced upward and over the colder air (frontal).*

available to set up convection cycles, which facilitate mixing. Seeding is a technique that is receiving increasing attention in initiating the precipitation of fog since fog frequently hinders airport operations.

Precipitation requires that water vapor condense, a process which results when the vapor in ascending air cools. Three mechanisms by which air rises, thus allowing for precipitation, are convectional, orographic, and frontal. These are illustrated in Fig. 23-3b.

Convectional precipitation is a result of convection cycles. This type of precipitation predominantly occurs in the summer, since localized heating is required to initiate the convection cycle. Condensation may occur quickly due to the convectional updrafts, and the precipitation is usually confined to the local area. The sudden summer shower is an example.

Orographic precipitation arises when air is forced to rise due to land forms such as mountain ranges. The wind blows along the surface of the earth and ascends along geographical variations. The ascending wind may give rise to orographic precipitation on the windward side of the mountain. (See Fig. 22-2b.)

Frontal precipitation results from the meeting of air masses of different temperatures. Moving warm air flows up and over the cooler air since the warm air is lighter. This ascent is associated with horizontal motion and the warm air travels upward at an angle.

As a result, the cooling is usually less rapid than in the vertical convection motion.

The boundary between two air masses is called a front and is characterized by varying degrees of precipitation and storms. Fronts will be considered in detail in Chapter 24.

The type of precipitation depends on atmospheric conditions, and can be in the form of rain, snow, sleet, hail, dew, or frost. **Rain** is the most common form of precipitation in the lower and middle latitudes. The formation of large water droplets which fall as rain has been described previously. If the dew point is below 32°F, the water vapor freezes on condensing, and ice crystals result which fall as **snow.** In cold regions, these ice crystals may fall individually, but in warmer regions, the crystals become stuck together, forming a snowflake which may be as large as one inch across. Since ice crystallizes in a hexagonal pattern, most snowflakes are six-sided, or hexagonal. Frozen rain, or pellets of ice in the form of **sleet,** occurs when rain falls through a cold surface layer of air and freezes or, more likely, when the ice pellets fall directly from the clouds without melting before striking the earth. Large pellets of ice, or **hail,** result from successive vertical descents and ascents in vigorous convection cycles associated with thunderstorms. Additional condensation on successive cycles into supercooled regions that are below freezing may produce layered, structured hailstones the size of baseballs. See Fig. 23-3c.

Dew is formed by atmospheric water vapor condensing on the earth's surface. The land surfaces cool quickly at night and the temperature near the surface may fall below the dew point. Water vapor condenses on the available surfaces such as blades of grass, giving rise to the "early morning dew." If the dew point is below freezing, the water vapor condenses on objects in the form of frost. The white frost, sometimes referred to as hoar frost, is composed of ice crystals. **Frost** is not frozen dew, but results from the direct change of water vapor into ice. Should the air near the earth be supersaturated, heavy ice formation

Figure 23-3c *A hailstorm* (right) *and hailstones* (below) *that fell in Oklahoma. The successive vertical ascents of ice pellets into super-cooled air and regions of condensation produce large, layered "stones" of ice. (Courtesy NOAA.)*

forms on objects when the temperature is below 32°F. This heavy frost is called **rime.**

One of the most frequently observed effects accompanying precipitation in the form of rain is the rainbow. The colorful arc of a **rainbow** across the sky is the result of refraction and internal reflection of sunlight by water droplets. As a result of dispersive refraction, the sunlight is separated into its component wavelengths, or colors, by the water droplets in much the same way as a prism. Often, two rainbows are observed, a primary and secondary, as shown in Fig. 23-3d. The primary rainbow results from a single internal reflection in the raindrops and the secondary rainbow from a double internal reflection as illustrated in Fig. 23-3e. The colors of the primary rain-

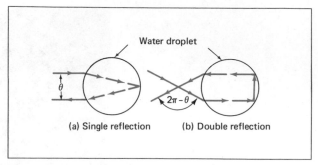

Figure 23-3e *A diagram showing the internal reflections in water droplets. A single reflection gives rise to the primary rainbow and a double reflection produces the secondary rainbow, whose colors are inverted.*

Figure 23-3d *A double rainbow. The fainter, secondary rainbow lies above the primary rainbow. (Courtesy J. Phinney.)*

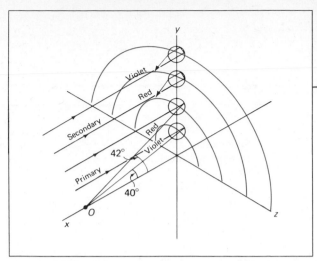

Figure 23-3f *Rainbow geometry. An observer at O sees the red light in the primary rainbow from water droplets on an arc at 42° altitude and the violet light on an arc at 40° altitude. The intermediate colors come from intermediate arcs. The secondary rainbow is formed at a higher altitude.*

Figure 23-3g *An illustration of a mirage. The refraction of light by layers of warm air near the earth produces the illusion of a water surface that is actually the image of the sky.*

bow run vertically from violet to red. The secondary rainbow's colors are reversed, running vertically from red to violet.

The observation and length of arc of a rainbow depends on the position of the sun and the elevation of the observer. Consider an observer at *O*, as shown in Fig. 23-3f, with the sun's rays being parallel to the earth's surface. It can be shown from optical principles that the angle θ between an incident and reflected ray of red light for a single internal reflection in Fig. 23-3e is 42°. Hence, for sunlight parallel to the earth's surface, an observer at *O* sees a reflected red ray above the horizon at an altitude of 42°. Similarly, violet light has an angle θ of 40° and the observer sees the reflected violet ray at an altitude of 40°. Thus, all the raindrops which lie on a circle subtending 42° reflect red light, and all those on the circle subtending 40° reflect violet light. The reflection of other colors results from raindrops which lie on circles subtended by angles between 40° and 42°. The secondary rainbow's reflections occur between 50.5° and

54°. As the sun rises, the arcs are lowered in altitude and less of the rainbow is seen. A moment's thought should reveal that an observer on the ground cannot see a primary rainbow when the sun is at an altitude greater than 42°. Conversely, if the observer is elevated, more of the arc is seen. It is common to see a completely circular rainbow from an airplane, thus eliminating the possibility of finding a pot of gold at the rainbow's end.

Another phenomenon which results from the reflection and refraction of light is a mirage. A **mirage** is the image of an object appearing where it is not. Mirages are caused by the refraction or bending of light by air layers having different temperatures. The image of the sky appearing near a hot surface Fig. 23-3g) is often mistaken for a body of water. This water mirage is quite common in the desert or near a black-top road in the summer. Conversely, light from objects on the earth's surface may be reflected back to earth by air layers at higher elevations, giving rise to mirages such as trees or ships appearing in the sky.

Clouds
Cirrus
Cumulus
Stratus
Nimbus
High clouds
Middle clouds
Low clouds
Clouds of vertical development
Cirrocumulus
Cirrostratus
Parhelia
Contrails
Altrostratus
Altocumulus
Advection fog
Radiation fog

Stratocumulus
Nimbostratus
Cumulonimbus
Fracto
Congestus
Humilis
Mammatus
Ceilometer
Lapse rate
Dry adiabatic lapse rate
Wet adiabatic lapse rate
Adiabatic
Stable
Stable layer
Temperature inversion
Smog
Supersaturated

Supercooled
Hygroscopic nuclei
Coalescence
Bergeron process
Rainmaking
Convectional
Orographic
Frontal
Rain
Snow
Sleet
Hail
Dew
Frost
Rime
Rainbow
Mirage

LEARNING OBJECTIVES

After reading and studying this chapter, you should be able to do the following without referring to the text:

1. Tell how clouds are classified.
2. Name and describe the various types of common clouds.
3. Explain the mechanisms by which clouds are formed.
4. Describe the conditions necessary for condensation in the atmosphere.
5. Distinguish between coalescence and the Bergeron process and discuss their relative importance in the formation of precipitation.
6. State the three classifications of precipitation based on air ascension.
7. Describe the different common forms of precipitation.
8. Explain how rainbows are formed.

QUESTIONS

1. What is a cloud?
2. According to what criteria are clouds classified?
3. Name the cloud family for each of the following: (a) nimbostratus, (b) cirrostratus, (c) altocumulus, (d) cumulonimbus.

4. Name the cloud type that is associated with each of the following: (a) mackerel sky, (b) parhelia, (c) solar or lunar corona, (d) the hazy shade of winter, (e) thunderhead.
5. How is cloud height or the "cloud ceiling" determined?
6. What condition is necessary for cloud formation?
7. Describe what happens when an air mass is locally heated, i.e., does a cloud form?
8. What determines the thickness of a cloud of vertical development?
9. Distinguish between dry and wet adiabatic lapse rates.
10. Why do clouds dissipate?
11. Which is heavier, a cubic foot of dry air or a cubic foot of moist air? (Hint: Consider the fact that some of the molecules of the gases in dry air are replaced with molecules of water.)
12. How is an inverted lapse rate associated with air pollution conditions?
13. Explain how the temperature dependence of the concentration of water vapor in air is analogous to the concentration of the solute in an aqueous solution.
14. Describe two processes of precipitation formation. Which process is more important?
15. What are the principles and methods of modern rainmaking?
16. What are three types of precipitation based on mechanisms by which air rises?
17. Distinguish between dew and frost, i.e., is frost frozen dew?
18. How are large hailstones formed?
19. What are the colors of the rainbow and how are rainbows formed?

PROBLEMS

1. Sketch typical clouds from each of the four cloud families as a function of altitude.

2. If the stationary air temperature near the ground is 60°F, what is the temperature of the air at 4000 ft? Assume a normal lapse rate.

   Answer: 46°F

3. The stationary ground air temperature is 70°F. (a) What is the approximate air temperature at an altitude of one mile? (b) What is the air temperature at the top of Mt. Everest ($h \simeq 29,000$ ft)? Assume a normal lapse rate.

4. If the stationary ground air temperature is 65°F, what would be the air temperature outside (a) a commercial aircraft at 30,000 ft and (b) an SST (supersonic transport) at 50,000 ft? Assume a normal lapse rate.

   Answer: (a) −40°F (b) −110°F

5. If the stationary ground air temperature is 74°F, above what altitude will the clouds be in the form of ice crystals?

   Answer: 12,000 ft

6. How many 10-micron droplets would be needed to form a raindrop 2-mm in diameter?

   Answer: 8×10^6

7. How many 100-micron droplets would be needed to form a 6-mm diameter raindrop?

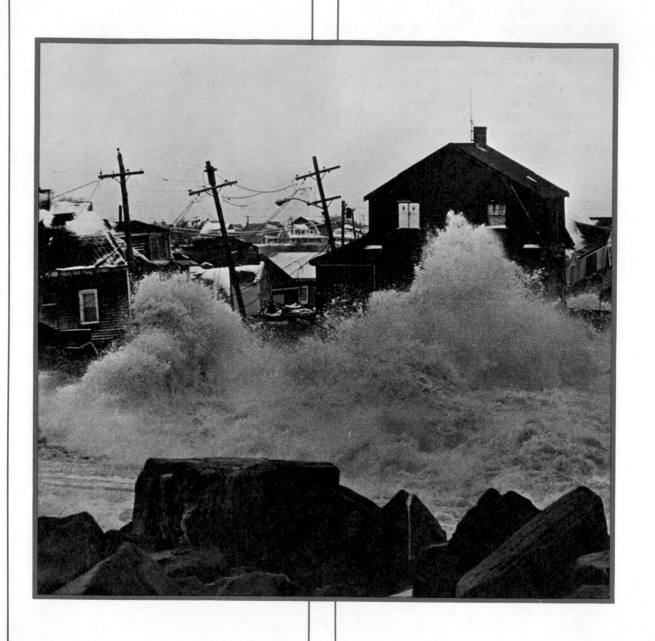

Chapter 24

AIR MASSES AND STORMS

And pleas'd the Almighty's orders to perform,
Rides in the whirlwind and directs the storm.

Addison, The Campaign

If the air of the troposphere were static, there would be little change in the local atmospheric conditions which constitute the weather. The dynamics of air movement in the form of winds have been discussed in Chapter 22. In this chapter, we shall consider the movements of large masses of air which have distinguishing physical characteristics. The air we now breathe may have been far out over the Pacific Ocean a week ago. As air moves into a region, it brings with it the temperature and humidity that are mementos of its origin and travel. Cold, dry arctic air may cause sudden drops in the temperatures of the regions in its path. Warm, moist air from the Gulf of Mexico may bring heat and humidity to some regions and make the summer seem unbearable. Thus, moving air transports the physical characteristics that produce changes and influences the weather. A large mass of air can influence the weather of a region for a considerable period of time or have only a brief effect. The movement of air masses depends a great deal on the earth's air circulation structure and its seasonal variations.

When air masses meet, the variations of their properties may trigger storms along their boundary. The types of storms depend on the properties of the air masses. Also, local variations within an air mass can give rise to storms. Storms can be violent and sometimes destructive. They remind us of the vast amount of energy contained in the atmosphere and also of its capability. The variations of our weather are therefore closely associated with air masses and their movements and interactions.

24-1
AIR MASSES

When a body of air takes on physical characteristics which distinguish it from the other air surrounding the earth, it is referred to as an **air mass.** The main distinguishing characteristics are temperature and moisture content. Air that remains over a large surface area, which has uniform characteristics for a sufficient period of time, comes to equilibrium with the surface, and a huge volume of air acquires relatively constant physical characteristics dependent on those of the surface. These characteristics define an air mass.

The underlying surface from which the air mass derives its characteristics is called its **source region.** The time required for an air mass to come to equilibrium with its source region depends on the characteristics of the surface. If the surface area is a warm body of water, convection currents in the air mix the overlying air, thus hastening equilibrium. The resulting air mass is warm and moist. If the surface area is a cold land mass, the air is more stable, and a longer time would be required to reach equilibrium. A cold, dry air mass would result.

An air mass eventually moves from its source region, bringing its characteristics and a change in weather to the regions in its path. As an air mass travels, its properties may become modified due to local variations. For example, if Canadian polar air masses did not become warmer as they traveled southward, Florida might experience extremely cold winters. Whether an air mass is termed cold or warm is relative to the surface over which it moves. Quite logically, if an air mass is warmer than the surface, it is referred to as a **warm air mass.** If colder, it is called a **cold air mass.** It should be remembered that this is a relative designation. The cold and warm prefixes do not always imply cold and warm weather. For example, a cold air mass in winter usually brings cold, frigid weather. However, the weather associated with a cold air mass of 70°F (21°C) traveling over an 80°F (27°C) surface may be quite pleasant. Such a cold air mass may bring welcome relief in the hot summer months.

Air masses are classified according to the surface and general latitude of their source regions:

Surface	*Latitude*
Maritime (m)	Arctic (A)
Continental (c)	Polar (P)
	Tropical (T)
	Equatorial (E)

The surface of the source region is abbreviated by a small letter and gives an indication of the moisture content of an air mass. An air mass forming over a body of water (maritime) would naturally be expected to have a greater moisture content than one forming over land (continental). The general latitude of the source region is abbreviated by a capital letter and gives an indication of the temperature of an air mass. The letter k or w is sometimes placed after the source region designation to indicate whether a moving air mass is respectively cold (German, *kalt*) or warm relative to the regions in its path. For example, mTw designates a warm maritime tropical air mass. A list of the air masses which affect the weather of the United States is given in Table 24-1a, along with their source regions, which are illustrated in Fig. 24-1a.

The movement of air masses is influenced to a great extent by the earth's general circulation patterns discussed in Chapter 22. Since the United States lies predominantly in the northern westerly zone, the general movement of air masses, and hence the weather, is from west to east across the continent. The circulation zones vary to some degree in latitude with

Table 24-1a *Air Masses That Affect the Weather of the United States*

Classification	Symbol	Source Region
Maritime Arctic	mA	Arctic regions
Continental Arctic	cA	Greenland
Maritime Polar	mP	Northern Atlantic and Pacific Oceans
Continental Polar	cP	Alaska and Canada
Maritime Tropical	mT	Caribbean Sea, Gulf of Mexico, and Pacific Ocean
Continental Tropical	cT	Northern Mexico, Southwestern United States

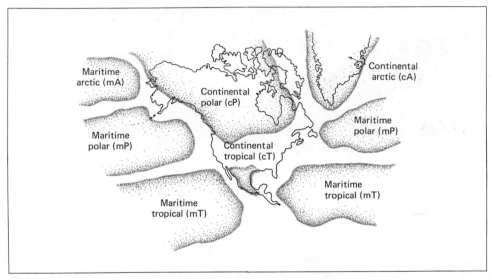

Figure 24-1a *Source regions of air masses for North America.*

the seasons, and the polar easterlies may also move air masses into the United States during the winter. Air masses are huge in magnitude and cover thousands of square miles. Their movement is the prime factor in the redistribution of moisture and heat energy. A maritime tropical air mass formed in the Gulf of Mexico, for example, may bring warm, wet weather along the eastern coast of the United States into New England.

24-2
FRONTS AND CYCLONIC DISTURBANCES

As defined previously, the boundary between two air masses is called a **front**. A **warm front** is the boundary of an advancing warm air mass over a colder surface and a **cold front** the boundary of a cold air mass moving over a warmer surface. These boundaries

Figure 24-2a (Above) *Squall line as seen from a satellite and* (below) *at the earth's surface. (Courtesy NOAA.)*

may vary in width from a few miles to a wide **frontal zone** of over 100 miles. It is along these fronts, which divide air masses of different physical characteristics, that drastic changes of weather occur. Turbulent weather and storms usually define a front. When this turbulent weather extends for some distance horizontally, it is referred to as a **squall line;** a squall line is shown in Fig. 24-2a. The series of storms which form this boundary along a front are referred to as **line squalls.**

The degree and rate of weather change depend on the difference in temperatures of the air masses and the vertical slope of the front. A cold front moving into a warmer region causes the lighter warm air to be displaced upward over the front. The lighter air of an advancing warm front cannot displace the heavier, colder air as readily and may move slowly over the colder air.

Cold air is associated with high pressure, and the downward divergent air flow in a high-pressure region in general gives a cold front a greater speed than a warm front. A cold front may have an average speed of 20 to 25 miles per hour, while a warm front moves 10 to 15 miles per hour. The features of cold and warm fronts are illustrated in Fig. 24-2b.

Cold fronts have sharper vertical boundaries than warm fronts, and warm air is displaced faster by an advancing cold front. As a result, cold fronts are accompanied by more violent or sudden changes in the weather. The sudden decrease in temperature is often described as a "cold snap." Dark altocumulus clouds mark the cold front's approach. The sudden cooling and rising warm air may set off rain- or snowstorm activity along the front. The effects of an advancing cold front are illustrated in the newspaper report at the top of the next page.

The warm front may also be characterized by precipitation and storms. Since its approach is more gradual it is usually heralded by a period of lower clouds. Cirrus and mackerel scales drift ahead of the front, followed by the alto clouds. As the front approaches, cumulus or cumulonimbus clouds resulting from the rising air produce precipitation and storms. Most of this occurs before the front passes.

Storms are also associated with cyclonic, or rotational, disturbances along active fronts. **Wave cyclones** are formed by air moving in opposite directions along a front. The development of a wave cyclone is illustrated in Fig. 24-2c. The graphic symbol for a cold front is △△△△△ and for a warm front ⌒⌒⌒⌒. The side of the line with the symbols indicates the direction of advance. Friction and shear forces between the moving air masses cause a variation from their parallel motions and start a circulation between them. This action may be demonstrated by holding a pencil between the palms of the hands. When the palms are moved in opposite directions, a rotational motion of the pencil is produced. The analogous frontal motion is shown in Fig. 24-2c, (a'), (b'), and (c'). Accompanying this frontal motion is the rising of warm air and a low-pressure area develops at the crest of the "wave" as illustrated in Fig. 24-2c, (d'). Air moving into this low-pressure region is deflected by the Coriolis force, producing cyclonic motion (cf. Chapter 22).

As the faster moving cold front advances, it may overtake the warm air mass and push it upward. The boundary between these two masses is referred to as an **occluded front** (△⌒△⌒△). The cold front occludes, or closes off, the warm air along the occluded front. Horizontal views of Fig. 24-2c, (d') and (e'), are shown in Fig. 24-2d. There are two types of occluded fronts. When the advancing cold air is

Figure 24-2b *Diagrams of cold and warm fronts. Notice in the upper diagram the sharp, steep boundary (in color) characteristic of a cold front. The boundary of a warm front, as shown in the lower diagram, is less steep. As a result of this difference, different cloud types are associated with the approach of the fronts.*

Figure 24-2d *Occluded front development. The drawings show ground-level views of parts (d') and (e') of Fig. 24-2c. The advancing colder air overtakes the retreating cold air and forces the intervening warm air upward, occluding it from the earth's surface.*

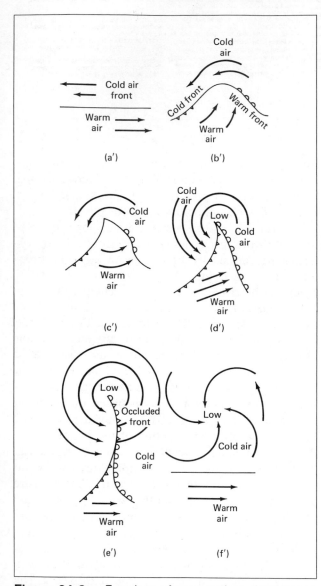

Figure 24-2c *Top views of wave cyclone development. Antiparallel wind motion produces the rotational motion for the development of wave cyclones (a'). An occluded front accurs when one front overtakes another (e'). As a result of the rotational action, a low-pressure wave cyclone spins off (f').*

colder than the air ahead of it as in Fig. 24-2d, the occluded front is called a **cold front occlusion.** That is, a cold front advances under a warm front. If the situation is reversed, and the cold air ahead is colder than the advancing air, the occluded front is referred to as a **warm front occlusion.** In this case, a warm front advances up and over a cold front.

As the cyclonic wave fully develops, a low-pressure area or storm center spins off. See Fig. 24-2c, (f'). The low moves away, carrying with it rising air currents, clouds, precipitation, and generally bad weather. Lows, or cyclones, are, therefore, generally associated with poor weather. Lows may be of a local nature or cover hundreds of square miles, depending on the extent and conditions of the frontal activity.

Highs, or anticyclones, on the other hand are generally associated with good weather. Developing in regions of divergent air flow, highs are relatively stable due to their descending air motion. As a result, dry air and a lack of precipitation usually characterize a high. There are exceptions, however, as the characteristics of a high depend on the conditions under which it develops. Because of their influence on the weather, the movements of highs and lows are closely observed. Opposing fronts may balance each other so that no movement occurs. This case is referred to as a **stationary front** (⌒⌣⌒⌣⌒⌣).

24-3
LOCAL STORMS

Storms are atmospheric disturbances associated with vertical air motion that may develop locally within a single air mass or due to frontal activity between two air masses. Several of these storms, which are distinguished by their intensity and violence, will be considered.

Rainstorms

The rate of rainfall is dependent upon the rate of the condensation and coalescence process discussed in Section 23-3. The duration of a rainfall is dependent on the amount of water vapor available. Should a rising air mass have rapid vertical motion, condensation and coalescence may occur quickly. The strong updraft will suspend the water droplets until they have coalesced into large drops. If there is a considerable amount of water vapor present, a heavy downpour of rain results. Otherwise, the rainfall may be just a sudden shower which begins and ends quickly. Rainfalls of one to two inches per hour are not uncommon. When wind motion accompanies a heavy rain, the storm may become violent with blowing "sheets" of rain. It beats against windows and quickly fills storm sewers. Wind damage in the form of bro-

ken tree limbs and downed power lines is common. When the convection currents cease or the cloud has an insufficient moisture content, the rainfall stops. However, if water vapor is continuously fed into the precipitating cloud, a prolonged heavy rainfall occurs. This condition is referred to as a **cloudburst.** When this type of rainfall occurs over dry hard ground, the water runoff may produce a **flash flood.**

Thunderstorms

The **thunderstorm** is a rainstorm distinguished by thunder and lightning and sometimes hail. Thunderstorms may result from frontal cyclonic disturbances or strong local heating. Strong updrafts of air which produce the rain cloud are essential for thunderstorm formation. In the upper portions of the resulting cumulonimbus cloud or thunderhead, the water vapor becomes supercooled. With proper nucleation, raindrops form and fall; however, some may evaporate before reaching the ground. This evaporation results in a cooling of the surrounding air, and the cool air descends, producing downdrafts. These spread out along the ground accounting for the cooling and the winds associated with thunderstorms. This development is illustrated in Fig. 24-3a. Hail may result from successive ascents and descents of raindrops if the supercooled region is below freezing.

The **lightning** of a thunderstorm is a discharge of electrical energy. The electrical nature of lightning is said to have been first demonstrated by Ben Franklin in his famous kite-flying key experiment. As he flew the kite with a key attached in a thunderstorm, lightning struck the key and was conducted toward the ground, where Franklin stored it in an electrical-storing device called a capacitor. He then demonstrated with the stored lightning that it had properties identical to those of ordinary electric charges.

Franklin was extremely lucky in the performance of this experiment in that it was a miracle that he was not electrocuted. Under no circumstances should du-

Figure 24-3a *Thunderstorm development. Strong vertical air motion is essential for thundercloud development. High-altitude winds often form an anvil cumulonimbus near the stable layer. (Courtesy NOAA.)*

plication of this experiment be attempted. Many people have been killed trying to duplicate it—not only from lightning but from a modern hazard not extant in Franklin's day, contact of the conducting kite string with high-voltage electric lines.

It is believed that a raindrop has an outer, negatively charged, electric sheath. When a raindrop is broken up or partially evaporates, negatively charged droplets become separated from the drop, leaving the remainder positively charged. This separation process of charged droplets, which produces the electric potential necessary for lightning, is not well understood.

However, when the potential is of sufficient magnitude, lightning is produced in the same manner that an electric spark arcs between two wires of different electric potential. The lightning flash usually occurs within a cloud or between clouds. Lightning occurring below the horizon or behind a cloud often illuminates the clouds with flickering flashes of light. This commonly occurs on a still summer night and is known as **heat lightning.**

Lightning may also arc between a cloud and the earth, which is a huge reservoir of charges (Fig. 24-3b). The shorter the distance between the separated

Figure 24-3b *Lightning arcs between clouds and the earth. (Courtesy NOAA.)*

charges, the more easily the electric discharge takes place. This is why lightning often strikes trees which extend upward toward the charged cloud. It is unadvisable, therefore, to take shelter from a thunderstorm under a tree, if the probability of lightning's striking in that area is great. A person in the vicinity of a lightning strike may experience an electric shock which causes respiration to fail. Mouth-to-mouth resuscitation or some other form of artificial respiration should be given immediately. Barns and other tall buildings are also occasional targets of lightning. A common practice to avoid lightning damage is the installation of **lightning rods.** A lightning rod consists of a metal rod which extends above the structure and is connected by a conductor to the earth. Jumping the shortest distance from a cloud, the lightning would strike the rod instead of the building and be conducted harmlessly to earth.

The old cliché that lightning never strikes twice in the same spot is practically true, as there is usually nothing left of that spot for the lightning to strike. Actually, lightning strikes more than once in the same spot. The Empire State Building has been struck many times, but it is protected by an elaborate lightning rod system.

Bolts of lightning have tremendous energy. It is estimated that the temperature in the vicinity of a lightning flash is of the order of 15,000–30,000°C. This is quite large compared to the sun's surface temperature of 6000°C. The energy expended by the lightning in a typical thunderstorm could supply the electric needs of a home for a considerable length of time. (The electric current of a large bolt of lightning may be as high as 100,000 Å.) The sudden release of energy of a lightning stroke explosively heats the air producing the compressions we hear as **thunder.** The sound compressions travel out from different parts of the lightning stroke. If an observer is centrally located such that the compressions from different portions of the stroke are received nearly simultaneously, a crash of thunder is heard. Rumbling thunder is partially due to the varying distances between the observer and the different portions of the lightning stroke. An initial report is heard followed by a period of rumbling sounds of decreasing intensity as the sound arrives from more distant portions of the lightning stroke. The sound waves are also reflected from thermal inversions in the air and from the surface of the ground and are modified by the turbulent conditions associated with a thunderstorm.

The association of thunder with lightning was postulated very early. Aristotle in his famous treatise, *Meteorologica,* presented the general idea, but in a somewhat incorrect, reversed fashion:

Heat when radiated is dispersed into the region above, but as much of the dry exhalation as is trapped in the change of the air's becoming cooler is radiated when the clouds condense, and moving violently and striking against the surrounding clouds delivers a blow, whose noise is called thunder. . . . This, then, is thunder, and this is the cause of it. Besides, the pressed out wind is usually consumed by a delicate and gentle flame, and this is what we call lightning, which we have whenever, as it were, the falling wind seems colored. It is produced after the blow and thus also after the thunder; but it seems to be produced before the thunder because seeing precedes hearing. We have clear proof that seeing precedes hearing when we observe the oarsmen of a trireme, for as they draw back their oars for a second strike we hear the sound of their first stroke.

Since lightning strokes occur near the storm center, the resultant thunder provides a method of easily calculating the distance to the storm center. Light travels at approximately 186,000 mi/s, and the lightning flash is seen without any appreciable time lapse. Sound, however, travels at approximately $\frac{1}{5}$ mi/s, and a time lapse occurs between seeing the lightning flash and hearing the thunder. This phenomenon is also observed when watching someone at a distance fire a

Figure 24-3c *Results of an ice storm. Rain freezes on objects when temperatures are below 32°F(0°C), causing damage due to the added weight of the ice. (Courtesy NOAA.)*

gun or hit a baseball. The report of the gun or the "crack" of the bat is always heard after the bullet reaches its mark or the baseball is well on its way. By counting the seconds between seeing the lightning flash and hearing the thunder, an estimate of the distance of the storm center may be obtained. For example, if 5 s elapsed, then the storm center would be approximately 1 mi away, taking the velocity of sound as $\frac{1}{5}$ mi/s; if 8 s elapsed, the storm would be $1\frac{3}{5}$ mi away.

Ice Storms and Snowstorms

When a warm air mass overrides a cold air mass as in Fig. 24-2b, rain may form. The rain falls to earth through the underlying cold air. If the temperature of the earth's surface is below 32°F and the raindrops do not freeze before striking the earth, the rain will freeze on the cold objects on which it falls. A layer of ice builds up on the objects exposed to the freezing rain. The resultant glaze is referred to as an **ice storm.** The ice layer may build up to over one-half inch in thickness, depending on the magnitude of the rainfall (Fig. 24-3c). Viewed in sunlight, the ice glaze produces a beautiful winter scene with the coated landscape glistening in the sun.

Other aspects of an ice storm are not so beautiful. Damage can be quite severe. Tree branches and power lines snap and fall from the weight of their ice coatings; animals may become frozen statues; transportation by foot or vehicle becomes extremely hazardous; and injuries usually result.

Snow is made up of ice crystals that fall from ice clouds. A **snowstorm** is an appreciable accumulation of snow. What may be considered a severe snowstorm in some regions might be thought of as a light snow in others where snow is more prevalent. For example, four to six inches of snow can paralyze areas like New York City and Washington, D.C.; however, in areas like upstate New York, Michigan, and Wisconsin, this accumulation would be just an incidental snowfall.

Figure 24-3d *A snowstorm often leaves huge quantities of snow, while accompanying high winds may blow the snow against objects forming deep drifts. (Courtesy North Dakota Highway Department.)*

When a snowstorm is accompanied by high winds and low temperatures, the storm is referred to as a **blizzard.** The winds whip the fallen snow into blinding swirls. Visibility may be reduced to a few inches. For this reason, a blizzard is often called a *blinding* snowstorm. The swirling snow causes a loss of one's sense of direction and people have become lost within a few feet of their homes. The wind may blow the snow across level terrain, forming huge drifts against some obstructing object as shown in Fig. 24-3d. Drifting is common on the flat prairies of western United States.

Snowstorms without appreciable winds may be equally severe due to huge quantities of snow. A single inundation of snow in western New York in 1966 measured nearly 100 inches. The region to the south and east of the Great Lakes is often termed a snow belt because of the large quantities of snow it receives, especially in the early part of winter. The lakes are still comparatively warm at this time of the year and dry continental polar air masses moving in from the northwest pick up large quantities of water from evaporation. Orographic ascension occurs as the air masses move eastward over the elevated land area around the lakes. Hence, the moist air masses are cooled and the conditions are favorable for snow. It may be recalled that during the particularly severe winter of 1977, Buffalo received nearly 200 inches of snow. As the winter progresses, the water of the lakes cools and evaporation is less, thus reducing the probability of snowstorms.

Tornadoes

The **tornado** is the most violent of storms. Although it may have less *total* energy than other storms (see Table 24-3a), the concentration of its energy in a relatively small region gives this storm its violent distinction. Characterized by a whirling, funnel-shaped cloud that hangs from a dark cloud mass, the tornado is commonly referred to as a **twister.** See Fig. 24-3e.

Tornadoes occur around the world, but are most prevalent in the United States and Australia. In the United States, most tornadoes occur in the Deep South and in the broad, relatively flat basin between the Rockies and the Appalachians. But no state is immune (Fig. 24-3f). Peak months of activity are April, May, and June with southern states usually hit hardest in winter and spring, northern states in spring and summer. However, tornadoes have occurred in every month at all times of the day or night. A typical time of occurrence is on an unseasonably warm, sultry spring afternoon between 4 and 6 P.M.

Most tornadoes travel from southwest to northeast. They average 5 to 10 minutes on the ground, with a path width about one-eighth of a mile wide and two to five miles long. Forward speed averages about 30 miles per hour, but may reach as much as 70. Wind speed in the rotating funnel is estimated to reach as much as 250 miles per hour in some of the larger, longer-lived tornadoes. Tornadoes are so violent that wind-measuring instruments do not survive a direct hit.

The complete mechanism of tornado formation is not known, due to its many variables. One essential component, however, is rising air, which occurs in thunderstorm formation and in the meeting of cold and warm air masses. This condition is frequently fulfilled during the spring and summer in the midwestern United States where tornadoes are common. In the case of air masses, warm air is forced upward when it collides with a cold air mass. The rapid rising of the warm air, coupled with the frictional and Coriolis forces acting on moving cooler air, produces a circulating spiral. Intersecting jet streams are also thought to contribute to the vortex formation as shown in Fig. 24-3g. As the ascending air cools, clouds are formed which are swept to the outer portions of the cyclonic motion and outline its funnel form. Since clouds form at certain heights (Section 23-2), the funnel may appear well above the ground. Under the proper conditions, a full-fledged tornado develops. The winds increase and the air pressure

Table 24-3a *Comparison of Average-Storm Characteristics*

	Cyclones	*Hurricanes*	*Tornadoes*	*Thunderstorms*
Wind speed	0–50 mph	74–200 mph	100–250 mph	20–30 mph
Width	500–1000 mi	300–600 mi	$\frac{1}{8}$ mi	1–2 mi
Duration	Week or more	Week	Few minutes	An hour or less
Approx. kinetic energy, based on wind velocity (joules)[a]	10^{14}	10^{15}	10^{10}	10^{9}

[a]Energy of atomic bomb $\approx 10^{14}$ joules

Figure 24-3e *Tornado funnel and damage. The high-speed winds and sudden pressure drop accompanying the awesome-looking tornado produces much damage along its path. (Courtesy NOAA.)*

near the center of the vortex is reduced as the air swirls upward. When the funnel is well developed, it may "touch down," or be seen extending up from the earth as a result of dust and debris picked up by the whirling winds.

The alerting system for tornadoes has two phases. A **tornado watch** is issued when atmospheric conditions are ripe for tornadoes, and a **tornado warning** is issued when a tornado has actually been sighted or indicated on radar. Despite radar, satellites, and other sophisticated instruments, most tornado reports are made by volunteer spotters who relay the word of sightings to official warning centers. One of the record years for tornadoes in the United States was 1973. During that year, 1109 tornadoes occurred on 208 days, killing 87 people, injuring 2481 others, and for the first time on record, caused property losses in excess of $500 million.

A great deal of tornado damage results from the sudden reduction in air pressure as it passes and the violent winds. The upward air movement produces a low-pressure region near the center of the tornado, and the air pressure may drop as much as 10 percent in a few seconds as a tornado passes over an area. It is this sudden pressure drop that causes houses and other structures literally to explode. Wind, rain, and thunderstorms usually precede a tornado as the result of the frontal activity associated with its formation. One's natural reaction is to close the windows of the house to keep out the rain. However, this helps to maintain the normal atmospheric pressure of 14.7 lb/in² inside. In the presence of a tornado, the outside pressure is suddenly reduced by 10 percent to approximately 13.2 lb/in². The resulting pressure difference of 1.5 lb/in² or 216 lb/ft² exerts a force outwardly on the walls and roof of the house. The

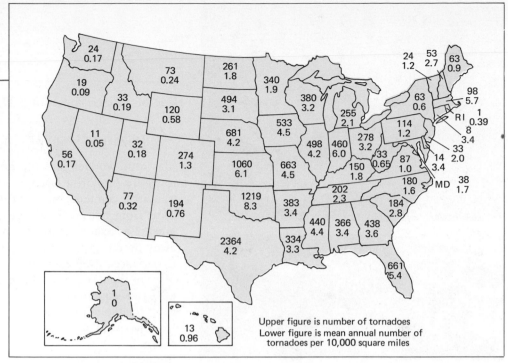

Figure 24-3f *The tornado incidence by state and month for the United States for the period 1953–1973. (Courtesy NOAA.)*

Figure 24-3g *Conditions favorable for tornado formation. Cyclonic motion associated with frontal movement and jet-stream activity combine to produce tornadoes. (Courtesy NOAA.)*

Figure 24-3h *The "lifting" effect of tornadoes. The roofs of buildings are forced upward, or "lifted," during a tornado because of the sudden reduction of pressure outside the house. Walls may also "explode" outward. (Courtesy NOAA.)*

side of a house measuring 30 ft × 10 ft would then feel a force of

$$(216 \text{ lb/ft}^2)(30 \times 10) \text{ ft}^2 = 64{,}800 \text{ lb} = 32.4 \text{ tons!}$$

It is little wonder that the house "explodes." Since roofs are primarily fastened to houses by their own weight, a common prank of a tornado is to lift the roofs from houses. The roof is lifted mainly by the greater pressure inside the house and then blown away by the external high speed winds (Fig. 24-3h). Experienced dwellers in tornado areas open the windows of their houses at the approach of a tornado in hopes of equalizing the air pressure, then quickly head for the storm cellar.

The wind damage accompanying a tornado is enormous. A wind speed of 250 miles per hour is frightening as well as almost inconceivable. Along with the havoc and destruction caused by the wind come stories of remarkable occurrences of a less serious nature. Chickens have been picked up by tornadoes, plucked clean, and redeposited otherwise unharmed. Livestock and barns have been lifted over fences into a neighbor's field. Newspapers have been driven deep into the wood of trees. Documented feats of tornadoes are endless.

A tornado occurring over water is known as a **waterspout.** Although the funnel appears to be a spout of water rising up from the water surface, the upper portions of the waterspout are chiefly composed of whirling clouds like that of a land tornado. Dust devils, or whirling columns of dust, that are common to dry areas are sometimes looked upon as mini-tornadoes.

24-4
TROPICAL STORMS

Tropical storms, or tropical cyclones, are more violent than the cyclones discussed previously (see again Section 24-2). A tropical storm becomes a **hurricane** when its wind speed exceeds 74 miles per hour. The hurricane is known by different names in different parts of the world. In southeast Asia, it is called a **typhoon.** In the region of the Indian Ocean **cyclone** is the term used, while off the Australian coast, the tropical cyclone is a **willy-willy** (Fig. 24-4a). Regardless of the name, the hurricane is characterized by high-speed rotating winds, the energy of which is spread over a large area. A hurricane may range from 300 to 600 miles in diameter and have wind speeds of 74–200 miles per hour.

Hurricanes form over the tropical oceanic regions where the sun heats huge masses of moist air. An ascending spiral motion results, in the same manner as described in tornado formation. When the moisture of the rising air condenses, the latent heat provides additional energy and more air rises up the column. This latent heat is a chief source of the hurricane's energy and is readily available from the condensation of the moist air of its source region. The growth and development of a hurricane is shown in

Figure 24-4a *Tropical storm regions of the world. Tropical storms are known by different names in different parts of the world. (Courtesy NOAA.).*

SEPT. 6	SEPT. 7	SEPT. 8	SEPT. 9	SEPT. 10
SEPT. 11	SEPT. 12	SEPT. 13	SEPT. 14	SEPT. 15
SEPT. 16	SEPT. 17	SEPT. 18	SEPT. 19	SEPT. 20

Figure 24-4b *The development of a hurricane as photographed by a satellite. (Courtesy NOAA.)*

the satellite photograph of Fig. 24-4b. Unlike the tornado, a hurricane gains energy from its source region. As more and more air rises, the hurricane grows, accompanied by clouds, increasing winds, and a central low-pressure area called the eye (Fig. 24-4c). The eye may be 20 to 30 miles wide, and ships sailing into this area have found that it is usually calm and clear with no indication of the surrounding storm. The air pressure is reduced 6–8% (to about 28 inches of Hg) near the eye. Hurricanes move rather

slowly at a few miles per hour. Hence, the pressure drop due to an approaching hurricane may take several hours. This is not as drastic as the pressure drop associated with a tornado. The destruction of a hurricane results chiefly from its winds, which also produce huge waves and resulting flooding along coastal areas. (See the two photographs of Fig. 24-4d.)

Tropical storms form in the regions between 5° and 20° latitude in the late summer (Fig. 24-4a). Below 20° latitude the stabilizing influence of the

Hurricane eye

Vertical height—thousands of feet

30

20

10

0

0   25   50   75

Horizontal distance in miles

Figure 24-4c *A hurricane's eye as seen by satellite and radar. The lower photo is a radar profile of a hurricane. Note the generally clear eye at about 35 to 55 miles from the radar station and the vertical build up of clouds (over 35,000 ft) at the near side of the eye. (Courtesy NOAA and NASA.)*

Figure 24-4d *Hurricane damage from flooding. The high winds of the hurricane produce waves that do much damage along coastal areas. (Courtesy NOAA.)*

high-pressure horse latitudes is small and large quantities of moisture are evaporated from the warm tropical oceans. Thus, in these regions the conditions for hurricane formation are optimum. The breeding grounds of the hurricanes which mostly affect the United States are in the Atlantic Ocean southeast of the Caribbean Sea. As the hurricane forms, it moves westward with the tradewinds, traveling about 100–200 miles a day. During the summer in the northern hemisphere, the northern shift of the doldrums region is accompanied by the exchange of air from the southern to the northern hemisphere over the Mid-Atlantic. The Coriolis force acts on this air movement, deflecting it toward the northwest. As a result, the hurricanes move away from the equator and into the Gulf of Mexico as they travel westward. Monsoon-type effects may also have some influence on the hurricane movement toward land. Those hurricanes which do strike the United States usually hit along the Gulf and south Atlantic coasts. These areas may be lashed by several hurricanes during the late summer hurricane season. Once over a land mass and deprived of its moist air source of energy, the hurricane dissipates. However, if the hurricane proceeds far enough northward, it may be blown back into the Atlantic by the prevailing Westerlies which lie above 30° N. As a result, the coastal area may be hit by a hurricane both coming and going. (See Fig. 24-4e.)

During the hurricane season, the area of their formation is constantly monitored by aircraft and satellite reconnaissance (Fig. 24-4f). These observations permit advanced hurricane warnings along coastal areas, which save lives and reduce property damage. The course of a hurricane is constantly observed, as it may suddenly change its path and strike elsewhere with seemingly malicious intent.

Figure 24-4e *Hurricane paths. Hurricanes that form in the Atlantic travel northwest until they are blown eastward by the prevailing westerlies. As a result, these hurricanes often strike the coastal areas of the United States, coming and going. The satellite photograph above shows a hurricane approaching the east coast of the United States. (Courtesy NOAA.)*

For several hundred years, many hurricanes in the West Indies were named after the particular saint's day on which the hurricane occurred. For the last 25 years, tropical storms and hurricanes have been designated by female names. This practice began in World War II when military personnel named typhoons in the western Pacific after their wives and girl friends. The National Weather Service adopted this means of storm identification in 1953. Under this method the storms are named in alphabetical order. The first hurricane of a season receives a name beginning with A such as Alma, the second begins with B, and so on. Names beginning with Q, U, X, Y, and Z are excluded because of their scarcity.

The practice of giving hurricanes exclusively female names has now been challenged and may be changed so that the yearly list of tropical storms includes male as well as female names.

Figure 24-4f *A satellite view of tropical storms around the world in the northern hemisphere. (Courtesy NOAA.)*

IMPORTANT WORDS AND TERMS

Air mass	Line squalls	Thunderstorm	Tornado watch
Source region	Wave cyclones	Lightning	Tornado warning
Warm air mass	Occluded front	Heat lightning	Twister
Cold air mass	Cold front occlusion	Lightning rods	Waterspout
Front	Warm front occlusion	Thunder	Hurricane
Warm front	Stationary front	Ice storm	Typhoon
Cold front	Rainstorm	Snowstorm	Cyclone
Frontal zone	Cloudburst	Blizzard	Willy-willy
Squall line	Flash flood	Tornado	

LEARNING OBJECTIVES

After reading and studying this chapter, you should be able to do the following without referring to the text:

1. Distinguish the different air masses, particularly those that affect the weather in the United States.
2. Describe the different kinds of fronts and their associated characteristics.
3. Explain the formation of a wave cyclone.
4. Describe the formation and characteristics of a thunderstorm.
5. Explain the cause of ice storms.

6. Discuss the formation and effects of tornadoes.
7. Distinguish between tropical storms and hurricanes and describe their formation and effects.

QUESTIONS

1. How are air masses classified? Explain the relationship between air mass characteristics and source regions.
2. What air masses affect the weather in the United States?
3. What would be the abbreviated designation of an air mass which developed over Siberia? over northern Africa? over the Arabian Sea?
4. What is a front? List the meteorological symbols for four types of fronts.
5. Describe the characteristics and weather associated with cold and warm fronts. What is the significance of the sharpness of their vertical boundaries?
6. Assume a low-pressure air mass approaches and advances over your location. Give the variations in barometric readings and winds you would expect to occur.
7. Explain why the household barometer often lists descriptive adjectives, such as stormy, unsettled, and fair, on the barometer's face instead of direct pressure readings.
8. How do conditions for lightning discharges occur?
9. What type of first aid should be given to someone suffering from the shock of a lightning stroke?
10. What is thunder and why is it sometimes heard as a crash and other times as a rumble?
11. An ice storm is likely to result along what type of front? Explain.
12. What is the most violent of storms?
13. Describe the formation and characteristics of a tornado.
14. Distinguish between a tornado watch and a tornado warning.
15. What is the major source of energy for a tropical storm? When does a tropical storm become a hurricane?

PROBLEMS

1. How far on the average does a cold front travel daily? A warm front?

   Answer: Cold front, 500 mi
   Warm front, 250 mi

2. Sketch top and horizontal views of a warm front occlusion.

3. Using the ideal gas law (Eq. 22.1), obtain an approximation of the increase of air pressure in the vicinity of a lightning flash. (Hint: Assume a constant volume and use a proportion.)

   Answer: A factor of 50

4. If thunder is heard 7 seconds after the observation of a lightning flash, (a) approximately how far away is the storm center? (b) after 12 seconds?

   Answer: (a) 1.4 mi

5. Calculate the resulting force on the ceiling of your classroom in the presence of a tornado. Assume normal atmospheric pressure is maintained within the room.

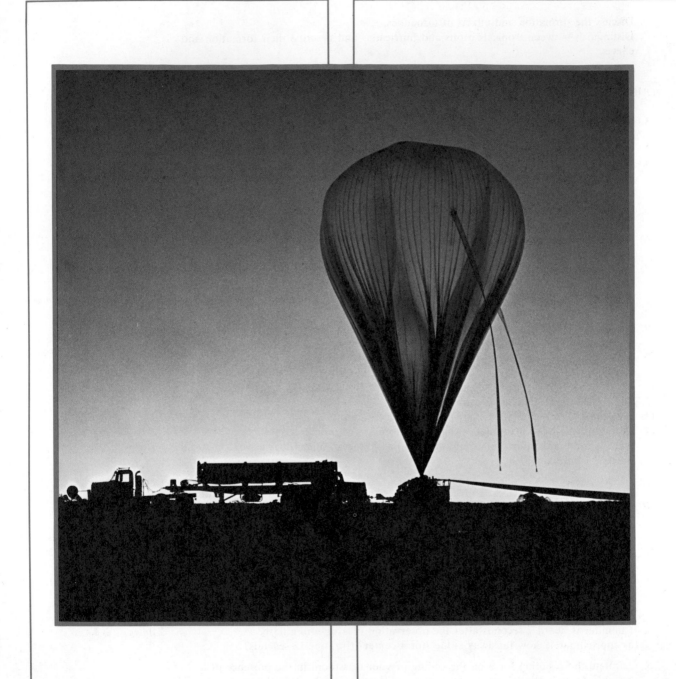

Chapter 25

WEATHER FORECASTING

When ye see a cloud rise out of the west,
  straight way ye say
There cometh a shower, and so it is
And when ye see the southwind blow, ye
  say
There will be heat; and it cometh to pass.

Luke 12:54–55

Human beings have long desired to control the weather. However, little progress has been made toward achieving this desire, and as the next best thing, the preoccupation of observing and attempting to predict the weather developed. The influence of the weather on everyday activity has grown to the extent that this preoccupation has become the sole occupation of many modern meteorologists who are known collectively as "the weatherman."

The weatherman is a much maligned person. When the weather fails to follow the forecast, he is the target of many unflattering remarks. Bad weather which spoils a planned outing is blamed on the weatherman's shortsightedness, if not on the weatherman himself. Although meteorologists are scapegoats, they do a commendable job and on the average are correct about 80 percent of the time. The accurate prediction of atmospheric behavior is a difficult task. Changes in the weather are governed by scientific principles. However, the atmosphere is complex and contains so many variables that meteorologists must combine empirical and scientific knowledge in making reasonable forecasts. Weather data are collected and processed by the most modern means. From the current weather conditions, future conditions are projected according to established behavior patterns. This chapter will deal with a few of the many facets involved in weather forecasting.

25-1

THE NATIONAL WEATHER SERVICE

The United States National Weather Service is the federal organization which provides national weather information. It forms part of the National Oceanic and Atmospheric Administration (**NOAA**—pronounced Noah), which was created within the U.S. Department of Commerce in 1970. Its formation brought together the functions of the Weather Bureau, Coast and Geodetic Survey, Environmental Data Service, National Environmental Satellite Center, Research Laboratories, and various oceanic and marine offices. The role and functions of NOAA are best described by the agency itself:

NOAA was created to improve man's comprehension and uses of the physical environment and its oceanic life.

NOAA will explore, map, and chart the global oceans, their geological cradles, their geophysical forces and fields, and their mineral and living resources. New physical and biological knowledge will be translated into systems capable of assessing the sea's potential yield, and into techniques which the Nation and its industries can employ to manage, use, and conserve these animal and mineral resources.

NOAA will monitor and predict the characteristics of the physical environment—the protean changes of atmosphere and ocean, sun and solid earth, gravity and geomagnetism—in real time, given sufficiently advanced knowledge and technology. It will warn against impending environmental hazards, and ease the human burden of hurricane, tornado, flood, tsunami, and other destructive natural events.

NOAA will monitor and predict such gradual and inexorable changes as those of climate, seismicity, marine-life distributions, earth tides, continental position, the planet's internal circulations, and the effects of human civilization and industry on the environment and oceanic life.

To accomplish these objectives, NOAA will draw upon the talent and experience of its personnel, the wide range of its facilities, and mutually important links between government, universities, and industry. NOAA and its institutional partners will develop the technology and the systems with which to comprehend this broad province of service and investigation—systems leading to effective resource assessment, utilization of environmental data, environmental monitoring and prediction, and, possibly, environmental modification and control. Here, the growing family of satellites, sensors, ships, data buoys, computers, and simulators, which have enriched scientific understanding and provided the base for essential environmental services in recent decades, will find their best achievement.

In these ways, NOAA will improve the safety and quality of life, the efficiency and timing of oceanic hunts and harvests, and man's comprehension, use, and preservation of his planetary home.*

The role and functions of NOAA just stated do indeed make for a sizeable task.

The **National Weather Service** (formerly the Weather Bureau) is primarily concerned with the meteorological aspect of this endeavor.

The U.S. Weather Bureau grew out of the Army Signal Corps, which maintained an early telegraph system that was used for weather reporting. It became a part of the Department of Agriculture in 1891, and then part of the Department of Commerce in 1940. The Weather Bureau's early activities were primarily directed toward weather forecasting as an aid to agriculture. However, the greatest impetus for the Bureau's growth came with the development of aviation,

*NOAA PA 70001 1971

for which up-to-date weather reports are vital. Today, the Weather Bureau is known as the National Weather Service, and its activities are concerned with every phase of the weather. This is expressed in the following excerpt:

Under the National Oceanic and Atmospheric Administration, the National Weather Service reports the weather of the United States and its possessions, provides weather forecasts to the general public, issues warnings against tornadoes, hurricanes, floods, and other weather hazards, and records the climate of the United States. In addition to these basic weather services, the National Weather Service develops and furnishes specialized weather services which support the needs of agricultural, aviation, maritime, space, and military operations. The services of the National Weather Service are supported by a national network of surface and upper-air observing stations, aircraft, satellite systems, communications, and computers. Some 300 National Weather Service offices in cities across the land maintain close contact with the general public, to ensure prompt and useful dissemination of weather information.*

The nerve center of the National Weather Service is the **National Meteorological Center** (NMC) located just outside Washington, D.C. in Suitland, Maryland. It is the NMC which receives and processes the raw weather data taken at numerous weather stations. Each day the NMC receives:

12,000	reports from surface stations
25,000	hourly surface aviation reports
1,500	reports from ships at sea
1,500	upper-air reports

*NOAA letter: A Brief History of the National Weather Service

400	weather reconnaissance reports
2,500	aircraft reports
500	radar reports

All available cloud and temperature data from weather satellites.

The operations of NMC are almost completely automated and data are handled by high-speed computers, printers, and communication systems. Fifty teletype circuits feed weather data directly into computers at rates up to 4500 words per minute. Analyses are made twice daily on observations taken over the northern hemisphere at 0000 hours and 1200 hours Greenwich Mean Time (7 A.M. and 7 P.M. EST). A preliminary analysis is made after an hour and a half of data reception. An operational analysis is made after three and a half hours of data collection, when 80 percent of all incoming data has been received. The computer takes a half hour to analyze the information and the analysis is completed before the data are four hours old. NMC estimates the same analysis would take five people working eight hours to complete. Manual analyses are made on certain portions of the data as a check on the computer as shown in Fig. 25-1a.

Once the analysis is completed, the forecasts are made. These may be for agricultural, marine, aviational, or local use. NMC prepares forecasts for the day, week, and month. This is done on a large computer, which takes an hour and a half to complete a 48-hour forecast for the northern hemisphere. A final analysis of the incoming data is made once a day after 10 hours collection and covers 10 atmospheric levels over the northern hemisphere from the surface to 53,000 ft. Only surface charts are prepared for the southern hemisphere and tropical regions (20° N–20° S latitudes). These are 12 hours old when completed, as they do not receive the immediate attention given to those prepared for the northern hemisphere. From the final analysis, a final forecast is also made, not for general distribution, but as an initial input for the next cycle. A research analysis is done once a day

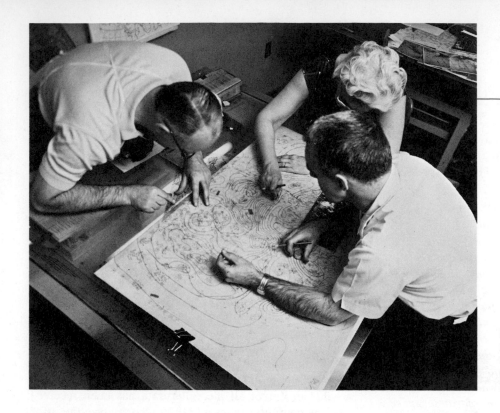

Figure 25-1a *Manual analyses are done on portions of the weather data as a check on the computer. (Courtesy NOAA.)*

for the region between 53,000 and 102,000 ft commercial supersonic aviation.

The analyses and general forecasts are then distributed to local and regional offices which prepare the forecasts for their specific areas. The teletype circuits which brought in the data are used for dissemination. Facsimile circuits transmit charts for immediate use and 2500 weather maps are also distributed daily (Fig. 25-1b). Weather information also is distributed to other nations, since NMC is part of the World Meteorological Organization (WMO) which promotes international meteorological efforts and cooperation. The four phases of collecting and processing, analysis, forecasting, and distribution are all completed within 12 hours, in time for the next cycle to begin.

Other central National Weather Service organizations deal with specialized weather conditions, such as hurricanes, tornadoes, and severe storms. From these central organizations, data go to Forecast Offices, which have the responsibility for warnings and

forecasts for states, or large portions of states, and assigned zones. State forecasts are issued twice daily for a time period up to 48 hours. An extended outlook, up to five days, is issued daily for these same areas. The forecast organizational structure is shown in Fig. 25-1c.

Weather Service Offices represent the third echelon of the forecast system. Local forecasts are adaptations of state forecasts. They are issued to meet local requirements and to provide general weather information to the public. The local forecasts are distributed by telephone services, direct radio and television broadcasts, including VHF-FM (very high frequency-frequency modulated) radio in some areas, and local teletype circuits (Fig. 25-1d).

A major task of the National Weather Service is getting the weather information to the general public. This is primarily accomplished by newspapers, radio, and television. To aid in this operation, the National Weather Service maintains VHF-FM broadcasting facilities near major population centers which trans-

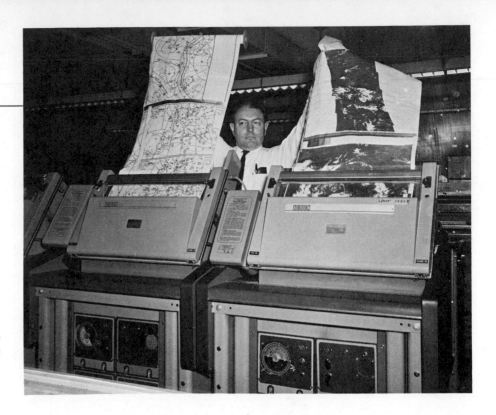

Figure 25-1b *Facsimile map transmission. Facsimile transmission allows fast distribution of weather maps. (Courtesy NOAA.)*

Figure 25-1c *Structure of the National Weather Service's forecast organization. (Courtesy NOAA.)*

| TROPICAL ANALYSIS CENTER | NATIONAL HURRICANE CENTER | NATIONAL METEOROLOGICAL CENTER | NATIONAL SEVERE STORMS FORECAST CENTER |

FORECAST OFFICES

ALBANY	FORT WORTH	OMAMA
ALBUQUERQUE	GREAT FALLS	PHILADELPHIA
ANCHORAGE	HONOLULU	PHOENIX
ATLANTA	INDIANAPOLIS	PITTSBURGH
BIRMINGHAM	JACKSON, MS	PORTLAND, ME
BOISE	KANSAS CITY	PORTLAND, OR
BOSTON	LITTLE ROCK	RALEIGH
BUFFALO	LOS ANGELES	SALT LAKE CITY
CHARLESTON, WV	LOUISVILLE	SAN ANTONIO
CHICAGO	LUBBOCK	SAN FRANCISCO
CLEVELAND	MEMPHIS	SAN JUAN
COLUMBIA	MIAMI	SEATTLE
DENVER	MINNEAPOLIS	SIOUX FALLS
DES MOINES	NEW ORLEANS	ST LOUIS
DETROIT	NEW YORK	WASHINGTON, DC
	OKLAHOMA CITY	

290 WEATHER SERVICE OFFICES

USERS

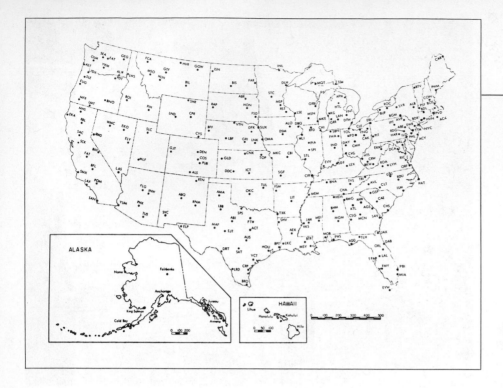

Figure 25-1d *The locations of offices in the United States issuing local weather forecasts. (Courtesy NOAA.)*

mit weather information continuously 24 hours a day. Through the efforts of the National Weather Service up-to-date local weather information readily available to anyone desiring it. Special weather conditions such as storm warnings are made available as soon as they are known. The National Weather Service is dedicated to getting the weather information to the public by the fastest means available. However, inquiries and calls to the local offices often hamper their operations, especially during a time of severe weather. Weather information should be obtained from the local news media, which receive the latest information on teletype circuits. General questions on weather information may be addressed to the NOAA's Office of Public Information in Silver Spring, Maryland.

There is an exception to the prompt distribution of weather information. In wartime, weather data are of great military value and become classified information. During World War II weather reports were not made public except in cases of severe weather, in which warnings were necessary to prevent property damage and loss of life. The news media were pro-

hibited from mentioning anything that pertained to the weather. Farmers were simply told it would be a good day to make hay or a rained-out ball game was called off due to conditions beyond control. As one sportscaster put it, "Well folks, I can't say anything about the weather, but that isn't perspiration on the pitcher's face."

As technology increases and facilities expand, better weather forecasts become available. It is estimated that the predictions of the National Weather Service meteorologists are about 80 percent accurate. This, of course, depends on the weather element and location. For example, predictions of no rain in parts of Arizona during the summer are almost 100 percent correct. The method of reporting the weather has also changed. In the past, forecasts of precipitation were given using such terms as probable, likely, or occasional. Precipitation forecasts are now expressed in "percent probabilities," such as a 70 percent chance of rain. The percent probability, given in such a forecast, results from a consideration of two quantities: (1) the probability that a precipitation-producing storm will develop in, or move into, the forecast area;

and (2) the percentage of the area which the storm is expected to cover. Thus, in summer when storms (convectional) tend to be more isolated, or scattered, the probability that your immediate area will get rain tends to be less than in the winter when frontal storms are more prevalent. The percentage probabilities give the public a better indication of what the weather might be.

Although many hours of work and endeavor go into making the daily weather forecasts as accurate as possible, they sometimes fail to hold true. But after all, if the weatherman were always correct, wouldn't it take a bit of the fun out of life?

25-2
DATA COLLECTION AND WEATHER OBSERVATION

In 1870 weather data were taken by the Army Signal Corps at 24 stations. Today, weather observations are collected from approximately 1000 land stations, six fixed ocean stations, and several hundred merchant vessels of all nationalities. Many of the land stations are associated with airport operations. Volunteer observers also supply climatology data from about 12,000 substations around the country. Weather stations are concerned with the basic meteorological measurements of temperature, pressure, humidity, precipitation, and wind direction and speed, as discussed in Chapter 21. The instruments used are also basically the same, but have official specifications and in some cases may be automated to supply continuous readings.

A typical installation for data collection is shown in Fig. 25-2a. The small hut to the left is an instrument shelter. It is constructed with louvered sides, a ventilated floor, and a double roof with an air space between. This construction permits air measurements to be taken free from insolation influences. The **instrument shelter** is designed to hold a thermograph, a maximum-minimum thermometer, and a sling psy-

Figure 25-2a *A typical installation for the collection of weather data. Shown are an instrument shelter (left), containing temperature and humidity measuring devices; a wind vane and anemometer mount; and a rain gauge. (Courtesy NOAA.)*

chrometer. A **thermograph** is an automatic temperature-recording device. A **hydrothermograph** is sometimes available and records the relative humidity by some hygrometric means as well as the temperature, thus eliminating the need for the psychrometer. The relative humidity and temperature measurements are taken at a specified distance of six feet above ground level.

The maximum and minimum daily temperatures are important weather observations. There are measured with a **maximum-minimum thermometer.** This

Figure 25-2b *Principles of maximum-minimum thermometers. (a) The constriction in the capillary bore prevents the mercury from returning to the bulb and the thermometer retains its highest reading. (b) Surface tension draws the index rod to the lowest descent of the liquid column surface. When the temperature increases and the alcohol expands, the liquid flows around the index, leaving it at the lowest temperature reading.*

consists of a set of two thermometers, one of which records the maximum temperature, and the other the minimum temperature.

The **maximum thermometer** is a mercury thermometer with a constriction in the lower part of its capillary bore (Fig. 25-2b). The mercury passes through the constriction as the temperature increases, due to the pressure of the expanding mercury in the bulb. When the temperature decreases, the mercury is unable to pass back through the narrow constriction. Thus, the column of mercury above the constriction indicates the highest temperature reached. The maximum thermometer is reset by shaking. Shaking forces the mercury back into the bulb and readies the thermometer for a new reading. Clinical thermometers are of similar construction and are maximum thermometers. A common error in taking one's temperature at home is forgetting to reset the thermometer.

The **minimum thermometer** is an alcohol thermometer that contains a thin-colored glass rod in its

bore. This dumbbell-shaped rod is called the index. The surface tension of the alcohol surface draws the index with it to the lowest point of descent of the liquid column surface. When the temperature increases and the alcohol expands, the liquid flows around the index, leaving it at the lowest position. Thus, the upper end of the index marks the lowest or minimum temperature reached. In operation, the minimum thermometer is positioned horizontally; vertically, the glass index would fall through the liquid to the bottom of the tube. The thermometer is reset by tilting it downward.

The tall cylindrical object on the right in Fig. 25-2a is an 8-inch **rain gauge.** The gauge automatically weighs and records the accumulated precipitation. The collection bucket of the rain gauge is mounted on a weighing mechanism with a scale that converts the weight of the rain to equivalent inches. The rain gauge is also used for snow measurements. The collection bucket is sprayed with special chemicals, which melts the snow so its water equivalent can be measured.

Unlike the rain gauge, open exposure and immediate locale is not critical for barometric readings, and barometers are usually kept indoors because of their intricate construction.

Radar (*RA*dio *D*etecting *A*nd *R*anging) is used to detect and monitor precipitation, especially that of severe storms. A radar scan of a rainstorm is shown in Fig. 25-2c. Radar operates by sending out electromagnetic waves and monitoring the returning waves which have been reflected back from some object. In this manner, the location of the object may be determined. The objects of interest in weather observations are storms and precipitation.

The most common type of radar used by the National Weather Service is a relatively high-powered (500 kW) system operating on a 10-cm wavelength. This type of radar has a range of about 250 miles. Radar installations are located mainly in the tornado belt of midwestern United States and along the Atlantic and Gulf coasts where hurricanes are

Figure 25-2c *Rainfall intensity as seen on a radar screen. (Courtesy NOAA.)*

probable. Additional radar information is obtained from traffic control radar systems at various airports. The NMC is linked to radar stations by telephone circuits, and its meteorologists can view the current radar scans around the country on television screens.

Upper-air observations are important in the collection of meteorological data. These observations are made chiefly with radiosondes (radio sounding). A **radiosonde** is a small package of meteorological instruments combined with a radio transmitter. These are carried aloft by balloons, and data are transmitted to ground receiving stations (Fig. 25-2d). The wind direction and speed may also be obtained by tracking the flight of the radiosonde. The carrier balloon will eventually burst, and the radiosonde descends slowly by means of a small parachute. This prevents damage to the instruments and anything in the radiosonde's path on landing. Directions on the radiosondes request the finder to return the radiosonde to the National Weather Service where they are reconditioned

and flown again. Approximately 150 National Weather Service offices engage in radiosonde observations.

Rockets are also used in upper-atmosphere data collection, but probably the greatest progress to general weather observation has come with the **weather satellite.** After much experimentation, the first fully operational weather satellite system was launched in 1966. The current weather satellites of the TIROS series have a "cartwheel" design (Fig. 25-2e). After the satellite is put into orbit, it is turned so as to spin on an axis perpendicular to its orbital plane. In this manner, the satellite rolls in orbit like a wheel several hundred miles above the earth. Cameras point out from the rim of the satellite and take pictures when facing the earth. This provides continuous photographs of the clouds over the earth at least once a day. Its two cameras have wide-angle lenses and are mounted 180° apart. Pointing downward once per satellite rotation, an automatic picture transmission

Figure 25-2d (Left) *A radiosonde about to be carried aloft by a weather balloon. The man to the left holds the radiosonde instrument package. (Courtesy U.S. Air Force.)*

Figure 25-2e (Below) *A wheel satellite. The wheel spacecraft for TOS (Trans Orbiting Satellite) rotates as its cameras scan the earth and adjacent atmosphere. (Courtesy NOAA.)*

Figure 25-2f *Satellite nephanalysis. Satellite scans are put together to give a wide view of the earth's weather picture. Note the opposite cyclonic circulations in the northern and southern hemispheres. (Courtesy NOAA.)*

(APT) camera automatically takes and transmits a picture at specified intervals while the satellite is in the daylight. The picture signals are projected on special television screens and photographed. Coastlines and longitude and latitude grids are prepared by a computer and electronically combined with the picture signal. The cloud cover maps prepared from these gridded photographs are called **nephanalyses.** A nephanalysis for the world is shown in Fig. 25-2f. Ground stations can acquire two or three pictures per orbit. To photograph the entire United States about three orbits are required (Fig. 25-2g). An average time for a satellite to complete a revolution of the earth is on the order of 100 minutes, depending on the orbit.

With satellite photographs, meteorologists have a panoramic view of the weather conditions. The dominant feature of the photograph is, of course, the cloud coverage. However, with the aid of radar, which uses a wavelength that picks up only precipitation, the storm areas are easily differentiated from regular cloud coverage. Successive photographs of these storm centers give an indication of their movement, growth, and behavior thus aiding weather forecasting.

Another important feature of the satellites is **infrared measurements.** As discussed in Chapter 21, the energy and wavelength of emitted radiation depends on the temperature of the radiator. Objects with normal temperatures emit radiation energy in the infrared region of the spectrum. Hence, using a spectrometer to determine the wavelength in this region provides a method of determining the temperature. See Fig. 25-2h. Normally, temperature measurements versus altitude are taken by conventional methods using radiosondes. However, recent advances in satellite infrared spectrometers now permit these measurements to be taken by satellite. Fig. 25-2i shows a comparison of the temperature profiles taken by a radiosonde and a Nimbus III satellite infrared spectrometer (SIRS). The SIRS measurements offer a great advantage since a satellite covers all areas of the earth, some of which have never been monitored by radiosonde due to their inaccessibility.

ACQUISITION RANGE
CIRCLE

3rd ORBIT

2nd ORBIT

1st ORBIT

Figure 25-2g *Nephanalysis of the United States and adjacent ocean areas (top); (below) the TIROS weather satellite makes three orbits to get a complete nephanalysis of these areas. (Courtesy NOAA.)*

TIROS II INFRARED MEASUREMENTS

1. TEMPERATURE, TOP OF WATER VAPOR LAYER

2. SURFACE TEMPERATURE NIGHT CLOUDY AREAS CLOUD TOP TEMPERATURE AND HEIGHT

3. REFLECTED SUNSHINE

4. EMITTED HEAT RADIATION

5. LOW RESOLUTION CLOUD PICTURES

61-95

Figure 25-2h *Some satellites monitor the infrared radiation of objects, thereby measuring their temperatures which is proportional to the wavelength of the radiation. (Courtesy NOAA.)*

This represents a major step in making the meteorologist's global weather picture more complete.

Special weather observations are made to provide advance notice of such things as severe storms, floods, tornadoes, and hurricanes. The National Weather Service issues notices of varying degrees. For example, when a tropical storm moves within a few hundred miles of the coast a small-craft warning is issued. Advisories warn small-craft operators to take precautions and not to venture into the open sea. When wind of 38–55 miles per hour are expected, a gale warning is added to the advisory message. When winds of 55–74 miles per hour are expected, a storm warning is added. If the tropical storm continues to gain strength and its advance threatens coastal and inland areas, a hurricane watch is added to the advisory. A **hurricane watch** means hurricane conditions are a definite possibility. It does not mean the conditions are a certainty, but it does alert people who would be affected to be prepared to act quickly, should the hurricane watch be changed to a warning. A **hurricane warning** indicates that hurricane conditions are expected within 24 hours and appropriate precautionary actions should be taken. Because of the

Figure 25-2i *A comparison of radiosonde temperature data (dotted line) and that of Nimbus III "SIRS" infrared data (solid line). (Courtesy NOAA.)*

28 EPISODES WEST

59 EPISODES EAST

0 200 400
MILES

Figure 25-2j *Air pollution potential. (Courtesy NOAA.)*

Figure 25-2k *Air pollution. A modern problem. (Courtesy NOAA.)*

long time of development and migration, the hurricane requires a great deal of observation. Radar-equipped airplanes, or **"hurricane hunters,"** seek out and track tropical storms and hurricanes. Prior to the advent of satellites, hurricanes were almost exclusively monitored by aircraft and surface ships. This is a dangerous business; however, to be able to predict a hurricane's behavior, on-the-spot weather data not furnished by satellites, such as air pressure and wind velocity, are essential.

One of the more recently initiated atmospheric observations is the **air pollution potential.** As discussed in Chapter 23, the pollution problem becomes severe when certain atmospheric conditions prohibit the dispersal of pollutants. Using upper-air observations, the NMC prepares wind and air current data which are analyzed for potential pollution conditions. Advisory forecasts are issued to areas in which stagnation conditions are expected to persist for at least 36 hours (Figs. 25-2j and 25-2k). The air pollution potential provides an opportunity for air pollution control and research, particularly in studying the influence of meteorological factors on the pollution conditions observed (see Chapter 26).

Another increasingly common atmospheric observation not related to weather forecasting is the pollen count. These counts are made by health officials and are issued to give hay-fever sufferers an indication of the amount of airborne pollen which may aggravate their condition.

Today, questions on pollution, its extent, and how it may be abated are increasingly common. To provide data needed to answer these questions, several government services have been reorganized to better monitor our environment. Several such NOAA organizations are: (1) The National Environmental Satellite Service. This organization plans and operates environmental satellite systems, gathers and analyzes satellite data, and develops new methods of using satellites to obtain environmental data. As environmental satellite technology matures, sensors will be added to measure additional atmospheric charac-

teristics, and to provide data on solar, ionospheric, and other geophysical phenomena. (2) The Environmental Research Laboratories. Headquartered in Boulder, Colorado, the laboratories conduct fundamental investigations needed to improve man's understanding of the physical environment. (3) The Atmospheric Physics and Chemistry Laboratory. Also located in Boulder, Colorado, the emphasis of this facility is on the development of methods of practical, beneficial weather modification. Atmospheric observations by these three organizations will not only help answer questions related to ecology, but also provide knowledge for better weather predictions.

25-3
WEATHER MAPS

To help analyze the data taken at the various weather stations, maps and charts are prepared which display the weather picture. Since these charts present a synopsis of the weather data, they are referred to as **synoptic weather charts.** There are a great variety of weather maps with various presentations. However, the most common are the daily weather maps issued by the U.S. National Weather Service. A set consists of a surface weather map, a 500-millibar height contour map, a map of the highest and lowest temperatures, and one showing the precipitation areas and amounts. Examine those of Fig. 25-3a carefully.

The Surface Weather map presents station data and analysis for 7 A.M. EST, for a particular day. This map shows the major frontal systems and high- and low-pressure areas at the constant surface level. Isobars are represented by solid lines, and prevalent isotherms (the lines of constant temperature) are indicated by dashed lines (in degrees Fahrenheit). The height contour map gives a representation of a surface of constant pressure, in this case, 500 millibars. Contours of 300- and 700-millibar pressure are also common. The temperature map gives the highest and

Figure 25-3a *Daily weather maps. Weather maps are issued daily by the National Weather Service. The maps from which these drawings were made gave many more locations and data than are shown here. (Redrawn courtesy NOAA.)*

HIGHEST AND LOWEST TEMPERATURES

PRECIPITATION AREAS AND AMOUNTS

500- MILLIBAR HEIGHT CONTOURS
AT 7:00 A.M., E.S.T.

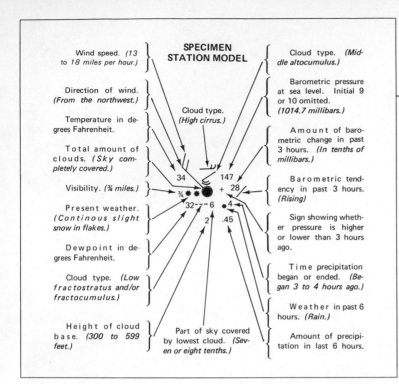

Figure 25-3b *The weather data from each reporting station are arranged in this model format. (Redrawn courtesy NOAA.)*

lowest temperatures for weather stations throughout the country. The complete precipitation picture is given by the Precipitation Areas and Amounts map. Major precipitation areas are shaded. On the Surface Weather map a station model, as shown in Fig. 25-3b, gives the pertinent weather data at the station's location. The stations shown in Fig. 25-3a are only a fraction of those included in the National Weather Service's operational weather maps. Also, the station model information has been abbreviated.

Figure 25-3c shows how contour maps are prepared. A surface map shows isobars at surface level, whereas a height contour map shows height contours at a constant pressure. The contours indicate the heights of the designated pressure in feet above sea level. This is analogous to a topographical map which indicates the contour of the earth's surface. If constant pressure maps were three-dimensional and viewed from the side, the hills and valleys represented by the contours would be the surfaces of equal barometric readings. On the height contour map, the wind symbols show the wind directions and speeds at the 500-millibar level. Isotherms also appear on the

height contour maps. Since many weather conditions and changes depend on the dynamics of the upper atmosphere, these maps are important in forecasting future conditions. Note the correspondence of the low-pressure regions on the height contour map and areas of precipitation on the precipitation map in Fig. 25-3a.

From a knowledge of the general behavior of fronts, cyclonic motion, and air masses studied in the previous chapters, reasonable weather forecasts may be made from surface maps. This is not always easy, as the weather conditions may be quite complex. For example, note the complexity of the frontal systems over the northwestern United States and Canada shown in Fig. 25-3d. The student may find it difficult to predict the weather changes for that region. At the National Meteorological Center, a computer projects the weather conditions based on stored empirical knowledge, statistical methods, and continuous data. A **prognosis chart** (prog chart) is then printed out, based on the computer's analysis of what the weather conditions will be at some time in the future. However, some unpredictable atmospheric occurrence may take place at any time and make the projection incorrect. One should now begin to realize how difficult it is to accurately forecast the weather.

Figure 25-3a contains a classical cyclonic formation, with a stationary front extending toward the northwest into Canada. In the low-pressure occluded front region, the conditions are favorable for convective and frontal precipitation, and its occurrence is indicated by the colored region over Nebraska, Iowa, and Minnesota. Moreover, the counterclockwise circulation around the low may draw in moist air from the Great Lakes region. The clockwise rotation of the adjacent highs to the west carries the precipitation

farther west into Wyoming, as shown on the precipitation areas map. The cyclonic disturbances off the east coast contribute to the precipitation in that area.

To forecast the next day's weather, one must consider the probable movement of the fronts and pressure areas. Localized forecasts are more concerned with the internal dynamics of these phenomena. Because of the general west-to-east movement of weather across the United States and the circulation around a low-pressure area, the low-pressure system over the Midwest in Fig. 25-3a should move northeast carrying its precipitation with it. The high-pressure areas would be expected to move over the middle

Figure 25-3c *The preparation of a pressure contour map. Such contour maps are prepared by automatic curve traces. (Courtesy NOAA.)*

Figure 25-3d *Weather maps for the day following that of Fig. 25-3a. Notice the complex array of frontal systems on the surface map. Complicated atmospheric conditions make weather forecasting a difficult task. (Redrawn courtesy NOAA.)*

Figure 25-3e (Above and p. 510) *A series of three surface maps showing the advance of a cold, high-pressure cP air mass across Canada and the United States. Notice the temperature and precipitation variations as the air mass moves across the country. (Redrawn courtesy NOAA.)*

United States, and the low-pressure region off the California coast would move inland. The speeds of the high- and low-pressure areas vary. Generally, highs and lows cross the United States in three to five days.

The northeastern United States would experience a warm front, which usually indicates an advancing low. The center of a low-pressure region is normally 0.5 inch of mercury or more below normal atmospheric pressure and increases radially outward. Hence, aside from the small deflection of the Coriolis force, the winds are generally toward the center of the low. As a low from the west approaches an observer, the barometer falls and the winds are generally from the east. As the low passes, the barometer begins to rise and the winds shift, coming from the west. In Fig. 25-3a, these conditions should occur in the Great

Lakes region over which the low should pass. The wind conditions are reversed for the passage of a high and would be expected in the northern Plains states. The southeastern United States should feel the effect of the advancing cold front. The upward displacement of the warm air by the advancing cold front will produce a slight fall in the barometric pressure, followed by a sharp rise as the front passes, due to the incoming high. Little change would be expected for the northwest states, which will remain under the influence of high pressure. California may expect precipitation from the advancing low.

Figure 25-3d shows the next day's weather maps which prove these predictions to be fairly accurate. The entire central United States is now dominated by high pressure. Highs are characterized by descending air. The descending air becomes warm and can hold more moisture, so that precipitation will not occur. In the summer, these conditions are responsible for the high humdiity and frequent droughts in the central United States. Note also that high-pressure areas are usually larger than low-pressure areas, with widely spaced and less streamlined isobars. In the winter, a high may enter and move across the United States in the form of a cold, cP air mass. This occurrence is evident in the maps of Fig. 25-3e. (The nephanalysis for the third surface map appears in Fig. 25-2g.) Hence, it can be seen that weather forecasting is, to a large part, the projection of the movement of fronts, pressure areas, and air masses. This process is complicated by a multitude of local variations and influences.

25-4
FOLKLORE AND THE WEATHER

Folklore and legends have been associated with the weather from the beginning of mankind. People in the early civilizations worshipped deities whose actions were manifested by weather phenomena. Myths were created to account for those occurrences which could not be otherwise explained. Thunder and lightning, for example, indicated the presence of the violent god Thor. Clouds were the cattle of the sun god Apollo, grazing in the meadows of heaven.

Men judge by the complexion of the sky
The state and inclination of the day.

Shakespeare (*Richard II*)

As human beings progressed and gave up the mythological interpretations of the weather, they began associating its behavior with things they observed. In this manner a similar observation might allow a prediction of future weather conditions. It is these observations that have come down to us as folklore. Usually in the forms of sayings or verse, they have been handed down by word of mouth from generation to generation and often appear in literature. Some of these sayings are well founded and can be explained scientifically. Others, seemingly without any valid explanation, live on, much the same as superstitions. A few of these well-known sayings will be examined in this section. Those which are related to the meteorological explanations of the previous chapters will be pointed out. The others are left to the imagination of the reader.

Many amateur weather forecasters closely observe the activities of birds and animals. The antics and instincts of these creatures are often without explanation. However, they form an essential part of folklore. The braying of a donkey or ducks flapping their wings, for example, are thought by some to be indicative of a rainstorm. Other observations are of a seasonal nature. Marks on a caterpillar or the amount of food stored by squirrels are used to predict the severity of the coming winter. The southward flight of birds may indicate an early winter, while wild geese flying north is taken as a sure sign that warm weather is coming soon.

Probably the most celebrated weather predicting animal is the ground hog. Legend has it that the ground hog emerges from his winter hibernation each year on February 2 to check on the approach of spring. If he sees his shadow, he returns to his burrow and six more weeks of winter are imminent. It is somewhat surprising that his shadow, which results from the sun, should warn the ground hog of continuing winter, but that's the legend.

More dependable observations result from the sun and its effects.

Above the rest, the sun who never lies
Foretells the change of weather in the skies.

Vergil

There are several sayings about the sun and the weather indications it gives. Among these are the following:

The weary sun hath made a golden set
And by the bright tracks of his fiery car
Gives token of a goodly tomorrow.

Shakespeare (*Richard II*)

If the red sun begins his race
Be sure the rain will fall apace.

The Pharisees also with the Sadducees came
and tempting desired him that he would show
them a sign from heaven. He answered and
said unto them, When it is evening ye say, It
will be fair weather for the sky is red. And in
the morning, It will be foul weather today for
the sky is red and lowering. Oh ye hypocrites,
ye can discern the face of the sky, but can ye
not discern the signs of the times.

Matthew 16:1–4

The weather predictions alluded to in the Bible are more commonly stated:

Red sky at night, sailors delight
Red sky in the morning, sailors take warning.

"Rainbow" is often substituted for "red sky" in the above saying. All these sayings involve the red sky, which we often see at sunrise and sunset. What then causes the sun and sky to appear red? When the sun is on the horizon, the sunlight travels further through the atmosphere to reach us than when the sun is overhead. The blue portion of the sunlight is normally scattered (Section 21-4). However, if the air contains impurities such as dust, more of the longer wavelengths are scattered as the sunlight travels near the surface of the earth, with the result that only the red portion of the spectrum may reach an observer. This, of course, makes the sky and sun appear red. This condition is enhanced if there is a stable high-pressure region between the observer and the sun, as the high pressure holds air contaminants near the earth and scattering is increased. As discussed in

Section 24-2, highs are generally associated with good weather so if one sees a red sky in the evening, it is quite probable that there is a high-pressure area to the west, and good weather will accompany it as it moves eastward, delighting sailors in the westerly wind zone where this weather movement applies. Should the red sky occur at sunrise in the east, the high has probably passed and will usually be followed by a low-pressure system which is generally associated with poor weather.

However, one must temper this prediction with a knowledge of the location of rain-producing clouds. When the sun sets, clouds in the east often appear pink, due to the reflection of the red portion of the sunlight, and some people consider this to be a red sky. At sunrise the reflecting pink clouds may be in the west. Since our general weather movement is from west to east, the clouds in the east have already passed us and should not affect the future weather. However, should there be clouds in the west in the evening, there is a good chance that it may rain during the night or early in the morning. If there are clouds in the west at sunrise, they may produce precipitation as they move eastward, bringing poor weather. These conditions may occur with or without an accompanying red sky. Hence, the above saying deserves some merit, but must be carefully interpreted.

Another saying describing an observed phenomenon of sunlight is as follows:

When the sun draws water, rain will follow.

The sun's rays are often blocked by dense clouds. Occasionally a cloud may be thin enough in a small area to allow the sunlight to shine through. In the humid air associated with the clouds, the sunlight may be scattered by fine water droplets or other particles, giving the effect of a fanlike ray of sunlight extending from the earth to the cloud. This effect may also be observed when sunlight shines through dense leaf coverage in a thickly wooded area. There is no drawing up of water, but clouds *are* available for the production of rain.

Mackerel sky and mare's tails
Make lofty ships carry low sails.

Trace in the sky the painter's brush
The winds around you soon will rush.

The cirrus and cirrocumulus clouds referred to in the above sayings usually precede the approach of a warm front (Fig. 24-2b). As the front approaches, the warm air rising over the cold air mass is likely to produce the winds predicted by these verses.

The wind in the West
Suits everyone best.

In general our good weather comes from the west as opposed to the north and east, so the saying is to some degree valid. It must be remembered, however, that some locations on the earth receive very poor weather from the west. Hence the saying is not universal.

When the morn is dry
The rain is nigh
When the morn is wet
No rain you get.

Or in different form and meter.

When the grass is dry at morning light
Look for rain before the night;
When the dew is on the grass
Rain will never come to pass.

The reasoning behind these rhymes is obvious. If condensation in the form of dew has occurred, the air will have a lower relative humidity and rain will be unlikely. However, a better last line to the second poem might read "Rain will *seldom* come to pass," as the air at higher elevations may have sufficient moisture to produce rain. Also, the lack of dew in the morning is not a positive indication of rain. The general validity of these sayings is, therefore, somewhat questionable.

The moon also shares considerable prominence in weather folklore. Many of its related sayings are unfounded. Among these are the indication of rain when "the new moon holds the old moon in its arms." This refers to the time when the crescent new moon is near the lower portion of the moon and appears to hold the upper darkened portion. The rain supposedly results from water being spilled from the saucer-shaped new moon. Such sayings are best answered by the countersaying,

Moon and weather may change together
But a change of the moon does not change the
weather.

More reasonable predictions are given by the following:

Clear moon, frost soon.

Or:

A ring around the moon is a sure sign of rain.

Or:

When the stars begin to huddle
The Earth will soon become a puddle.

The first of the above sayings refers to the lack of cloud coverage which acts like an insulation to keep the earth warm. In the absence of clouds, the moon is clearly seen and the land masses cool quickly, making frost likely. The latter two sayings refer to the appearance of the moon and stars as viewed through cirrostratus clouds. The moon appears with a diffuse halo, while the indistinct stars appear to be closer together. Cirrostratus clouds normally precede an approaching warm front, which is accompanied by turbulent weather, usually in the form of rain.

Although not directly related to meteorology, the moon and its phases are often referred to when planting crops. Many gardeners follow the *Farmer's Almanac*, which gives the periods of the proper phases of the moon for planting the proper crops. In general, crops which produce above ground are to be planted in the "light of the moon," or in the waxing phase. Crops which produce below the surface are to be planted in the "dark of the moon," or in the waning phase. For example, it has been reported that a certain gardener had difficulty in keeping dirt on potatoes, which were planted in the light of the moon. Wood is also said to be affected by the phases of the moon. A board lying on the ground will supposedly curl up at the ends in the light of the moon and stick firmly to the ground in the dark of the moon. Whether or not this is true is subject to doubt, but some builders will put on roofs of wooden shingles only during the waning phase or dark of the moon.

Other weather sayings include:

Sound travelling far and wide
A stormy day will betide.

Sound waves may be reflected by air layers of different temperatures in the atmosphere. As a result, the sound is heard at a considerable distance where it is reflected back to earth. A reflecting cold air layer may also be the source of precipitation, accounting for the stormy day.

Rain before seven, stops before eleven.

This is a reasonably safe prediction for convectional precipitation. If it were raining prior to 7 A.M., there would probably be little moisture left in the precipitating cloud by 11 A.M. Moreover, the sun would be rising high in the sky and cause the temperature to rise above the dew point temperature. The clouds would then dissipate and the rain cease, which

would bear out the prediction. However, for frontal precipitation, which is controlled by huge air masses, the prediction may not prove to be so accurate.

Leaves turn silver before a rain.

When the wind blows the leaves so that the shiny underside is exposed, they take on a silvery appearance. This turning up of the leaves may result from vertical air motion which may cause cloud formation. If a nimbus cloud develops, the leaves' prediction of rain will be fulfilled.

It smells like rain.

Before a rain there may be a musty, or earthy, smell in the air. Since convectional precipitation is associated with low pressure and rising air, air and gases in the ground may diffuse out, giving rise to the "smell of rain."

Several common sayings which are without scientific foundations are

Rain on the first Sunday,
rain every Sunday [of the month].

Rain on Easter Sunday,
rain for seven straight Sundays.

Rain on Good Friday,
the Saint is pouring water on a flat rock.

The latter demonstrates how the meanings of folklore saying may be disguised. Its interpretation is that if it rains on Good Friday, the following summer will be dry, such that the summer rains will run off the hard, dry ground as though it were a rock.

Another special day for rain is St. Swithin's Day, which is July 15. Should it rain on St. Swithin's Day, legend has it, 40 days of rain will follow. St. Swithin was an English bishop whose wish it was to be buried in the open church yard. When sainted and moved into the church, his spirit protested with 40 days of rain. A rain occurring on the anniversary of this event will supposedly arouse St. Swithin's spirit for a repeat performance.

Certain weather occurrences are given special names. A period of warm weather in October or November is sometimes referred to as **Indian Summer.** The seasonal cooling at this time of year gives rise to low-lying fogs. It is said that thin fogs lying near the tops of corn shocks reminded the early settlers of smoke coming from Indian tepees, hence the name Indian Summer. There is nothing uncommon about a brief warm "spell" this late in the year, as the season changes from autumn to winter. Its special significance is chiefly psychological. People are aware of the coming harsh winter and are sentimental toward this last warm reminder of the pleasant summer. The fondness for summer prompts a special name for this last trace. A wintry period in April is not so honored as Indian Winter, but is deplored and quickly forgotten. More significance is given to the break in winter called the January thaw which often occurs in February.

Some people are able to forecast a change in the weather or rain by an ache in the knee or some other joint. This was thought to be a joke for some time; however, it is now believed that changes in the pressure and humidity affect the aches and pains of rheumatic joints. Others who are not afflicted may rely on folklore to help predict weather behavior. The date of the first snow is sometimes taken as the number of snows that will occur during the winter. Those who believe in this prediction become very discriminating between an actual snow and a noncountable flurry, as the number of snows approaches the predicted number.

Folklore, whether true or unfounded, will exist as long as people talk about the weather, and weather is one of our most talked-about subjects. When we are without anything to say or at a loss for words, it is a favorite topic that comes to our aid. Everyone comments on the weather while an old English proverb surmises:

Weather is the discourse of fools.

IMPORTANT WORDS AND TERMS

NOAA	Hydrothermograph	Radar	Hurricane warning
National Weather Service	Maximum-minimum	Radiosonde	Hurricane hunters
National Meteorological	thermometer	Weather satellite	Air pollution potential
Center	Maximum thermometer	Nephanalysis	Synoptic weather charts
Instrument shelter	Minimum thermometer	Infrared measurements	Prognosis chart
Thermograph	Rain gauges	Hurricane watch	Indian Summer

LEARNING OBJECTIVES

After reading and studying this chapter, you should be able to do the following without referring to the text:

1. Describe the organization and functions of NOAA.
2. Explain how weather data is processed and how forecasts are distributed.

3. State the instruments, their operation, and the data collected in a typical weather data station.
4. Describe the operation and data collected by radar, radiosondes, and weather satellites.
5. Explain the different types of synoptic weather charts.
6. Describe how forecasts are made from weather maps.
7. Distinguish between folklore sayings with scientific merit and those without.

QUESTIONS

1. What is NOAA, and what is its purpose?
2. Has the United States always had a National Weather Service or a similar organization?
3. Where are weather data processed and forecasts made? How often is this done?
4. What are the locations of the Forecast Office and the Office of the National Weather Service nearest your hometown? your college or university?
5. What type of data are taken at a typical weather data station?
6. Explain the operation of a maximum-minimum thermometer.
7. For what does the word "radar" stand and how is radar used in weather observations?
8. What is a radiosonde?
9. What is a nephanalysis?
10. How are temperature profiles obtained via satellite?
11. Distinguish between a hurricane watch and a hurricane warning?
12. What regions of the United States have high air pollution potentials?
13. What is a synoptic weather map and what information is given in the station models?
14. Explain a 500-millibar height contour map.
15. How can weather forecasts be made from weather maps?
16. Are folklore sayings concerning the weather reliable? Do they have any scientific merit?

PROBLEMS

1. Given the following data, plot a weather station model:
   Temperature: 68°F
   Sky: One-half covered with cirrus, cirrocumulus, and a few stratocumulus clouds
   Winds: 15 miles per hour out of the northeast
   Visibility: 3 miles
   Psychrometer wet bulb: 66°F
   Barometer: 1012.7 mb, falling from 1013.4 in the last 3 hours
   Started raining 2 hours ago with current accumulation of 0.15 inch

Map for Problem 4.

Map for Problem 6.

2. From Fig. 25-3e, calculate the approximate average speed of the high-pressure cP air mass moving from western Canada into the central United States.

Answer: 500 to 1000 mi/day

3. Prepare a local weather forecast for your area from Fig. 25-3a. Use Fig. 25-3d to check your predictions.

4. From the surface map in the accompanying map (p. 518) what are the weather conditions across the country? Give particular attention to weather changes and variations. Check your answer with the newspaper report given in Section 24-2.

5. Prepare a general forecast for the United States from the last surface map in Fig. 25-3e. Use the nephanalysis of Fig. 25-2g to help.

6. Prepare a 24-hour forecast for the United States east coast from the accompanying surface map (p. 518). Give the reasons for your predictions.

7. Examine the following weather sayings for meteorological merit and meaning:
   (a) A red sun has water in his eye.
   (b) Two full moons in a calendar month bring on a flood.
   (c) Mackerel clouds in the sky
       Expect more wet than dry.
   (d) The moon with a circle brings water in her beak.
   (e) Expect stormy weather when ants travel in lines, and fair weather when they scatter.
   (f) February rain is only good to fill ditches.
   (g) Candles burn dim before rain.
   (h) March comes in like a lion and goes out like a lamb.
   (i) A year of snow, a year of plenty.
   (j) Once in a blue moon.

Answer: (j) There are two interpretations of this saying which is used to refer to events that occur very seldom. (1) Under special conditions, the red end of the visible spectrum of moonlight passing through thin clouds is preferentially scattered. The moon therefore appears to have a bluish hue. (2) Although not directly related to meteorology, the occurrence of two full moons in one calendar month is a relatively infrequent event. The second full moon is sometimes called a "blue" moon.

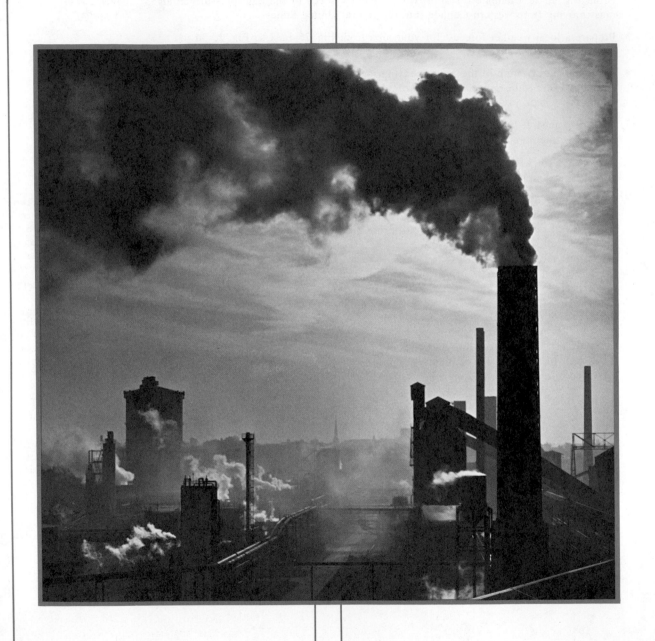

Chapter 26

AIR POLLUTION AND CLIMATE

*". . . this most excellent canopy, the air,
look you, this brave o'erhanging firmament,
this majestical roof fretted with golden fire,
why, it appears no other thing to me but a
foul and pestilent congregation of
vapours."*

(*Shakespeare—Hamlet*)

To make a complete survey of the atmosphere one cannot neglect the current problem of air pollution. The various pollutants being expelled into the troposphere may have direct effects on our health and living conditions, as well as influencing the local weather conditions and possibly the global climate. What then, are the nature and sources of air pollution? Before answering this question, we would do well to explain what we mean by air pollution. An obvious definition might be any unnatural constituent of the atmosphere. But we must then decide if the gases and particulate matter from volcano eruptions and forest fires (started by lightning) are unnatural contributions. To limit and focus our discussion, let us simply define **air pollution** as any atypical contributions to the atmosphere resulting from the activities of human beings.

The air pollution resulting from these activities has received a great deal of attention in the past decade, and will probably continue to do so in the foreseeable future. The concentrations of air pollutants have become critical in many localities and preventive actions have been implemented. The belching smokestacks, which were once a symbol of industrial prosperity, are now frowned on as sources of pollution. It is becoming increasingly evident that the atmosphere is not limitless in the disposal of our combustive wastes, and that air pollution may have long-term effects in addition to direct and immediate effects.

In this chapter we shall consider the sources of air pollution and the effects of pollution on meteorological processes which may result in the modification of our weather and climate.

26-1
AIR POLLUTANTS

Air pollution results primarily from the products of combustion and industrial processes which are released into the atmosphere. It has long been a common practice to vent these wastes into the atmosphere, and the resulting problems are not new, particularly in areas of population concentrations. Smoke and soot from the burning of coal plagued England over 700 years ago. London recorded air pollution problems in the late 1200s, and particular smoky types of coal were taxed and even banned. The problem was not alleviated, and in the middle 1600s Charles II was prompted to commission one of the outstanding scholars of the day, Sir John Evelyn, to make a study of the situation. The degree of London's air pollution at that time is described in his report, *Fumifugium* (Latin, generally meaning *On Dispelling of Smoke*, see Fig. 26-1a).

. . . the inhabitants breathe nothing but impure thick mist, accompanied with a fuliginous and filthy vapor, corrupting the lungs. Coughs and consumption rage more in this one city (London) than in the whole world. When in all other places the aer is most serene and pure, it is here eclipsed with such a cloud . . . as the sun itself is hardly able to penetrate. The traveler, at miles distance, sooner smells than sees the City.

However, the Industrial Revolution was about to begin, and Sir John's report was ignored and so gathered dust (and soot). An indication of the pollution of

Figure 26-1a *The title page of Sir John Evelyn's 1661 book on air pollution. The Latin quotation from the Roman poet Lucretius (97–53 B.C.) near the bottom of the page may be translated* How easily the heavy potency and odor of the carbons sneak into the brain. *(Courtesy British Museum.)*

17th-century London air may also be gained by the quotation from Shakespeare at the beginning of this chapter.

As a result of this air pollution, London has experienced several disasters involving the loss of life. Thick fogs are quite common in this island nation, and the combination of smoke and fog forms a particularly noxious mixture known by the contraction of *smoke—fog*, or **smog**. The presence of fog indicates the temperature of the air near the ground is below

Figure 26-1b *A scene in Donora, Pennsylvania, taken in 1949. During a five-day smog episode in 1948 in this Monongahela River Valley town, located 20 miles southeast of Pittsburgh, hundreds of people became ill and at least 20 died. (Courtesy EPA-DOCUMERICA.)*

the dew point and the existence of a temperature inversion (cf. Section 23-2). The gases and smoke are thus held near the ground. Continued combustion causes the air to become polluted with smog.

In London's worst smog episode in 1952, over 4000 deaths were attributed to air pollution. Such pollution conditions are particularly hazardous for persons with heart and lung ailments. Smog episodes are not confined to London. Others have occurred in Donora, Pennsylvania (Fig. 26-1b), the Meuse Valley in Belgium, and to lesser degrees in New York City and Los Angeles.

It should be obvious that the major source of air pollution is the combustion of fuels, which are primarily **fossil fuels**—coal, gas, and oil. More accurately, air pollution results from the **incomplete combustion** of **impure fuels.** Technically, combustion is

the chemical combination (burning) of certain substances with oxygen. Fossil fuels are the remains of plant and animal life and are composed chiefly of carbon and compounds of carbon and hydrogen, or **hydrocarbons.** If a fuel is pure and combustion is complete, the products are CO_2 and H_2O and are not usually considered pollutants, since they are part of the natural atmospheric cycles. For example, if carbon (coal) or methane (natural gas), CH_4, are burned completely, the reactions are

$$C + O_2 \longrightarrow CO_2$$

$$CH_4 + 2\,O_2 \longrightarrow CO_2 + 2\,H_2O$$

However, if fuel combustion is incomplete, the products may include carbon (soot), various hydrocarbons, and carbon monoxide (CO), all of which

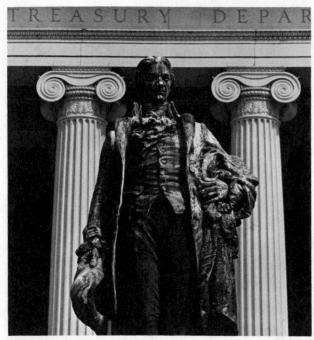

Figure 26-1c *An effect of air pollution. The statue is that of Alexander Hamilton at the U.S. Treasury Building in Washington, D.C. (Courtesy EPA-DOCUMERICA.)*

may contribute to air pollution. **Carbon monoxide** results from the incomplete combustion (oxidation) of carbon. That is,

$$2\,C + O_2 \longrightarrow 2\,CO$$

But, even increased concentrations of CO_2 can affect our environment. Carbon dioxide combines with water to form carbonic acid.

$$CO_2 + H_2O \longrightarrow H_2CO_3$$

Carbonic acid is an agent of chemical weathering in geologic processes (see Chapter 30). But, as an indirect product of air pollution, may also aid in the corrosion of metals and react with certain materials causing decomposition (Fig. 26-1c). Also, there is some concern about the increased CO_2 content of the atmosphere causing a change in global climate through the greenhouse effect. This aspect will be considered later in the chapter.

Oddly enough, some by-products of complete combustion contribute to air pollution, viz., nitrogen oxides (NO_x). **Nitrogen oxides** are formed when combustion temperatures are high enough to cause a reaction between nitrogen and the oxygen of the air. This typically occurs when combustion is nearly complete, a condition which produces the high temperature, or when combustion takes place at high pressure, e.g., in the cylinders of automobile engines. These oxides, normally NO (nitric oxide) and NO_2 (nitrogen dioxide), can cause lung irritation and are a key substance in the chemical reactions producing "Los Angeles" smog.

This is not the classical London smoke-fog variety, but a smog resulting from the chemical reactions of hydrocarbons with oxygen in the air and other pollutants in the presence of sunlight called **photochemical smog.** Since it was first identified in Los Angeles, it is often referred to as Los Angeles smog. Some 13 million people live in the Los Angeles area, which is in the form of a basin with the Pacific Ocean on one side and mountains on the other (east). This topography makes air pollution and temperature inversions a particularly hazardous combination. A temperature inversion can essentially put a "lid" on the city, which then becomes engulfed in its own fumes and exhaustive wastes.

Los Angeles has had its share of temperature inversions, with occurrences as frequent as 320 days per year. The California coast is usually under the influence of the edge of a semipermanent high-pressure center which extends over the Pacific Ocean. The descending air, coupled with the Los Angeles topography, gives rise to frequent (sometimes daily) subsidence temperature inversions. These atmospheric conditions, a generous amount of air pollution, and an abundance of sunshine, set the stage for the production of photochemical smog (see Fig. 26-1d).

Figure 26-1d *Scenes of Los Angeles on a clear day and on days with inversion layers at approximately 300 feet* (middle photo) *and 1500 feet* (bottom photo) *above the ground. (Courtesy Los Angeles County Air Pollution Control District.)*

In contrast to the smoke fogs of London, photochemical smog produces eye irritation and contains many more dangerous contaminants. One of the products in the photoxidation of hydrocarbons is a family of organic compounds called **aldehydes.** These result from the reaction of certain hydrocarbons with oxygen (Section 16-1).

Two of the major aldehydes in photochemical-polluted air are formaldehyde (HCHO) and acrolein (CH_2CHCHO). Formaldehyde usually accounts for about 50 percent of the total aldehydes in polluted air and acrolein for perhaps 5 percent. The identities of the remaining aldehydes are unknown, but these two are problem enough. Their undesirability in our atmosphere can be inferred from the fact that formaldehyde in liquid form is used as an embalming fluid, and acrolein was used as a poison gas in World War I. The cause of eye irritation in photochemical smog may be attributed, in part, to nitrogen-hydrocarbon compounds. The main culprit is peroxyacetal nitrate, or PAN for short. Another category of photochemically produced pollutants, nitro-olefins, also cause eye irritation. Although not photochemically produced but nonetheless dangerous is the hydrocarbon from motor exhausts called benzo(a)pyrene, or more simply, benzpyrene. Benzpryene is classified as a **carcinogen** because of its cancer-producing capabilities. There is some evidence that at least six to eight other hydrocarbons found in exhaust emissions are carcinogens.

One of the best indicators of photochemical reactions, but also a pollutant itself, is **ozone** (O_3), which is found in relatively large quantities in polluted air. In Los Angeles, air pollution warnings of various degrees are given on the basis of ozone concentrations in the air. To generate the amount of ozone in some polluted areas, a process other than dissociation of O_2 and the combination of O_2 and O, which occurs in the ozonosphere, must be involved (see Section 21-3). The energetic photons responsible for ozone production in the stratosphere do not reach the earth's surface, so they are not responsible. Scientists, however, are believed to have at least part of the answer, as we shall see.

Nitrogen dioxide (NO_2), formed by the oxidation of nitric oxide (NO), does absorb energy from the insolation and dissociates. This leads to a series of reactions in which ozone is an intermediate product.

$$NO_2 \longrightarrow NO + O$$

$$O + O_2 \longrightarrow O_3$$

$$O_3 + NO \longrightarrow NO_2 + O_2$$

However, the NO_2 cycle shown in these equations only explains the initial formation of O_3 in polluted atmospheres; it does not explain the development of concentrations as large as those measured, because (as the equations show) NO and O_3 are formed and destroyed in equal quantities. Laboratory experiments indicate that hydrocarbons provide the necessary added reactants. Certain types of hydrocarbons from exhaust emissions enter the NO_2 cycle. Studies suggest that oxygen atoms attack the hydrocarbons and the resultant oxidized compounds react with NO to form more NO_2. Thus, the balance of O_2 consumption is upset so that O_3 and NO_2 levels build up, while NO is depleted.

Both O_3 and NO_2 are potentially dangerous to plants and animals. NO_2 has a pungent, sweet odor and is yellow-brown in color. During the peak rush hour traffic, it is often evident in a whiskey-brown haze over large cities. NO_2 reacts with water vapor in the air to form nitric acid (HNO_3), which is very corrosive.

Fuel impurities occur in a variety of forms. Probably the most common impurity in fossil fuels and the most critical to air pollution is **sulfur.** Sulfur is present in fossil fuels in different concentrations. A low sulfur fuel has less than one percent sulfur content, and a high sulfur fuel greater than two percent. When fuels containing sulfur are burned, the sulfur combines with oxygen to form sulfur oxides (SO_x), the most common being **sulfur dioxide** (SO_2).

$$S + O_2 \longrightarrow SO_2$$

Figure 26-1e *A future advertisement? (Courtesy American Petroleum Institute.)*

However, in the atmosphere these sulfur compounds can cause considerable damage to practically all forms of life and property. Anyone familiar with sulfuric acid will be able to appreciate its undesirability as an air pollutant. Sulfuric acid is the electrolyte used in car batteries.

Thus, air pollution can be quite insidious and can consist of a great deal more than the common particulate matter in the form of smoke, soot, and fly-ash, which blackens buildings and the weekly wash if it is hung outside. Pollutants may be in the forms of mists and aerosols. Some pollutants are metals, such as lead and arsenic. Approximately 100 atmospheric pollutants have been identified and 20 of these are metals which come primarily from industrial processes. The extent of pollution varies a great deal and depends on the conditions, sources, and types of pollutants. Hopefully, it will not reach the extent characterized by the ad in Fig. 26-1e.

About 60 percent of the SO_2 emissions comes from burning coal and 14 percent from burning fuel oils. These are the major fuels used in the generation of electricity. Almost one-half of the SO_2 pollution in the United States occurs in seven northeast industrial states.

By either one of the following double reactions, SO_2 may combine with water and oxygen to form **sulfuric acid** (H_2SO_4):

$$SO_2 + H_2O \longrightarrow H_2SO_3 \quad \text{(Sulfurous acid)}$$

$$2\,H_2SO_3 + O_2 \longrightarrow 2\,H_2SO_4$$

or $\quad 2\,SO_2 + O_2 \longrightarrow 2\,SO_3 \quad \text{(Sulfur trioxide)}$

$$SO_3 + H_2O \longrightarrow H_2SO_4$$

Sulfurous acid (H_2SO_3) is mildly corrosive and is used as an industrial bleaching agent. Sulfuric acid, a very corrosive acid, is a widely used industrial chemical.

26-2
SOURCES OF AIR POLLUTION

Table 26-2a and Fig. 26-2a show the sources and magnitudes of atmospheric pollutants. As can be seen from the figure, transportation is the major source of total air pollutants. The United States has a mobile society, with over 100 million registered vehicles powered by internal combustion engines.

There are various types of internal combustion engines, but most common is the gasoline engine used in the automobile. The completeness of combustion depends primarily on the air-fuel mixture. The proper **air-fuel ratio** for near-complete combustion is 15:1. When accelerating, the air-fuel ratio is on the order of 12:1; while cruising 13:1; and while idling and decelerating, as in stop-and-go city traffic, 11:1. When the ratio is less than 15:1, substantial amounts of CO and hydrocarbons are formed as shown in Fig.

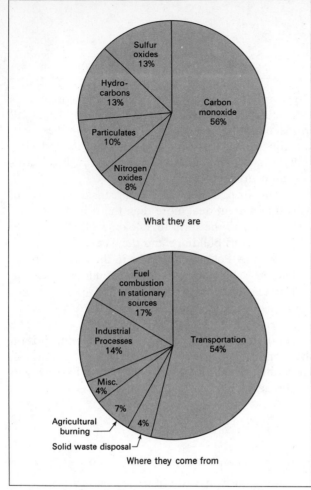

What they are

Where they come from

Figure 26-2a *Graphic representations of data in Table 26-2a showing sources and relative magnitudes of air pollution emissions.*

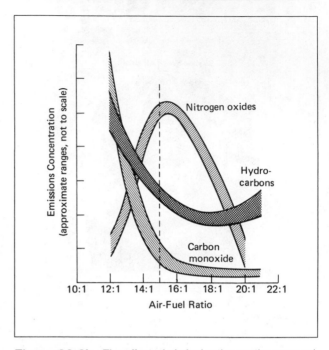

Figure 26-2b *The effect of air-fuel ratios on the composition of exhaust emissions of an automobile engine. The lower the air-fuel ratio is below 15:1 (complete combustion), the greater the carbon monoxide and hydrocarbon emissions. More nitrogen oxides are formed when combustion is nearly complete, due to higher temperatures. (Courtesy U.S. DHEW. Source: D.A. Trayser et al., Battelle Memorial Institute, Columbus, Ohio.)*

26-2b. Note, however, that the emission of nitrogen oxides is near maximum when the air-fuel ratio is 15:1.

Considerable efforts are being made by automobile manufacturers to reduce these emissions through engine modifications and special devices. It should now be readily understood why one is urged to keep their automobile engine properly tuned. A properly tuned engine operates nearer the correct air-fuel ratio, which minimizes emissions (see Fig. 25-2c). However, even with the current efforts to reduce

pollution emissions from automobiles, greater pollution is predicted in the future due to the increasing number of motor vehicles as shown in Fig. 25-2d.

Another source of pollution comes from the lead additive used in the gasoline as an "anti-knock" agent. You may recall the big emphasis in the early 1970s to use lead-free gasoline. The engines of most new automobiles are designed to use this type of gasoline.

Other internal combustion engines are the diesel and jet engines. The diesel engine operates with an

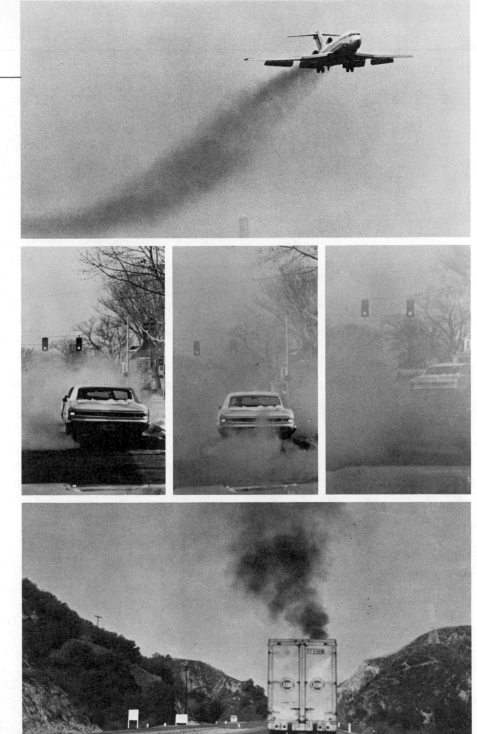

Figure 26-2c *Air pollution from automobile, jet, and diesel engines. (Courtesy EPA. Truck photo, Los Angeles County APCD; auto photo, Ames Iowa Daily Tribune.)*

Figure 26-2d *Graphs of estimated motor vehicle emissions. As a result of control devices and greater engine efficiency, CO and HC emissions are expected to decrease until the early 1980s. The increased number of motor vehicles then will produce greater amounts of emissions. The more complete combustion that reduces CO and HC emissions increases NO_x emissions. (Courtesy U.S. DHEW.)*

unrestricted air supply (no carburetor), and the fuel is injected directly into the combustion chamber; hence, the air-fuel ratio is usually higher than that for the gasoline engine. The diesel engine has no spark plug for ignition, using the heat from the high compression of the air-fuel mixture to cause ignition. Due to excess air there should be little unburned fuel, theoretically. However, improper operation and overloading, particularly in trucks, make diesels worse polluters than

gasoline engines (Fig. 25-2c). With high operating temperatures, diesels also form nitrogen oxides more readily.

Jet engines operate on somewhat different principles than the diesels. Combustion in a jet engine is used to heat compressed air, which expands and escapes out of the rear of the engine. By the conservation of momentum (Section 3-4), a plane powered by jet engines receives a forward thrust from the high-speed jet of escaping gases. In some cases, the exhaust gases are used to rotate a propeller or fan, and such engines are called turboprops and turbofans, respectively. A jet engine is, of course, larger and more powerful than an automobile engine, and with its larger size comes increased pollution. It is estimated that a jet engine produces roughly 10,000 times the emissions of an automobile gasoline engine. In this respect, we are fortunate that there are fewer jet aircraft than automobiles.

From Fig. 26-2a, it can be seen that the other major sources of air pollution are from stationary sources and industrial processes. Stationary sources refer mainly to electrical generating facilities. As can be seen from Table 26-2a, these sources account for the majority of the sulfur oxide (SO_x) pollution. This results primarily from the burning of coal, which always has some sulfur content.

Table 26-2a *Nationwide Emission Estimates[a] (10^6 tons/year)*

Source Category	SO_x	Part.	CO	HC	NO_x
Transportation	1.0	0.7	111.0	19.5	11.7
Fuel combustion in stationary sources	26.5	6.8	0.8	0.6	10.0
Industrial process losses	6.0	13.3	11.4	5.5	0.2
Solid waste disposal	0.1	1.4	7.2	2.0	0.4
Agricultural burning	Neg[b]	2.4	13.8	2.8	0.3
Miscellaneous	0.3	1.5	4.5	4.5	0.2
Total	33.9	26.1	149.0	34.9	22.8

[a](Courtesy EPA)
[b]Negligible (less than 0.05×10^6 tons/year).

26-3
THE COST OF AIR POLLUTION

The costs of air pollution are difficult to access. Visible evidence of the costs resulting from air pollution comes from lowering real estate values, repairs and replacements from corrosion, cleaning costs, animal and agricultural damages, and many more sources, not excluding the effects on human health. Some examples of air pollution damage are shown in Fig. 26-3a.

The corrosion of metals and other building materials is 3 to 30 times greater in cities than in rural areas, depending on the type of pollutant. Corrosive action also depends on the amount of moisture in the air. The rate of corrosion increases sharply when the relative humidity is over 50 percent.

Pollution effects range from damage to seemingly permanent stone structures to electrical insulators. Concentrations of ozone in photochemical smog can cause the deterioration and cracking of rubber, a common electrical insulator. This can cause electrical

Figure 26-3a *Examples of air pollution damage. (Above, left) District Building in Washington, D.C. being cleaned shows the accumulation of dirt due to air pollution. (Courtesy EPA. Photo by Washington, D.C. Health Department.) (Right) House paint damage resulting from sulfur gases from a burning coal pile. The gases react with the lead in the paint to produce lead sulfide which is black. (Courtesy Pennsylvania State Department of Health.) (Above, right) Copper Basin, Tennessee, as photographed in 1943. SO_2 emissions from the copper smelting process killed virtually all plant life in the surrounding area, leading to erosion. Although the SO_2 emissions have been controlled for over 50 years, the injury to the environment is still very evident. (Courtesy EPA. Photo by C.R. Hursh, U.S. Forest Service.)*

problems and damage. In some cities where photo-chemical smog is severe, car owners can purchase special anti-ozone tires for their automobiles to help alleviate the problem. Also, dust and particulate matter soil our clothes and home furnishings and can cause damage to delicate instruments. To alleviate some of this problem, we pay for the cost of filtering the air.

Crops, shrubs, trees, and other plants in general suffer from air pollution. There are many cases of industrial pollution having effects on the flora of an area. The classic example is Copper Basin, Tennessee, where a copper smelting plant was built shortly after the Civil War. Its production increased in the late 1800s, and consequently its SO_2 emissions. In time, these deadly fumes killed the plant life of some 30,000 acres of the surrounding timberland (Fig. 26-3a). Flowers for the Rose Bowl Parade used to come from Pasadena (the location of the Rose Bowl) and the nearby Los Angeles area, but now they must come from other places too, because the effects of smog on the floral industry of that area. Plants are particularly vulnerable to gaseous pollutants. Gases enter the microscopic openings, called stomata, on the underside of the plant leaves and attack the plant cells.

The preceding are but a few of the effects of air pollution which cause the government to estimate that the average yearly cost of air pollution in the United States is about $65 per person. With a population of over 200 million, simple arithmetic shows the total yearly cost estimate to exceed $13 billion. This includes estimates of $100 million for paint damage, $800 million for laundering, $240 million for car washing, and $500 million in damages to livestock and agriculture.

Not only does air pollution have severe effects on the objects of our physical environment, but also on ourselves. The cost of air pollution on human health is difficult to estimate. How much is your health worth? Rather than conjecture about the monetary costs of air pollution on health, let's consider and become aware of the effects on the respiratory system. Humans inhale about five quarts of air per minute. Air is normally taken in through the nose and is warmed and moistened in the nasal cavity. On its route to the lungs the air passes through the pharynx, which is also connected to the mouth as illustrated in Fig. 26-3b. Below the pharynx is the funnel-shaped larynx, or voice-box, the rim of which projects outward forming the Adam's apple. Two muscular tissues across the larynx, the vocal cords, are capable of producing sound when air is exhaled. Passing through the larynx the inhaled air enters the trachea, or wind pipe, which divides into two bronchial tubes. Each of these bronchi enters one of the lungs and divides again and again into fine bronchioles. At the end of the bronchioles are millions of tiny air spaces, or sacs, called **alveoli,** which are surrounded by capillaries. (See the blowup in Fig. 26-3b.) The oxygen of the air is exchanged through the walls of the alveoli for the CO_2 waste from the cells which is transported to the lungs by the blood.

The respiratory system has natural protection mechanisms against foreign materials. The hair within the nostrils filters out large particles in the air. To guard the lungs against finer particles, the upper respiratory tracts—the nasal cavity, pharynx, larynx and bronchial tubes—are lined with mucous membranes that excrete liquid mucus. The fine particles are caught in the mucus and are swept out of the respiratory system by the action of tiny hairlike projections called **cilia,** Fig. 26-3b.

This natural defense is effective for particulate matter, which is nature's major contaminant in air, but the respiratory tract is vulnerable to gaseous pollutants. The cilia when attacked by SO_2 and other smog constituents become less efficient, allowing particulate matter to penetrate deeper into the lungs. The irritation caused by pollutants causes coughing, increased mucus flow, and muscle spasms which limit the air flow to and from the lungs. This condition is symptomatic of inflamed bronchi and is referred to as bronchitis. Irritated tissue in the bronchiole may

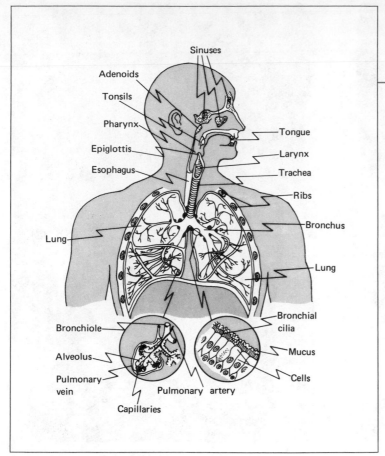

Figure 26-3b *The anatomy of the human respiratory system. (From the Air Pollution Primer, The National Tuberculosis and Respiratory Disease Association.)*

Figure 26-3c *An obstructed bronchiole, or small bronchial tube. This condition, which restricts normal breathing, can arise from irritation of the respiratory tract by air pollutants. The lower drawing is a cross section of an obstructed bronchiole. (Courtesy J.D. Wilson, An Environmental Approach to Physical Science, Heath, Lexington, Mass., 1974.)*

swell, and with the other conditions, the small bronchiole may be obstructed as illustrated in Fig. 26-3c.

The air passages to the alveoli normally expand and contract with breathing. However, if this action is impaired by pollution irritation, some of the air may be retained as new air is introduced. As a result of the increase in air volume, the air sacs get larger, and the walls of the alveoli may eventually rupture, forming a larger sac. Large pouches may form as shown in the drawing of Fig. 26-3d. This condition, characterized by a shortness of breath, is called **emphysema,** a respiratory disease that is becoming increasingly common. The air pouches may enlarge, as the process

continues, further destroying the oxygen-CO_2 exchange capability of the lungs. Personal air pollution through smoking may give rise to respiratory conditions such as these, as well as aggravating existing ones.

One thing common to all respiratory problems from pollution is the decrease of oxygen exchange to the blood. When there is an oxygen deficiency, the body tries to rectify the situation by increased breathing and blood flow. The latter requires the heart to work faster, and so a secondary effect of air pollution is heart disease. It can be easily understood why the health problems associated with air pollution

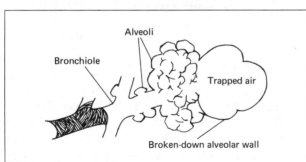

Figure 26-3d *A diagram of the lung condition emphysema (above). Irritation of the lungs may cause air to be trapped in the alveoli. With the in-take of more air, the increased volume results in the break down of the alveolar walls, forming large pouches of trapped air, which restricts breathing and reduces the oxygen-CO_2 exchange capacity of the lungs. The photo shows a section of normal lung (left) and an emphysematic lung section (right). (Photo courtesy EPA. Photo by Webb-Waring Institute for Medical Research.)*

are the most critical for the sick and elderly. Figure 26-3e illustrates the effect of age in a comparison of emphysema cases between two cities with different degrees of pollution.

We have all experienced the lack of oxygen, if only by holding our breath. Its effects are dizziness and lightheadedness. Another air pollutant, carbon monoxide, can also account for oxygen deficiencies in the body which impose extra physical burdens. Inhaled CO can displace the combined oxygen in the blood and reduce the amount carried to the tissues. The oxygen normally combines with a protein in the blood called hemoglobin (Hb).* Oxygen and CO react with Hb in a similar manner, but the hemoglobin's affinity for CO is 200 times greater than for O_2. Carbon monoxide reacts with Hb to form carboxyhe-

*Hb is not the chemical formula for hemoglobin. Hb is used for simplicity.

moglobin, thus reducing the oxygen-carrying capacity of the blood.

$$CO + Hb \longrightarrow COHb \text{ (Carboxyhemoglobin)}$$

Small concentrations of contaminants are measured by the volume ratio expressed as parts per million (ppm). A concentration of 1 ppm means there is one volume of contaminant in a million such volumes. Exposure to air with 10 ppm of CO for eight hours produces mental dullness. CO concentrations of 100 ppm can produce headaches and dizziness in a short time, so there is great cause for concern when CO concentrations greater than 300 ppm have been measured in idling cars in a traffic jam. The concentration of CO in the mainstream of cigarette smoke is over 2500 times greater than the recommended maximum atmospheric concentration (15 ppm). Consequently, smokers have 5–10 percent less hemoglobin available for oxygen transfer due to increased CO concentrations in the blood.

The other main gaseous air contaminants, SO_2, NO_2, and O_3, are also quite dangerous in small concentrations. Sulfur dioxide (SO_2) dissolves in the mucous lining of the upper air tract, while NO_2, and O_3, being less soluble, travel further. However, soot, which has a very chemically active surface, can carry absorbed SO_2 deeper into the respiratory tract. Carcinogens may also use soot and particulate matter as a vehicle into the lungs. Aerosols may also be inhaled deeply into the lungs. You should now begin to realize the forbidding dangers of a smoky, photochemical smog. An additional hazard of many pollutants is **synergism,** or their combined effect. For example, the combinations of CO and nitrogen oxides, or CO, CO_2, and H_2O, are several times worse than CO alone.

There are many other air pollutants which create health hazards. These are usually associated with particular industrial processes and are confined to the ambient air of the factory or the atmosphere of the

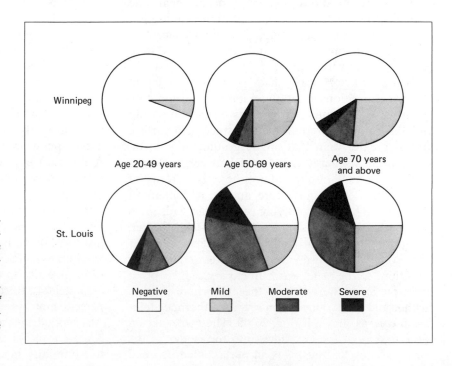

Figure 26-3e *A comparison of emphysema cases in two cities with contrasting levels of air pollution. The graphs show the results of a survey of 300 residents of heavily industrialized St. Louis, Missouri, and relatively unpolluted Winnipeg, Canada. Notice the prevalence of emphysema with age. (From the Air Pollution Primer, The National Tuberculosis and Respiratory Disease Association.)*

local region. Processes involving metallic chemicals are particularly dangerous due to metal poisoning by such metals as beryllium and arsenic. Beryllium is now being used in rocket fuels, so its pollution may be more widespread. Lead, the antiknock agent used in gasoline, may also cause metal poisoning. Like most metals, the lead accumulates in certain parts of the body, causing blood disorders and other complications.

26-4
WEATHER AND CLIMATIC EFFECTS OF AIR POLLUTION

Although we sometimes purposely attempt to modify the weather, e.g., rainmaking or fog dissipation, the effects resulting from air pollution are inadvertent modifications of weather and climate. The most immediate modifications naturally occur near the sources of pollution. Since air pollution results from the activities of human beings, the major concentrations of pollutants are in and around population centers, or industrial areas.

Cities have always had somewhat different weather conditions than those of the surrounding countryside. The buildings of a city obstruct the wind and absorb much more insolation, since their surface area is greater than that of the ground they cover. Also, the heat energy is radiated between buildings rather than back into the atmosphere. As a result of such effects, the temperature and heat content of cities are higher than outlying rural areas. This gives rise to what is known as the urban **heat-island effect.** Because of the concentrated heat of the city, warm air rises from it, carrying combustive wastes and other air pollution. As the rising warm air expands and cools, it flows out over the edges of the city where it cools further and sinks. This action sets up a thermal circulation cell, as shown in Fig. 26-4a. The cooler air on the outskirts of the city flows into the center in the convection cycle, but this is, in part, cooled polluted city air. Thus, the heat-island effects sets up a self-contained thermal circulation system in which the air is continually polluted.

In the absence of winds, e.g., frontal air movement, to break up the circulation system and sweep the pollution away, serious pollution conditions can arise in cities. This is particularly true should a high-pressure air mass move over the city and become stationary. In this windless condition, along with possible topographical effects, such as a city being in a river valley, the air of a city becomes continually polluted by its own activities, which may eventually have to be curtailed. This condition is not uncommon in several U.S. cities, typically in the summer as a result of a "stagnant" air mass.

It might be thought that additional insolation on an urban area would supply sufficient heating and convection to break up the self-contained circulatory system of the heat-island effect. This is a possibility, but it is counterbalanced by the formation of a **dust dome** over the city (Fig. 26-4a). City pollutants contain large amounts of aerosols and particulate matter. As the polluted air moves up and spreads out over the city, it forms an often-discernible (dust) dome. The dust concentration may be several hundred to a thousand times that of rural air. This polluted ceiling decreases the amount of insolation that penetrates the city atmosphere and prevents the sun from efficiently heating the metropolitan area so as to break up the circulation cell of the urban heat island. A city must be content to wait for outside air movements to sweep away its pollution.

Thus, air pollution causes cities to have modified weather conditions. This is manifested not only in a temperature increase and attenuated insolation, as just mentioned, but also in increased cloudiness and precipitation. The updrafts and concentration of particles give rise to cloud formation and precipitation (Chapter 24). Records show that cities receive several percent more precipitation than nearby rural areas. Also, with drains and sewer systems, the water runoff is more rapid in cities, which results in a lower relative humidity near the surface.

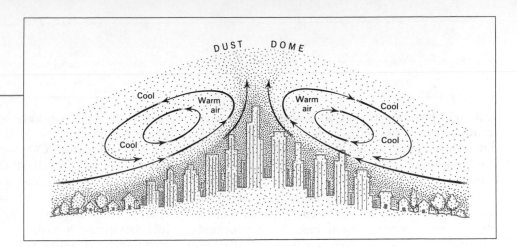

Figure 26-4a *An illustration of the heat-island effect, showing the air circulation and dust dome of a city.*

Climate

Although not as immediately obvious, there is a general belief that there are (and will be) changes in the global climate brought about by atmospheric pollution. **Climate** is the long-term average weather conditions of a region. Some regions are identified by their climates. For example, when Florida or California is mentioned, one usually thinks of a warm climate, and Arizona is known for its dryness and low humidity. Because of such favorable conditions, the climate of a region often attracts people to live there, and thus the distribution of population (and pollution) is affected.

Dramatic climate changes have occurred throughout the earth's history. Probably the most familiar is that of the Ice Age, when glacial ice sheets advanced as far south as the northern United States. Some evidence now exists from ocean sediment cores to support the theory that dramatic changes in global climate result from subtle, regular variations in the earth's orbit around the sun. For example, one cyclic variation is the advance of the earth in its elliptical orbit. That is, the earth's closest approach to the sun occurs at different times of the year in a cycle of 23,000 years. The earth and the sun are now closest in January; in 10,000 years it will be closest in July. This would result in cooler summer temperatures, less snow melting, and a growth of the polar ice caps. Such a change could slowly lead into a new ice age.

However, climatic fluctuations have been continually occurring, but of much smaller magnitude than that of the Ice Age. For example, in the last several decades there has been a noticeable southward shift of world climate. The Sahara Desert is reportedly advancing southward in places as much as 30 miles a year, and rainfall in some regions is barely one-half of what it was 20 years ago, giving rise to great bands of land with drought conditions. The ice borders of the northern polar cap appear to be paralleling the shifts of wind patterns and climatic zones, producing new record low temperatures. It appears that while the earth's average total rainfall shows little variation, more rain is now falling in the southern hemisphere.

The question now being asked is whether or not air pollution may be responsible for some of the observed climate variations. It is reasonably well established that the climates of cities can differ from that of the surrounding countryside as previously noted, however, the extent of the effects of human activities on global climate is not clear. For example, from 1880 to 1940, the average annual temperature of the earth's surface increased by about 0.6°C (1.1°F). Since 1940, the average temperature has decreased by about 0.3°C (0.55°F). Associated with this lowering temperature has been the shift of the frost and ice boundaries, a weakening of zonal wind circulation, and marked variations in the world's rainfall pattern.

Global climate is sensitive to atmospheric contributions which affect the radiation balance of the atmosphere, such as the concentration of CO_2, the particulate concentration, the extent of cloud cover, and the albedo of the earth's surface. Air pollution and other human activities do contribute to changes in these conditions, and scientists are now trying to determine whether or not this accounts for any substantial portion of the observed variations in global climate.

Alterations of the climate are brought about primarily through changes in the energy content of the atmosphere, which is maintained by the methods discussed in Chapter 21. Any activity which alters the mechanisms of these methods may produce changes in the atmosphere's energy balance. For example, the 1815 eruption of the volcano Tambora, located on an island just east of Java, vented an estimated 35 cubic miles of particulate debris into the air. Fine volcanic dust was circulated around the earth by global wind patterns to the extent that the normal insolation was decreased and the year of 1816 was uncommonly cold (see Fig. 26-4b).

Air pollution affects the balance of atmospheric constituents which are critical in maintaining the normal energy balance. For example, concern has been expressed over the effects of aerosol spray can propellants on the ozone layer. The common gas propellants in question were the chlorofluorocarbons $CFCl_3$ and CF_2Cl_2. It was theorized that when these gases were released, they would be carried up to the stratosphere by air currents. In the stratosphere, the chlorofluorocarbon atoms would be broken apart by ultraviolet radiation, releasing reactive chlorine atoms. These, in turn, would react with and destroy ozone molecules in a repeating cycle:

$$Chlorofluorocarbon \xrightarrow{uv} Cl$$

$$Cl + O_3 \longrightarrow ClO + O_2$$

$$ClO + O \longrightarrow Cl + O_2$$

The depletion of the protective, ultraviolet-absorbing ozone layer in the stratosphere would have serious environmental consequences. The predicted effects on the ozone layer by chlorofluorocarbons has not been verified, and hopefully will not be. Fortunately, as a result of environmental concern, manufacturers are now using alternative, environmentally

safe propellants, or have switched to mechanical pump systems.

Vast amounts of CO_2 are being expelled into the atmosphere as a result of the combustion of fossil fuels (Table 26-2a). As discussed in Section 21-4, CO_2 and water vapor play important roles in the earth's energy balance through the greenhouse effect. An increase in the atmospheric concentration of CO_2 could alter the amount of radiation absorbed from the earth's surface and produce an increase in the earth's temperature. An early calculation of the effect of such an increase was made in 1896 by Svante Arrhenius. Arrhenius' calculations showed that a doubling of the atmospheric **CO_2 concentration** would increase the earth's surface temperature by 5–6°C. It might be pointed out that Arrhenius' concern arose from the CO_2 escaping from volcanoes, not from automobile exhaust emissions.

During the 19th century the atmospheric CO_2 concentration increased by about 10 percent, as determined from old records. More recent calculations show an annual increase of about 0.2 percent (0.7 ppm), a considerable portion of which results from combustion. Using simplified atmospheric models, scientist's calculations predict that a 5 percent change in the CO_2 atmospheric content would lead to a global warming of the troposphere of 0.15°C (at fixed relative humidity) and that a doubling of the CO_2 concentration would produce an average global temperature change of about 2°C.

Is it then possible that an increase in atmospheric CO_2 concentration led to the increase in global temperature (and subsequent changes in the climate) observed from 1880 to 1940? It is generally accepted that this is a possibility. However, it should be noted that a substantial portion of the change in the atmospheric CO_2 concentration has occurred in the last several decades, during which the contribution due to the activities of human beings almost doubled. At the same time (since 1940), the average global temperature has decreased. Evidently, air pollution also affects other mechanisms of the atmosphere's energy balance.

Figure 26-4b *Particulates from the volcano on Mt. Redoubt in Alaska.*

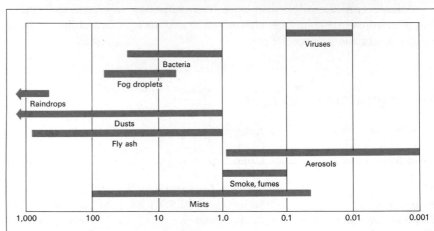

				Viruses				
		Bacteria						
	Fog droplets							
Raindrops								
	Dusts							
	Fly ash			Aerosols				
				Smoke, fumes				
1,000	100	10	1.0	Mists	0.1	0.01	0.001	

Figure 26-4c *The sizes of several different particles expressed in microns. (One micron is $\frac{1}{1000}$ mm, or $\frac{1}{25,000}$ in.)*

The opposing mechanism of the atmosphere's increasing temperature due to increased CO_2 concentrations most commonly cited is that which results from the presence of particles. **Particle pollution** contributes to changes in the earth's thermal balance by decreasing the transparency of the atmosphere to insolation; that is, there is a decrease in the amount of solar radiation the earth receives, through a change in the atmospheric albedo, which results in a lower surface temperature. Such changes resulting from volcano eruptions are well documented, as in the case of the previously cited example of Tambora. It should be mentioned, however, the effect of particulate pollution depends on the number and size of the particles. Small particles can also cause increased absorption of the outgoing, infrared terrestial radiation (greenhouse effect). See Fig. 26-4c.

Contributions to an **albedo** change could also be from an increase in cloud coverage that resulted from an abundance of particle nuclei. Recall from Section

21-4 that clouds are good reflectors of solar radiation, and thus they play a major role in the earth's albedo. It has been suggested that increased aircraft activity in the upper troposphere has resulted in an increase in cirrus clouds. Supersonic transport aircraft operating in the lower stratosphere have also caused some concern because of the persistent contrails they make. Also, moisture from the jet engine's combustion causes an increase in the relative humidity. In the troposphere, precipitation processes act to "wash" out particle and gaseous pollutants, but there are no cloud or rain-washout mechanisms in the stratosphere. Also, the stratosphere is a region of high chemical activity, and chemical pollutants (e.g., NO_x and hydrocarbons) could possibly give rise to climate-changing reactions.

Albedo considerations also arise from activities that affect the earth's surface. Urbanization and agriculture affect the surface albedo.

Thus, we see that air pollution may contribute to a variety of atmospheric effects—both local and global. Effects of pollution on local climatic conditions are somewhat understood, but there is too little data to understand or predict global climatic effects. We are inclined to believe that increased atmospheric CO_2 concentrations give rise to an increase in global temperature. This would cause more water evaporation and increase in the relative humidity. Particulate pollution could then give rise to increased cloud formation, which would increase the albedo and cause a decrease in global temperature. Certainly little comfort can be found in this speculative, counterbalancing pollution cycle. There are too many unanswered questions on the effects of man's pollution, which has occurred over a relatively short time, on the natural interacting cycles of the atmosphere and biosphere, which have taken millions of years to become established.

IMPORTANT WORDS AND TERMS

Air pollution
Smog
Fossil fuels
Incomplete combustion
Impure fuels
Hydrocarbons
Carbon monoxide
Nitrogen oxides
Photochemical smog

Aldehydes
Carcinogen
Ozone
Sulfur
Sulfur dioxide (SO_2)
Sulfuric acid (H_2SO_4)
Air-fuel ratio
Alveoli
Cilia

Emphysema
Synergism
Heat-island effect
Dust dome
Climate
CO_2 concentration
Particle pollution
Albedo

LEARNING OBJECTIVES

After reading and studying this chapter, you should be able to do the following without referring to the text:

1. Discuss and describe the major air pollutants.
2. Distinguish between regular smog and photochemical smog.
3. State the major sources of the various air pollutants.

4. Discuss the effects of the air-fuel ratio of the gasoline internal combustion.
5. Explain some of the damaging effects of the various air pollutants to property.
6. Describe some of the effects of air pollutants on the human respiratory system.
7. Discuss the effects of air pollution on local weather conditions.
8. Discuss the effects of air pollution on global climate.

QUESTIONS

1. Define *air pollution*.
2. Is air pollution a relatively new problem?
3. What gives rise to the majority of air pollution?
4. What are the products of complete combustion? incomplete combustion?
5. Are nitrogen oxides products of complete or incomplete combustion? Explain.
6. Distinguish between plain smog and photochemical smog.
7. What is one of the best indicators of photochemical smog?
8. What is the major fossil fuel impurity?
9. What is the sulfur content of a low sulfur fuel? a high sulfur fuel?
10. The air-fuel mixture is important in an automobile engine. What is the proper air-fuel ratio and what are the effects of an improper air-fuel ratio?
11. Name the major source of each of the following pollutants: (a) carbon monoxide, (b) sulfur dioxide, (c) particulate matter, (d) nitrogen oxides.
12. What is the estimated yearly expense due to the effects of air pollution in the United States?
13. The human respiratory system's natural defense mechanism is primarily for what type of pollutant?
14. What is the cause and effects of emphysema?
15. Why is carbon monoxide a dangerous air pollutant?
16. What is meant by synergism?
17. What are the urban heat-island effect and dust dome, and what are their effects on urban weather conditions?
18. How has the earth's average temperature varied over the last 100 years?
19. What is the possible effect(s) of increased atmospheric CO_2 concentrations?
20. What is a possible explanation for the decrease in the earth's average temperature since 1940?
21. What effect could chlorofluorocarbon propellants of aerosol spray cans possibly have on the earth's climate?
22. What is the concern about air pollution in the stratosphere?

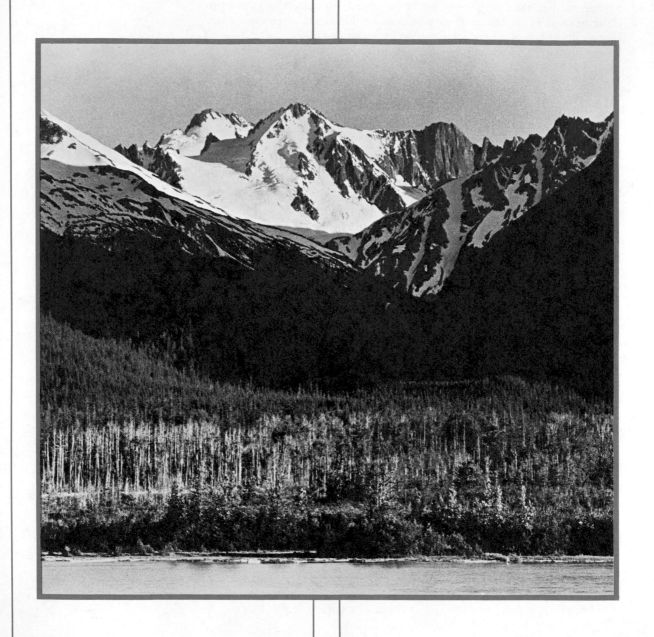

Chapter 27

GEOLOGY
AND TIME

In addition to the atmosphere, a major part of our physical environment is the earth upon which we live. Geology is the study of the earth, its processes and history. With the increasing emphasis on natural resources obtained from the earth, geology has taken on a new sense of importance. An understanding of the various geologic processes is critical in locating and developing these natural resources. Also, such an understanding may allow the prediction of catastrophic events, such as volcanic eruptions and earthquakes, with the result of saving lives and property. In addition, geological studies of meteorites and the moon have helped us understand the origin of the solar system.

The earth is indeed a dynamic place, both externally and internally. However, the rate of most earth processes gives geology a unique time scale known as geologic time. Rather than an actual time scale, geologic time is more of a concept, in which it is assumed that the earth developed gradually over an incredibly long time span. The several thousand years of the recorded history of human beings is only a "tick" on the geologic clock.

The history of the earth is recorded in geologic time and events. Geologists must consider processes that may have taken place over millions of years, a time interval that is very difficult to comprehend. However, if natural processes are constant and the same processes operate today on and within the earth, then the present may be considered the key to

This earth, a spot, a grain, an atom.

Milton, Paradise Lost

the past. This is an important geologic concept called the **principle of uniformity** (or uniformitarianism). Changes occurring on and within the earth today are clues to the long-term total picture.

Within the last 25 years, there has been a revolution in the science of geology that gives us a better insight into the processes occurring near the earth's surface. This has come about with the development of the theory of plate tectonics, which considers the earth's surface to be made up of movable "plates." The assumptions of the theory of plate tectonics are surprisingly simple, but the consequences are far-reaching. With this new set of concepts, a new understanding is gained of the formation of mountains and ocean basins, volcanic activity, earthquakes, and the mechanisms of operation of many geologic processes.

In an effort to give the student a contemporary view of geology, the concepts of plate tectonics will be presented in this chapter and used to explain various geologic phenomena in the following chapters. It should be kept in mind, however, that although the current enthusiasm for plate tectonics is grounded by recent experimental evidence, it must stand the test of time—perhaps geologic time. Other theories may someday replace that of plate tectonics just as it is currently replacing previous ones.

27-1
THE EARTH'S STRUCTURE

Is the earth hollow? If not, is its interior solid or liquid? These are fundamental questions of geology. Some ancient beliefs considered the earth to be hollow like a basketball. However, today we have experimental evidence to the contrary. Only the earth's surface and a rather insignificant part of its interior in mines and drilling projects can be observed directly. To date, we have drilled about five miles into the earth, but it is unlikely that a hole can be drilled into the earth's deep interior. Materials extruded in volcanic eruptions obviously come from the earth's interior, but these are from relatively shallow depths compared to the earth's radius of approximately 4000 miles. How then do geologists study the earth's deep interior structure?

Our knowledge of the earth's interior structure comes almost exclusively from the monitoring of shock waves generated by earthquakes. The enormous releases of energy associated with movements along fractures in the solid exterior, which result in "earthquakes", generate both transverse and longitudinal waves (Chapter 5) that propagate through the earth. By monitoring these waves at different locations on the earth and from a knowledge of wave properties in various types of earth materials, such as wave velocity and refraction, information about the earth's interior structure is gained. The monitoring of these so-called seismic waves will be considered in detail in Chapter 29. For the present discussion, only the resulting theory of the earth's internal structure will be presented.

From these indirect observations of the earth's interior, it is believed that the earth is made up of a series of concentric shells as illustrated in Fig. 27-1a. There are three major shells: (1) the core, which is partly molten, (2) the mantle, and (3) the crust. The different shells are characterized by different composition and physical properties.

The two innermost regions are together called the **core** and have an average density of over 10 g/cm³. This density suggests a metallic composition, which is believed to be chiefly iron and nickel. The core is believed to be composed of about 80 percent iron. This estimate is based on the behavior of seismic waves passing through the core, the known abundance of iron in meteorites, and the measured proportion of iron in the sun and stars. The solid inner core has a radius of approximately 750 miles, while the outer core, some 1400 miles thick, is believed to be composed of molten, highly viscous, "liquid" ma-

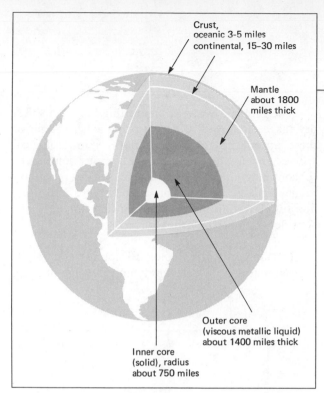

Crust,
oceanic 3-5 miles
continental, 15-30 miles

Mantle
about 1800
miles thick

Outer core
(viscous metallic liquid)
about 1400 miles thick

Inner core
(solid), radius
about 750 miles

Figure 27-1a *The interior structure of the earth. The crustal thickness is greatly exaggerated.*

terial. The magnetic field of the earth is thought to be related to the liquid nature of the outer core. As pointed out in Chapter 6, the magnetic field of the earth resembles that of a huge bar magnet within the earth. However, the interior temperatures of the earth are probably too high to have permanently magnetized ferromagnetic materials with the interior temperatures exceeding the Curie temperatures* of these materials. The slow change of the positions of the earth's magnetic poles suggests that the magnetic field is due to currents, which must arise somehow from motions of the molten metals in the outer core.

Around the core is the **mantle,** which is on the average about 1800 miles thick. The composition of the rocky mantle differs sharply from the metallic core, and their boundary is distinct. The average density of the mantle is of the order of 4.5 g/cm³,

*The atomic temperature where the atomic coupling in a ferromagnetic material disappears and the material becomes paramagnetic.

which indicates that its composition is iron magnesium rock-type material.

Around the mantle is a thin, rocky, outer layer upon which we live, called the **crust.** It ranges in thickness from about 3-5 miles beneath the ocean basins to about 15-30 miles under the continents. About 65 percent of the earth's crust is oceanic crust. That is, about 65 percent of the earth's surface is made up of ocean basins.

The boundary between the crust and the mantle was discovered by A. Mohorovičić, a Yugoslav seismologist, while studying a Balkan earthquake in 1909. He found that the speed of the shock waves increased suddenly several miles below the earth's surface, indicating a change in the constituent material at this depth. This boundary is called the Mohorovičić discontinuity after its discoverer, or simply the **Moho discontinuity** for short.

Thus, we see that the internal structure of the earth is somewhat analogous to that of an egg with the yolk, white, and shell corresponding to the core, mantle, and crust, respectively. How did this layered structure of the earth come about? Perhaps when the earth was formed in an original molten state, the denser material settled to the bottom, in a fashion similar to the gases in the heterosphere. The liquid outer core could not be due to remanent heat from the earth's formation, since the interior would have long since had time to cool and solidify. Although the outer core is subjected to enormous pressures that would result in high temperatures, calculations show that these temperatures are not high enough to melt the iron-nickel material of the outer core. Rather, it is believed that the heat is generated from the decay of radioactive materials.

It should be remembered that our picture of the general structure and composition of the earth comes chiefly from indirect observations. Although materials from the upper mantle are observed in volcanic eruptions and perhaps someday exploratory drillings may be made into the mantle, scientists will probably never directly observe the earth's interior as fantasized in Jules Verne's novel *Journey to the Center of the Earth*. However, as science and technology progress, geologists search for methods to provide a more detailed picture of the earth's anatomy. Recently, major advances have been made in the understanding of the makeup and dynamics of the earth's outer structure as will be described in the following sections.

27-2
CONTINENTAL DRIFT AND SEA FLOOR SPREADING

When looking at a map of the world, we are tempted to speculate that the Atlantic coasts of Africa and North and South America could fit nicely together as though they were pieces of a jigsaw puzzle (Fig. 27-2a). This observation has lead scientists at various times in history to suggest that these continents, and perhaps the other continents, were once a single, giant supercontinent that broke and drifted apart. However, there was no evidence to support this theory other than the shapes of the continents.

In the early 1900s, Alfred Wegener (1880–1930), a German meteorologist and geophysicist, revived the theory of **continental drift** and brought together various geological evidences for its support. Wegener's theory gave rise to considerable controversy and only relatively recently has reasonably conclusive evidence been found that supports some aspects of the theory of continental drift.

Wegener's assumption was that the continents were once part of a single giant continent, which he called **Pangaea** (Greek, pronounced pan-jee-ah and meaning "all lands"). According to his theory, this hypothetical supercontinent somehow broke apart about 200 million years ago and its fragments drifted to their present positions and became today's continents (Fig. 27-2b).

The geologic evidence supporting Wegener's theory takes on several different approaches. Some of these approaches are (a) similarities in biological species and fossils found on the various continents, (b) continuity of geologic structures such as mountain ranges and the distribution of rock types and ages, and (c) glaciation in the southern hemisphere. Let's consider each of these briefly:

(a) Similarities in biological species and fossils suggest that there was formerly an exchange of these

Figure 27-2a *Continental jigsaw puzzle. The coasts of Africa and North and South America appear as though these continents could fit together nicely as if they were pieces of a jigsaw puzzle. The dashed lines show the paths the continents might have followed if they were once together and rifted and drifted apart. The heavy line indicating where the continents might have been once joined marks a present mid-oceanic ridge that runs the length of the Atlantic Ocean.*

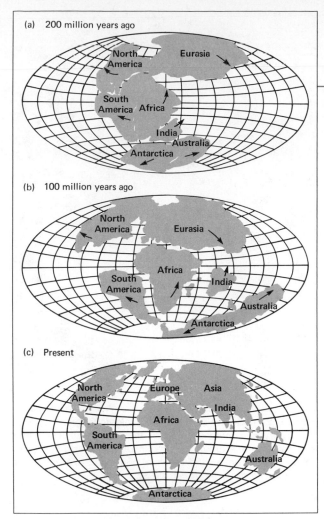

(a) 200 million years ago

(b) 100 million years ago

(c) Present

Figure 27-2b *An illustration of the sequence of events in the theory of continental drift. (a) 200 million years ago the giant supercontinent Pangaea rifted and the continental fragments began to drift apart. (b) Continental positions 100 million years ago. Notice India was drifting northward on a collision course with the Asian continent. (c) Present position of the continents.*

young genus of earthworm is found in the same latitudes of Japan and the Asian and European continents, while it is found in similar latitudes on the east coast, but not the west coast of North America. Similarly, fossils of identical reptiles are found in South America and Africa, and identical plant fossils have been found in South America, Africa, India, Australia, and Antarctica.

(b) As has been noted, it was the roughly interlocking shapes of the coastlines of the African and American continents that inspired the theory of continental drift. Imagine that the continental shapes are cut from a printed page like jigsaw-puzzle pieces and an attempt is made to put the separated pieces together again. When putting the page pieces back together, it is evident that the continuity of the printed lines would be common to the fitted pieces as shown in Fig. 27-2c. If indeed the continents had rifted and drifted apart, we might expect some similar "printed lines" common to the pieces. Such evidence does occur in the form of geologic features. If the continents were put back together, the Cape mountain range in southern Africa would line up with the Sierra range near Buenos Aires, and these mountains are strikingly similar in geologic structure and rock composition (Fig. 27-2d). In the northern hemisphere, the Hebrides mountains in northern Scotland match up with similar formations in Labrador, and the Caledonian mountains in Norway and Scotland have a logical extension in the Canadian Appalachians. Various other similarities in the plateau rock formations of Africa and South America have been found. More recently, with the development of radioactive dating techniques, transatlantic areas of rocks of similar ages have been found.

(c) There is solid geological evidence that a glacial ice sheet covered the southern parts of South America, Africa, India, and Australia about 300 million years ago, similar to the one that covers Antarctica today. Hence, it is reasonable to conclude that the

forms between continental regions when they were together as Pangaea. It would not be expected that these biological species could traverse the present-day oceans. For example, a certain variety of garden snail is found only in the western part of Europe and the eastern part of North America. Also, a relatively

Figure 27-2c *An illustration of the continuity of continental features. Continental jigsaw pieces cut from a printed page would show the continuity of printed lines when put back together. If the continents were once together and drifted apart, we might expect some similar "printed lines" common to the separated continents in the form of continuities of geologic features such as mountains.*

Figure 27-2d *The continuity of geologic features supports the theory of continental drift. If the continents were fitted together, various mountain ranges of similar structure and rock composition on the different continents would line up analogous to the print in Fig. 27-2c.*

southern portions of these continents must have been under the influence of a polar climate at this time. There are no traces of this ice age found in Europe and North America. In fact, fossil evidence indicates that a tropical climate prevailed in these regions during that period. Yet, evidence of glaciation is found in India and in Africa near the equator. How could this be? It seems unreasonable that an ice sheet would cover the southern oceans and extend northward to the equator.

Wegener's theory suggests an answer and derives support from these observations. The directions of the glacier flow is easily determined by marks of erosion on rock floors and by moraines (deposits of rock and soil debris transported by glaciers). If the continents were once grouped together as Wegener's theory indicates, then the glaciation area was common to the various continents, as illustrated in Fig. 27-2e. The glacial movements, as indicated by the arrows in the second drawing support the idea of a single ice cap and subsequent continental drift.

Although there was evidence supporting Wegener's theory, it was not generally accepted, primarily because the proposed mechanism for continental drift was unsatisfactory. Wegener depicted the continents as giant rafts moving through the oceanic crust due to the earth's rotation. This mechanism is unacceptable because the force associated with the earth's rotation is not strong enough to overcome the measured strength of the rocks, which would be disintegrated if the continental crust would slide over or move through the oceanic crust.

A more satisfactory theory for the mechanism behind continental drift was suggested in 1960 by H. H. Hess, the American geologist. It was known that a mid-oceanic ridge system stretches through the major oceans of the world. In particular, there is the Mid-Atlantic Ridge that runs along the center of the Atlantic Ocean between the continents. This and other mid-oceanic ridges run along large fissures in the earth's crust, as evidenced by volcanic and earthquake activity along the ridges. Hess suggested a theory of **sea floor spreading** where the sea floor spreads slowly and moves sideways away from the mid-oceanic ridges. This movement is believed to be accounted for by convection currents of subterranean molten materials that causes the formation of mid-oceanic ridges and surface motions in lateral directions from the fissure, as illustrated in Fig. 27-2f.

Support for this theory has come from remanent magnetism studies and the determination of the ages of the rocks on each side of the mid-oceanic ridges. **Remanent magnetism** refers to the magnetism of rocks due to a special group of minerals called ferrites, one of which is ferrous (iron) ferrite, Fe_3O_4, commonly called magnetite or lodestone. When molten material containing ferrites is extruded upward from the earth's mantle, e.g., in a volcanic eruption, and solidifies in the earth's magnetic field, it becomes magnetized. The direction of the magnetization indicates the direction of the earth's magnetic field at the time.

These solidified rocks, called igneous rocks (Section 28-1) are worn down by erosion. The fragments are carried away by water and they eventually settle in bodies of water, where they become layers in future sedimentary rocks (Section 28-2). In the settling process, the magnetized particles, which are in fact small magnets, become generally aligned with the earth's magnetic field (Fig. 27-2g). By studying the remanent magnetization of geologic rock formations and layers, scientists have learned about changes that have taken place in the earth's magnetic field.

Measurements of the remanent magnetism of rocks on the ocean floor revealed long, narrow, symmetric bands of **magnetic anomalies** on both sides of the Mid-Atlantic Ridge (Fig. 27-2f). That is, the direction of the magnetization was reversed in adjacent parallel regions. Along with data on land rocks, the

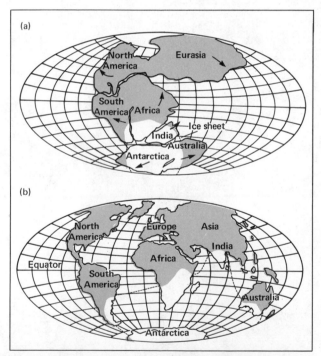

Figure 27-2e *Glaciation evidence of continental drift. Geologic evidence shows that a glacial ice sheet covered parts of South America, Africa, India, and Australia some 300 million years ago, but there is no evidence of this glaciation in Europe and North America. Also, evidence of this glaciation is found near the equator in India and in Africa. This is explained if a single ice sheet covered the southern polar region of Pangaea and the continents subsequently drifted apart.*

Figure 27-2f *Sea floor spreading and the Mid-Atlantic Ridge. The Mid-Atlantic Ridge runs along the ocean floor between the continents. It is believed that sea floor spreading takes place along the ridge due to convection currents of molten material rising from the upper mantle and spreading outward from the ridge. This is evidenced by magnetic anomalies in the ocean floor rock (shown as strippled areas in the drawing) caused by reversals of the earth's magnetic poles, which have occurred relatively frequently and regularly throughout geologic time. (Photo courtesy National Geographic Society.)*

magnetic anomalies indicate that the earth's magnetic field has abruptly reversed fairly frequently and regularly throughout recent geologic time. The most recent reversal occurred about 700,000 years ago. Why this reversal should occur is not known. We are evidently living in a period between pole reversals. However, the symmetry of the anomaly bands on either side of the Ridge indicates a movement away from the Ridge at the rate of a few centimeters per year and provides evidence for sea floor spreading.

Remanent magnetism also provides other support for continental drift. The remanent magnetism of a rock remains fixed throughout the history of the rock, even if forces within the earth move the rock. If the rock is moved, its remanent magnetism is no longer aligned with the earth's magnetic field. This is similar to the nonalignment of printed lines on moved puzzle pieces as described in the previous analogy on the continuity of geologic features. Scientists have studied the remanent magnetism of rocks of varying ages and have found evidence of such changes of alignment. Moreover, it was found that rocks of the same age on a continent have the same misalignment, but that the misalignment is different for different continents. This suggests that the entire continent must have moved as a unit and that each

continent moved or "drifted" in its own separate direction.

In recent years, investigative drilling into the ocean floor has been done from the oceanographic research vessel Glomar Challenger (Fig. 27-2h). These drillings have shown that the ocean floor between Africa and South America is covered by rela-

Figure 27-2g *An illustration of remanent magnetism deposition in sedimentary layers. Magnetized particles from eroded rock become generally aligned with the earth's magnetic field as they settle in bodies of water. The sediment layers eventually become sedimentary rock with remanent magnetism.*

Figure 27-2h *The Glomar Challenger. This 10,500-ton vessel is capable of conducting drilling operations in the open sea using dynamic positioning to maintain position over a bore hole. The drilling derrick amidship stands 194 feet above the waterline. Forward of the derrick (partially visible) is the automatic pipe racker containing 24,000 feet of 5-inch drill pipe. Scientific laboratories and crew's quarters are located aft. (Courtesy Deep Sea Drilling Project/SIO, Scripps Institution of Oceanography.)*

tively young sediment strata. Also, the strata thicknesses increase away from the Mid-Atlantic Ridge, which implies that the older parts of the ocean floor are farther away from the Ridge. The older parts would be covered by a greater thickness of sediment because there would have been a longer time for the sediment to accumulate.

These observations support the idea of sea floor spreading as a mechanism for the theory of continental drift, which has culminated into the modern theory of plate tectonics.

27-3
PLATE TECTONICS

The theory of **plate tectonics*** was fully realized in the late 1960s. In this theory, the outer layer of the earth is viewed as a series of rigid **plates** or huge slabs of rock, which are in (very slow) relative motion. The

---

*Tectonics (Greek, *tekton*, builder) is the study of the earth's general structural features and their changes.

basic assumptions of this theory are quite simple, but its explanations of basic geologic phenomena are far-reaching.

An immediate question might be: How can these plates move as a whole as though they are floating on something? Actually, they are floating in a sense. The geological term for this flotation is isostasy. Suppose the outer portions of the earth were not rigid enough to support crustal features of different weight placed on it. These features of different weights would sink to different depths until they came to equilibrium like the blocks in Fig. 27-3a. It is found that there is a similar occurrence for the earth's crust. That is, mountains have deeper "roots" than plateau areas as illustrated in the figure. This is the concept of **isostasy**—that crustal material is "floating" in gravitational equilibrium with other crustal materials on a "fluid" substratum. In terms of plate tectonics, it is convenient to divide the outer portion of the earth as shown in Fig. 27-3b. The outermost solid portion is called the **lithosphere** (Greek, *lithos,* rock). It includes the crust and some of the upper mantle, extending to a depth of about 50 miles. The material below the lithosphere, which is considered to be rocky material

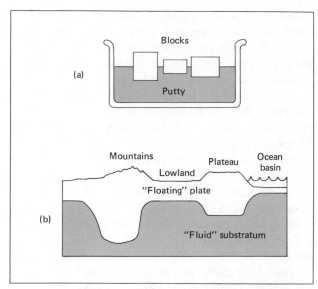

Figure 27-3a *An illustration of the concept of isostasy. (a) Blocks of some heavy material will sink to different depths until they float in equilibrium in some viscous "fluid" material like putty. (b) Similarly, the solid outer portion of the earth "floats" on a "fluid" substratum with different crustal features floating at different depths.*

fill the gap. The rising material cools near the surface and becomes part of the lithosphere material forming the mid-oceanic ridges and becoming part of the plates on either side of the boundary. The average rate of sea floor spreading is of the order of 4–6 centimeters per year.

If the surface area of the earth remains constant, as it is believed to be, then as new lithospheric material is generated along divergent plate boundaries, the plates must be driven together and material "destroyed" at convergent plate boundaries. The types of convergent plate boundaries are illustrated in Fig. 27-3d. The first two processes in which one plate is deflected downward beneath the other into the asthenosphere is called **subduction.** When two oceanic plates collide, one is deflected downward and returns to the mantle. The subduction zone of oceanic plates is marked by a deep oceanic trench. An example of such a trench is the Marina Trench in the western Pacific, which has the greatest known depth below sea level on the earth's surface—about 6.8 miles. The

that is hot enough to be deformed and capable of internal "flow," is called the **asthenosphere** (Greek, *astheno,* weak). The rigid plates of the lithosphere move about on the fluid substratum of the asthenosphere. The continents are like passengers on the moving plates.

If indeed plates of crustal material move relative to each other, then the relative motion of the plates is an important feature. Obviously this relative motion can be of three general types as illustrated in Fig. 27-3c—divergent, convergent, and sliding motions. Dynamic geologic processes occur at the plate boundaries.

In the case of a divergent plate boundary, sea floor spreading occurs. As the plates recede from each other, hot, molten material from the mantle rises to

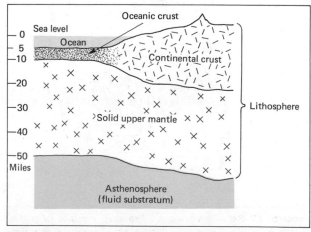

Figure 27-3b *A cross section of the lithosphere. The lithosphere is made up of the oceanic and continental crusts and the solid upper portion of the mantle. The material below the lithosphere, which is considered to be hot enough to be easily deformed and capable of internal flow, is called the asthenosphere.*

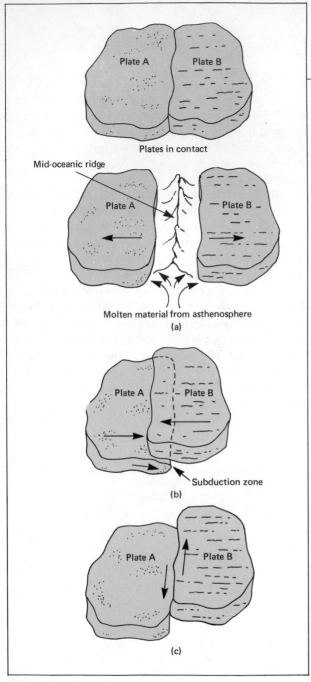

Mid-oceanic ridge

Plate A

Plate B

Molten material from asthenosphere

(a)

Plate A

Plate B

Subduction zone

(b)

Plate A

Plate B

(c)

Figure 27-3c *Three types of relative motions of plates. Plates in contact can (a) diverge as a result of sea floor spreading, (b) converge where plates are driven together and one plate is deflected beneath the other (subduction zone), or (c) slide past each other while in contact.*

material of the oceanic crust is denser than that of the continental crust. Hence, when oceanic and continental plates collide, the oceanic plate is deflected below the continental plate. In each of these cases, part of the descending oceanic crust in contact with the mantle begins to melt. The molten material rises into the overlying plate, and on breaking through the surface gives rise to volcanoes. In the case of the convergence of oceanic plates, the volcanoes build island arcs, of which Japan is a good example. In the case of the convergence of oceanic and continental plates, a volcanic mountain chain may be formed. The Andes mountains on the west coast of South America are a good example of such a formation. Friction between the descending and overlying plates causes stresses in the plates. The release of energy accompanying movement of the plates gives rise to earthquakes in the region of subduction zones.

When two continental plates collide, material is pushed upward to form mountain ranges. It is believed that the Himalayas were formed in this manner when the Indian plate originally broke loose from Pangaea and collided with the Eurasian plate. (See Fig. 27-2b.)

Finally, the third type of relative plate motion shown in Fig. 27-3c is where adjacent plates slide horizontally past each other without a gain or loss in area. This occurs along faults, which mark the plate boundaries. Movements and the release of energy along these boundaries give rise to earthquakes. Examples of such fault zones are along the San Andreas fault in California and the Anatolian fault in Turkey.

With the resulting geologic features of earthquakes and volcanoes along plate boundaries, it should be easy to identify the various plates. If one plots the earthquake centers, as in Fig. 27-3e, a general idea of the plate boundaries is obtained (volcanoes are less frequent than earthquakes but occur in the same general regions). In 1968, the determination of the six major plates of the earth's surface and their relative motions was made. These six major plates, the African, American, Pacific, Eurasian, Indian

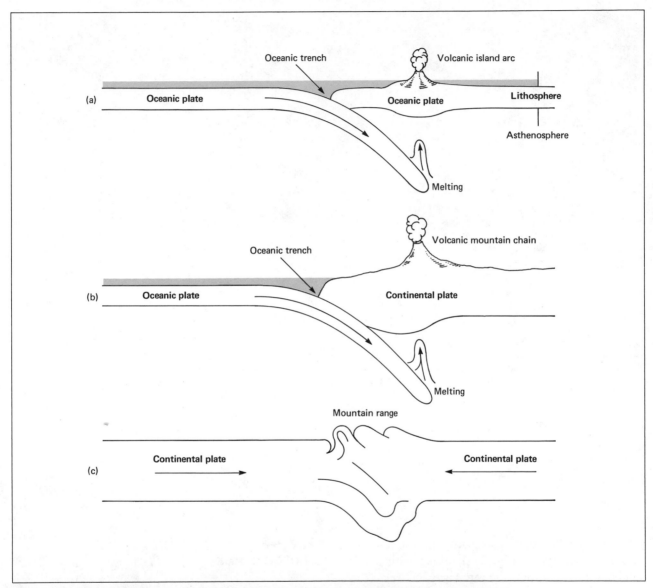

Figure 27-3d *Types of convergent plate boundaries. (a) Collision of two oceanic plates. One is deflected below the other and the melted material in the subduction zone rises to form volcanic island arcs. (b) Collision of oceanic and continental plates. The less dense oceanic plate is deflected below the continental plate and the melted material in the subduction zone rises to form volcanic mountain chains. (c) Collision of two continental plates. Material is pushed upward forming mountain ranges.*

Figure 27-3e *World map of earthquake locations from 1961–1969. As can be seen, the earthquake locations outline general boundaries. (Courtesy NOAA.)*

(sometimes called the Australian), and the Antarctic Plates are shown in Fig. 27-3f along with several smaller plates. Notice the similarity between Figs. 27-3e and 27-3f. The consistency of the theory of plate tectonics with geological phenomena has given rise to new explanations of geological processes which will be presented in greater detail in the following chapters.

27-4
GEOLOGIC TIME

In the previous sections, references were made to geologic events that happened millions of years ago. For example, Pangaea is believed to have broken up about 200 million years ago. It is fair to ask how we know, or what evidence we have, that certain geologic

Figure 27-3f *The six major plates of the world and several smaller ones. The relative motions along the plate boundaries are indicated. Compare this with Fig. 27-3e. Notice the rifting or divergence on the interior of the African Plate in East Africa. This East African Rift marks the place of possible further plate breakup. It is uncertain whether this rifting will lead to the eventual division of the African Plate and the creation of a new ocean basin.*

events took place so long ago. Early geologists had a great deal of difficulty in establishing a geologic time scale of events. As you know, the measurement of a time interval requires reference points or events, in particular a zero reference or an initial time or event.

The first "geologic clock" to be systematically used in the interpretation of earth history was based upon the study of fossils, the preserved remains of

former living plants and animals. Using the so-called **law of superposition,** which is the simple observation that in a succession of stratified deposits the younger layers lie on top of the older layers, geologists established a *relative* **geologic time scale** based on the characteristic fossils of each stratum. The major unit of geologic time is called an **era,** which in turn is divided into **periods** and **epochs.** The particular units

Figure 27-4a Simplified geologic time scale. (Courtesy U.S. Geological Survey.)

in geologic time are given names as indicated in Fig. 27-4a.

Relative time simply identifies an object or event as being younger or older than something else. From the law of superposition, we know that a set of fossils and the associated events in an upper stratum are younger or happened later in time than those in a lower stratum. But, it is difficult to assign times within a stratum. Since no accurate rate of deposition can be determined for most rock strata, the actual length of geologic time represented by a given stratum is at best an educated guess. As a result, events are assigned to a particular geologic time unit. For example, a geologist might say that dinosaurs became extinct (as evidenced by fossil remains) near the end of the Mesozoic Era, which precedes the Cenozoic Era. The boundary between these eras is *estimated* to be 70 million years ago, so dinosaurs became extinct over 70 million years ago.

Modern scientific methods have allowed the development of an **atomic time scale** based on natural radioactivity. As presented in Chapter 9, the rate of decay of a radioactive isotope is conveniently expressed in terms of the isotope's half-life, or the time it takes for one-half of the nuclei in a sample to decay. Many isotopes have relatively short half-lives a few days or years. However, some isotopes decay slowly and several of these are used as "atomic clocks" in measuring the ages of geologic events. The parent and daughter nuclei of the decay schemes most commonly used in geologic radioactive dating are listed in Table 27-4a.

Radioactive dating is simple in theory as described in Chapter 9 for carbon-14 dating. However, the actual laboratory procedures are complex. The principal difficulty with geologic samples lies in the accurate measurement of very small amounts of radioactive or associated elements.

Carbon-14 decay is a useful tool in dating relatively recent events. Because of its relatively short half-life, carbon-14 dating techniques are accurate for dating events that have taken place within the last

Table 27-4a *Radioactive Decay Schemes in Geologic Dating*

Parent Nucleus	Daughter Nucleus	Half-Life
Carbon-14	Nitrogen-14	5730 years
Potassium-40	Argon-40	1.3 billion years
Rubidium-87	Strontium-87	50 billion years
Thorium-232	Lead-208	14.1 billion years
Uranium-235	Lead-207	713 million years
Uranium-238	Lead-206	4.5 billion years

50,000 years, which is a very short interval in geologic time. Fortunately, other radioactive isotopes associated with geologic events have longer half-lives.

The potassium-argon decay scheme is one of the most useful to the geologist because it can be used on rocks as young as a few thousand years as well as on the oldest rocks known. Potassium is a constituent of several common minerals. The half-life of potassium-40 is such that measurable quantities of argon-40 (the daughter isotope) have accumulated in potassium-bearing rocks of nearly all ages; and even in very small quantities the amounts of potassium and argon isotopes can be measured accurately. However, the daughter product, argon, is a gas and can escape from the system, thereby causing the computed date of rock to be younger than it actually is. Fortunately, some rocks retain the argon quite well.

Each decay scheme dating method finds different geologic applications, but they must be applied carefully. Whenever possible, two or more methods of analysis are used on the same specimen of rock to confirm and verify the results. The correlation of the relative geologic time scale deduced from fossil evidence with atomic time poses certain problems, chiefly because rocks solidified from molten material can be dated by atomic methods more reliably than other kinds and these rocks do not ordinarily contain fossils. Many solidified igneous rocks retain the parent and daughter isotopes within them, thereby allowing accurate determination of the relative amounts of the isotopes. Other rocks, such as sandstone and limestone (sedimentary rock formed from sediments), are related to the atomic time scale by correlating them to time intervals that are determined from the dating of appropriate igneous rocks. However, even with these problems, the "atomic clocks" of the radioactive decay processes have helped geologists relate the relative geologic time scale to one that is more absolute.

IMPORTANT WORDS AND TERMS

Geology	Sea floor spreading	Subduction
Principle of uniformity	Remanent magnetism	Law of superposition
Core	Magnetic anomalies	Geologic time scale
Mantle	Plate tectonics	Era
Crust	Plates	Period
Moho discontinuity	Isostasy	Epoch
Continental drift	Lithosphere	Atomic time scale
Pangaea	Asthenosphere	

LEARNING OBJECTIVES

After reading and studying this chapter, you should be able to do the following without referring to the text:

1. Describe the earth's interior structure and composition.
2. Describe the early formulation of the theory of continental drift and the geologic evidence offered for its support.
3. Explain the mechanism and evidence for sea floor spreading and its relationship to continental drift.
4. Explain the theory of plate tectonics.
5. List the general relative motions of plates and the resulting geologic implications.
6. Distinguish between the relative geologic time scale and the atomic time scale.
7. Describe several radioactive decay schemes used in geologic dating.

QUESTIONS

1. What is the principle of uniformity and how is it used in the study of geologic processes?
2. State the compositions and structural divisions of the earth's interior. Are they all solid?

3. Why can there not be a permanent magnet within the earth that gives rise to the earth's magnetic field?
4. What is the name of the boundary between the crust and the mantle?
5. How is the earth's internal structure analogous to an egg?
6. What is Pangaea?
7. Describe the geologic evidence that supported Wegener's theory of continental drift.
8. What is the theory of the mechanism for continental drift? Why was Wegener's explanation unacceptable?
9. What is meant by remanent magnetism?
10. What evidence supports the theory of sea floor spreading?
11. What is the average rate of sea floor spreading?
12. What is the concept of isostasy?
13. What are the lithosphere and the asthenosphere?
14. What is a "plate" in the context of plate tectonics?
15. Describe the three general types of relative plate motions and the geologic results of each.
16. What is a subduction zone?
17. How are volcanic and earthquake activity explained in the theory of plate tectonics?
18. What are the names of the six major plates?
19. What is the law of superposition?
20. On what is the relative geologic time scale based?
21. What is the atomic time scale and what "atomic clocks" are used in this scale?
22. How are the relative geologic and atomic time scales correlated?
23. What is the average length of an era, a period, an epoch?

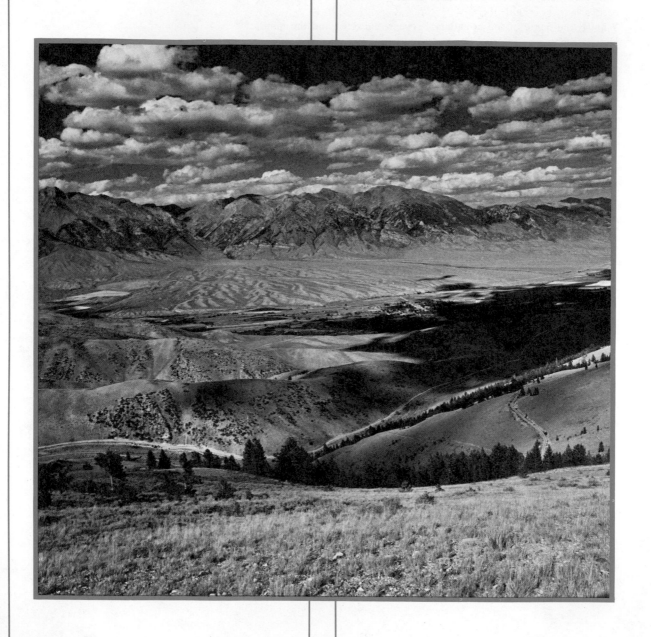

Chapter 28

ROCKS AND MINERALS

Touch the earth, love the earth, honour the
  earth,
her plains, her valleys, her hills and her
  seas;
rest your spirit in her solitary places.

Henry Beston

What is the earth made of? From our vantage point on its surface, "rocks" would be a typical answer. Even the sand and soil were once part of rocks. Since rocks are such an intimate part of the earth, an important question in geology is, How are rocks formed and what are they made of?

The processes of rock formation provide a means of classifying different types of rock, and, as will be discussed in this chapter, there are three types or families of rock: igneous, sedimentary, and metamorphic. We tend to think of rocks as permanent things; however, in the context of geologic time they are not. Geologic processes result in the creation, destruction, and alteration rocks—the so-called rock cycle. This was alluded to in the previous chapter, where in the theory of plate tectonics parts of the lithosphere are created and others destroyed.

A common definition of **rock** is any natural occurring, solid, mineral mass that makes up part of the earth. What are rocks made of? Chemical elements and compounds, of course. However, the geologist finds it more convenient to say most rocks are made up of one or more minerals, which are defined to be inorganic, crystalline substances. With the large number of available chemical elements, it would seem that the number of such substances that could be formed would be multitudinous. But, the number of minerals is limited because only a few chemical elements make up the bulk of the earth's crust, as will be learned in this chapter.

Igneous (Latin, *ignis*, fire) **rocks** are those formed by the cooling and solidification of hot, molten rock material called **magma.** If we accept the idea that the earth was originally a molten mass (magma), we would expect it to have cooled and solidified externally, so that the earth's crust would be composed largely of igneous rock. In fact, igneous rock is the most abundant of the three rock categories of the earth's crust, comprising as much as 80 percent.

The formation of magma is a continuing process. As was learned in the last chapter, plates are deflected downward toward the earth's interior, where they melt. Investigations in deep oil wells and mines have shown the temperature of the outer portion of the earth's crust increases about one degree Fahrenheit per 150 feet of depth. This variation in temperature with depth is called the **geothermal gradient.** If this rate of temperature increase were extrapolated to the center of the earth, the central interior temperature would be greater than 200,000°F, which is unreasonable. The temperature of the surface of the sun is only about 10,000°F. A more reasonable estimate of 4000°F to 5000°F for the temperature of the earth's interior comes from the study of melting points of materials on the surface that are believed to be the same as those in the central interior. Although heat is produced by the enormous pressure in the earth's interior, the major source of the earth's interior heat is believed to arise from the decay of radioactive materials.

Since the formation of magma takes place below the earth's crust, it is impossible to observe the natural formation of magma. But one can observe magma and the formation of one type of igneous rock. The formation of igneous rock takes place both on and within the earth's surface. Igneous rock is classified as being either extrusive or intrusive, depending on whether it solidified from molten magma above or below the earth's surface, respectively.

Extrusive igneous rock is formed from the cooling and solidification of magma poured out onto the earth's surface by volcanic eruptions. Magma that reaches the earth's surface through a volcanic vent is called **lava.** The lava cools and solidifies, becoming igneous rock (Fig. 28-1a). When exposed to the atmosphere (and often the ocean), lava cools and solidifies relatively quickly compared to magma cooling below the earth's surface. Ordinarily, lava cools slowly enough for crystalline grains to form. However, some lava cools and solidifies too quickly for this to happen and glassy materials—those without orderly crystalline structures—are formed. Obsidian is such a glassy volcanic rock. (See Table 28-1a and Fig. 28-1e). Such glassy substances tell the geologist that molten material underwent rapid cooling. Most often, volcanic eruptions or explosions are accompanied by the escape of gases. These gases may be originally held in solution in the molten subterranean magma by high pressure, much like gas is held in a capped soft drink. When the pressure is released on a soft drink by removing the cap, gas can be seen escaping from solution. Similarly, gases are released from magma reaching the earth's surface (lava) in volcanic eruptions. However, in the case of highly viscous lava, the gas bubbles cannot escape quickly and may be "frozen in" as the lava solidifies. Extrusive igneous rocks formed with "frozen in" gas bubbles have a frothy texture such as scoria and pumice shown in Fig. 28-1e.

Intrusive igneous rock is formed from the cooling and solidification of magma below the earth's surface. A distinguishing characteristic between intrusive rock and extrusive rock is their relative texture, or grain size. When an igneous mass cools and crystallizes, the size of the mineral particles depends on the rate of cooling. Extrusive rock cools relatively quickly, and is characterized by fine grains. Large crystalline grains do not have time to grow. As pointed out previously, the lava rock of volcanoes is fine grained and even glassy in some cases. Intrusive rock cools slowly below the earth's surface, and such rock is character-

Figure 28-1a *A solidified lava flow west of Pisgah Volcano, California. Magma reaching the earth's surface through a volcanic vent is called lava. The lava cools and solidifies, becoming igneous rock. (Courtesy U.S. Geological Survey.)*

ized by relatively larger grains and coarser texture. Granite is an example of intrusive rock (Fig. 28-1e). The grain size is dependent on the rate of cooling, which involves several factors, such as depth, location, size or mass, and the specific heat of the material.

Large bodies of intrusive igneous rock are called **plutons,** after Pluto, the Greek god of the underworld. These plutonic bodies are classified according to their size, shape, and orientation in surrounding rock. Intrusive rock bodies are referred to as being **concordant** when they lie more or less parallel to older formations, and as being **discordant** when they cut across older formations. The four major plutonic forms are illustrated in Fig. 28-1b. Large discordant bodies are called **batholiths.** Batholiths are enormous in size, and to be so classified, must occupy an area of at least 40 square miles. Protruding from a batholith may be an irregular dome formation called a **stock,** which may have once supplied magma to a volcanic vent. The locations of some batholith formations are

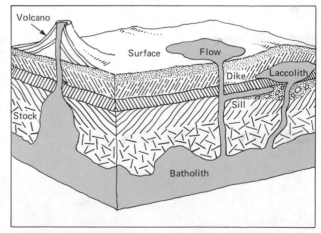

Figure 28-1b *An illustration of plutonic bodies. Magma solidifying within the earth forms intrusive igneous bodies. The batholith is the largest intrusive body. Sills and laccoliths are concordant bodies that lie parallel to existing rock formations. The discordant dike cuts across existing rock formations. A volcanic stock is shown rising from the batholith.*

Table 28-1a *Examples of Types of Rocks[a]*

Type	Characteristic
Igneous	
Intrusive	*Texture*
Granite	Coarse-grained
Extrusive	
Obsidian	Glassy
Scoria	Porous
Pumice	Glassy and porous
Sedimentary	
Clastic	*Sediment*
Conglomerate	Gravel
Sandstone	Sand
Shale (mudstone)	Mud
Breccia	Volcanic fragments
Nonclastic, chemical	*Chemical composition*
Limestone	$CaCO_3$, calcium carbonate
Gypsum	$CaSO_4 \cdot 2 H_2O$, hydrous calcium sulfate
Rock salt	NaCl, sodium chloride
Flint and chert	SiO_2, silicon dioxide
Nonclastic, organic	
Shell limestone (e.g., Coquina)	Compacted or cemented calcareous plant or animal remains
Coal	Compacted plant remains
Metamorphic	*Parent rock*
Marble	Limestone
Quartzite	Sandstone
Slate	Shale (low grade metamorphism)
Gneiss	Shale (high grade metamorphism) or granite

[a] See photographs in Fig. 28-1e.

Source: J. D. Wilson, *An Environmental Approach to Physical Science.* Heath, Lexington, Mass., 1974.

Figure 28-1c *The locations of large batholiths in the western part of North America. To be classified as a batholith, the intrusive body must have an area of at least 40 square miles.*

shown in Fig. 28-1c. The Coastal Range batholith along the western Canadian coast is a classic example.

Another discordant plutonic body is a **dike,** which is formed when magma fills a nearly vertical fracture in rock layers. A dike is tabular with its width or

Figure 28-1d (Top) *A discordant dike formation. This exposed dike shows how the forming magma intruded across existing rock formations. (Bottom) A concordant still formation. The intrusion of magma that formed this sill on Unalaska Island, Aleutian Islands, Alaska, was parallel to existing rock formations. (Courtesy U.S. Geological Survey. Sill photo by G.L. Snyder.)*

thickness much smaller than its other dimensions, as shown in Fig. 28-1d. Although originally formed below the earth's surface, some intrusive bodies are now visible, due to erosion and other geologic processes.

Sills and laccoliths are intrusive concordant bodies. **Sills** are similar to dikes, but form between and parallel to existing rock beds (Fig. 28-1d). A **laccolith** is a blisterlike intrusion that has pushed up overlying rock layers.

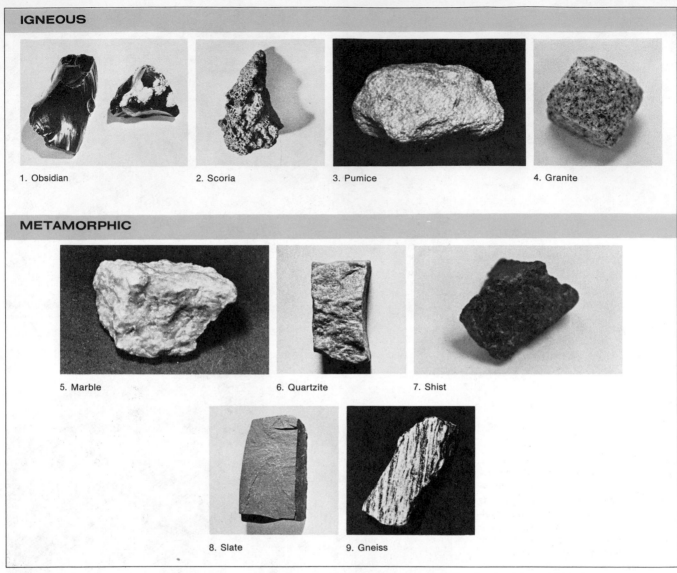

IGNEOUS

1. Obsidian 2. Scoria 3. Pumice 4. Granite

METAMORPHIC

5. Marble 6. Quartzite 7. Shist

8. Slate 9. Gneiss

Figure 28-1e *Examples of igneous, sedimentary, and metamorphic rocks. (Courtesy Ron Docie, Athens, Ohio.)*

28-2
SEDIMENTARY ROCK

In the previous section on intrusive rock formations, it was stated that magma intrudes between layers of preexisting rock. In general these rocks near the surface of the earth are **sedimentary rocks.** These are rocks formed from sediment composed of particles of older rocks and other material. Rocks on the earth's surface are continually being both physically and chemically worn down by environmental processes. The resulting rock particles make up the sediment that is transported by streams and rivers and eventually deposited in large bodies of water as sedimentary layers. The consolidation of this sediment forms sedimentary rock.

Sedimentary rocks, which occur in many varieties depending on the type of rock fragments making up the sediment, are the most common rocks in our

10. Sandstone

11. Conglomerate

12. Shale (Mudstone)

13. Breccia

14a. Limestone (chemical)

14b. Limestone (organic)

15a. Flint (black) with Chert

15b. Chert

16. Coquina (shell limestone)

17. Gypsum

18. Rock salt (Halite)

environment. Although sedimentary rocks make up only 5 percent of the earth's crust, they cover about 75 percent of the earth's surface. The stratified layers of sedimentary rock are commonly observed in the environment as shown in Fig. 28-2a for both natural and artificial reasons.

Because of their abundance and variety of content, sedimentary rocks are important in our everyday life. Petroleum and coal are found in sedimentary rock formations and sedimentary rock supplies many

essential building materials. The transformation of sediment into rock is referred to as **lithification** or **consolidation.** For example, when relatively fine-grained mud is subjected to high pressure such that a substantial portion of water is forced out, the mud is compacted into the sedimentary rock called shale (sometimes called mudstone, Fig. 28-1e). In some cases of compaction the grains may recrystallize to form an interlocking structure. Since the fine grains of compacted rocks are held together by forces that are

Figure 28-2a *Examples of exposed sedimentary rock layers in the environment due to natural (erosion) and human (road-cut) causes. (Top photo courtesy U.S. Geological Survey.)*

not very strong, the rocks are relatively soft and break easily. If the grains of sediment are coarse, such as sand or gravel, compaction will not form a stable rock. The particles of coarse-grained sediment may be cemented together by materials dissolved in the water which permeates the sediment. Common cementing agents are silica (SiO_2), calcium carbonate ($CaCO_3$), and iron oxides. Regardless of the method of lithification, sedimentary rocks are not as hard as the fused, crystallized igneous rocks.

Sedimentary rocks are classified into two subcategories: clastic and nonclastic. **Clastic** (Greek, *klastos,* broken) **sedimentary rocks** are formed from sediment composed of fragments of preexisting rocks. Common

examples are sandstone and shale (Fig. 28-1e). **Nonclastic sedimentary rocks** are formed primarily from chemical and organic sediments. For example, dissolved chemicals in water may precipitate to form nonclastic chemical sedimentary rocks.

Many caverns contain dripstone, which is formed by material precipitated from dripping water. Dripstone takes on a variety of forms, but most comon are the icicle-shaped **stalactites** and **stalagmites** (Fig. 28-2b). As drops of water seep through the cavern roof and fall on its floor, minute amounts of dissolved calcium carbonate ($CaCO_3$) and other materials precipitate and cause the formations to grow. To distinguish between which formation hangs from the ceiling and which formation protrudes upward from the floor, simply remember stalactites or "tites" come from the "top." If the stalactites and stalagmites are joined, the combined formations are referred to as **columns.**

Organic sediments result from living organisms, either animal or plant. The shells of marine animals may be cemented together to form organic sedimentary rocks such as coquina shown in Fig. 28-1e. Chalk is another sedimentary rock formed from accumulations of shells of microscopic animals. Probably the best known organic rock of plant origin is coal. Coal is a black sedimentary rock consisting chiefly (over 60 percent) of partly decomposed fossil plant matter. In what was once low swampy areas, sediment layers rich in vegetation were deposited and later became coal beds. Initially, bacterial action decomposed the plant matter with the release of gases; and the plant residue gradually became concentrated in carbon content. This carbon-rich decayed plant matter is called **peat.** Existing peat beds are an important fuel source in some countries.

As peat was buried by overlying strata of sediment deposits of sand, silt, or clay, it was compressed and water was squeezed out. Volatile organic compounds such as methane (CH_4) escaped, leaving a material of increased carbon content. The peat was successively compacted into **lignite, or brown coal** (65 to 70 percent carbon), **sub-bituminous coal** (70 to 75 percent carbon), and **bituminous, or soft coal** (75 to 80 percent carbon). These coals are sedimentary rocks. A further change results in **anthracite, or hard coal** (approximately 90 percent carbon). Anthracite coal is a metamorphic rock (see Section 28-3).

The ignition of coal depends on its content of volatile materials. With a relatively high volatile con-

Figure 28-2b *Nonclastic sedimentary cavern dripstone. Stalactites (tite—top) form from the ceiling down and stalagmites build from the floor up as a result of the precipitation of dissolved chemicals from dripping water. When the formations join together they are referred to as a column. Scientists estimate the formations in this cave have been building for more than 10 million years and the formations are still building at the very slow rate of one cubic inch per 120 years. (Courtesy Luray Caverns, Virginia.)*

Figure 28-2c *Coal deposits in the conterminous United States. (Courtesy U.S. Geological Survey.)*

Figure 28-2d *Bedding and folding. (Top) Cross-bedding in sandstone formations in Navajo County, Arizona. Sand deposited by winds that shifted directions formed these cross-bedded formations. (Bottom) Folded rock strata. Stresses in the earth's crust are evidenced by folded rock strata, such as the ones shown here in a weathered shale formation in Johnson County, Tennessee. (Courtesy U.S. Geological Survey. Photos by H.E. Gregory and W.B. Hamilton.)*

EXPLANATION

Anthracite and semianthracite

Low-volatile bituminous coal

Medium- and high-volatile bituminous coal

Subbituminous coal

Lignite

0   200   400   600 MILES

tent, lignite and sub-bituminous coals ignite easily, but burn with a great deal of smoke. As a result, these coals are sometimes referred to as smoky coals, and can contribute to air pollution (Chapter 26). Anthracite coal, on the other hand, is difficult to ignite and burns with almost no smoke. Deposits of lignite and sub-bituminous coals are found predominantly in the western United States, and deposits of bituminous and anthracite coals are found predominantly in the eastern United States (Fig. 28-2c). Coal is an important fuel resource for the highly industrialized eastern part of the country, however, the eastern coals have a relative high sulfur content compared to western coals—another major consideration in air pollution.

The decay of organic matter, both plants and animals, also produce the petroleum and natural gas resources upon which we depend so much today. The majority of the petroleum used today comes from liquid oil pools below the surface of the earth. However, another source of petroleum, which may become increasingly important as our energy demands increase, is tar sands and oil shale. The hydrocarbons in these reserves occur in the solid state, filling the pores of sedimentary strata. The tar sands and oil shale must be mined and heated to remove the solid hydrocarbons. Important tar sands and oil shale deposits occur in the western United States and Canada.

Perhaps the most characteristic feature of sedimentary rock formation is the stratification, or bedding, which results from the depositing of the layers of sediment. The sedimentary layers are not always horizontal, however. They may be either tilted, as a result of the deposition of sediment on an inclined surface, or folded as a result of pressures from within the earth's crust (Fig. 28-2d). The layered feature of sedimentary rocks gives insight into the recent geologic history of the earth's surface. Each layer characterizes a certain time period, much the same as the rings of a tree do on a smaller time scale. In the sedimentary layers the preserved remains of plants and animals, or fossils, are sometimes found as shown in Fig. 28-2e. Fossils supply important clues to the history of life and to past features of the earth's surface.

Figure 28-2e *A clam fossil. This 570 million-year-old clam, 70 million years older than any previously known, was discovered among rocks collected near Albany, New York. (Courtesy U.S. Geological Survey.)*

28-3
METAMORPHIC ROCK

As the name implies, metamorphic rock results from a change or metamorphism (from the Greek word to change or transform) of preexisting rock. The metamorphism, which takes place below the earth's surface, is a change in the composition and texture of the rock caused primarily by heat, pressure, and moisture.

When rocks of any type are formed, they are stable in the surrounding conditions of formation. However, if these conditions change, the rock may become unstable and changes so as to conform to its new environmental conditions. For example, limestone with coral and clam shells is found in Pennsylvania and New York (indicating that this area was once covered by a shallow sea). As such, this sedimentary limestone was formed under conditions of little heat or pressure. However, if the limestone should experience increased heat and pressure because of some geologic conditions, it may undergo a change, and under proper conditions become the metamorphic rock marble, as is found in western New England, Georgia, and Alabama.

Having undergone a metamorphic process, the resulting metamorphic rock is different from the original rock. In most cases, the bulk chemical composition of the rock remains unchanged, but the new rock usually contains different chemical compounds or minerals. (Minerals or particular rock-forming chemical compounds will be discussed in Section 28-4). In some instances, the existing rock-mineral grains are recrystallized and a new texture develops. Either or both of these processes may take place in metamorphism.

Metamorphism occurs for sedimentary, igneous, and even metamorphic rocks. Due to geologic changes, rocks formed at the surface of the earth may become buried to depths of several thousand feet. The rocks are then subject to high temperatures and pressures. Certain rock components are unstable at these temperatures and pressures and react with other rock components to form new compounds or minerals. These metamorphic reactions occur within the solid rock. Should the temperature be great enough to cause the rock to become molten, it is then magma and its solidification would result in an igneous rock. Also, as pointed out previously, during a metamorphic process the original rock grains may recrystallize becoming larger with perhaps slightly different forms. It is these changes in the features of rock composition, texture, and structure that distinguishes metamorphic rock from sedimentary and igneous rock.

Moisture is also an important factor in metamorphism. Sedimentary rock has open spaces between its grains that are accessible to moisture. Sandstone is a good example. When metamorphism takes place, the water between the grains tends to speed up the chemical reactions. Sandstone is metamorphized to quartzite. (See Fig. 28-1e.) Igneous rocks, on the other hand, are usually more dense and have relatively little open space between the rock grains and hence contain little water. To metamorphize a "dry" rock, much higher temperatures and pressures are required to initiate chemical reactions. In general, then, the metamorphism of igneous rocks require greater temperatures and pressures than do sedimentary rocks.

As has been mentioned, metamorphic rocks may be further metamorphized into other metamorphic rocks. That is, increased temperature and pressure can cause a change in an existing metamorphic rock. An example of such a sequence begins with the metamorphism of sedimentary shale or mudstone into metamorphic slate. This platy rock was once commonly used for blackboards and as a roofing material. Under the proper conditions, slate may be changed to schist; and schist in turn may be changed to gneiss (Fig. 28-1e). The examples of these types of rocks in the figure are not metamorphisms of the same rock, but are examples with different chemical contents. A type of gneiss also results from the metamorphism of granite.

Metamorphic rocks provide geologists a means to study the forces and temperatures in the earth's crust. The pressure for metamorphism of rock materials may arise from the weight of materials deposited on top of the rock formation or from a shift in the earth's crust that produces a more horizontal pressure. Pressure generates heat; however, the heat associated with metamorphism may also be supplied by nearby igneous bodies. Metamorphism should not be thought of as an overnight process, rather it is a geologic process that in general requires a considerable time interval. If a change is brought about primarily as a result of pressure it is referred to as **dynamic metamorphism**. During the formation of some metamorphic rocks, an applied directional pressure causes the rock grains to be oriented, or formed in the same direction. Such metamorphic rocks are said to be foliated. As a result of this orientation or **foliation**, the rocks may be split along a smooth plane and have a sheetlike, or platy, characteristic. Slate is an example of foliated metamorphic rock. Other metamorphic rocks may not be foliated and not exhibit this characteristic, particularly if they were formed by **thermal metamorphism** (change by heat with little pressure). Most metamorphic rocks have experienced both high temperatures and pressure. This type of change is called **regional metamorphism.** Such metamorphic rocks have been affected by both chemical recrystallization from the heat and mechanical deformations from the pressure.

Under sufficient heat and pressure a metamorphic rock may become molten and therefore be classified as magma. This process essentially completes a cycle, which you may have noticed emerging. Magma forms igneous rocks which are worn by weathering. The particles are lithified into sedimentary rocks that are most susceptible to metamorphism. Both sedimentary and metamorphic rocks are worn away by weathering, and their particles again form sedimentary rocks. These interrelated processes by which rock is destroyed, created, and altered form what is known as the **rock cycle** and is illustrated in Fig. 28-3a.

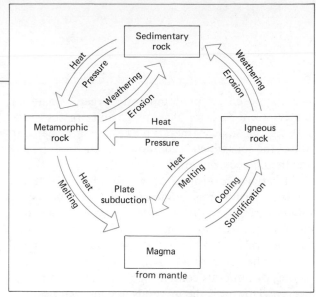

Figure 28-3a *An illustration of the rock cycle. Rock material in the lithosphere follows various routes in the cycle. Plate additions along mid-oceanic ridges and plate destruction in subduction zones correspond to an entrance and exit to the rock cycle, respectively.*

Within the framework of plate tectonics, the additions to the plates along the mid-oceanic ridges and the destruction of the plates in the subduction zones correspond to an entrance and exit to the rock cycle of the lithosphere.

28-4
MINERALS

As has been discussed, rocks can be classified into three broad categories based on the method of their formation. However, the chemical and physical classification of rocks depends on the composition of their particles. Certainly rock particles are made up of one or more chemical compounds; but rather than using specific compounds (or elements), geologists refer to minerals as being the building blocks of rocks. A mineral is any naturally occurring, inorganic, crystalline substance. *Most* rocks are made up of an aggregate of one or more minerals. *Some* rocks, e.g., coal, are composed of predominantly organic material. Others are a mixture of minerals and organic material, e.g., oil shale. Minerals make up a large percentage of the soil, which is composed for the most part of weathered rock material.

Everywhere around us we see minerals or the products of minerals. Some are quite valuable, while others are essentially worthless. For example, precious stones such as diamonds and rubies are valuable minerals. Also, we speak of a nation's mineral wealth when referring to natural raw materials such as ores or minerals containing iron, gold, silver, copper, etc. However, the minerals of common rock, like sandstone, have little monetary value. The word mineral has also taken on popular meanings. For example, foods are said to contain vitamins and "minerals." In this case, the term mineral refers to compounds in food which contain elements needed in small quantities by the human body, such as Fe, Na, I, Mn, Mg, and Cu. The chemical composition of these compounds may be the same as those naturally occurring minerals that make up the earth's crust, or they may actually be the naturally occurring minerals themselves. The names of minerals, like those of chemical elements, have historical connotations and many reflect the names of localities.

Mineral classification is advantageous because it is based on the physical and chemical properties of substances, which distinguishes between different forms of minerals composed of the same element or compound. For example, graphite—a soft black, slippery substance commonly used as a lubricant—and diamond are both composed of carbon. But, because of different crystalline structures their properties are quite different. (Graphite mixed with other substances to obtain degrees of hardness is the "lead" in lead pencils.) Also, a mineral may have a series of related compounds in which one metallic element is replaced by another. The mineral feldspar may have the compositions $NaAlSi_3O_8$, $CaAl_2Si_2O_8$, or $KAlSi_3O_8$. In this case, distinction is made between different types of feldspar.

Let's take a look at the three prerequisites of a mineral as given in the previous definition: (1) Naturally occurring. This means that a synthetic compound, such as a plastic, is not a mineral. (2) Inorganic. The implication here is in the biological sense—that a mineral was not previously part of a living thing. For example, a fragment of plant fossil is not a mineral. Coal is conveniently considered to be a rock, although it has less than 40 percent inorganic—mineral—content. The energy value of coal comes from its nonmineral (organic) content. (3) Crystalline. The atoms of minerals are organized in orderly arrangements, and each mineral has a unique internal structure. Substances lacking a systematic arrangement of atoms and not having a definite composition are said to be amorphous. Amorphous (noncrystalline) substances are not minerals. There is an important class of geologic materials that falls into this nonmineral category, viz., glasses. Recall that glasses are substances that cooled so rapidly from molten material that there was not time for orderly internal structures to form before solidification. Naturally occurring glasses are relatively rare in everyday activities. The common glass materials, e.g., window glass, is not a mineral on two counts— they are neither naturally occurring nor crystalline.

With the large number of known inorganic elements, the number of possible minerals that might form in the earth's crust would seem enormous. However, the number of inorganic compounds that occur naturally in any substantial proportion is limited, because only a few elements make up the bulk of the earth's crust. As can be seen from Table 28-4a, only eight elements make up about 98 percent of the earth's crust. Notice that only two elements, oxygen and silicon, make up over 70 percent.

There are over 2000 known minerals, but only a few minerals—less than 20—make up over 95 percent of the rocks in the earth's crust. Most mineral compounds are composed of two or more elements which are held together by ionic bonding. Diamond is an example of one of the few covalent mineral com-

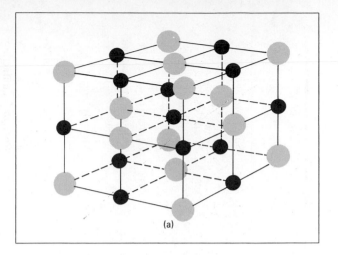

Figure 28-4a *The relative sizes and charges of some of the common mineral ions. The actual radii of the ions in angstroms (1Å = 10^{-10} m) are listed below the circles.*

pounds. In general, the ability of an ion to combine with other ions or atoms to form the crystalline structure of minerals depends on their electric charges and their relative sizes. Figure 28-4a shows the relative sizes of several common ions.

As was learned in Chapter 13, sodium and chlorine ions form NaCl (mineral name, halite) with the internal arrangement of the ions in a crystal structure of a simple cubic lattice (Fig. 28-4b). This structure depends on the electric charge and the relative sizes of the ions. As another example, let's consider the group of minerals called silicates (silicon-oxygen compounds), which are the most abundant rock-forming minerals. The positive silicon ion is Si^{4+} and

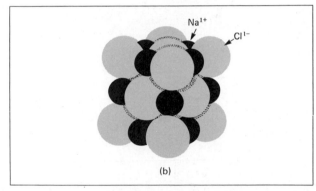

(b)

Figure 28-4b *An illustration of the cubic lattice structure of NaCl. (a) A simplified expanded view. (b) A close-packed view showing the relative sizes of the sodium and chlorine ions.*

Table 28-4a *Relative Abundances of Elements in the Earth's Crust*

Element	Approx. Percentage (weight)
Oxygen (O)	46.5
Silicon (Si)	27.5
Aluminum (Al)	8.1
Iron (Fe)	5.3
Calcium (Ca)	4.0
Magnesium (Mg)	2.7
Sodium (Na)	2.4
Potassium (K)	1.9
All Others	1.6
Total	100.0

the negative oxygen ion is O^{2-}. The simplest silicate compound is quartz (SiO_2, silicon dioxide).

All other important silicate minerals are combinations of silicon and oxygen and one or more other elements. The fundamental unit in these silicate crystalline structures is the **silicon tetrahedron** (Fig. 28-4c). This is a pyramid-shaped complex ion $(SiO_4)^{4-}$. How the tetrahedral units are bound together define the various silicate crystalline types. Several of these structures are illustrated in Fig. 28-4d. Olivine is a mineral with a single independent tetrahedron that is combined with either Mg^{2+} or Fe^{2+}. These metal ions can substitute for each other, depending on the conditions of mineral formation.

Table 28-4b *Some Common Minerals and Their Properties*

Minerals	Chemical Composition	Color	Luster	Hardness (Mohs' Scale)	Specific Gravity (approx.)	Cleavage	General Use or Occurrence
Argentite	Silver sulfide, Ag_2S	Black	Metallic	2–2.5	7.3	Yes	Silver ore
Asbestos (chrysotile)	Hydrous magnesium silicate, $H_4Mg_3Si_2O_9$	Green (different shades)	Silky	Low	2.5	Yes	Fireproofing; insulation
Calcite	Calcium carbonate, $CaCO_3$	White or colorless	Glassy to earthy	3	2.7	Yes	Cement manufacture; optical application; lime
Cinnabar	Mercury sulfide, HgS	Red-brown	Dull, earthy	2–2.5	8.1	Yes	Mercury ore
Clay	Hydrous aluminum silicates containing Na, K, Fe, Mg, etc.	Light	Earthy	Very low	2.6	Yes	Chinaware, pottery
Dolomite	Calcium (Mg, Fe, Mn, Ca) carbonate e.g., $CaMg(CO_3)_2$	White-gray	Pearly	3.5–4.5	2.9	Yes	Cement manufacture; building materials; lime
Feldspar plagioclase	Calcium & Sodium aluminum silicate, $CaAl_2Si_2O_8$ $NaAlSi_3O_8$	White-green	Pearly	6–6.5	2.7	Yes	Ceramic glazes
orthoclase	Potassium aluminum silicate $KAlSi_3O_8$	White-pink	Pearly	6	2.5	Yes	
Galena	Lead Sulfide, PbS	Gray	Metallic	2.5	7.4	Yes	Lead ore
Gypsum	Hydrous calcium sulfate, $CaSO_4 \cdot 2H_2O$	White-gray	Silky-dull	2	2.3	Yes	Plaster of Paris; neutralizer of alkaline soils

Mineral	Composition	Color	Luster	Hardness	Density	Cleavage	Uses
Halite	Sodium chloride, $NaCl$	Colorless or white	Glassy	2.5	2.2	Yes	Table salt
Hematite	Iron oxide, Fe_2O_3	Gray-reddish	Metallic, dull	5.5-6.5	5.1	No	Iron ore
Hornblende	Silicate mineral containing Mg, Fe, Al, etc.	Black-green	Silky	5-6	3.2	Yes	Igneous rock mineral
Magnetite	Iron oxide, Fe_3O_4	Black	Metallic, dull	6	5.2	No	Magnetic iron ore
Mica biotite chloriet muscovite	Group of hydrous metallic aluminum silicates	black-green; clear; reddish-brown	Pearly; Pearly	2.5-3; 2-2.5; 2.5-3	3.0; 2.8; 2.9	Yes; Yes; Yes	"Isinglass" heat proof windows; electrical insulator
Olivine	Metallic silicate, $(Mg, Fe, Mn)_2SiO_4$	Green	Glassy	6.5-7	3.2-4.4	No	Igneous rock mineral
Pyrite	Iron Sulfide FeS_2	Pale yellow	Metallic	6-6.5	5.0	No	Source of sulfuric acid (mine acid); fool's gold
Pyroxene	Metallic aluminum silicate (Ca, Mg, Fe, Na)	Black-green	Glassy	5-6	3.4	Yes	Igneous rock mineral
Quartz	Silicon dioxide, SiO_2	Colorless or white (many colors)	Glassy	7	2.65	No	Optical applications
Sphalerite	Zinc sulfide, ZnS	Black, Yellow-brown	Resinous	3.5-4	4.0	Yes	Zinc ore
Talc	Hydrous magnesium silicate, $Mg_3 (Si_4O_{10}) (OH)_2$	Green-white Silvery	Pearly, greasy	1-2.5	2.7	Yes	Cosmetics

Source: J. D. Wilson, *An Environmental Approach to Physical Science.* Heath, Lexington, Mass., 1974.

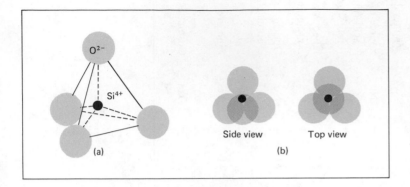

Figure 28-4c *An illustration of the silicon tetrahedron. (a) An expanded view of the complex ion (SiO₄)⁴⁻ showing the larger oxygen ions (O²⁻) at the four corners of the tetrahedron (three-sided pyramid) equidistant from the smaller silicon ion (Si⁴⁺) at the center. (b) Close-packed side and top views.*

The silicon tetrahedra can form single and double chain structures as shown in Fig. 28-4d. Examples of minerals with these structures are pyroxene and hornblende, respectively. In the pyroxene chain, each tetrahedron shares two oxygens; while in the double hornblende chain, half the tetrahedra share two oxygens and the other half share three oxygens. These structures have various metallic ion components as in the case of olivine. (See Table 24-4b.) The two-dimensional sheet structure of tetrahedra shown in Fig. 28-4d is the structure of mica and clay. The three-dimensional silicate structures are too complex to be shown in a simple illustration.

Minerals can be identified by chemical analysis, but most of these methods are detailed and costly, and are not available to the average person. More commonly, physical properties are used as the key to mineral identification. These properties are well known to all serious rock and mineral collectors. Some of the physical properties used in mineral identification are as follows. **Crystal form.** Crystals have definite geometric arrays of atoms or ions, that are often recognizable from the external crystal configurations of some minerals (Fig. 28-4e). Crystalline forms in minerals can be studied in detail by means of X-ray analysis. **Cleavage** and **fracture.** Cleavage refers to the splitting of a mineral along directions of weak bonding between planes of atoms (Fig. 28-4f). Fracture is the breaking up of a mineral other than

along a cleavage plane. **Specific gravity.** This is the ratio of the mass of body to the mass of an equal volume of water. Specific gravity indicates the compactness of the mineral structure. **Hardness.** This property is comparative and so is indicated by a harder mineral being able to scratch a less-hard mineral. The varying degrees of hardness are represented on **Mohs' scale of hardness,** which runs from 1 to 10, soft to hard. This arbitrary scale is expressed by the 10 minerals listed in Table 28-4c. Talc is the softest and diamond is the hardest. A particular mineral on the scale is harder than (can scratch) all those with lower numbers. Using these minerals as standards, one finds the following on the hardness scale: fingernail, 2–3; a penny, 3; window glass, 5–6.

Luster. The characteristics of light on the surface of the mineral due to absorption, reflection, and/or refraction give rise to a mineral's look or luster. Common descriptions of a mineral's luster are metallic, glassy, greasy, and dull. **Color** and **streak.** Color is

Table 28-4c *Mohs' Scale of Hardness*

1. Talc	6. Feldspar (orthoclase)
2. Gypsum	7. Quartz
3. Calcite	8. Topaz
4. Fluorite	9. Corundum
5. Apatite	10. Diamond

Silicate structure	Arrangement of tetrahedra (top view)	Typical mineral
Single independent tetrahedron		Olivine
Single chain		Pyroxene
Double chain		Hornblende
Continuous sheet		Mica
Three-dimensional network	Too complex to be shown by simple two-dimensional sketch	Quartz and feldspar

Figure 28-4d *Some silicate structures.*

Figure 28-4e *Mineral crystal forms. Certain minerals are characterized by natural crystal forms. Quartz (SiO_2) crystals are shown below, fluorite (CaF_2) crystals above. (Courtesy Ron Docie, Athens, Ohio.)*

Figure 28-4f *Cleavage of an NaCl crystal. Some minerals are characterized by cleavage planes along which they may be easily cleaved. (Courtesy Ron Docie, Athens, Ohio.)*

just what it implies, the color of a mineral. Color depends a great deal on the type of metal atoms a mineral contains. Streak refers to the color of the powdered material on a streak plate (unglazed porcelain). The streak of some metallic ores differ from the color of the unpowdered mineral. Other properties used to identify minerals include magnetism, radioactivity, and fluorescence.

Minerals are the natural source of many commercially important chemicals. Metal-rich rocks (ores) and mineral deposits are mined and processed for our use. Because of the vast extent of the earth's crust, it was once thought that the supply of many common minerals was virtually inexhaustible. However, at the current rates of use, some mineral resources may last only a few more generations.

IMPORTANT WORDS AND TERMS

Rock
Igneous rock
Magma
Geothermal gradient
Extrusive igneous rock
Lava
Intrusive igneous rock
Pluton
Concordant
Discordant
Batholith
Stock
Dike
Sill
Laccolith
Sedimentary rock

Lithification
Consolidation
Clastic sedimentary rock
Nonclastic sedimentary rock
Stalactite
Stalagmite
Column
Peat
Lignite coal
Sub-bituminous coal
Bituminous coal
Anthracite coal
Bedding - layers
Metamorphic rock
Dynamic metamorphism pressure
Foliation - directional pressure

Thermal metamorphism - heat
Regional metamorphism - pressure + heat
Rock cycle
Mineral
Silicon tetrahedron
Crystal form
Cleavage
Fracture
Specific gravity - weight in water
Hardness
Mohs' scale of hardness
Luster
Color
Streak

LEARNING OBJECTIVES

After reading and studying this chapter, you should be able to do the following without referring to the text:

1. State the three types of rocks and how they are formed.
2. Give several examples of the three types of rocks.
3. Describe the major intrusive rock formations.
4. State the difference between clastic and nonclastic rocks.
5. Explain the different types of metamorphism.
6. Explain the rock cycle.
7. Discuss minerals and mineral structures.
8. Describe the methods of mineral identification.

QUESTIONS

1. How are igneous rocks formed?
2. What is the geothermal gradient? Does this gradient extend to the center of the earth?
3. What is the source of the earth's internal heat? *pressure & decay of radioactive subs.*
4. Distinguish between intrusive and extrusive igneous rocks.
5. What is the difference between concordant and discordant igneous formations?
6. Describe each of the following: (a) batholith, (b) laccolith, (c) dike, (d) sill. *between 4 11*
7. What is lithification and how does it take place? *push up*
8. Distinguish between clastic and nonclastic sedimentary rocks and give examples of each.
9. What are stalactites, stalagmites, and columns? How are they formed?
10. Give the four general types of coal and their differences. Are all coals sedimentary rocks?
11. What is meant by "bedding"?
12. Describe how metamorphic rocks are formed and the types of metamorphism.
13. What types of rocks are metamorphized? Give an example of each.
14. What is foliation?
15. Distinguish between dynamic, thermal, and regional metamorphism.
16. What is the rock cycle and what geologic processes correspond to an entrance and exit to the cycle?
17. How is mineral defined?
18. What two elements make up the majority of the earth's crust and in what percentages?
19. On what does the formation of the crystalline structure of minerals depend?
20. What is the silicon tetrahedron? Sketch its structure.
21. The mineral beryl has a ring arrangement of six silica tetrahedra. Sketch this arrangement.
22. Give six physical properties used in mineral identification. ——————— *color texture chemical crystaline hardness, water weight luster*
23. What are the hardest and softest substances on the Mohs scale? *talc diamond*

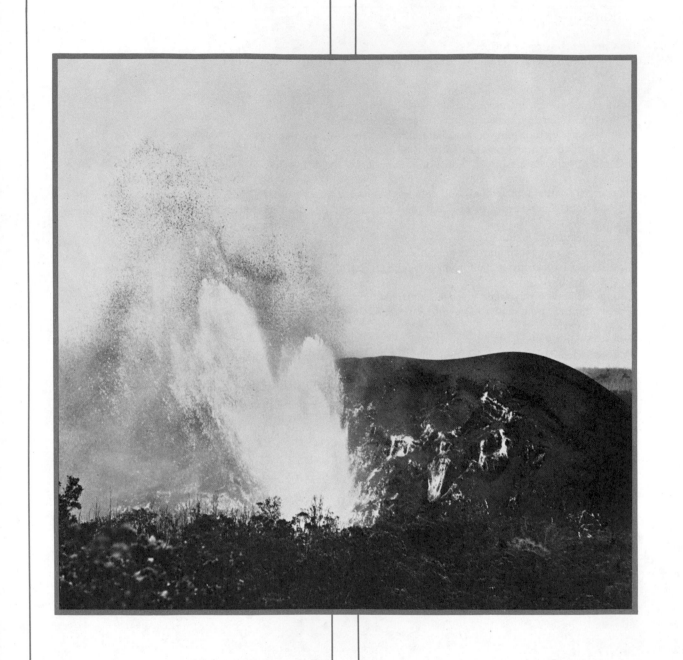

Chapter 29

INTERNAL PROCESSES

It is useful to be assured that the heavings of the earth are not the work of angry deities. These phenomena have causes of their own.

Seneca

We have explored the atmosphere and beyond—landing on the moon and probing the nearby planets. Yet, the direct exploration of the earth has gone no more than a few miles below its surface. The vast majority of our knowledge about the earth's interior composition and structure is the result of indirect observations. That is, we must be content to observe the manifestations of internal processes from our surface vantage point and then form models of the earth below us. The theory of plate tectonics discussed in Chapter 27 is an example of such a model.

Occasionally materials of the upper mantle can be observed through the eruptions of volcanoes when mantle magma is vented to the earth's surface. The dynamics of the earth's internal processes is also evidenced on the surface in the form of earthquakes, which are the result of movements in the earth's lithosphere. It is the shock waves from these enormous releases of energy traveling through the earth's interior that gives us our internal picture of the earth's structure. Land forms, such as mountains, also provide clues as to what has happened within the earth. By the principle of uniformity, these processes are still taking place.

In this chapter we shall examine these occurrences and features that give us our current "view" of the earth's interior structure and processes.

29-1

The word **volcano** comes from the volcanic island of Vulcano in the Mediterranean Sea. It was once believed that Vulcano was the chimney of the forge of Vulcan, the blacksmith of the Roman gods. Eruptions were thought to be the by-products of the forge of Vulcan as he beat out thunderbolts for Jupiter, king of the gods, and weapons for Mars, the god of war. However, we now know that a volcano is formed when magma from deep within the earth's crust or the upper part of the mantle is vented to the surface. Above the surface, the molten **lava** solidifies into extrusive igneous rocks.

There have been several hundred recorded volcanic eruptions, and geologic evidence exists for many other cases of volcanic activity. Some volcanic eruptions make their mark in history. A classic example is the eruption of the dormant volcano Mount Vesuvius in 79 A.D., that buried the Roman cities of Pompeii and Herculaneum. Other violent eruptions include the volcano Tamboro located on a small island in the East Indies. This most violent volcanic explosion in recorded history occurred in 1815 and blew as much as 35 cubic miles of particle debris into the air. As discussed in Chapter 26, the large volume of fine dust carried around by the global atmospheric circulation affected the earth's albedo and 1816 was an uncommonly cold year. In 1883, a violent eruption on the island of Krakatoa, located between Java and Sumatra in the East Indies, literally blew apart this and a neighboring island. Fortunately these eruptions occurred in a relatively sparsely populated area.

In this century, the volcano hall of fame includes Mount Pelee on the island of Martinique in the West Indies. This volcano was known to have been previously active, but without serious incident. However, in April of 1902, activity was observed by gases emitting from **fumaroles**—volcanic vents from which chiefly gases are emitted. The activity increased and in May the volcano exploded, emitting a huge cloud of gas and dust that engulfed the city of St. Pierre on the volcano's lower slopes. In a matter of minutes, an estimated 30,000 to 40,000 people lost their lives.

In 1912, Mount Katmai, a dormant volcano on the Alaskan peninsula, erupted with a similar explosion that sent great quantities of dust and glassy volcanic ash into the air. This eruption occurred in an uninhabited area, so there was no loss of human life as in the case of Mt. Pelee. The fine dust of the ejected material was carried around the world, but the larger particles dropped out within a few hundred miles. A hot blanket of material collected around the volcano almost 100 feet thick. Areas of Kodiak Island, some 60 miles away, were covered with several feet of dust. It took many years for this blanket to cool. Steam from precipitation and gaseous emissions from millions of fumaroles (gas vents) led to the name of "Valley of Ten Thousand Smokes" being applied to the area. Studies of this region have given much information about volcanic gaseous emissions and their contents. A great deal of the gas from Katmai was hydrogen chloride, which gave rise to raindrops of hydrochloric acid.

In more recent memory are the eruptions of Paricutin and Surtsey. In 1943, Paricutin erupted in a Mexican farmer's cornfield. In the first year, the volcanic cone rose to a height of over 1000 feet. After about 10 years of subsiding activity, Paricutin is now a dormant volcano with a cone of a height over 1300 feet (Fig. 29-1a). In 1963, volcano Surtsey (after *Surtr,* the subterranean god of fire in Icelandic mythology) boiled up from the ocean floor off the coast of Iceland, forming a permanent volcanic island, which now may be found on the maps of the world. Geologists were able to study Surtsey from its "birth."

Not all volcanoes are so dramatic, however, and there are various types and degrees of volcanic eruptions. Frequently magma reaches the earth's surface through cracks or fissures. Such **fissure eruptions** are typically nonexplosive and erupt relatively quietly with an outflow of lava. One of the most noted exam-

Figure 29-1a *Volcanoes Paricutin and Surtsey. (Lower left) The volcano Paricutin, in Mexico, which erupted in a farmer's cornfield. (Top and right) The volcano and island Surtsey off the coast of Iceland after several months of eruption. Notice the lava fountains to the left. The molten lava flows to the right into the ocean. (Courtesy U.S. Geological Survey. Paricutin photo by F. O. Jones.)*

ples of such eruptions is in the chain of volcanoes that forms the foundation of the Hawaiian Islands (Fig. 29-1b). Eruptions of huge quantities of lava from cracks in the floor of the Pacific Ocean have caused the formation of one of the most extensive mountain chains on earth, most of which is beneath the ocean. The Hawaiian chain is nearly 1600 miles long and the volcanic formations on the ocean's floor, nearly three miles below sea level, rise to form island projections above sea level.

The occurrence of volcanic activity is for the most part unpredictable. New volcanoes may be formed unexpectedly while existing volcanoes lying dormant may suddenly erupt with practically no warning. However, the locations of eruptions and potential eruptions are known. From the theory of plate tectonics (Chapter 27), volcanic activity takes place predominantly at plate boundaries above subduction zones, where one plate is deflected downward into the earth's interior. (See Fig. 27-3d.) For example, the area of the Pacific Ocean is one of widespread volcanic activity. The outer rim of the Pacific Ocean is marked by a ring of volcanoes, which is known as the **"Ring of Fire"** (Fig. 29-1c). Comparing Fig. 29-1c with Fig. 27-3f, the correlation with plate boundaries is easily seen.

The formation of the Hawaiian Islands in the central Pacific Ocean and on the central portion of the Pacific Plate is explained by the so-called **hot spot theory.** This theory visualizes that as one of the earth's crustal plates moves over a fixed source of heat beneath it, a "hot spot," a succession of volcanoes are generated. Each volcano slides away or moves with the plate, leaving room for the next. According to the theory, volcanic island chains trace out the plates' movement over a hot spot. The Hawaiian Islands are cited as a classic example of this process, and studies have yielded data in support of the theory (age dating of sea floor samples). However, not every island chain appears to fit the theory, so geologists have more work to do in explaining the earth's internal processes.

Volcanism in the United States has been confined mainly to the Hawaiian Islands, Alaska and the Aleutian Islands, and the Cascade Mountains in Oregon and Washington. Most of these volcanoes, particularly in the latter region, are now dormant. However, an active volcano can be seen in Hawaii. The volcano Kilauea, on the island of Hawaii (the largest island in the Hawaiian chain) is one of the most active in the world. (See Fig. 29-1b.) Because of its frequent, yet mild eruptions, the U.S. Geological Survey maintains a Volcano Observatory on the summit of Kilauea. It is of special interest to geologists because the erupted material comes from great depths and thus provides clues to the geochemical composition of the earth's interior. The Observatory is located in the Hawaii Volcanoes National Park where over 500,000 visitors come each year to view an active volcano.

When a volcano erupts, a cone is built up around its central vent. The shape and size of a volcanic cone gives an indication of the type of volcanic activity that took place and the form of the emitted material. The two basic types of materials are lava and pyroclastic debris. Lava, as we know, is molten rock. How fast lava flows depends on its **viscosity**—the internal property of a substance that offers resistance to flow. A highly viscous liquid does not flow readily, for example, molasses on a cold day is highly viscous. The viscosity of lava depends on its composition, primarily its silica (SiO_2) content, and on its temperature. Low-viscosity lava is characterized by a low silica content and is called **basaltic lava.** In rare instances, basaltic lava can flow faster than a person can run.

More commonly flow rates vary from a few meters per hour to a few meters per day. **Felsitic lava,** with a SiO_2 content of 70 percent or more, flows so slowly that its movement is not immediately evident. Highly viscous lavas tend to plug the volcano vent, and the internal build up of pressure may result in a violent explosive eruption. A volcano may also emit large volumes of solid material called **pyroclastic debris,** ranging in size from fine dust to huge boulders. Large masses of lava thrown from a volcano harden in mid-air, forming large rocks called **volcanic bombs.** These boulder-size rocks can be heard whistling through the air as they descend.

As was pointed out previously, many volcanoes do not erupt violently, but relatively calmly eject liquid magma from fissures in the earth's surface. If the lava has a relatively low viscosity such that it

Figure 29-1c *The Pacific "Ring of Fire." The outer rim of the Pacific Ocean is circumscribed by a ring of volcanoes which tend to have violent eruptions, hence the name "Ring of Fire." (Courtesy U.S. Geological Survey.)*

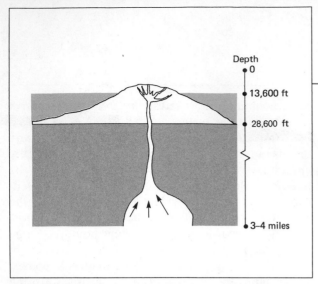

Figure 29-1d *An idealized profile of Mauna Loa. The magma from depths of three to four miles rises to a shallow reservoir, from which it erupts onto the surface. Mauna Loa's gently-sloping profile is characteristic of shield volcanoes.*

flows easily, a gently-sloping low profile **shield volcano** is formed by frequently repeated lava flows. The classic example of a shield volcano is Mauna Loa in the Hawaiian island chain. In fact, Mauna Loa is a huge volcanic mountain—the largest single mountain on earth in sheer bulk. Although not as tall as Mt. Everest (slightly over 29,000 feet above sea level), Mauna Loa rises 15,000 feet from the ocean floor to sea level and protrudes an additional 13,600 feet above sea level for a total height of about 28,600 feet (Fig. 29-1d). Its bulk comes from the fact that this partially submerged mountain has a base of almost 100 miles in diameter.

Volcanic eruptions of both lava and pyroclastic debris form a more steeply sloping, layered composite

Figure 29-1e *Mount Fuji in Japan. Mount Fuji is a dormant stratovolcano that was formed from eruptions of both lava and pyroclastic debris. Seen in the foreground of the photograph is Lake Kawaguchi, one of the five lakes around the base. (Japan Air Lines Photo.)*

cone that is called a **stratovolcano** (also called a **composite volcano**). The lava of stratovolcanoes has a relatively high viscosity and eruptions are more violent and generally less frequent than those of shield volcanoes. Many stratovolcanoes have a build up of material up to 6000 to 8000 feet above their base and have a characteristic symmetrical profile. Dormant stratovolcanoes include Mt. Fuji in Japan (Fig. 29-1e), and Mts. Shasta, Hood, and Rainier in the Cascade mountain rage in the western United States (Washington and Oregon) and Canada.

A volcanic eruption may also consist primarily of pyroclastic debris. In this case, steeply sloped **cinder cones** are formed, which rarely exceed 1000 feet in height. A cinder cone is shown in Fig. 29-1f rising above a previous lava flow. Sunset Crater near Flagstaff, Arizona, is another example of these small, steep symmetric cones of pyroclastic debris.

Figure 29-1f *Cerro Negro Volcano, near Leon, Nicaragua (1968). Notice the initial lava flows to the side and in back of the steep symmetrical cinder cone. The shape or profile of a volcano gives a good indication of the type of volcanic activity. Also notice the huge volume of smoke and particulate matter being vented into the atmosphere. (Courtesy U.S. Geological Survey.)*

Figure 29-1g *Crater of Mauna Loa in the Hawaiian Islands. A funnel-shaped crater exists near the summit of most volcanoes from which gases and material are ejected. (Courtesy U.S. Geological Survey.)*

Figure 29-1h *Crater Lake caldera. Large depressions or caldera are formed when an evacuated volcano chamber collapses. This occurred to a stratovolcano in Oregon (now called Mount Mazama) over 6000 years ago. The caldera is now filled with water and forms Crater Lake with a small protrusion called Wizard Island (small flat-topped mound near the center of lake foreground in photo). The lake is approximately 5 miles in diameter. (Courtesy Oregon State Highway Department.)*

Volcanic activity is usually not confined to the region of the central vent of a volcano. Fractures may split the cone with volcanic material being emitted along the flanks of the cone. Also, material and gases may emerge from small auxiliary vents forming small cones on the slope of the main central vent. A funnel-shaped depression called a **volcanic crater** exists near the summit of most volcanoes from which material and gases are ejected (Fig. 29-1g). Many volcanoes are marked by a much larger depression called a **caldera.** These roughly circular, steep-walled depressions may be up to several miles in diameter. Calderas result primarily from the collapse of the chamber at the volcano's summit from which lava and ash were emitted. The weight of the ejected material of the partially empty chamber causes its roof to collapse much like the collapse of the snow-ladened roof of a building. Crater Lake on top of Mt. Mazama in Oregon occupies the caldera formed by the collapse of the volcanic chamber of this once active stratovolcano (Fig. 29-1h). In some instances, calderas are formed by a violent explosion, rather than a collapse. Felsitic lava may clog the volcanic vent, and the resulting build up of pressure causes the volcano to "blow its top," so to speak.

29-2
EARTHQUAKES

As was learned in Chapter 5, the sudden release or transfer of energy gives rise to the propagation of waves in elastic media. Sound waves and water are commonly observed examples. The earth's internal processes involve transfers of energy, which may occur either quickly or slowly. Although the earth is not ordinarily thought of as a wave-propagating medium, when a sudden release or transfer of energy occurs, the earth vibrates and waves propagate, resulting in what we call **earthquakes.** An earthquake,

as everyone knows who has experienced a substantial one, is manifested by the vibrating and sometimes violent movement of the earth's surface. But waves are also propagated through the earth. It is these waves that provide one of the most useful ways of studying the earth's interior. **Seismology** is the branch of geophysics that studies these waves and uses them as probes to "see" below the earth's surface to determine its internal structure.

Earthquakes may be caused by explosive volcanic eruptions or even man-made explosions, but most earthquakes appear to be associated with movements in the earth's crust along a fault or fracture. According to the theory of plate tectonics, the optimum place for such movement is at the plate boundaries, and the major earthquake belts of the world are observed in these regions. (See Fig. 29-2a as well as Figs. 27-3e and 3f.) Notice how the earthquake region around the Pacific Ocean is similar to that of the volcanic "Ring of Fire" (Fig. 29-1c).

Movements of the earth's plates would certainly put strains on rock formations in the earth's crust near the plate boundaries. Rocks possess elastic properties and energy is stored in the elastic deformations of the lithosphere. If the forces causing this deformation are great enough to overcome the force of friction along a plate boundary or some nearby fault, then the fault walls move suddenly and the stored energy in the rocks is released, causing an earthquake.

If the elastic limit of a stressed rock formation is not exceeded, the rock may rebound elastically and snap back to its original shape after a shift along a fault has relieved the stress—much like a spring. However, if the elastic limit of the rock is exceeded, the rock cannot fully recover when the strain is removed. If the elastic limit is greatly exceeded by the deformation force, the rock formation may even rupture.

The relative motions along fault boundaries give rise to three general types of faulting—normal, reverse, and strike-slip. These three types of faulting

EARTHQUAKE "BELTS" OF THE WORLD

REYKJAVIK
MADRID
ALGIERS
TEHRAN
BANGKOK
MANILA
TOKYO
ANCHORAGE
SEATTLE
HONOLULU
ATLANTIC OCEAN
SAN JUAN
PACIFIC OCEAN
LIMA
SANTIAGO
INDIAN OCEAN
CHRISTCHURCH

Figure 29-2a *The earthquake belts of the world. Notice the belt that circumscribes the Pacific Ocean in a manner similar to that of the volcanic "Ring of Fire" in Fig. 29-1c. (Courtesy U.S. Geological Survey.)*

correspond to the three relative motions of plate boundaries (cf. Fig. 27-3c).

Normal faulting occurs along a nonvertical fault as the result of expansive forces that causes the overlying side of the fault to move downward relative to the side beneath it (Fig. 29-2b). In this case, the stress forces are in opposite directions and the faulting tends to expand or pull the material apart. Normal faults occur near diverging plate boundaries in the regions of sea floor spreading. (See the plate boundaries in Fig. 27-3f.)

Reverse faulting also occurs along a nonvertical fault and is the result of compressional stress forces that cause the overlying side of the fault to move upward relative to the side beneath it. A special case of reverse faulting, called **thrust faulting,** describes the faulting when the fault plane is at a small angle to the horizontal. As might be expected, this type of faulting occurs in subduction zones where one plate slides over a descending plate, or along convergent plate boundaries.

Finally, **strike-slip faulting** occurs when the stresses are parallel to the fault boundary, such that the fault slip is horizontal. This type of faulting takes

place along the boundary of two plates that *strike* and *slip* by each other without an appreciable gain or loss of surface area. Strike-slip faulting occurs along what are commonly called **transform faults.** Transform faults occur near plate boundaries that do not move directly away from or toward each other. An example of a transform fault is the famous San Andreas fault in California. It is the master fault of an intricate network of faults that runs along the coastal regions of California (Fig. 29-2c). This huge fracture in the earth's crust is 600 or more miles long and extends vertically to depths of at least 20 miles. Over much of its length, a linear trough of narrow ridges reveal the fault's presence, as shown in Fig. 29-2c.

As can be seen in Fig. 29-2c, the San Andreas fault system lies on the boundary of the Pacific Plate and the American Plate (cf. Fig. 27-3f). Movement along the transform fault arises from the relative motion between these plates. The Pacific Plate is moving northward relative to the American Plate at a rate of several centimeters per year. At this current rate and direction, in about 10 million years Los Angeles will have moved northward to the same latitude as San Francisco. In another 50 million years,

Figure 29-2b *Block diagrams of general types of faulting. (a) Normal faulting occurs along a nonvertical fault as a result of expansive forces that cause the overlying side of the fault to move downward relative to the side beneath it and tends to expand or pull the material apart. Normal faulting occurs near diverging plate boundaries. (b) Reverse faulting also occurs along a nonvertical fault, but as a result of compressive forces that cause the overlying side of the fault to move upward* *relative to the side beneath it. This type of faulting occurs near subduction zones where one plate slides over a descending plate. (c) Strike-slip faulting occurs when the stress forces are parallel to the fault boundary such that the resulting fault slip is horizontal. This type of faulting takes place along the boundary of two plates that slip by each other—along what is commonly called transform faults.*

(a)

(b)

Figure 29-2c (a) *A map of the San Andreas fault system in California. The San Andreas fault forms part of the boundary between the Pacific and American plates. (b) An aerial view of the San Andreas fault in the Carrizo Plain area of central California. (Courtesy U.S. Geological Survey.)*

the segment of continental crust on the Pacific Plate will have become completely separated from the continental land mass of North America.

The horizontal movement along the San Andreas fault is a cause for considerable concern for the populated San Francisco Bay area through which it runs (Fig. 29-2d). The famous San Francisco earthquake of April 18, 1906, which measured 8.3 on the Richter scale (earthquake scales will be discussed shortly), resulted in the loss of approximately 700 lives and in millions of dollars of property damage. There have been many other milder earthquakes since then along the San Andreas and other branch faults.

In 1971, an earthquake in the San Fernando Valley near Los Angeles resulted in the deaths of over 60 people. Although this fault region had been dormant for hundreds of years, the occurrence of this recent earthquake serves as a warning and reminder that another quake as strong as that of 1906 could happen again.

There is little that can be done about the San Andreas fault except to learn to live with it. Geologists measure the stresses along the fault in hope of predicting future earthquakes. Also, building codes in the area are strict and are concerned with construction that will lessen earthquake damage. However, increased population and an ever-increasing need for housing have caused land developers to use any available land, with the result that rows of houses straddle the trace fault of 1906 and sit on hills where earthquake-induced landslides could occur. The people in these houses live there with the knowledge of the fracture in the earth below and the hope that any future movement along the fault will occur elsewhere.

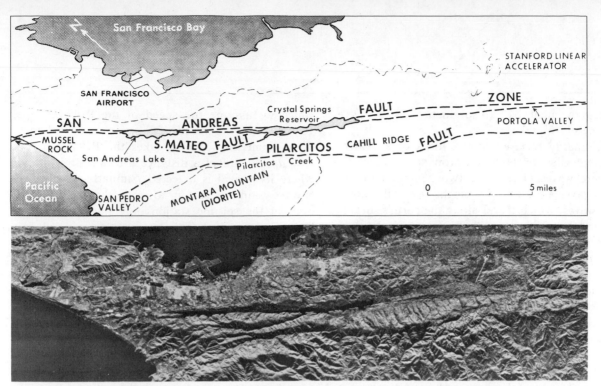

Figure 29-2d *The San Andreas fault and the City by the Bay. The image shown here of the San Francisco Peninsula is not a regular photograph, but is a radar image produced with equipment mounted on an airplane. In this image, the geological structural elements, especially those of the San Andreas fault system, are clearly delineated. (Courtesy NASA and USDA-SCS.)*

When an earthquake does occur, the point or region of the initial energy release or slippage is called the **focus** of the "quake." The vast majority of earthquakes originate in the crust or upper mantle, so an earthquake's focus generally lies at some depth— from a few miles to several hundred miles. Because of this, it is convenient to designate the location on the earth's surface directly above the focus. This point is called the **epicenter** of the earthquake.

Within the framework of the theory of plate tectonics, earthquake activity occurs predominantly in **subduction zones.** In general, geologists classify earthquakes as originating in four depth regions or subzones associated with plate activity. The *outer zone* lies near the surface near the trench boundary of the subduction zone. The earthquakes in this region are thought to result from stresses in the descending

plate along normal faults, predominantly in the oceanic crust. The *shallow zone* of earthquake activity extends to a depth of 100 km (62 miles). Thrust faulting is the major mechanism in this zone, and earthquakes result from the frictional forces between the descending and overlying plate. Some of the greatest earthquakes on record have occurred in the shallow zone. The region from 100–300 km (62–186 miles) in depth is called the *intermediate zone*. Earthquakes originating in this zone appear to occur as a result of stresses within the descending slab. Presumably, these stresses are associated with the descent of the slab into lithosphere. The *deep zone* extends from 300 km–700 km (186–434 miles). Seismic indications are that earthquakes in this zone also result from compressions within the descending slab. At these depths, the slab may be encountering materials that resist its movement. Not all subduction regions have deep zone earthquakes.

The energy released at the focus of an earthquake, regardless of its depth, propagates outwardly as **seismic waves.** There are two general types of seismic waves produced by earthquake vibrations: **surface waves,** which travel along the earth's surface or a boundary within it; and **body waves,** which travel through the earth. Surface waves cause most earthquake damage as they move along the earth's surface. Body waves propagate through the earth's interior and so do less damage. There are two types of body waves, P (for primary) or compressional waves and S (for secondary) or shear waves. The **P waves** are longitudinal compressional waves that are propagated by particles in the propagating material moving longitudinally back and forth in the same direction as the wave is traveling (cf. Chapter 5). Sound waves are an example of longitudinal compressional waves. In the case of **S waves,** the particles move at right angles to the direction of the wave travel, and hence are transverse waves. An example of transverse shear waves is the vibrations in a plucked guitar string.

There are two important differences between P and S waves. S waves can travel only through solids. Liquids and gases cannot support a shear stress, that is, they have no elasticity in this direction and therefore their particles will not oscillate in the direction of a shearing force. For example, little or no resistance is felt when one shears a knife through a liquid or gas. The compressional P waves, on the other hand, can travel through any kind of material—solid, liquid, or gas. The other important difference is the velocity of the waves. P waves derive their name "primary" because they always travel faster than the "secondary" S waves in any particular solid material and hence arrive earlier at a seismic station. It is these differences between P and S waves that allow seismologists to locate the focus on an earthquake and learn about the earth's internal structure.

To illustrate how the epicenter of an earthquake might be located, let us take a simplified example in which we assume the speed of the P waves $v_p = 10$ km/s and the speed of S waves is $v_s = 4$ km/s. Then, from the focus, it takes the P wave a time $t_1 = d/v_p$ to travel a distance d to a seismic station where its arrival is detected. Similarly, it takes the S wave a time $t_2 = d/v_s$ to reach the station. However, at the seismic station only the time interval between the arrival of the waves is determined. But,

$$\Delta t = t_2 - t_1$$

$$= \frac{d}{v_s} - \frac{d}{v_p}$$

$$= d\left(\frac{1}{v_s} - \frac{1}{v_p}\right)$$

Knowing the wave speeds and Δt, the distance d can be calculated. Then, the earthquake must have occurred somewhere on a circle of radius d from the seismic station. (See Fig. 29-2e.) Suppose the time delay between the arrival of the waves was 150 seconds, i.e., $\Delta t = 150$ s, we have

$$\Delta t = d\left(\frac{1}{v_s} - \frac{1}{v_p}\right)$$

or

$$d = \frac{\Delta t}{(1/v_s) - (1/v_p)}$$

$$= \frac{150}{\frac{1}{4} - \frac{1}{10}}$$

$$= 150/(0.25 - 0.1)$$

$$= 1000 \text{ km} \quad (620 \text{ miles})$$

Using similar data from two other seismic stations, the epicenter of the quake must then lie near the intersection of the three distance circles as illustrated in Fig. 29-2e. Of course, this example is oversimplified and in practice more complicated methods are used to locate both epicenters and focuses.

The speeds of the body waves depend on the density of the material, which generally increases with depth. As a result, the waves are curved or refracted. Also, the waves are refracted when they

Figure 29-2e *An illustration of the method of determining the location of an epicenter of an earthquake. P waves travel faster than S waves, as illustrated in the graph, and hence arrive earlier at a seismic station. Knowing the time interval between the arrivals of the P and S waves at a seismograph station and the speeds of the waves, the distance d of the earthquake epicenter from the station can be calculated. The location of the epicenter is somewhere on a circle of radius d from the station. Using data from three stations, the epicenter lies near the intersection of the three distance circles. In this example, seismic stations in Hawaii; Pasadena, California; and Puerto Rico locate the earthquake epicenter to be in northern Chile. (Courtesy U.S. Geological Survery.)*

composition of the earth's crust. These measurements have shown the crust to be thicker below mountain ranges than the ocean basins, which supports the concept of isostasy (Section 27-3). The refraction of seismic waves and the fact that S waves cannot travel through liquid media provide our present view of the earth's structure as shown in Fig. 29-2g. Because of the so-called *shadow zones* of the S and P waves, the outer core of the earth is believed to be a highly viscous liquid.

The seismic waves of an earthquake are monitored by an instrument called a **seismograph,** the principle of which is illustrated in Fig. 29-2h. The recorded seismogram gives the time delay of the arrival of different types of waves, as well as an indication of their energy. (The greater the energy, the greater the amplitude of the traces on the seismograph.)

The severity of earthquakes is represented on different scales. The two most common are the **Richter scale,** which gives an absolute measure of the

cross the boundary or discontinuity between different media—in the same manner light waves are refracted (Chapter 5). A knowledge of these properties led to the discovery of the Moho discontinuity. When monitoring the seismic waves from an earthquake, the Yugoslav seismologist Mohorovičić observed two distinct sets of P and S waves. He concluded that one set must have traveled directly from the earthquake's focus, while the other arrived later because they were refracted through a deeper zone, as illustrated in Fig. 29-2f. By using such seismic measurements, scientists can determine the thickness and infer the

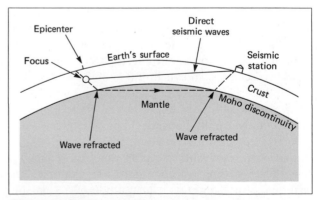

Figure 29-2f *An illustration of the seismic detection of the Moho discontinuity. Two distinct sets of P and S waves are received at a seismic station from a shallow-focus earthquake. One set is the waves traveling directly from the focus; the other set, which arrives later, must be refracted at some boundary through a deeper zone. The boundary is between the crust and the mantle—the Moho discontinuity.*

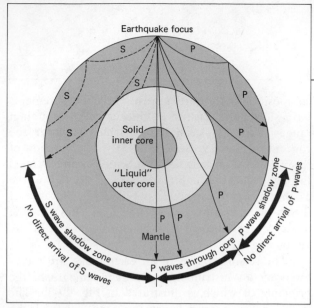

Figure 29-2g *An illustration of seismic waves traveling through the earth's interior. Because of the S and P wave shadow zones, the earth is believed to have a liquid outer core. The S wave shadow zone arises because the liquid outer core will not transmit the S shear waves. The P wave shadow zone results from the refraction the compressional P waves have upon entering the liquid outer core. The S and P waves are separated on opposite sides of the earth in the drawing for simplicity.*

energy released by calculating the energy of seismic waves at a standard distance, and the modified **Mercalli scale,** which describes the severity of an earthquake by its observed effects. A comparison of these two scales is shown in Table 29-2a.

Of these two scales the Richter scale is used more often, with magnitudes expressed in whole numbers and decimals usually between 3 and 9. However, the scale is a logarithmic function, that is, each whole number step represents about 31 times more energy than the preceding number step. For example, an earthquake that registers a magnitude of 5.5 on the Richter scale indicates the measuring seismograph received about 31 times as much energy as for an earthquake that registers a magnitude of 4.5. Similarly, an earthquake with a magnitude of 8 does not represent twice as much energy as one with a magnitude of 4, but about one million times as much! (See

Fig. 29-2i). As can be seen from Table 29-2a, an earthquake with a magnitude of 2 to 3 is the smallest tremor felt by human beings. The largest recorded earthquakes are in the magnitude range of 8.7 to 8.9.

The Richter scale gives no indication of the damage caused by an earthquake, only its potential for damage. Earthquake damage depends not only on the magnitude of the quake but also on the location of its focus and epicenter and the environment of that region, specifically, the local geologic conditions, the density of population, and the construction designs of buildings. The modified Mercalli scale gives a better indication of earthquake effects, since the scale is based on actual observations. The Mercalli scale was developed in the 1890s before the advent of seismographs. This scale has a range of values from I, which indicates that the earthquake is so slight that it is not felt by human beings (except only by a few under especially favorable circumstances), to XII, which designates extensive damage with practically all works of construction damaged greatly or destroyed. Waves are seen on the ground surface and objects are

Table 29-2a *Scales of Earthquake Activity*

Richter Scale	Modified Mercalli Scale	
Magnitude	*Maximum Expected Intensity* *(at epicenter)*	
1–2	I.	Usually detected only by instruments
3–4	II–III:	Can hardly be felt
4–5	IV–VI:	Generally felt; slight damage
6–7	VI–VII:	Moderately destructive
7–8	IX–X:	Major earthquake
8+	XI–XII:	Great earthquake

Specific Earthquakes

1906 San Francisco	8.3 Richter	XI Mercalli
1964 Alaska	8.4	X

Figure 29-2h *An illustration of the principle of the seismograph. The rotating spool (anchored in bedrock) vibrates during the quake. A light beam from the relatively motionless source on a spring traces out a record or seismograph of the earthquake's energy on the light sensitive photographic paper. (Redrawn courtesy U.S. Geological Survey.)*

thrown upward into the air for an earthquake with a modified Mercalli rating of XII. The general indications of this scale are given in Table 29-2a. The modified Mercalli scale is more illustrative with its classifying descriptions. For example,

VI. Felt by all, many frightened and run outdoors. Some heavy furniture moved; a few instances of fallen plaster or damaged chimneys. Damage slight.

VII. Everybody runs outdoors. Damage negligible in buildings of proper design and construction; slight to moderate in well-built ordinary structures; considerable in poorly built or badly designed structures; some chimneys broken. Noticed by persons driving motor cars.

Earthquake damage may result directly from the vibrational tremors or indirectly from landslides and subsidence, as shown in Fig. 29-2i. In populated areas, a great deal of the property damage is caused by fires because of a lack of ability to fight them, due to disrupted water mains and so on. When the energy release of a quake occurs in the vicinity of, or beneath the ocean floor, huge waves called **tsunamis** are sometimes generated. These waves travel across the

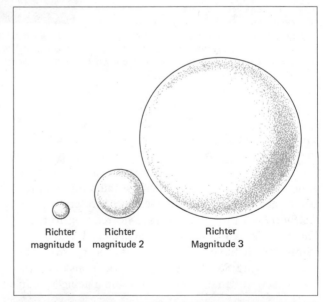

Figure 29-2i *An illustration of the relationship between earthquake magnitude on the Richter scale and energy. The volumes of the spheres are roughly proportional to the amount of energy released by earthquakes of the magnitudes given, and illustrate the exponential relationship between magnitude and energy. At the same scale, the energy released by the San Francisco earthquake of 1906 (Richter magnitude 8.3) would be represented by a sphere with a diameter of 220 feet. (Redrawn courtesy U.S. Geological Survey.)*

Figure 29-2j *Earthquake and tsunami damage. The photo at the left shows subsidence in Anchorage, Alaska, resulting from the "Good Friday" earthquake of March 27, 1964. The quake generated a tsunami that carried the fishing boat shown in the right-hand photo several hundred feet inland. (Courtesy U.S. Geological Survey.)*

oceans at speeds up to 600 miles per hour. In the open ocean, the waves may be 100 miles long and only 3 feet high. However, as they travel in shallower water, the tsunamis grow in height and may be 50 feet high or higher when they smash into shore with immense force (See Fig. 29-2j). Tsunamis coming ashore are commonly and incorrectly referred to as *tidal waves,* although they have no relation to tides. The Hawaiian Islands have experienced many tsunamis, which have raced across the ocean from the Alaska–Bering Strait area or other locations of earthquake activity that circumscribe the Pacific Ocean.

29-3
MOUNTAIN BUILDING

The sheer massiveness and heights of mountains have always fascinated and awed observers. To ancient cultures their lofty summits were the dwelling places of gods. Throughout history, and even today, superstitions surround mountains. To some, mountains present a challenge that must be met by scaling the highest peak. However, to the geologist, mountains are a key to the earth's geologic history. In an effort to know this history, geologists study and theorize how mountains were built or formed.

Mountains differ greatly. Some are singular masses rising majestically above plains such as the volcanic cone of Mt. Kilimanjaro in Africa. Other mountains belong to a geologic unit or series of mountains that make up a **mountain range** such as the Sierra Nevada range in the western United States. A group of similar ranges form a **mountain system.** The Sierra Nevada range along with several other ranges form the Rocky Mountain system. An elongated unit of ranges is referred to as a **mountain chain.**

Mountains are generally classified into three principal kinds based on characteristic features: (a) volcanic mountains, (b) fault-block mountains, and (c) folded mountains.

Volcanic mountains are those that have been built by volcanic eruptions. As has been noted, most volcanoes and hence volcanic mountains are located above the subduction zones of plate boundaries. If the colliding plates are both oceanic, volcanic mountain chains are formed on the ocean floor of the overlying plate. New mountains are being formed beneath the sea today by this process. Chains of oceanic volcanic mountains are evidenced above sea level by island arcs. Japan and the islands of the West Indies are good examples of island arcs. Island arcs lie adjacent to ocean trenches which mark the boundaries of oceanic plates.

Continental volcanic mountain chains occur when the overlying plate of the subduction zone is continental crust. The Andes mountains along the western coast of South America is an example of such a continental mountain range (Fig. 29-3a). The Cascade mountains in Washington and Oregon is the only volcanic mountain range in the United States.

Fault-block mountains are believed to have been built by normal faulting in which giant pieces of the earth's crust were tilted and uplifted (Fig. 29-3b). These mountains evidence great stresses within the earth's crustal plates. Fault-block mountains rise sharply above the surrounding plains. In the United States, the Sierra Nevada range in California, the Grand Teton mountains in Wyoming, and the Wasatch range in Utah are examples of fault-block mountains.

Folded mountains, as the name implies, are characterized by folded rock strata. Examples of folded mountains include the Alps, Himalayas, and the Appalachian mountains (Fig. 29-3c). Although folding is the main feature of folded mountains, these complex structures also contain external evidence of faulting and central evidence of igneous metamorphic activity. Folded mountains are also characterized by exceptionally thick sedimentary strata, which indicates that the material of these mountains was once associated with ocean basins. Indeed, marine fossils have been found at high elevations in the Himalayas and other folded mountain systems.

Figure 29-3a *Volcanic mountains. A view of the Chile-Argentina lake district of the Andes mountains. The Andes are continental volcanic mountains that were formed from continental crust overlying a subduction zone. (Courtesy Lan-Chile Airlines.)*

Two eminent American geologists, James D. Dana (1813–1895) and his contemporary James Hall (1811–1898) advanced many of the original concepts of how such mountains were formed. These concepts considered the mountains on the earth's surface to be a result of the cooling and contracting of the earth during its formation—much like the wrinkling of the skin of an apple as it dries out. A good statement of this theory is presented in the conclusion of an article

Figure 29-3b *Fault-block mountains. The diagram illustrates how fault-block mountains are formed due to the uplifting of normal faulting. (Courtesy J .D. Wilson,* An Environmental Approach to Physical Science, *Heath, Lexington, Mass., 1974.) The Grand Teton mountains in Wyoming shown in the photograph are examples of fault-block mountains. (Courtesy U.S. Geological Survey.)*

by Dana that appeared in the *American Journal of Science and Arts* in 1873 (Series 3, Vol. 106).

The views on mountain building now sustained suppose the existence, through a large part of geologic time, of a thin crust, and of liquid rock beneath that crust so as to make its oscillations possible and refers the chief oscillations, whether of elevation or subsidence (sinking), to lateral pressure from the contraction of that crust; and this accords with my former view, and with that earlier presented by the clear-sighted French geologist, Prevost.

I hold also, as before, that the prevailing position of mountains on the borders of continents, with the like location of volcanoes and of the greater earthquakes, is due to the fact that the oceanic areas were much the largest, and were the areas of greatest subsidence under the continued general contraction of the globe.

However, Dana and Hall disagreed on how the subsidence of ocean basins occurred. From a study of fossils once contained in **geosynclines,** which are long, narrow, subsiding ocean troughs containing large accumulations of sediment, Hall found evidence that over geologic time periods a basin sank at approximately the same rate as the sediment accumulated. He believed that the basin sank because of the accumulating burden of sediment. Dana, on the other hand, thought that the sinking was the result of **diastrophism,** which is a term meaning the movement of the earth's crust. Most of the present-day geological data support Dana's view. Certainly, the weight of the sediment contributes to subsidence, but the fundamental mechanism of mountain building appears to be diastrophism.

The theory of plate tectonics offers an explanation of the mechanism of diastrophism and the building of folded mountains. Oceanic geosynclines containing large accumulations of sediment occur along the margin of continents, which supply the eroded sediment. If these occur in regions of descending ocean plates, the downward movement of a plate pushes the stratified layers in the geosyncline against the continental crust (Fig. 29-3d). The deep-water strata are dragged downward by the moving plate, where they are heated and subjected to metamorphism. Some of the deepest strata may melt and rise

Figure 29-3c *Folded mountains. A view of the Alps near Brechtesgaden in Germany. In the theory of plate tectonics, the Alps are believed to have formed when the African and Eurasian plates collided. (Courtesy Lufthansa German Airlines.)*

to form igneous intrusions in the strata above, becoming the core of a new chain of folded mountains. The shallow water strata are pushed aside, and are folded and faulted along the marginal regions of the folded chain. Such mountain building is active today along the western coasts of North and South America. Large geosynclines lie along the Atlantic coasts of North and South America, but these regions are not along plate boundaries. If perhaps the growing accumulations of sediment in the geosynclines causes a downward plunge into the lithosphere at some future geologic time, then a folded mountain chain may form along the Atlantic coast.

Let's consider the formation of some of the folded mountain ranges of the earth within the framework of plate tectonics. The Himalayas are believed to have formed after the supercontinent Pangaea broke up some 200 million years ago. Essentially in terms of surface features, India broke away from Africa and ran into Asia with the collision resulting in the Himalayas. In more detail, the north-

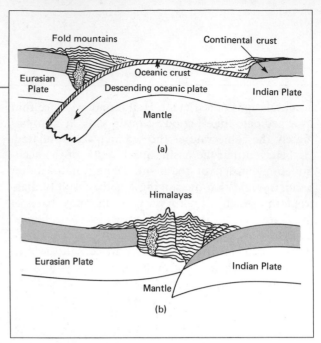

Figure 29-3e *An illustration of the formation of the Himalayas. (a) After its separation from Africa, the northward-moving Indian Plate descended under the Eurasian Plate and the geosynclines along the Eurasian Plate were folded into mountains. (b) When the two continental crusts met, the edge of the Eurasian Plate was lifted, giving rise to the lofty Himalayas.*

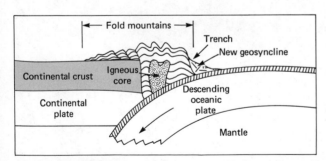

Figure 29-3d *An illustration of the formation of folded mountains. The downward movement of the descending oceanic plate pushes the geosyncline strata against the continental plate. The deep water strata are dragged downward by the moving plate, where it is heated. The deepest strata melts and rises to form the core of the folded mountain chain, while the shallow water strata are folded and faulted along the marginal regions of the chain. Accumulations of sediment eventually form a new geosyncline in the oceanic trench.*

ward movement of the Indian Plate after its separation from Africa resulted in the loss of oceanic lithosphere formerly separating India and Asia, with the eventual collision of these two continental masses. The Indian Plate descended under the Eurasian Plate and the geosynclines along the Eurasian Plate were folded into mountains, as illustrated in Fig. 29-3e. When the two continental crusts met, the edge of the Eurasian Plate was lifted with the spectacular result of the highest mountain range on earth. Present-day analysis of plate movements indicate that the Indian Plate is still moving northward relative to the Eurasian Plate causing the slow, continuing uplift of the Himalayas.

The Alps are another example of what is believed to be the result of plate collision. In this case, the colliding plates were the African Plate and the Eurasian Plate.

The Appalachian mountains are quite old compared to the Himalayas and the Alps, and were apparently once joined to mountains in northern Europe (See Fig. 27-2d). How did these folded mountains form? That is, where was the geosyncline? It is speculated that before Pangaea an ancient ocean existed between North America and Europe. The ocean disappeared as the plates of these land masses converged to form Pangaea. In the collision, the Appalachians were formed. Thus, the Appalachians were once an interior mountain range of Pangaea. When Pangaea broke up, a plate boundary cut across the mountain range and part went with the new American Plate and part with the new Eurasian Plate.

A similar explanation of the collision of pre-Pangaea plates is given for the Ural mountain range, which runs north-south between Europe and Asia. In this case, the European and Asian plates remained in tact as the Eurasian Plate when Pangaea broke up.

Thus we see that the theory of plate tectonics provides a beautiful and simplified explanation of mountain building. However, it must be kept in mind that this theory is relatively new; and although there is a great deal of supporting evidence, it has not really been examined and tested fully. There are other theories of mountain building that also give acceptable explanations. One such theory suggests that mountains can arise independently of horizontal plate movements. According to this theory, gravity causes heavier rocks to sink in the lithosphere while lighter rocks rise to form mountains such as the Alps. Any theory must stand the test of time and evidence; and because of the complex nature of the earth, geologic evidence is slow in coming. There is a great deal of work to be done before it is fully established whether or not the sequence of events described by the theory of plate tectonics happened in this manner.

IMPORTANT WORDS AND TERMS

Volcano	Volcanic crater	P waves
Lava	Caldera	S waves
Fumaroles	Earthquake	Seismograph
Fissure eruptions	Seismology	Richter scale
"Ring of Fire"	Normal faulting	Mercalli scale
Hot spot theory	Reverse faulting	Tsunami
Viscosity	Thrust faulting	Mountain range
Basaltic lava	Strike-slip faulting	Mountain system
Felsitic lava	Transform fault	Mountain chain
Pyroclastic debris	Focus	Volcanic mountains
Volcanic bombs	Epicenter	Fault-block mountains
Shield volcano	Subduction zones	Folded mountains
Stratovolcano	Seismic waves	Geosyncline
Composite volcano	Surface waves	Diastrophism
Cinder cone	Body waves	

LEARNING OBJECTIVES

After reading and studying this chapter, you should be able to do the following without referring to the text:

1. Explain the cause and general locations of volcanic activity.
2. Describe the characteristics and different types of volcanic structures.
3. Explain the causes of earthquakes.
4. Distinguish the different types of faulting.
5. State the difference between S and P waves and explain how these waves provide information about the earth's interior.
6. Define the *Richter scale* and the modified *Mercalli scale*.
7. Describe the three principal types of mountains and give an example of each.
8. Explain mountain building in the framework of plate tectonics.

QUESTIONS

1. What is the difference between lava and magma?
2. Give the locations of the following volcanoes: (a) Mt. Vesuvius, (b) Mt. Pelee, (c) Katmai, (d) Krakatoa, (e) Paricutin, (f) Surtsey.
3. What is the "Ring of Fire" and how is it explained by plate tectonics?
4. Describe the Hawaiian island chain and what is the "hot spot" theory?
5. Are there any volcanoes in the United States? Explain.
6. What is the difference between basaltic and felsitic lavas?
7. Distinguish between (a) shield volcanoes, (b) stratovolcanoes, and (c) cinder cones.
8. What is a caldera and how are they formed?
9. What causes an earthquake?
10. Explain each of the following: (a) normal faulting, (b) reverse faulting, (c) thrust faulting, and (d) strike-slip faulting.
11. What types of relative plate motions correspond to the different types of faulting?
12. What type of faulting is associated with a transform fault? Give an example of an active transform fault.
13. What are the focus and epicenter of an earthquake?
14. Describe the four depth regions of earthquakes associated with subduction zones.
15. Give the two general types of seismic waves and the subdivision of one of these types.
16. What is the difference between S and P waves? How do these waves allow seismologists to locate the focus of an earthquake and to investigate the earth's interior structure?
17. If the speeds of P and S waves from an earthquake are 10 km/s and 4 km/s, respectively, and the time delay between the arrival of the waves at a seismic station is 200 seconds, how far away from the station is the epicenter of the earthquake?

18. What is the basis of (a) the Richter scale, and (b) the modified Mercalli scale?
19. Why is *tidal wave* an incorrect and misleading term for a huge ocean wave generated by an earthquake? What is the correct term?
20. What are the three principal kinds of mountains and how are they distinguished? Give an example of each.
21. How is the formation of each kind of mountains explained by the theory of plate tectonics?
22. What is a geosyncline?
23. What is diastrophism?
24. Describe how the formation of each of the following mountain ranges is explained by the theory of plate tectonics: (a) the Himalayas, (b) the Alps, (c) the Appalachians.

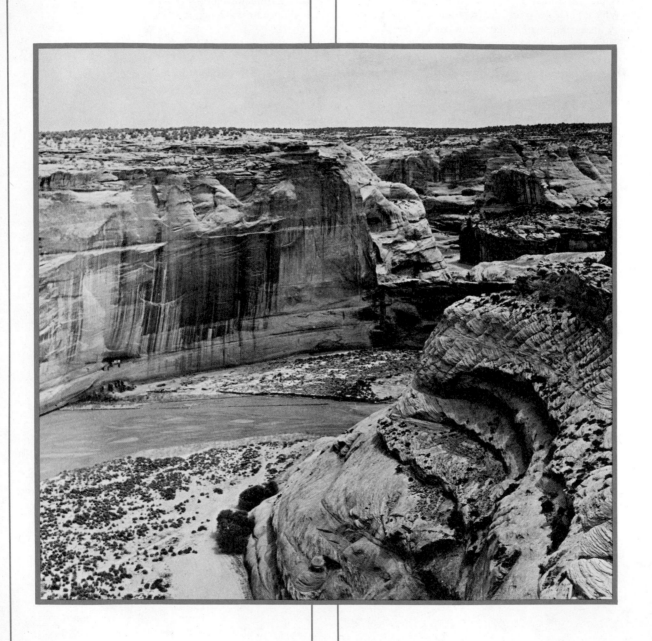

Chapter 30

SURFACE PROCESSES

Nothing is permanent on the earth's surface—in the context of geologic time. Since early times, people have erected edifices and monuments as memorials to persons, peoples, and as symbols of various cultures. In doing so, the most durable rock materials were used; however, even recent erections show the inevitable signs of deterioration. Buildings, statues, and tombstones will eventually crumble when exposed to the elements, and, on a larger scale, nature's mountains are continually being leveled.

An integral part of the rock cycle, as discussed in Chapter 28, is the decomposition of rocks by weathering and other natural processes. As soon as rock is exposed at the earth's surface, its destruction begins. Many of the resulting particles of rock are washed away by surface water and are transported in streams and rivers as sediment toward the oceans. Over geologic time, this constitutes a leveling of the earth by the wearing away of high places and transporting of sediment to lower elevations.

We say that the wearing away and leveling of the earth's surface is due to **erosion,** which is a group of interrelated processes by which rock is broken down and the products are removed. While volcanism and diastrophism account for the uplifting of land masses, erosion accounts for their leveling. In this chapter we shall consider some of the processes and agents of erosion—including "unnatural" agents—human beings.

The same regions do not remain always seas or always land, but all change their condition in the course of time.

Aristotle

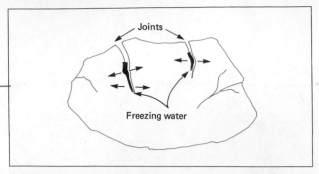

Also, the oceans and ocean currents, along with sea floor topography will be considered. Important geologic processes take place on the ocean floor which account for about 65 percent of the earth's surface.

Figure 30-1a *Frost wedging. The expansive forces of freezing water in rock joints cause the rock to fracture. The photo shows an example of this type of physical weathering in a glacial cirque or depression in the Big Horn Mountains of Wyoming. (Courtesy U.S. Geological Survey. Photo by N. H. Darton.)*

30-1
WEATHERING AND MASS WASTING

Weathering

The erosion process begins with weathering, which is the physical disintegration and chemical decomposition of rock. Weathering depends on a number of factors, such as the type of rock, moisture, temperature, and overall climate.

Physical weathering involves the mechanical disintegration or fracture of rock, primarily as a result of pressure. For example, a common type of physical weathering in some regions is **frost wedging.** When rocks are formed by solidification, lithification, or metamorphism, internal stresses are produced in the rocks. One common result of these stresses is cracks or crevices in the rock, called joints. Joints provide an access route for water to penetrate into the rock. If the water freezes, the less dense ice requires more space than the water and exerts a strong pressure on the surrounding rock. As a result, the rock may break apart, such as those illustrated in Fig. 30-1a. The expansive force of freezing water is readily observed by freezing water in a glass container, which usually cracks. Frost wedging is most effective in regions where freezing and thawing occur daily.

The action of freezing water is also effective in loosening the soil, which promotes its erosion. In the winter, water in the soil freezes as the ground cools. An accumulation of freezing water from the upward migration of water by capillary action or from rain water or melting snow can give rise to an upward expansion that heaves up the material above and creates a bulge at the surface. This so-called **frost heaving** can result in fractures and bulges in street pavement. Highway joints and cracks are usually "sealed" with waterproofing materials to prevent accumulations of water and frost heaving. Also building foundations are laid below the surface frost zone to prevent damage from frost heaving.

In cold upper latitudes, the subsurface soil may remain frozen permanently, giving rise to a permafrost layer. During a few weeks in the summer the top soil may thaw to a depth of a few inches to a few feet. The subsurface permafrost provides a stable base and prohibits the melted water from draining. As a result, the ground surface becomes wet and spongy (Fig. 30-1b). The permafrost in Alaska caused many problems in the construction of the Alaskan oil pipe-

Figure 30-1b *A "roller-coaster" railroad near Strelna, Alaska. This condition was caused by differential subsidence stemming from the thawing of top soil over subsurface permafrost. (Courtesy U.S. Geological Survey.)*

line. On the North Slope where the oil fields were discovered, the thawed surface is especially unstable because of a relatively high moisture content. The thawed surface does not recover easily if disturbed and, if removed or compressed, the subsurface permafrost is susceptible to progressive thawing and erosion.

Frost wedging and frost heaving account for little if any physical weathering in hot, arid climates, but a similar process can. This arises from the growth of salt crystals rather than water crystals. Ground water with high salt concentrations seeps into porous rock—sandstone is a prime example. When the water evaporates and the salt crystallizes, the salt crystals exert pressure on the surrounding rock. This process

of **salt wedging** can loosen and fracture the cementing material holding the sand grains together. The loosened sand grains can then be carried away by wind and water from rainstorms.

Plants and animals play a relatively small role in physical weathering. Most notable is the fracture of rock by plant root systems when they invade and grow in rock crevices. Burrowing animals, earthworms in particular, loosen and bring soil to the surface. This promotes aeration and access to moisture, which are important factors in chemical weathering. Thus one type of weathering promotes another. The activities of human beings also give rise to weathering and erosion that is often unwanted, as we will learn later in the chapter.

In all cases of physical weathering the disintegrated rock still has the same chemical composition. **Chemical weathering,** however, involves a chemical change in the rock's composition. Since heat and moisture are two important factors in chemical reactions, this type of weathering is most prevalent in hot, moist climates. One of the most common types of chemical weathering involves limestone, which is made up of the mineral calcite ($CaCO_3$). Rain can absorb and combine with carbon dioxide (CO_2) in the atmosphere to form a weak solution of carbonic acid.

$$H_2O + CO_2 \longrightarrow H_2CO_3$$
$$\text{Carbonic acid}$$

Also, as water moves downward through soil, it can take up even more carbon dioxide that is released by soil bacteria involved in plant decay. Recall that carbonic acid (carbonated water) is the weak acid in carbonated drinks.

When carbonic acid comes in contact with limestone, it reacts with the limestone to produce calcium hydrogen carbonate or calcium bicarbonate.

$$H_2CO_3 + CaCO_3 \longrightarrow Ca(HCO_3)_2$$

Carbonic acid   Limestone   Calcium bicarbonate

Figure 30-1c *A sinkhole collapse in Barstow, Florida. Water flowing through limestone formations can carve out large underground caverns. A cavern ceiling can collapse, causing a sinkhole to appear on the land surface. In this case, the collapse caused considerable property damage. (Courtesy U.S. Geological Survey.)*

Calcium bicarbonate dissolves readily in water and is carried away in solution. Since limestone is generally impermeable to water (and dilute carbonic acid), this type of chemical weathering acts primarily on the surfaces of limestone rocks along which water flows. Water flowing through underground limestone formations can carve out large caverns over millions of years. The cavern ceilings may collapse, causing **sinkholes** to appear on the land surface (Fig. 30-1c). In many stable caverns, as water seeps through the cavern ceiling and drips to its floor, it loses its carbon dioxide and minute amounts of calcium carbonate are precipitated. These precipitations build up to form icicle shaped stalactite, stalagmite, and column dripstone (see Fig. 28-2b). Do you recall which formation extends down from the ceiling and which protrudes up from the floor?

The rate of chemical weathering depends on many factors. The principal factors are climate and the mineral content of rock, with humidity and temperature being their chief climate controls. Chemical weathering is relatively rapid in hot, humid climates as compared to chemical decomposition in polar regions. On the other hand, chemical weathering is quite slow in hot, dry climates. Egyptian pyramids and statues have stood for centuries with relatively minimal weathering. Some rock minerals are more susceptible to chemical weathering than others. This is evidenced in the weathering of tombstones (Fig. 30-1d). Marble tombstones, which consist of soluble calcite, may show a great deal of chemical weathering as compared to sandstone tombstones, which consist of relatively insoluble quartz. However, the weathering of sandstone tombstones can cause a loosening of the cementing materials holding the sand grains together, and give rise to the erosion of large blocks of rock.

As mentioned previously, burrowing animals, both large and small, bring soil or rock particles to the surface where the exposed particles are more susceptible to chemical weathering. The mounds of soil particles built up by ants and earthworms are common sights.

Mass Wasting

The weathered material that accumulates on base rock is called **overburden.** This general term refers to rock and rock fragments or soil that is free of the base rock and so may start its descent toward lower elevations. The downslope movement of overburden

Figure 30-1d *Examples of chemical weathering. Marble tombstones, which consist of soluble calcite, are susceptible to chemical weathering. The weathering of sandstone, which consists of relatively insoluble quartz, is primarily due to the loosening of cementing materials holding the sand grains together. As a result, large blocks of material are eroded.*

under the influence of gravity is called **mass wasting.** If you drop a rock from a cliff, it is, in the geological sense, mass wasted. However, mass movement usually occurs with the aid of a transport medium such as running water. It is convenient to divide mass wasting into two categories, fast and slow, which are based on the comparative times involved in mass movement process. As might be expected, the angle of the land slope is a critical factor in the distinction between fast and slow mass wasting.

Two important types of *fast mass wasting* are landslides and mudflows. **Landslides** involve the downslope movement of large blocks of weathered materials. Spectacular landslides occur in mountainous areas when large quantities of rock break off and move rapidly down the steep slopes. This type of landslide is termed a **rockslide.** A comparative slower form of landslide is a **slump.** This is the downslope movement of an unbroken block of overburden, which leaves a curved depression on the slope (Fig. 30-1e). Slumps are commonly accompanied by debris flows that consist of a mixture of rock fragments, mud, and water that flows downslope as a viscous liquid. Small slumps are commonly observed on the bare slopes of new road constructions.

Mudflows are the movements of large masses of soil that have accumulated on steep slopes and become unstable with the absorption of large quantities of water from melting snows and heavy rains. Mudflows are likely to occur on slopes lacking vegetation, which hinders mass movement. Consequently, mudflows are common in hilly and mountainous regions where land is cleared for development without regard to soil conservation.

Mudflows are also associated with volcanoes. Volcanic mudflows result from loose ash on the slopes of the volcano that absorb water from thunderstorms commonly associated with volcanic activity. The mudflows of ash can flow long distances and build up thick deposits of sediment. In some instances, volcanic mudflows have buried nearby towns.

Fast mass wasting is quite dramatic, but *slow mass wasting* is a more effective geological transport process. In contrast to fast mass wasting, the rates of slow mass wasting processes are generally imperceptible. An important type of slow mass wasting is called creep. **Creep** is the slow particle-by-particle movement of weathered debris down a slope, taking place year after year. It cannot be seen happening, but the manifestations of creep are evident such as

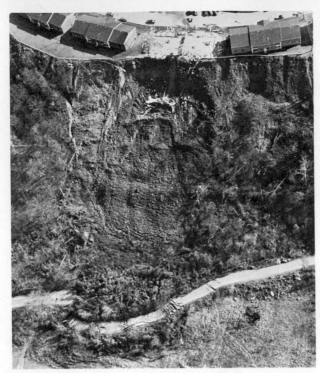

Figure 30-1e *A large slump that threatens an apartment complex and debris flow. The missing building was torn down a short time before the landslide occurred and the other buildings were evacuated. Notice the damage to the road by the debris flow in the foreground. (Courtesy David Kantner, Marietta, Ohio.)*

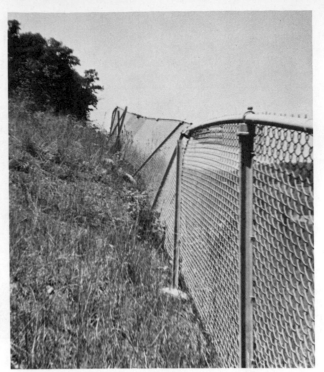

Figure 30-1f *A result of creep. Tilted fence posts indicate the imperceptibly slow mass wasting due to creep. (Courtesy J. D. Wilson,* An Environmental Approach to Physical Science, *Heath, Lexington, Mass., 1974.)*

that shown in Fig. 30-1f. Although the spectacular landslides and slumps involve large quantities of mass movement, this is but a fraction of the cumulative total of mass movement by creep over a period of time.

Another type of slow mass wasting is **solifluction** or soil flow, which is the "flow" of weathered material over a solid, impermeable base. This type of mass wasting is common in cold climates where permanently frozen ground (permafrost) forms a solid base for the solifluction of a surface layer that thaws during the short summers. Moisture in the thawing surface and from summer rains is trapped in the surface layer due to the frozen ground below. As a result, the surface layer becomes saturated and the water-soaked soil "flows" slowly downhill.

30-2
AGENTS OF EROSION

The interacting factors in erosion processes are quite complex. To simplify matters, it is convenient to talk about the **agents of erosion,** that is, the major physical phenomena and mechanisms that supply the energy for erosion. These are running water, ice, wind, and waves.

■ *Running water* refers primarily to the waters of streams and rivers that erode the land surface and transport and deposit eroded materials. Rainfall that is not returned to the atmosphere by evaporation either sinks into the earth as ground water or flows over the surface as runoff. Runoff or overland flow occurs whenever rainfall exceeds the amount of water that can be immediately absorbed in the ground. Overland flow usually occurs only for short distances before the water ends up in a stream. The erosion performed by overland flow is referred to as **sheet erosion.** The flowing "sheets" of water can move a surprisingly large amount of sediment, particularly from unirrigated slopes and cultivated fields (Fig. 30–2a).

Stream flow is the flow of water occurring between well-defined banks. The eroded material in a stream is referred to as the stream's **load.** The load of a stream varies from dissolved minerals and fine particles to large rocks, and the transportation process, as well as erosion, depends on the volume and swiftness (discharge) of the stream's current. A stream's load is divided into three components: a dissolved load, a suspended load, and a bed load.

The *dissolved load* consists of dissolved water-soluble minerals that are carried along by a stream in solution. As much as 20 percent of the material reaching the oceans is transported in solution. Fine

particles not heavy enough to sink to the bottom are carried along in suspension. The *suspended load* is quite evident after a heavy rain when the stream appears muddy (Fig. 30-2b). Coarse particles and rocks along or close to the bed of the stream constitutes its *bed load.* These rocks and particles are rolled and bounced along by the current. This transport mechanism of bed load is referred to as **traction.**

The coarser load material near the bottom of a stream is further broken into smaller rocks and particles by **abrasion,** as it is moved along by traction. Stream abrasion is evidenced by smooth rounded rocks and pebbles (Fig. 30-2b). During the transport process the bed load is worn finer and finer and eventually may be transported by suspension. The final products of a long journey are sand (rock particles) or a mixture of finer particles and organic material picked up along the way which is collectively referred to as **silt** and clay.

The action of the flowing water in a river causes the erosion of its bed. By studying the development of this process, the age of a river may be determined.

Figure 30-2a *Sheet erosion. The overland flow of water removes a large amount of sediment. This field, which was plowed in the fall and left without any cover winter crops, shows the signs of sheet erosion. (Courtesy USDA, Soil Conservation Service.)*

Figure 30-2b *Stream erosion and transport. The upper photo shows the cutting of a road embankment that occurred when the stream flooded due to heavy rains. Fine soil particles are carried away as suspended load. The lower photo shows the result of such erosion in an aerial view of the Loosahatchie River pouring suspended sediment into the Mississippi River one mile north of Memphis, Tenn. (Courtesy EPA and USDA, Soil Conservation Service.)*

Figure 30-2c *Meandering. (Top) An aerial view of a classic case of river meanders in the Shenandoah River near Woodstock, Virginia. The 1 to 1½-mile-wide meander belt cuts into the adjoining shale to depth up to 150 feet. (Bottom) When erosion along two sharp meanders causes a stream to meet itself, it may abandon the meander loop, leaving behind an oxbow lake as occurred here in the White River in Arkansas. (Courtesy U.S. Geological Survey. Bottom photo by J. R. Balsley.)*

Over geologic time a river goes through three general stages: youth, maturity, and old age. In its youth, it is characterized by a V-shaped valley or canyon which reflects downcutting erosion. The depth of an eroded valley is limited by the level of the body of water into which the river flows. The limiting level below which a stream cannot erode the land is called its **base level.**

In general, the ultimate base level for rivers is sea level.

As a stream or river flows overland, it twists and turns, following the path of least resistance. A loop-like bend in a river channel is called a **meander** (Fig. 30-2c). Meanders shift or migrate due to greater erosion on the side of the stream bed on the outside

Figure 30-2d *A satellite photograph of the Nile Delta. The dark areas are cultivated lands. The Mediterranean Sea is in the left foreground and the Suez Canal and the Red Sea are above the dark delta region. The rich sediment load deposited at the mouth of the river makes the delta area an important agricultural region. (Courtesy NASA.)*

of the curved loops. The speed of the stream flow is greater in this region than near the inside bank of the meander where sediment can deposit. When the erosion along two sharp meanders cause the stream to meet itself, it may abandon the water-filled meander, which is then known as an oxbow lake.

As the river matures it becomes **graded,** which is the condition when its erosion and transport capabilities are in balance. A mature graded river no longer has much tendency to deepen its channel. Instead, it begins to widen its valley, forming a **flood plain.** This widening continues until in old age the river has a very wide flood plain.

The amount of material eroded and transported by streams and rivers is enormous. The oceans receive billions of tons of sediment each year as a result of the action of running water. The Mississippi River alone discharges approximately 500 million tons of sediment yearly into the Gulf of Mexico. A river's suspended and bed loads may accumulate at its mouth and form a **delta,** such as the Nile River Delta (Fig. 30-2d). The rich sediment makes delta areas important for agriculture. However, some sediment deposits are not always to our benefit as shown in Fig. 30-2e. Many harbors must be dredged to remove sediment deposits to keep them navigable, and the sediment build up behind dams poses operational

problems. Our activities which remove erosion-preventing vegetation from the land may give rise to sediment pollution in rivers and streams and result in environmental problems. Some of these activities and problems will be discussed shortly.

■ *Ice* One of the eroding actions of ice, frost wedging, has been mentioned previously. Many of us are familiar with the ice of winter, but parts of the earth are covered with large masses of ice the year round. These large masses of ice are called **glaciers.** To most of us, the term glacier usually brings to mind the thought of the Ice Age. Indeed, large areas of Greenland and Antarctica are presently covered with glacial ice sheets or **continental glaciers** similar in size to those which covered Europe and North America during the last Ice Age—over 10,000 years ago. But are there any glaciers in the United States today? The answer is yes. In fact, you may be surprised to learn that there are about 1100 glaciers in the western part of the conterminous (48 states) United States and that 3 percent of the land area of Alaska is covered by glaciers (17,000 square miles). However, these glaciers differ greatly in size and form from ice sheets of Greenland and Antarctica.

Glaciers are formed when, over a number of years, more snow falls than melts. As the snow accu-

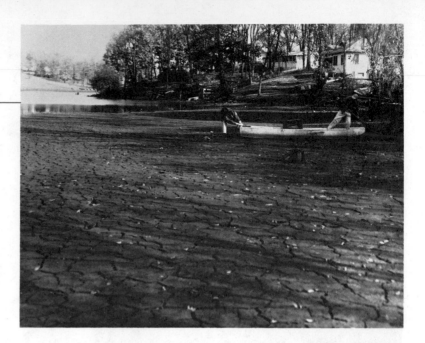

Figure 30-2e *A sediment problem. To reach the water, the men must drag their boat over a large area of silt deposits in Lake Calhoun near Lafayette, Ill. The erosion of land fills harbors and lakes with unwanted sediment. (Courtesy EPA.)*

mulates and becomes deeper, it is compressed by its own weight into solid ice. When enough ice accumulates, the glacier "flows" downhill or out from its center if on a flat region. The icebergs commonly found in the North Atlantic and Antarctic oceans are huge chunks of ice that have broken off from the edges of the glacial ice sheets of Greenland and Antarctica, respectively.

Small glaciers, called **cirque glaciers,** form in hollow depressions along mountains that are protected from the sun. A majority of the glaciers in the United States are of this variety. The ice movement further erodes the land and forms an amphitheatre-like depression called a *cirque* (see Fig. 30-1a). When the ice melts these glacier-eroded cirques often become lakes.

If snow accumulates in a valley, the valley floor may be covered with compressed glacial ice, and a **valley glacier** or mountain glacier is formed, which flows down the valley. At lower and warmer elevations the ice melts, and where the melting rate equals the glacier's flow rate, the glacier becomes stationary (Fig. 30-2f). The flow rate of a glacier may be a few inches to over 100 feet per day, depending on the glacier's size and other conditions. The end of a glacier is called its *snout,* which may advance and recede with the seasons. The erosion action of a valley gla-

cier is not unlike that of a stream. As the ice flows, it loosens and carries away material, or bed load, that will be ground fine by abrasion. Although much slower than a stream, a glacier can pick up huge boulders and gouge deep holes in the valley. The paths of vigorous preexisting mountain glaciers are well-marked by the deep U-shaped valleys they leave (Fig. 30-2g).

Glaciers, like streams, also deposit the material they carry. The general term **drift** is applied to any type of glacial sediment deposit. Material that is transported and deposited by ice, in contrast to meltwater, is called **till.** Till deposits are not layered or sorted as would be sediment carried away by the melted water of a glacier. Near the end and sides of a glacier the till may form ridges known as **moraines.** The terminal moraine marks the farthest advance of the glacier. Terminal moraines give us an indication of the extent and advance of the glacial ice sheet in North America which retreated about 10,000 years ago. These moraines lie as far south as Indiana, Ohio, and on Long Island, New York.

Not only do glaciers erode the land surface, but they supply an estimated 560 billion gallons of water to stream flow during the summer months in the conterminous United States alone. These huge fresh water reservoirs are being eyed as water sources for

Figure 30-2f *Glaciers. The top photo shows the Blue Glacier in Olympic National Park, Washington. The lower photo shows the snout of Wolverine Glacier on the Kenai Peninsula near Seward, Alaska. A glacier advances until the melting rate equals its flow rate. (Courtesy U.S. Geological Survey.)*

the heavily populated western areas. The glacier cycle fits well into our needs. Glaciers accumulate and store water in the winter when most areas have sufficient water supplies and release it in the summer when many city reservoirs are low.

■ *Wind* The action of wind as an agent of erosion is a slow process, but nevertheless wind contributes significantly to the leveling of the earth. Dust particles that are small enough are transported great distances by the wind, while larger particles are moved short distances by rolling or bouncing along the surface. The transport action of the wind is not unlike that of traction in a stream, only a broader and less confined scale. This action is quite evident in areas with large quantities of loose weathered debris.

Dust storms may darken the sky and be of such intensity that visibility is reduced to almost zero. During the 1930s, drought conditions created areas known as *dust bowls,* in which layers of fertile topsoil were blown away by the wind.

Because sand grains are relatively heavy, the transport action of sand by the wind is usually within a few feet of the ground, similar to the coarse sediment transportation near the bed of a stream. On the ground the sand forms drifts, as shown in Fig. 30-2h. The resulting abrasion from such sandblast action is evident in nature.

■ *Waves* The waves of large bodies of water also erode the land along their shorelines. This is most evident in the oceans where the surf pounds the

Figure 30-2g (*Right*) *A U-shaped glacial valley with moraine in the foreground in Glacier National Park, Montana. The valleys formed by glacial action mark the paths of preexisting mountain glaciers. (Courtesy U.S. Geological Survey.)*

Figure 30-2h (*Below, left*) *A sand and dust storm. At the height of the storm the sand drifts have almost covered the railroad tracks. (Below, right) Isolated sandhills are evidence of the eroding action of the wind. (Courtesy USDA, Soil Conservation Service.)*

Figure 30-2i *Marine terraces. The eroding action of waves is evident in this view along the California coast near Los Angeles. (Courtesy U.S. Geological Survey. Photo by K. S. W. Kew.)*

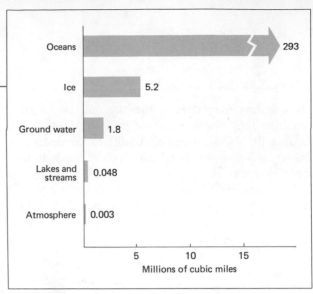

Figure 30-3a *Distribution of the earth's water supply. (Courtesy U.S. Geological Survey.)*

shoreline. Some coasts are rocky and jagged, which evidences that only the hardest materials can withstand the unrelenting wave action over long periods of time. Along other coastlines, cliffs are formed on the shoreline, terraced from the eroding action of waves (Fig. 30-2i).

30-3
THE EARTH'S WATER SUPPLY

All rivers run into the sea, yet the sea is not full; into the place whence the rivers come, thither they return again.

Eccles. 1:7

Water is often referred to as the basis of life. The human body is composed of 65–70 percent water by weight, and water is necessary to maintain our body functions. This common chemical compound is an essential part of our physical environment, not only in life processes, but also in other areas, such as agriculture, industry, sanitation, fire-fighting, and even religious ceremonies. Early civilizations developed in valleys where water was in abundance, and even today the distribution of water is a critical issue. Consider how your life would be affected without an adequate water supply.

The earth's water supply, some 300 million cubic miles, may seem inexhaustible, since it is one of our most abundant natural resources. Approximately 70 percent of the earth's surface is covered with water. However, about 98 percent of the water on the earth is salt water, and only about two percent is fresh water (Fig. 30-3a). Most of the fresh water is frozen in the glacial ice sheets of Greenland and Antarctica. Even so, there are about 10^{15} gallons of fresh water available each year. The oceans contain another 10^{20} gallons of water which is easily accessible, but this water has 3–4 percent salt content—principally sodium chloride—which makes it unfit for human consumption without a costly desalination process.

Although the daily use of fresh water in the United States is about 360 billion gallons, the supply of fresh water is generally adequate in most areas. Regional problems arise due to population concentrations, but these are human problems that involve water management.

Figure 30-3b *An illustration of the hydrologic cycle. Evaporated moisture from oceans, rivers, lakes, and soil is distributed by atmospheric processes. It eventually falls as precipitation and returns to the soil and bodies of water. The ground water percolates downward through the zone of aeration until it reaches the level (water table) where the ground is saturated with water. (Courtesy USDA.)*

The Hydrologic Cycle

The earth's water supply is a reusable resource that is constantly being redistributed over the earth. There are many factors that enter into this redistribution, but in general it is a movement of moisture from large reservoirs of water, such as oceans and seas, to the higher elevated inland regions. As discussed in the preceding section, the water flows back to the sea, eroding as it goes. This gigantic cyclic process is known as the **hydrologic cycle** (Fig. 30-3b). Moisture evaporated from the oceans moves over the continents through atmospheric processes and falls as precipitation. Some of this water evaporates and returns to the oceans via atmospheric processes, but a large part of it soaks into the earth to become ground water.

Some of the precipitation falling on the land may be lost in direct runoff. How much water goes into the ground depends on the permeability of the soil and rocks near the earth's surface. **Permeability** is a measure of a material's capacity to transmit fluids. Naturally, loosely packed soil components, such as sand and gravel, permit a greater movement of water. Clay, on the other hand, has fine openings and relatively low permeability. **Porosity** is closely related to permeability. This is the percentage volume of unoccupied space in the total volume of a substance. The porosity of rocks and soil near the earth's surface determines the capacity of the ground for water storage. In total, ground porosity is so great that, except for glaciers, the ground is our largest reservoir of fresh water.

Under the influence of gravity, water percolates downward through the soil until at some level the ground becomes saturated with water. The upper boundary of this **zone of saturation** is called the **water table.** The unsaturated zone above the water table is called the **zone of aeration** (Fig. 30-3b).

In the zone of aeration the pores of the soil and rocks are partially or completely filled with air. However, in the zone of saturation all the voids are saturated with ground water, forming a reservoir from which we obtain part of our water supply by drilling wells to depths below the surface of the water table. Lakes, rivers, and springs occur where the water table is at the same level as the earth's surface. Springs may also form above the water table as a result of impervious rock layers, but these tend to dry up in seasons of light precipitation. Also, the level of the water table shows seasonal variations, and shallow wells may go dry in late summer.

A body of permeable rock through which ground water moves is called an **aquifer** (Latin, water carrier). Sand, gravel, and loose sedimentary rock are good aquifer materials. Aquifers are found under more than half of the area of the conterminous United States, and it is in these aquifers that we find wells and springs. Ordinary wells and springs fill with water because they intersect the water table. A *spring* is

Figure 30-3c *An illustration of wells, streams, and springs relative to the water table. As shown in the blowup, a special geometry of impermeable rock layers gives rise to artesian wells and springs. If the ground water is not recharged by precipitation, the level of the water table will fall; then wells, springs, and streams can go dry. This often occurs during a hot, dry summer.*

simply a flow of ground water that emerges naturally on the earth's surface. Ordinary springs result from the gravitational flow of water and are hence sometimes called gravity springs (Fig. 30-3c).

Special geometry of impermeable rock layers give rise to what are called **artesian wells** and **artesian springs.** These are wells and springs in which water rises above the aquifer. As illustrated in Fig. 30-3c, this can occur when an aquifer is sandwiched between sloping impermeable rock strata. The pressure due to gravity can cause water to spurt or bubble onto the surface above the aquifer. The name artesian comes from the French province of Artois where such wells and springs are common.

Thermal or hot springs are found in many areas. These involve heat energy from the earth's interior. Ground water seeping several thousand feet underground is heated as a result of the earth's geothermal gradient (1°F per 150 feet of depth). Also at these depths, the water can come in contact with hot igneous rocks and may be heated to 400°F or higher (superheated) without boiling because of the great underground pressures. The superheated water expands and rises to the earth's surface through fissures, where it forms thermal springs. Heat energy is lost during the water's journey to the surface, but some springs are boiling hot. The reduction of pressure near the surface can cause the water to boil. The resulting stream causes some hot springs to erupt in the form of a **geyser** (Icelandic, *geysir,* gush or rage). Geysers from time-to-time spurt water into the air to heights of 150 feet or more. Most geysers have very irregular eruptions, but a few, such as Old Faithful in Yellowstone National Park, are regular enough to satisfy the impatient tourists (Fig. 30-3d). Old Faithful's eruptions vary between 30 and 90 minutes apart with an average time interval of 65 minutes.

Geothermal areas are found around the world, principally in volcanic regions associated with plate boundaries (Fig. 30-3e). Hot springs have been used as baths or spas down through history. In Iceland, hot water from geyser fields is used for domestic heating. A significant commercial application occurred in 1905 with the building of the first geothermal electric generation plant in Larderello, Italy. The first geothermal generation facility in the United States was built by the Pacific Gas and Electric Company in 1960 at The Geysers, California, some 90 miles north of San Francisco (Fig. 30-3f). Other geothermal areas in the western United States are being eyed as possible energy sources.

Geothermal fields are of two general types—hot springs at the surface and deep, insulated, superheated reservoirs with little surface leakage. It is the latter type that is important in electrical generation application. When such reservoirs are tapped by drillings, the water boils and the expanding stream causes the hot water to be propelled to the surface. The drilled well acts as a continuously erupting geyser, so there is very little pumping cost. In some wells "flash" boiling occurs and only steam erupts with very little water discharge. This type of well is suitable for electrical generation, and is the case with the wells at The Geysers and Larderello.

Although these geothermal sources eliminate the air pollution associated with combustive electrical generation, they are not without problems. Gases in the steam and dissolved minerals in the hot water cause equipment to erode rapidly. The cooling and disposal of the salty, mineral-rich water from the spring may also present a problem. This water could and possibly should, be returned to the water table of the area rather than allowing it to run off. There is evidence that the ground water in areas where geothermal energy is used must be replaced to prevent subsidence.

Thus, we see that our water supply, within the context of the hydrologic cycle, is an intricate system that depends on many factors. The quality of the

Figure 30-3d *Old Faithful geyser. Eruptions of hot water vary between 30 and 90 minutes apart with an average time interval of 65 minutes. (Courtesy National Park Service.)*

GEOTHERMAL AREAS IN WESTERN UNITED STATES

EXPLANATION

★ Developed Geothermal fields
1 The Geysers
2 Cerro Prieto (Mexico)

● Promising Geothermal areas
3 Boise
4 Klamath Falls
5 Pyramid Lake
6 Bradys
7 Beowawe
8 Steamboat
9 Casa Diablo
10 Valles
11 Salton Sea

● Hot Springs

EXPLANATION

● Producing geothermal fields

● Promising geothermal areas

⊥ Mid-ocean rifts-centers of ocean floor spreading along which heat flow is high local water supply depends to a great deal on the mineral composition of the soil and rocks through which it percolates in the zone of aeration. Water soluble minerals are dissolved and carried along with the water in solution. In some cases, water containing bicarbonates (hydrogen carbonates) and chlorides is sold commercially as "mineral water." However, dissolved minerals in "hard" water have undesirable effects. **Hard water** is due to dissolved calcium and magnesium salts (bicarbonates, chlorides, and sulfates). Iron salts also contribute to hard water. Such dissolved minerals not only affect the taste of the water, but also cause usage problems. The salts combine with the organic acids in soaps to form insoluble compounds, thereby reducing the lather and cleansing quality of the soap. It is because of these insoluble compounds from soap that clothing does not come out "whiter white" on wash day and also that there is a ring around the bathtub. At high temperatures, the metal ions present may form insoluble compounds (e.g., carbonates and sulfates) and precipitate. This precipitation tends to clog pipes and forms scale in teakettles and boilers. Boiler scale is a poor conductor of heat and reduces the boiler efficiency and wastes fuel. To avoid these conditions, the hard water may be "softened" by removing the mineral salts prior to use. Water softening is an active business in many parts of the country.

Figure 30-3f *The Geysers, California. A commercial geothermal electric generation plant is located at The Geysers in Sonoma County, 90 miles north of San Francisco. Production of electricity was started in 1960 with a capacity of 12,500 kW and has since been increased to over 83,000 kW. (Courtesy U.S. Geological Survey.)*

30-4

THE OCEANS AND SEA FLOOR TOPOGRAPHY

The Oceans

The vastness of the restless oceans imparts an awesome and humbling feeling to most observers. And rightly so, since the oceans cover about 70 percent of the earth's surface. There are five major oceans. In order of decreasing size, these are the Pacific, the Atlantic, the Indian, the Antarctic, and the Arctic oceans. The average depth of the oceans is about 2.5 miles, with the greatest measured depth of about 7 miles in the western Pacific.

To the novice beachgoer, one of the first things noticed on an initial swim is the saltiness of the sea water. About 3.5 percent of the average sea water by weight consists of dissolved mineral salts. Through the hydrologic cycle, the dissolved salts of over 2 billion years of erosion have found their way into the ocean waters. Some ions of the dissolved salts are effectively removed from the oceans by biological processes and some by precipitation or absorption by other minerals. (Recall from Chapter 15 that dissolved chemical salts dissociate into metallic and nonmetallic ions.) These salts become part of the ocean sediment and reenter the rock cycle from which they originated. It is believed that during the course of geologic time these processes have reached a steady state, such that the present salt composition of sea water has been maintained for millions of years. The quantities of salts in the seas may vary at some locations due to variations in the size of the mineral deposits of the drainage regions whose rivers feed the oceans. However, the same salts are found everywhere, generally in the same proportions.

The saltiness of the sea is measured in terms of its **salinity.** This is commonly expressed in parts per thousand (ppt) rather than percent or parts per hundred. Parts per thousand is the same as the grams of dissolved salt per kilogram of sea water (g/kg). The salinity of the average sea water is therefore 35 ppt or

Table 30-4a *Some Chemical Constituents of Sea Water* (average concentrations)

Chemical	Concentration (ppt)	Chemical	Concentration (ppt)
Chloride	19.3	Nitrate	3.5×10^{-4}
Sodium	10.7	Iodine	5×10^{-5}
Sulfate	2.7	Iron	1×10^{-5}
Magnesium	1.3	Copper	5×10^{-6}
Calcium	0.4	Zinc	5×10^{-6}
Potassium	0.39	Lead	4×10^{-6}
Bicarbonate	0.14	Silver	3×10^{-7}
Bromine	0.070	Nickel	1×10^{-7}
Strontium	0.013	Mercury	3×10^{-8}
Fluorine	0.001	Gold	6×10^{-9}

*Over 60 chemical elements have been found in sea water.

35 g/kg (equivalent to 3.5 percent). The range of sea water salinity in the open sea is from 33 ppt to 38 ppt. As can be seen from Table 30-4a, the principal elements that contribute to the salinity of sea water are sodium and chloride.

The sea is sometimes referred to as a mineral storehouse from which we might extract minerals. This is true in part, but one must not overlook the concentrations and the difficulties of extracting these minerals from solution. It is a fact that NaCl has been taken from the ocean since ancient times, and that needs prompted the extraction of magnesium and bromine from sea water during World War II. Even today much of our output of these substances still comes from sea water. The extraction of metals like gold, however, is a different matter. There are 10^{10} tons of gold in the oceans, but the concentration is only about 6×10^{-11} oz/gal. This means that with 100 percent recovery about 17,000 million gallons of sea water would have to be processed to obtain one ounce of gold. This would hardly pay for the effort.

Another feature noted when at the beach is the restless motion of the ocean waters. There are three types of sea water movements that are quite noticea-

Figure 30-4a *An illustration of a surface wave approaching the shore. The water particles move in more or less circular paths, the diameters of which decrease with increasing depth. In shallow water near the shore, the water particles are forced into more elliptical paths. This causes the wave to grow higher and steeper and to eventually break, forming a surf.*

ble at the beach or along the coast. These are surface waves, long shore currents, and tidal currents. **Surface waves** continually lap the shore. In general, the waves near the surface are a combination of longitudinal and transverse wave motions. It is tempting to think of water waves as being transverse since the sinusoidal profile is clearly evident. However, the water "particles" move in more or less circular paths as illustrated in Fig. 30-4a. Circular particle motion is a combination of longitudinal and transverse motions.

The diameter of the circular paths of the water particles decreases rapidly with depth and the wave motion is hardly observed when one dives several feet below the surface. As a wave approaches shallower water near the shore, the water particles experience difficulty in completing their circular paths and are forced into more elliptical paths. The surface wave then grows higher and steeper. Finally, when the depth becomes too shallow, the water particles can no longer move through the bottom part of their paths and the wave breaks, with the crest of the wave falling forward to form a **surf.**

When at the beach, you may have noticed debris moving along the shore as it bobs up and down. This movement is an indication of a **long shore current** flowing along the shore. These currents arise from incoming ocean waves that break at an angle to the shore. The component of water motion along the shore causes a current in that direction. Waves and their resulting long shore currents are important agents of erosion along coastlines.

The periodic rise and fall of the tides are also quite evident at the beach. These **tidal currents** result from the two tidal bulges that "move" around the earth daily as a result of the gravitational attractions of the moon and sun and the rotation of the earth (cf. Chapter 19). The water level may rise as much as 40 feet in some regions at high tide. In the open ocean, tidal currents are of little significance, but then as they approach the shore, the rise of the water level is evident, particularly when they are confined within the boundaries of a bay or a river outlet. The water level can rise rapidly and sea water can back up into rivers that open to the sea. In France, tidal currents are used to generate electrical power on the Rance River. Not only is power generated from the incoming tidal current as the sea water flows rapidly into the Rance River estuary, but also from the outgoing tidal current as the sea water flows back into the Gulf of St. Malo.

In the open ocean the major movement of sea water results from two general types of currents: surface currents and density currents. **Surface currents** in the ocean are broad drifts of surface water that are set in motion by the prevailing surface winds. These currents rarely extend more than 100 meters in depth (~100 yards—one football field length). As might be expected, surface currents are influenced primarily by the prevailing winds of the earth's atmospheric semipermanent circulation structure as discussed in Section 22-3. A comparison of the global surface currents shown in Fig. 30-4b with the prevailing wind zones in Fig. 22-3b should make this evident. Surface currents, like air movements, are affected by the Coriolis force (cf. Section 22-1). As a result, the general circulations of surface currents are

Figure 30-4b *Major ocean surface currents. Notice that the general circulations of the surface currents are clockwise in the northern hemisphere and counterclockwise in the southern hemisphere. Like atmospheric circulations, this results from the Coriolis force. (Courtesy Hydrographic Center, DOD.)*

clockwise in the northern hemisphere (deflection to the right) and counterclockwise in the southern hemisphere (deflection to the left). The surface current circulations are larger in the southern hemisphere since there is more open ocean—two-thirds of the earth's land surface is north of the equator. The lack of land forms in high southern latitudes makes it possible for a global circulating surface current—the West-Wind Drift. Water can move from one ocean to another in this drift.

The surface current circulation patterns are important climatic factors. As can be seen from Fig. 30-4b, the circulations form sea water convection cycles and the currents carry warm water from equatorial regions to higher latitudes. For example, in the Atlantic Ocean the North Equatorial Current carries warm water westward into the Gulf of Mexico. It emerges as the Florida Current and flows northward as the Gulf Stream. The circulation continues as

the North Atlantic Current flowing toward Europe, where it is diverted to the north and south. The warm water the northward-deflected current brings to the British Isles and to the northern European coast is an important factor in moderating the climate of this northern region. The current deflected to the south cools and returns to the equatorial regions via the Canary Current, where it is warmed and starts on another cycle. In a similar manner, the surface current circulation in the northern Pacific brings a warming influence to Japan via the Kuroshio Current, and the returning cool California Current moderates the climate along the coast of Southern California.

Deep ocean currents are called density currents. A **density current** is the motion of sea water caused by dense water sinking through less dense water. The density of sea water is dependent on its temperature, salinity, and sediment load. The cooler the water and the greater its salinity, the greater its density.* In the polar regions, the surface water in contact with the atmosphere is cooled and its density increases. Also, when sea water freezes, the dissolved salts are not incorporated in the ice, which is fresh water. The residual salts from the frozen water increase the salinity of the surrounding sea water. By these methods the polar water becomes denser and sinks, giving rise to deep density currents that may extend to the tropics. Such a deep-water current starts in the North Atlantic with the surface water brought northward by the surface North Atlantic Current. This water already has a high salinity due to evaporation in the equatorial region where it began its journey. This North Atlantic Deep Water Current flows southward along the ocean bottom far past the equator before mixing decreases its density and halts its flow.

Deep-water currents flow northward from Antarctica, where cooling and freezing give rise to some of the densest water in the oceans. These currents flow northward into the Atlantic, Pacific, and the Indian oceans. However, there is no counterpart of the south-bound North Atlantic Deep Water Current in the Pacific. This is because there is no large, deep, cold-water source in the northern Pacific. A shallow barrier in the Bering Strait prevents the cold Arctic water from entering this region.

Sea Floor Topography

It was once thought that the surface features or topography of the ocean basins consisted of an occasional mountainous island arc on a relatively smooth

---

*The density of water increases with decreasing temperatures to a maximum of 1.0 g/cm^3 at 4°C. Above and below 4°C, the water is slightly less dense. For this reason, lakes freeze at the top rather than at the bottom. As the water near the surface loses heat to the atmosphere, it becomes denser and sinks (causing a density current), displacing the warm water below. This continues until the air and surface water temperature is below 4°C. The cool surface water is then less dense than the water below and hence remains at the surface where freezing begins when the temperature goes below 0°C.

sediment-covered floor. This incorrect view resulted from the lack of direct observation. Surprisingly enough, the surface of a major portion of the earth —the oceanic crust—was not explored in great detail until after World War II. With advent of modern technology, sounding and drilling operations revealed that the ocean floor is about, if not more, irregular than the surfaces of the continents (Fig. 30-4c).

We now know that the sea floor has a system of mid-oceanic ridges, which mark divergent plate boundaries. These rocky submarine mountain chains are along fracture zones through which magma rises from below to form new oceanic crust. Oceanic volcanism is widespread. Large volcanic mountains also rise from the ocean floor in the midst of plates, such as those in the Hawaiian island chain. As pointed out in Chapter 29, these are believed to arise from plates moving over internal "hot spots."

Many isolated submarine volcanic mountains have also been discovered. These are known as **seamounts.** They are individual mountains that may extend to heights of over a mile above the sea floor. Some seamounts are found to have flat tops, and are given the special name of **guyots*** (Fig. 30-4c). Their shapes suggest that the tops were once islands that were eroded away by wave action. However, many of the guyot tops are several thousand feet below sea level. Evidently, the eroded seamounts must have subsided and sunk below sea level. This subsidence in the oceanic crust perhaps occurs as the oceanic crust moves away from a spreading ridge or results from an isostatic adjustment due to the weight of the large volcanic seamount.

Another marked feature of sea floor topography are **sea floor trenches,** which mark the locations of plunging plate boundaries. These trenches are as much as 150 miles in width and 15,000 miles or more

---

*Named in honor of Arnold Guyot, the first geologist at Princeton University, by Professor Harry Hess, a geologist at Princeton University who discovered the first flat-topped seamounts in the 1950s.

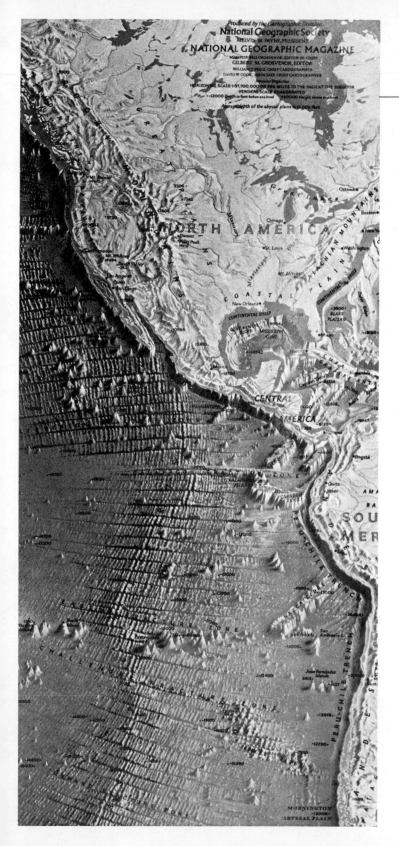

Produced by the Cartographic Division
National Geographic Society
MELVIN M. PAYNE, PRESIDENT
NATIONAL GEOGRAPHIC MAGAZINE
MELVILLE BELL GROSVENOR, EDITOR-IN-CHIEF
GILBERT M. GROSVENOR, EDITOR
WILLIAM T. PEELE, CHIEF CARTOGRAPHER
DAVID W. COOK, ASSOCIATE CHIEF CARTOGRAPHER

HORIZONTAL SCALE 1:37,700,000 OR 595 MILES TO THE INCH AT THE EQUATOR
VERTICAL SCALE EXAGGERATED
-12000 Depth in feet below sea level   +40000 Height above sea level
Average depth of the abyssal plains is 15,000 feet.

NORTH AMERICA

SOUTH AMERICA

CENTRAL AMERICA

Figure 30-4c *The Pacific Ocean Floor. The sea floor topography was once thought to be smooth due to the accumulation of sediment deposits. However, relatively recent explorations have shown the sea floor surfaces to be as irregular as the surfaces of the continents. There are many isolated submarine volcanic seamounts, some of which are flat-topped guyots. See also Fig. 27-2f for the topography of the Atlantic Ocean floor. (Courtesy National Geographic Society.)*

in length. The deepest trenches show little evidence of sediment accumulation, whereas other trenches near land areas are partially filled with sediment. The huge volumes of sediment flowing into the oceans from continental regions do have an affect on sea floor topography. Distributed by ocean currents, sediment accumulates in some regions such that a sedimentary layer covers and masks the irregular features of the rocky ocean floor. The resulting large flat areas are called **abyssal plains.** Abyssal plains are most common near the continents, which supply the sediment.

Although 70 percent of the earth's surface is covered with water, the oceanic crust basins account for only about 65 percent of the surface area. This means that the continental crust comprises 35 percent of the earth's surface area; but since only about 30 percent of the earth's surface is land area, 5 percent of the continental crust must be submerged. This occurs along the continental margins, and the shallowly submerged borders of the continental masses are called **continental shelves** (Fig. 30-4d). The widths of these shelves vary greatly. Along the Pacific coast of South America, there is almost no continental shelf—only a relatively sharp, abrupt continental slope. However, off the north coast of Siberia, the continental shelf extends outwardly into the ocean for about 800 miles. The average width of the continental shelves is of the order of 40–50 miles.

The continental shelves have recently become a point of international interest and dispute. The ma-

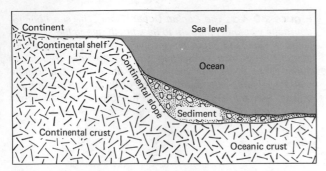

Figure 30-4d *A cross-sectional illustration of a continental shelf. Continental shelves are the shallowly submerged borders of the continents. Continental slopes define the true edges of the continents. Sediment from continental erosion collects at the bases of the continental slopes.*

jority of commercial fishing is done in the waters above the continental shelves. Also, the continental shelves are the locations of oil deposits that are now being tapped by off-shore drilling. As a result, many countries including the United States have extended their territorial claims to an off-shore 200-mile limit.

Beyond a continental shelf, the surface of the continental land mass slopes downward to the floor of the ocean basin. The **continental slopes** define the true edges of the continental land masses. Erosion along these slopes gives rise to deep submarine canyons that extend downward toward the ocean basins. And near the edges of the ocean basins, the sediment collects in geosyncline troughs. Thus, we come back to one of the factors in continental mountain building.

Undersea exploration is a relatively new phase of scientific investigation. Indeed, there is a great deal more to be learned about this vast region that makes up almost 70 percent of the earth's surface. As advances in technology provide more data on this previously inaccessable part of the earth's surface, we may expect our knowledge of the earth and its geologic processes to grow.

IMPORTANT WORDS AND TERMS

Erosion
Physical weathering
Frost wedging
Frost heaving
Permafrost
Salt wedging
Chemical weathering
Sinkhole
Overburden
Mass wasting
Landslides
Rockslides
Slump
Mudflows
Creep
Solifluction
Agents of erosion
Sheet erosion
Stream load
Traction

Abrasion
Silt
Base level
Meander
Graded river
Flood plain
Delta
Glacier
Continental glacier
Cirque glacier
Valley glacier
Drift
Till
Moraine
Hydrologic cycle
Permeability
Porosity
Zone of saturation
Water table

Zone of aeration
Aquifer
Artesian wells
Artesian springs
Geyser
Hard water
Salinity
Surface wave
Surf
Long shore current
Tidal current
Surface current
Density current
Seamount
Guyot
Sea floor trench
Abyssal plains
Continental shelves
Continental slopes

LEARNING OBJECTIVES

After reading and studying this chapter, you should be able to do the following without referring to the text:

1. Define *erosion*.
2. Distinguish between physical and chemical weathering and give examples of each.
3. Define *mass wasting* and give two important types of mass wasting with examples of each.
4. Describe the agents of erosion.
5. Distinguish between the three components of a stream's load.
6. Explain how glaciers are formed and distinguish between the different types of glaciers.
7. Discuss the distribution of the earth's water supply.
8. Describe the hydrologic cycle.
9. Explain why the sea is salty.
10. State and explain the types of ocean currents.
11. Explain how ocean currents affect climate.
12. Describe the major features of sea floor topography.

QUESTIONS

1. What is erosion?
2. Distinguish between physical and chemical weathering.
3. What are frost wedging and frost heaving?
4. What type of weathering takes place in hot, arid climates that is similar to frost wedging?
5. How do plants and animals contribute to weathering?
6. On what factors does chemical weathering depend?
7. Describe in detail the chemical weathering process of limestone and the formation of caverns and dripstone.
8. What is a sinkhole?
9. What is mass wasting?
10. Explain each of the following and state whether it is a fast or slow type of mass wasting. (a) rockslide, (b) creep, (c) slump, (d) solifluction, (e) mudflow.
11. State the four major agents of erosion.
12. What is sheet erosion?
13. State and explain the three components of a stream's load.
14. Explain stream abrasion and traction.
15. What is silt?
16. State and describe the three general stages of river development.
17. Why does a river meander and what is a graded river?
18. Distinguish between continental and valley glaciers.
19. Are there any glaciers in the United States? Explain.
20. Describe each of the following: (a) drift, (b) till, and (c) a moraine.
21. How much of the earth's water supply is fresh water and where is most of it located?
22. Explain the hydrologic cycle.
23. Explain each of the following: (a) permeability, (b) porosity, (c) the zone of aeration, (d) the zone of saturation, and (e) the water table?
24. What is an aquifer?
25. Distinguish between gravity springs and artesian wells and springs.
26. What causes hot springs and geysers?
27. What are the two general types of geothermal fields? Which of the types is more important in the generation of electricity?
28. What is the cause of "hard" water and what are some of its effects?
29. What is salinity and in what units is it measured? What is the average salinity of the oceans?
30. Describe the three types of sea water movements along a coast.
31. Why do waves form a surf?
32. What causes surface currents in the open ocean and what are some of the major surface currents?
33. Why is the surface current circulation larger in the Southern Hemisphere? Are there any global circulating surface currents?
34. Explain how the climate of the British Isles and northern Europe are moderated by surface currents.

35. What are density currents? Describe their effect on ocean circulation.
36. Why is there no counterpart of the North Atlantic Deep Water Current in the Pacific Ocean?
37. Define and explain the formation of each of the following: (a) seamounts, (b) guyots, (c) sea floor trenches, (d) abyssal plains.
38. What are continental shelves and continental slopes? What defines the true edges of the continental land masses?
39. Why are the continental shelves the focus of current international interest?

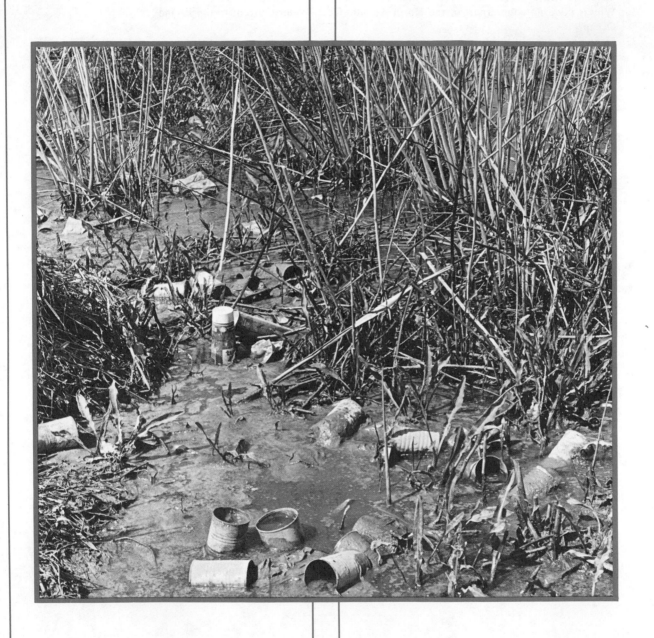

Chapter 31

LAND AND WATER POLLUTION

As in the case of air pollution, land and water pollution result primarily from the indiscriminate disposal of waste products. Land and water pollution are closely related through the hydrologic cycle. Through decomposition and erosion processes many materials contributing to land pollution, or their by-products, eventually find their way into the earth's water supply. Many of the waste products were once natural parts of the environment, but were processed by human beings into different forms for their use. Other pollution results directly from our activities.

Having studied the physical and chemical processes of the environment, you should now realize that what may not be a problem in a particular situation or location may contribute to or give rise to environmental problems elsewhere. Water is used to carry away waste as is the air, but such pollution is better defined within waterway boundaries. That is, waste dispersion in water is in general not as great as in the atmosphere and the path of waste in surface waterways is well charted. Ground water pollution is less easily observed, but its effects can be obvious and dangerous. In this chapter some of the sources and effects of land and water pollution will be examined.

You can't get pure water and put sewage into the lake. I say this on behalf of your children.

President Theodore Roosevelt
1910, at Astabula, Ohio (on Lake Erie)

31-1
LAND POLLUTION

As was mentioned, it is difficult to separate land and water pollution because of the interacting hydrologic cycle. However, we shall first look at some of the **land pollution** or land surface problems that may give rise to water pollution problems. Later, we will consider water pollution in more detail.

One of the chief sources of pollution of our countryside is improper solid waste disposal. **Solid waste** is a general term that is applied to any normally solid material resulting from human or animal activities that is useless or unwanted. More than 12 million tons of solid waste is produced each day in the United States. The major sources of solid wastes are listed in Table 31-1a.

Modern packaging technology and disposable items have contributed greatly to urban wastes. "Waste not" is no longer a popular adage. The domestic solid waste per capita in this country has more than doubled from its 1920 value of 2.9 pounds per person per day. If you have had to empty the garbage and trash for an average family, you are aware of the magnitude of the solid waste generated.

The terms refuse, garbage, trash, and rubbish are used to describe solid waste. Their definitions vary slightly in different regions. Refuse is a rather general term, while garbage usually refers to food wastes,

Figure 31-1a *Solid waste can be a form of land pollution.*

rubbish, trash, and combustible (paper cartons, etc.) and noncombustible (junk) items. Regardless of the term used to describe solid waste, the problem is what to do with it (Fig. 31-1a).

Solid waste management begins with proper disposal. However, when situations occur such as those shown in Fig. 31-1b, the pollution is in the form of litter. **Littering** is the indiscriminate disposal of solid waste on public and private property. The litterbug's disrespect is quite costly. The cost of litter removal from public property alone exceeds an estimated $500 million of tax money annually. All states have litter laws, but evidence of their violation can be seen everywhere. Tons of litter are strewn along streets and

Table 31-1a *Major Sources of Solid Waste*

Source	Production (millions of tons)
Residential, municipal, and commercial	260
Industrial	110
Agriculture	2280
Mineral (mining)	1700
Federal	45
Total	4395

Source: Bureau of Solid Waste Management, HEW.

highways and in parks and along beaches. Some wastes such as food scraps and paper items are **biodegradable** (capable of being decomposed by bacteria) and will relatively quickly decompose and return to the soil. However, other items such as plastics are nonbiodegradable and are semipermanent. Some show little or no sign of weathering as in the case of chemical weathering of metal wastes and plastics.

Solid waste disposal is a serious problem, particularly in metropolitan areas where the concentrations of solid waste are enormous. At one time, the **open dump** was the common method of solid waste disposal (Fig. 31-1c). Dump sites were usually land of little economic value and at some distance from the center of population that used it. However, open dumps are now forbidden by law for obvious reasons. Solid waste disposal now takes place in **sanitary landfills.** In a sanitary landfill, refuse is covered with a layer of soil within a day after being deposited. This action eliminates some of the nuisances and hazards of the open dump. The landfill method is more sightly, and odor-breeding places for vermin and air pollution from burning are eliminated.

However, sanitary landfills still have potential pollution danger. As with the open dump, the seepage of water through the wastes can leach out undesirable components that pollute the ground water. The extent of pollution from leaching is largely dependent on the geologic features of the area in which the wastes are deposited.

Some cities along the ocean use the convenience of this body of water to dispose of their solid waste. Solid wastes from New York City are hauled offshore by barge and dumped into the ocean. Incineration is also a common method of combustible solid waste disposal, but air pollution is associated with this method.

The activities of human beings often lead to erosion and water pollution. Housing and land developments are necessary to accommodate our growing population. Many prefer country settings, so housing developments are frequently made in wooded areas.

Figure 31-1b Littering, the indiscriminate disposal of solid waste on public and private property. (Courtesy USDA-SCS.)

Figure 31-1c *Past and present. Open dumps were once a common method of solid waste disposal. Today, most solid wastes are disposed of in sanitary landfills. (Courtesy EPA and USDA-SCS.)*

Figure 31-1d *Land development and a result of poor planning in the removal of erosion-preventing trees and vegetation. (Courtesy California State Dept. of Conservation, Division of Soil Conservation.)*

Poor planning and the removal of trees and vegetation that prevent erosion can give rise to landslide situations as shown in Fig. 31-1d. Another likely result is sediment or silt concentrations in streams and rivers. Silting is a natural process, as was learned in the last chapter. However, increased silt concentrations from unnatural silting can have a profound effect on the environments of lakes and ponds into which streams drain (Fig. 31-1e). The suspended sediment reduces the extent of sunlight penetration into the water. This reduces the photosynthesis of submerged vegetation; thus reducing the oxygen production and the food supply for fish that eat the vegetation. Also, silt can foul the spawning beds of fish.

Mining operations can give rise to land and water pollution. The minerals and fuels needed by society are mined from the earth. Many such deposits lie well below the surface and are reached by deep mines. Other deposits lying near the surface are surface

Figure 31-1e *Silting from people's activities can foul streams and lakes, causing severe damage to animal and plant life. (Courtesy California State Dept. of Conservation, Division of Soil Conservation.)*

Figure 31-1f *Surface mine operation. The large dragline shovel in the background removes the overburden and the smaller shovel removes the coal. This strip mine is near Oak Creek, Colorado. (Courtesy USDA-SCS. Photo by N. P. Mc-Kinstry.)*

mining. Sediment yields from surface-mined areas average nearly 300,000 tons per square mile annually. This is as much as 60 times the amount of sediment removed from agricultural lands.

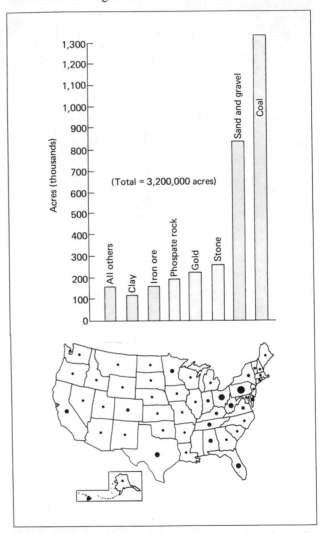

Figure 31-1g *Acreage disturbed by surface mining in the United States. The dots on the map show the proportionate surface-mined acreage in each state. About only one-third of the mined area has been reclaimed. The bar graph shows the extent of surface mining by commodities. (Courtesy USDI and USDA.)*

mined. Large excavating equipment of the type shown in Fig. 31-1f remove the overlying earth (overburden) to expose the commercially valuable materials. Surface mining is faster and cheaper than deep mining. Properly treated and managed, the land after being surface mined can be returned to its natural condition. Unclaimed, it remains an ugly scar on the earth's surface and a source of stream-fouling sediment and leached pollutants.

The extent of surface mining in the United States is illustrated in Fig. 31-1g, along with the proportionate acreage of particular mining operations. Erosion is the most obvious problem associated with surface

Another pollution problem arising from mining is **acid mine drainage.** Acid formations occur when oxygen (from the air) and water react with sulfur-bearing minerals to form sulfuric acid and iron compounds. For example,

$$2\ FeS_2 + 7\ O_2 + H_2O \longrightarrow 2\ FeSO_4 + 2\ H_2SO_4$$

Pyrite   Ferrous   Sulfuric
sulfate   acid

About 70 percent of the acid mine drainage comes from deep coal mines—chiefly from inactive mines. The major problems of acid drainage occur in the coal regions of Appalachia and in some specific western regions. Coal formation in these regions have a high sulfur content ($>2\%$) primarily due to the sulfur-bearing mineral iron pyrite (FeS_2—a yellowish mineral sometimes called fool's gold).

Acid mine waste may directly affect the plant life that absorbs it or have an indirect effect on the flora of a region by affecting the soil minerals and microorganisms. The acid wastes that drain off into ponds and streams can dramatically reduce the aquatic plant and animal life. Changes in the pH of water may cause fish to grow slowly or die. To reduce pollution from acid mine drainage, reclamation projects include sealing abandoned mine shafts and other accesses to the mine. Sealing prevents the circulation of air and so reduces the oxidation of the sulfur minerals.

31-2
WATER POLLUTION

I counted two-and-seventy stenches,
All well defined, and several stinks.
The river Rhine, it is well known,
Doth wash your city of Cologne;
But tell me, nymphs! what power divine
Shall henceforth wash the river Rhine?

(From the poem *Cologne* by
Samuel Taylor Coleridge, 1828)

Table 31-2a *Sources and Types of Water Pollution*

Agricultural	Industrial	Domestic
Pesticides	Chemical	Detergents
Fertilizers	Thermal	Organic wastes
Animal wastes	Radioactive	

Water pollution may occur in many different forms and originate from many different sources. To simplify matters, the major sources and types of pollution are summarized in Table 31-2a.

Let's consider each of these briefly.

Agriculture

Water pollution from agricultural sources is usually not direct, but enters the water supply via runoff and ground water absorption in areas of agricultural activity.

■ *Pesticides* **Pesticides** may be subdivided into insecticides, herbicides, and fungicides. Their purpose is to kill some unwanted insect and plant species, usually for the promotion of an agricultural crop. There are a variety of other uses. For example, in the home, insecticides have all but replaced the pollutionless fly-swatter.

In modern agriculture, pesticides are essential to food production and other crop products to produce the yield required by our population. Without them, tons of agricultural products would be lost each year to pests. Prior to World War II mostly natural organic pesticides were used, which were easily decomposed in the environment. In 1939, DDT (dichlordiphenyl-trichlorethane) was discovered and used extensively during the war to combat insects in tropical areas. Afterward, this effective agent was expanded to general agricultural use, but in 1972 the government banned the use of DDT in the United States. DDT is a chlorinated hydrocarbon which, unlike natural pesticides, is not biodegradable. As a result of its exten-

Figure 31-2a *A cattle feed lot. Commercial beef cattle no longer roam the open range. Instead, they are raised in feed lots such as the one shown here near Omaha, Nebraska, which has facilities to feed about 6000 cattle. Such concentrations of livestock create a problem with animal wastes and water pollution. (Courtesy USDA.)*

sive usage (worldwide), DDT accumulated in the environment. DDT residue was found in appreciable concentrations in animals, entering through the biological food chains. Even penguins in Antarctica were found to have DDT concentrations. DDT is only one of several chlorinated hydrocarbons used as pesticides. Other pesticides include arsenic compounds of calcium, lead, and mercury.

Pesticide pollution of water in appreciable concentrations poses critical health hazards. There are some 34,500 registered pesticide products composed of one or more of 900 chemical compounds. Although pesticides are usually sold for a specific pest, they often kill nonpest species and have side effects on the growth and reproduction of birds and fish.

■ *Fertilizers* Like pesticides, **fertilizers** are used to increase food production. Farmers in the United States apply millions of tons of fertilizers to their fields each year. Water pollution results from the **phosphates** (PO_3^{4-}) and **nitrates** (NO_3^-) present in the fertilizers. These substances enter the water supply primarily through water runoff and through the erosion of topsoil. In sufficient concentration, nitrates are toxic to animals and humans. Nitrates can be reduced to nitrites (NO_2^-) which interfere with the transport of oxygen by hemoglobin in the blood. Normal water purifying procedures do not remove the nitrate contaminants. Nitrates and phosphates are both believed to contribute to the excessive growth of

microscopic plant algae in lakes, a condition that affects the local ecosystem. This effect will be considered in the discussion of domestic detergents, the major contributor of phosphates to water pollution.

■ *Animal wastes* Animal wastes are a source of water pollution with potential health hazards. With the trend toward raising animals in feed lots, disposal of these wastes becomes a problem (Fig. 31-2a). Table 31-2b gives an indication of the magnitude of the problem. From the data in the table, we see that a feed lot handling 1000 head of cattle would have the same weight-equivalent waste-disposal problem as that of a city with a population of 157,000. The city would probably have a sewage treatment plant, the feed lot would probably not.

Traditionally, the disposal of animal waste has been as a fertilizer. However, studies show that com-

Table 31-2b *Animal Waste Production*

Source	Pounds/Day
Human being	0.33
Cow	52.0*
Pig	6.0
Sheep	2.5
Chicken	0.4

*Perhaps you, like the author, thought this was an inordinate amount, but a county agricultural agent said it is a good figure.
Source: Soil Conservation Service, U.S. Dept. of Agriculture.

Figure 31-2b *A case of water pollution. The photographs show the discharge of paper pulp wastes being discharged at its source (not to mention the air pollution) and its effects downstream. (Courtesy Bureau of Sport Fisheries and Wildlife.)*

pared to modern chemical fertilizers the benefits of soil fertilization from animal wastes do not justify the costs of hauling and application. Thus the livestock owner must contend with a daily supply of a product which he cannot sell, give away, or burn, and which may cause water pollution in water runoff. The effects of this pollutant will be discussed with those of domestic waste.

Industrial

Industrial pollution, one of the more serious forms of water pollution, arises from the disposal of noxious industrial wastes into rivers and streams (Fig. 31-2b). This has been a common industrial practice for many years, and at one time streamflows were adequate to dilute and carry away the wastes without environmental damage. But with the increase of industry and population, industrial pollution has become serious. Most responsible companies are aware of the problems and have instituted programs to reduce or to eliminate their pollution activities. Just as the pollution problems did not develop overnight, neither can the solutions be developed and applied overnight. While solutions are being sought, industry continues to be the largest user of water with some 240,000 industries in the United States using this resource. The amount of water usage, of course, depends on the nature of the industry. A comparison of amounts used is shown in Fig. 31-2c.

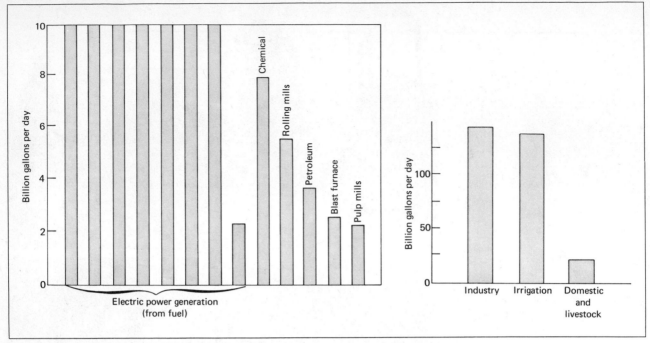

Figure 31-2c *The use of water in the United States. The greatest part of industry's water is for cooling in the generation of electric power. (Courtesy USDI.)*

■ *Chemical* The number of chemical compounds discharged by industry are legion, and it would be hopeless to attempt to list and discuss them here. Instead, let us consider only one as an example, mercury. Mercury (Hg) is a readily available element obtained from the abundant ore cinnabar (HgS). Mercury has many industrial uses (as many as 3000), and compounds of mercury are used extensively as pesticides.

Metallic mercury in liquid form is not overly dangerous unless ingested in large amounts. However, mercury vapor is highly toxic and readily absorbed through the lungs. Also, organic compounds of mercury, such as methylmercury, are quite dangerous. The action of microorganisms in water on metallic mercury can produce methylmercury, which in sufficient quantities harms the central nervous system of animals, including human beings. Methylmercury is retained in the body for long periods of time so there can be a cumulative effect from otherwise small harmless doses. Mercury compounds were used to treat the beaver pelts used in the early hat

industry, and the early symptoms of mercury poisoning (giddiness, silliness, and strange behavior) were common among the workers; hence the term "mad as a hatter." It is said that the madness of the Mad Hatter in Lewis Carroll's *Alice in Wonderland* was the result of mercury poisoning.

As a form of water pollution, mercury compounds find their way into the food chains of wildlife, particularly fish. Fish are a food supply for many species, including human beings. **Mercury pollution** poses a severe health threat to the inhabitants of areas in which fish forms a substantial part of the diet. Such a case occurred in the coastal town of Minamata, Japan, where a plastics manufacturing plant discharged quantities of wastes containing mercury into Minamata Bay. Between 1953 and 1960, 110 people, mostly from the families of fishermen, died or were severely disabled after eating fish caught in the mercury-polluted waters.

■ *Thermal* A great deal of the water required by industry is used in cooling processes. In fact, 94 percent of the water used by industry is for this purpose. The major cooling application of water is in thermoelectric power generation, which explains the disproportionate share of this user shown in Fig. 31-2c. Other industries, such as oil refining and steelmaking, also use an appreciable amount of water as a coolant.

In thermoelectric power generation, steam is used to drive turbines in both conventional fuel and nuclear methods. Water is used to cool the steam from the turbine. The cooled steam condenses, and so releases the back pressure on the turbine and increases its efficiency. The water for cooling does not have to be of high quality. In some locations, salt water is used as the coolant, but this requires that parts of the equipment be made of special materials because of the salt water's corrosive properties.

The temperature rise of water used in electric generation cooling is between 10° and 30°, depending on plant design and operation. Nuclear power plants require more cooling than conventional (fossil fuel) plants of equivalent size. Either the temperature of the effluent water from nuclear plants is greater or a greater volume of water must be used, as much as 50 percent more. The **thermal pollution** problem associated with power generation is due to the return of the heated water to rivers and lakes. The heated water causes the normal temperature of the lake or a portion of the river to rise, which is a type of pollution. One might not think that a few degrees rise in the temperature of a body of water would create an environmental problem. However, the thermal aspect of an ecologic balance is very delicate. A temperature rise in a natural water formation causes adverse conditions for some aquatic life, reduces the dissolved oxygen content of the water, and accelerates plant growth and other biologic processes.

The most observable effect of thermal pollution is that on aquatic life, particularly fish. The amount of dissolved oxygen in water decreases with increasing temperature. Fish absorb oxygen from the water passing through their gills and may die in water with a reduced oxygen content. This possibility is particularly likely in streams with thermal pollution in the summertime when the water temperature rises naturally (Fig. 31-2d). The environmental effects of thermal pollution can be quite severe and far-reaching, however, thermal pollution can be readily prevented by cooling processes, as shown in Fig. 31-2d.

■ *Radiation* Natural radioactivity occurs in the environment, however, **radiation pollution** arises from the use of radioactive materials. The alpha, beta, and gamma particles emitted from radioactive decay, as well as the neutrons emitted from fission reactions, fall into the general category called **nuclear radiation.** Other particles, such as protons in cosmic rays, and quanta of energy such as X-rays, are also referred to as radiation. The hazard of radiation is the effect it may have on living cells. A cell may receive slight radiation damage and, apparently repaired, retain its normal functions. However, radiation damage may also be critical.

In general, there are two types of **radiation cell damage,** direct and indirect. In the direct type, biologically important molecules, such as those in the cell nucleus, may be ionized directly by radiation and split into biologically useless fragments. This results in a cell that can live but not reproduce itself properly. Such progressive cell death without replacement soon leads to the malfunction and eventual death of the irradiated tissue. In other instances, damage to the cell nucleus may cause the cell to lose its identity; but the cell still has the capability to reproduce at an uncontrolled rate. This gives rise to **cancer,** a condition of unregulated cell growth rate. Cancer cells may grow slowly with little effect on the surrounding normal cells, forming a benign tumor; or they may grow at the expense of surrounding cells, producing a malignant tumor. Skin cancer and leukemia, or "cancer of the blood," are possible results of radiation exposure. **Leukemia** is characterized by an abnormal increase of white blood cells (leucocytes).

In the indirect type of damage, a less critical molecule, such as a water molecule in the cytoplasm

Figure 31-2d *An effect and prevention of thermal pollution. (Right) Thousands of dead shad, victims of hot weather and river pollution, washed ashore on the Anacosta River near Washington, D.C. The oxygen content of the river water was reduced to the point where the fish could not survive. (Upper left) Although hot weather cannot be prevented, thermal pollution can be minimized by the spray air cooling of heated water before it is discharged into nearby waterways. (Lower left) Cooling towers may be employed in locations where bodies of water are not accessible. The cooled water is recycled and used again. The towers are emitting water vapor, not smoke. (Courtesy USDA, Bureau of Sport Fisheries and Wildlife, EPA, and ERDA.)*

(that part of the cell exterior to the nucleus), may be ionized by radiation into reactive complex ions. The ions and the removed electrons may simply recombine, but some of the reactive fragments may also drift within the cell to react with critical molecules of the nucleus. The damage will be much the same as if the nuclear molecules had been affected directly. On the other hand, complete cell destruction by radiation may also have a useful effect. Radioisotopes are used in medicine to kill diseased cells, such as the uncontrolled cancer cells.

Three main sources of radiation pollution are coal-fired power plants, nuclear power plants, and nuclear explosions. Radiation from coal-fired power plants results from natural radioactive materials in coal. The radiation escaping from nuclear power plants is less than that from coal-fired plants and is continuously monitored to be sure it is under the maximum permissible levels. However, permissible levels are the subject of much debate and disagreement. Some persons feel that any radiation is too much. The possibility of radiation pollution resulting from accidents in nuclear power plants is extremely low, but it is often a topic of discussion.

Direct sources of radiation pollution are nuclear explosions. In the past, test explosions were usually done in the atmosphere. These are now banned by the Nuclear Test Ban Treaty, which was signed by the major nuclear powers in 1963. Most test explosions are presently carried out underground, which reduces the risk of radiation pollution. But, not all nations signed the Test Ban Treaty. Several exceptions are France, India, and China.

Domestic

Pollution from domestic sources increases as our population increases. No one wants to be labeled a polluter, particularly at home. But we all contribute, inadvertently and unknowingly. We are now finding that even with accepted disposal methods there are long-term effects of water pollution that are evident only after several years. Of course, there can be im-

mediate health hazards with improper and unsanitary disposal practices. In the case of domestic pollution it may be said, quite truthfully, that pollution prevention begins at home. Two major considerations in domestic pollution are detergents and human organic wastes.

■ *Detergents* A detergent in the general sense is a cleansing agent. Ordinary soap, which is a cleansing agent and detergent, is made by combining either fats or oils with alkalies such as sodium or potassium hydroxide. The complex soap molecules in water can effectively interact with grease and oil, so that both are carried away with dirt in this solution. This is the detergent action, or detergency, of soap.

Few problems occur with ordinary soap in wastes, since it is normally degraded into harmless substances by bacterial action in sewage. However, the detergent action of soap is impaired by dissolved minerals in water. To accommodate the consumer, synthetic detergents were invented that react with the dissolved minerals without greatly impairing the cleansing action. But it was quickly found that the synthetic detergents were nonbiodegradable and not removed by accepted waste treatment methods. As a result, the synthetic detergents persisted when discharged into streams, sometimes to reappear as shown in Fig. 31-2e. The manufacturing of these "hard" detergents was voluntarily stopped in the mid-1960s. Through chemical modification of the molecular structures, other synthetic detergents were made that were biodegradable. After sufficient time, these "soft" detergents are completely broken down by bacterial action. There is still some concern, however, about possible yet unknown effects of the decomposition products on the environment. Detergent sales in the United States are in excess of $1 billion.

Another more obvious pollution problem arises from **detergent builders.** These are substances added to the detergent to make its cleaning action more efficient. In some cases, detergent builders account for as much as 40 percent of the detergent weight. Phosphate compounds are commonly used as detergent

Figure 31-2e *Suds from detergents. The use of "hard" detergents caused many such scenes as the one shown here in Montgomery County, Pennsylvania. The use of hard detergents has been discontinued and replaced by "soft" biodegradable detergents, since this is obviously not the way to "clean up" our rivers and streams. (Courtesy USDA.)*

builders. The phosphates form compounds with the Ca, Mg, and Fe ions in hard water. This prevents the ions from reacting with the soap and reducing its detergency. As expected, the waste water has high phosphate content. Unfortunately, the phosphates are not removed by ordinary waste treatment processes.

Environmental problems arise from phosphate waste water in lakes, where the phosphates act as a nutrient for microscopic plant algae. A normal amount of algae is good for a lake, because oxygen is added to the water by algae through photosynthesis. Algae also serves as food for fish, which means the algae population is kept in a natural balance. However, the phosphates from detergents cause the algae to multiply rapidly, or "bloom." The excess algae bloom dies and in time decays as the result of bacterial action. Decay reduces the oxygen content of the water, and so the aquatic animal life of the lake is affected. Thus, with the natural balance upset, the lake begins to "die." It becomes covered with algae mats and is so deficient in oxygen that only low animal forms can survive. The process by which this occurs is called **eutrophication** (Greek, *eutrophos*, well

nourished). Eutrophication is also a natural process and takes place over thousands of years, gradually turning lakes into marshes. However, human beings hasten the natural aging process of lakes by introducing phosphates, nitrates (fertilizers), and other pollutants. The classic example is Lake Erie which has aged as much in the past 50 years as it might have normally aged in 15,000 years.

■ *Organic wastes* Large quantities of human wastes along with the other unwanted liquids of human living are generally referred to as sewage. Sewage is composed of over 99 percent water and only 0.02 to 0.04 percent organic solids. Any concentration of people creates a sewage disposal problem, and rivers and streams are usually relied upon to carry away waste products (Fig. 31-2f). Ancient civilizations were not exempt from these problems. Rome had an adequate sewage disposal system because of the ample water supply provided by aqueducts. Surplus water was used to continually flush the city sewers into the Tiber River. Some 1500 to 1700 years later the practice of washing away human and animal

Figure 31-2f *Sewage pollution. (Left) Raw sewage being discharged into the Brandywine River near Wilmington, Delaware. (Right) An isolated case of direct water pollution. (Courtesy EPA and USDA-SCS.)*

wastes from city streets was reinstituted in Europe and was considered a major breakthrough in sewage disposal.

The flushing of raw sewage into rivers and streams, however, causes water pollution and contamination problems as would be expected—perhaps not in the location of the sewage dumping, but somewhere further downstream. Rivers may be able to handle a certain amount of sewage effectively, but with high concentrations of population, problems arise. Even if sewage disposal is attempted by underground storage and drainoff, the water supply may easily become contaminated if there are large volumes of waste.

When sewage is flushed into rivers and streams the organic matter is broken down by microorganisms, which use up oxygen in the process. Streams usually replenish the oxygen supply by aeration, but a high degree of organic pollution causes the depletion of the water's oxygen supply. This creates problems for most aquatic life.

No doubt the most critical effect of water pollution on human beings is the possibility of waterborne bacterial diseases such as cholera, typhoid fever, and dysentery. A healthy human being lives in balance with bacteria, since intestinal processes depend on certain bacterial action. However, some strains of bacteria are pathogenic and affect some people but not others. As a result, healthy individuals may be carriers of pathogenic (disease producing) bacteria. Although there may be no known cases of a disease in a locality its bacterial agent can be introduced into sewage along with normal waste. This could create a health hazard should a water supply become polluted by the sewage waste. In a public system, the water is treated and its quality checked prior to consumption, but this is usually not the case in private water supplies.

Other Sources

Another source of water pollution has come about through the demand for petroleum. Our nation runs on oil, which accounts for 96 percent of the fuel used for transportation and 85 percent of the organic chemicals produced in this country. Domestic production does not keep up with demand. As a result, about 50 percent of the oil currently used in our country is imported.

The transportation of oil poses an environmental problem. Nearly half of the world's oil production is transported by ocean-going tankers. Some 4000 tankers throughout the world form a petroleum supply chain. The potential hazard to the environment is not from the transportation of the oil, but from oil spills. Oil spills may occur accidentally—or purposefully, since it has been common practice in the past for ships to discharge unwanted oil or oil residues at sea. In 1971, the U.S. Coast Guard reported approximately 8700 known oil spills in the United States waters.

The spilled oil is not readily degradable and floats on the water, being carried by currents. Aquatic life is stifled by the black crude oil, which coats practically everything with which it comes into contact. Should the currents carry the oil spill ashore, beaches are coated with the black ooze and are useless for recreation. An oil spill's toll of sea birds can be enormous (Fig. 31-2g). One of the most notorious of all oil spills happened in 1967 when the tanker Torrey Canyon went aground off the shore of southern England and discharged some 100,000 tons of oil into the sea. The oil washed ashore and ruined the many recreational beaches and boating facilities along England's southern coast, as well as laying waste the sea life in the vicinity.

To satisfy the demand for oil, wells are being drilled offshore, even though it is two to three times more costly. Since 1960, over 9700 wells have been drilled on the continental shelf of the United States. Offshore oil wells increase oil production, but they also increase the possibility of oil spills. In 1969, a well off the coast of California near Santa Barbara discharged into the ocean some 200,000 gallons of oil, which later drifted ashore (Fig. 31-2h). The consequences were particularly severe for animal life. Peo-

Figure 31-2g *Oil-soaked birds due to an oil spill. (Courtesy USDI.)*

Figure 31-2h *Oil spills. (Top) The Santa Barbara oil spill, January 1969. Before the break in the well was repaired, an estimated 200,000 gallons of oil had escaped. Currents carried the oil ashore in the Santa Barbara area. (Bottom) Infrared photo of an oil spill in the Gulf of Mexico, March 1970, just after an accompanying fire was extinguished. A line of booms and barges are in position to attempt to contain the spill, some of which had already floated past the barges. Fortunately, the oil did not reach the Gulf coast. (Courtesy U.S. Geological Survey.)*

ple tried to clean the oil-covered birds, but only 10 percent of 1700 birds treated survived. In 1970, fire broke out in eight offshore wells in the Gulf of Mexico, and 800,000 gallons of oil were released into the water. Fortunately, in this instance the wind and currents kept the oil from the beaches and wildlife refuges along the shore.

31-3
THE COST OF CLEAN WATER

As in the case of determining the cost of clean air (Chapter 26), determining the cost of clean water is also an insurmountable task. Millions of dollars are spent each year in the treatment of municipal water supplies to make the water safe for human consumption. However, this amount is minor compared to the monetary expenditures needed to alleviate the water pollution problems discussed in the previous section. These problems are interwoven with land management and waste disposal. The sheer magnitude of these factors alone makes the analysis of even the indirect costs of clean water enormously complex. Monetary evaluations cannot be placed on the effects of water pollution on the environment and on human health.

Because of the large variety of water pollutants, many measures are required to ensure clean water. Specific treatment may be required for some pollutants, particularly toxic industrial wastes. Space limitations in this text do not permit a detailed consideration of the various methods of pollution prevention and water treatment, so we will consider only the general aspects of two major problems: land management and sewage treatment.

■ *Land Management* With proper planning, the unnatural loads of sediment and silt in our rivers and streams can be prevented. The material from natural erosion processes is part of a balanced environment, but the activities of human beings often upset this balance, as previously pointed out. Human beings disturb the land for their agricultural and other domestic needs, such as construction, and obtaining natural resources.

Proper land management can virtually eliminate stream-fouling sediment and other by-products. For example, **contour farming** prevents the erosion of valuable top soil and increases food production without causing environmental problems (Fig. 31-3a). Perhaps it would be somewhat easier to prepare a hillside for cultivation by going up and down the hill instead of around as in the case of contour farming. However, the farmer would increase erosion resulting from the natural water runoff. Planted in contours the crops hold the soil in place and prevent it from being washed away into streams and rivers. In housing developments, similar considerations of proper land management prevent erosion.

A great deal of water pollution comes from coal-mining operations. Nearby streams receive an overload of sediment from surface strip mines. Some 32 million acres of land in the United States have been disturbed by surface mining operations. Of this figure, one-half of the acreage needs no treatment to prevent sediment and other damage to adjacent land and water. However, the remainder is subject to various degrees of erosion and does need treatment.

Strip mined land can be put back into productive usefulness through **reclamation.** For reclamation to be effective the mounds of spoil need to be leveled and graded and sometimes treated so that trees, shrubs, and grass can take root (Fig. 31-3b). A recent government project which properly reclaimed 650 acres of stripped land cost an average of $1650 per acre.

Sewage Treatment
Sewage constitutes a large part of water pollution and is an important consideration because of potential health hazards from organic wastes. How is sew-

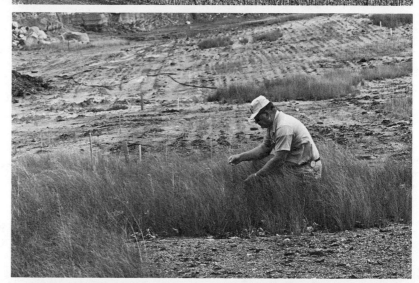

Figure 31-3a *Land management. Through proper land management, such as the contour strip farming shown here, soil erosion from water runoff can be prevented. (Courtesy USDA.)*

Figure 31-3b *Reclamation. (Top) The checking of weeping lovegrass on a heavily limed test plot on a strip mine spoil in White Oak Mountain, West Virginia. The highly acid soil originally had a pH of 2.8. The bare spots were seeded, but not limed. Vegetation is needed to prevent erosion as shown in the background. (Bottom) Properly reclaimed land can be put back into productive use. (Courtesy USDA and Consolidated Coal Co.)*

age handled and how is it treated to avoid pollution problems? In rural areas vaults and septic tanks are the normal means of waste disposal and treatment in rural homes. These consist of storage containers in the ground to which water and human wastes flow or are deposited. Bacterial action breaks down the organic matter and the water seeps into the ground directly or through subsurface leach beds. However, the lifetimes of these systems are finite since the soil eventually becomes clogged with nonbiodegradable particulate matter. Also, the septic tank-leach bed is not a good system in areas of heavy clay soils.

As the water percolates through the ground toward the water table it is further purified by natural processes; however, proper planning is required so that the effluent does not seep into and contaminate nearby sources of drinking water.

In cities and municipalities, sewers are used to collect waste water (and organic contents) and deliver it to treatment plants for processing before it is discharged into oceans, lakes or rivers. Some city dwellers may be unaware of how this is done, but all the householders are intimately aware of the sewage assessments in terms of the bills they must pay. There are two general types of sewer systems in use—combined and separated systems. The **combined sewer system** uses the same network as both a sanitary and storm sewer. Many older cities and towns have a combined system because the original sewer installation was for storm and rain water runoff. Human waste disposal was usually an individual home problem and handled as described previously. However, when the population became sufficiently large that water supplies were threatened with contamination, the original storm sewer system was pressed into combined service. With small populations, the receiving streams sufficiently diluted the sewage, and microorganisms consumed or converted the sewage into less noxious products. Large populations, however, produced excessive sewage loads and upset and impaired this process. Today, sewage treatment plants have been installed in most cities, but some raw sewage still finds its way into our waterways.

Combined sewers usually have overflow bypasses, which divert part of the load during a rainstorm when the water volume is much greater than usual. This safeguards treatment plants from overload, but the part of the water bypassed usually flows directly into the receiving streams and, of course, includes amounts of human wastes.

Separated sewer systems provide separate networks for sanitation and storm water. The storm sewer water does not require treatment and may be run directly into the receiving system. The sanitary sewer runs to the treatment plant.

Through applied technology, waste treatment plants essentially speed up the natural processes by which water purifies itself. The two basic methods of treating ordinary municipal wastes are called primary and secondary treatments. In the **primary sewage treatment** there are four main steps, as illustrated in Fig. 31-3c. First, the large objects are removed by passing the effluent through screens with openings of a fraction of an inch. This debris is usually removed from the screens and discarded, but some plants have a disintegrator that grinds or breaks the coarse material into smaller pieces. Beyond the first stage screening process is a grit chamber where sand, gravel, and heavy coarse materials settle out. The sediment of the grit chamber may be used for landfill or construction purposes. The grit chamber stage is particularly important in combined sewer systems where the grit load is heavy.

The third step of primary treatment is a sedimentation tank in which suspended solids are removed. This sediment is known as raw sludge and has a high organic content. In some plants, the raw sludge is dried and sold or given away as fertilizer. In more sophisticated plants the sludge is put into a heated digestion tank where bacteria digest the solids, producing gaseous by-products. About two-thirds of the gas is methane which may be sold or used as fuel in generating electrical power for the treatment plant. The final step is the chlorination of the remaining liquid. Chlorination kills disease-causing bacteria and also reduces odors. The product of this primary treat-

Figure 31-3c *A diagram of the four steps of primary sewage treatment. (1) Large objects are removed by screens. (2) Heavier materials settle out in a grit chamber. (3) Suspended solids are removed in a sedimentation tank. (4) Bacteria and odors are destroyed by chlorination. (Courtesy EPA.)*

ment is then placed into the receiving steam which dilutes and carries it away.

About 30 percent of the cities in the United States give only primary treatment to their sewage. However, in dealing with sewage in large amounts and with a variety of wastes, secondary treatment is recommended. **Secondary sewage treatment** removes up to 90 percent of the suspended organic matter. This is accomplished by adding an additional step to the primary treatment between the sedimentation tank and the chlorination. This secondary step is done by either of two processes—trickling filters and activated sludge.

In the **trickling filter process** the sewage effluent is trickled through a bed of stones several feet deep. A trickle filter bed is shown in Fig. 31-3d. Bacteria gather and multiply on the stones until they have the capacity to consume most of the organic matter in the sewage effluent.

The current trend favors the use of an **activated sludge process** for the secondary treatment step. In this process, the sewage effluent passes into an aeration tank and then into a large container where it mixes with sludge heavily laden with bacteria (Fig. 31-3e). The aeration oxygenates the effluent and speeds up the bacterial breakdown of the organic

Figure 31-3d *Sewage treatment plant with trickle bed filter. Bacteria gather and multiply on the stones until they have the capacity to consume most of the organic matter in the sewage effluent. (Courtesy USDA-SCS.)*

Figure 31-3e *A diagram of the activated sludge process for secondary sewage treatment. Bacteria in the sludge break down the organic matter in the sewage effluent not removed by the primary treatment. The aeration oxygenates the effluent-sludge mixture to replace the oxygen used in the bacterial action and thus speeds up the process. (Redrawn courtesy EPA.)*

Figure 31-3f *The ultimate goal—unpolluted water. (Courtesy EPA.)*

matter. Though more expensive, the activated sludge process is more efficient and is a closed system which reduces odors.

The sewage treatment methods just described are generally adequate for normal sewage. However, other treatments are necessary for sewage with complex industrial wastes. Some treatments may be specific for a particular type of chemical wastes that cannot be broken down by bacterial action (nonbiodegradable) of conventional methods. Precipitation and coagulation methods have been used for years to treat industrial wastes. Chemicals are added to precipitate dissolved materials or to coagulate fine suspended particles. The precipitate or coagulate is then allowed to settle out in a sedimentation tank. Another special treatment for organic matter that resists normal bacterial action is adsorption. The waste effluent is passed through beds of activated charcoal which adsorbs as much as 98 percent of the organic material. It is obvious that all of these treatment methods add to the cost of clean water (Fig. 31-3f).

IMPORTANT WORDS AND TERMS

Land pollution
Solid waste
Litter
Biodegradable
Open dump
Sanitary landfill
Acid mine drainage
Water pollution
Pesticides
Fertilizers

Phosphates
Nitrates
Mercury pollution
Thermal pollution
Radiation pollution
Nuclear radiation
Radiation cell damage
Cancer
Leukemia
Detergent

Detergent builders
Eutrophication
Contour farming
Reclamation
Combined sewer system
Separated sewer system
Primary sewage treatment
Secondary sewage treatment
Trickling filter process
Activated sludge process

LEARNING OBJECTIVES

After reading and studying this chapter, you should be able to do the following without referring to the text:

1. Describe the different types of land pollution.
2. State the major sources of solid waste production in the United States.
3. Define biodegradable.
4. Distinguish between deep and surface mining.
5. Explain acid mine drainage.
6. State the sources and types of water pollution.
7. Explain eutrophication and the effects of pollution.
8. Discuss land management.
9. Distinguish between (a) combined and separated sewer systems, and (b) primary and secondary sewage treatments.

QUESTIONS

1. What is one of the chief sources of land pollution?
2. What is the greatest source of solid waste in the United States?
3. Approximately how much domestic solid waste is generated per capita in the United States?
4. Distinguish between open dumps and sanitary landfills.
5. What is the potential pollution danger of sanitary landfills?
6. How have human activities given rise to land and water pollution?
7. What is acid mine drainage? Describe its cause, source, and effects.
8. State the sources and types of water pollution.
9. What are the effects of pesticide pollution?
10. How do fertilizers contribute to water pollution?

11. Why are animal wastes a current potential pollution problem?
12. What are the effects of mercury poisoning?
13. What is thermal pollution, and what are its sources and effects?
14. Describe the effects of radiation damage on living cells.
15. What are the major sources of radiation pollution?
16. What are detergents and how do they contribute to water pollution?
17. What are detergent builders?
18. Describe the causes and effects of eutrophication due to pollution.
19. What is the composition of sewage?
20. Describe the sources and effects of oil spills.
21. Give some examples of land management that prevent pollution.
22. What is the difference between combined and separated sewer systems?
23. What are the four main steps in primary sewage treatment?
24. How does secondary sewage treatment differ from primary sewage treatment?
25. Describe the operation of trickling filters and the activated sludge process.

Appendices

Appendix I

THE SEVEN BASE UNITS OF THE INTERNATIONAL SYSTEM OF UNITS (SI)

Meter-m (length)

The meter is defined as 1,650,763.73 wavelengths of the orange-red line of the spectrum of krypton-86.

Kilogram-kg (mass)

The kilogram is a cylinder of platinum-iridium alloy kept by the International Bureau of Weights and Measures, at Paris. A duplicate in the custody of the National Bureau of Standards serves as the mass standard for the United States. This is the only base unit still defined by an artifact.

Second-s (time)

The second is defined as the duration of 9,192,631,770 cycles of the radiation associated with a specified transition of the cesium-133 atom.

Ampere-A (electric current)

The ampere is defined as that current which if maintained in each of two long parallel wires separated by one meter in free space, would produce a force between the two wires (due to their magnetic fields) of 2×10^{-7} newton for each meter of length.

Kelvin-K (temperature)

The kelvin is defined as the fraction 1/273.16 of the thermodynamic temperature of the triple point of water. The temperature 0°K is called "absolute zero."

Mole-mol (amount of substance)

The mole is the amount of substance of a system that contains as many elementary entities as there are atoms in 0.012 kilogram of carbon-12.

Candela-cd (luminous intensity)

The candela is defined as the luminous intensity of 1/600,000 of a square meter of a blackbody at the temperature of freezing platinum (2045°K).

Appendix II

FUNDAMENTAL CONSTANTS

$g = 9.8 \text{ m/s}^2$
$\quad = 980 \text{ cm/s}^2$
$\quad = 32 \text{ ft/s}^2$

$G = 6.67 \times 10^{-11} \text{ N} \cdot \text{m}^2/\text{kg}^2$
$\quad = 6.67 \times 10^{-8} \text{ dyne} \cdot \text{cm}^2/\text{g}^2$

$q = 1.6 \times 10^{-19} \text{ coulomb}$

$0°\text{K} = -273°\text{C}$

$c = 3.0 \times 10^8 \text{ m/s}$
$\quad = 3.0 \times 10^{10} \text{ cm/s}$
$\quad = 1.86 \times 10^5 \text{ mi/s}$

$h = 6.625 \times 10^{-27} \text{ erg} \cdot \text{s}$

Avogadro's number $= 6.02 \times 10^{23}$
molecules/mole

$m_{proton} = 1.677 \times 10^{-27} \text{ kg}$
$\qquad = 1.677 \times 10^{-24} \text{ g}$
$\qquad = 1.00783 \text{ amu}$

$m_{neutron} = 1.674 \times 10^{-27} \text{ kg}$
$\qquad = 1.674 \times 10^{-24} \text{ g}$
$\qquad = 1.00867 \text{ amu}$

$m_{electron} = 9.11 \times 10^{-31} \text{ kg}$
$\qquad = 9.11 \times 10^{-28} \text{ g}$

CONVERSION FACTORS

Mass
1 gram $= 10^{-3} \text{ kg} = 6.85 \times 10^{-5}$ slug
1 kg $= 10^3 \text{ g} = 6.85 \times 10^{-2}$ slug
1 slug $= 1.46 \times 10^4 \text{ g} = 14.6$ kg
1 amu $= 1.66 \times 10^{-24} \text{ g} = 1.66 \times 10^{-27}$ kg

Length
1 cm $= 10^{-2} \text{ m} = 0.394$ in
1 m $= 10^{-3} \text{ km} = 3.28 \text{ ft} = 39.4$ in
1 km $= 10^3 \text{ m} = 0.62$ mi
1 in $= 2.54 \text{ cm} = 2.54 \times 10^{-2}$ m
1 ft $= 12 \text{ in} = 30.48 \text{ cm} = 0.3048$ m
1 mi $= 5280 \text{ ft} = 1609 \text{ m} = 1.609$ km
1 Å $= 10^{-10} \text{ m} = 10^{-8}$ cm
1 parsec $= 3.26$ light years $= 205,265$ a.u.

Time
1 h $= 60 \text{ min} = 3600$ s
1 day $= 24 \text{ h} = 1440 \text{ min} = 8.64 \times 10^4$ s
1 year $= 365 \text{ days} = 8.76 \times 10^3$ h
$\quad = 5.26 \times 10^5 \text{ min} = 3.16 \times 10^7$ s

Speed
1 m/s $= 3.6 \text{ km/h} = 3.28 \text{ ft/s} = 2.24$ mi/h
1 km/h $= 0.278 \text{ m/s} = 0.621 \text{ mi/h} = 0.911$ ft/s
1 ft/s $= 0.682 \text{ mi/h} = 0.305 \text{ m/s} = 1.10$ km/h
1 mi/h $= 1.467 \text{ ft/s} = 1.609 \text{ km/h} = 0.447$ m/s
60 mi/h $= 88$ ft/s

Force
1 newton $= 10^5 \text{ dynes} = 0.225$ lb
1 dyne $= 10^{-5} \text{ N} = 2.25 \times 10^{-6}$ lb
1 lb $= 4.45 \times 10^5 \text{ dynes} = 4.45$ N
Equivalent weight of 1 kg mass $= 2.2$ lb
$\qquad\qquad\qquad\qquad\quad = 9.8$ N

Pressure
1 atm $= 14.7 \text{ lb/in}^2 = 1.013 \times 10^5 \text{ N/m}^2$
$\quad = 1.013 \times 10^6 \text{ dyne/cm}^2$
$\quad = 30 \text{ in Hg} = 76 \text{ cm Hg}$
1 bar $= 10^6 \text{ dyne/cm}^2$
1 millibar $= 10^3 \text{ dyne/cm}^2$

Energy   1 joule $= 10^7$ erg $= 0.738$ ft $\cdot$ lb
$= 0.239$ cal $= 9.48 \times 10^{-4}$ Btu
$= 6.24 \times 10^{18}$ eV
1 kcal $= 4186$ J $= 4.186 \times 10^{10}$ erg $= 3.968$ Btu
1 Btu $= 1055$ J $= 1.055 \times 10^{10}$ erg
$= 778$ ft $\cdot$ lb $= 0.252$ kcal
1 cal $= 4.186$ J $= 3.97 \times 10^{-3}$ Btu
$= 3.09$ ft $\cdot$ lb
1 ft $\cdot$ lb $= 1.36$ J $= 1.36 \times 10^7$ erg
$= 1.29 \times 10^{-3}$ Btu
1 eV $= 1.60 \times 10^{-19}$ J $= 1.60 \times 10^{-12}$ erg

Power   1 watt $= 0.738$ ft $\cdot$ lb/s $= 1.34 \times 10^{-3}$ hp
$= 3.41$ Btu/h
1 ft $\cdot$ lb/s $= 1.36$ W $= 1.82 \times 10^{-3}$ hp
1 hp $= 550$ ft $\cdot$ lb/s $= 745.7$ watt
$= 2545$ Btu/h

SOLAR SYSTEM DATA

Radius of earth		Mass of moon	7.4×10^{22} kg $= \frac{1}{81}$ mass of earth
equatorial	3963 mi $= 6.378 \times 10^6$ m	Mass of sun	2.0×10^{30} kg
polar	3950 mi $= 6.357 \times 10^6$ m	Average distance	
Radius of moon	1080 mi $= 2.689 \times 10^6$ m	of earth from sun	93×10^6 mi $= 1.5 \times 10^8$ km
Radius of sun	432,500 mi $= 6.951 \times 10^8$ m	Average distance	
Mass of earth	6.0×10^{24} kg	of moon from earth	2.4×10^5 mi $= 3.8 \times 10^5$ km

Appendix III

PSYCHROMETRIC TABLES (Pressure: 30 in Hg)

Table 1 *Relative Humidity (%) and Maximum Moisture Capacity**

Air Temp. (°F) (dry bulb)	Max. Moisture Capacity (gr/ft³)	Degrees Depression of Wet-Bulb Thermometer (°F)													
		1	2	3	4	5	6	7	8	9	10	15	20	25	30
25	1.6	87	74	62	49	37	25	13	1						
30	1.9	89	78	67	56	46	36	26	16	6					
35	2.4	91	81	72	63	54	45	36	27	19	10				
40	2.8	92	83	75	68	60	52	45	37	29	22				
45	3.4	93	86	78	71	64	57	51	44	38	31				
50	4.1	93	87	80	74	67	61	55	49	43	38	10			
55	4.8	94	88	82	76	70	65	59	54	49	43	19			
60	5.7	94	89	83	78	73	68	63	58	53	48	26	5		
65	6.8	95	90	85	80	75	70	66	61	56	52	31	12		
70	7.8	95	90	86	81	77	72	68	64	59	55	36	19	3	
75	9.4	96	91	86	82	78	74	70	66	62	58	40	24	9	
80	10.9	96	91	87	83	79	75	72	68	64	61	44	29	15	3
85	12.7	96	92	88	84	80	76	73	69	66	62	46	32	20	8
90	14.8	96	92	89	85	81	78	74	71	68	65	49	36	24	13
95	17.1	96	93	89	85	82	79	75	72	69	66	51	38	27	17
100	19.8	96	93	89	86	83	80	77	73	70	68	54	41	30	21
105	23.4	97	93	90	87	83	80	77	74	71	69	55	43	33	23
110	26.0	97	93	90	87	84	81	78	75	73	70	57	46	36	26

*To use table, determine air temperature with dry-bulb thermometer and degrees depressed on wet-bulb thermometer. Read maximum capacity directly. Read relative humidity (in percent) opposite and below these values.

Table 2 *Dew Point (°F)**

Air Temp. (°F) (dry bulb)	Degrees Depression of Wet-Bulb Thermometer (°F)													
	1	*2*	*3*	*4*	*5*	*6*	*7*	*8*	*9*	*10*	*15*	*20*	*25*	*30*
25	22	19	15	10	5	−3	−15	−51						
30	27	25	21	18	14	8	2	−7	−25					
35	33	30	28	25	21	17	13	7	0	−11				
40	38	35	33	30	28	25	21	18	13	7				
45	43	41	38	36	34	31	28	25	22	18				
50	48	46	44	42	40	37	34	32	29	26	0			
55	53	51	50	48	45	43	41	38	36	33	15			
60	58	57	55	53	51	49	47	45	43	40	25	−8		
65	63	62	60	59	57	55	53	51	49	47	34	14		
70	69	67	65	64	62	61	59	57	55	53	42	26	−11	
75	74	72	71	69	68	66	64	63	61	59	49	36	15	
80	79	77	76	74	73	72	70	68	67	65	56	44	28	−7
85	84	82	81	80	78	77	75	74	72	71	62	52	39	19
90	89	87	86	85	83	82	81	79	78	76	69	59	48	32
95	94	93	91	90	89	87	86	85	83	82	74	66	56	43
100	99	98	96	95	94	93	91	90	89	87	80	72	63	52
105	104	103	101	100	99	98	96	95	94	93	86	78	70	61
110	109	108	106	105	104	103	102	100	99	98	91	84	77	68

*To use table, determine air temperature with dry-bulb thermometer and degrees depressed on wet-bulb thermometer. Find dew point opposite and below these values.

Appendix IV

RULES AND EXAMPLES OF THE FOUR BASIC MATHEMATICAL OPERATIONS USING POWERS OF TEN NOTATION

1. In addition or subtraction, only like terms can be used; that is, the exponent for the number 10 must be of the same value. Thus,

$$\begin{array}{r} 4.6 \times 10^{-8} \\ +1.2 \times 10^{-8} \\ \hline 5.8 \times 10^{-8} \end{array} \quad \text{and} \quad \begin{array}{r} 4.8 \times 10^{-8} \\ -2.5 \times 10^{-8} \\ \hline 2.3 \times 10^{-8} \end{array}$$

2. In multiplication, the exponents are added. Thus,

$$2 \times 10^4 \text{ multiplied by } 4 \times 10^3 = 8 \times 10^7$$
$$1.2 \times 10^{-2} \text{ multiplied by } 3 \times 10^6 = 3.6 \times 10^4$$

3. In division, the exponents are subtracted. Thus,

$$\frac{4.8 \times 10^8}{2.4 \times 10^2} = 2.0 \times 10^6$$

and

$$\frac{3.4 \times 10^{-8}}{1.7 \times 10^{-2}} = 2.0 \times 10^{-6}$$

4. A power of ten may be transferred from the numerator to the denominator, or vice versa, by changing the sign of the exponent. Thus,

$$\frac{3.4 \times 10^{-8}}{1.7 \times 10^{-2}} = \frac{3.4 \times 10^{-8} \times 10^2}{1.7} = 2.0 \times 10^{-6}$$

Appendix V

PROBLEM SOLVING: FIVE MAJOR STEPS

1. Read the problem carefully to find out what is given and what is inferred.
2. Determine the unknown or unknowns; that is, what are you looking for?
3. Apply the mathematical relationship that connects the unknown quantity or quantities with the given data.
4. Check data for correct units.
5. Solve mathematically for the unknown value or values.

  ▪ *Example 1* A ball is thrown upward from the ground with an original velocity of 128 ft/s. How long does the ball remain in the air? What was the maximum height attained by the ball?

Step 1. Write down the given data:

$$v_o = 128 \text{ ft/s}$$

and inferred data:
  (acceleration) $a = -32$ ft/s^2 (The minus sign is used because v_o and the acceleration are in opposite directions.)
  (final velocity) $v_f = 0$ at maximum height.

Step 2. Write down the unknown term or terms:
$t = ?$ seconds (total time the ball is in the air)
$s = ?$ feet (maximum height or displacement of ball).
  Since the original velocity is in ft/s, find the unknown time in seconds and the unknown height in feet.

Step 3. The terms v_o, v_f, and a are known; therefore, use the relation

$$a = \frac{v_f - v_o}{t}$$

which contains these terms, plus the time, which is unknown.

Step 4. All the quantities used in solving a problem should be in the same system of units. The units of the result will then also be in that system. In this example, the units are in the British system, hence the height will be in feet (Step 5). Units are often mixed in a problem. For example, velocity may be given in feet per second while an acceleration may be given in meters per second². One of these must then be converted before being used in an equation.

Step 5. Solve the equation for t, and obtain

$$t = \frac{v_f - v_o}{a}$$

Substitute the given data and solve to obtain

$$t = \frac{0 - 128 \text{ ft/s}}{-32 \text{ ft/s}^2}$$

$t = 4$ s (time for ball to reach maximum height)

The total time in the air is twice this value since the ball will require the same amount of time to return to the ground.

After you know the value of t, v_o, v_f, and a, the unknown height can be determined by

$$s = v_o t + \frac{at^2}{2}$$

Since v_o at the top of the ball's flight $= 0$

$$s = \frac{at^2}{2}$$

Since a in this case is equal to g, which is 32 ft/s², we can substitute the data and find

$$s = \frac{32 \text{ ft/s}^2 \times 4^2 \text{ s}^2}{2}$$

therefore

$$s = \frac{512}{2}$$

or

$$s = 256 \text{ ft}$$

In solving problems in grade school, you learned not to add apples to oranges, because doing so gives an answer that has no meaning. In physical science, you must remember that only like terms can be added or subtracted. When you have a mathematical relationship (i.e., an equation), terms on the left side of the equals sign must be equal to the terms on the right side, not only in numerical value, but also in units. For example, in the equation

$$a = \frac{v_f - v_o}{t}$$

both sides of the equation must have like dimensions. A **dimensional analysis** of an equation is done by substituting the fundamental units of length, mass, and time in the equation for the given terms.

■ *Example 2* We can use dimensional analysis to show the above is an equality. Although it has no mass term, the expression has terms that can be traced back to the fundamental units of length and time. Therefore, by substituting

$$\frac{L}{t^2} = \frac{\frac{L}{t} - \frac{L}{t}}{t}$$

Solving just for the numerator on the right, and remembering that we are subtracting units rather than numerical or algebraic values, we obtain

$$\frac{L}{t} - \frac{L}{t} = \frac{L}{t}$$

The top expression now can be written

$$\frac{L}{t^2} = \frac{\dfrac{L}{t}}{t}$$

Using negative exponents to bring our denominators above the line, we find

$$Lt^{-2} = Lt^{-1} \times t^{-1}$$

or

$$Lt^{-2} = Lt^{-2}$$

The left side of the expression has the same dimensions as the right side, showing the original to be an equality. If the two sides were not the same dimensionally, it would not be an equation that would give a correct solution.

Data found in physics problems often must be converted to different units in order to give the answer desired. For instance, measurements for rates of travel may be given in miles per hour, and the answer requested in feet per second. How can we obtain the proper result?

■ *Example 3* What is the distance in feet traveled by an automobile in 10 s if the average speed of the car is 60 miles/hour? Since we want to know the distance traveled in feet, we must convert the mi/h to ft/s. This can be done as follows:

1. Write down the term you want to convert and set it equal to itself. At the same time, make a mental or written note of the terms to which you wish to convert.

$$\frac{60 \text{ mi}}{h} = \frac{60 \text{ mi}}{h}$$

2. Multiply the right side of the equation by 1 or 1/1, which is the same thing.

$$\frac{60 \text{ mi}}{h} = \frac{60 \text{ mi}}{h} \times \frac{1}{1}$$

3. Place a unit in the numerator or denominator of the 1/1 which will cancel out an unwanted unit in the original term. In this case, we would place the unit "mile" in the denominator. At the same time, place an equal term above

our newly introduced denominator unit, in order to keep the value equal to 1/1. Since we are after feet in our answer for this problem, it would be well in this case to place 5280 ft in the numerator of the 1/1, since that is equal to 1 mi. Then cancel like units.

$$\frac{60 \text{ mi}}{h} = \frac{60 \cancel{\text{mi}}}{h} \times \frac{5280 \text{ ft}}{1 \cancel{\text{mi}}}$$

4. The next step is to eliminate the unit hour in the denominator. This can be accomplished by multiplying again by 1/1 and placing the unit hour in the numerator of the 1/1 term. At the same time, an equal unit should be placed in the denominator of the 1/1 term in order to keep the value equal to 1/1. Since 60 minutes equal 1 hour, we can use it, then cancel both "hour" terms.

$$\frac{60 \text{ mi}}{h} = \frac{60 \cancel{\text{mi}}}{\cancel{h}} \times \frac{5280 \text{ ft}}{1 \cancel{\text{mi}}} \times \frac{1 \cancel{h}}{60 \text{ min}}$$

5. At this point, we have feet per minute, but we want feet per second; therefore we must convert the minutes to seconds. Again we multiply by 1 or 1/1 and add the appropriate units for conversion. Since 60 seconds equal 1 minute, we obtain

$$\frac{60 \text{ mi}}{h} = \frac{60 \cancel{\text{mi}}}{\cancel{h}} \times \frac{5280 \text{ ft}}{1 \cancel{\text{mi}}} \times \frac{1 \cancel{h}}{60 \cancel{\text{min}}} \times \frac{1 \cancel{\text{min}}}{60 \text{ s}}$$

6. After all like units are canceled, the numbers are canceled insofar as is possible; we arrive at

$$\frac{60 \text{ mi}}{h} = \frac{528 \text{ ft}}{6 \text{ s}}$$

or, after dividing, $\dfrac{60 \text{ mi}}{h} = \dfrac{88 \text{ ft}}{s}$

The problem can now be solved using 88 ft/s for 60 mi/h. Knowing that

$$s = \bar{v}t$$

and substituting $\quad s = \dfrac{88 \text{ ft}}{s} \times 10 \text{ s}$

we obtain $\quad s = 880 \text{ ft}$

DERIVATION OF DISPLACEMENT EQUATION

The displacement of an object moving with constant acceleration can be determined by

$$s = v_o t + \frac{at^2}{2}$$

This relationship is derived as follows:

$$\bar{v} = \frac{s}{t}$$

therefore,

$$s = \bar{v} t$$

We know that

$$\bar{v} = \frac{v_f + v_o}{2}$$

when the acceleration is constant. Substituting, we obtain

$$s = \left(\frac{v_f + v_o}{2}\right) t$$

Knowing that $v_f = v_o + at$ we may substitute for v_f and obtain

$$s = \left(\frac{v_o + at + v_o}{2}\right) t$$

or

$$s = \left(\frac{2v_o + at}{2}\right) t$$

or

$$s = v_o t + \frac{at^2}{2} \qquad \text{Eq. A.1}$$

The simplest way to obtain an equation of motion which does not include time is to take two equations that include time and eliminate the time factor by cancellation. By definition

$$a = \frac{v_f - v_o}{t}$$

From the derivation of Eq. A.1 we know that

$$s = \left(\frac{v_f + v_o}{2}\right) t$$

Multiplying these equations

$$\left(a = \frac{v_f - v_o}{t}\right) \times \left[s = \left(\frac{v_f + v_o}{2}\right) t\right]$$

and canceling the time, we obtain

$$as = (v_f - v_o)\left(\frac{v_f + v_o}{2}\right)$$

or

$$2as = v_f^2 - v_o^2 \qquad \text{Eq. A.2}$$

DERIVATION OF WORK–KINETIC ENERGY RELATIONSHIP

In Chapter 4 we use the fact that a change in kinetic energy is related to work done against inertia. This equation can be derived as follows: First, we write the equation for the work done by an unbalanced force in accelerating a mass through a horizontal distance.

$$W = Fs \qquad \text{Eq. A.3}$$

The unbalanced force F, obtained from Newton's second law, is

$$F = ma$$

Substituting in Eq. A.3 for F, we have

$$W = mas \qquad \text{Eq. A.4}$$

Acceleration has been defined as

$$a = \frac{v_f - v_o}{t}$$

Substituting in Eq. A.4 for a, we have

$$W = m\left(\frac{v_f - v_o}{t}\right)s \qquad \text{Eq. A.5}$$

The distance traveled can be obtained from

$$s = \bar{v}t$$

or

$$s = \left(\frac{v_f + v_o}{2}\right)t$$

Substituting in Eq. A.5 for s, we obtain

$$W = m\left(\frac{v_f - v_o}{t}\right) \times \left(\frac{v_f + v_o}{2}\right)t$$

Canceling the t, we have

$$W = m\left(\frac{v_f - v_o}{1}\right) \times \left(\frac{v_f + v_o}{2}\right)$$

Multiplying $(v_f - v_o)$ by $(v_f + v_o)$ we have

$$W = \left(\frac{m}{2}\right)(v_f^2 - v_f v_o + v_f v_o - v_o^2)$$

or

$$W = m\left(\frac{v_f^2 - v_o^2}{2}\right)$$

or

$$W = \left(\frac{mv_f^2}{2}\right) - \left(\frac{mv_o^2}{2}\right)$$

which is the equation used in Chapter 4.

Appendix VI

DERIVATION OF BOHR'S EQUATIONS

Niels Bohr's quantization of angular momentum led to two useful equations (see Chapter 10, Eqs. 10.2 and 10.3). These equations are derived as follows:

Bohr's set of equations were

$$(1) \quad \frac{kq^2}{r^2} = \frac{mv^2}{r}$$

$$(2) \quad E = -\frac{kq^2}{2r}$$

and

$$(3) \quad mvr = \frac{nh}{2\pi}$$

The third equation can be restated as

$$v = \frac{nh}{2\pi mr}$$

Then, substituting this in the first yields

$$k\frac{q^2}{r^2} = \frac{m}{r}(v^2)$$

$$k\frac{q^2}{r^2} = \frac{m}{r}\left(\frac{n^2h^2}{4\pi^2 m^2 r^2}\right)$$

or

$$r = \left[\frac{h^2}{kq^2 4\pi^2 m}\right]n^2 \qquad \text{Eq. A.6}$$

We can now solve for E

$$E = -\frac{kq^2}{2}\left(\frac{1}{r}\right)$$

$$E = -\frac{kq^2}{2}\left(\frac{kq^2 4\pi^2 m}{h^2 n^2}\right)$$

$$E = -\left[\frac{k^2 q^4 4\pi^2 m}{2h^2}\right]\frac{1}{n^2} \qquad \text{Eq. A.7}$$

The quantities in brackets in Eq. A.6 and A.7 are all constants and can be evaluated. The results are Eqs. 10.2 and 10.3:

$$r = 0.529n^2 \text{ angstrom}$$

$$E = \frac{-13.6}{n^2} \text{ eV}$$

Glossary

Abyssal plain a large flat area on the ocean floor where layers of sediment have covered the original sea floor topography.

Acceleration the change in velocity divided by the change in time: $a = \Delta v / \Delta t$.

Acceleration of gravity usually given as the symbol g; equal to 32 ft/s^2, 980 cm/s^2, or 9.8 m/s^2.

Acid a substance which acts as a proton donor.

Acid mine drainage water drainage from mining areas that contain sulfuric acid due to the reaction of air (oxygen) and water with sulfur-bearing minerals.

Actinides the 14 elements following actinium in the periodic table—the second inner transition series.

Activated sludge process a sewage treatment process in which aerated sewage effluent is mixed with bacteria-laden sludge for the removal of organic matter.

Activation energy the energy necessary to get a chemical reaction started.

Adiabatic a process in which no heat is added or removed from the system.

Advection fog a fog formed as a result of convectional air movement.

Air current vertical air movement.

Air-fuel ratio the ratio of air to fuel mixture that is important in the operation of the gasoline internal combustion engine.

Air mass a mass of air with physical characteristics that distinguish it from other air.

Air pollution any atypical contributions to the atmosphere resulting from human activities.

Air pollution potential weather conditions that are favorable for potential pollution conditions, e.g., stagnation conditions as a result of a temperature inversion.

Albedo the reflectivity or average fraction of light a body reflects.

Alcohol any compound that contains an [OH] group and a hydrocarbon group.

Aldehydes any compound that contains a —C=O group.
 |
 H

Alkanes hydrocarbons which have a composition that satisfies the general formula C_nH_{2n+2}.

Alkenes a hydrocarbon containing a double bond. The general formula for the alkenes is C_nH_{2n}.

Alkynes a hydrocarbon containing a triple bond. The general formula for the alkynes is C_nH_{2n-2}.

Alpha decay the disintegration of a nucleus into an alpha particle ($_2^4$He nucleus) and the nucleus of another element.

Alpha particle the nucleus of a helium-4 atom ($_2^4$He).

Altitude the angle measured from the horizon to a celestial object.

Alto the prefix associated with the middle cloud family.

Alveoli tiny air spaces, or sacs, in the lungs.

Amino acids compounds that contain both —NH$_2$ and —COOH groups.

Ampere the unit of electric current defined as that current which, if maintained in each of two long parallel wires separated by one meter in free space, would produce a magnetic force between the two wires of 2×10^{-7} newton for each meter of length.

Amplitude the maximum displacement of a wave from its equilibrium position.

amu one atomic mass unit—$\frac{1}{12}$ the mass of the ^{12}C isotope.

Anemometer an instrument used to measure wind speed.

Angstrom a unit of length equal to 10^{-8} cm.

Angular momentum mvr for a mass m going at a speed v in a circle of radius r.

Anion a negatively charged ion.

Annular eclipse an eclipse of the sun in which the moon blocks out all of the sun except for a ring around the outer edge of the sun.

Anticyclone a high-pressure area characterized by clockwise air circulation (in the northern hemisphere).

Aquifer a body of permeable rock through which ground water moves.

Artesian wells and springs wells and springs in which water rises above an aquifer. The pressure due to gravity can cause the water to bubble or spurt onto the surface.

Asteroids large chunks of matter that orbit the sun (usually between Mars and Jupiter) and which are too small to be labeled as planets.

Asthenosphere the rocky substratum below the lithosphere that is hot enough to be deformed and capable of internal flow.

Astronomical unit the mean distance between the earth and the sun, 93,000,000 mi.

Atmosphere the gases that surround a planet or celestial object.

Atom the smallest particle of an element that can enter into a chemical combination.

Atomic number the number of protons in the atom.

Atomic time scale a geologic time scale based on radioactive dating.

Atomic weight the weight of an average atom of an element with reference to the ^{12}C isotope. ^{12}C is 12 exactly.

Autumnal equinox the time (near September 21) when the sun's declination crosses the equator moving south (for the northern hemisphere).

Avogadro's law equal volumes of gases at the same temperature and pressure contain equal number of molecules.

Avogadro's number the number of molecules in a mole of gas, liquid or solid—6.02×10^{23} molecules per mole.

Btu the amount of heat required to raise one pound of water one degree Fahrenheit at normal atmospheric pressure.

Barometer a device used to measure pressure, viz., atmospheric pressure.

Basaltic lava lava with a low silica content and low viscosity.

Base a substance that acts as a proton acceptor.

Batholith a large intrusive igneous rock formation that has an area of at least 40 mi^2.

Bedding the stratification of sedimentary rock formations.

Beta decay the disintegration of a nucleus into a beta particle (electron) and the nucleus of another element.

Beta particle an electron.

Big Bang Theory theory of the universe that states that the known universe was concentrated in a massive glob of extremely dense material which exploded approximately 15 billion years ago.

Binding energy total binding energy—the amount of energy necessary to completely separate the protons and neutrons of the nucleus of an atom.

Biodegradable capable of being broken down or degraded by bacterial action.

Black hole a very dense collapsed star from which no light can escape.

Blizzard a snow storm accompanied by high winds that whip the snow into blinding swirls and drifts.

Boiling point the temperature at which a substance changes from the liquid to the gas phase.

Boyle's gas law the volume of a perfect gas varies inversely as the absolute pressure, if the temperature remains constant.

British system the system of measurement used in the United States with the foot as the length standard, the pound as the weight standard, the second as the time standard, and the coulomb as the electric charge standard.

Builders substances added to a detergent to make its cleansing action more efficient. Phosphates are common detergent builders.

cgs system the metric system that has the centimeter, gram, and second as the standard units of length, mass, and time, respectively, and the coulomb as the standard of electric charge.

Caldera a roughly circular, steep-walled depression formed as a result of the collapse of a volcanic chamber.

Calorie the amount of heat necessary to raise one gram of pure liquid water one degree Celsius at normal atmospheric pressure.

Cancer a condition of unregulated cell growth rate.

Carbohydrate compounds composed of carbon, hydrogen, and oxygen with the hydrogen-to-oxygen ratio usually 2 to 1; sugars and starches are typical.

Carcinogen a cancer-producing agent.

Catalyst a substance that changes the rate of a chemical reaction without undergoing a permanent change itself.

Cation a positively charged ion.

Ceilometer an instrument used to measure the heights of clouds by light reflection.

Celestial sphere the imaginary sphere on which all the stars seem to be.

Centripetal force a "center-seeking" force that causes an object to travel in a circle.

Cepheid variables stars that vary in magnitude with a fixed period of between 1 and 100 days.

Chain reaction a self-propagating process in which product neutrons from one fission are used to induce fission in other nuclei.

Charles' gas law the volume of a perfect gas varies in diect proportion to the absolute temperature, if the pressure remains constant.

Chemical properties the properties involved in the transformation of one substance into another.

Chinook the name applied (literally, snow eater) to the warm leeward winds on the eastern slopes of the Rocky Mountains that give rise to the rapid melting of snow. In Europe the term *föhn* is used.

Chromosphere an outer layer of the sun which lies just outside the photosphere.

Cilia tiny, hairlike projections in the respiratory tract.

Cinder cone a volcano with a steeply sloped cinder cone formed by eruptions of pyroclastic debris.

Cirque glacier a small glacier formed in a hollow depression along a mountain.

Cirrus a root name used to describe wispy, fibrous cloud forms.

Clastic rocks sedimentary rocks formed from sediment composed of fragments of preexisting rocks.

Cleavage the splitting of a mineral along an internal molecular plane.

Climate the long-term average weather conditions of a region or the world.

Cloud buoyant masses of visible droplets of water vapor and ice crystals in the lower troposphere.

Coalescence the combining of small droplets of water vapor to make larger drops.

Column the cavern dripstone formation consisting of a joined stalactite and stalagmite.

Combined sewer system a system that uses the same network as both a sanitary and storm sewer.

Comet a chunk of matter that displays a long tail as it passes near the sun and has a highly elliptical orbit.

Composite volcano a volcano with a steeply sloping symmetric cone formed by eruptions of high viscosity lava and pyroclastic debris. Also called a stratovolcano.

Compound a substance composed of two or more elements chemically combined in a definite proportion.

Concave mirror a mirror shaped like the inside of a small section of a sphere.

Conduction the transfer of heat energy by molecular transfer.

Conductor a material that easily conducts an electric current because some electrons in the material are free to move.

Conjunction the time at which a planet and sun occur on the same meridian.

Conservation of angular momentum, law of the angular momentum of an object remains constant unless acted upon by an external torque.

Conservation of energy, law of the total energy of an isolated system remains constant.

Conservation of linear momentum, law of the total linear momentum of a system remains constant if there are no external unbalanced forces acting on the system.

Conservation of mass, law of there is no detectable change in the total mass during a chemical process.

Consolidation the process of forming sedimentary rock from sediment. Also called lithification.

Continental drift the theory that continents move or drift apart.

Continental glacier (ice sheet) a large mass of ice that covers a large surface region and flows outwardly.

Continental shelf the shallowly submerged margin of a continental land mass.

Continental slope the seaward slope beyond the continental shelf that extends downward to the ocean basin.

Convection the transfer of heat through the movement of a substance.

Convection cycle the cyclic movement of matter, e.g., air, due to localized heating and convectional heat transfer.

Convex mirror a mirror shaped like the outside of a small section of a sphere.

Core the innermost region of the earth, which is composed of two parts, a solid inner core and a molten, highly viscous, "liquid" outer core.

Coriolis force a pseudoforce arising in an accelerated reference frame on the rotating (accelerating) earth. The apparent deflection of objects is attributed to the Coriolis force.

Coulomb the unit of electric charge equal to one ampere-second (A · s).

Coulomb's law the force of attraction or repulsion between two charged bodies is directly proportional to the product of the two charges and inversely proportional to the square of the distance between them.

Covalent bond a chemical bond in which electron pairs are shared by atoms.

Covalent compounds compounds formed by the electron-sharing process.

Crater the funnel-shaped depression of the summit of a volcano.

Creep a type of slow mass wasting involving the slow particle-by-particle movement of weathered debris down a slope which takes place year after year.

Crust the thin outer layer of the earth.

Crystal an orderly arrangement of atoms.

Cumulus a root name used to describe billowy, round cloud forms.

Current rate of flow of electric charge.

Cyclone a low-pressure area characterized by counterclockwise air circulation (in the northern hemisphere).

DNA deoxyribonucleic acid, a nucleic acid located primarily in the nucleus of the cell. DNA has the molecular structure of a double-stranded helix.

Declination the overhead position of a celestial object such as the sun measured in degrees latitude.

Definite proportions, law of different samples of a pure compound always contain the same elements in the same proportions by weight.

Degree a unit of angle. There are 360 degrees in a circle.

Density current a deep ocean current that is due to dense water sinking through less dense water.

Detergent a general term for a cleansing agent.

Deuteron the nucleus of a deuterium atom, ($_1^2$H).

Dew point the temperature at which a sample of air becomes saturated, i.e., has a relative humidity of 100 percent.

Diastrophism a geologic term meaning the movement of the earth's crust.

Diffraction the bending of waves when an opening or obstacle has a size smaller than or equal to the wavelength.

Dike a discordant pluton formation that is formed when magma fills a nearly vertical fracture in rock layers.

Dispersion the fact that different frequencies are refracted at slightly different angles.

Doldrums the low-pressure region near the equator.

Doppler effect an apparent change in frequency resulting from the relative motion of the source or the observer.

Drift glacial deposits of eroded material that are deposited either by ice or meltwater.

Dust dome a concentration of dust over a city due to the self-contained thermal circulation cell set up as a result of the heat-island effect.

Dyne a unit of force, 1 g · cm/s^2.

Earthquake the sudden release or transfer of energy resulting from sudden movement due to stresses in the earth's lithosphere.

Ecliptic plane the plane of the earth's orbit around the sun.

Electric charge a fundamental property of matter that can be either positive or negative and gives rise to electrical forces.

Electric field a set of imaginary lines that indicate the direction a small positive charge would move if it were placed at a particular spot.

Electricity the effects produced by moving charges.

Electrolysis the production of an oxidation-reduction reaction by means of an electric current.

Electrolyte a compound in the molten or dissolved form that conducts an electric current.

Electromagnet a current-carrying coil of insulated wire wrapped around a piece of soft iron that creates a magnetic field inside the iron only when the wire conducts a current.

Electromagnetic wave a wave caused by oscillations of electric and magnetic fields.

Electromagnetism the interaction of electric and magnetic effects.

Electromotive series a list of standard oxidation potentials at a given temperature.

Electron an elementary subatomic particle with a very small mass of 9.01×10^{-31} kilograms and a negative charge of 1.602×10^{-19} coulombs which goes in orbit around the atomic nucleus.

Electronegativity the measure of the ability of an atom to attract electrons in the presence of another atom.

Electron period a set of energy levels all of which have approximately the same energy.

Electron shell consists of all the orbits of electrons with the same principal quantum number (n).

Electron subshell similar to an energy level, all electrons in the same subshell have the same energy.

Electron volt the amount of kinetic energy an electron (or proton) acquires when it is accelerated through an electric potential of one volt.

Element a substance that has the same number of protons in all of its atoms.

Emphysema a lung condition characterized by large air sacs due to the breakdown of the walls of the alveoli.

Endothermic reaction a reaction where energy is absorbed as in the melting of ice.

Energy the capacity to do work.

Entropy a measure of the disorder of a system.

Enzymes organic substances of high molecular weight that catalize reactions in living organisms.

Epicenter the point on the surface of the earth directly above the focus of an earthquake.

Epoch an interval of geologic time that is a subdivision of a period.

Era an interval of geologic time made up of periods and epochs.

Erg a unit of energy, 1 dyne · cm or 1 g · cm²/s².

Erosion the group of interrelated processes by which rock is broken down and the products removed.

Ester any compound that conforms to the general formula R—C—O—R′.

$$\underset{\text{O}}{\overset{\parallel}{}}$$

Eutrophication the natural aging process of lakes that is accelerated by phosphate pollution.

Excited state a state of the atom with energies above the ground state. (*See* ground state.)

Exothermic reaction a reaction in which energy is released, as in the burning of a candle.

Fault a fracture along which a relative displacement of the sides has occurred.

Fault-block mountains mountains that were built by normal faulting in which giant pieces of the earth's crust were uplifted.

Faulting the relative motion along a fracture or fault that results in the displacement of rock masses on one side of the fracture relative to those of the other side.

Felsitic lava lava with a high silica content and high viscosity.

First quarter moon the phase of the moon between the new and full moon in which an observer in the United States sees the right half of the moon bright and overhead at 6 P.M. local solar time.

Fission the splitting of the nucleus of an atom into two nuclei of approximately equal size with the release of energy.

Focal length the point at which light rays from a distant source will converge after being reflected from a mirror or refracted by a lens.

Focus (earthquake) the point within the earth where the initial energy release of an earthquake occurs.

Fog a maze of visible droplets of water vapor near the earth's surface.

Föhn the name applied to the warm leeward winds on mountain slopes that give rise to the rapid melting of snow. In the Rocky Mountain region, the term *chinook* is used.

Folded mountains mountains characterized by folded rock strata, with external evidence of faulting and central evidence of igneous metamorphic activity. Folded mountains are believed to be formed at convergent plate boundaries.

Foliation the orientation characteristic of some metamorphic rocks due to formational directional pressures.

Formula weight the sum of the atomic weights given in the formula of the compound. If the formula is the molecular formula, the formula weight is also its molecular weight.

Fossil fuels fuels of organic origin, viz., coal, gas, and oil.

Foucault pendulum a pendulum with a very long length that swings for a long period of time and demonstrates the rotation of the earth.

Frequency the number of oscillations of a wave in a given period of time.

Front the boundary between two air masses.

Frost heaving the uplifting of rock and soil due to accumulations of freezing water.

Frost wedging a form of physical weathering by which rocks are pushed apart and fractured due to the pressure of freezing water within the rocks.

Full moon the phase of the moon that occurs when the moon is on the opposite side of the earth from the sun.

Fundamental properties the four fundamental or basic properties of nature are length, time, mass, and electric charge.

Fusion the transmutation of lightweight atomic nuclei into heavier nuclei with the release of energy.

G the universal gravitational constant—$G = 6.67 \times 10^{-11}$ N·m²/kg².

Galaxy an extremely large collection of stars occupying an extremely large volume of space.

Gamma decay the emission of electromagnetic energy from the nucleus of an atom. The atomic number, mass number, and neutron number of the atom remain the same, but the nucleus decreases in energy.

Gamma ray high-energy electromagnetic radiation with wavelength ranging from 3×10^{-14} to 3×10^{-12} m.

Gas matter that has no definite volume or shape.

Gay-Lussac's law when gases combine to form new gaseous compounds, all at the same temperature and pressure, the volumes of the initial and final gases are in the ratio of small whole numbers.

General theory of relativity a theory of relativity true for systems that are accelerating with respect to one another.

Generator a device that converts mechanical work or energy into electrical energy.

Geocentric theory the old false theory of the solar system which placed the earth at the center of the universe.

Geologic time scale a relative time scale based on the fossil index of rock strata.

Geology the study of the earth, its processes and history.

Geosyncline long, narrow ocean troughs containing large accumulations of sediment.

Geothermal gradient the increase of temperature with depth that occurs in the outer portion of the earth's crust. The rate of increase is about 1°F per 150 feet of depth.

Geyser a surface opening through which steam and boiling water erupt intermittently.

Glacier a large ice mass, consisting of recrystallized snow, that flows on a land surface under the influence of gravity.

Graded river the condition of a river when its erosion and transport capabilities are in balance.

Gram atomic weight the weight of an atom of an element relative to ¹²C, expressed in grams.

Gravitational collapse the collapse of a very massive body because of its attraction for itself.

Greenhouse effect the heat-retaining process of atmospheric gases, viz., water vapor and CO_2, due to the selective absorption of long-wavelength terrestrial radiation.

Greenwich meridian the meridian or half-circle on the earth's surface which has zero longitude.

Ground state the lowest energy level of an atom.

Group a number of elements appearing in any one column of the periodic table.

Guyot a seamount with a flat top.

H-R diagram a plot of absolute magnitude versus temperature of stars.

Half-life the time required for half of any sample of radioactive element to disintegrate.

Hard water water containing dissolved calcium and magnesium salts.

Heat a form of energy; energy in transit.

Heat capacity the amount of heat energy in calories required to raise the temperature of a substance one degree Celsius.

Heat engine a device that uses heat energy to perform useful work.

Heat-island effect the condition of the temperature and heat content of a city being higher than the surrounding rural areas due to greater radiation absorption, etc.

Heat lightning lightning occurring below the horizon or behind a cloud which illuminates the cloud with flickering flashes of light.

Heat pump a device used to transfer heat from a low-temperature reservoir to a high-temperature reservoir.

Heisenberg uncertainty principle it is impossible to know simultaneously the exact velocity and position of a particle.

Heliocentric theory the current theory of the solar system which places the sun in the center.

Hertz one cycle per second.

Heterogeneous matter that is of nonuniform composition.

Heterosphere a region of the atmosphere based on the heterogeneity of the atmospheric gases between approximately 60 and several hundred miles in altitude.

Homogeneous matter that is of uniform composition throughout.

Homosphere a region of the atmosphere based on the homogeneity of the atmospheric gases between approximately 0 to 60 miles in altitude.

Horse latitudes the high-pressure region near 30° latitude.

Horsepower a unit of power, 550 ft · lb/s.

Hot spot theory the theory that explains central plate volcanic chains as being formed as a result of plate movement over a hot spot beneath it.

Hubble's law the recessional speed of a distant galaxy is proportional to its distance away.

Humidity a measure of the water vapor in the air.

Hurricane a tropical storm with winds in excess of 74 mi/h.

Hurricane warning an alert that hurricane conditions are expected within 24 hours.

Hurricane watch an advisory alert that hurricane conditions are a definite possibility.

Hydrocarbons compounds composed of carbon and hydrogen, e.g., methane, CH_4.

Hydrologic cycle the cyclic movement of the earth's water supply from the oceans to the mountains and back again to the oceans.

Hydronium ion H_3O^+

Hygroscopic nuclei particulate matter that acts as nuclei in the condensation process.

Ice point the temperature of a mixture of ice and air-saturated water at normal atmospheric pressure.

Igneous rock rock formed by the cooling and solidification of hot, molten material.

Incomplete combustion combustion in which the carbon element of the fuel is not completely reacted to form CO_2, but rather CO.

Index of refraction the ratio of the speed of light in a vacuum and the speed of light in a medium.

Indian summer a period of warm weather in the late fall.

Inertia the property of matter to resist any change of motion.

Inner planets the four planets closest to the sun (Mercury, Venus, Earth, and Mars).

Insolation the solar radiation received by the earth and its atmosphere—incoming solar radiation.

Insulator a material that does not conduct an electric current because the electrons in the material are not free to move.

International date line the meridian that is 180° E or W of the prime meridian.

Inter-transition elements elements in which the third shell from the outer shell is increasing from 18 to 32 electrons.

Ion an atom or group of atoms with a net electric charge.

Ionic bond a chemical bond in which electrons have been transferred from atoms of low ionization potential to atoms of high electron affinity.

Ionic compounds compounds that are formed by the process of electron transfer.

Ionization potential the electrical voltage required to remove an electron from an atom.

Ionosphere a region of the atmosphere characterized by ion concentrations between 50 to several hundred miles in altitude.

Isobar a line drawn through points of equal pressure.

Isomers molecules with the same molecular formula but a different structure or arrangement of atoms, hence slightly different physical properties.

Isostasy the concept that the earth's crustal material "floats" in gravitational equilibrium on a "fluid" substratum.

Isothermal a constant-temperature process.

Isotope atoms whose nuclei have the same number of protons but different number of neutrons.

Jet streams fast-moving "rivers" of air in the upper troposphere.

Joule a unit of energy, 1 N · m or kg · m²/s².

Kepler's harmonic law the ratio of the square of the period to the cube of the semimajor axis (one-half the larger axis of an ellipse), is the same for all the planets.

Kepler's law of elliptical paths all planets (asteroids, comets, etc.) go in elliptical orbits around the sun.

Kepler's law of equal areas as a planet (or asteroid or comet) goes around the sun, an imaginary line joining the planet to the sun sweeps out equal areas in equal periods of time.

Kilo prefix that means 10^3 or one thousand.

Kilogram the unit of mass in the mks system. One kilogram has an equivalent weight of 2.2 pounds.

Kinetic energy energy of motion equal to $\frac{1}{2}mv^2$.

Laccolith a concordant pluton formation formed from a blisterlike intrusion that has pushed up the overlying rock layers.

Landslide a type of fast mass wasting that involves the downslope movement of large blocks of weathered material.

Lathanides the 14 elements following lanthanum in the periodic table—the first inner-transition series.

Lapse rate the rate of temperature change with altitude. In the troposphere the normal lapse rate is $-3\frac{1}{2}°F$ per 1000 ft.

Last quarter moon (or third quarter moon) the phase of the moon between the full and new moon in which an observer in the United States sees the left half of the moon bright and overhead at 6 A.M. local solar time.

Latent heat of fusion the amount of heat required to change one gram of a substance from the solid to the liquid phase at the same temperature.

Latent heat of vaporization the amount of heat required to change one gram of a substance from the liquid to the gas phase at the same temperature.

Latitude for a point on the surface of the earth, the angular measurement, in degrees, north or south of the equator.

Lava magma that reaches the earth's surface through a volcanic vent.

Leukemia a cancerous condition characterized by an abnormal increase in the white cells (leucocytes) of the blood.

Lightning an electric discharge in the atmosphere.

Linear momentum mass × velocity.

Linearly polarized transverse waves that only vibrate in one direction.

Line squalls a series of storms along a front.

Lipid a general term that includes such substances as fats, oils, and waxes.

Liquid matter that has a definite volume but no definite shape.

Lithification the process of forming sedimentary rock from sediment. Also called consolidation.

Lithosphere the outermost solid portion of the earth which includes the crust and part of the upper mantle.

Load (stream) the eroded material transported by a stream.

Local Group the cluster of galaxies including our own Milky Way galaxy.

Longitude for a point on the surface of the earth, the angular measurement, in degrees, east or west of the prime meridian.

Longitudinal wave a wave in which the vibrations are in the same direction as the wave velocity.

Long shore current a current along a shore due to the waves that break at an angle to the shore line.

Lorentz-Fitzgerald contraction the length of a moving object appears to be shorter to an observer who is at rest than to an observer moving with the object.

Lunar eclipse an eclipse of the moon caused by the earth's blocking of the sun's rays to the moon.

MeV million electron volts, a unit of energy.

mks system the metric system that has the meter, kilogram, and second as the standard units of length, mass, and time and the coulomb as the standard of electric charge.

Magma hot, molten rock material.

Magnetic anomolies adjacent regions of rocks with remanent magnetism of opposite polarities. That is, the directions of the magnetism are reversed.

Magnetic declination the angular variation of a compass from geographic north.

Magnetic field a set of imaginary lines that indicate the direction a small compass needle would point if it were placed at a particular spot.

Magnetic monopole a single magnetic north or south pole without the other—as yet undiscovered.

Main sequence a narrow band on the H-R diagram on which most stars fall.

Mantle the interior region of the earth between the core and the crust.

Maria the large dark areas on the moon.

Mass a quantity of matter and a measurement of the amount of inertia that a body possesses.

Mass number the number of protons plus the number of neutrons in an atom.

Mass wasting the downslope movement of overburden under the influence of gravity.

Matter anything that exists in time, occupies space, and has mass.

Matter waves the waves produced by moving particles.

Meandering the looping, ribbonlike path of a river channel due to accumulated deposits of eroded material that diverts the stream flow.

Mean solar day the average length of a solar day. One solar day is the elapsed time between two successive crossings of the same meridian by the sun.

Mega prefix that means 10^6 or one million.

Mercalli scale a scale of earthquake severity based on the physical effects produced by an earthquake.

Mesosphere a region of the atmosphere based on temperature between approximately 35 to 60 miles in altitude.

Metal an element that tends to lose its valence electrons.

Metalloid an element that exhibits the properties of both metals and nonmetals.

Melting point the temperature of a substance at which the substance changes from a solid to a liquid.

Metamorphic rock rock that results from a change or metamorphism in pre-existing rock due to heat and pressure.

Meteor a small chunk of matter that burns up as it goes through the earth's atmosphere and appears to be a shooting star.

Meteorites chunks of matter from the solar system that fall through the atmosphere and strike the earth's surface.

Meteorology the study of atmospheric phenomena.

Meter the standard of length in the mks system. It is equal to 39.37 inches or 3.28 feet.

Micro prefix that means 10^{-6} or one one-millionth.

Milky Way the galaxy in which our solar system is.

Milli prefix that means 10^{-3} or one one-thousandth.

Millibar a unit of pressure. 1 mb = 10^3 dyne/cm^2.

Mineral any naturally occurring, inorganic, crystalline substances.

Mixture a nonchemical combination of two or more substances of varying proportions.

Moho discontinuity the boundary between the earth's crust and mantle.

Moh's scale a 10-point scale of mineral hardness based on mineral standards—diamond being the hardest and talc being the softest.

Molarity the number of moles of solute in one liter of solution. Molarity is designated by the letter M.

Mole a gram formula weight.

Molecule an uncharged particle of an element or compound.

Molecular weight the weight of one molecule of a substance relative to ^{12}C, expressed in grams.

Monomer the fundamental repeating unit of a long chain molecule.

Monsoons winds associated with seasonal convectional cycles set up between continents and oceans.

Moon illusion the optical illusion in which the moon appears to be larger on the horizon than when high in the sky.

Moraine a ridge of glacial till.

Motor a device that converts electrical energy into mechanical energy.

Mountain range a geologic unit or series of mountains.

Mountain system a group of similar mountain ranges.

Mudflow the movement of large masses of soil that have accumulated on steep slopes and become unstable with the absorption of large quantities of water from melting snows and heavy rains.

Multiple proportions, law of whenever two elements, say E_1 and E_2, combine to form different compounds, the various amounts of E_2 combined with the same amount of E_1 are in a ratio of small whole numbers.

Muons mu mesons which are particles with properties like the electron but are more massive than the electron—1 mu meson mass = 207 electron masses.

NOAA the National Oceanic and Atmospheric Administration of the U. S. Department of Commerce, which is responsible for the study and monitoring of oceanic and atmospheric phenomena.

National Meteorological Center the nerve center of the National Weather Service where weather data are received and processed and forecasts are made.

National Weather Service a federal organization under NOAA which provides national weather information.

Neap tide moderate tides with the least variation between high and low.

Nephanalysis a satellite cloud-cover photograph with a map grid.

Neutralization the mutual disappearance of the H⁺ and the OH⁻ ions. They combine to form water (HOH).

Neutrino a subatomic particle that has no rest mass or electric charge but does possess energy and momentum.

Neutron number the number of neutrons in the atom.

Neutrons elementary particles that have approximately the same mass (1.6×10^{-27} kg) as protons but have no charge. They are one constituent of the atomic nucleus.

New moon the phase of the moon that occurs when the moon is between the earth and the sun.

Newton $1 \text{ kg} \times \text{m/s}^2$.

Newton's first law of motion a body will move at a constant velocity unless acted upon by an external unbalanced force.

Newton's law of universal gravitation $F = Gm_1m_2/r^2$.

Newton's second law of motion the acceleration on an object is equal to the unbalanced force on the object divided by the mass of the object ($a = F/m$).

Newton's third law of motion whenever one mass exerts a force upon a second mass, the second mass exerts an equal and opposite force upon the first mass.

Node a point where the moon's path crosses the ecliptic.

Nonclastic rocks sedimentary rocks formed primarily from chemical and organic sediments.

Nonmetal an element that tends to gain electrons to complete its outer shell.

Normal faulting movement along a nonvertical fault in which the overlying side moves downward relative to the side beneath it.

Novae stars that suddenly increase dramatically in brightness for a brief period of time.

Nucleic acid very high molecular weight polymers that are present in living cells.

Nucleus the central core of the atom.

Occluded front a front that is occluded, or closed off, from the earth's surface due to forced ascension.

Octet rule the tendency of atoms to have eight electrons in their outer shell when forming molecules or ions. There are many exceptions to the rule.

Ohm the unit of resistance equal to one volt per ampere.

Ohm's law the voltage across two points is equal to the current flowing between the points times the resistance between the points.

Opposition the time at which a planet is on the opposite side of the earth from the sun.

Organic acid any compound that contains the molecular arrangement $-C\underset{\text{OH}}{\overset{\text{O}}{\big<}}$ called the carboxyl group.

Organic compounds compounds of carbon.

Outer planets the five planets farthest from the sun (Jupiter, Saturn, Uranus, Neptune, and Pluto).

Overburden the weathered material that accumulates on base rock.

Oxidation the process in which electrons are lost.

Oxidation number a positive or negative number assigned to an atom as a measure of the electric charge that an atom in a molecule would have if it were ionically bound.

Oxidation state a term used to indicate the oxidation number.

Oxide a compound containing oxygen and one or more other elements.

Ozone the compound O_3. It is found naturally in the atmosphere in the ozonosphere and is also a constituent of photochemical smog.

Ozonosphere a region of the atmosphere characterized by an ozone concentration between approximately 0 to 50 miles in altitude.

pH the exponent of the negative power to which 10 is raised when used to express H⁺ ion concentration.

P waves primary (P) waves, so-called because they reach a seismic station before the S waves. P waves are longitudinal or compressional waves, i.e., their particle oscillations are in the direction of propagation and are transmitted by solids, liquids, and gases.

Pangaea the single giant supercontinent that is believed to have existed over 200 million years ago.

Parhelia two bright-colored patches that appear on each side of the sun due to scattering by the ice crystals of cirrostratus clouds. (Sometimes called side suns.)

Parallax the apparent motion, or shift, that occurs between two fixed objects when the observer changes position.

Parsec the distance to a star where the star exhibits a parallax of one second. This distance is equal to 3.26 light years or 206,265 astronomical units.

Pauli exclusion principle no two electrons can have the same set of quantum numbers.

Penumbra a region of partial shadow. During an eclipse, an observer in the penumbra sees only a partial eclipse.

Perfect gas law the relationship that exists between the volume, absolute pressure, and temperature of a gas. $PV/T = K$.

Period an interval of geologic time that is a subdivision of an era and made up of epochs; the time for a complete cycle of motion; a horizontal row of the periodic table, elements that have approximately the same energy.

Periodic law the properties of elements are periodic functions of the atomic number.

Permafrost ground that is permanently frozen.

Permeability a measure of a material's capacity to transmit fluids.

Pesticides chemicals used to kill unwanted insect and plant species.

Photochemical smog air pollution conditions resulting from the photochemical reactions of hydrocarbons with oxygen in the air and other pollutants in the presence of sunlight.

Photon a "particle" of electromagnetic radiation.

Photosphere the sun's outer surface that is visible to the eye.

Photosynthesis the process by which plants convert CO_2 and H_2O to sugars with the release of oxygen.

Physical properties those that do not involve a change in the chemical composition of the substance.

Pi (π) the ratio of the circumference of a circle to its diameter. It is equal to 3.14159. . . .

Pitch the highness or lowness of a sound. It is a consequence of the frequency of the sound waves received by the ear.

Planck's constant a constant of proportionality relating the energy and frequency of a photon. The constant has the value of 6.625×10^{-27} erg $\cdot$ s.

Plasma matter that is a disordered population of ionized atoms where the ion cores and the free electrons are not in thermal equilibrium—a state in which the temperature is so high that electrons and atomic nuclei are separated and move very rapidly, constantly colliding with each other.

Plate a huge slab of rock that makes up a portion of the outer layer of the earth and is in relative motion with respect to other plates.

Plate tectonics the theory that the outer layer of the earth is made up of rigid plates that are in relative motion with respect to each other.

Pluton a large body of intrusive igneous rock.

Polar easterlies the prevailing winds in the latitudes from 60° to 90°.

Polymer the resulting chainlike structure produced by many monomers.

Porosity the percentage of unoccupied space in the total volume of a substance.

Potential energy the energy a body possesses due to its position in a force field.

Potential well a term used to indicate a negative potential energy.

Power work per unit time.

Precession the slow rotation of the axis of spin of the earth around an axis perpendicular to the ecliptic plane.

Pressure force per unit area.

Pressure gradient a variation of pressure with position.

Pressure trough a region with a pressure minimum.

Primary sewage treatment the basic method of sewage treatment consisting of four main steps—screening, grit removal, sedimentation, and chlorination.

Principal quantum number the numbers (1, 2, 3, . . .) used to designate the various principal energy levels that an electron may occupy in an atom.

Protein the complex combination of amino acids. Proteins are composed mainly of carbon, oxygen, nitrogen, and hydrogen.

Protons elementary particles that are 1836 times as massive (1.67×10^{-27} kg) as electrons and have a positive charge (1.6×10^{-19} coulomb). They are one constituent of the atomic nucleus.

Psychrometer an instrument used to measure relative humidity.

Pulsar a star that has a radio signal that pulses regularly and very rapidly.

Pyroclastic debris solid material emitted by volcanoes that ranges in size from fine dust to large boulders.

Quantum a discrete amount.

Quantum mechanics (or wave mechanics) a field of physics used to solve problems when geometrical sizes are comparable with the wavelengths of particles. Schrödinger's equation forms the basis of wave mechanics.

Quantum number a number assigned to one of the various values of a quantized quantity.

Quasar a shortened term for "quasi-stellar radio source."

RNA ribonucleic acid, a nucleic acid located primarily in the cytoplasm or outer structure of the cell. RNA has the structure of a double-stranded helix.

Radar an instrument that sends out electromagnetic (radio) waves, monitors the returning wave that is reflected by some object, and thereby locates the object. Radar stands for RAdio Detecting And Ranging. Radar is used to detect and monitor precipitation and severe storms.

Radian a unit of angle equal to 57.4°. There are 2π radians in a circle.

Radiation the transfer of energy by means of electromagnetic waves.

Radiation fog a fog formed as a result of radiative heat loss (sometimes called a valley fog).

Radioactive the spontaneous disintegration of certain atomic nuclei, generating one or more of three types of radiation: alpha, beta, and/or gamma.

Radiosonde a small package of meteorological instruments with a radio transmitter that is carried aloft by a balloon.

Rays streaks of light-colored material extending outward from craters on the moon.

Rayleigh scattering the preferential scattering of light by air molecules and particles that accounts for the blueness of the sky. The scattering is proportional to $1/\lambda^4$.

Real image an image from a mirror or lens which can be brought to focus on a screen.

Red giant a red star that has a much larger diameter than average.

Red-shift a Doppler effect caused when a light source, such as a galaxy, moves away from the observer and shifts the light frequencies lower or toward the red.

Reduction the process in which electrons are gained.

Refraction the bending of light waves caused by a velocity change as light goes from one medium to another.

Remanent magnetism the magnetism retained in rocks containing ferrite minerals after solidifying in the earth's magnetic field.

Representative elements elements in which the added electron enters the outermost shell in which the outermost shell is incomplete—the A groups in the periodic table.

Resistance the opposition to the flow of electric current.

Resonance a wave effect that occurs when an object has a natural frequency that corresponds to an external frequency.

Rest mass the mass of an object at zero velocity.

Retina the part of the human eye onto which light is focused by the lens of the eye.

Reverse faulting movement along a nonvertical fault in which the overlying side moves downward relative to the side beneath it.

Revolution the movement of one mass around another.

Richter scale a scale of earthquake severity based on the amplitude or intensity of seismic waves.

Rill a narrow trench or valley on the moon.

Ring of Fire the area generally circumscribing the Pacific Ocean that is characterized by volcanic activity.

Rock any naturally occurring, solid, mineral mass that makes up part of the earth's lithosphere.

Rock cycle the cyclic movement of rock, during which the rock is created, destroyed, and metamorphized by the earth's internal and external geologic processes.

Rockslide a type of landslide occurring in mountain areas when large quantities of rock break off and move rapidly down steep slopes.

Rotation a spinning motion.

S waves secondary (S) seismic waves, so-called because they reach a seismic station after the P waves. S waves are transverse or shear waves, i.e., their particle oscillations are at right angles to the direction of propagation, and are transmitted only by solids.

Salinity a measure of the saltiness of water.

Salt an ionic compound containing the cation of a base and the anion of an acid.

Sanitary landfill a method of solid waste disposal in which refuse is covered with a layer of soil within a day after being deposited at the landfill site.

Saturated hydrocarbons contain only single bonds and their hydrogen content is at a maximum.

Saturated solution when the maximum amount of solute possible is dissolved in the solvent.

Scalar a quantity that has a magnitude and units but no direction is associated with it.

Sea floor spreading the theory that the sea floor slowly spreads and moves sideways away from mid-ocean ridges. The spreading is believed to be due to convection cycles of subterranean molten material that causes the formation of the ridges and a surface motion in a lateral direction from the ridges.

Seamount an isolated submarine volcanic structure.

Second the standard unit of time. It is now defined in terms of the frequency of a certain transition in the cesium atom.

Secondary sewage treatment the treatment that adds an additional step of either a trickling filter or an activated sludge process to the primary treatment for the removal of suspended organic matter.

Sedimentary rock rock formed from the consolidation of layers of sediment.

Seismic waves the waves generated by the energy release of an earthquake.

Seismograph an instrument that records the intensity of seismic waves.

Separated sewer system a system with separate networks for sanitation and storm water transportation.

Sheet erosion erosion resulting from runoff or the overland flow of water.

Shield volcano a volcano with a low, gently sloping profile formed by a fissure eruption of low-viscosity lava.

Silicon tetrahedron the pyramid-shaped structure of the complex silicate ion $(SiO_4)^{4-}$.

Sill a pluton formation that lies between and parallel to existing rock layers.

Silt a mixture of fine particles and organic material delivered by a stream.

Slug a unit of mass, $1\ lb/ft/s^2$.

Slump a type of landslide involving the downslope movement of an unbroken block of overburden that leaves a curved depression on the slope.

Smog a contraction of smoke-fog used to describe the combination of these conditions.

Solar constant the average amount of solar energy received per area at the top of the atmosphere per time, $2.0\ cal/cm^2 \cdot min$.

Solar eclipse an eclipse of the sun caused by the moon's blocking of the sun's rays to an observer on earth.

Solid matter that has a definite volume and a definite shape.

Solid waste any normally solid material resulting from human or animal activities that is useless or unwanted.

Solifluction (soil flow) a type of slow mass wasting involving the "flow" of weathered material over a solid, impermeable base, permafrost.

Solubility the amount of solute that will dissolve in a specified volume of solvent (at a given temperature) to produce a saturated solution.

Solute the dissolved substance in a solution.

Solution a homogeneous mixture.

Solvent the substance in excess in a solution.

Sound a wave phenomenon caused by variations in pressure in a medium such as air.

Source region the region or surface from which an air mass derives its physical characteristics.

Special theory of relativity a theory of relativity true for systems that are moving at a constant velocity with respect to one another.

Specific heat the amount of heat energy in calories necessary to raise the temperature of one gram of the substance one degree Celsius.

Spectroscope an instrument used to separate electromagnetic waves into their component wavelengths.

Spectrum an ordered arrangement of various frequencies or wavelengths of electromagnetic radiation.

Speed of light 186,000 mi/s or 3×10^{10} cm/s or 3×10^8 m/s.

Spring tide the tides of greatest variation between high and low.

Squall line a region along a front characterized by storms and turbulent weather.

Stable layer a layer of air with uniform temperature and density.

Stalactite an icicle-shaped dripstone formation extending downward from a cavern roof.

Stalagmite a blunt icicle-shaped dripstone formation extending upward from a cavern floor.

Standard conditions $0°$ C and sea level atmospheric pressure.

Standard unit a fixed and reproducible reference value used for the purpose of taking accurate measurements.

Steam point the temperature at which pure water, at normal atmospheric pressure, boils.

Stratosphere a region of the atmosphere based on temperature between approximately 10 to 35 miles in altitude.

Stratovolcano a volcano with a steeply sloping symmetric cone formed by eruptions of high viscosity lava and pyroclastic debris. A stratovolcano is also called a composite volcano.

Stratus a root name used to describe stratified or layered cloud forms.

Streak the color of a powdered mineral on a streak plate (unglazed porcelain).

Strike-slip faulting movement along a horizontal transform fault in which the two sides of the fault strike and slip by each other.

Subduction the process in which one plate is deflected downward beneath another plate into the asthenosphere.

Substance a homogeneous sample of matter all specimens of which have identical properties and identical composition.

Summer solstice the farthest point of the sun's declination north of the equator (for the northern hemisphere).

Sunspots patches of cooler, darker material on the surface of the sun.

Supercooled same as supersaturated—see below.

Supernova an exploding star.

Supersaturated the condition of air when its temperature is lowered below the dew point without condensation.

Supersaturated solution a solution that contains more than the normal maximum amount of dissolved solute.

Surf the breaking of surface waves along a shore.

Surface current a broad drift of surface water in the ocean that is set in motion by prevailing winds.

Surface wave (oceanic) a wave on the surface of the ocean near a shore line.

Surface wave (seismic) a seismic wave that travels along the earth's surface or a boundary within it.

Superposition, law of the principle that in a succession of stratified deposits the younger layers lie on the older layers.

Synergism the combined effect of two or more pollutants.

Synoptic weather charts charts or maps that present a synopsis of weather data.

Tectonics the study of the earth's general structural features and their changes.

Temperature a measure of the average kinetic energy of the molecules.

Temperature inversion a condition characterized by an inverted lapse rate.

Ten-degree rule a 10-degree Celsius increase in temperature leads to a doubling of the chemical reaction rate.

Thermal circulation the cyclic movement of matter (e.g., air) due to localized heating and convectional heat transfer.

Thermal conductivity a measure of the ability of a substance to conduct heat energy.

Thermal pollution pollution resulting from warm water, chiefly from industrial cooling processes, that is discharged into natural waterways.

Thermodynamics, first law of the heat energy added to a system must go into increasing the internal energy of the system, or any work done by the system, or both. The law also states that heat energy removed from a system must produce a decrease in the internal energy of the system, or any work done on the system, or both. The law is based upon the law of conservation of energy.

Thermodynamics, second law of it is impossible for heat to flow spontaneously from an object having a lower temperature to an object having a higher temperature.

Thermodynamics, third law of a temperature of absolute zero can never be attained.

Thermosphere a region of the atmosphere based on temperature between approximately 60 to several hundred miles in altitude.

Thrust faulting a special case of reverse faulting in which the fault plane is at a small angle to the horizontal.

Thunder the sound associated with lightning that arises from the explosive release of electrical energy.

Tidal current a current due to the tidal movement of sea water.

Till eroded material that is transported and deposited by glacial ice rather than meltwater.

Time the continuous forward flowing of events.

Time dilation (the stretching out of time) the passage of time for a moving object appears to be longer to an observer who is at rest than to an observer moving with the object. All motions appear to be slower than to an observer at rest.

Tornado a violent storm characterized by a funnel-shaped cloud and high winds.

Tornado warning the alert issued when a tornado has actually been sighted or indicated on radar.

Tornado watch the alert issued when conditions are favorable for tornadoes.

Total internal reflection a phenomenon in which light is totally reflected because refraction is impossible.

Traction the transport of a stream's bed load as a result of rocks and particles being rolled and bounced along by the stream's current.

Trade winds the prevailing winds in the latitudes from 0° to 30°.

Transformer a device that increases or decreases the voltage of alternating current.

Transform fault a fault with a horizontal fault plane along which strike-slip faulting occurs.

Transition elements elements in which the second shell from the outer shell is increasing from 8 to 18 electrons—the B groups in the periodic table.

Transverse wave a wave in which the vibrations are perpendicular to the wave velocity.

Trickling filter a bed of bacteria-laden stones through which sewage effluent trickles for the removal of organic matter.

Tropical year the time interval from one vernal equinox to the next vernal equinox.

Troposphere a region of the atmosphere based on temperature between the earth's surface and 10 miles in altitude.

Umbra a region of total darkness in a shadow. During an eclipse, an observer in the umbra sees a total eclipse.

Uniformitarianism *see* Uniformity.

Uniformity, principle of the principle that the same processes operate today on and within the earth as in the past. Hence, the present is considered the key to the past.

Unsaturated hydrocarbons compounds that can take on hydrogen atoms to form saturated hydrocarbons or atoms of substances other than hydrogen to form derivatives.

Valence the net electric charge of an atom or the number of electrons an atom can give up (or acquire) to achieve a filled outer shell.

Valence electrons the electrons that are involved in bond formation, usually the outermost electrons.

Valley glacier a glacier that flows downward creating a valley.

Vector a quantity that has not only a magnitude and units but also a direction associated with it.

Velocity, average the change in displacement divided by the change in time, $v = \Delta s / \Delta t$.

Velocity, instantaneous the velocity at a particular instant of time.

Vernal equinox the time (near March 21) when the sun's declination crosses the equator moving north (for the northern hemisphere).

Virtual image an image from a lens or mirror that cannot be brought to focus on a screen.

Viscosity the internal property of a substance that offers resistance to flow.

Vitamins organic substances that are needed in minute amounts to perform specific functions for normal growth and nutritional needs of the human body.

Volcanic mountains mountains that have been built by volcanic eruptions.

Volt the unit of voltage equal to one joule per coulomb.

Voltage the amount of work it would take to move an electric charge between two points, divided by the value of the charge, i.e., work per unit charge.

Waterspout a tornado over water.

Water table the boundary between the zone of aeration and zone of saturation.

Watt a unit of power, $1 \, kg \cdot m^2/s^3$ or $1 \, J/s$.

Wave cyclone cyclonic or rotational disturbances along a front that are formed by air motion in opposite directions.

Wavelength the distance from any point on a wave to an identical point on the adjacent wave.

Weather the atmospheric conditions of the lower troposphere.

Weathering the physical disintegration and chemical decomposition of rock.

Weight the force of gravity on the earth's surface.

Westerlies the prevailing winds in the latitudes from 30° to 60°.

White dwarf a white star that has a much smaller diameter than average. It is believed to be the end stage of small and average mass stars.

Wind horizontal air motion.

Winter solstice the farthest point of the sun's declination south of the equator (for the northern hemisphere).

Work the product of a force and the parallel distance through which it acts.

Zenith the position directly overhead of an observer on earth.

Zenith angle the angle between the zenith and the sun at noon.

Zodiac a section of the sky extending around the ecliptic 8° above and below the ecliptic plane.

Zone of aeration the zone of rock and soil above the water table whose open spaces are filled mainly with air.

Zone of saturation the subsurface zone of rock and soil in which all the openings are completely filled or saturated with water.

Index

Ante meridiem, 339
Antibiotic, 299
Anticyclone, 433, 470
Apogee (moon), 359
Apparent solar day, 338
Appleton layer, 407
Aqueous solution, 254
Aquifier, 626
Archimedes, xv
Aristarchus, xv, 308
Aristotle, xv, 308
Arrhenius acid-base concept, 273
Arrhenius, Svante, 273
Artesian
  spring, 626
  well, 626
Ascending node (moon), 365
Aspirin, 298
Asteroids, 320
Asthenosphere, 553
Astronomical unit, 314, 377
Atmosphere, 401–425
  composition, 402
  energy content, 409–417
  humidity of planets, 421–424
  origin, 403
  of planets, 403–404
  pressure, 404, 418–421
  temperature, 404–405
Atom, 152
Atomic
  nucleus, 154–155
  number, 154
  quantum numbers, 193–197
  weight, 157, 207
Aurora
  australis, 408
  borealis, 408
Autumnal equinox, 343
Average
  speed, 21
  velocity, 23
Avogadro's
  law, 251
  number, 253

B

Balmer series, 182
Barnard, Edward M., 389
Barnard's star, 389
Barometer
  aneroid, 420
  mercury, 418
Base, 273
  conjugate, 274
Batholith, 565
Battery, lead storage, 270
Becher, Johann J., 245
Bedding, 573
Bed load, 617
Benzene ring, 240
Bergeron process, 457–458
Berzelius, Jons Jacob, 207
Bessel, Friedrich W., 332
Beta decay, 159
Big Bang Theory, 397
Binary star, 381
Binding energy
  per nucleon, 157
  total, 157
Biodegradable, 643
Black hole, 385
Blizzard, 476
Bode's law, 314
Bohr, Niels, xviii, 173, 175
Bohr's equation for the hydro-
      gen atom, 175
Boiling point, 120
Bombs, volcanic, 589
Bond
  covalent, 229
  ionic, 229
  nonpolar, 232
  polar, 232
Boyle, Robert, xviii, 206
Boyle's law, 126
Brahe, Tycho, xvi, 309
Breeder reactor, 166
Breeze
  land, 436

mountain, 436
sea, 436
valley, 436
British system, 5
Brönsted, J. N., 273
Brönsted-Lowry theory, 273
Btu, 118
Buys Ballot's law, 433–434

Caldera, 593
Calendar, 348–349
  Gregorian, 348
  Julian, 348
  Roman, 348
Calorie, 118
Cancer, 651
Candela, 7
Carbohydrates, 287
Carbon dioxide
  in greenhouse effect, 414–415
  pollution, 538–540
Carbonic acid, 524, 613
Carbon monoxide, 524, 534–535
Carboxyl, 286
Carcinogen, 526
Cartesian coordinate system, 336
Catalyst, 266
Catastrophe theories, 327
Cathode, 269
Cation, 228
Caverns, 614
Ceilometer, 451
Celestial
  prime meridian, 377
  sphere, 376
Cellulose, 288
Celsius scale, 116
Centripetal
  acceleration, 175
  force, 42
Cepheid variables, 381

H

Hahn, Otto, xix
Hail, 458
Half-life, 160
Half-reaction, 228
Hall, James, 603–605
Halley's comet, 325
Halogens, 216
Hardness (mineral), 580
Hard water, 629
Heat, 115–131
  capacity, 120
  engine, 122
  of fusion, 120
  -island effect, 536
  pump, 122–123
  transfer, 129–130
  of vaporization, 120
Heaviside layer, 407
Heisenberg, Werner, 191
Heisenberg uncertainty principle,
    191
Heliocentric theory, 308
Hemoglobin, 534–535
Henry, Joseph, 94
Heracleides, Ponticus, 308
Heroin, 299
Hertz (unit), 70
Hess, H. H., xx, 549
Heterogeneous, 246
Heterosphere, 405
High clouds, 444–447
Highs (anticyclone), 470
Hipparchus, xvi
Homogeneous, 246
Homosphere, 406
Horizon, 344
Horse latitudes, 438
Horsepower, 53
Hot spot theory, 588
H-R diagram, 380
Hubble's law, 394
Humidity, 421–424
  absolute, 421
  relative, 421, 423

Humilis clouds, 449
Hurricane, 480–486
  hunters, 503
  warning, 501
  watch, 501
Hutton, James, xix
Huygens, Christian, xviii, 49
Hydrated, 277
Hydrocarbons, 237
Hydrogen
  burning, 384
  isotopes of, 207
Hydrogenation, 289
Hydrologic cycle, 625
Hydronium ion, 274
Hydrothermograph, 495
Hygrometer, 421
  adsorption, 421
  hair, 421
Hygroscopic nuclei, 456

Ice
  point, 116
  storm, 475
Igneous rocks, 564–567
  extrusive, 564
  intrusive, 564
Index of refraction, 78
Indian summer, 516
Inertia, 34
Inertial mass, 34
Infrared
  measurements (satellite), 499
  rays, 71
Inner planets, 312
Inner transition elements, 212
Instantaneous
  speed, 22
  velocity, 23
Instrument shelter, 495
Insulation, 409
Insulator, 97
Intensity of sound wave, 74

Interference effects, 81
International Date Line, 340
International system of units,
    7
Ion(s), 218, 223
  complex, 231
  layers, 407–408
Ionic
  bond, 229
  compounds, 225, 227
Ionization potential, 178,
    219
Ionosphere, 406
Island arcs, 603
Isomers, 241
Isostasy, 552
Isothermal, 126
Isotope, 154

Jet stream, 440
Joule, James Prescott, 115
Joule (unit), 51
Julian calendar, 348
Jupiter (planet), 321

Kelvin, 7
  scale, 117
Kepler, Johannes, xvi, 309
Kepler's
  harmonic law, 310
  law of elliptical paths, 309
  law of equal areas, 310
Kilo, 14
Kilocalorie, 118
Kilogram, 6
Kinematics, 19
Kinetic
  energy, 54
  theory, 127–129
  theory of gas, 124
Krypton-86, 6

1 2 3 4 5 6 7 8 9 10